URBAN DYNAMICS AND GROWTH
Advances in Urban Economics

CONTRIBUTIONS
TO
ECONOMIC ANALYSIS

266

URBAN DYNAMICS AND GROWTH
Advances in Urban Economics

Edited by

Roberta Capello
Dept. of Management, Economics & Industrial Engineering
Politecnico di Milano
Milano, Italy

and

Peter Nijkamp
Dept. of Spatial Economics
Free University
Amsterdam, The Netherlands

United Kingdom – North America – Japan –
India – Malaysia – China

Emerald Group Publishing Limited
Howard House, Wagon Lane, Bingley BD16 1WA, UK

First edition 2004.

British Library Cataloguing in Publication Data
A catalogue record for this book is available from the British Library

ISBN: 978-0-444-51481-3

Printed and bound in Great Britain by
CPI Antony Rowe, Chippenham and Eastbourne

LIST OF CONTRIBUTORS

Numbers in paranthesis indicate the pages where the authors' contributions can be found.

Hesham M. Abdel-Rahman (443), Department of Economics and Finance, University of New Orleans, Lakefront, New Orleans, LA 70148, USA

Zoltan J. Acs (635), Max Plank Institute, Jena, Germany; University of Baltimore, Merrick School of Business, 1420 N. Charles Street, Baltimore, MD 21201, USA

Åke E. Andersson (837), Department of Economics, Jönköping International Business School, P.O. Box 1026, SE-551 11 Jönköping, Sweden

Marcus Berliant (533), Department of Economics, Washington University, St Louis, MO 63130-4899, USA

Johannes Bröcker (609), Department of Economics, University of Kiel, Olshausenstrasse 40, 24098 Kiel, Germany

K.J. Button (153), School of Public Policy, George Mason University, Fairfax, VA, USA

Meagan Cahill (729), University of Arizona, Tucson, AZ 85721, USA

Roberto Camagni (121, 495), Department of Management, Economics and Industrial Engineering, Politecnico di Milano, Milan, Italy

Roberta Capello (1, 57, 495), Department of Management, Economics and Industrial Engineering, Politecnico di Milano, Milan, Italy

John Carruthers (729), Mundy Associates LLC, Seattle, WA 98109, USA

Lata Chatterjee (837), Centre for Transportation Studies and Department of Geography and Environment, Boston University, 675 Commonwealth Ave., Boston, MA 02215, USA

Thomas de Graaff (381), Department of Spatial Economics, Free University Amsterdam, De Boelelaan 1105, 1081 HV, Amsterdam, The Netherlands

Henri L.F. de Groot (381), Department of Spatial Economics, Free University Amsterdam, De Boelelaan 1105, 1081 HV, Amsterdam, The Netherlands; Tinbergen Institute, De Boelelaan 1105, 1081 HV, Amsterdam, The Netherlands

Kieran P. Donaghy (583), Department of Urban and Regional Planning, University of Illinois at Urbana-Champaign, 111 Temple Hoyne Buell Hall, 611 East Lorado Taft Drive, Champaign, IL 61820, USA

Manfred M. Fischer (319), Department of Economic Geography and Geoinformatics, Vienna University of Economics and Business Administration, Vienna, Austria

Peter Friedrich (691), University of the Federal Armed Forces Munich, Werner-Heisenberg-Weg 39, 85579 Neubiberg, Germany

Manie Geyer (803), Urban and Regional Planning, University of Potchefstroom, Potchefstroom, South Africa

Joanna Gwiazda (691), University of Agriculture, UI. Orlat Lwowkowskich 38/51, 02-495, Warsaw, Poland

Geoffrey J.D. Hewings (213), Regional Economics Applications Laboratory, University of Illinois, Urbana, IL 61801, USA

Tschangho John Kim (213), Department of Urban and Regional Planning, and Department of Civil and Environmental Engineering, University of Illinois, 210 Temple Buell Hall, 611 Taft Drive, Champaign, IL 61820, USA

Rajendra Kulkarni (659), School of Public Policy, George Mason University, Fairfax, VA, USA

T.R. Lakshmanan (837), Department of Geography and Environmental Studies, Centre for Transportation Studies, Department of Geography and Environment, Boston University, 675 Commonwealth Ave., Boston, MA 02215, USA

Lars Lundqvist (181), Unit for Transport and Location Analysis, Department of Infrastructure, KTH-Royal Institute of Technology, SE-100 44 Stockholm, Sweden

Philip McCann (31), Department of Economics, University of Reading, RG6 6AW, UK

Gordon Mulligan (729), University of Arizona, Tucson, AZ 85721, USA

Chang Woon Nam (691), Ifo Institute for Economic Research, Poschinger Str. 5, 81679 Munich, Germany

Peter Nijkamp (1, 87, 153), Department of Spatial Economics, Free University Amsterdam, De Boelelaan 1105, 1081 HV, Amsterdam, The Netherlands; Department of Regional Economics, Free University of Amsterdam, De Boelelaan 1105, 1081 HV, Amsterdam, The Netherlands

Jos van Ommeren (347), Free University, FEWEB, De Boelelaan, 1081 HV Amsterdam, The Netherlands

Shin-Kun Peng (413), Academia Sinica and National Taiwan University, Taipei 11529, Taiwan, ROC

Aura Reggiani (319), Department of Economics, Faculty of Statistics, University of Bologna, Bologna, Italy

Piet Rietveld (153), Department of Regional Economics, Free University of Amsterdam, De Boelelaan 1105, 1081 HV, Amsterdam, The Netherlands

Jan Rouwendal (249), Department of Spatial Economics, Free University, Amsterdam, The Netherlands; Economics of Consumers and Households Group, Wageningen University, Wageningen, The Netherlands

Stephen Sheppard (285), Department of Economics, Williams College, Williamstown, MA 01267, USA

Jungyul Sohn (213), National Center for Smart Growth Research and Education, University of Maryland, College Park, MD 20742, USA; Regional Economics Applications Laboratory, University of Illinois, Urbana, IL 61801, USA

Roger R. Stough (659), School of Public Policy, George Mason University, Fairfax, VA, USA

J. Willemijn Van Der Straaten (249), Department of Spatial Economics, Free University, Amsterdam, The Netherlands

Erik T. Verhoef (87), Department of Spatial Economics, Free University Amsterdam, De Boelelaan 1105, 1081 HV, Amsterdam, The Netherlands

Ping Wang (533), Vanderbilt University, Nashville, TN, USA; NBER, Cambridge, MA, USA

CONTENTS

PART 2 ACCESSIBILITY

PART 3 SPATIAL INTERACTION, MIGRATION AND COMMUTING

Urban Dynamics and Growth, CEA, vol. 266
Roberta Capello and Peter Nijkamp (editors)

CHAPTER 1

The Theoretical and Methodological Toolbox of Urban Economics: From and Towards Where?

Roberta Capello[a] and Peter Nijkamp[b]

[a]Department of Management, Economics and Industrial Engineering, Politecnico di Milano, Milan, Italy
[b]Department of Economics, Free University, Amsterdam, The Netherlands

Abstract

Many attempts have been made in the past decade, with more or less success, to provide comprehensive or critical reviews on the state of the art in Regional Science. The reason behind these conceptual and methodological exercises has often been referred to as a "mid-life crisis", reached after forty years since the establishment of regional science as a discipline; the assessment of the path that led from there to here, a comparison of the aims achieved with those expected, and the exploration of new possibilities for the future were the main aims of the various reflections and evaluations that from different perspectives were addressed to regional science.
The interest in editing a book like the one presented here is mainly to underline theoretical and methodological advances in urban science from the perspective of economics, in order to highlight the scientific achievements obtained so far and the theoretical or methodological gaps which still need to be filled out.
The overview of theories, models, methodologies and scientific frameworks presented in the overview witnesses the richness of this discipline, and explains the interest in editing a book in advances urban economics like the present one.

Keywords: urban dynamics and growth, advances in urban economics

JEL classifications: R0, R20, R30, R40

1.1. Pathways in regional science and urban economics

Many attempts have been made in the past decade, with more or less success, to provide comprehensive or critical reviews on the state of the art in regional science. The reason behind these conceptual and methodological exercises has often been referred to as a 'mid-life crisis', reached after 40 years since the establishment of regional science as a discipline; the assessment of the path that led from there to here, a comparison of the aims achieved with those expected, and the exploration of new possibilities for the future were the main aims of the various reflections and evaluations that from different perspectives were addressed to regional science (Funck, 1991; Bailly, 1992; Isserman, 1993, 1995; Bailly and Coffey, 1994; van Geenhuizen and Nijkamp, 1996).

In the various reviews and reflections offered, the tendency was to analyse regional science as a unique and appealing discipline, underlining positive, negative, successful and problematic – theoretical as well as empirical and practical – trends in its life cycle. In the overall attempt to identify success and failure of theoretical and methodological advances in spatial science, the regional dimension was prominently present; regional economics, regional planning, or methods and modelling in regional science are principally highlighted in these retrospective contributions, treating the urban dimension often as a by-product of regional science. The urban scale was called for attention when dealing with location theory, or land use and mobility patterns, but it is somehow astonishing that hardly anybody felt the need to unfold regional science methodologies into consistent sub-disciplines (including the urban one), in order to highlight the role they played during the evolution of regional science. In the light of the importance of urban modes of living and working, it would have been plausible to expect a proposal to rename the discipline 'Regional and Urban Science'.

The interest in editing a book like the one presented here is mainly to underline theoretical and methodological *advances in urban science from the perspective of economics*, in order to highlight the scientific achievements obtained so far and the theoretical or methodological gaps which still need to be filled out.

The reasons behind the interest in urban economic studies are manifold. The first reason concerns the fact that urban economics is at the core of regional science; it is a strategic discipline whose future trends and developments in theoretical and methodological contributions will be decisive for the future of regional science as a whole, as the basic models in urban location theory of Von Thünen, Alonso, Christaller and Lösch did in the past.

The second reason lies in the fact that the city – or the urban area – is more and more the location or heartland of the major part of the world population, in both the developed and developing countries. The city is the cradle of ancient civilization, the origin of culture and science, the source of industrial development, the node of any information and communication system, and the command centre of a modern network society. But it is also the source of many evils (congestion, criminality, social deprivation and social inequality). Therefore, all negative and positive effects associated with the presence of a high geographical population density are concentrated in metropolis and urban areas, and call for specific spatial-economic analysis to be offered to practitioners and policy makers.

Moreover, the main tendencies generated by the rhythm and by the profound changes in the world economy are exacerbated at the urban level. Cities in developed countries play both the role of gatekeepers towards world markets being the nodes of international infrastructure networks, and the role of loci where competition creates the greatest market tensions (both in input markets, like local labour markets, and in output markets, with strong product competition). Cities in developing countries are both important and problematic realities, being the recipient of rural unemployment for a long time, and thus the locus where the rural crisis generates its negative effects: poverty, social tensions and social diseases, high income inequality, natural resource scarcity, environmental decay; they all mirror unprecedented and dramatic appearances, concentrated in particular territorial settings, and call for particular attention in spatial economic analysis (Glaeser *et al.*, 1992).

And last but not least, the city is by its very nature the locus where the socio-economic effects caused by a high territorial density of productive and residential activities, manifest all their strengths, and where space plays a fundamental role in generating efficient resource-based production systems. Innovation and learning processes, increasing returns in knowledge and other production factors, and economies of scale in services and infrastructure provision, generated by the simple geographical concentration of activities in space, are all key factors explaining a cumulative self-reinforcing endogenous growth.

With a view to the prominent position of modern cities in a global network economy, the focus of this introductory chapter is on urban economics as a sub-discipline of regional science. The general reflections we present in this chapter will mostly have the aim:

- to highlight the role urban economics has played – and will continue to play – in regional science (Section 1.2);
- to provide an overview of recent developments in both theoretical and methodological reflections in the field of urban economics (Section 1.3);
- to explore the role these reflections in urban economics may play in attempts in the regional science community to cope with its so-called crisis (Section 1.4);
- to identify future development patterns and the limits that hamper at present urban economic development (Section 1.5).

The overview of theories, models, methodologies and scientific frameworks presented here witnesses the richness of this discipline, and explains the interest in editing a book in advance is in urban economics like the present one.

1.2. The role of urban economics in regional science

Urban economics has always played a central role in the development of regional science. Various path finding contributions to spatial economic analysis can be found in the work of Von Thünen, Alonso, Christaller and Lösch, all dealing with location and choice behaviour of firms and residents mainly at the urban level. Also when one envisages contributions in the broader field of spatial development, it appears that many seminal works were conducted at the urban level, like the Hoyt (1954) model, born as an urban planning tool and, therefore, as a spatial model of physical urban growth. Once again, considering methodological tools, the same kind of conclusion emerges; the first applications of gravity (and more recently entropy) models have mainly taken place at the urban level and were developed in order to solve practical urban problems (e.g. spatial interaction within cities, and consequent infrastructure planning).

One can identify the reasons for this strategic role of urban economics as follows. The first argument lies in the nature of the city itself, being a complex system where social, economic and environmental aspects interact and define urban physical and economic growth patterns. Therefore, if the field of regional science is conceived as a cooperative venture among distinct spatially oriented disciplines, this is even more true for urban phenomena; a single discipline such as geography, economics or political science cannot provide the basis for a comprehensive understanding of a city, with all its social and economic complexity.

The second reason for the strategic role played by urban economics within regional science is that, given the economies of density of residential and productive activities, the territorial principles governing the spatial organisation of activities are strongly manifested at the urban level. In particular, in a textbook published 12 years ago, urban economics theory and models have very efficiently been organised around five principles governing the activities on territory, namely (Camagni, 1992):

- the *agglomeration principle*. The high density of population and productive activity prompts all positive (and negative) phenomena stemming from physical proximity; agglomeration economies, in the form of both urbanisation and localisation economies, are recognised to be one of the genetic elements in the existence of cities;
- the *accessibility* principle. The understanding of mutual interaction between transport costs and land use finds its first immediate and more rational application at the urban level;
- the *spatial interaction principle*. The high density of residential and productive activities present in cities facilitates the needs for contacts, and consequently the spatial interaction mechanisms, with all positive and negative effects associated with them;
- the *urban hierarchy principle*. The spatial division of labour is clearly reflected in the socio-economic disparity patterns among different cities;
- the *competitiveness principle*. As cities are the major location of productive activities, competitiveness is very important at

the urban level, and calls for specific provisions in favour of urban efficiency mechanisms.

Clearly, these principles may be mutually complementary, but they may also be mutually conflicting. The recognition of the importance and development of proper theories and methods in the field of urban economics is, therefore, an important means to understand both as yet existing bottlenecks and future possible directions in spatial economics.

1.3. Advances in urban economics: recent theoretical and methodological directions

1.3.1. Prefatory remarks

Although regional science is a relatively young discipline, in its 50 years of existence a surprisingly large variety of theories, methods and models have been developed which provide a relatively comprehensive theoretical and methodological toolbox for spatial analysis. Urban economics is not an exception in this respect; contemporary urban economics records, in fact, many advances and even breakthrough achievements, which enrich and reinforce both the theoretical and empirical frameworks of spatial analysis.

A great deal of our present understanding of the fundamental interaction between space and local economic behaviours originates from the fields of location theory and urban economics; the great number of relatively new and advanced contributions in this field does not allow for a detailed review on all individual achievements made; in addition, a disaggregated analysis of all novelties would probably not be so stimulating. We feel that an attempt to highlight general tendencies, at both a theoretical and methodological level, will turn out to be more fruitful for a debate on present weaknesses and on possible future directions of urban economics (see also Table 1.1). Inevitably, the set of 'tendencies' that follows is both selective and incomplete, primarily reflecting our own views and research interests.

1.3.2. Tendencies in theory

By looking at the theoretical trajectories followed in urban economics, one of the major tendencies which has accompanied the theoretical development in the field is the *need for more realism*

Table 1.1. Main tendencies in theories and methods of urban economics

	Spatial Principles				
	Agglomeration	Accessibility	Spatial Interaction	Urban Hierarchy	Competitiveness
Tendencies in Theories					
More realism in theoretical approaches	Efficient city size rather than optimal city size Spatial equilibrium models considering both environmental and agglomeration externalities Relational rather than physical proximity as a source of urban externality	Endogenous bid rent functions Inter-city location models Absolute vs. differential urban rent Income differences in location choices Externalities in residential location Randomly distributed idiosyncratic tastes Non-uniform generalised cost of travel with respect to location Externalities in land use and social optimum in land use	Search for economic rationale of gravity models Spatial interaction models considering both competing destinations and intervening opportunity factors Computational intelligence approaches to spatial interaction modelling	Aspatial logics behind urban systems	Endogenous growth determinants

(continued)

Table 1.1. Continued

	Spatial Principles				
	Agglomeration	Accessibility	Spatial Interaction	Urban Hierarchy	Competitiveness
Dynamic rather than static approaches	Dynamic urbanisation economies	Dynamic locational choice decisions		Dynamic urban hierarchy models	Cumulative and circular effects in urban growth
Tendencies in modelling and methods					
More refined and advanced techniques	(Bio)ecologically based models	Discrete models of choice (logit and probit models)	Entropy models Neural networks		Multiple equilibria models Non-deterministic growth models Path-dependent growth models
Interest in quantitative measures	Measurement of dynamic urban economies measured	Measurement of differential vs. absolute rent measured		Measurement of dynamic urban network externalities measured	Measurement of knowledge spillover measured Measurement of endogenous growth determinants measured

in sometimes rather abstract conceptual approaches, by relaxing most of the glaring unrealistic assumptions of the basic theoretical models. This tendency is justified by the need to broaden the interpretative capacity of the theoretical toolbox in this research field by searching for theories that are better able to reflect the real world.

In the context of the agglomeration principle, the need for more realism has led to the recognition that city size cannot be interpreted on the basis of an 'optimal city size', but of an 'efficient size', which depends on the functional characteristics of the city and on the spatial organisation within the urban system. Economies of scale exist up to a certain city size. However, urban development generates conditions leading to structural readjustments which may create new economic advantages. These structural adjustments may either be sectoral transformations towards higher order functions, or the increase of external linkages with other cities. Therefore, these new perspectives were inclined to accept what Richardson (1972) had already emphasised some years ago, namely that, beyond size, in the real world most cities differ in terms of functional specialisation and spatial organisation. Moreover, decisive forward steps have been made by accepting that environmental aspects (both positive and negative) are intrinsic and intertwined elements of agglomeration economies, contributing to the definition of urban attractiveness, urban growth and degree of competitiveness (Roback, 1982). An important conceptual step forward has been provided by the acceptance that physical proximity cannot be the source for all advantages of an urban location, and that relational proximity, i.e. the degree of social interaction, and sense of belonging (called 'social capital' in the social sciences), can sometimes have a greater interpretative power on urban dynamics than the advantages obtained by the mere physical proximity.

The area where the need of realism has strongly been felt is in land use and in location choice models, explaining the competition that derives among activities to obtain the most central location in a city. The analysis of economic behaviour in space represents the core of urban economics; extensions and refinements of the basic Von Thünen–Alonso–Muth work, in which at equilibrium a marginal reduction in rent from further decentralisation was exactly offset by a marginal increase in travel costs, defining a condition of indifference among locations (the famous 'Muth condition'), led to the birth of an established particular sub-discipline; all advanced models in this direction can be interpreted under the label 'New Urban Economics', and more recently 'Analytical Urban Economics'.[1] The development trajectory in this branch of urban economics has been the relaxation of the simple assumptions made in the basic models; the introduction of income differences in location choices, randomly distributed idiosyncratic tastes, heterogeneous urban space and the existence of externalities in the use of land (congestion, zoning, segregation, fiscal jurisdictions) are some examples in this respect.[2] A higher degree of realism was achieved in the models at the expense of a higher level of analytical sophistication, highly criticised when giving birth to a pure 'l'art pour l'art' attitude so detrimental to further acceptance and advances in this branch of urban economics.

In spatial interaction models, a great deal of effort has also been devoted to the introduction of various more realistic assumptions. Recent analyses and models contain both competing destination and intervening opportunity factors (Fotheringham, 1983). Attempts to make interaction models more realistic are also developed by considering possible alternative paths between nodes. When congestion requires a path different from the off-peal ideal, the intervening opportunities along the alternative path are taken into consideration (Fischer and Getis, 1999). An important breakthrough has been the establishment of a consistent link between spatial interaction models and behavioural discrete choice models (see Nijkamp and Reggiani, 1999).

In the study of urban hierarchy, two main directions have been followed in new theoretical contributions. The first attempt is to insert more realism into the two path finding models of Christaller and Lösch, by relaxing strong assumptions regarding the homogeneous demand distribution (Beckmann and McPherson 1970) and non-existence of location and production choice interdependencies (Long, 1971; Beguin, 1988). In this respect, the pioneeristic attempt of Long to introduce in the Christaller model the interdependence of goods demonstrates that the honeycomb structure achieved by

---

[1] See Richardson *et al.* (1996).

[2] The volume edited by Richardson *et al.* (1996) contains a very comprehensive set of papers on this issue.

Christaller strongly depends on the assumption of no interdependence in production and demand, although the mathematical complexity of the Long model does not allow for analytical solutions. More recently, the interpretation of new economic relationships among cities, primarily based on cooperative and horizontal relationships, has required to break the conceptual approach of urban hierarchy, and generated a new interpretative paradigm, that of 'city networks' (Camagni, 1993). The most important theoretical novelty provided by this paradigm is the break of the link between urban size and urban functions imposed by the Christallerian logic. With Christaller's approach, it is in fact impossible to explain why a city like Zurich, with only 300,000 inhabitants, is specialised in international finance in the same way as the city of New York or Tokyo. In the real world, the urban size is not always characteristic of core functions.

Last but not least, in theories dealing with urban competitiveness, decisive developments have been made with regard to the understanding of endogenous determinants of urban growth. The question of whether a city (or a region) is intrinsically capable of growing as a result of endogenous forces has been a source of debate for decades; industrial specialisation, infrastructure endowment, central location, production factor endowment or agglomeration economies have alternatively been emphasised in the academic arena as driving forces of local economic success. The decisive step forward in this field has been the focus on economies of scale in production which, together with non-linear transportation costs, are introduced into a (quantitative) interregional growth model; the final spatial distribution of activities critically depends on initial conditions including the starting distribution of activities and the nature of the non-linearities embedded in the activity–transportation interactions, which give rise to multiple equilibria (Krugman, 1991). The additional value of Krugman's approach resides in skilfully modelling the interaction between transportation costs and economies of scale in production, although the determinants of endogenous growth have already since long been emphasised, starting from the Myrdal–Kaldor model (increasing returns, cumulative self-reinforcing growth patterns). In parallel to Krugman's efforts, in the field of endogenous determinants a great emphasis has recently been put on knowledge as a driving force to development, and, what is really new, on the endogenous self-reinforcing mechanisms of knowledge creation. Macroeconomic models of endogenous growth (where knowledge is generally embedded in human capital) (Romer, 1986; Lucas, 1988), as well as in microeconomic models, where knowledge creation is analysed in terms of learning processes (at institutional, at territorial and at firms' level) have widely dominated the academic arena in the last decade.

A second clear tendency in theoretical developments in urban economics has been the attempts to move towards *dynamic approaches*. Time matters as well as space in regional science, and this also holds in urban economics. The effort to encapsulate time in spatial analyses has taken place in two different ways, according to two different meanings of time: a more traditional chronological time, and time as rhythm of innovative phenomena which occur in the territory. The introduction of a chronological time within spatial analysis is not at all a simple task, since it requires a mathematical and methodological toolbox, only recently available to regional scientists, with which we will deal in Section 1.3.3. Theories on non-linear urban dynamics – framed in the context of chaos theory, synergetics theory or predator–prey analysis – may be mentioned here (Nijkamp and Reggiani, 1999).

Conceptually speaking, time has also entered theoretical reflections on cities through the concept of innovation; time à la Bergson–Heidegger is interpreted as duration and a continuous process of creation, characterised by discontinuity, irreversibility, sequentiality and cumulativity. Time has thus been conceived by an important part of urban studies as the pace of learning, innovation and creation processes. Cities are, by definition, the loci where learning and cumulative learning processes take place; the identification of the sources and of the endogenous determinants of such processes, besides simple physical proximity, represents a great challenge for urban economists. Knowledge spillovers, collective learning, learning regions (or learning space) are all theories that embrace the most advanced perspectives in this direction.[3]

[3] In these fields of research, see among others Aydalot (1986), Jaffe (1989), Camagni (1991, 1999), Jaffe *et al.* (1993), Maillat *et al.* (1993), Rallet (1993), Feldman (1994), Audretsch and Feldman (1996), Anselin *et al.* (1997, 2000), Ratti *et al.* (1997), Capello (1999, 2001), Feldman and Audretsch (1999), Maskell and Malmberg (1999), RERU (1999) and Crevoisier and Camagni (2000).

1.3.3. Tendencies in models and methods

Parallel to the theoretical reflections underlined above, a series of advances can be highlighted in the fields of models and methods, which are here grouped in two major tendencies: refinements and advances in operational models and techniques, accompanied by a clear tendency to prove the theoretical reflections through quantitative measures.

In terms of *refinements of models and methods*, in the sphere of physical city growth, a particular paradigm has received a great deal of attention in the modelling literature of the 1980s, based on the competition (substitution/complementarity) among populations in a space–economy network, generally formalised by means of (bio)ecologically-based models. An important property of these models is that they allow oscillating and chaotic behaviour, like the previous non-linear models, with which they are strongly connected (van Geenhuizen and Nijkamp, 1996). More recently, the Lotka–Volterra (or prey–predator) model has been reformulated in order to explain urban dynamics through the relative dynamics of land rents (Camagni, 1992). Urban rent is interpreted as a share of total income; the substitution link between production profits and urban rent (the former decrease when the latter increases) generates, as a consequence, a decrease in investments, limiting economic growth in the urban area concerned. In this version of the model, urban rent therefore plays the role of spatial resource allocator, since it influences location choices: an increase in urban rent pushes residential and production activities towards the periphery, which is characterised by lower land prices.[4]

The area of location decisions has witnessed more theoretical advances than new applications – being the greatest steps forward made in theoretical modelling – with a high degree of abstraction and analytical complexity. However, recent locational analysis is increasingly based on disaggregated models of choice: logit and probit models have widely been applied in this field, which give rise to a much more refined analysis, since they allow to take into account various individual locational determinants, including

---

[4] Recently Capello and Faggian (2002) applied this model to the Italian urban system.

qualitative factors. In this context, a new interest in spatial computable general equilibrium models can be observed.

In spatial interaction, a great deal of effort has been devoted to the explanation of the reasons behind the strong interpretative power of gravity models on urban phenomena. At least, three theoretical foundations have been developed for this family of models: entropy maximisation, random utility, and finally models based on neurocomputing principles that represent a recent innovation in the design of spatial interaction models (Griffith, 1999, in this volume). Wilson (1970) introduced the entropy maximising theory supporting spatial interaction models, later extended by many others (see for example, Snickars and Weibull, 1977; Roy and Lesse, 1981; Smith, 1988). The fundamental assumption is that, at the outset, all outcomes are equally likely. The number of outcomes is a combinatorial problem, counting the number of ways of assigning the total number of flows to all possible origin–destination pairs. Maximising this function identifies the most probably geographic patterns that are consistent with origin–destinations, and/or average distance travel. The entropy maximizing approach was followed by a host of alternative derivations. The most important is the choice-theoretic approach that was first proposed by Niedercorn and Bechdolt (1969) and has generated a great deal of interest since then.[5] The essential idea of this approach is to model spatial interaction behaviour within the microeconomic paradigm of random utility maximizing choice behaviour. More recently, the emergence of GeoComputation as a subject (see Longley *et al.*, 1998; Fischer and Leung, 2001) and the powerful and fast computing environment has inspired many scholars to apply neurocomputing principles and techniques to revisit old and to solve new spatial interaction problems. The interaction models derived are given a very general formulation represented in the form of specific neural networks and viewed as universal function approximators(Fischer and Reggiani, in this volume).

[5] See Golob and Beckman (1971), Choukroun (1975), Nijkamp (1975), Smith (1975, 1978), Batten and Boyce (1986), Fotheringham and O'Kelly (1989); Fotheringham (1983), Fotheringham *et al.* (2000) and Wilson (2000).

A clear new evolution in Christaller and Lösch's basic models is linked to attempts to present a dynamic picture of an urban hierarchy, and the work of Parr is a breakthrough in this respect (Parr, 1978, 1981, 1985). Starting from the honeycomb structure of Christaller, Parr analyses the evolution of the spatial organisation of the urban hierarchy when some external effects occur, like the change in the allocation of economic functions at different hierarchical levels, or the creation of different lower order levels in the hierarchical structure. The result achieved is that the hexagonal structure of Christaller modifies into rectangular, triangular or varying hexagonal forms along the urban hierarchy.

In growth models, until a few years ago, the large majority of experiments and applications has taken for granted the existence of linear – and thus regular – growth processes. Linear models are certainly able to generate unstable solutions, but the solutions of such models are restricted to certain regular standard types. Such models may provide approximate replications of short- and medium-run changes, but fail to encapsulate long-term developments characterised by structural shifts of an irregular nature. This limit has recently been overcome with the adoption of non-linear models, which allow for a change in the dynamics of a system generated even by small perturbations in structural forms; structural instability means the possible existence of significant qualitative changes in the behaviour of the system (i.e. in the state variables) that are closely connected with bifurcation and catastrophe phenomena that can occur if the parameter values (i.e. the control variable) are changing. The application of non-linear models to the well-known neoclassical and Keynesian models has shown that the deterministic and unique results achieved by the dynamic linear models are no longer guaranteed: interregional income convergence determined by the traditional neoclassical model collapses and opens the way to alternative possible trajectories, and equilibria solutions; non-linear Keynesian Myrdal–Kaldor models substitute the deterministic result of continuous growth or decline with new and opposite development trajectories, after a catastrophe phenomenon occurs (Miyao, 1984, 1987a,b).

A second clear tendency in models and methods is the *interest in quantitative measures*. Econometric and statistical tools have in fact exerted a dominant influence on regional and urban economics.

In the past several statistical methods have been developed for dealing with regional and urban data, such as cluster techniques, principal component analysis, spatial autocorrelation analysis, spectral analysis, and so on (Nijkamp and Mills, 1986). The 'quantitative' revolution in economics has no doubt exerted a significant impact on the methodology of regional science as a whole. In urban economics this has led to the possibility to 'measure the unmeasurable': examples are dynamic urban economies, urban milieu effects, environmental externalities in cities, social costs of alternative land use patterns, city network advantages, knowledge spillovers and collective learning processes; the results achieved provide robust empirical evidence for policy makers and practitioners. Developments in the field of geographic information systems – also in the urban field – are complementing this development.

1.4. Urban economics and regional science transition

The so-called regional science crisis of the 1990s was mainly interpreted as the result of two sources of difficulties, widely recognised by many regional scientists: the lack of relevance on practical problems, on one side, and the loss of interdisciplinarity, on the other. The first was signalled as the result of the tendency of that period to develop descriptive or analytical tools and models, which "had the sweet and intoxicating flavour of l'art pour l'art" (Bolton and Jensen, 1995, p. 137). The second source of malaise was related to the somewhat ironic recognition that, despite openness and breadth – in terms of disciplines, methods and objects of analysis – were the major goals to which the field aspired in its early days, in the 1990s the major weakness of regional science was its narrowness of perspective (Bailly and Coffey, 1994). We may now appreciate that these phenomena may not be regarded as 'crisis signs', but as normal transition phenomena reflecting a sound dynamics of the discipline. Science – including Regional and Urban Science – goes through the normal upswings and downswings of a 'scientific product life cycle'.

Urban economics, as an important branch of regional science, played certainly at that time a crucial role in that critical reorientation. One of the major divergences between regional science and practice is in the field of behavioural choices and location models, where the degree of analytical complexity is in some models so high that the role theoretical models should have to provide a coherent framework within which to think about empirical issues was lost. Moreover, the attitude to provide classic mainstream economic models to the analysis of location choices and spatial behaviour has heavily provided a divergence from traditionally more interdisciplinary oriented models and approaches of regional science.

Nowadays, we recognise that the field of regional science is changing, especially for what concerns the need to reduce the unfortunate discrepancy between regional science and practice, mainly due to urban economics. The reason behind this feeling stems from the emphasis put on some recent and updated research fields in urban economics; two examples in this respect are, on one side, the important consideration of urban environmental problems, and on the other, the interest in capturing sources of endogenous urban competitiveness, with the aim to guarantee economic and social growth.

Urban sustainability is nowadays a major field of research, dealing with problems of efficient natural resource consumption, with negative environmental externalities, but also with a larger concern on urban quality of life. Social, economic and environmental problems are taken into consideration in the analysis of quality of life. After the avalanche of interest in global environmental issues (see, e.g. the Bruntland Report or the Report of the World Commission on Environment and Development (—WCED, 1987), the awareness has grown that many environmental problems have a local origin, while also global environmental decay often manifests itself at a local level. Thus, there is a simultaneous need for local action and global reflection. Consequently, cities may act as focal points for creative environmental strategies (see also Stanners and Bourdeau, 1995). This holds for both the industrial and the developing world.

Urban sustainability as a field of research can also offer an opportunity to overcome the second malaise encountered in regional science, that of loss of multidisciplinarity: in fact, environmental problems are not only a subject for economic reflections, but they also call for a multidisciplinary approach, from economics to urban planning, from biology to ecology. There are, in fact, many ways for a simultaneous analytical treatment of economics and environment. Since the 1960s a great many attempts has been made to link the economy to the ecology (Costanza *et al.*, 1997). An important contribution to the integration of economics and ecology began simply with a reflection on the principles of the materials balance for resources (extracted or collected, transformed, consumed and emitted) and on the need to take account of an economic viewpoint of such processes (Ayres *et al.*, 1999). Several attempts have also been made to build economic and social accounting systems that could incorporate the measurement of economic welfare and performance together with the measurement of environmental indicators and performances. The integration of economics with ecology has also been approached from the viewpoint of land-use – where economic and ecological processes have the most disruptive effects – and of urban environments. In addition, the interaction between economic and ecology has been dealt with for situations with global risks and uncertainties.

Even the second field which is more and more in the agenda of urban scientists, that of the identification of the determinants of urban competitiveness and growth, presents in some respects a good opportunity for regional science to recover from the diseases emphasised in the 1990s. It replies to the more and more stringent need of practitioners and policy makers to build efficient urban systems. As mentioned in the ESPD document, first written on the occasion of the European Ministers Council meeting in Noordwijk in 1997, and revised at the meeting in Glasgow in June 1999, "The development of Europe's cities and the relations between them constitutes the most important factor affecting the spatial balance of the territory of Europe" (ESDP, 1998, p. 47) and, moreover, "regions as a whole can become competitive only if their towns and cities are motors of economic growth" (ESDP, 1998, p. 51).

Interestingly enough, while it seems to us that in the field of urban economics, regional scientists are spontaneously moving towards much more practical problems, it also seems that practitioners and policy makers at different governmental levels are calling for more and more attention at the local, and particularly urban, level. It is a great chance offered to urban scientists, a chance to recover from previous 'diseases' and to relaunch the field of regional science as a strategic area of research, not only in academic, but also in policy making, arenas. Whether this is the case depends strongly on the way regional and urban scientists will react to this opportunity.

1.5. Hurdles to be crossed

Our impression on the future of regional science, and urban economics in particular, is optimistic. After a period of reflection, regional science shows clear signs of recovery, such as a deep interest in practical problems, and the recognition that an 'art pour l'art' approach is detrimental to further acceptance and advances in this field.

However, we still envisage some risks for what concerns the lack of interdisciplinarity. Since the time this problem has been underlined (Bailly and Coffey, 1994), hardly any signs of recovery can be identified, and we feel that the situation has become even more problematic. This pessimistic interpretation is based on some clear tendencies encountered in some recent theoretical developments, where some wide fields of unexplored interdisciplinarity still exist and no tendency to fill them seems to show up.

Some examples are useful in this respect. The theory on 'social capital' developed by quantitative sociology is an example in this respect: the concept could take advantage and provide advantage to all reflections on local synergies and milieu effects developed by regional and urban economists, and by the strategic planning studies in the field of urban planning (Camagni, 2003). The reflections in the field of knowledge spillovers developed by industrial economists could take advantage from the concepts of collective learning and relational proximity of regional scientists, in which the endogenous spatial development patterns of knowledge are not left to simple probabilistic contacts, but explained through territorial processes (Camagni and Capello, 2002). Last but not least, the theoretical reflections characterising the 'new economic geography' seem to be the result of a skilful effort of a group of mainstream economists driven, however, by a somehow unexplainable attitude to deny the importance of well-known spatial concepts (i.e. technological spatial externalities), or to (re-)invent important spatial concepts (i.e. cumulative self-reinforcing processes of growth; transportation costs vs. agglomeration economies in location choices). The inevitable consequence of such attitude is to mix the important and undeniable steps forward made by the 'new economic geography' school with already well-known knowledge in the field of regional science.

Some risks of disciplinary barriers and closeness to interdisciplinary views on strategic problems are still there. They are the result of a regional scientists' narrow perspective, as mentioned by Bailly and Coffey (1994), but also on some idiosyncratic approaches of mainstream disciplines towards a clearly multidisciplinary science like regional science. Especially in the case of economics, we hope that after the (re-)discovered interest by mainstream economists of space, and of spatial phenomena, the attitude towards regional science changes in favour of a more cooperative attitude and pronounced interest.

1.6. Reasons and structure of the volume

After the overview of theoretical and methodological approaches to urban economics, the reasons to edit a book on 'Advances in Urban Economics' become straightforward. The discipline is vast and in a mature phase; its richness is witnessed by many advanced contributions in theoretical as well as methodological and applied fields of research. The book aims to bring together a wide variety of advanced contributions located at the frontier of the discipline in order to witness multiplicity of scientific approaches, and to underline the important role the discipline has in solving practical problems encountered in urban life.

In terms of richness of approaches, urban economics is nowadays developing in two main directions: (a) a conventional urban economics direction, in which a traditional macroeconomic growth approach on one side, and a neoclassical (mostly microeconomic) location behavioural approach, on the other, develop more and more sophisticated and advanced formalised models and (b) a less conventional approach, more of an applied, mostly qualitative, nature, characterised by the undeniable advantage of grasping the tangible and intangible aspects of city growth and dynamics (i.e. elements like social capital, collective learning, environmental disease accompanying city growth, etc.), at the expense of synthetic (economic) indicators of urban structure, growth and dynamics. The present book provides advanced contributions which develop in both the directions, with the aim to present a panorama of advances in urban economics.

The book is structured around the five principles that theoretically explain the organisation of activities in space; each part of the book is devoted to one principle, and a final part to normative elements. Within each part, the first chapter is devoted to a state-of-the-art contribution on advanced theories, models and techniques developed so far in the literature, while the remaining contributions witness the tendencies that emerge in advanced urban economics, reflecting the attempt to achieve more realism in theories and to move towards dynamic conceptual and methodological frameworks (Table 1.2).

For what concerns the agglomeration principle (Part 1), the first chapter witnesses the steps forward made in conceptualising static and dynamic agglomeration economies (McCann). In this context, contributions on the idea of an efficient rather than optimal city size (Capello), on the importance of considering environmental elements in looking for equilibrium solutions (Nijkamp and Verhoef), and on dynamic approaches in the explanation of agglomeration effects, conceptualised in dynamic urbanisation economies (Camagni), demonstrate the capacity in inserting more realism in theoretical approaches.

The international literature on the accessibility principle (Part 2) contains a richness of contributions aiming to achieve more realism in the theoretical models. After the review chapters on the strategic and important linkages between land use, transportation and urban development (Button, Nijkamp and Rietveld; Lundqvist), the book presents chapters dealing with polycentric rather than monocentric cities (Hewings, Kim and Sohn), and with location choices in the presence of dual earners (Rouwental and van der Straaten). Moreover, the linkage between land use regulation and social welfare is taken into account (Sheppard).

Part 3 is devoted to the principle of spatial interaction. For this principle, the book provides a state of the art on models and techniques for the measurement of spatial interaction flows of people, from well-known gravity models to entropy maximizing and random utility maximizing models and finally to models based on neurocomputing principles that represent the most recent innovation in the design of spatial interaction models. Moreover, in this part,

Table 1.2. Structure of the book

Chapters' characteristics	Spatial principles					
	Part 1 *Agglomeration*	Part 2 *Accessibility*	Part 3 *Spatial Interaction*	Part 4 *Urban Hierarchy*	Part 5 *Competitiveness*	Part 6 *Policy Issues*
Review chapters on advances in theories/models/ techniques	Static vs. dynamic urbanisation economies (McCann)	Land use, transportation and urban development (Button, Nijkamp and Rietveld; Lundqvist)	Entropy models and neural networks (Fischer and Reggiani)	Advanced models of the central place theory (Peng)	Endogenous growth determinants (Berliant and Wang; Donaghy)	Local public finance (Friedrich and Nam)
Chapters dealing with more realism in theoretical approaches and dynamic rather than static approaches	Efficient city size rather than optimal city size (Capello) Spatial equilibrium models considering both environmental and agglomeration externalities (Nijkamp and Verhoef) Dynamic urbanisation economies (Camagni)	Polycentric urban development (Sohn, Hewings and Kim) Dual earners and housing demand (Rouwendal and Van der Straaten) Externalities in land use and social optimum in land use (Sheppard)	Different commuting costs on spatial interaction (van Ommeren) Presence of externalities like ethnic concentration and human capital formation on migration flows (de Graaff and de Groot)	Dynamic urban hierarchy models (Abdel-Rahman) Aspatial logics behind urban systems (Camagni and Capello)	Economies of scale and different degrees of knowledge diffusion and of labour and capital mobility (Bröcker) Cumulative and circular effects in urban growth (Acs) Changes in urban forms and in urban functions (Stough)	Urban quality life and public policy (Mullingan, Carruthers and Cahill) Policy issues in the urban South (Geyer) Urban policy in a global economy (Andersson, Chatterjee and Lakshmanan)

the book provides contributions which witness the attempt to insert more realism in theories and models of spatial interaction, like the presence of different commuting costs in individual's location choice behaviour (van Ommeren), and of externalities like ethnic concentration and human capital formation on migration flows (de Graaff and de Groot).

For what concerns the urban hierarchy principle (Part 4), the review chapter is devoted to advanced models of the traditional 'central place theory', of the rank size rule and urban systems and city growth (Peng). In terms of advanced theories and models, dynamics in urban hierarchy systems are presented within a general equilibrium model (Abdel-Rahman). Moreover, the book provides advanced contributions witnessing that aspatial organisational logics can explain urban systems, moving away from the traditional and conventional approaches to the study of urban systems based on Christaller and Lösch's contributions (Camagni and Capello).

In the urban competitiveness principle (Part 5), the book provides an overview of the decisive developments that have been made with regard to the understanding of endogenous determinants of urban growth we discussed in Section 1.3.2. Non-linear urban dynamic and endogenous models explain urban growth (Berliant and Wang) on the basis of skilful models of interaction between transportation costs and economies of scale in production à la Krugman (Donaghy), and between economies of scale and different degrees of knowledge diffusion and labour and capital mobility (Bröcker). Moreover, urban growth is also studied on the basis of the concept of time à la Bergson-Heidegger, interpreted as duration and a continuous process of creation, characterised by discontinuity, irreversibility, sequentiality and cumulativity on which city growth depends (Acs). Finally, changes in urbanisation and urban form towards more widespread settlements (sprawl) give rise to radical quantitative and qualitative changes in the nature of traditional urban functions which have to be taken into consideration at the normative side (Stough).

In their scientific agendas, urban scientists present a particular interest in problems that have a strong practical contour. This is witnessed by the contributions present in the last part (Part 6), which are related to urban policy issues; from local public finance issues (Friedrich and Nam), to urban environmental problems (Mulligan, Carruthers and Cahill), and North–South divides (Geyer). Last but not least, urban policy issues in the globalisation process of the economy are debated (Andersson, Chatterjee and Lakshmanan).

References

Anselin, L., A. Varga and Z. Acs (1997), "Local geographic spillovers between university research and high technology innovations", *Journal of Urban Economics*, Vol. 42, pp. 422–448.

Anselin, L., A. Varga and Z. Acs (2000), "Geographic and sectoral characteristics of academic knowledge externalities", *Papers in Regional Science*, Vol. 79(4), pp. 435–443.

Audretsch, D. and M. Feldman (1996), "R&D spillovers and the geography of innovation and production", *American Economic Review*, Vol. 86(3), pp. 630–640.

Aydalot, Ph. (ed.) (1986), *Milieux Innovateurs en Europe*, Paris: GREMI.

Ayres, R.U., K. Button and P. Nijkamp (eds.) (1999), *Global Aspects of the Environment*, Vols. 1 and 2, Cheltenham, UK: Edward Elgar.

Bailly, A. (1992), "Representation et analyse des territoires: une epistemologie de la science regionale", pp. 3–23, in: P.H. Derycke, editor, *Espace et Dynamique Territoriale*, Paris: Economica.

Bailly, H. and W. Coffey (1994), "Regional science in crisis: a plea for a more open and relevant approach", *Papers in Regional Science*, Vol. 73(1), pp. 3–14.

Batten, D.F. and D.E. Boyce (1986), "Spatial interaction, transportation, and interregional commodity flow models", pp. 357–406, in: P. Nijkamp, editor, *Handbook of Regional and Urban Economics*, Vol. I, Amsterdam: North-Holland.

Beckmann, M.J. and J. McPherson (1970), "City size distribution in a central place hierarchy: an alternative approach", *Journal of Regional Science*, Vol. 10, pp. 25–33.

Beguin, H. (1988), "La Région et les Lieux Centraux", in: C. Ponsard, editor, *Analyse Economique Spatiale*, pp. 231–275, Paris: Universitaire de France.

Bolton, R. and R.C. Jensen (1995), "Regional science and regional practice", *International Regional Science Review*, Vol. 18(2), pp. 133–145.

Camagni, R. (1991), "Local milieu, uncertainty and innovation networks: towards a new dynamic theory of economic space", pp. 121–144, in: R. Camagni, editor, *Innovation Networks: Spatial Perspectives*, London: Belhaven-Pinter.

Camagni, R. (1992), *Economia Urbana*, Rome: Nuova Italia Scientifica.

Camagni, R. (1993), "From city hierarchy to city networks: reflections about the an emerging paradigm", pp. 66–87, in: T.R. Lakshmanan and P. Nijkamp, editors, *Structure and Change in the Space Economy: Festschrift in Honor of Martin Beckmann*, Berlin: Springer.

Camagni, R. (1999), "The city as a milieu: applying the Gremi approach to urban evolution", *Révue d'Economie Régionale et Urbaine*, Vol. 3, pp. 591–606.

Camagni, R. (2003), "Interdisciplinarietà e multidisciplinarietà nelle scienze regionali", *Scienze Regionali*, Vol. 2, pp. 103–109.

Camagni, R. and R. Capello (2002), "Apprendimento collettivo, innovazione e contesto locale", pp. 11–26, in: R. Camagni and R. Capello, editors, *Apprendimento Collettivo e Competitività Territoriale*, Milano: Franco Angeli.

Capello, R. (1999), "Spatial transfer of knowledge in high-technology milieux: learning vs. collective learning processes", *Regional Studies*, Vol. 33(4), pp. 353–365.

Capello, R. (2001), "Urban innovation and collective learning: theory and evidence from five metropolitan cities in Europe", pp. 181–208, in: M.M. Fischer and J. Froehlich, editors, *Knowledge, Complexity and Innovation Systems*, Berlin: Springer.

Capello, R. and A. Faggian (2002), "An economic-ecological model of urban growth and urban externalities: empirical evidence from Italy", *Ecological Economics*, Vol. 40(2), pp. 181–198.

Choukroun, J.M. (1975), "A general framework for the development of gravity-type distribution models", *Regional Science and Urban Economics*, Vol. 5, pp. 177–202.

Costanza, R., C. Perrings and C.J. Cleveland (eds.) (1997), *The Development of Ecological Economics*, Cheltenham, UK: Edward Elgar.

Crevoisier, O. and R. Camagni (eds.) (2000), *Les Milieux Urbains: Innovation, Systèmes de Production et Ancrage*, Neuchâtel: EDES.

ESDP (1998), *European Spatial Development Perspective*, complete version, Glasgow, 8 June.

Feldman, M. (1994), *The Geography of Innovation*, Boston: Kluver.

Feldman, M. and D. Audretsch (1999), "Innovation in cities: science-based diversity, specialisation and localised competition", *European Economic Review*, Vol. 43, pp. 409–429.

Fischer, M.M. and A. Getis (1999), "New advances in spatial interaction theory", *Papers in Regional Science*, Vol. 78, p. 2, Special issue.

Fischer, M.M. and Y. Leung (eds.) (2001), *GeoComputational Modelling: Techniques and Applications*, Berlin: Springer.

Fotheringham, A.S. (1983), "A new set of spatial interaction models: the theory of competing destinations", *Environment and Planning A*, Vol. 15, pp. 15–56.

Fotheringham, A.S. and M.E. O'Kelly (1989), *Spatial Interaction Models: Formulations and Applications*, Dordrecht: Kluwer.

Fotheringham, A.S., C. Brunsdon and M. Charlton (2000), "Quantitative geography. Perspectives on spatial data analysis", in: P.A. Sag Longley, S.M. Brocks, R. McDonnell and B. MacMillan, editors, *GeoComputation: A Primer*, Chichester: Wiley.

Funck, R. (1991), "Regional science in transition", *Papers in Regional Science*, Vol. 70, pp. 1–8.

Golob, T.F. and M.J. Beckman (1971), "A utility model for travel forecasting", *Transportation Science*, Vol. 5, pp. 79–90.

Glaeser, E., H. Kallal, J. Scheikman and A. Shleifer (1992), "Growth in cities", *Journal of Political Economy*, Vol. 100, pp. 1126–1152.

Griffith, D.A. (1999), "Statistical and mathematical sources of regional science theory: map pattern analysis as an example", *Papers in Regional Science*, Vol. 78(1), pp. 21–46.

Hoyt, H. (1954), "Homer Hoyt on the development of economic base concept", *Land Economics*, May, pp. 182–187.

Isserman, A.M. (1993), "Lost in space? On the history, status and future of regional science", *Review of Regional Studies*, Vol. 23, pp. 1–50.

Isserman, A.M. (1995), "The history, status and future of regional science: an American perspective", *International Regional Science Review*, Vol. 17(3), pp. 249–296.

Jaffe, A. (1989), "Real effects of academic research", *American Economic Review*, Vol. 79, pp. 957–970.

Jaffe, A., M. Trajtenberg and R. Henderson (1993), "Geographic localisation of knowledge spillovers as evidenced by patent citations", *Quarterly Journal of Economics*, Vol. 63, pp. 577–598.

Krugman, P. (1991), *Geography and Trade*, Cambridge, MA: MIT Press.

Long, W. (1971), "Demand in space: some neglected aspects", *Papers and Proceedings of the Regional Science Association*, Vol. 27, pp. 45–62.

Longley, P.A., S.M. Brocks, R. McDonnell and B. MacMillan (eds.) (1998), *GeoComputation: A Primer*, Chichester: Wiley.

Lucas, R. (1988), "On the mechanics of economic development", *Journal of Monetary Economics*, Vol. 22, pp. 3–42.

Maillat, D., M. Quévit and L. Senn (eds.) (1993), *Réseaux d'Innovation et Milieux Innovateurs: un Pari pour le Développement Régional*, Neuchâtel: EDES.

Maskell, P. and A. Malmberg (1999), "Localised learning and industrial competitiveness", *Cambridge Journal of Economics*, Vol. 23, pp. 167–185.

Miyao, T. (1984), "Dynamic models of urban growth and decay: a survey and extensions", paper presented to the *Second World Conference of Arts and Sciences*, Rotterdam, 4–15 June.

Miyao, T. (1987a), "Dynamic urban models", pp. 877–925, in: E. Mills, editor, *Urban Economics: Handbook of Regional and Urban Economics*, Vol. 2, Amsterdam: North-Holland.

Miyao, T. (1987b), "Urban growth and dynamics", pp. 1–41, in: E. Mills, editor, *Urban Economics. Handbook of Regional and Urban Economics*, Amsterdam: North-Holland.

Niedercorn, J.H. and B.V. Bechdolt (1969), "An economic derivation of the gravity law of spatial interaction", *Journal of Regional Science*, Vol. 9, pp. 273–281.

Nijkamp, P. (1975), "Reflections on gravity and entropy models", *Regional Science and Urban Economics*, Vol. 5, pp. 203–225.

Nijkamp, P. and E. Mills (eds.) (1986), *Handbook of Regional and Urban Economics*, Amsterdam: Elsevier.

Nijkamp, P. and A. Reggiani (1999), *The Economics of Complex Spatial Systems*, Amsterdam: Elsevier.

Parr, J. (1978), "Models of the central place system: a more general approach", *Urban Studies*, Vol. 15, pp. 35–49.

Parr, J. (1981), "Temporal change in a central-place system", *Environment and Planning A*, Vol. 13, pp. 97–118.

Parr, J. (1985), "A note on the size distribution of cities over time", *Journal of Urban Economics*, Vol. 18, pp. 199–212.

Rallet, A. (1993), "Choix de proximité et processus d'innovation technologique", *Revue d'Economie Régionale et Urbaine*, Vol. 3, pp. 365–386.

Ratti, R., A. Bramanti and R. Gordon (eds.) (1997), *The Dynamics of Innovative Regions*, Ashgate: Aldershot.

RERU (1999), "Le paradigme de milieu innovateur dans l'économie contemporaine", *Revue d'Economie Régionale et Urbaine*, Vol. 3.

Richardson, H. (1972), "Optimality in city size, systems of cities and urban policy: a sceptic's view", *Urban Studies*, pp. 29–47.

Richardson, H., K. Button, P. Nijkamp (with H. Park) (eds.) (1996), *Analytical Urban Economics*, Cheltenham: Edward Elgar.

Roback, J. (1982), "Wages, rents and the quality of life", *Journal of Political Economy*, Vol. 90(6), pp. 1257–1278.

Romer, P. (1986), "Increasing returns and long-run growth", *Journal of Political Economy*, Vol. 94(5), pp. 1002–1037.

Roy, J.R. and P.F. Lesse (1981), "On appropriate microstate descriptions in entropy modelling", *Transportation Research*, Vol. 15B, pp. 85–96.

Smith, T.E. (1975), "A choice theory of spatial interaction", *Regional Science and Urban Economics*, Vol. 5, pp. 137–176.

Smith, T.E. (1978), "A cost-efficiency principle of spatial interaction behaviour", *Regional Science and Urban Economics*, Vol. 8, pp. 313–337.

Smith, T.E. (1988), "A cost-efficiency theory of dispersed network equilibria", *Environment and Planning A*, Vol. 20, pp. 231–266.

Snickars, F. and J.W. Weibull (1977), "A minimum information principle", *Regional Science and Urban Economics*, Vol. 7, pp. 137–168.

Stanners, D. and P. Bourdeau (eds.) (1995), *Europe's Environment (The Dobris Report)*, Brussels: European Environmental Agency.

van Geenhuizen, M. and P. Nijkamp (1996), "Progress in regional science: a European perspective", *International Regional Science Review*, Vol. 19(3), pp. 223–246.

WCED (1987), *Our Common Future*, New York: Oxford University Press.

Wilson, A.G. (1970), *Entropy in Urban and Regional Planning*, London: Pion.

Wilson, A.G. (2000), *Complex Spatial Systems: The Modelling Foundations of Urban and Regional Analysis*, Harlow: Pearson Education.

PART 1

Agglomeration

Urban Dynamics and Growth, CEA, vol. 266
Roberta Capello and Peter Nijkamp (editors)

CHAPTER 2

Urban Scale Economies: Statics and Dynamics

Philip McCann
Department of Economics, University of Reading, RG6 6AW, UK

Abstract

This chapter explores the various theoretical and empirical approaches to understanding the static and dynamic aspects of the urban economy. The early work on the subject is related to more recent models in order to clarify the ways in which the analysis of these issues has progressed. An important finding here, is that when we compare the traditional analysis with the newer research, the empirical evaluation of the dynamics of the urban economy is a much more complex and subtle issue than it initially appears. The reason for this is that there are many different interpretations of agglomeration and industrial clustering. Therefore, it is necessary for us to consider the types of inter-firm and inter-agent transactions which take place within cities, and to specify exactly how such linkages can be captured empirically.

Keywords: agglomeration, clusters, externalities, transactions costs, information, innovation

JEL classifications: R120, R340, R380

2.1. Introduction

Cities have played a central role in the last 4000 years of human history. The cultural, political and anthropological roles played by cities are well documented by urban historians and urban geographers, whereas the economic role played by cities within the process of human wealth generation is rather less well understood by historians (Hall, 1998). Ironically, the economic role played by cities has also been much less well understood by economists than other aspects of the economy. Until the works of Porter (1990) and Krugman (1991) focussed attention on the role played by geography, cities and clusters in the determination of national and international trade patterns and growth, most economists outside the regional science tradition were largely unaware of the importance of these issues. Over the last decade, however, the situation has changed dramatically. Researchers are now queuing up to demonstrate the essential role played by cities in fostering localised increasing returns to scale within national growth processes. In part, this new interest may be because of the availability of new analytical techniques (Krugman, 1998). It may also be simply because of changes in academic fashions and debates (Martin, 1999). Whatever the reasons for this growth in interest, these are welcome developments from the perspective of urban and regional economists, who inherently see cities as one of the central economic phenomena.

There are two reasons why urban and regional economists perceive cities to be so important. Firstly, a generally observed phenomenon is that most industrial and commercial activities tend to be clustered together in space. These clusters may take the form of industrial parks, small towns or major cities, but the general observation appears to be true. This observation, therefore, raises the obvious question of why it is that activities are generally grouped together geographically and what role this clustering plays in economic growth. Secondly, in most countries there is generally observed to be a size and activity distribution of cities (Gabaix, 1999a,b), with different ranges of activities taking place in different centres. In particular, within an individual country or market area, there will usually be a single largest city cluster which exhibits almost all the types of industrial and commercial activities, followed by larger numbers of other smaller clusters which increase in number as their individual size falls, and which exhibit smaller individual ranges of activities. What role these scale and diversity distributions play in economic development are also of central importance to urban and regional economists.

Apart from the centrality of roles played by cities, and city-size or city-activity distributions, in economic development, a further key issue for urban and regional economists is the dynamic aspects of these roles. In other words, how do changes in these phenomena impact on the economy as a whole, and how rapid are any such changes in the urban economic system?

This chapter will provide an overview of our current understanding of the economic nature of the city, the major features of the urban hierarchy, and the key issues underlying the dynamic aspects of the urban system. In order to discuss these various issues, the chapter is organised as follows. In Section 2.2 we will review the various traditional approaches to discuss the economic aspects of cities and industrial clustering. This discussion will focus initially on our present understanding of the nature of agglomeration economies. As we will see, however, identifying agglomeration economies in practice is extremely difficult, not least because this is a rather static concept, but also because analytically there are so many alternative interpretations of the nature of industrial clustering. Therefore, in Section 2.3 we will provide a brief discussion of alternative dynamic descriptions of industrial clustering which have often been employed in the literature. As we will see, these descriptions focus on the nature of information generation, innovation and firm foundation, and attempt to relate these issues to city economies. In Section 2.4 we will employ a transactions-costs framework in order to identify the essential spatial–industrial–organisational features underlying our various descriptions of cities and industrial clusters. In Section 2.5 we will review the latest evidence on spatial transactions costs, information spillovers and innovation, in order to determine whether the role played by cities is becoming more or less important over time.

2.2. *Cities, returns to scale and agglomeration*

In attempting to explain the observation that economic activities and people are generally clustered together in space, it is necessary to employ the notion that economies of scale can be place-specific. This assumption is essential, because the clustering of activities in space increases competition for land. The resulting increases in local land and labour prices[1] will increase the costs experienced by firms, and reduce their profitability ceteris paribus, unless there are some more than compensating benefits associated with clustering. Such location-specific economies of scale are generally known as agglomeration economies. It was Alfred Marshall who is generally credited[2] with first providing a detailed description of the sources of these economies. In Marshall's (1920) discussion these economies are generally understood to be external economies, which are independent of a single firm, but which accrue to all of the firms located in the same area. Marshall provided three reasons why such localised economies of scale might exist, namely information spillovers, local non-traded inputs produced under scale economies, and a local skilled labour pool.

Firstly, if many firms in the same industry are grouped together in the same location, it implies that the employees of any one particular firm have relatively easy access to employees from other local firms. This easy access allows for frequent direct informal face-to-face contact between individuals, which may allow for tacit information to be shared between the individuals. Tacit information is information which is incomplete and which is shared on a non-market basis. This process of the mutual informal trading of information allows each of the market participants to build up a more coherent picture of the overall market environment, thereby improving their ability to compete in the market. The more such participants in the local area, the more complete a picture can be assembled by each participant. The advantage of spatial clustering in this case is, therefore, that proximity maximises the mutual accessibility of all individuals within the cluster, thereby improving the information spillovers available to all local participants. In market environments characterised by rapidly changing information, such clustering affords the clustered and co-located firms an information advantage relative to all other firms. The extent of this advantage would, therefore, appear to

[1] Increases in nominal local labour prices are required in order to maintain real wages.
[2] In particular by Krugman (1991). However, the language and nature of Marshall's discussion suggests that he was almost certainly aware of the 1851 British parliamentary inquiry into the Expansion of the Railway System to Manchester. This documents in detail the advantages of 'manufacturing districts' in a manner which is strikingly similar to Marshall's textbook description.

depend directly on the number of such firms which are located in the same area. Secondly, at the same time, in situations where many firms in the same industry are grouped together in the same area, there will also be possibilities for certain specialist inputs to be provided to the group, in a more efficient manner than would be the case if all of the firms were dispersed. In many cases, the costs of providing such services would normally be prohibitive to most market participants. However, the fact that a large number of potential customer firms within the same sector are also located at the same place allows the costs to be spread across the group, and the costs of the input to each firm within the group will fall as more firms join the group. Thirdly, the spatial grouping of firms allows for the creation of a local specialised labour pool. This allows all firms within the group to reduce their labour acquisition costs, and there are two aspects to these cost reductions. Firstly, a local skill-base reduces the general hiring and re-training costs incurred by the local firm which are associated with any job changes within the group. Labour can move between firms relatively easily and the firms are also able to employ new workers relatively easily. Moreover, this ease of local labour adjustment also acts as a risk reduction mechanism on the part of both the firm and the labour in the face of demand fluctuations, provided these fluctuations are firm specific.

While these three sources of agglomeration economies can be distinguished, at least in principle, observing them in reality can be much more difficult. In order to do this, urban and regional economists generally employ a classification which was first employed by Ohlin (1933) and Hoover (1937, 1948). This classification splits agglomeration economies into three types, namely internal returns to scale, localisation economies and urbanisation economies. The first Ohlin–Hoover category of agglomeration of internal returns to scale does not entirely concur with the Marshallian notion of agglomeration. However, the fact that some firms achieve significant location-specific economies of scale means that large levels of both capital investment and people are concentrated at one particular location, rather than across a range of different locations. This can then contribute to the development of the other two forms of agglomeration economies. If the ensuing agglomeration benefits accrue to activities in the same sector located at the same place these are termed localisation economies, whereas if the benefits accrue to a range of local sectors, they are termed urbanisation economies. In the case of localisation economies, the mechanisms by which such externality benefits are generated correspond largely to Marshall's tripartite description. These Marshallian externalities are, therefore, generally associated with the development of industrial clusters of particular sectors. In the case of urbanisation economies, however, the externality benefits accrue to firms across different sectors (Jacobs, 1960) located in the same place. These various activities, although not directly related to the sector experiencing internal returns to scale and localisation economies, will still cluster in the local economy in order to provide services for the firms and employees of this sector. It is these overall urbanisation economies which are normally regarded as the typical agglomeration benefits associated with major cities (Duranton and Puga, 2000), and these are generally assumed to comprise a plethora of lower order localisation economies (Lichtenberg, 1960). In these situations, the economy of the city as a whole will tend to dominate that of any particular sector. In small cities, on the other hand, it may be that particular localisation economies tend to dominate. In these cases, a small number of individual sectors will tend to dominate the city, rather than the city dominating the sectors (Duranton and Puga, 2000).

New economic geography models (Fujita *et al.*, 1999) give a central role to the idea of diversity, rather than specialisation, in promoting local growth. In particular, it is argued that a greater diversity of local products provides increased welfare gains via both the consumption process of final consumers and also goods producers. However, capturing these effects empirically is rather complex. Although there is evidence concerning the relative importance of localisation vs. urbanisation economies which appears to provide limited support to the new economic geography argument, at least as far as employment growth is concerned (Glaeser *et al.*, 1992), the evidence is still far from conclusive (Henderson *et al.*, 1995). Moreover, the balance between specialisation and diversity appear to have no effect on real wage growth (Glaeser *et al.*, 1992) or on labour productivity (Henderson, 2003).[3]

[3] The empirical literature is surveyed in the forthcoming (2004) North-Holland *Handbook of Regional and Urban Economics Vol IV*.

It is important to note here that in terms of the mechanisms by which agglomeration externalities are mediated within cities, each of Marshall's sources of agglomeration can contribute to the development of either localisation economies or urbanisation economies. As such, the distinction between the Ohlin–Hoover agglomeration categories initially appears to be primarily related to the definitional boundaries of the firms and sectors in which the externality benefits accrue locally. As we will see in the following sections, however, it can also be argued that the various Ohlin–Hoover agglomeration categories do exhibit some differences in the transmission mechanisms by which the agglomeration benefits are generated.

The discussion so far has focussed on a rather static notion of cities, in that cities are perceived to represent location-specific economies scale. While these assumptions are useful from analytical point of view, and in particular when used within in new economic geography frameworks, it is also possible, however, to introduce dynamic elements into the discussion of cities. This is achieved by incorporating notions of learning and the evolution in information within the standard spillovers-type framework. In order to do this, it is necessary to consider the role played by cities in the process of information generation and firm creation.

2.3. Cities, innovation and firm creation

Within the academic literature, the clarity of the tripartite Marshallian sources and Ohlin–Hoover classifications of agglomeration economies are somewhat complicated by a range of alternative notions of the ways in which cities can generate economies of scale and growth. In particular, although the simple Marshallian description allows for information spillovers, no discussion of the types of information or the role of the information spillovers is provided. In response to this, a series of models have been developed which attempt to relate the role of the city to the generation of particular types of information and information spillovers which have very specific functions. In particular, these models concern the local generation within the city of information related to new ideas and techniques, under the broad heading of innovation. There are five principal models relating to these issues:

the growth pole model, the incubator model, the product cycle model, the clusters model and the new industrial areas model. The first three models are the traditional models which emerged during the 1960s, whereas the latter two models reflect rather more recent developments.

The growth pole model of Perroux (1950) employed some of the ideas of Schumpeter (1939), and was extended to geographical space by Boudeville (1966). In this framework economic relationships are assumed to exhibit certain polarities regarding financial transactions via particular network structures. In other words, the decisions made by key large firms have major financial implications for other firms which are linked to the key firm through particular structures of customer–supplier relationships. As such, the innovation behaviour of a firm alone may not be of such significance, but rather when it is combined with large firm size it may engender significant local growth effects. Meanwhile, at the other end of the spectrum is the incubator model of Chinitz (1961, 1964). The argument of Chinitz is that cities which are highly diversified, and which contain a range of different types of industries and firm sizes, will act as superior 'incubators' for the development and growth of new and small firms. The reason for this is that in such an environment, there will be a variety of local business services available to these small firms which will facilitate their growth. On the other hand, in cities dominated primarily by large firms, many of these requisite services will not be available, because the large firms will be able to supply such activities internally. Once again, issues of firm ownership and structure appear to play an independent role in affecting the growth of the city. Recent work on so-called nursery cities (Duranton and Puga, 2001) employs these arguments, and suggests that the size distribution of firms within the city may, therefore, be important for promoting growth via innovation.

An alternative approach is to assume that firm size, size distribution and organisational boundaries are not central features determining the innovation behaviour of geographical areas, but rather the structure of the urban system is itself what is essential. The major model adopting this approach is that of the product cycle model (Vernon, 1960, 1966). Vernon's argument is that firms will separate activities by location according to the stage in the life cycle of the product. In dominant central cities, firms will tend to locate information intensive activities such as R&D and high-level decision-making, all of which relate to the early stages of the product life cycle. On the other hand, more geographically peripheral areas exhibiting lower labour costs and skills, will tend to have plants producing more mature, less novel, and more standardised products. The result is that there will tend to be a clear separation of activity types between central city–regions and more peripheral areas. The important observation of the product-cycle model is that there may, therefore, emerge a qualitative distinction between the types of activities which take place at the economic centre or at the economic periphery of any geographical area.

During the last 15 years, two other approaches have been assimilated into the traditional city-innovation literature which had previously been dominated by the three approaches described above. These two new models are the clusters model of Porter (1990) and the new industrial spaces model (Scott, 1988), both of which draw heavily on observations of the development of particular types of industrial areas over the last two decades. In terms of the cluster model, Porter's (1990) argument is that proximity engenders mutual visibility between competitors. In others words, firms are able to observe the competitive developments of each other, and this visibility itself acts as a spur to all firms to continue to improve their own individual competitiveness. The result of this process of localised competition is that the competitiveness of the cluster as a whole is increased. Porter argues that spatial clustering is particularly important in the case of small firms which rely mainly on external sources of information and technology.

As well as the Porter clusters model an alternative set of relationships between spatial industrial concentration and innovation is also provided by the literature on new industrial areas (Scott, 1988). The growth of regions dominated by large numbers of small firms such as Silicon Valley and the Emilia-Romagna region of Italy (Scott, 1988), appears to be related crucially to innovation. As such, these observations have led to suggestions from many observers that industries which are made up of spatial clusters of small firms, tend to be more highly innovative than industries comprised mainly of large firms (Saxenian, 1994), because such environments provide the appropriate 'milieux' for such innovations to take place (Aydalot and Keeble, 1988). As well as Marshall's classic sources of the agglomeration economies, it is also argued that in these types of regions, networks of important business relationships operate between local decision-makers thereby allowing for mutual risk-taking in collaborative and cooperative activities.

These various dynamic models of innovation, learning and cities are often regarded as being complementary to the standard static descriptions of agglomeration discussed in Section 2.2. However, the exact ways in which the city-innovation models discussed here relate to the city-agglomeration models discussed above are still rather unclear. There are two reasons for this. Firstly, the empirical literature on innovation (Acs, 2002) tends to focus primarily on manufacturing innovations, particularly as measured by patent citations (Jaffe *et al.*, 1993), with very little evidence being available concerning the service industries. This is an area for future research. Secondly, the analytical issues involved in relating cities and clusters to innovation are rather more complex than at first they might appear. This is because each of these various models is predicated on a series of assumptions as to the nature of the inter-firm relations, firm structures and inter-firm competition which exists within the city. These are the issues to which we now turn.

2.4. Clusters, firm types and the nature of transactions

From the arguments outlined in Sections 2.2 and 2.3 we see that all geographical clusterings of industrial activities, whether within a single city or spread across a metropolitan city–region, are maintained in the face of rising local factor prices, by the existence of crucial inter-firm linkages and transactions. Engaging in these local inter-firm transactions and linkages implies transactions costs are incurred by the locally clustered participating firms. For our purposes, it is important to examine the nature of these transactions and the local inter-firm relations, because this may help us to understand the relationship between the static and dynamic conceptions of city agglomeration economies.

In order to understand these relations, it is necessary for us to focus on the characteristics of firms which exist in the cluster, and the transactions which take place within the cluster. Adopting this approach, we see from the literature that there are three broad typologies of spatial industrial clusters, as defined in terms of the features they exhibit (Gordon and McCann, 2000). These are the *pure agglomeration*, the *industrial complex*, and the *social network*. The key feature which distinguishes each of these different ideal types of spatial industrial cluster is the nature of the relations between the firms within the cluster. The characteristics of each of the cluster types are listed in Table 2.1, and as we see, the three ideal types of clusters are all quite different. In reality, all cities and spatial clusters will contain characteristics of one or more of these ideal types, although one type will tend to be dominant in each city or cluster.

In the model of pure agglomeration, inter-firm relations are inherently transient. Firms are essentially atomistic, in the sense of having no market power, and they will continuously change their relations with other firms and customers in response to market arbitrage opportunities, thereby leading to intense local competition. As such, there is no loyalty between firms, nor are any particular long-term relations. The external benefits of clustering accrue to all local firms simply by reason of their local presence, the price of which is the local real estate market rent. There are no free riders,

Table 2.1. Industrial clusters

Characteristics	Pure Agglomeration	Industrial Complex	Social Network
Firm size	Atomistic	Some firms are large	Variable
Characteristics of relations	Non-identifiable Fragmented Unstable	Identifiable Stable trading	Trust Loyalty Joint ventures Joint lobbying Non-opportunistic
Membership	Open	Closed	Partially open
Access to cluster	Rental payments Location necessary	Internal investment Location necessary	History Experience Location necessary but not sufficient
Space outcomes	Rent appreciation	No effect on rents	Partial rental Capitalisation
Notion of space	Urban	Local but not urban	Local but not urban
Example of cluste	Competitive urban economy	Steel or chemicals Production Complex	New industrial areas
Analytical approaches	Models of pure agglomeration	Location–production theory Input–output analysis	Social network theory (Granovetter)

access to the cluster is open, and consequently it is the growth in the local real estate rents which is the indicator of the cluster's performance. This idealised type is best represented by the pure Marshall model, and the localisation and urbanisation economies of Hoover. The industrial structure represented by these models is that of perfect or monopolistic competition, and the notion of space is essentially urban space. In other words, this type of clustering only exists within individual cities. The empirical verification of this pure agglomeration phenomenon relies on evidence of localised productivity growth, associated with growth in local real estate prices and real wages.

The industrial complex is characterised primarily by long-term stable and predictable relations between the firms in the cluster. This type of cluster is the type of spatial cluster typically discussed by classical (Weber, 1909) and neo-classical (Moses, 1958) location–production models, representing a fusion of locational analysis with input–output analysis. Component firms within the spatial grouping each undertake significant long-term investments, particularly in terms of physical capital and local real estate, in order to become part of the grouping. Access to the group is, therefore, severely restricted both by high entry and exit costs, and the rationale for spatial clustering in these types of industries is that proximity is required primarily in order to minimise inter-firm transport transactions costs. Rental appreciation is not a feature of the cluster, because the land which has already been purchased by the firms is not for sale. This ideal type of cluster more closely reflects the internal returns to scale and localisation arguments of Hoover as well as aspects of the growth pole model of Perroux. The literature on these types of inter-form relations has typically focussed on traditional commodities-based sectors (Isard and Kuenne, 1953) such as steel, chemical and pharmaceuticals. However, it is also possible to argue that this notion of inter-firm relations is equally applicable to certain high-technology sectors (McCann, 1997) and even some elements of the financial services industry such as international merchant banking. The key issue here is that the industrial structure within the complex demonstrates oligopoly (or occasionally monopoly) characteristics. Within this generally oligopolistic framework the notion of space in the industrial complex model is local, but not necessarily urban. In other words, these types of local industrial clusterings can exist either within, outside, or beyond the limits of an individual city. The empirical verification of this industrial complex phenomenon relies on evidence of long-term and stable inter-firm transactions (McCann, 1997).

The third type of spatial industrial cluster is the social network model. This is associated primarily with the work of Granovetter (1973, 1985, 1991, 1992), and is a response to the markets and hierarchies model of Williamson (1975). Whereas the pure agglomeration model and the industrial complex described above represent the clustered spatial equivalents of the market and the hierarchy alternative modes of coordination, the social network model argues that mutual trust relations between key decision-making agents in different organisations may be at least as important as decision-making hierarchies within individual organisations. These trust relations will be manifested by a variety of features, such as joint lobbying, joint ventures, informal alliances, and reciprocal arrangements regarding trading relationships. However, the key feature of such trust relations is an absence of opportunism in that individual firms will not fear reprisals after any reorganisation of inter-firm relations. Inter-firm cooperative relations may, therefore, differ significantly from the organisational boundaries associated with individual firms, and these relations may be continually reconstituted. All of these behavioural features rely on a common culture of mutual trust, the development of which depends largely on a shared history and experience of the decision-making agents. This social network model is essentially aspatial, but from the point of view of geography, it can be argued that spatial proximity will tend to foster such trust relations, thereby leading to a local business environment of confidence, risk-taking and cooperation. Spatial proximity is necessary but not sufficient to acquire access to the network. As such, membership of the network is only partially open, in that local rental payments will not guarantee access, although they will improve the chances of access. The social network model, therefore, contains elements of both the Porter model (1990, 1998) and the new industrial areas model (Scott, 1988), and has been employed to describe the characteristics and performance of areas such as Silicon Valley and the Emilia-Romagna area of Italy. In this model space is once again local, but not necessarily urban, and may extend across a broader definition of a city–region. In terms of the industrial structure, the literature on these new industrial areas generally assumes that the area is dominated by many small firms (Saxenian, 1994) rather than large hierarchically organised firms. As such, as with the model of pure agglomeration, the new industrial areas model corresponds most closely to a model of perfect or monopolistic competition. The empirical verification of this pure agglomeration phenomenon relies on evidence of collaborative behaviour such as joint lobbying (Bennett, 1998).

In reality all spatial concentrations of economic activity will exhibit at least one of these cluster types, and in the case of large cities, will exhibit aspects of all three models. However, the important point is to identify exactly what are the dominant transactions-costs features of the industrial cluster in question, by assessing the stability, longevity and loyalty of inter-firm transactions and relations. Only by doing this can we clearly identify the economic rationale for the existence of each cluster.

All three of these industrial clustering types exhibit economies of scale which can compensate for either local factor price appreciation or for the costs involved in the overcoming of geographical space. However, if we assume that one of the major functions of cities is not only to reduce the costs of spatial inter-firm transactions but also to generate new ideas and new firms, then each of these three possible models of city-clustering should also be associated empirically with increased levels of innovation and new firm foundation, relative to other locations.

On this point, it is relatively easy to reconcile the model of pure agglomeration model with local innovation tendencies by including elements of the Porter, Chinitz and Vernon models in the pure agglomeration model. In order to do this we simply assume that the information spillovers which are assumed to operate within the city are also assumed to be spillovers of new information. The apparent ease with which dynamic innovation behaviour and static city-agglomeration models can be reconciled in theory means that in the literature these different concepts are regularly treated as being synonymous. Similarly, one of the major features of the new industrial areas model is that industrial clustering is perceived to be beneficial because firms are assumed to be willing to undertake joint highly innovative and risky activities without the fear of opportunistic behaviour[4] on the part of other local firms. Once again, as with the model of pure agglomeration, in the new industrial areas model, industrial clustering and innovation are generally perceived to be synonymous (Saxenian, 1994; Castells and Hall, 1994). In the case of the industrial complex, however, the theoretical situation is rather more complex, in that it is not so easy to reconcile the location behaviour of these types of industrial organisational arrangements to cities alone. Nor is it so obvious why such large hierarchically organised firms would benefit from informal information spillovers, in a manner analogous to the pure agglomeration or the social network (Arita and McCann, 2002a,b; McCann *et al.*, 2002). As such, in the case of the many sectors and firm-types which do not approximate to being either perfectly or monopolistically competitive, the straightforward link between innovation, clustering and cities which is assumed by naïve interpretations of the agglomeration models, is far from obvious. In Section 2.5 we will, therefore, review some of the empirical evidence regarding spatial transactions costs, information spillovers and innovation, in order to assess whether cities and urban economies of scale really are becoming more important than previously.

2.5. *The empirics of cities*

In order to identify the links between cities and dynamic growth, there are four major issues on which empirical evidence may be able to throw some light. These four issues are; firstly, the role of specialisation, secondly, the link between cities and innovation, thirdly, the critical distances over which city agglomeration externalities operate, and fourthly, changes in the nature and levels of spatial transactions costs.

On the first point, it is a well-known observation that larger cities are generally more industrially diversified than smaller cities which tend to be more specialised (Duranton and Puga, 2000). This observation would tend to suggest that urbanisation economies tend to dominate in larger cities and localisation economies tend to

---

[4] In these models, trust relations are assumed to be enforced because any individuals entering into opportunistic behaviour will be simply expelled from the social network. However, how such an enforcement mechanism is related to clustering is not clear.

dominate in smaller cities. While this is the generally held view, from an empirical point of view, however, distinguishing between urbanisation economies and localisation economies can be notoriously difficult (Glaeser *et al.*, 1992; Henderson *et al.*, 1995), because many cities will contain all three types of industrial clusters described above both within and across a range of local sectors. Moreover, the distinction between the three Ohlin–Hoover agglomeration classifications is rather arbitrary in may cases, given that mergers and acquisitions means that firms are frequently changing ownership and sectors without necessarily changing either their locations or the nature of many of their transactions. This latter issue is particularly pertinent in the case of many service industries (Cohen, 1998). Therefore, while we generally accept that aggregate urbanisation economies tend to dominate in larger cities, we must still remember that large cities contain many disaggregated localisation economies.

On the second point, in terms of the link between innovation and cities, there is much evidence to suggest that this link can be strong in certain sectors (Acs, 2002). Yet, the evidence on these issues is also not always conclusive. Cities do not always appear to be centres of innovation, and nor does innovation necessarily appear to be centred on cities (Simmie and Sennett, 1999; Simmie, 2002). Part of the reason for this is that a large proportion of patents and innovations actually take place within multinational oligopoly industries, whose spatial industrial-complex type clustering and locational behaviour may or may not be related to the assumed benefits of cities, as described in the pure agglomeration model. Indeed, here is strong evidence to suggest that that many highly innovative oligopoly firms do not locate in cities in order to avoid any outward information spillovers (Simmie, 1998; Cantwell and Kosmopoulou, 2002). The implication here is that the assumed link between Marshallian inter-firm information spillovers and innovation in cities is probably most closely associated with city-sectors dominated by many small firms, whereas cities with larger firms will tend to benefit more from the Marshallian labour pooling arguments rather than from any information–innovation links. Therefore, whil[e] we generally accept that innovation and spatial concentration tend t[o] be correlated, we must still remember that industry structure an[d] ownership independently plays a very significant role in the generation of innovation (Cantwell and Iammarino, 2003).

On the third point, in the models of pure agglomeration, the critical distances over which agglomeration externalities are assumed to operate is that of the individual city-metropolitan area. In other words, locations outside this area are assumed not to benefit from any city-externalities, whereas all areas within this spatial boundary are assumed to benefit. On the other hand, in both the industrial complex and new industrial areas models, the critical spatial dimension is never explicitly mentioned. In both the cases, it is assumed that clustering-externality benefits may operate over a rather larger scale than that of a single city, and may include hinterland areas. In neither case, however, are we assuming that such spillovers operate at the national level. While there is much evidence to suggest that in many cases the critical distance over which urban agglomeration externalities operate may be that of the city-metropolitan area (Gordon and McCann, 2000), there is also much evidence to suggest that for many firm-types and industries, the critical distances over which urban agglomeration externalities operate may be very much larger than that of the city–region (Cantwell and Iammarino, 2000, 2003; Caniels, 2000). Moreover, this appears to be particularly so in many of the high-technology and information-intensive sectors which are assumed to benefit from urban clustering (Simmie, 1998; Suarez-Villa and Walrod, 1997; Arita and McCann, 2000). As such, information spillovers would appear to be localised within the urban setting for small firms, whereas for large firms such arguments would appear to be much less relevant.

On the fourth issue concerning the changes in modern spatial transactions costs, the evidence is rather difficult to interpret, because spatial transactions costs are of two major types, namely spatial information costs and transportation costs.

In the case of spatial information costs, since the 1980s we have seen dramatic improvements in the ability of decision-makers and planners coordinate activities across space. The primary reasons for these improvements have been the enormous technological developments in information and communications technology (ICT), and also the advent of widespread usage of these technologies. These developments have meant that complex operations can now be managed both more efficiently and effectively than was previously possible. There are two aspects to these developments. Firstly, the new information technologies have reduced the real costs of communicating across distance, allowing us to more efficiently control existing spatial arrangements of activities (The Economist, 1999a). This is a common observation in industrial sectors and activities where physical commodities are being moved across large distances, such as in the management of international importing and exporting supply chains (Financial Times, 1999) or the coordination of multinational manufacturing activities (The Economist, 1999b). Analogous arguments also exist for the case of the service sectors, in situations where information rather than physical goods is being transferred across space. In many situations, information technologies employing satellite and fibre-optical technology allow for greater quantities of information to be transmitted at a much lower costs than was previously possible. Secondly, the existence of these new information technologies also allows decision-makers to undertake the coordination of spatial arrangements of activities which were previously not possible. Meanwhile, for service industries such as finance and marketing, the new possibilities provided by information technologies for the supply of information-based services across global space appear almost unlimited (The Economist, 1999b). On the other hand, however, there are some other arguments which suggest that over time the development of these information technologies is actually leading to increases in the costs of transmitting information across space, thereby increasing the relative importance of geographical centrality. The argument here is that an increase in the quantity, variety and complexity of information produced itself increases the costs associated with transmitting this information across space. This is because much of the information will be of a non-standardised tacit nature, and the transmission of this type of information essentially requires face-to-face contact. The opportunity costs involved in not having face-to-face contact will consequently increase with the quantity, variety and complexity of the information produced. The effects of this will be to increase the costs of doing business across large geographical distances.

In the case of transportation costs, the picture is also rather complex. Transportation technologies have improved dramatically over recent years. Obvious examples of this include the growth in roll-on roll-off trucking, containerization, rapid-turnaround shipping, and the increased efficiency and frequency of airline services. On the other hand, the quantity, variety and complexity of market information generated in the modern economy is increasing. This also implies that in many industries which involve the production or shipping of goods across space, the variety and complexity of the logistics operations being undertaken will also increase. The reason for this is that as modern consumer demand requirements become more sophisticated, there is an increasing preference for goods shipments characterised by speed, reliability and timeliness. The most extreme example of this trend towards more frequent shipments, is the application of Just-In-Time (JIT) manufacturing and distribution techniques, the influence of which has pervaded all areas of modern production, distribution and retailing (Nishiguchi, 1994; Schonberger, 1996). Yet, these technological developments have also led to a change in consumer behaviour. Both household and industrial consumers now expect goods to be delivered JIT. As such, the nature of demand for transactions across space has changed dramatically. Customers require much shorter lead-times than was previously possible. As such, there is a direct parallel with the argument regarding information costs, only in this case, the opportunity (time) costs of goods shipments are tied up in the levels of inventory being held, rather than the opportunity (time) costs of not having face-to-face contact.

There is a range of empirical evidence which suggests that the spatial transaction costs involved in shipping of goods have indeed increased over the last two decades because of this demand for more frequent deliveries. Firstly, the average inventory levels for almost all manufacturing and distribution sectors in the developed world have fallen dramatically since the 1980s, relative to the value of output (Schonberger, 1996; Financial Times, 1998). This implies that the average lead times of goods-shipments have fallen over recent years, with a concomitant increase in goods-shipment frequencies. Secondly, by carefully disentangling the various components of transport costs it becomes clear that the proportion of global output which is accounted for by logistics and transportation activities in the economy has not fallen over recent decades (Financial Times, 1997; Hummels, 1999). Thirdly, while the transportation cost component of bulk materials has indeed generally fallen, in the case of manufactured goods, there is evidence that this proportion has actually increased over the recent decades, in spite of the improvement in transportation and logistics technologies (Hummels, 1999). Fourthly, industries which are very dependent on JIT shipments have tended to reorganise their trade patterns in favour of geographically close suppliers and customers (Reid, 1995; McCann, 1998). Moreover, this behaviour is even evident in industries in which the product value–weight ratios are extremely high (McCann and Fingleton, 1996). In other words, such localisation behaviour is present in the very industries which many economic geographers and traditional trade theorists would have ruled out.

In addition to observations of changes in spatial transactions costs described above, there are also two other sources of evidence which support the argument that spatial information transactions costs have increased over recent decades, thereby increasing the importance of the urban area as the potential source of economies of scale. The first source of evidence comes from observations of telephone usage patterns (Gaspar and Glaeser, 1998). Using data from Japan and the US, they observe the relationship between the density and frequency of telephone usage and the location of the users. Firstly, they find that users who are geographically closer together, and for whom greater face-to-face contact is therefore easier, spend more time talking to each other on the telephone, than do users who are at greater distances from each other. Secondly, the same result also holds for urban size, in that users in larger urban areas talk to each other relatively more frequently than users in smaller urban centres. Thirdly, the frequency of airline business travel has increased more or less in line with the growth in telecommunications usage. While this evidence of the relationship between face-to-face contact, ICT usage and proximity, is only in terms of observed correlation and not causation, it is certainly supportive of spatial interaction arguments based on the importance of information transfer via face-to-face communication.

Meanwhile, the second source of evidence suggesting that the individual urban area has become progressively more important as a source of economies of scale involves an assessment of the rates of global urbanisation. Over the last three decades, the proportion of people living in urban areas has increased in all parts of both the developed and developing world (United Nations, 1997). While the reasons for this are complex, and particularly in relation to the out-migration of labour from rural areas in developing economies, the ubiquitous urbanisation phenomenon in the developed parts of the world where information technologies are mostly applied, also suggests that the geographical proximity of firms and people within individual urban areas is becoming relatively more important over time.

These apparently conflicting pieces of evidence which suggest that in some cases spatial transactions costs have fallen over time and in other cases have increased over time can be reconciled in that the decreasing transportation and information costs have increased income, profits and growth and these increases in turn have changed all kinds of transportation demands. These in turn have then had an impact on spatial interaction costs, such as a higher demand for more frequent deliveries. However, different types of changes in transactions costs have tended to take place in different types of sectors and activities. The sectors in which spatial transactions costs have fallen significantly over recent decades, appear to be the sectors in which the nature of the spatial transactions undertaken have not changed fundamentally over time, in terms of the required frequency of interaction. This is typically the case in many raw material, agricultural or extraction industries, and in industries producing manufactured products at a mature stage within their product cycles (Vernon, 1966). This is also the case in service sector industries in which the nature of the information being transacted is rather standardised, such as retail banking. On the other hand, in production sectors in which the demand lead-times have fallen dramatically, or in industries in which the variety and complexity of information generated has increased significantly, spatial transactions costs would appear not to have fallen over recent decades, and in some cases will actually have increased. In these cases, the requirement for geographic proximity would appear to have increased.

Glaeser (1998) argues that taking a broad view of all the empirical evidence indicates that the aggregate share of total output accounted for by transportation costs has fallen markedly over time. These reduced costs of doing business across geographical space would alone appear to imply that the range of activities supplied across all spatial areas will tend to converge, thereby obviating the role of cities as centres of economic growth, unless spatial information costs have increased over recent decades. Yet, the evidence on changes in the nature of spatial information costs (Gaspar and Glaeser, 1998) and delivery modes (McCann and Fingleton, 1996; Hummels, 1999) suggest that while the overall effect is indeed that transportation costs in aggregate have fallen as a proportion of global output, at the same time, the relative differences in the importance of proximity for different sectors appears to have increased. For a majority of sectors and activities spatial transactions costs have fallen, while for some sectors they have actually increased. For certain types of activities, cities appear to have become even more important as centres of production.

2.6. Conclusions

The implication of these theoretical and empirical observations is that the individual urban industrial area is, if anything, therefore becoming even more important nowadays as a determinant of domestic scale economies than it was previously. The reason for this is that while spatial transactions costs are generally decreasing, the opportunity costs associated with spatial interaction are actually increasing for many activities and sectors. This is because information and communications technologies and face-to-face contact, are not necessarily substitutes for each other, but are often complements for each other. In other words, a general increased usage of information and communications technologies often leads to an increase in the quantity, variety and complexity produced, which itself leads to an increase in spatial information transactions costs, and an associated increased need for spatial proximity to facilitate face-to-face contact. At the same time, an increase in the levels of spatial proximity encourages a greater usage of information and communications technologies, and the production of more varied and complex information, such that the process becomes cumulative. Glaeser's arguments (Glaeser, 1998; Gaspar and Glaeser, 1998), therefore, suggest that in the modern world, the Marshallian foundations of agglomeration externalities may be becoming an ever-more significant determinant of domestic economies of scale for many types of firms and activities. This would appear to be particularly relevant for sectors dominated by small firms. On the other hand, in sectors dominated by oligopolistic structures, many of the information spillover arguments in favour of cities and agglomerations appear rather weak. This is because the organisation boundaries of the firms are designed specifically to internalise information and to avoid information spillovers. In these cases, where such firms are observed to be clustered together in cities, it would appear that the reasons for this behaviour are rather more connected with either labour pooling arguments or local non-traded inputs, rather than informal information spillovers. As such, when considering the relationship between clustering and agglomeration, it is necessary to consider specifically the nature of the local industrial structure, as this will provide clues as to the nature of the local transactions and inter-firm relationships and the rationale for such clustering behaviour.

References

Acs, Z. (2002), *Innovation and the Growth of Cities*, Cheltenham: Edward Elgar.

Arita, T. and P. McCann (2000), "Industrial alliances and firm location behaviour: some evidence from the US semiconductor industry", *Applied Economics*, Vol. 32, pp. 1391–1403.

Arita, T. and P. McCann (2002a), "The spatial and hierarchical organization of Japanese and US multinational semiconductor firms", *Journal of International Management*, Vol. 8(1), pp. 121–139.

Arita, T. and P. McCann (2002b), "The location of technological innovations within the Japanese semiconductor industry", in: Z.J Acs, H.L.F. de Groot and P. Nijkamp, editors, *The Emergence of the Knowledge Economy: A Regional Perspective*, Heidelberg: Springer.

Aydalot, P. and D. Keeble (1988), *Milieux Innovateurs en Europe*, Paris: GREMI.

Bennett, R.J. (1998), "Business associations and their potential to contribute to economic development: re-exploring an interface between the state and the market", *Environment and Planning A*, Vol. 30(8), pp. 1367–1387.

Boudeville, J.R. (1966), *Problems of Regional Planning*, Edinburgh: Edinburgh University Press.

Caniels, M.C.J. (2000), *Knowledge Spillovers and Economic Growth*, Cheltenham: Edward Elgar.

Cantwell, J.A. and S. Iammarino (2000), "Multinational corporations and the location of technological innovation in the UK regions", *Regional Studies*, Vol. 34(4), pp. 317–332.

Cantwell, J.A. and S. Iammarino (2003), *Multinational Corporations and European Regional Systems of Innovation*, London: Routledge.

Cantwell, J.A. and E. Kosmopoulou (2002), "What determines the internationalization of corporate technology?", in: M. Forsgren, H. Hakanson and V. Havila, editors, *Critical Perspectives on Internationalization*, London: Pergamon Press.

Castells, M. and P. Hall (1994), *Technopoles of the World: The making of 21st Century Industrial Complexes*, London: Routledge.

Chinitz, B. (1961), "Contrasts in agglomeration: New York and Pittsburgh", *American Economic Review*, Vol. 51, pp. 279–289.

Chinitz, B. (1964), "City and suburb", in: B. Chinitz, editor, *City and Suburb: The Economics of Metropolitan Growth*, Englewood-Cliffs, NJ: Prentice-Hall.

Cohen, B. (1998), *The Geography of Money*, Ithaca, NY: Cornell University Press.

Duranton, G. and D. Puga (2000), "Diversity and specialisation in cities: where and when does it matter?", *Urban Studies*, Vol. 37(3), pp. 533–555.

Duranton, G. and D. Puga (2001), "Nursery cities: urban diversity, process innovation, and the life cycle of products", *American Economic Review*, Vol. 91(5), pp. 1454–1477.

Financial Times (1997), *Survey: Logistics*, October 7.

Financial Times (1998), *Survey: Supply Chain Logistics*, December 1.

Financial Times (1999), *Survey: Supply Chain Logistics*, June 17.

Fujita, M., P. Krugman and A.J. Venables (1999), *The Spatial Economy: Cities, Regions and International Trade*, Cambridge, MA: MIT Press.

Gabaix, Z. (1999a), "Zipf's law and the growth of cities", *American Economic Review: Papers and Proceedings*, Vol. 89(2), pp. 129–132.

Gabaix, Z. (1999b), "Zipf's law for cities: an explanation", *Quarterly Journal of Economics*, Vol. 114(3), pp. 739–767.

Gaspar, J. and E.L. Glaeser (1998), "Information technology and the future of cities", *Journal of Urban Economics*, Vol. 43, pp. 136–156.

Glaeser, E.L. (1998), "Are cities dying", *Journal of Economic Perspectives*, Vol. 12(2), pp. 139–160.

Glaeser, E., H.D. Kallal, J.A. Schinkmann and A. Shleifer (1992), "Growth in cities", *Journal of Political Economy*, Vol. 100, pp. 1126–1152.

Gordon, I.R. and P. McCann (2000), "Industrial clusters, complexes, agglomeration and/or social networks?", *Urban Studies*, Vol. 37, pp. 513–532.

Granovetter, M. (1973), "The strength of weak ties", *American Journal of Sociology*, Vol. 78, pp. 1360–1389.

Granovetter, M. (1985), "Economic action and social structure: the problem of embeddedness", *American Journal of Sociology*, Vol. 91, pp. 481–510.

Granovetter, M. (1991), "The social construction of economic institutions", pp. 75–81, in: A. Etzoni and R. Lawrence, editors, *Socio-economics: Towards a New Synthesis*, New York: Armonk.

Granovetter, M. (1992), "Problems of explanation in economic sociology", pp. 25–56, in: N. Nohria and R. Eccles, editors, *Networks and Organisations: Form and Action*, Cambridge, MA: Harvard Business School Press.

Hall, P.G. (1998), *Cities in Civilization: Culture, Innovation and the Urban Order*, London: Weidenfeld and Nicolson.

Henderson, J.V. (2003), "The urbanization process and economic growth: the so-what question", *Journal of Economic Growth*, Vol. 8(1), pp. 47–71.

Henderson, J.V., A. Kuncoro and M. Turner (1995), "Industrial development in cities", *Journal of Political Economy*, Vol. 103, pp. 1067–1085.

Hoover, E.M. (1937), *Location Theory and the Shoe and Leather Industries*, Cambridge, MA: Harvard University Press.

Hoover, E.M. (1948), *The Location of Economic Activity*, New York: McGraw-Hill.

Hummels, D. (1999), *Have International Transportation Costs Declined?, Department of Economics Working Paper*, West Lafayette, IN: Purdue University.

Isard, W. and R.E. Kuenne (1953), "The impact of steel upon the Greater New York–Philadelphia industrial region", *Review of Economics and Statistics*, Vol. 35, pp. 289–301.

Jacobs, J. (1960), *The Economy of Cities*, New York: Random House.

Jaffe, A.B., M. Trajtenberg and R. Henderson (1993), "Geographic localization of knowledge spillovers as evidenced by patent citations", *Quarterly Journal of Economics*, Vol. 108, pp. 577–598.

Krugman, P. (1991), *Geography and Trade*, Cambridge, MA: MIT Press.

Krugman, P. (1998), "Space: the final frontier", *Journal of Economic Perspectives*, Vol. 12, pp. 161–174.

Lichtenberg, R.M. (1960), *One Tenth of a Nation*, Cambridge, MA: Harvard University Press.

Marshall, A. (1920), *Principles of Economics*, 8th edition, London: Macmillan.

Martin, R. (1999), "The new geographical turn in economics: some critical reflections", *Cambridge Journal of Economics*, Vol. 23, pp. 65–91.

McCann, P. (1997), "Hoe deeply embedded is 'Silicon Glen'? A cautionary note", *Regional Studies*, Vol. 31(7), pp. 695–703.

McCann, P. (1998), *The Economics of Industrial Location: A Logistics-Costs Approach*, Heidelberg: Springer.

McCann, P. and B. Fingleton (1996), "The regional agglomeration impacts of just-in-time input linkages: evidence from the Scottish electronics industry", *Scottish Journal of Political Economy*, Vol. 43(5), pp. 493–518.

McCann, P., T. Arita and I.R. Gordon (2002), "Industrial clusters, transactions costs and the institutional determinants of MNE behaviour", *International Business Review*, Vol. 11(6), pp. 647–663.

Moses, L. (1958), "Location and the theory of production", *Quarterly Journal of Economics*, Vol. 78, pp. 259–272.

Nishiguchi, T. (1994), *Strategic Industrial Sourcing: The Japanese Advantage*, Oxford: Oxford University Press.

Ohlin, B. (1933), *Interregional and International Trade*, Cambridge, MA: Harvard University Press.

Perroux, F. (1950), "Economic space, theory and applications", *Quarterly Journal of Economics*, Vol. 64, pp. 89–104.

Porter, M. (1990), *The Competitive Advantage of Nations*, New York: Free Press.

Porter, M.E. (1998), "Clusters and the new economics of competition", *Harvard Business Review*, Vol. 76(6), pp. 77–90.

Reid, N. (1995), "Just-in-time inventory control and the economic integration of Japanese-owned manufacturing plants with the county, state and national economies of the United States", *Regional Studies*, Vol. 29(4), pp. 345–355.

Saxenian, A. (1994), *Regional Advantage*, Cambridge, MA: Harvard University Press.

Schonberger, R.J. (1996), *World Class Manufacturing: The Next Decade*, New York: Free Press.

Scott, A.J. (1988), *New Industrial Spaces*, London: Pion.

Schumpeter, J.A. (1939), *The Theory of Economic Development*, Cambridge, MA: Harvard University Press.

Simmie, J. (1998), "Reasons for the development of 'islands of innovation': evidence from Hertfordshire", *Urban Studies*, Vol. 35(8), pp. 1261–1289.

Simmie, J. (ed.) (2002), *Innovative Cities*, London: Spon.

Simmie, J. and J. Sennett (1999), "Innovative clusters: global or local linkages?", *National Institute Economic Review*, Vol. 170, pp. 87–98.

Suarez-Villa, L. and W. Walrod (1997), "Operational strategy, R&D and intrametropolitan clustering in a polycentric structure: the advanced electronics industries of the Los Angeles basis", *Urban Studies*, Vol. 34(9), pp. 1343–1380.

The Economist (1999a), *The World in your Pocket: A Survey of Telecommunications*, October 9.

The Economist (1999b), *The Net Imperative: A Survey of Business and the Internet*, June 26.

United Nations Centre for Human Settlements (1997), *Human Settlements Basic Statistics*, Nairobi, ISBN: 92-1-0031003-9.

Vernon, R. (1960), *Metropolis 1985*, Cambridge, MA: Harvard University Press.

Vernon, R. (1966), "International investment and international trade in the product cycle", *Quarterly Journal of Economics*, Vol. 80(2), pp. 190–207.

Weber, A. (1909), *Uber den Standort der Industrien, Alfred Weber's Theory of the Location of Industries*, Chicago: University of Chicago Press, translated by Friedrich, C.J. (1929).

Williamson, O.E. (1975), *Markets and Hierarchies*, New York: Free Press.

Urban Dynamics and Growth, CEA, vol. 266
Roberta Capello and Peter Nijkamp (editors)

CHAPTER 3

Beyond Optimal City Size: Theory and Evidence Reconsidered

Roberta Capello

Department of Management, Economics and Industrial Engineering, Politecnico di Milano, Milan, Italy

Abstract

The aim of the chapter is to present a critical review of theoretical and empirical works undertaken on the concept of the optimal city size, with the aim to highlight the achievements made from the theoretical and empirical point of view. We begin with the consideration that, during the 1960s and 1970s, the question of optimal city-size tended to be expressed in a misleading way. The real issue underlined by the more advanced theory is that one has to deal not with an 'optimal city size' but with an 'efficient size', which depends on the functional characteristics of the city and on the spatial organisation within the urban system. Economies of scale exist up to a certain city size. However, urban development generates conditions leading to structural readjustments which may create new economic advantages. These structural adjustments may either be industry transformations towards higher order functions, or the increase of external linkages with other cities. The paper also provides a critical review of all empirical methods developed to measure the old and new theoretical insights in this field.

Keywords: urban size, city networks, alternative urban growth patterns

JEL classification: R0

3.1. Introduction

In the real world, the number of people living in cities is growing in all countries and continents. The urbanisation process is a phenomenon which, in the last decade, has been increasingly intense in the developing countries as well. The share of urban population in the more developed continents, such as Europe and North America, is extremely high, and at the world scale is nearly 50%. This percentage, according to the forecasts of the World Resources Institute (1996), is expected to rise further in future decades. As a consequence of increasing population, cities physically expand through processes which have been defined as 'ville éclatée' (disseminated town), 'ville éparpillée' (sparsely populated town), 'ubiquitous city'. The population of large cities is continuing to grow, though sometimes more slowly than previously (Camagni, 1999).

The dynamics of cities encountered in the real world is in contrast with the famous 'optimal city size' theory, which envisages a size above which an increase in physical dimension decreases the advantages of agglomeration. The reason for the contradiction lies in the fact that the declining rate of urban population growth recorded in the last decade in most developing countries appears to be common to all cities, independently of the physical size, and represents a general slowing down, rather than a specific crisis in the largest cities. During the 1970s, negative population growth rates in the urban system of the Po Valley in Northern Italy were not registered only in the major cities, but also in a number of secondary centres of 75,000 to 150,000 inhabitants (8 out of 19) and even some smaller towns of 20,000 to 75,000 inhabitants (27 out of 113) (Camagni *et al.*, 1985, 1986).

This seemingly mistaken interpretation of the real world by the 'optimal city size theory' has already been pointed out by various authors. Richardson was the first to present a 'sceptic's view', by underlining that an apparent paradox existed between the theoretical acceptance of an 'optimal city size' and the contradictory development patterns of urban systems in the real world. According to Richardson (1972), this paradox could be explained by the existence of other determinants influencing urban agglomeration economies, not merely physical size. Since Richardson's paper, other interpretations have been given to this apparent paradox, through the 'urban life cycle' theory,[1] and the integration of

[1] On this theory see, among others, van den Berg *et al.* (1983) and Camagni *et al.* (1985).

dynamic elements, such as innovation, continuous information and knowledge acquisition, into the static framework of 'optimal city size theory'.

In this chapter, the aim is to go through the different theoretical steps that have been made since the idea of the 'optimal city size' theory and to present the most recent and advanced (theoretical and empirical) studies in this field.

The structure of the chapter is as follows. The first seminal ideas on the concept of the optimal city size are presented in Section 3.2; Section 3.3 presents the new theoretical paradigms that overcome the limits of the traditional theory. Section 3.4 presents an overview of the traditional empirical analyses in the field, while Sections 3.5 and 3.6 contain the most recent empirical investigations. Section 3.7 highlights some concluding remarks.

3.2. *Optimal city size: an old and still unsolved issue*

Since the 1960s, urban economists and geographers have put their attention to the problem of the optimal city size. In particular, the main theoretical question raised in this theory is whether increasing returns exist with the increase in urban size. The reply to this question has been provided many years ago: with an increase in urban size, the average (and marginal) advantages related to an urban location increase, and at the same time the average (and marginal) costs associated to an urban location decrease (Alonso, 1971).

A larger urban size generates positive externalities related to the concentration of public intervention, to a greater market for output and to more qualified labour and input markets. All these externalities are generally known as 'agglomeration economies'. At the same time, from a certain city size average location costs decrease, thanks to size effects; once achieved a certain indivisibility level, per capita investments in social overhead capital and in the implementation of public services decrease with the increasing number of people living in city (Richardson, 1978).

A general and large consensus exists in the literature on the fact that the positive advantages exist up to a certain urban size: beyond that size, opposite mechanisms are at work, which translate the positive externalities into negative, the economies into diseconomies, while location costs start to increase, with an additional negative effect on the net location benefit; congestion, prohibitive urban rent and environmental costs are all elements which explain the increase in the costs of an urban location. The city, as all resources used in an intensive way, shows, from a certain urban size, decreasing returns to scale.

Therefore, according to this theory, urban average location advantages and costs have both an U-shaped form: the former grow and then decline, the latter decrease and then grow. The size at which the difference between the curves of average benefits and costs is maximum corresponds to the (per capita) optimal city size, an optimal situation from the point of view of the population already located in the city.[2]

During the 1970s, some additional refinements to the theory have been developed:

1. the difference between optimal city size for existing inhabitants, and for potential inhabitants has been introduced: the former stems from the maximum difference between average benefits and costs curves, the latter is reached when marginal benefits and marginal costs curves are equal (Richardson, 1972);
2. a more precise definition of location cost has been introduced in the theory, defined as the urban rent which families and firms have to pay for an urban location (Alonso, 1971);
3. costs related to the natural environment are introduced in the concept of urban location (Anderson and Crocker, 1972): already Duncan (1956) stresses the importance of the level of criminality and air pollution in the definition of an optimal city size. Richardson underlines the difference between private and social costs, the latter also containing the environmental costs associated to urban size, and on the basis of this difference introduces the concepts of private and social optimal city size (Richardson, 1972).

---

[2] Alonso stressed the mistaken tendency of many authors to look for 'optimal city size' only by minimising the location cost function. As he argued, this would be sensible only if output per capita were constant (Alonso, 1971, p. 70).

The simplicity of the theory, its strong empirical validation provided by empirical analyses witnessing the U-shaped form of average costs and benefits curves, and the importance that the subject has for normative interventions have generated a particular 'allure' around this theory. An in-depth analysis of the literature on this issue, however, shows that little theoretical background exists beyond the concept of optimal city size: no economic rules, theories or models explain the law of urban increasing/decreasing returns. All mechanisms mentioned above stem from empirical analyses which support them; in the case the reality did not show increasing/decreasing returns to scale, the existence of economies of agglomeration would be impossible to be theorised. In this field, empirical analysis supports theory. In the words of Mills (1993): "this is one of the few fields in which econometrics is ahead of theory".

Over time, many criticisms have been made of the neoclassical approach to optimal city size theory. These include the observations that:

- cities are different from one another. They are characterised by different functions and perform different specialisations (Henderson, 1985, 1996). The use of the same urban production function for all cities in econometric analyses estimating optimal city size is extremely restrictive. In the words of Richardson: "we may expect the efficient range of city sizes to vary, possibly dramatically, according to the functions and the structure of the cities in question" (1972, p. 30);
- if cities are different from one another, the optimal city size may be different, depending on the specific characteristics. Richardson elegantly compares the 'optimal city size' theory with the theory of the behaviour of firms. In the real world, we would never expect the optimal position for each and every firm to occur at the same level of output, so why should we expect the optimal point in each city to be identified at the same population level?
- cities exist in an inter-urban environment. The optimal city size theory, on the contrary, does not consider the spatial context in which cities operate;
- cities generate a large variety of externalities as a result of the qualitative characteristics of the urban production environment.

Already Chinitz (1961) expressed some doubts about the fact that urban factor productivity depends mainly on the physical size of cities. He emphasised, on the contrary, the importance of a diversified and competitive urban production system as a source of urban productivity. Such a system is able to provide a far larger variety of externalities for small firms than an oligopolistic and specialised urban structure, in which the internalisation processes of service functions put in place by large firms reduce urbanisation economies. Chinitz supported his thesis with an empirical analysis of New York, a large and diversified urban area, and Pittsburgh, a highly specialised city; his findings supported the idea that in the more diversified urban area urbanisation advantages have an impact on urban productivity, while in the more specialised city economies of scale play a role in definition of urban productivity.[3]

The necessity to overcome the limits of the theory on the optimal city size has increased in recent years, when the urbanisation process has drastically been affected by a rapid growth.

3.3. New paradigms for an old problem

The theories which have superseded the above limitations of the optimal city size theory can be grouped into two different conceptual paradigms.[4] We refer to these two paradigms as the 'neoclassical city interpreted within a logic based on the Christaller model' and the 'network city paradigm' (Table 3.1).

3.3.1. The neoclassical and Christallerian city

The first paradigm deals with some of the limitations of the optimal city size theory by stating that: (a) urban size is defined as the

---

[3] Carlino (1980) provides a criticism of Chinitz' analysis, and demonstrates on a sample of 65 American towns that economies of scale, both internal and external to the firm, play a role in the definition of urban productivity.

[4] These theories supersede the previous ones in conceptual terms, rather than chronologically, in that the Christaller model (Christaller, 1933) dates before the optimal city size theory.

Table 3.1. A comparison of three theoretical paradigms

Elements	Paradigms		
	Optimal City Size	The Neoclassical and Christallerian city	The Network City
Characteristics of the approach	Empirical	Theoretical	Theoretical and empirical
Characteristics of the city	Undefined city (aggregated)	Despecialised city	Specialised city linked with a large urban system
Characteristics of the urban system	Not considered	Hierarchical	Networked
Characterising element	Urban size	Urban size interpreted through urban functions	Distinction between size and urban function. Analysis developed in a spatial context
Urban efficiency	Agglomeration economies	Functional upgrading of economies	Coexistence of network externalities, economies of agglomeration and functional upgrading
Result of the analysis	An intra-urban equilibrium exists which has to be reached	An intra and inter-urban equilibrium exists by definition	An intra-urban equilibrium exists which can be reached through inter-urban system relationships
Urban policy aims	Achievement of an intra urban equilibrium between costs and benefits obtainable through the urban dimension	None. The system is in equilibrium by definition	Achievement of a cost/benefit equilibrium through specialisation policies and/or network integration

Source: Capello, 1998a.

equilibrium between marginal production benefits and marginal location costs, and (b) cities are not all the same, but produce different goods according to their size. In the neoclassical approach to urban location theory, developed in the 'New Urban Economics' paradigm, urban equilibrium is achieved when marginal location benefits and costs are equal. This is true in an intra-urban equilibrium logic considered in extensions and refinements of the basic Von Thünen–Alonso–Muth work; all claim that at equilibrium a marginal reduction in rent from further decentralisation is exactly offset by a marginal increase in travel costs.[5] The result of the model is an indifferent location choice among all possible locations (the famous 'Muth condition'), i.e. lower accessibility to the centre is compensated for by lower rents and higher environmental quality. The same holds for an inter-urban equilibrium; in an equilibrium solution, the same profits and utility levels have to be guaranteed in each city. If this is not the case, ceteris paribus, a city offering higher rents but lower agglomeration benefits (with the hypothesis of non-existent transport costs) would lose both residents and firms. Urban size is, in this case, the result of market forces, pushing towards the maximisation of utility levels for residents and profits for firms (Table 3.1).

In the neoclassical approach of the 'New Urban Economics', the use of the same production function for all cities inevitably generates cities of the same size (Camagni, 1992). This evident paradox can be overcome either through the hypothesis of different production functions for each city (and thus a single production function for each city) as suggested by Henderson (1985), or by expressing the neoclassical logic through the use of the Christaller model. This famous model on urban hierarchy suggests that the urban system organises itself around a limited number of urban ranks; each higher order centre produces goods/services typical of its rank and all goods/services of lower order centres. In this reasoning, the importance of the functional specialisation of a city is emphasised and even becomes a proxy of the size of the city. When the neoclassical logic is applied to the Christaller model, it leads to the definition of a hierarchical urban system – by assumption in equilibrium, thanks to market forces – where differences in city size can be interpreted as the compensation between agglomeration advantages, on one hand, and higher urban rents and diseconomies of congestion, on the other.

More recently, urban hierarchical systems have been studied within the 'New Economic Geography' approach. Based on a pure economic reasoning, the (endogenous) development of hierarchical urban systems is the result of an equilibrium between different forces:

[5] Alonso (1964) and Fujita (1985).

transport costs, product variety competition and economies of scale, the first pushing towards dispersed activity location (under the assumption of a spatially distributed demand), the second and third pushing towards agglomeration of activities in large cities, because of lower cost of living (because of greater competition among products) and greater economies of scale. Under the logic of this approach, small cities (and, therefore, cities with low agglomeration economies), or cities which are geographically peripheral, with low product competition (and, therefore, high cost of living) can still capture small market areas if low transport costs to the final market counterbalance the two other sources of locational advantages. While these models suggest how Christaller-type urban hierarchies can be a natural response to economic development over time, being able to endogenise the critical distance up to which the product is sold efficiently, they do not specify whether location forces differ among urban functions, and thus whether cities of the same size can host different economic functions.[6]

The neoclassical city within the Christaller approach, though elegant and fascinating in its theoretical interpretation, overcomes perhaps rather too simply the problem of optimal city size. The problem simply does not exist, thanks to the ability of an urban system of any size, to equalise costs and benefits, and find by definition an equilibrium solution. In the 'New Urban Economics' approach an indifferent location choice emerges over the whole geographical space, since the agglomeration advantages are perfectly capitalised into urban rents; in the 'New economic geography' approach indifference in location choice emerges over geographical space when agglomeration advantages and lower cost of living are counterbalanced by lower transportation costs.

While optimal city size theory gives the impression that it is an empirical exercise without a theory behind it, we could argue that the 'neoclassical city interpreted within the logic of Christaller' is a theory without empirical application. The result achieved by the model – a general equilibrium of cities, which denies the existence of

---

[6] See Fujita and Krugman (1995) and Fujita *et al.* (1999).

an excessive city size or urban growth – is undoubtedly unrealistic. This approach leaves no chance to normative interventions.

3.3.2. The network city

An interesting paradigm which has emerged recently, which overcomes some of the limits of both the 'optimal city size' and of the 'neoclassical/Christallerian approach' (Table 3.1), is the so-called 'network city paradigm'. The most important theoretical novelty provided by this paradigm is the break of the link between urban size and urban function imposed by the Christallerian logic. With Christaller's approach, it is in fact impossible to explain why a city like Zurich, with only 300,000 inhabitants, is specialised in international finance in the same way as the city of New York or Tokyo. In the real world, the urban size is not always characteristic of the function in which the city is specialised.

The break between urban size and function is one of the main characteristics of the SOUDY (Supply Oriented Dynamic Approach) model (Camagni *et al.*, 1986), which puts forward the new hypothesis that an 'efficient' city size interval exists separately for each hierarchical city rank, associated with its specific economic functions. In other words, for each economic function, characterised by a specific demand threshold and a minimum production size, a maximum city size also exists beyond which urban location diseconomies overcome production benefits; the minimum and the maximum sizes define the city size interval in which the output is produced under efficiency conditions (positive net gains). The model is based on some assumptions:

- higher order functions require higher thresholds for the level of appearance in the city (in terms of urban population) ($d_1, d_2, d_3, \ldots$ in Figure 3.1) to be produced under efficiency conditions, as was the case in the Christaller model;
- each average (aggregate) production benefit curve corresponds to a function developed by the city; higher order functions are represented by higher average benefits curves (ABC_1; $\mathrm{ABC}_2, \ldots$ in Figure 3.1) because of (a) growing entry barriers, (b) decreasing elasticity of demand, due to a lower number of substitutes for higher order goods/services, which allows extra profits to be gained in all market conditions far from the long run equilibrium,

Figure 3.1. Efficient urban size for different urban functions

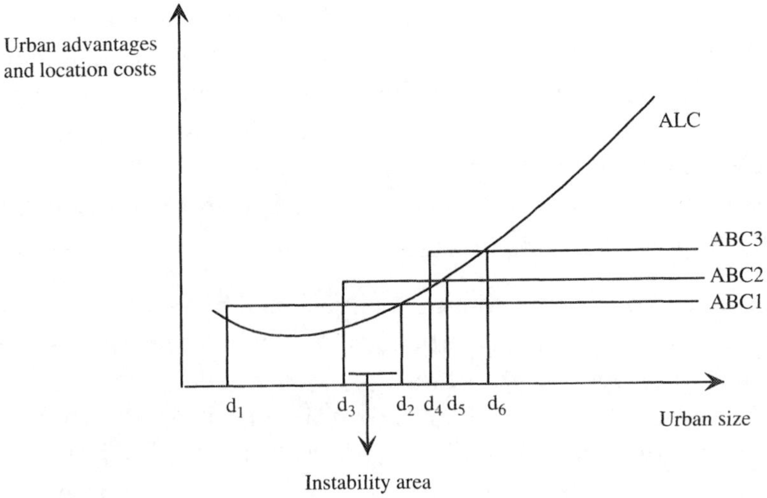

Source: Camagni *et al.* (1986).

and (c) increasing possibility of obtaining monopolistic revenues due to the use of scarce, qualified factors;

- the location cost curve has the traditional form suggested by Alonso (ALC in Figure 3.1).

As Figure 3.1 shows, under these conditions for each economic function and each associated urban rank, it is possible to define a minimum and a maximum city size in which the city operates under efficiency conditions (i.e. with net positive gains, with extra-profits to be gained) (d_1-d_2 for the function and centre of rank 1, d_3-d_5 for the function and centre of rank 2,...). The efficient size interval takes place at greater urban sizes, the higher the urban function and rank are (Figure 3.1).

As each centre grows, it becomes a suitable location for higher order functions, thanks to the achievement of a critical demand and of sufficient economies of scale in the production. In dynamic terms, each city's long-term growth possibilities depend on its ability to move to ever higher urban rank, developing or attracting new and higher order functions. Its ability is not mechanically attained, but depends on the innovativeness of the private and public urban sectors, to be treated as a stochastic variable in a model. A city size interval, called 'instability area', can be identified in Figure 3.1

$(d_3 - d_2)$; having achieved the minimum threshold level for a higher rank function in this city size interval the city has the chance to develop and attract higher order functions, and therefore continue to grow under efficiency conditions, or can choose to remain specialised in the previous function, achieving within the short run inefficiency conditions (Figure 3.1).

The interest of this model is that it overcomes some of the limits of the 'optimal city size' theory, by suggesting:

- the need to replace 'optimal city size' by an interval within which the city size is efficient,[7] i.e. where average benefits exceed average location costs;
- the possibility to separate urban function from urban size. Differently from Christaller's approach, two cities of the same size (for example d_2 in Figure 3.1) can be specialised in two different functions, depending on their capacity to attract/develop higher functions;
- the important role played by the economic functions in the definition of the efficient city size.

The interpretations of the city size provided by the SOUDY model are enriched if analysed together with another theoretical approach, defined as the 'network city' paradigm.[8] The logic underlying the paradigm is that the spatial organisation in which cities operate is fundamental to the understanding of their efficiency, growth, factor productivity and sometimes their specialisation. While the organisational logic underlying the central place model is a territorial logic, emphasising a gravity-type control of market areas, in the city network model, another logic prevails, referring to long distance competition and cooperation regardless of the distance barrier. While

[7] Richardson (1972) has already suggested replacing the concept of optimal city size with an efficient interval of urban size in which urban marginal benefits are greater than marginal location costs.

[8] Camagni (1993) theorised the concept and applied it to urban systems. The same concept has already been applied to many fields, such as the behaviour of the firm, and macroeconomic organisational behaviour. For a review of the concept, see Capello (1996); for the policy implications, see Capello and Rietveld (1998). For the most advanced studies and reasoning in this theoretical paradigm, see Chapter 16 by Camagni and Capello in this volume.

transport costs and economies of scale were the main forces shaping the spatial organisation of functions and urban areas, in the network logic other kinds of economies come to the fore: economies of vertical or horizontal integration and network externalities similar to those emerging from 'club goods' à la Buchanan.[9] By network behaviour a set of privileged synergetic relationships among centres that cooperate or interact in the same fields or functions are intended; these relationships spontaneously provide externalities to partners which cooperate on the basis of both vertical and horizontal linkages. When horizontal relationships are put in place, cities cooperate even if they are specialised in the same function. Thus, the city network concept mainly consists of three elements:

(a) the *network* element, meaning that the relationships among centres do no longer occur only on the basis of territorial hierarchy-type relationships, driven by non-overlapping market logics, à la Christaller. Other new types of non-territorial and long-distance relationships emerge, among cities of the same size, of different or similar specialisation patterns;
(b) the *network externality* element, which represents the main economic advantage explaining network behaviour. It is in fact no longer a matter of minimisation of transport cost and of maximisation of control on non-overlapping market areas, but a matter of exploiting scale economies in complementary relationships and synergic effects in cooperative activities, achieved through the participation to network. In this sense, network advantage is a real club good, achieved only by those economic actors who are partners of the economic and spatial network and distributed among partners despite the private marginal costs each partner bears to participate to the network. In this sense, private marginal costs of network participation differ from private marginal benefits, and network advantages turn out to be network externalities, in the real economic sense given to the concept;
(c) the *cooperation* element. City relationships are no longer governed by hierarchy among centres, or competition among specialised

---

[9] On the concept of 'club good', see Buchanan (1965), Berglas, (1980), and Cornes and Sandler (1986).

centres, where location economies à la Hoover and input–output relationships à la Perroux reinforce each single centre's growth against the others. Cooperative relationships are at the basis of the city network paradigm, leading to the achievement of urban economies of scale in a cooperative way, without implying a physical growth of each single centre, and distributing the consequent advantage among partners, generating economic development.

The city network paradigm provides a successfully theoretical framework to overcome the limiting interpretative power of the traditional central-place model; it provides a theoretical understanding of why cities of the same size can cooperate, a situation that is denied in a Christallerian logic.

Moreover, the joint application of the SOUDY model and the 'network city' theory implies something very important for the definition of economies of agglomeration: city size is not the only determinant of factor productivity and economies of agglomeration of large centres. The presence of higher urban functions and integration in the network of urban systems are also extremely important in explaining the size of the city. Both these elements may permit the achievement of economies of scale, even in small cities.

3.4. Traditional empirical analyses: the aspatial city

The strength of the theory of the 'optimal city size' is its vast and supporting empirical evidence. Many empirical studies have in fact been developed since the 1970s which successfully measured increasing returns of urban size. The methodologies applied for these studies may be grouped into three main approaches (Table 3.2); though all of them have limits in their application, the results achieved confirm both the existence of urbanisation and localisation economies in an urban environment, and the U-shape of the (average and marginal) costs and advantages curves, witnessing the possible existence of an 'optimal city size'.

The first approach deals with the estimate of an aggregate urban production function, thanks to which the existence of a multiplicative constant linked to the urban size variable is envisaged (Table 3.2). The hypothesis that is behind all these econometric

Table 3.2. A comparison of different methodological approaches

Characteristics	Methods		
	Aggregate Urban Production Function	Sectoral Production Function at the Urban Level	Analyses of Rent and Wage Differentials
Level of analysis	Urban (aggregate)	Sectoral	Urban (aggregate)
Methodology	Estimate of a production function at the urban level	Estimate of a sectoral production function at the urban level	Estimate of the reasons for rent and wage differentials
Limits of the methodology	All cities have the same production function	The sectoral mix is left aside	High urban rents and wages do not only reflect higher productivity levels
Critics to the methodology	Cities of different size cannot have the same production function	The sectoral analysis does not contain urbanisation economies	High wages and rent can compensate for high social and environmental costs
Results	U-shaped costs curve. Higher labour productivity in larger cities	Significant economies of scale in different sectors	Higher wages and rents in larger cities
Bibliographic references	Hirsch (1968), Alonso (1971), Mera (1973), Henderson (1974), Kawashima (1975), Segal (1976), Marelli (1981), Catin (1991), Ladd (1992), Rousseaux and Proud'homme (1992), Capello (1996)	Mills (1970), Shefer (1973), Sweikauskas (1975), Carlino (1980), Moomaw (1983), Sveiskauskas *et al.* (1988)	Fuchs (1967), Hoch (1972), Getz and Huang (1978), Rosen (1979), Cropper (1981), Henderson (1982), Roback (1982, 1988), Izraeli (1987), Clark and Kahn (1988, 1989), Clark and Cosgrave (1991), Burnell and Galster (1992), Boyer and Savageau (1985), Herzog and Schlottmann (1993)

Source: Capello, 1998b.

exercises is that the same production function can be applied to each city, despite the size and the functional specialisation that characterise each city; but this hypothesis is simply non-realistic.

In the first part of the 1970s, the empirical studies in this field have concentrated on the analysis of per capita expenses for public services.[10] Alonso (1971) and Mera (1973) estimate, on a sample of, respectively, American and Japanese cities, that per capita public expenses are greater for cities with more than one million inhabitants.[11] Beyond that size, per capita expenses increase again, witnessing a U-shaped curve for average urban costs.[12] Hirsch (1968) shows that this rule holds only for specific services, like fireworks, while the average cost curve has either a linear function with respect to urban size for some services, like education, or an increasing function, for others, like water, gas, electricity, etc. Alonso shows that the average labour productivity is greater in American cities which have more than 5 million inhabitants, and demonstrates, like other scholars do later, that the minimum of the location cost curve is achieved for an urban size which is smaller than the size which guarantees the maximum of location advantages.

The estimates of urban production function at the aggregate urban level show that increasing returns to scale exist. Through the estimate of an aggregate urban Cobb–Douglas production function on a sample of 58 American cities, Segal demonstrates that the parameter of the urban size variable is a significant one: metropolitan areas with more than 3 million inhabitants show a factor productivity which is 8% higher than the other cities (Segal, 1976). In a study on 230 American cities, in cross-section, Marelli achieves similar results: larger cities have a greater factor productivity than smaller cities, but this holds up to a certain urban size, after which factor productivity shows again decreasing

---

[10] In the first studies, the minimum of the cost curve was interpreted as the optimal city size, without considering that also advantages change with city size.

[11] Data presented by Alonso were previously analysed by Douglas (1967).

[12] A doubt remains with these results: in larger cities higher per capita expenses may be due to a higher willingness to pay for public services than to economies of scale. Moreover, the difference in income between large and small cities exceeds the difference in average costs; therefore, if an optimal dimension exists, this is characterised more by higher productivity than by lower average costs.

returns (Marelli, 1981). Other empirical studies in France consider the difference between the positive effects of size and of modernisation and innovation processes:[13] Catin demonstrates that the greater industrial labour productivity registered in the Île de France in the period 1974–1984 with respect to other French cities, largely depends on the innovation process that characterises the city, despite its urban size (Catin, 1991). Rousseaux and Proud'homme find out that the productivity is 30% greater in the Île de France and 12% greater in Marseille, Lyon and Nice than in the rest of the French cities (Rousseaux and Proud'homme, 1992; Rousseaux, 1995). The conclusion they achieve is that a correlation does not exist between urban productivity and urban size: the functional structure probably explains the vast majority the differences encountered (Table 3.2).

For Italy, Capello shows through the estimate of an urban translog production function on a sample of 58 Italian cities, in cross-section, that economies of agglomeration decrease with the increase of size: the positive income elasticity to urban size decreases while urban size increases (Capello, 1996).

The second approach for the measurement of factor productivity and increasing returns at the urban level is based on the estimate of a production function disaggregated at sectoral level (Table 3.2). In order to overcome the limits encountered with an aggregate and similar production function for all cities, despite their sectoral specificity, these studies estimate the size effects at the sectoral level. Through a CES production function, Shefer (1973) witnesses the existence of wide economies of scale in 10 sectors located in American cities; Carlino (1980) has divided the index used by Shefer into three parts, in order to capture economies of scale, economies of localisation and urbanisation in 19 manufacturing sectors, and has found significative results for both localisation and urbanisation economies in 12 sectors over 19. Sweikauskas (1975) has estimated industrial labour productivity in 14 sectors, and found that productivity increases of 6.4% every time the city doubles in size. Moomaw (1983) achieves the same conclusions:

[13] Similar studies have been developed in other Countries. See, for example, Beeson (1992).

he demonstrates that if the population doubles in size, labour productivity increases by 6%.

Although these analyses overcome some limits of the aggregate approach analysed before, they still contain some limitations: these analyses are unable to take into account that most of the advantages of an urban location stem from the great sectoral mix present in large urban areas, and that this mix is usually made of tertiary activities, not encountered in the previous analysis for the well-known problem of measuring the product of the tertiary sector.

Sveiskauskas, Gowdy and Funk have stressed the importance of a distinction between urban size effects and industry size effects, i.e. between urbanisation and localisation economies, when industrial location policies are concerned. In fact, if increasing returns depend on urban size, a concentration of firms in urban areas is legitimate; on the contrary, if industry size effects explain increasing returns, it seems reasonable to develop a process of decentralisation of some sectors. The analysis, developed on 174 metropolitan areas in the food industry, shows that the greater productivity in large urban areas is certainly due to urbanisation economies, while some doubts remain for what concerns localisation economies (Sveiskauskas *et al.*, 1988).

The third approach to the measurement of increasing returns in urban areas is related to income and wage differentials, adjusted for the different life costs, between large and small cities (Table 3.2). The large city, in this approach, should show higher wages which compensate for higher productivity: the empirical results confirm this hypothesis (Fuchs, 1967; Hoch, 1972). The criticism moved to this hypothesis is the major wages paid in large cities may in reality be the monetary compensation for the greater social and environmental diseases that workers in urban areas have to bear, more than the signs of greater productivity levels.[14]

3.5. Recent empirical analyses: city size and environmental aspects

Based on the interest for environmental problems and for the emerging environmental sustainability paradigm, during the 1990s

[14] Against this criticism it might be claimed that if higher wages can be paid, nevertheless demonstrates that labour productivity must be higher.

the studies on the optimal city size have shifted the attention towards the relationship between urban size and environmental quality in urban areas: the question addressed in these studies is whether cities of major size are the most polluted and polluting.

Two methodologies are applied in this field. The first methodology refers to the approach based on the differences between wages in cities of different size; the major idea behind this methodological approach is that negative environmental externalities are registered in the urban price system. The higher wages and house prices (urban rent) in large towns are justified by the necessity to compensate people for the greater environmental costs and social diseases they face by living in large cities.[15] For example, on a sample of 249 American metropolitan areas, Herzog and Schlottmann (1993) find out that wage differentials are significantly and largely explained by the quality of the labour force, and moreover, by the size of the city and by the quality of the urban environment, like the level of criminality and of unemployment. However, they also find out that wage differentials decrease with urban size up to an urban size of 4.5 million inhabitants; they explain this phenomenon with the presence of increasing positive economies of agglomeration which compensate for the loss in the wage differentials between that city size and cities of smaller sizes. After that threshold, economies of agglomeration turn out to be diseconomies, and wage differentials between that city size and cities of larger sizes increase.

The second methodology is based on the attempt to measure a 'quality of life' index in cities of different size.[16] Burnell and Galster use a quality of life index calculated by Boyer and Savageau on a sample of 249 American cities, and find out that 41% of the quality of life index variance is explained by city size. However, the square quality of life index has a negative sign, which prevails for city sizes of more than 4.5 million people; thus, the quality of life curve is a U-shaped curve, with a maximum

[15] See, among others, Berger *et al.* (1987), Blomqvist *et al.* (1988) and Corielli *et al.* (1996).

[16] For a comparative analysis among quality of life index, see Liu (1976), Bowman *et al.* (1981) and Conway and Liston (1981).

obtained for 4.5 million inhabitants (Boyer and Savageau, 1985; Burnell and Galster, 1992).[17]

Recently, through the estimate of a translog production function on 58 Italian cities, Capello demonstrates that if it is true that the largest cities are also the most polluted, they are also the most efficient in the use of environmental resources: income elasticity of environmental resources increases with city size, witnessing the existence of increasing returns in natural resource consumption in large towns.[18]

Recent empirical analyses have been developed with the aim to look for the 'optimal urban form' (more than the 'optimal urban size'); in these studies the optimal urban form is the one which allows the city to grow, in physical terms, with the lowest social (including environmental) costs involved. The necessity to put in place these studies is due to the strong qualitative change which is going on in the urbanisation process in European cities:[19] following the example of American metropolitan areas, the physical growth of cities in Europe is taking place through an enlargement of the physical boundaries of the city, through the indiscriminate and fragmented use by industrial and residential activities of agricultural and natural land, without any drastic increase in the number of people or industrial activity located in urban areas. Land consumption index calculated in 1992 by the French Agence d'Urbanisme shows that between 1950 and 1975 in 22 French urban areas, population has doubled while the territory occupied has increased by 20–30%; however, between 1975 and 1990 population has increased by 25% while the territory occupied by urban activities has doubled (Camagni, 1999).

The phenomenon described above, of urban sprawl, does not characterise only French cities. Other studies exist which witness

---

[17] Burnell and Galster demonstrate that the results are highly dependent on the way in which the quality of life index is calculated; the same regression analysis developed on the basis of a quality of life index obtained as the difference between wages in the different cities shows different results, although in this case the econometric result is less robust.

[18] In the same study, the results are confirmed by a simple linear regression analysis between the consumption of environmental resources and urban size (or urban density); when the urban size (or urban density) increases, the per capita use of energy decreases.

[19] See the debate on the sprawl effect, firstly introduced by Breheny (1992) and Owens (1992).

these urban patterns; in the Lombardy region, in the North of Italy, for example, an analysis of 186 cities shows the 'wasteful' character of sprawling development patterns in terms of land consumption (Camagni *et al.*, 2002).

3.6. Recent empirical analyses: the specialised city in an urban system

A recent empirical analysis has the clear aim to prove empirically the 'network city' paradigm, by measuring the role that other variables, more than the physical size of a city, can explain urban advantages and costs. As the 'network city' paradigm underlines, these variables have been envisaged in the functional specialisation and in the degree of interaction that the city has within its urban system (Capello and Camagni, 2000).

Through the estimate of a translog function of urban advantages (containing all positive social, economic and environmental advantages associated to an urban location, called 'city effects'), and of a cost location function (capturing all negative externalities associated to a urban location, labelled 'urban overload effects'), on a sample of 58 Italian cities, the results confirm what was theoretically envisaged by the 'network city paradigm'. The 'city effect' describes a situation in which agglomeration economies should be associated with positive environmental externalities and social network externalities. The most important difference with the traditional neoclassical view on 'optimal city size' is that the advantages and disadvantages referred to relate not only to the 'economic environment', but also the interaction of the three environments constituting the city (Table 3.3). The 'city effect' is the result of not only a pure economic efficiency goal, but also responds to wider policy objectives, defined as environmental equity, long-term allocation efficiency and distributive efficiency. On the contrary, an urban overload takes place when economic location costs are associated with negative social and environmental externalities. On the basis of the traditional view, we incorporate in the cost curve all negative externalities stemming from an urban location. In other words, in the case of costs, we widen the definition to all negative externalities which are encountered at the urban level and which have not been explicitly mentioned by Alonso (Table 3.3).

Table 3.3. City effect and urban overload

	Interaction Between Economic and Physical Environment	Interaction Between Economic and Social Environment	Interaction Between Social and Physical Environment
City effect	Efficient energy use Efficient use of non-renewable natural resources Economies of scale in the use of urban environmental amenities	Accessibility to: good housing facilities, skilled jobs, social amenities, social contacts, education facilities, health services	Green areas for social amenities Residential facilities in green areas Accessibility to urban environmental amenities
Urban overload	Depletion of natural resources Intensive energy use Water, air pollution Depletion of green areas Traffic congestion noise	Suburbanisation forced by high urban rents Social friction in the labour market New poverty	Urban health problems Depletion of historical buildings Loss in cultural heritage

Source: Capello, 1998b.

City effects and urban overload indicators are expressed by different variables belonging to the different aspects characterising the social, environmental and economic environments which constitute a city and have been calculated as the un-weighted sum of the different indices in each group of variables presented in Table 3.4.[20]

The results obtained are presented in Figure 3.2, where the city effects and urban overload are calculated for different city size (measured in terms of population), for the presence of higher urban functions (measured as the share of private tertiary value-added produced by the city), and for the level of network integration of the city with the rest of the world, measured through the stock of per capita telephone subscribers.[21] The results confirm what was

[20] The first group of indices, relating to the interaction between the economic and the natural environment, enter the sum with their 'complement to one' value, reflecting their negative correlation with city size.

[21] The lack of statistical information on the flows of interaction between our sample cities (duration of phones calls or number of phone calls) for these groups of cities has obliged to choose a variable representing the stock of per capita telephone subscribers. However, the share of flows of international phone calls (both duration and number of phone calls) and the per capita telephone subscribers available for a different group of cities (municipalities) in the metropolitan areas of Milan have shown a correlation equal to 0.80.

Table 3.4. City effect and urban overload indicators

	Interaction Between Economic and Physical Environment	Interaction Between Economic and Social Environment	Interaction Between Social and Physical Environment
City effect indicator (ALB)	Per capita energy use Per capita petrol use Per capita water use	Number of graduates/population Number of schools/population Number of banks/population Supply of public services/population Urban rent per square metre	Square meters of green areas in city per capita
Urban overload indicator (ALC)	Per capita NO_x emissions Per capita kg of urban waste Number of vehicles per square metre	Unemployment/ population	Number of murders/population

theoretically envisaged by the network city paradigm (Figure 3.2) since:

- the estimate of the average urban overload and city effect curves associated to urban size confirms their traditional U-shape. The maximum city effect is reached with an urban size of 361,000 inhabitants, while the minimum of urban overload is obtained with 55,000 inhabitants (Figure 3.2a,b). As already claimed by Alonso, the minimum of overload costs is associated with an urban size which is smaller than the one associated with the maximum of city effects;
- the development of decreasing returns to scale can be postponed to larger size of the city if the city develops higher order functions; higher city effects are reached, ceteris paribus, through the presence of a high share of tertiary activities in the city (Figure 3.2c). Moreover, urban overload increases but at decreasing rates while tertiary activities increase their presence in the city (Figure 3.2d);
- as suggested by the theory, increasing city effects are also obtained thanks to a strong network integration of the city

Figure 3.2. Estimated city effect and urban overload

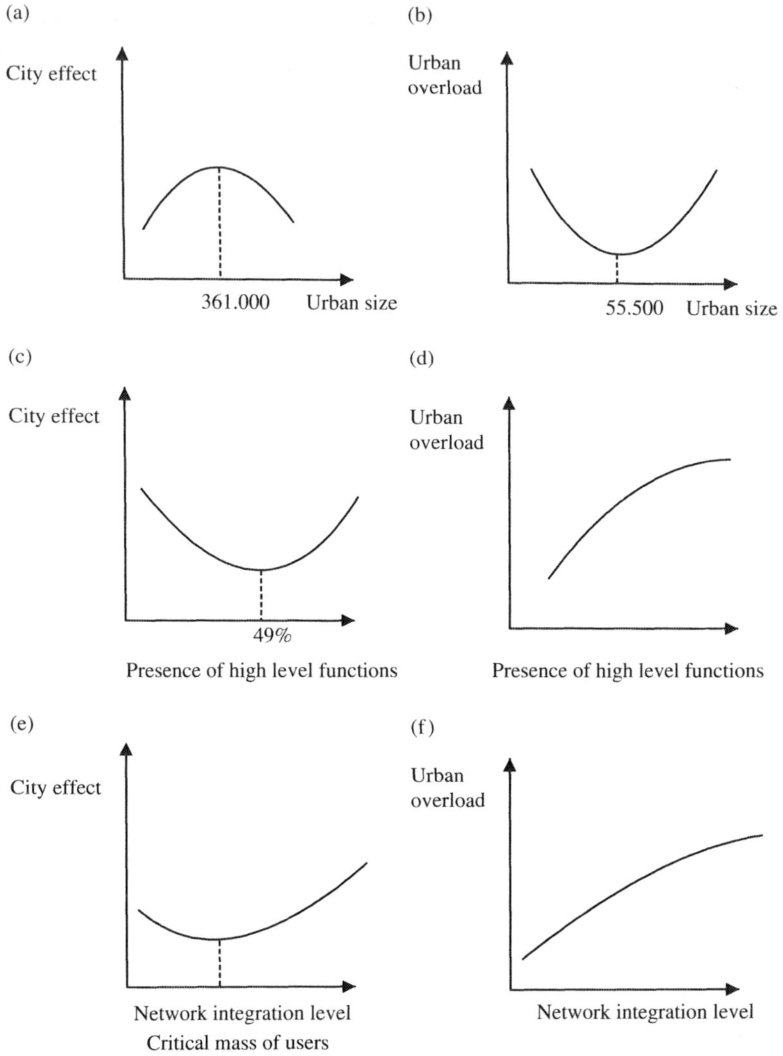

Source: Capello and Camagni, 2002.

(Figure 3.2e).[22] Also in this case, ceteris paribus, urban overload effects increase with the integration of the city in an urban system, but at decreasing rates (Figure 3.2f).

---

[22] On this concept, see also Chapter 16 in this volume.

Given the results obtained, the hypotheses on strong agglomeration economies in the presence of large urban size, of high-order function, of high-network linkages, are confirmed. If this is true, a large range of alternative urban policies exist for planners and local policy makers; it is not only a matter of 'optimal city size', but also of processes of functional upgrading towards higher order function (more dynamic, richer and more innovative) and of strong linkages with the urban system in which the city lies.

3.7. Concluding remarks

A critical approach to the theory of 'optimal city size' has produced the following findings. The influence of urban size exists and is important, but cannot be efficiently assessed without overcoming some of the limitations imposed by the theory. It is not a problem of *optimal* city size, but of *efficient* size, which largely depends on what the city produces, how it produces and the way in which it cooperates within the urban system. Urban size inevitably influences location costs and benefits; however, the same also holds for its level of specialisation and integration with the urban system. Economies of scale exist, ceteris paribus, but turn into diseconomies after a certain urban dimension. However, with increasing size, the preconditions for developing structural changes allowing a greater mix of higher urban functions increase.

The most advanced empirical analyses have confirmed these hypotheses. In particular, the type of economic function and the spatial organisation within which the city is integrated appear to be strategic elements for the definition of location benefits and costs, analysed in relation to all aspects constituting the city, i.e. the social, environmental and economic aspects.

Our analysis has important normative consequences. Since it is difficult to envisage a large city having a strong city effect without facing high overload costs, local urban policies are absolutely vital and play a significant role in the definition of the growth potential of our cities. These policies should focus, among other things, on upgrading the economic functions within the city, as well as the development of linkages outside the city, such as alliances, cooperation agreements, advanced international transport and telecommunications infrastructure. All these elements are undoubtedly important elements for guaranteeing the survival of a modern city.

References

Alonso, W. (1964), *Location and Land Use: Towards a General Theory of Land Rents*, Cambridge, MA: Harvard University Press.

Alonso, W. (1971), *The Economics of Urban Size, Papers and Proceedings of the Regional Science Association*, Vol. 26, pp. 67–83.

Anderson, R. and T. Crocker (1972), "Air pollution and residential property values", *Urban Studies*, pp. 171–180.

Beeson, P. (1992), "Agglomeration economies and productivity growth", pp. 19–35, in: E. Mills and F. McDonald, editors, *Sources of Metropolitan Growth*, New Brunswick, NJ: Center for Urban Policy Research.

Berger, M., G. Blomquist and W. Waldner (1987), "A revealed-preference ranking of quality of life for metropolitan areas", *Social Science Quarterly*, Vol. 68, pp. 761–778.

Berglas, E. (1980), "The market provision of club goods once again", *Journal of Public Economics*, Vol. 15, pp. 389–393.

Blomqvist, G., M. Berger and J. Hoehn (1988), "New estimates of the quality of life in urban areas", *American Economic Review*, Vol. 78, pp. 89–107.

Bowman, T., G. Giuliani and M. Minge (1981), *Findings Your Best Place to Live in America, West Babylon*, New York: Red Lion Books.

Boyer, R. and D. Savageau (1985), *Places Rated Almanac*, New York: Rand McNally.

Breheny, M. (1992), "Sustainable development and urban form: an introduction", in: M. Breheny and S. Owens, editors, *Sustainable Development and Urban Form*, Londra: Pion.

Buchanan, J. (1965), "An economic theory of clubs", *Economica*, February, pp. 1–14.

Burnell, J. and G. Galster (1992), "Quality-of-life measurements and urban size: an empirical note", *Urban Studies*, Vol. 29(5), pp. 727–735.

Camagni, R. (1992), *Economia Urbana*, Roma: Nuova Italia Scientifica.

Camagni, R. (1993), "From city hierarchy to city networks: reflection about an emerging paradigm", pp. 66–87, in: T. Lakshmanan and P. Nijkamp, editors, *Structure and Change in the Space Economy: Festschrifts in Honor of Martin Beckmann*, Berlin: Springer.

Camagni, R. (1999), "La qualità della vita nelle città", *Consumatori, Diritti e Mercato*, Vol. 3, pp. 6–26.

Camagni, R., F. Curti and M.C. Gibelli (1985), "Ciclo urbano: le città tra sviluppo e declino", in: G. Bianchi and I. Magnani, editors, *Sviluppo Multiregionale: Teorie, Problemi, Metodi*, Milan: Franco Angeli.

Camagni, R., L. Diappi and G. Leonardi (1986), "Urban growth and decline in a hierarchical system", *Regional Science and Urban Economics*, Vol. 16, pp. 145–160.

Camagni, R., M.C. Gibelli and P. Rigamonti (2002), "Urban mobility and urban form: the social and environmental costs of different patterns of urban expansion", *Ecological Economics*, Vol. 20, pp. 199–216.

Capello, R. (1996), "Rendimenti Urbani e Risorse Ambientali: una Stima delle Esternalità Ambientali nella Funzione di Produzione Urbana", pp. 53–82, in: R. Camagni, editor, *Economia e Pianificazione della Città Sostenibile*, Bologna: Il Mulino.

Capello, R. (1998a), "Economies d'echelle et taille urbaine: théorie et etudes empiriques révisités", *Révue d'Economie Régionale et Urbaine*, Vol. 1, pp. 43–62.

Capello, R. (1998b), "Urban return to scale and environmental resources: an estimate of environmental externalities in an urban production function", *International Journal of Environment and Pollution*, Vol. 10(1), pp. 28–46.

Capello, R. and R. Camagni (2000), "Bejond optimal city size: an evaluation of alternative urban growth patterns", *Urban Studies*, Vol. 37(9), pp. 1479–1497.

Capello, R. and P. Rietveld (1998), "The concept of network synergy in economic theory: policy implications", pp. 57–83, in: K. Button, P. Nijkamp and H. Priemus, editors, *Transport Networks in Europe*, Cheltenham: Edward Elgar.

Carlino, G. (1980), "Constrast in agglomeration: New York and Pittsburgh reconsidered", *Urban Studies*, Vol. 17, pp. 343–351.

Catin, M. (1991), "Économies d'agglomération et gains de productivité", *Revue d'Économie Régionale et Urbaine*, Vol. 5, pp. 565–598.

Chinitz, B. (1961), "Constrast in agglomeration: New York and Pittsburgh", *American Economic Review Papers*, Vol. 51, pp. 279–289.

Christaller, W. (1933), *Die Zentralen Orte in Suddeuschland*, Jena: Gustav Fischer Verlag.

Clark, D. and J. Cosgrave (1991), "Amenities versus labour market opportunities: choosing the optimal distance to move", *Journal of Regional Science*, Vol. 31, pp. 311–328.

Clark, D. and J. Kahn (1988), "The social benefits of urban cultural amenities", *Journal of Regional Science*, Vol. 28, pp. 363–377.

Clark, D. and J. Kahn (1989), "The two stage hedonic wage approach: a methodology for the valuation of environmental amenities", *Journal of Environmental Economics and Management*, Vol. 16, pp. 106–120.

Conway, H. and L. Liston (1981), *The Good Life Index*, Atlanta: Conway Publications.

Corielli, F., P. Frigieri, A. Messori and P. Tedeschi (1996), "Applicazione della teoria dei prezzi edonici al mercato immobiliare milanese", pp. 123–144, in: R. Camagni, editor, *Economia e Pianificazione della Città Sostenibile*, Bologna: Il Mulino.

Cornes, R. and T. Sandler (1986), *The Theory of Externalities, Public Goods and Club Goods*, Cambridge: Cambridge University Press.

Cropper, M. (1981), "The value of urban amenities", *Journal of Regional Science*, Vol. 21, pp. 359–374.

Douglas, R. (1967), *Selected Indices of Industrial Characteristics for US: Metropolitan Statistical Areas, 1963, Discussion Paper No. 20*, Philadelphia: Regional Science Research Institute.

Duncan, O. (1956), "The optimum size of cities", in: J. Spengler and O. Duncan, editors, *Demographic Analysis*, New York: The Free Press.

Fuchs, V. (1967), Differentials in hourly earnings by regions and city size, 1959, *NBER Occasional Papers No.101*, New York.

Fujita, M. (1985), *Urban Economic Theory: Land Use and City Size*, Cambridge, MA: Cambridge University Press.

Fujita, M. and P. Krugman (1995), "When is the economy moncentric? von Thünen and Chamberlin unified", *Regional Science and Urban Economics*, Vol. 18, pp. 87–124.

Fujita, M., P. Krugman and T. Mori (1999), "On the evolution of hierarchical urban systems", *European Economic Review*, Vol. 43, pp. 209–251.

Getz, M. and Y. Huang (1978), "Consumer revealed preference for environmental goods", *Review of Economics and Statistics*, Vol. 60, pp. 449–458.

Henderson, J. (1974), "The sizes and types of cities", *The American Economic Review*, Vol. 64, pp. 640–656.

Henderson, J. (1982), "Evaluating consumer amenities and interregional welfare differences", *Journal of Urban Economics*, Vol. 11, pp. 32–59.

Henderson, J. (1985), *Economic Theory and the Cities*, Orlando: Academic Press.

Henderson, J. (1996), "Ways to think about urban concentration: neoclassical urban systems vs. the new economic geography", *International Regional Science Review*, Vol. 19(1/2), pp. 31–36.

Herzog, H. and A. Schlottmann (1993), "Valuing amenities and disamenities of urban scale: can bigger be better?", *Journal of Regional Science*, Vol. 33(2), pp. 145–165.

Hirsch, W.Z. (1968), "The supply of urban public services", in: H. Perloff and L. Wingo, editors, *Issues in Urban Economics*, Baltimore: John Hopkins Press.

Hoch, I. (1972), "Income and city size", *Urban Studies*, Vol. 9, pp. 299–328.

Izraeli, O. (1987), "The effect of environmental attributes on earnings and housing values across SMSAs", *Journal of Urban Economics*, Vol. 22, pp. 361–376.

Kawashima, T. (1975), "Urban agglomeration economies in manufacturing industries", *Papers of the Regional Science Association*, p. 34.

Ladd, H. (1992), "Population growth, density and the costs of providing public services", *Urban Studies*, Vol. 29(2), pp. 237–295.

Liu, B. (1976), *Quality of Life Indicators in US Metropolitan Areas*, New York: Praeger.

Marelli, E. (1981), "Optimal city size, the productivity of cities and urban production functions", *Sistemi Urbani*, Vol. 1/2, pp. 149–163.

Mera, K. (1973), "On the urban agglomeration and economic efficiency", *Economic Development and Cultural Change*, Vol. 21, pp. 309–324.

Mills, E. (1970), "Urban density functions", *Urban Studies*, Vol. 7, pp. 5–20.

Mills, E. (1993), "What makes metropolitan areas grow?", pp. 193–216, in: A. Summers, P. Cheshire and L. Senn, editors, *Urban Change in the United States and Western Europe*, Washington: The Urban Institute.

Moomaw, R. (1983), "Is population scale worthless surrogate for business agglomeration economies?", *Regional Science and Urban Economics*, Vol. 13, pp. 525–545.

Owens, S. (1992), "Energy, environmental sustainability and land use planning", in: M. Breheny and S. Owens, editors, *Sustainable Development and Urban Form*, Londra: Pion.

Richardson, H. (1972), "Optimality in city size, systems of cities and urban policy: a sceptic's view", *Urban Studies*, pp. 29–47.

Richardson, H. (1978), *Regional and Urban Economics*, Harmondsworth: Penguin Books.

Roback, J. (1982), "Wages, rents and the quality of life", *Journal of Political Economy*, Vol. 90, pp. 1257–1278.

Roback, J. (1988), "Wages, rents and amenities: differences among workers and regions", *Economic Inquiry*, Vol. 26, pp. 23–41.

Rosen, S. (1979), "Wage-based indices of urban quality of life", pp. 74–104, in: P. Mieszkowski and M. Straszheim, editors, *Current Issues in Urban Economics*, Baltimore: Johns Hopkins University Press.

Rousseaux, M.-P. (1995), "Y a-t-il une surproductivité de l'Île de France?", pp. 157–167, in: M. Savy and P. Veltz, editors, *Économie globale et Réinvention du Local*, Paris: DATAR/éditions de l'aube.

Rousseaux, M.-P. and R. Proud'homme (1992), *Les Bénéfis de la Concentration Parisienne*, Paris: L'OEIL-IAURIF.

Segal, D. (1976), "Are there returns to scale in city size?", *Review of Economics and Statistics*, Vol. 58, pp. 339–350.

Shefer, D. (1973), "Localization economies in SMSA'S: a production function analysis", *Journal of Regional Science*, Vol. 13, pp. 55–64.

Sveiskauskas, L., J. Gowdy and M. Funk (1988), "Urban productivity: city size or industry size", *Journal of Regional Science*, Vol. 28(2), pp. 185–202.

Sweikauskas, L. (1975), "The productivity of city size", *Quarterly Journal of Economics*, Vol. 89, pp. 393–413.

Berg van den, L., R. Drewett, L. Klaassen, A. Rossi and C.H.T. Vijverberg (1983), *Urban Europe: A Study of Growth and Decline*, Oxford: Pergamon Press.

World Resources (1996), *World Resources 1996–97, The Urban Environment, World Resources Institute, United Nations Environment Programme*, United Nations Development Programme/The World Bank.

Urban Dynamics and Growth, CEA, vol. 266
Roberta Capello and Peter Nijkamp (editors)

CHAPTER 4

Spatial Externalities and the Urban Economy

Erik T. Verhoef and Peter Nijkamp

Department of Spatial Economics, Free University Amsterdam, De Boelelaan 1105, 1081 HV, Amsterdam, The Netherlands

Abstract

This chapter is concerned with the economics of urban externalities. We start by reviewing the literature on urban externalities, and observe that although many interesting contributions have been made, there seems to be sufficient scope and need for further research, both theoretically and empirically. We identify what we believe to be important advances to be pursued in future research on urban externalities. These include (1) the explicit consideration of mutual interactions between externalities; (2) a thorough analysis of the relationship between these externalities and urban form; and (3) a clear focus on (realistic) second-best policies. The importance of these issues is illustrated by developing a simple urban general equilibrium model in which we study the interactions between agglomeration externalities and pollution from commuting. Our results show that what may appear impossible from a non-spatial perspective, namely a simultaneous stimulation of agglomeration externalities and a reduction of environmental externalities, is in fact the outcome of first-best policies in our spatial model. Moreover, while the incentives from road pricing and labour subsidies would seem to be perfectly opposite in a non-spatial setting, leaving one of the two instruments redundant, our results show that their welfare effects may, in contrast, turn out to be strongly super-additive when a spatial perspective is taken.

Both authors are affiliated to the Tinbergen Institute, Roetersstraat 31, 1018 WB Amsterdam. We thank two referees for helpful comments on an earlier draft, but of course we assume full responsibility for any remaining deficiencies.

Keywords: urban equilibrium, environmental and agglomeration externalities, second-best regulation

JEL classifications: D62, R13, R14

4.1. Cities in perspective

Over the past hundred years cities have developed into engines for the economic development of our globe and its regions. At present, the city mirrors part of a global network society by acting as a nodal point in an interlinked information and communication configuration (see Castells, 1996). But whatever appearance a city may have had in the history of mankind, it has always formed the cradle of civilisation. The key role of the city in ancient times is eloquently presented in a fascinating study of Tulleken (1988), when he writes:

> Yet by 3000 BC, an astonishingly different panorama was unfolding. Along the length of the valley, magnificent cities sprawled on the riverbanks. Around them, fields of grain spread like a tide of fecundity across the once desolated flatlands. Groves of date palms swayed in the wind, offering fruit and shade. Within the massive walls that ringed the cities, temples towered over both streetscape and plain. There were brick places and mansions and street after street of comfortable houses. People thronged the avenues and marketplaces; in hundreds of workshops artisans turned out all manner of goods, from pottery to sparkling jewelry. On holy days, processions of the worshipful wound through the streets to the temples. What had happened in this land the Greeks later called Mesopotamia, 'between the rivers', was the most crucial event in human history: the birth of civilization (p. 1).

Cities are a centre of socio-economic interplay, human confrontation, political dialectics, birthplaces of civilisation, centres of science and art, and a melting pot of cultures. According to Jacobs (1969), cities generate economic growth inter alia from the disordered order of human interaction. In the urban economics literature, the concept of agglomeration advantages reflects that a spatial clustering of economic activities (firms, industries, households, public services) leads to various types of economies of scale, which cannot be generated outside agglomerations. Indeed, a geographic juxtaposition may lead to win-win situations for all actors involved.

But of course, city life does not only have positive benefits, but also several disadvantages. A discussion on city life often witnesses uneasy feelings. O'Sullivan (2000) presents two quotes that nicely express the contrasting views one may have on the merits of the city: "Cities have always been the fireplace of civilisation, whence light and heat radiated out into the dark" (Theodore Parker) and "I'd rather wake up in the middle of nowhere than in any city on earth" (Steve McQueen) (p. 1). Despite the existence of mixed feelings about the city, the idea that the city is a 'blessing in disguise' is still prevalent. This has also to do with the great variety of roles cities are able to play. We will mention here a few of such important roles of cities, without striving to be exhaustive.

Shelter role: The city offers settlement facilities for numerous people, based on its scale advantages in housing many citizens. Shelter has even become a human right, and cities facilitate the housing needs of many people. From this perspective, cities offer a significant contribution to a sustainable human habitat.

Religious role: Cities have for long played an important role as centres of religious activities. Nevertheless, in the early biblical history the city was often regarded as a source of evil (Babylon, Nineveh). But in the later history we observe a more positive appreciation of the city. Jerusalem was the seat of King David and the New Jerusalem became even a metaphor for a total re-birth of mankind.

Cultural role: Historically, the city was the place where arts and sciences were flourishing. Venice, Bologna, Padua, Paris, Augsburg, Amsterdam and many other places offer an overwhelming evidence of the favourable seedbed conditions of an urban way of life for the advancement of culture.

Political role: Democracy was a new type of governance which found its seedbed in the city. And still nowadays political power is largely concentrated in cities and governments have established their premises in cities. Deconcentration of physical government facilities (e.g. premises) has never become very successful. Administrative functions are usually executed in capital cities of countries, or at least in cities with a critical political mass.

Economic role: The city is the market place for economic activity. It is also the place where usually products are designed and often manufactured. Furthermore, it is a marketplace where capital is supplied and advisory services are offered. In addition, the city is – as a result of various types of agglomeration advantages – a very efficient way of organising production and consumption.

Social role: Cities house thousands of people who are through the associative nature of city life able to communicate with a great number of others, intensively or less intensively. But they have a social contact and communication spectrum which far exceeds that of a uniform distribution of people.

Engineering role: The city is also the cradle of technological inventions and innovations. It brings together craftsmanship, technical expertise, hardware, software and orgware. As a result, cities are still the breeding places for the genesis of new products and services.

Network role: In an emerging network society cities become more and more the virtual centres of global network forces. The city brings together a *triple-C* potential: communication, competence and creativeness. Despite doomsday scenarios on the 'death of distance' and on the threats to city life, it is more plausible that cities continue to reinforce their role in local, regional, national and international networks.

The manifold strategic functions of the city have also induced many negative forces which might erode city life. Congestion, pollution, poor health conditions and criminality are examples of phenomena which exert a threat for survival of the modern city. The present chapter will, within limitations, focus attention in particular on the non-market dimensions of the urban economy. We will address in particular the externalities involved with a modern urban system as a nucleus of a network economy. In this context, not only the transport sector, but also urban land use (including residential sites, job places and public or private facilities) deserves a prominent place in such a discussion. In evaluating urban polices for such a complex urban system, we have to recognise the feasibility of first-best policy instruments *vis-à-vis* second-best policy instruments in order to care for some realism in our analysis. This sense of political realities is also prompted by the need to consider explicitly socio-economic and spatial equity impacts of externalities in an urban setting, particularly as unregulated urban markets may create many impediments to a balanced urban economic development.

4.2. Urban externalities

The spatial-economic architecture of the city can still powerfully be described by means of the well-known Von Thünen's land rent theory. In a monocentric world, distance friction together with competition among land users will lead to concentric rings of economic activity reflecting the users' bidding power. Clearly, this model has many limitations: it is static, it is based on uniform land use, it draws on equal spatial accessibility and it neglects (positive and negative) externalities. Several of these limitations have thoroughly been analysed in the urban economics literature (see e.g. Miyao, 1981; Fujita, 1989), while more recently insights from the new economic geography have been added to the body of knowledge on the economics of urban development (e.g. Fujita *et al.*, 1999).

Agglomeration economies are still at the heart of modern theories on urban growth. Geographical clustering often offers a great variety of economies of scale and scope (including transaction benefits, contact opportunities and search advantages), so that we observe at a global scale a continual urbanisation process. The agglomeration forces are apparently so strong that the shadow sides of urban areas are exceeded by positive density economies. The urban economy is full of externalities, both positive and negative, and the question is how the existence of such 'market failures' affects urban development. Examination of such externalities (ranging from social costs like waste or criminality to social benefits like increased contact potentials) requires a thoughtful economic framework that addresses, for instance, the urban activity market, the property conditions of real estate and land, and the supply of public and private goods. The role of governments in the case of market failures has extensively been discussed in the history of economic thinking by Pigou (1920) and Coase (1960). Market organisation, as well as (the distribution of) property rights, are main socio-economic arrangements affecting the effectiveness of government interventions (Webster and Wai-Chung Lai, 2003).

Thus, the modern urban scene offers many perspectives, of both a policy and a theoretical economic nature. It is widely recognised that the emerging network economy is prominently shaped by an urban force field. Cities exhibit a wide array of attractive (e.g. the proximity to a wide variety of goods, services and jobs) and unattractive (e.g. expensive housing, lack of open space and environmental quality) features, and are hence governed by centripetal and centrifugal forces, often leading to complex spatial dynamic processes. Although in the long run market forces may lead to some equilibrium between positive and negative aspects of city life, it is doubtful whether the result would represent an efficient spatial-economic configuration. Many positive and negative features of city life involve what economists call externalities: unpriced effects that economic agents impose upon one another. Important examples from the urban arena include traffic congestion, noise and smell, pollution, agglomeration advantages, and ethnic segregation and/or concentration. Free markets involving externalities typically do not yield the most efficient outcome. And this may have serious implications for the city, and for urban policies.

Given the high degree of urbanisation in most contemporary societies, and given the importance of the aforementioned externalities in urban firms' and citizens' daily lives, and given most citizens' concerns with the quality of urban life as well as the economic health and general well-being of the cities they live in, the economic analysis of these phenomena and the evaluation of policy options to deal with them are of great social importance, in addition to the academic interest that this type of analysis may provoke.

If the urban economy is fraught with externalities, a closer look at this phenomenon is no doubt warranted. The definition of externalities – unpriced effects that actors impose upon other actors – implies that externalities are often more important in urban areas than elsewhere, both absolutely and relatively. Urban areas are denser, and proximity makes the occurrence of unpriced spill-overs more likely. Nevertheless, most economic studies of externalities take a non-urban perspective, and the spatial dimension is typically lacking from the analysis. But this dimension may often be crucial in assessing the implications of externalities and policies, as both receptors and creators of externalities may often respond to these primarily in terms of spatial behaviour. For instance, spatial segregation may result from the desire to live close to some groups or activities and less close to others; potential visitors may avoid central city smell and noise by shopping elsewhere; and firms may seek a location near other firms in the hope of benefiting from knowledge spill-overs or labour market pooling. In aggregate, such behaviour may have important consequences for urban densities and urban form. And given the discrepancies that often exist between the private and social cost and benefits of individuals' location decisions, the resulting aggregate urban configuration may be far from optimal from the social perspective, indicating a clear potential and need for welfare enhancing policies.

Urban externalities have been studied from various perspectives in the prior literature. This broad literature encompasses contributions from various disciplines, including urban economics, geography, planning and sociology. It is therefore relevant to identify what we believe to be important advances to be pursued in future economic research on urban externalities. These include (1) the explicit consideration of mutual interactions between externalities, and (2) between these externalities and urban form, and (3) the focus on (realistic) second-best policies. It is important to motivate these priorities.

Interactions between externalities are important because these may strongly affect economic policy guidelines for (optimally) coping with external effects. To give a simple example, the transport economics literature has provided substantial underpinning for the use of pricing instruments in coping with traffic congestion externalities ('road pricing'). By setting charges equal to the travel time delay costs that road users impose on others (the 'marginal external costs'), unregulated traffic volumes will be reduced to optimal levels, and substantial social cost savings can be realised, outweighing the social benefits foregone for those priced off the road. Society as a whole would thus benefit. True as this conclusion may be in the context of a transport model that ignores relations between the transport system and the urban economy, it may require reconsideration when those travelling during peak hours are mainly workers who contribute to positive agglomeration economies during their workdays. These externalities may be reduced, or even eliminated, when the charge induces these workers to change their job location, or to work at home. The interaction between externalities would, in this example, call for a downward adjustment in congestion charges. In contrast, an upward adjustment might be in order if negative environmental externalities (pollution) result either from the commute itself, or from production activities following the commute.

Evident as these qualitative observations may seem, and despite the potentially far-reaching policy implications in an urban context where multiple externalities typically prevail simultaneously, such issues have only recently been investigated in the economics literature, usually applying non-spatial modelling approaches (see, for instance, Parry and Bento, 1999, on the interactions between inefficient labour and transport markets in the context of congestion pricing; and Verhoef *et al.*, 1997, on the policy implications arising from interactions between pollution from transport and from production, from a spatial perspective). There is certainly a need for significant contributions to this important and growing literature, in particular by exploring such interactions while explicitly accounting for the spatial urban dimension.

Indeed, it may often be important not only to consider interactions between externalities as such, but also interactions between externalities and urban form. These interactions may run in both directions: the existence of externalities may affect urban form (e.g. when citizens choose to live further from the centre to avoid noise or crowding), and urban form may affect the severity of externalities (e.g. traffic congestion will depend on the spatial arrangement of residential and job locations). Such interactions may affect the optimal design of policies directly, may also affect the consequences of these policies, and may in the third place have an indirect impact on the above-mentioned interactions between externalities (e.g. if congestion pricing would lead to a relocation of firms to less dense areas, the above-mentioned negative effect of congestion pricing on agglomeration externalities may become even larger insofar as these agglomeration externalities decrease with the distance between firms).

Thirdly, it is also necessary to explicitly consider what economists call ‘second-best’ policies. These are policies that are imperfect from the strict economic perspective, but that are often much more realistic from a practical viewpoint. An example, again from the transport economics literature, is when a road operator can only put charges on a limited number of links in a network, instead of all links as is required for an optimal, ‘first-best’, control (Verhoef, 2002). Economic policy rules under such second-best circumstances become much more complicated than under hypothetical first-best conditions. The reason is that economic distortions caused by the instrument itself should be taken into account when using that instrument. In the network example, prices equal to marginal external costs on those links that can be charged are not (second-best) optimal. It is even conceivable that such 'quasi first-best pricing' would lead to welfare losses, instead of the gains that are hoped for. Second-best policies are therefore certainly not an academic *curiosum*, but should instead be seen as of much greater relevance than the first-best pricing that has traditionally been considered in the literature. Given the complex nature of cities, where multiple externalities, and other market failures, exist simultaneously, and given the practical unavailability of first-best instruments, the development of analytical frameworks that allow consideration of second-best policies in urban economic models deserves high priority.

Thus, the field of urban externalities offers a rich research spectrum which ought to be explored more thoroughly. In this endeavour, several important distinct fields may be distinguished, such as traffic congestion in or around the city, environmental externalities and urban environmental quality, agglomeration externalities and industrial and services clustering, and neighbourhood externalities in relation to social and ethnic segregation and urban housing markets. It is noteworthy that all these externalities do not only refer to inefficiently operating urban markets, but involve also various types of distributive effects. In the practice of policy making and analysis, equity aspects tend to play as important a role as efficiency aspects, and this is certainly not different for the urban policy arena.

The above-mentioned range of urban economic and urban policy issues has intensively been discussed in the literature. In the next section, we will offer a concise and selected overview of various types of research, separated into theoretical and applied studies.

4.3. An overview of urban externalities studies

As mentioned, the economic literature on (urban) externalities is vast and covers many aspects of the urban economy. A very selective presentation of such studies with a major emphasis on policy is given here, while making a distinction between theoretical and empirical contributions.

4.3.1. Theoretical studies

Recent years have demonstrated an avalanche of theoretical urban policy evaluation studies. The long-run impacts of such policies – on externalities as well as on urban form – can typically best be identified using spatial general equilibrium approaches. Partial equilibrium approaches would often miss exactly those (spatial) interactions that were identified above as important in an urban context. For example, an analysis that would take into account the relocation of households due to the existence of environmental externalities in a city centre, or a policy designed to reduce these, but would fail to account for the implied change in labour supply in the city centre and the associated change in demand for goods supplied in this centre, would give a biased picture of the effects of the policy, and of how it should be designed in a (second-best) optimal way. The evaluation would namely be based on the characteristics of an urban system out-of-equilibrium, which therefore cannot be a long-run stable configuration.

No single analytical and integrated framework is currently available that can readily be used to analyse all or even most key research questions of urban economic policy. The development of proper analytical modelling tools is still a fertile field, and is among the greatest challenges to be met in urban economics.

The theoretical work along these lines should address fundamental questions surrounding the conceptually sound economic modelling of (multiple) externalities in a spatial equilibrium setting. It may build upon the contemporary urban and spatial economic tradition (for overviews, see for instance Fujita, 1989; Anas *et al.*, 1998; Fujita *et al.*, 1999), and then seek to extend this tradition where appropriate. It need not be wise to focus on one single stream of modelling approaches, or indeed to pursue one single general analytical methodology. In contrast, it seems of great importance to design and test different analytical model specifications in studying urban externalities in the context of policy analysis, as these may yield qualitatively different and sometimes contrasting insights.

One example concerns the modelling of agglomeration externalities. Various approaches have been proposed in the literature (Duranton and Puga, 2004, provide an excellent overview), varying from spatial modifications of the Dixit–Stiglitz (Dixit and Stiglitz,

1976) model of monopolistic competition where agglomeration effects arise through consumers' preferences for heterogeneity, as in Fujita *et al.* (1999), to reduced-form formulations where production efficiency increases in the aggregate urban production level (e.g. Sullivan, 1986) or labour supply (e.g. Arnott, 1979).

Another example concerns alternative ways of modelling traffic congestion, which may take the form of static stationary-state congestion (e.g. as in the spatial urban model by Anas and Kim (1996)), dynamic flow congestion (as in Chu (1995)), or dynamic bottleneck congestion (as in Vickrey (1969)).

For both examples, it is not unlikely that conclusions on the relation between the externality and urban form, as well as on the spatial impacts of externality regulation, may vary across different model specifications (as in fact demonstrated elegantly in Arnott (1998)). Such differences are of course important to identify – and conversely, if similarities are to be identified, this would often be a welcome establishment of robustness of results over alternative analytical formulations, too. Not surprisingly, similar differences in modelling approaches can be distinguished for other urban externalities of interest.

Likewise, diverging insights may be derived for models in which, for example, the city is 'open' or 'closed' (i.e. whether or not in- and outmigration is considered), a single city vs. a system of cities is considered, labour supply is fixed vs. endogenous, a single good or multiple goods are produced in the city, etc. Given the potentially decisive effects of such modelling characteristics on the insights generated, it is a research challenge to explicitly consider different, 'competing' formulations in investigating the issues at hand.

New analytical economic studies on urban externalities can nevertheless be expected to have a number of fundamental characteristics in common. They will typically be based on principles of utility maximisation by consumers and profit maximisation by producers. The models will have to describe spatial equilibria, where possible in continuous space and otherwise in discrete space, so that urban land use patterns are endogenous. Analyses will typically first have to be performed in the context of linear-space, static, often monocentric urban models in the tradition of Alonso (1964) and Muth (1969), before possible explorations are made into, for instance, two-dimensional (ground) space, dynamics, or multi-centric configurations.

Relatively simple urban externalities models – like the one we will present in Section 4.4 – will clearly suffer from a higher level of abstraction, and hence a more limited degree of realism. The reason that such models nevertheless have been – and still are – extensively used in the urban economics literature is that they allow the identification of fundamental economic insights in a relatively manageable setting. The next step then typically concerns the investigation of whether these insights carry over to more complex, more realistic, but less transparent settings. In other words, the insights from the simpler models help explain the results of the more elaborate models, which otherwise will be harder or even impossible to understand due to the complex interactions occurring.

For the same reason, new types of desired extensions will typically not be implemented simultaneously, but rather one by one, and depending on whether the type of problem studied would call for that extension. For example, two-dimensional ground space is relevant when the objective of study concerns traffic congestion on a network. Dynamics become relevant when path-dependency issues may play an important role, for instance, when studying segregation issues. And polycentric configurations may be relevant when studying, again, transport networks, or agglomeration externalities in the context of sub-centres.

For such models, it is to be expected that closed-form analytical solutions will typically not exist – or if they do, they will be too complex to offer any clear insight – so that numerical simulation modelling will form an important tool in identifying the comparative static or dynamic properties of various possible equilibria. Some examples of how such approaches can be used are, among many others, Anas and Kim, 1996; Tabuchi, 1998; Fujita *et al.*, 1999; Verhoef and Nijkamp, 2002.

4.3.2. Towards empirical studies

Next to theoretical model development, there have also been many advances in empirical studies. Nevertheless, substantial further research will be needed to better understand the causes, nature and consequences of urban externalities, and to identify policy implications. This part of the chapter again outlines some general principles of such research.

A first goal of empirical research studies on the urban economic system will, at least for some externalities, be to identify the very nature of these externalities. This is in particular relevant for the 'container-notion' of agglomeration externalities. Practical definitions of agglomeration externalities may encompass a plethora of market-internal and external relations. A market-internal relation would, for instance, be a firm's desire to economise on transport costs for its inputs and outputs, from and to other firms, and thus to locate near these other firms. Insofar as these costs are market internal (and congestion would be priced optimally), it is to be expected that firms would face economically optimal incentives to agglomerate, and no market intervention would be called for. On the other hand, agglomeration effects such as caused by knowledge spill-overs would reflect genuine externalities, that are not optimally reflected in market prices; at least not if the land market is competitive (following Henderson, 1985, one might argue that a city developer would have the incentive to optimally internalise such external benefits). These effects would therefore potentially ask for government intervention. It is therefore important to have a clear picture not only of the overall importance of agglomeration advantages in the urban economy, but also of which part of these reflect market failures that should be accounted for in (second-best) policies, and which part should be left to the market.

A second goal is to assess the empirical relevance of the various externalities studied, and thus to provide inputs for the calibration of spatial equilibrium models to be developed. The very nature of externalities – implying that these are unpriced in free markets – often makes their empirical measurement and valuation a difficult task. Nevertheless, various attempts to measure the empirical relevance of the external effects considered in this study have been made. For example, the relevance of traffic congestion externalities in urban areas can be assessed by using estimates of total (and marginal) travel time losses and uncertainty, in combination with estimates of what is known as the 'value of time' and 'value of unreliability' (see, for instance, Small, 1992). Agglomeration externalities have, for instance, been investigated empirically by analysing the impacts of the size of a sector (for 'localization economies'), or of the urban economy as a whole (for 'urbanization economies'), upon a sector's productivity (e.g. Henderson, 1986; Mun and Hutchinson, 1995). Ciccone and Hall (1996) consider the relation between urban density and productivity. Environmental externalities in urban areas have, for example, been monetized using hedonic price methods (see Freeman, 1993, for an overview). And the empirical economic effects of segregation have, for instance, been assessed by looking at the relation between the degree of segregation and indicators such as high school dropouts, employment and single motherhood (Cutler and Glaeser, 1997).

It is noteworthy that also outside the field of (urban) economics, a wealth of literature is available that deals with similar phenomena. One can think of traffic engineering in the context of congestion, geography and planning for agglomeration, environmental studies for environmental externalities, and sociology and cultural anthropology for segregation. Such studies will guide us in further understanding the necessary inputs for the modelling exercises, but also more general empirical patterns of economic development in the city. One of the main challenges in an applied context appears to be the question of how people would respond to urban economic policy instruments.

An important methodological choice in empirical valuation studies is between revealed preference studies, studying actual market behaviour of the relevant agents, and stated preference studies, using questionnaire type of approaches. Both techniques have been used widely in valuation studies, and both have their respective advantages and disadvantages (see for instance Perman *et al.* (1999), for a discussion in the context of environmental externalities). There have been many recent advances, especially in the field of contingent valuation studies, and such studies may assume a more prominent methodological place in the study of urban externalities.

4.4. A modelling framework for urban externalities: analysing first-best and second-best policies for multiple externalities

It is helpful, at this point, to present an illustrative modelling framework for analysing first-best and second-best interventions when multiple externalities exist simultaneously in an urban economy. For reasons of simplicity, we have chosen the example of a traditional monocentric city, and assume that only one single product is produced in its (assumed spaceless) CBD. Two externalities are present in our city. First, agglomeration effects that are external to the individual firm explain why all production is concentrated in the CBD – and perhaps why the city exists in the first place. Agglomeration externalities are modelled in reduced form (i.e. no micro-foundation is spelled out explicitly), and cause marginal productivity to increase with total labour supplied, for example, because of knowledge spill-overs. This aggregate positive relation between labour supply and productivity is consistent with most structural models of agglomeration benefits (Duranton and Puga, 2004). The second externality is caused by the fact that commuters travelling to the CBD, using the automobile as the single mode available, pollute the urban environment.

As Tinbergen (1952) has argued, with two externalities present one would generally need two independent policy instruments to achieve the target of maximising social surplus. In the case studied here, these would be a road tax to control the environmental externality and a labour subsidy for stimulating labour supply and hence knowledge spill-overs. However, a spaceless analysis of the problem might easily suggest that these two 'sub-objectives' are in fact both directly and in a monotonous fashion dependent on aggregate labour supply, so that one instrument would suffice. There would then be a simple trade-off, stipulating that depending on whether the marginal agglomeration externality or the marginal environmental externality dominates, a first-best policy mix would either mean that aggregate labour supply and hence commuting should both be increased, or that both should be decreased. Under fixed vehicle technologies, a simultaneous stimulation of agglomeration externalities and reduction of pollution from commuting would seem impossible. We have chosen this example to highlight the importance of taking a spatial perspective on analysing urban externalities, and will present the result that when doing so, first-best policies indeed can result in such an attractive – but counter-intuitive – 'win-win' outcome. Moreover, while the incentives from road pricing and labour subsidies would seem to be perfectly opposite in a non-spatial setting, leaving one of the two instruments redundant, our results will show that their welfare effects may, in contrast, turn out to be strongly super-additive when a spatial perspective is taken.

The model deployed here is inspired by Verhoef and Nijkamp (2002), who consider the trade-offs between agglomeration externalities and pollution from production in a mono-centric city; the presentation of the model in Section 4.4.1 draws heavily from their earlier work and model. The main features of our model are the following. We consider a general spatial economic equilibrium: the land market, the labour market and the market for the industrial product in the city are simultaneously in equilibrium. Although we have a 'closed city', with a given population, labour supply is flexible and endogenous because a household can vary the hours of labour supplied. The firms are assumed to cluster in a spaceless CBD, and only use labour in their production process. The perfectly competitive firms have identical linear production functions. Agglomeration economies in the city are represented by 'Marshallian' externalities: marginal and average productivity increases in the aggregate labour supply. Because the city has only one sector, these agglomeration externalities are of the localisation (not urbanisation) type. Pollution from commuting arises in a fixed proportion of total kilometres driven, and disperses over the entire city and hence affects the local environmental quality in a spatially non-differentiated manner. Therefore, proximity to the CBD yields a benefit in terms of lower commuting costs, but no relative disbenefit due to a poorer environmental quality. We aim to investigate the market equilibrium that results in such a configuration, and compare it with equilibria that would arise under various policies: first-best policies, and two second-best policies that only internalise the congestion externality and the agglomeration externality, respectively.

4.4.1. The analytical model[1]

In this section, we present the details of the analytical model. Before turning to a detailed description of consumers' behaviour, firms, and a characterisation of general equilibrium, some introductory remarks are in order. First, z will be used to denote a one-dimensional continuous urban space. The location of the spaceless industrial area is at $z = 0$, and the residential area stretches from $z = 0$ to $z = z^*$,

[1] This section draws heavily from Verhoef and Nijkamp (2002).

with z^* being the endogenous city boundary. Commuting time increases linearly with z at a rate t: there is no traffic congestion. At the boundary of the city, the equilibrium residential bid-rent must be equalised to the exogenous and constant agricultural bid-rent r_A, because land should go to the highest bidder.

It is assumed that all excess land rents above r_A are redistributed among the city's population. Alternatively, we could have used the 'absentee land-lord assumption', which seems less realistic, as it assumes that none of the land rents generated in the city would be used for consumption in the city. At the same time, it is less plausible to assume that *all* land rents generated in the city are redistributed among the population, as this would imply that the endogenous city size could be expanded costlessly. The present representation would correspond to the situation where the public authority of the city buys the urban land against the relatively low rural land price, implying an equivalent per-unit-of time price of r_A, and redistributes all excess rents generated in the city among its population. It is a convenient assumption in the sense that it easily allows us to consider households with similar initial endowments. As a result of these assumptions, some share of the urban production will not be consumed in the urban area, but will be exported in exchange for the purchase of land against the agricultural rent.

Some final assumptions and remarks are to be made. All consumers and producers are assumed to be price-takers. Households are identical, and so are firms. The industrial product can be transported costlessly, and the world-market price of the industrial good p is used as the *numéraire*. We now turn to the various actors in the city and the resulting equilibrium issues.

4.4.1.1. Consumers

The closed city has N households, which we treat as a continuum of single economic entities. A household's utility depends on the consumption of the industrial good y, on the consumption of space or the size of the residence s, on the consumption of free time or leisure T_f, and on the environmental quality, Eq. A household's financial budget then consists of the wage rate w times the amount of hours worked T_w, plus the redistributed excess land rents (R in total, R/N per household), plus the share in the government's surplus or deficit as it arises from taxing road use and/or subsidising labour supply

(B in total, B/N per household). In equilibrium, the household's budget is fully spent on the consumption of y and s, and – if levied – on a road toll for commuters (τ_R per unit of distance). A household's given time budget is denoted by T, and can be spent on leisure (T_f), work (T_w) and commuting (T_c). All prices are treated parametrically by the (price-taking) households; w denotes the wage rate, p the price of the industrial good and r the rent. Commuting does not require financial outlays other than possibly a total toll (over the full trip) $\tau_R z$, but costs time at a given rate t: $T_c = tz$. The number of commuting trips made by an individual is assumed to be proportional with the amount of effective working time supplied (T_w).

A household's simultaneous labour supply and consumption decisions can be modelled by using the 'gross budget', that would be available under the maximum possible amount of time worked, and to let the household 'buy back' leisure time against the prevailing wage rate w. Observing that the household's optimisation problem is dependent on the residential location z, it can then be written as

$$\max_{y(z),s(z),T_f(z)} U(y(z), s(z), T_f(z), Eq(z))$$
$$\text{s.t.} \frac{R+B}{N} + (w - \tau_R z)(T - tz - T_f(z)) - py(z) - r(z)s(z) = 0 \tag{4.1}$$

with

$$R = \int_0^{z^*} r(z) - r_A \mathrm{d}z \tag{4.2a}$$

and

$$B = \int_0^{z^*} n(z)(T - tz - T_f(z))\tau_R z \mathrm{d}z - \int_0^{z^*} n(z)(T - tz - T_f(z))\sigma_L \mathrm{d}z \tag{4.2b}$$

where $n(z)$ gives the density of households at z, and σ_L is the second possible policy instrument of labour subsidies. The gross budget available at location z is thus defined as

$$M(z) = \frac{R+B}{N} + (w - \tau_R z)(T - tz) \tag{4.3}$$

A spatial equilibrium requires that utility $U(z)$ be constant over z for all $0 < z \leq z^*$ (and exceeds $U(z)$ for $z > z^*$): if a higher utility could be reached at any z, the rent $r(z)$ would be bid up, as households would like to move from their original location to z in order to enjoy the higher utility level prevailing there. This implies a particular equilibrium pattern of land rents. We can be more explicit about this when postulating a specific form for the utility function. We will be using a simple Cobb–Douglas structure:

$$\begin{aligned} &U(z) = y(z)^{\alpha_y} s(z)^{\alpha_s} T_f(z)^{\alpha_f} Eq^{\alpha_e} \qquad \text{with} \\ &\alpha_y + \alpha_s + \alpha_f = 1 \end{aligned} \tag{4.4}$$

Because utility is ordinal and any monotonic transformation of a given utility function represents the same preferences, the constraint on the parameters can be added without loss of generality. This utility function has the specific property of a unitary elasticity of substitution, implying that the gross income shares spent on y, s and T_f will be constant and given by the relative sizes of the parameters α. Specifically, the conditional demands for y, s and T_f are

$$y(z) = \frac{\alpha_y M(z)}{p} \tag{4.5a}$$

$$s(z) = \frac{\alpha_s M(z)}{r(z)} \tag{4.5b}$$

$$T_f(z) = \frac{\alpha_f M(z)}{w - \tau_R z} \tag{4.5c}$$

and the indirect utility – for analytical convenience defined as the logarithm of the maximum utility achievable under given prices and wage – can be written as

$$V(z) = \alpha_y \ln \alpha_y + \alpha_s \ln \alpha_s + \alpha_f \ln \alpha_f + \ln\left(\frac{R+B}{N} + (w - \tau_R z)(T - tz)\right) - \alpha_y \ln p - \alpha_s \ln r(z) - \alpha_f \ln(w - \tau_R z) + \alpha_e \ln Eq \tag{4.6}$$

The condition that V in Equation (4.6) be constant over space implies:

$$V'(z) = \frac{-wt - \tau_R T + 2z\tau_R t}{\dfrac{R+B}{N} + (w - \tau_R z)(T - tz)} - \alpha_s \frac{r'(z)}{r(z)} + \alpha_f \frac{\tau_R}{w - \tau_R z} = 0 \tag{4.7}$$

where a prime denotes a 'space derivative' (with respect to location). Equation (4.7) gives a first-order differential equation for $r(z)$ that can be solved to yield:

$$r(z) = K(w - \tau_R z)^{-\alpha_f/\alpha_s}(B + R + N(T - tz)(w - \tau_R z))^{1/\alpha_s} \tag{4.8}$$

where K is a constant of integration. Invoking the equilibrium condition that $r(z^*) = r_A$, we can solve for K:

$$K = \frac{r_A}{(w - \tau_R z^*)^{-\alpha_f/\alpha_s}(B + R + N(T - tz^*)(w - \tau_R z^*))^{1/\alpha_s}} \tag{4.9}$$

To derive the equilibrium level of utility, we must be explicit about the production structure. This issue will be addressed below. We first conclude this analysis with a few identities. We can find the local population density $n(z)$ as the inverse of the 'lot-size' $s(z)$:

$$n(z) = \frac{1}{s(z)} \tag{4.10}$$

The total population is given, so that

$$\int_0^{z^*} n(z)\mathrm{d}z = \int_0^{z^*} \frac{1}{s(z)}\mathrm{d}z = N \tag{4.11}$$

Total labour supplied equals:

$$L = \int_0^{z^*} n(z)(T - tz - T_f(z))dz \tag{4.12}$$

Total local consumption of the city's product equals:

$$Y = \int_0^{z^*} n(z)y(z)dz \tag{4.13}$$

The total amount of land consumed in the city must be equal to z^*, which is by definition true:

$$z^* = \int_0^{z^*} n(z)s(z)dz = \int_0^{z^*} 1\,dz = z^* \tag{4.14}$$

Finally, we define environmental quality such that a virgin state corresponds to $Eq = 1$, while the worst possible state, where an increase of consumption of other goods cannot even increase utility, occurs when $Eq = 0$. Eq is assumed to decrease in a simple linear fashion in total emissions from commuting E, which in turn depends in a linear fashion on the total amount of commuting (weighted by distance), Km, in a fixed proportion ε (i.e. ε gives the emissions per unit of distance driven):

$$Eq = 1 - E = 1 - \text{Km}\varepsilon = 1 - \int_0^{z^*} n(z)(T - tz - T_f(z))\varepsilon z\,dz \tag{4.15}$$

4.4.1.2. Producers

There is a continuum of firms, each of which is infinitesimally small relative to the market and takes all prices as given. The industrial output is homogeneous, and agglomeration externalities in our model thus arise from a more efficient production when aggregate labour supply L increases. These agglomeration effects are summarised in an efficiency measure A, which individual firms take as given, but that is endogenous on the city level to represent agglomeration economics. Firms have a simple linear production technology with one input (labour). A firm's production function thus exhibits constant returns to scale, and therefore qualifies for application of Euler's theorem. Therefore, also when the urban aggregate production function exhibits increasing returns to scale due to agglomeration externalities, we can model the firms' aggregate behaviour using a 'derived aggregate production function', in which the efficiency parameter A is treated parametrically. The following derived aggregate production function applies:

$$Q = AL \tag{4.16}$$

We will use the following functional form for A in the numerical model:

$$A = 1 + \alpha_L L^{\mathrm{aggl}_L} \tag{4.17}$$

Therefore, we use a simple relation, where $aggl_L$ in reality is likely to be (and will restricted to be) smaller than one. The fixed unitary factor can be interpreted as the efficiency level for the first firm to locate in the CBD, and might reflect the presence of local infrastructure, etc. A more than sufficient, but drastic condition for maintaining a monocentric configuration would be to set $A = 0$ for locations outside the CBD. As we wish to focus attention on monocentric configurations, we will make this assumption.

Perfect competition drives profits to zero, with the result that the following equality holds:

$$pA = w - \sigma_L \tag{4.18}$$

4.4.1.3. *General spatial equilibrium*

The model described above has 19 unknowns, some of which are functions of z. These unknowns are $V(z)$, w, $M(z)$, $r(z)$, R, B, K, $y(z)$, Y, $s(z)$, $n(z)$, $T_f(z)$, L, z^*, Eq, E, Km, Q and A (recall that r_A, p and N are given; all other scalars are clearly parameters). The 19 equations needed to solve this system are (4.2a), (4.2b), (4.3), (4.5a)–(4.5c), (4.6), (4.8)–(4.13) and (4.15)–(4.18); note that Equation (4.15) represents three equations. For other types of utility and production functions, as long as they imply unique conditional (factor) demands, a similar equality of numbers of equations and unknowns should in principle hold. We refrain from a formal analysis of existence, uniqueness and stability of equilibria and optima in our model.

In our list of equations, we did not include the 'aggregate demand equals aggregate supply' relation, which in our partly open system reads:

$$p(Q - Y) = r_A z^* \tag{4.19}$$

Equation (4.19) states that the value of the city's production in excess of its local consumption should be just sufficient to pay for the purchase of land against the exogenous terms of trade r_A/p. The share of local production not exported is consumed locally. The reason for not including this equilibrium condition explicitly is that it will be automatically satisfied under the zero profit condition and exhaustion of consumers' financial budgets – as in fact dictated by Walras' Law. To see why, first observe that zero profits imply that:

$$pQ = (w - \sigma_L)L \tag{4.20}$$

The exhaustion of consumers' total financial income implies (in aggregate terms) that the sum of redistributed land rents, the redistributed government surplus and wage income should be equal to the sum of expenses on the local product, rents and the road toll:

$$R + (\tau_R \mathrm{Km} - \sigma_L L) + wL = pY + (r_A z^* + R) + \tau_R \mathrm{Km}$$

$$\Rightarrow (w - \sigma_L)L = pY + r_A z^* \tag{4.21}$$

Substitution of Equation (4.21) into Equation (4.20) immediately yields Equation (4.19).

It is not possible to obtain any further analytical (equilibrium) results for our model. We therefore now move to the results of a numerical illustration, to study the comparative static properties of the free-market and some second-best and first-best equilibria.

4.4.2. A numerical example: base-case equilibrium

The numerical model is fully consistent with the analytical model just presented. Given that the model was built solely for the purpose

Table 4.1. Parameters and equilibrium values of some key variables in the base-case scenario

Parameters		Base-case Equilibrium Values of some Key Variables	
$\alpha_L = 0.03$	$N = 1000$	$Y = 298$	$pQ = 399$
$aggl_L = 0.5$	$T = 1$	$Q = 399$	$wL = 399$
$\alpha_s = 0.15$	$t = 1 \times 10^{-5}$	$A = 1.49$	$p(Q - Y) = 101$
$\alpha_y = 0.2$	$p = 1$	$w = 1.49$	$r_A z^* = 101$
$\alpha_f = 0.65$	$r_A = 0.005$	$L = 268$	$R = 122$
$\alpha_e = 0.1$		$Km = 2.09 \times 10^6$	$B = 0$
$\varepsilon = 1 \times 10^{-7}$	*Taxes and subsidies*	$E = 0.209$	
	$\tau_R = 0$	$Eq = 0.791$	
	$\sigma_L = 0$	$z^* = 20,256.5$	

of this paper, no particular attention was paid to its computational speed, and an algorithm was used that can most briefly be described as one that successively equilibrates the various markets, repeating the procedure until convergence (according to the criteria set) is reached.[2] Given the nature of the problem, a mathematical software package (*Mathematica* 4.1) was used that is capable of numerical solution of differential equations.

The left two columns of Table 4.1 show the parameter values we assumed for the base-case of our simulation model. The city has 1000 inhabitants, with a time endowment that is normalised to unity. Some 30% of the time available after commuting will be spent working (which appears reasonable when excluding 8 h sleeping time from the daily time budget: it would then mean an average of 33 h per week over the whole year), and 42% (100 × 0.15/(0.15 + 0.2)) of the monetary budget will be allocated to rent, the remaining share being used for the consumption of the locally produced good. The price p is the *numéraire* in the price dimension. The right two columns show that in equilibrium, the average productivity is 49% higher due to agglomeration benefits than would be the case without them. Around 25% of the urban product is exported for the purchase of land. The maximum commuting time at

[2] Once the starting values are set appropriately, the model reaches convergence within a minute on a modern PC.

z^* is 0.2; which implies an average of 3.2 h (for a round trip) per day (of 16 h) over a full year. The environmental quality has reduced to 79% of the virgin state. Because α_e is set at 0.1, this means that an inhabitant would be indifferent between obtaining a virgin state environmental quality or a 2.4% increase in the availability of all other goods (including leisure time) in the utility function, keeping everything else constant. These figures should give an adequate overall impression of the assumed base-case.

Figure 4.1 shows the spatial patterns of some variables of interest in the base equilibrium. The upper-left panel shows the equilibrium land rents, which have the expected convex shape. Land rents near the CBD are around four times as high as the agricultural rent applying at the fringe. As a result, households near the fringe will occupy nearly four times as much space as households near the CBD, as shown in the upper-right panel. Total expenses on space are somewhat lower near the fringe, because the monetary income is lower than in the centre due to a smaller number of hours worked (as shown in the lower-left panel). As a result, households near the fringe also consume less of the local product. The smaller

Figure 4.1. Equilibrium land rents (upper-left panel), consumption (upper-right panel), labour supply (lower-left panel) and utility (lower-right panel)

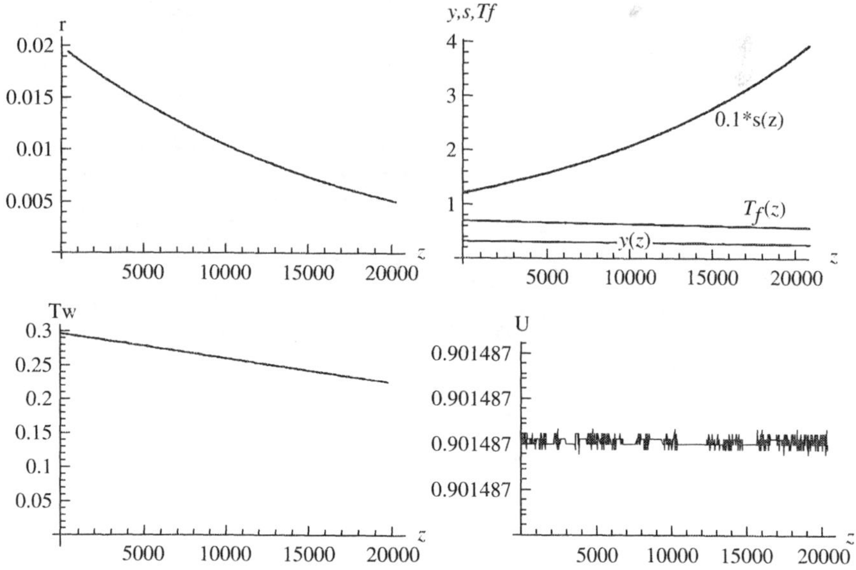

consumption of leisure time near the fringe, despite the smaller amount of hours worked, is explained by the extra time spent commuting (not shown in Figure 4.1). Finally, the lower-right panel shows that indeed a spatial equilibrium is obtained: utility – calculated directly according to Equation (4.4), with the endogenously determined equilibrium values of $y(z)$, $s(z)$, $T_f(z)$ and Eq substituted in – is constant over space, and the fluctuations shown are evident due to numerical imprecision only.

4.4.3. First-best regulation

Given the simultaneous existence of two externalities in our model, it should be no surprise that the free-market equilibrium ('base equilibrium' in the sequel) is not efficient. Given that environmental externalities arise in a fixed proportion to kilometres driven, and agglomeration externalities depend exclusively on the amount of labour supplied, first-best policy instruments are easily identified. These consist of the combination of a road tax τ_R for every kilometre driven, and a subsidy σ_L on every unit of labour supplied. In the optimum, the levels of these instruments should be equal to the marginal externality. These levels were determined numerically, and the first column of Table 4.2 shows the corresponding values, as well as the resulting values for some

Table 4.2. Key characteristics of first-best and second-best optima (% relative to base equilibrium)

	First-best	Second-best Road Pricing	Second-best Labour Subsidy
τ_R	2×10^{-5}	7.5×10^{-6}	0
σ_L	0.25	0	0.05
B	−30.9	14.3	−13.7
Y	107.97%	97.55%	102.76%
Q	103.53%	96.72%	102.31%
A	100.49%	99.53%	100.32%
w	117.26%	99.53%	103.68%
L	103.02%	97.17%	101.98%
Km	91.31%	91.45%	102.58%
E	91.31%	91.45%	102.58%
Eq	102.29%	102.25%	99.32%
z^*	90.48%	94.26%	100.98%
R	122.44%	100.28%	104.24%

key (non-spatial) endogenous variables as a proportion of their values in the base equilibrium.

The results clearly reject the hypotheses that one might infer from a non-spatial analysis. These would be that there is a simple trade-off between the two externalities, stipulating that depending on whether the marginal agglomeration externality or the marginal environmental externality dominates, a first-best policy mix would either mean that aggregate labour supply and hence commuting should be increased, or that both should be decreased; and that one of the two instruments would be redundant as they both affect the same margin of behaviour (labour supply, and in a fixed proportion to this, commuting). Under the first-best policy mix, labour supply increases by 3.0% and total commuting decreases by 9.7%. As a result, environmental quality improves by 2.3%, and marginal productivity increases by 0.49%.[3] This favourable combination can be realised because average commuting distances decrease: z^* falls by 9.5% compared to the base equilibrium. Instead of giving conflicting incentives, the road tax in an indirect way in fact contributes to the goal of stimulating labour supply. It gives a direct incentive to shorten commuting distances, which in turn gives an incentive to increase labour supply T_w, which – as was shown in Figure 4.1 – increases with proximity to the CBD.

Figure 4.2 compares the spatial patterns of two endogenous variables under first-best regulation to the base equilibrium. The left panel shows that land rents rise over most of the city, and more strongly so for locations closer to the CBD (near the new fringe, land rents of course must have fallen as $r(z^*) = r_A$ continues to hold). This is a direct consequence of the imposition of the road tax – increasing the per-unit-of-distance total commuting costs and hence making the equilibrium bid-rent steeper – and is indirectly aggravated by the substitution away from land consumption. Recall that excess land rents are redistributed over the city's population, so that the city's net land costs ($r_A z^*$) nevertheless decrease. The right panel shows that labour supply increases over most of the city, but

---

[3] The implied elasticity of average productivity (A) with respect to aggregate production (Q) of 0.14 (= 0.49/3.53) 0.14 is well in line with the empirical estimates reviewed by O'Sullivan (2002) for localisation economies, which range between 0.02 and 0.27 (p. 50).

Figure 4.2. ***First-best land rents (left panel) and labour supply (right panel); base equilibrium dashed***

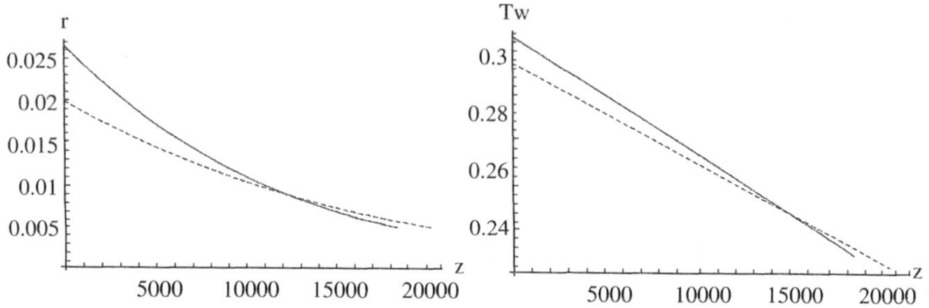

again more strongly so for locations closer to the CBD. These spatial differences are again explained by the increase in per-unit-of-distance total commuting costs, which decreases the shadow price of leisure time more strongly for more distant locations; compare Equation (4.5c).

4.4.4. Second-best regulation

First-best policies are, tautologically, the best choice when the city authority seeks to maximise the citizens' welfare. At the same time, the practical relevance of such idealised interventions may be smaller than most economists would hope, for a variety of reasons. First-best policies may be considered too complex, may require 'too much' cooperation between different governmental authorities (in our example the city's department of transport and the department of economic affairs), etc. Second-best policies then come into the picture, and deserve – we believe – much more attention from the urban economist and policy maker. We will therefore consider two possible second-best policies, namely the use of a road tax τ_R in isolation, and the use of a labour subsidy σ_L in isolation. Again we have refrained from the (notoriously difficult) task of deriving analytical expressions for the second-best optimal levels of these instruments, and have instead derived these numerically for our model. The second and third column in Table 4.2 provide the detailed impacts of these second-best policies for the non-spatial endogenous variables of interest, while Figures 4.4 and 4.5 show the resulting impacts on the spatial patterns of land rents and labour supply.

Figure 4.3. Second-best regulation: equivalent variation as a proportion of first-best equivalent variation as a function of tax/subsidy levels

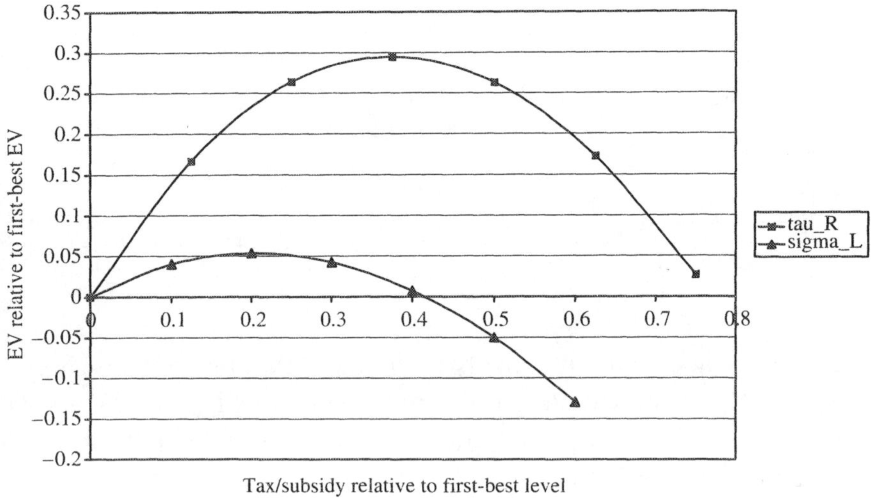

First, however, Figure 4.3 shows the relative welfare gain, expressed as the proportion of the first-best efficiency gain that can be achieved with these second-best instruments, as a function of their levels. Efficiency gains are pragmatically measured as the equivalent variation for the median-location household in the base equilibrium, while Figure 4.3 expresses tax and subsidy levels as proportions of first-best levels. Figure 4.3 conveys a number of messages. The first is that second-best policies are considerably less efficient than first-best policies, with the relative efficiency gain peaking just below 0.3 for the road tax, and just above 0.05 for the labour subsidy. Because the sum of these relative gains is around 0.35, it appears that the joint use of these instruments yields a gain nearly three times as high as the sum of the gains when the instruments are used in isolation. In other words, the efficiency gains from the two instruments are (strongly) super-additive.

The results in the second and third column in Table 4.2 show why: when used in isolation, the road tax reduces agglomeration benefits, while the labour subsidy increases pollution. The explanation for these negative side-effects is simple, and similar to what one would infer from a non-spatial analysis: a road tax discourages commuting and hence labour supply (as shown also in the right panel of

Figure 4.4. ***Second-best road pricing: land rents (left panel) and labour supply (right panel); base equilibrium dashed***

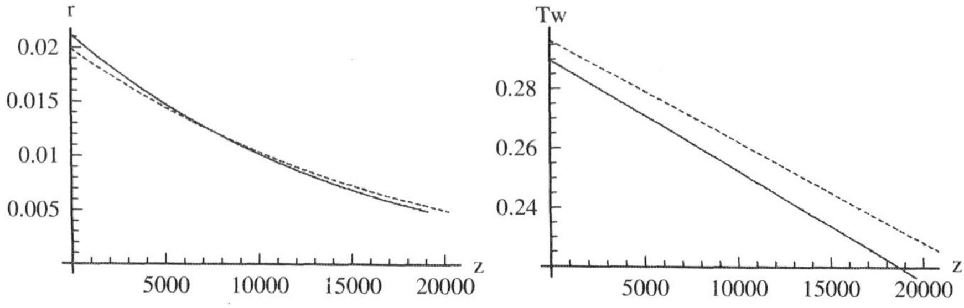

Figure 4.4), and a labour subsidy stimulates labour supply and hence commuting (as shown also in the right panel of Figure 4.5). As we discussed above, these unfavourable side-effects of the policies when used in isolation can be avoided when using them in combination.

But there is another lesson to be drawn from Figure 4.3, i.e. that the second-best optimal use of the two instruments is at a level significantly below their first-best levels (at around 20% for the labour subsidy and at around 40% for the road tax). The explanation is that the second-best optimal levels trade-off the 'good news' from the two instruments (enhancing agglomeration benefits for the labour subsidy and reducing pollution for the road tax) against the bad news (reducing agglomeration benefits for the road tax and enhancing pollution for the labour subsidy). The consequence is that a naïve use of the instruments, set at their first-best levels despite the fact that the other instrument is not used, would lead to efficiency gains below the second-best gains.

Figure 4.5. ***Second-best labour subsidies: land rents (left panel) and labour supply (right panel); base equilibrium dashed***

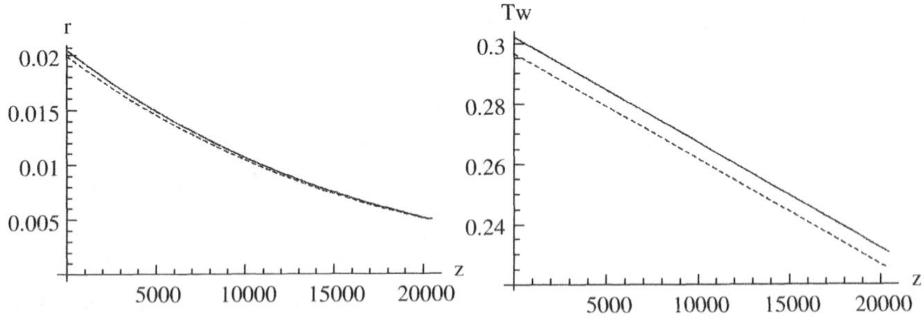

In fact, in our numerical example, for both instruments an efficiency loss would result from such naïve policies.

This is of course a nice illustration of our earlier claim that for an urban economy, expectedly fraught with many positive and negative externalities, it seems of great importance to develop modelling tools allowing for a simultaneous consideration of multiple externalities. The risk of performing 'less than second-best' is otherwise not inconceivable. Likewise, our simple model has illustrated why it is important to analyse urban externalities in a general spatial equilibrium setting: the optimal use of the two instruments in combination and the super-additivity of their efficiency gains could otherwise not have been identified. And the third priority for research that we mentioned – the importance of considering second-best issues when analysing inherently second-best instruments – is illustrated by the large discrepancy between 'truly' second-best and naïve first-best levels for the use of these instruments.

4.5. Conclusion

This chapter has addressed the economics of urban externalities. We started by reviewing the literature on urban externalities, and observed that although many important contributions of course have been made, there seems to be sufficient scope and need for further research, both theoretically and empirically. We identified what we believe to be important advances to be pursued in future research on urban externalities. These include (1) the explicit consideration of mutual interactions between externalities; (2) a thorough analysis of the relationship between these externalities and urban form; and (3) a clear focus on (realistic) second-best policies.

The importance of these issues was illustrated by developing a simple urban general equilibrium model, in which we examined the interactions between agglomeration externalities and pollution from commuting. Our results show that what seems impossible from a non-spatial perspective, namely a simultaneous stimulation of agglomeration externalities and reduction of environmental externalities, is in fact the result from first-best policies. The reason lies in spatial adaptations, in particular a shortening of commuting distances which directly reduces pollution, and indirectly – via the increase in labour supply that follows from a greater proximity to the CBD – stimulates agglomeration benefits. Moreover, while the incentives from road pricing and labour subsidies would seem to be perfectly opposite in a non-spatial setting, leaving one of the two instruments redundant, our results show that their welfare effects may, in contrast, turn out to be strongly super-additive when a spatial perspective is taken. The model thus underlines the importance of using spatial general equilibrium frameworks when analysing urban externalities.

Our results furthermore provided a nice illustration of our claim that for an urban economy, expectedly fraught with many positive and negative externalities, it seems of great importance to develop modelling tools allowing for a simultaneous consideration of multiple externalities. The risk of performing 'less than second-best' is otherwise not inconceivable. And finally, the large discrepancy between 'truly' second-best and naïve first-best levels for these instruments underlines the importance of seriously considering second-best issues when analysing inherently second-best instruments.

Evidently, the model we presented falls in the category of conceptual 'simple static monocentric models' as we called them in our review. It remains to be seen which additional insights can be obtained on the issues that we have studied from more elaborate modelling settings. But, we hope to have convinced the reader that this would be just one among many challenges facing the urban economist interested in the economics of urban externalities.

References

Alonso, W. (1964), *Location and Land Use*, Cambridge, MA: Harvard.

Anas, A. and I. Kim (1996), "General equilibrium models of polycentric land use with endogenous congestion and job agglomeration", *Journal of Urban Economics*, Vol. 40, pp. 232–256.

Anas, A., R. Arnott and K. Small (1998), "Urban spatial structure", *Journal of Economic Literature*, Vol. 36, pp. 1426–1464.

Arnott, R. (1979), "Optimal city size in a spatial economy", *Journal of Urban Economics*, Vol. 6, pp. 65–89.

Arnott, R. (1998), "Congestion tolling and urban spatial structure", *Journal of Regional Science*, Vol. 38, pp. 495–504.

Castells, M. (1996), *The Rise of the Network Society*, Oxford: Blackwell.

Chu, X. (1995), "Endogenous trip scheduling: the Henderson approach reformulated and compared with the Vickrey approach", *Journal of Urban Economics*, Vol. 37, pp. 324–343.

Ciccone, A. and R.E. Hall (1996), "Productivity and the density of economic activity", *American Economic Review*, Vol. 86, pp. 54–70.

Coase, R.H. (1960), "The problem of social cost", *Journal of Law and Economics*, Vol. 3, pp. 1–44.

Cutler, D.M. and E.L. Glaeser (1997), "Are ghettos good or bad?", *Quarterly Journal of Economics*, Vol. 112, pp. 827–872.

Dixit, A.K. and J.E. Stiglitz, (1976), Monopolistic competition and optimum product diversity, *American Economic Review*, Vol. 67, pp. 297–308.

Duranton, G. and D. Puga (2004), "Micro-foundations of urban agglomeration externalities", in: J.V. Henderson and J.-F. Thisse, editors, *Handbook of Regional and Urban Economics*, Vol. 4, Amsterdam: North-Holland.

Freeman, A.M., III (1993), *Property Value Models, The Measurement of Environmental and Resource Values: Theory and Methods Resources for the Future*, Washington, DC, pp. 367–420.

Fujita, M. (1989), *Urban Economic Theory*, Cambridge, MA: Cambridge University Press.

Fujita, M., P. Krugman and A. Venables (1999), *The Spatial Economy*, Cambridge, MA: MIT Press.

Henderson, J.V. (1985), "The Tiebout model: bring back the entrepreneurs", *Journal of Political Economy*, Vol. 93(2), pp. 248–264.

Henderson, J.V. (1986), "Efficiency of resource usage and city size", *Journal of Urban Economics*, Vol. 19, pp. 47–90.

Jacobs, J. (1969), *The Economy of Cities*, New York: Random House.

Miyao, T. (1981), *Dynamic Analysis of the Urban Economy*, New York: Academic Press.

Mun, S. and B.G. Hutchinson (1995), "Empirical analysis of office rent and agglomeration economies: a case study of Toronto", *Journal of Regional Science*, Vol. 35, pp. 437–455.

Muth, R.F. (1969), *Cities and Housing Chicago*, Chicago, IL: University of Chicago Press.

O'Sullivan, A. (2000), *Urban Economics*, 4th edition, Boston: Irwin/McGraw-Hill.

Parry, I.W.H. and A.M. Bento (1999), "Revenue recycling and the welfare effects of congestion pricing", *Scandinavian Journal of Economics*, Vol. 103, pp. 645–671.

Perman, R., Y. Ma, J. McGilvray and M. Common (1999), *Natural Resource and Environmental Economics*, 2nd edition, UK: Addison-Wesley, Longman Ltd.

Pigou, A. (1920), *The Economics of Welfare*, London: MacMillan.

Small, K.A. (1992), *Urban Transportation Economics. Fundamentals of Pure and Applied Economics*, Chur: Harwood.

Sullivan, A.M. (1986), "A general equilibrium model with agglomerative economies and decentralized employment", *Journal of Urban Economics*, Vol. 20, pp. 55–74.

Tabuchi, T. (1998), "Urban agglomeration and dispersion: a synthesis of Alonso and Krugman", *Journal of Urban Economics*, Vol. 44, pp. 333–351.

Tinbergen, J. (1952), *On the Theory of Economic Policy*, Amsterdam: North-Holland.

Tulleken, K. (1988), *The Age of God-Kings, 3000–1500 B.C.*, Amsterdam: Time-Life Books.

Verhoef, E.T. (2002), "Second-best congestion pricing in general networks: heuristic algorithms for finding second-best optimal toll levels and toll points", *Transportation Research*, Vol. 36B, pp. 707–729.

Verhoef, E.T. and P. Nijkamp (2002), "Externalities in urban sustainability: environmental versus localization-type agglomeration externalities in a general spatial equilibrium model of a single-sector monocentric industrial city", *Ecological Economics*, Vol. 40, pp. 157–179.

Verhoef, E.T., J.C.J.M. van den Bergh and K.J. Button (1997), "Transport, spatial economy and the global environment", *Environment and Planning*, Vol. 29A, pp. 1195–1213.

Vickrey, W.S. (1969), "Congestion theory and transport investment", *American Economic Review*, Vol. 59, pp. 251–260.

Webster, C. and L. Wai-Chung Lai (2003), *Property Rights, Planning and Markets*, Cheltenham: Edward Elgar.

Urban Dynamics and Growth, CEA, vol. 266
Roberta Capello and Peter Nijkamp (editors)

CHAPTER 5

Uncertainty, Social Capital and Community Governance: The City as a Milieu

Roberto Camagni

Department of Management, Economics and Industrial Engineering,
Politecnico di Milano, Milan, Italy

Abstract

In a condition of dynamic uncertainty, the adoption of a cognitive approach in regional science, considering the procedures through which decisions are taken at the micro-individual level, would greatly help throwing new light on territorial processes and phenomena of a macro character. Proximity space and above all the territory of local relations may be interpreted as powerful operators for reducing uncertainty and, through this, as tools for reducing the use costs of the market and the risks associated to future-oriented decision making.

In this chapter, this approach is followed in order to give an economic interpretation, and not only a geographical one, to the most important archetype of territorial organisation, namely the city, intended at the same time as a production system and a governance system. Two concepts are utilised: the concept of "innovative milieu", developed by the GREMI group and the concept of relational capital, similar to the social capital concept developed recently in political science. The thesis of this chapter is that, within the vast category of agglomeration economies which justify the existence of cities, the most interesting elements refer not only to indivisibilities or pecuniary externalities, but to "technological externalities" of a dynamic character and to elements such as synergies, interpersonal and inter-institutional relationships, collective action and shared development visions. All these externalities help actors in coping with an uncertain and globalising context.

Keywords: social capital, urban dynamics, urban milieu

JEL classification: R0

> *Especially remarkable is the explicit and complete exclusion from the theory of perfect competition of all personal relationships existing between the parties. In actual life the fact that our inadequate knowledge of the available commodities or services is made up for by our experience with the persons or the firms supplying them – that competition is in a large measure competition for reputation or good will – is one of the most important fact which enables us to solve our daily problems.*
>
> (*Friedrich August Von Hayeck, Individualism and Economic Order*, 1949)

5.1. Introduction: complexity and uncertainty

Modern economic and territorial systems are characterised by intrinsic complexity: in their internal structure, in their internal and external relationships, in the forces that determine their development, in the structure and evolution of preferences of actors and in governance systems.

At the macroscopic level, complexity means impossibility of perfect foresight about the evolutionary path of single systems and the changes of their internal structure as a response to a changing and turbulent external context. At the microscopic level, it manifests itself as uncertainty, and therefore as a limit to the possibility of fully rational action, generally assumed by economists: scarcity of relevant information, limited knowledge of causal links, and dependence of decision outcomes from the largely unpredictable behaviour of other actors. The cognitive context becomes one of bounded rationality, in which strategies and tools to limit risks in contracts become increasingly relevant, particularly within innovation processes.

Regional Science, in its recent history, has successfully treated the dimension I have called macroscopic, incorporating the complexity paradigm and many advanced approaches and methods proposed by other sciences (biology, mathematical ecology, physics of dissipative structures, etc.). Less attention has been paid, on the other hand, to the microscopic dimension, despite the relevant and fruitful results it allowed in other, related, disciplines like institutional economics, industrial economics and game theory.

The adoption of a cognitive approach in regional science, considering the procedures through which decisions are taken at the micro-individual level, would greatly help by throwing new light on territorial processes and phenomena of a macro character. Proximity space and above all the territory of local relations[1] may be interpreted as powerful operators for reducing uncertainty (Camagni, 1991a) and, through this, as tools for reducing the use costs of the market and the risks associated to future-oriented decision making.

In this chapter, this micro oriented approach is followed in order to give an economic interpretation, and not only a geographical one, to the most important archetype of territorial organisation, namely the city, intended at the same time as a production system and a governance system. Two concepts are utilised: the concept of 'innovative *milieu*', developed by the GREMI group (Aydalot, 1986; Camagni, 1991b), and the concept of relational capital, similar to the social capital concept developed recently in political science (Coleman, 1990; Putnam, 1993; Grootaert and Van Bastelaer, 2001). The thesis of this chapter is that, within the vast category of agglomeration economies which justify the existence of cities, the most interesting elements refer not only to indivisibilities or pecuniary externalities, but to 'technological externalities' of a dynamic character and to elements such as synergies, interpersonal and inter-institutional relationships, collective action and shared development visions.

All these elements and processes – which are not deterministically but only probabilistically linked to the reality of pure agglomeration – once empirically established may be seen as the heart of both the innovative nature of the *milieu* and the 'progressive' role of the city. Therefore it is argued that:[2]

---

[1] Reference here is to the local relational context and endogenous learning processes interpreted by the concepts of industrial district (Becattini, 1990), local *milieu* (Aydalot, 1986; Camagni, 1991b), proximity economies (Gilly and Torre, 2000), but also, in a more stylised way, by the endogenous growth theory (Romer, 1986). For similarities between the endogenous development approach and the endogenous growth models, see Capello (2004, chapters 8–10).

[2] Early reflections on the 'urban milieu' concepts are presented in Camagni (1999) and Crevoisier and Camagni (2000).

(a) under certain conditions, the comparison of the two concepts, innovative *milieu* (IM) and city, is legitimate;
(b) the two concepts, or theoretical archetypes, share many characteristics; the city is a more complex form of *milieu*, as it intrinsically encompasses economic differentiation (vs. the natural specialisation of the *milieu*) and the entire sphere of residential and life activities of population (which are only considered by the *milieu* concept when they generate synergy and learning effects directly useful for the innovation process);
(c) from a conceptual perspective, the relationships between city and *milieu* can take place in two distinct forms:
   - Urban Innovative Milieux: IM located in cities and exploiting the urban atmosphere;
   - the city as an innovative milieu: the entire city behaving as a *milieu* (the most interesting here).

As a consequence of this reorientation in the interpretation of the sources of local development, a parallel reorientation is needed in the approach to spatial policies: the central issue is no longer placed on the direct proposition of territorial schemes and projects, but rather on the – indirect – construction of the preconditions for the conception and implementation of these schemes; in a word, on the construction of relational or social capital. The policy tool, consistent with this new philosophy, is indicated in the new tradition of territorial strategic planning (Gibelli, 1996; Healey, 2004).

In Section 5.2 the characteristics of the *milieu* concept are presented, and its relationships with the relational capital concept (Section 5.3); in Section 5.4 the logical equation city-*milieu* is presented and in Section 5.5 the new policy tool is considered, namely strategic planning.

5.2. Uncertainty and the concept of local milieu

The intrinsic complexity of economic and territorial systems has relevant effects not only on the evolution, the predictability and control of their development path, but also on the decision making process of the single territorial actors, firms and individuals. This is so not only on the macroscopic level of aggregate development, but also on the microscopic level of individual behaviour under conditions of uncertainty.

The hypotheses and axioms of perfect information and absence of transaction costs, usual in mainstream economics, become less and less acceptable in the interpretation of the real economies. It is, therefore, necessary to pass from the abstract 'theory of choice' of economics textbooks to a 'theory of contracts' based on the observation of the constraints that the real market imposes to the single actors (Williamson, 2002). The same champions of the axiomatic approach in economics recognise that "instead of single transparent axioms there looms the likelihood of psychological, sociological and historical postulates" (Hahn, 1991, p. 47).

The cognitive context is one of bounded rationality which, in the traditional field of choice theory, means to abandon the maximisation criterion for a satisficing criterion à la Simon, but which, in the field of contract theory, means more stringently that we have to accept the incompleteness of all complex contracts, the possibility that the pay-off conditions of any transaction could not be clear ex ante to all partners, the possibility of opportunistic and free-rider behaviour; all elements which clash with the hypotheses of the traditional model of exchange and production. The necessity "to embed transactions in more protective governance structures" (Williamson, 2002, p. 439) becomes clear, in the same way as the necessity to dispose of operators and institutions, of a private (trust, cooperation) and public nature, that could force order into the incentive structure, reduce conflicts and allow the realisation of mutual advantages from the exchange.

For some simple typologies of contracts, the pure market is a sufficient governance tool – if supported by efficient institutions and legal frameworks, as the experience of 'transition' countries has widely shown, in negative terms. For the more complicated typologies of contracts, at the other extreme, the most efficient governance structure is direct control through hierarchy; for most intermediate cases, a cooperative form of governance is requested.

In this last case, social capital and *milieu* linkages in the local economy prove crucial: in fact, they work as facilitators of cooperative behaviours and reducers of uncertainty. In all cases, the existence of clear rules of the game, of reliable juridical institutions in charge of the control over the execution of contracts, of shares codes of economic behaviour, appears fundamental.

The concept of the *milieu innovateur* was created in order to understand innovative processes involving small enterprises and interprets spatial development as an effect of synergies developing over limited territorial areas (Aydalot, 1986; Camagni, 1991a; Ratti *et al.*, 1997). It is defined as a set of relations which bring together and integrate a local production system, a system of actors and representations and an industrial culture, and which generates a localised dynamic process of collective learning.

A fundamental element of a *milieu* is geographic proximity, which results in a reduction in production and transaction costs. But geographic proximity is also accompanied by *socio-cultural proximity* – which must inevitably be present for a *milieu* to exist. We can define it as the presence of shared models of behaviour, mutual trust, common language and representations, and common moral and cognitive codes. Geographic proximity and socio-cultural proximity mean a high probability of interaction and synergy between economic agents, density of informal repeated contracts, absence of opportunistic behaviours, high division of labour and cooperation within the *milieu*. This is what we have called local *relational capital* (Figure 5.1), and it is composed of cooperative attitudes, trust, cohesion and a sense of belonging to a community.[3]

The role of the local *milieu* in terms of economic theory is linked to three types of cognitive outcomes, supporting and completing the normal mechanisms of information circulation and coordination of agents performed through the market, namely: reduction of uncertainty in decisional and innovative processes; ex ante coordination among economic actors, facilitating collective action; and collective learning, a process occurring within the local labour market and industrial atmosphere, enhancing competencies, knowledge and professionalities. These outcomes, particularly the

---

[3] There is a clear relationship and similarity between the concepts of *milieu* and social or relational capital, as is shown below. We suggest, as is evident from Figure 5.1, that a local *milieu* is based on internal relational capital, which is used for the purpose of productive–innovative development.

***Figure 5.1. Basic elements and functions of a local* milieu**

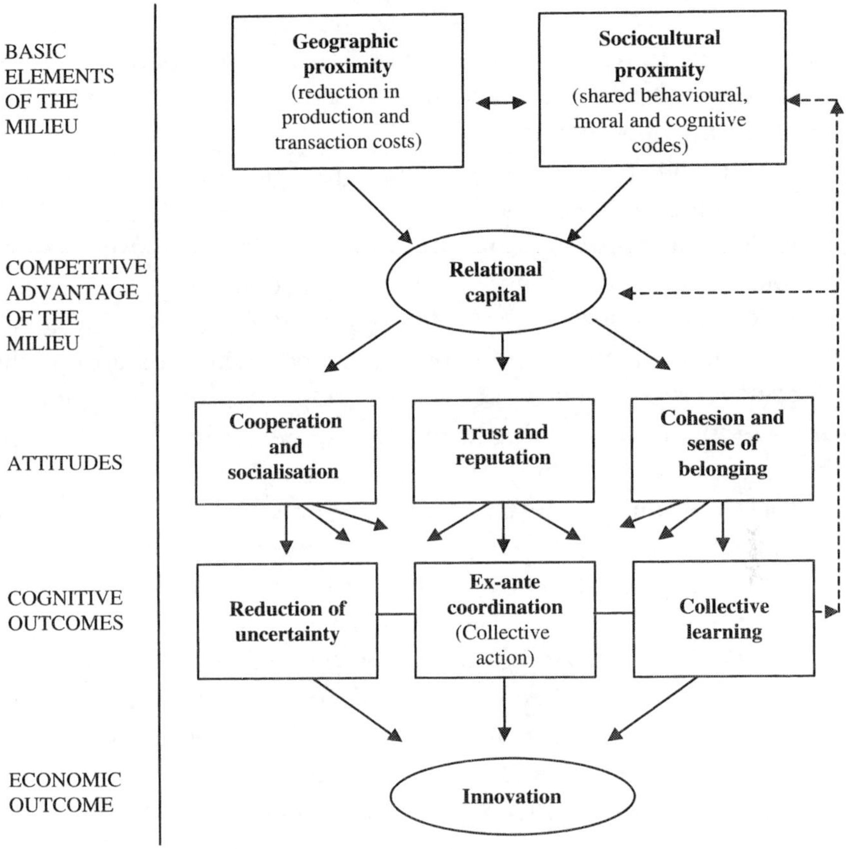

first two, are functions enabling well-known cases of 'market failure' to be overcome.

(A) The function of *reducing uncertainty in innovative processes.* Local relational space is seen as a means of reducing uncertainty, since – due to geographic and cultural proximity – collecting, evaluating and particularly transcoding information, selecting decisional routines, controlling and coordinating competition (all functions usually performed by research and development or strategic planning teams in large enterprises) are carried out collectively within the social context of the local *milieu* (Camagni, 1991a). In addition to this are the functions of external signalling, a kind of socialised territorial marketing and socialised quality control on local production. These functions are beneficial for both internal and external actors and are achieved through the creation of trust and reputation when dealings involve goods which are of intrinsically variable quality and difficult to evaluate.

We can also mention here the function of promoting informal guarantees for the *honouring of incomplete contracts*, which the *milieu* can provide due to its networks of interpersonal relations.[4] In modern economies – where an increasing importance is attached to product differentiation and ill-defined qualitative factors, and uncertainty regarding the properties of new, not yet existing, products is high – incomplete contracts gain importance. As a result, conditions of trust, respectability and reputation, on one hand, and sanctions and exclusion for opportunistic behaviour on the other, conditions which local communities of limited size can provide, gain significance and relevance in the economic sphere (Camagni and Rabellotti, 1997).

(B) A function favouring *ex ante coordination* of local actors and realisation of 'collective actions', due to the presence and development of conventions, norms of behaviour, shared codes of social inclusion and exclusion, mutual trust (Arrighetti and Seravalli, 1999). Once again it is evident how the characteristics of a 'pure' market, and the way information is assumed to operate within it, mean that it is not possible for complementary investment decisions to be taken concurrently, since the profitability of each decision is dependent on the concurrent decisions of other subjects (Richardson, 1960; Bruno and De Lellis, 1994). The presence of a *milieu* realises this important governance function of helping coordination of simultaneous decisions of different actors.

---

[4] Through models inspired by game theory, it has been shown that when there are interpersonal networks and effective mechanisms for punishment, social exclusion and reprisals, implying a reduction in the costs of monitoring and enforcement of contracts, it is possible to not only attain stable (cooperative) Nash equilibria which are not possible when costs are high, but it is also possible to achieve overall benefits for the partners which exceed the allocative costs of local contractual policies (or 'parochialism') (Bowles and Gintis, 2000b).

(C) A function favouring *collective learning*, which in the local *milieu*, and particularly in the local labour market, finds a permanent environment where it can be incorporated.[5] Thanks to the high mobility of skilled labour and professionals inside the local milieu, and thanks also to the wide cooperation agreements and synergies taking place, the learning processes that are at the base of innovation adoption and diffusion manifest themselves in a collective and socialised way, inside the local territory but outside the single firms.

5.3. Relational capital as a crucial constituent of the local **milieu**

In the last decade the concept of social capital has become established in the social sciences. It can be defined as the set of norms and values which govern interactions between people, the institutions where they are incorporated, the relationship networks set up among various social actors and the overall cohesion of society. In a word, social capital is the glue holding societies together.

For economists it includes the capital represented by the rules, habits and relationships which facilitate exchange and innovation – and consequently affects economic development. It is in fact almost unanimously accepted that if a market is to function properly, it needs shared norms as well as institutions and modes of behaviour that reduce the cost of transactions, that ensure contracts are observed and implemented, and that can rapidly resolve any disputes. If we add further factors – reciprocal trust, a sense of belonging to a community that shares values and behaviours and participation in public decisions – then a climate is created which encourages responsibility, cooperation and synergy. Such a climate enhances productivity, stimulates creativity and ensures more effective provision of public goods.

This is particularly noticeable in the case of development processes in backward areas or for transitions from planned economies to market economies or from traditional societies to

---

[5] See, for example, the theoretical considerations and empirical analyses in Camagni and Capello (2002).

modern societies. In fact these situations demonstrate, both in theory and in practice, the importance of social capital (Stiglitz, 1999; Grootaert and Van Bastelaer, 2001). But also in the case of advanced countries, most of the factors considered as components of social capital have for some time been identified as crucial in explaining the success of particular areas, such as industrial districts or *milieux innovateurs*.

However, the concept of social capital has some difficulties and ambiguities of an analytical and linguistic nature which are still limiting its full acceptance. The term 'capital' refers to the fact that it is an asset, or stock accumulated over time, which generates a flow of benefits, and is not just a set of values and social organisations. As a consequence, it should be possible to show that it is built up through a process implying costs or investments, at least in terms of individual and organisational time and effort. This is the rationale of research programmes attempting to measure social capital using suitable proxies (Putnam, 1993; Arrighetti *et al.*, 2001) so as to include it in an ideal production function along with human capital and physical capital.

On the other hand, it is possible to observe that it is created and accumulated through slow historical processes, and that its original function is not directly linked to economic goals, namely the increase in economic efficiency; therefore, it may be seen as "a by product of a pre-existent fabric of social relationships, oriented to other goals" (Bagnasco, 2002, p. 274).

Rather than being a measurable input to add to other factors of production, it can be considered as a public good that produces externalities for the entire economic system, increasing the efficiency of other factors. From this perspective it would be more appropriate to integrate it with another well-known economic variable, the level of technological knowledge which, in a production function, moves 'total productivity' of production factors upwards.[6]

---

[6] Solow (1995) notes that, in this case, a 'residual' should arise from the estimate of the growth function, but this residual does not seem to be significant even in the case of South-East Asian countries (where it was expected). However, Dasgupta (2000) attributes this result to the fact that social capital does not act directly on factor productivity, but rather on the process of accumulation; its effect on domestic product, therefore, occurs via more rapid accumulation of human capital and physical capital.

Whatever the outcome of these critical considerations, it seems justifiable to state that the concept of social capital contains interesting hypotheses – they are consistent with economic traditions dealing with institutions and evolutionary trends, and are supported by significant, though initial, empirical confirmation.

The main problem is that the definition of social capital should not be extended too far and used as a catchword for a range of phenomena. Therefore, it seems helpful to set out an initial classification of the different dimensions present in the concept of social capital, and the various component features indicated in the literature.

The dimensions, or better, the relevant dichotomies, are as follows (Figure 5.2):

- the micro–macro dichotomy, which distinguishes elements directly involving single individuals from those of the system, and
- the dichotomy between the formal and the informal dimension, distinguishing elements expressed through observable objects (roles, networks, norms or social structures), strengthened

Figure 5.2. Dimensions, forms and roles of social capital

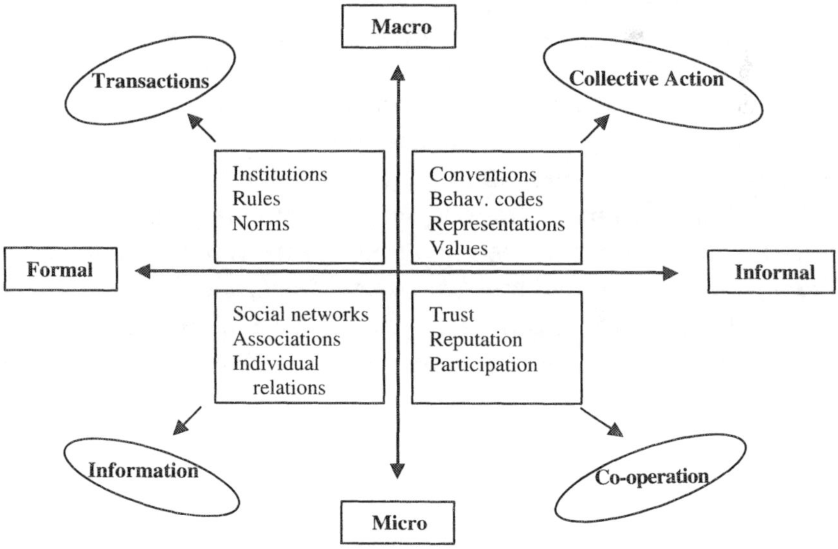

by rules and procedures, from more subjective elements such as values, representations, attitudes and codes of behaviour.[7]

In the macro dimension we can find institutions and rules in the sense of Williamson (1985), North (1990) and Coase (1991): "the rules of the game in a society or, more formally, the humanly devised constraints that shape human interaction" (North, 1990, p. 3). They may be formally expressed, objectively defined, or be informal, and here we refer to conventions, codes of behaviour, values and representations.

In the micro dimension we can find – among the formal elements – social networks and associations, the ability to focus and organise within organised structures (even loose structures), a large range of interactions among social actors; we can also find individual relationships, seen as the set of relations and contacts an individual possesses and which may be invested in economic and social activity. Among informal elements, however, we find trust and reputation and all the non-structured forms of individual participation in public or collective decisions.

There are many channels through which the different elements (or forms) of social capital may affect local development. At the risk of oversimplifying the theoretical framework, we may state that each case has a more direct role in a specific direction, indicated in Figure 5.2.

Institutions, rules and norms in fact fairly explicitly aim to reduce transaction costs, or the use costs of the market. They provide guarantees for contracts and obligations, efficiently manage problems of company law and governance, monitor for conflicts of interest and monopoly practice. Institutions and current regulations (and the organisations or administrations set up to implement and control them) always reduce the costs and time involved in formulating and drawing up contracts and transactions, the costs and time involved in settling disputes and legal cases, the costs and time involved in identifying and choosing commercial partners. All these elements are

[7] I prefer to define the second dichotomy in this way, compared to the definition given by Grootaert and van Bastelaer (2001), whose contributions have guided me, who speak of a 'structural/cognitive' dichotomy, with a similar meaning but logically less correct terminology.

focused on creating a favourable business climate, which benefits local firms and the attractiveness for external firms.

In the case of formal elements that operate at a micro level (individual and collective relationships), they aim to reduce the costs (and increase the availability) of information, particularly for current and potential commercial partners. All this increases the potential available to individuals operating on the professional market; makes it easier to identify opportunistic behaviours of local operators and their external partners, thereby reducing the probability that they will occur and increasing the speed and effectiveness of individual and collective sanctions. It also accelerates transmission of information regarding good practices – commercial, technological and organisational – facilitating their imitation and diffusion.

This component of social capital, operating at the formal and micro level, is usually accompanied by other informal elements, such as mutual trust and reputation which, in turn, strengthen the effectiveness of networks and formalised relations. The most significant effect is to reduce the incidence of free riding, to facilitate cooperation between individuals and actors, through different forms of partnership between private parties (strategic alliances, cooperation agreements, contracts – even incomplete – between customers and suppliers) or between public and private parties (given that trust can equally also involve local public administration). Public attitudes towards participation in collective decisions, enhancing individual commitment and responsibility, also improve the effectiveness of public decisions and promote synergies and complementarities.

Finally, the informal elements of social capital operating at a macro level, which are represented by shared values and codes of behaviour, allow collective action to be more easily undertaken, i.e. ex ante coordination of individual decisions in order to achieve the advantages of economies of scale, purpose and complementarity. In many cases it is only by taking decisions concurrently that costs can be reduced and a complex project made profitable and viable.

As stated above, just as different examples of social capital partly overlap and are simultaneously present in particular local situations, so also the outcomes which we have endeavoured to analyse often accumulate, are mutually beneficial and strengthened. One element in particular is common to all the various examples: the extent of relationality, a key factor affecting development. Relationality presents at the same time costs (psychological, informational and organisational), risks (of rewarding some opportunistic behaviours; of unevenly distributing benefits between partners), and advantages (in terms of the efficiency, effectiveness and innovative character of decisions). For all these reasons, it represents the most significant economic factor emerging from the analysis of social capital. This is why the expression 'relational capital' is preferred (Camagni, 1999), a concept that appears to be more selective, though almost synonymous with social capital and less exposed to the objection that wherever there is a society some form of social capital will also exist.[8]

Returning to the previous paragraph, it is evident that relational capital represents a relevant constituent of the local *milieu*, allowing it to perform its economic role of complement to the market and facilitator of collective action. Nevertheless it is noteworthy that many roles of the *milieu* do not imply an explicit co-operative intent by local actors: the functions of information collection and transcoding, external signalling and collective learning are the outcome of 'socialised' processes taking place spontaneously thanks to geographical and organisational proximity of competing firms, and sometimes even the outcome of opportunistic and imitative behaviour.

5.4. The city as a **milieu**

5.4.1. The conditions for a comparison

The central aim of this chapter is to present a theoretical reflection on the relationships between the concept of IM and that of the city interpreted in economic and spatial terms.

---

[8] In a recent study on the effects of social capital on the performance of small and medium-sized enterprises in the United Kingdom (Cooke and Clifton, 2002) it was found that – 'surprisingly' or perhaps not – "the use of social capital by SMEs was present everywhere" (p. 20). It seems to me that also a sociologist like Bagnasco explicitly prefers a 'relational' interpretation instead of a 'systemic' interpretation of social capital (Bagnasco, 2002, pp. 271–72).

At first glance, the concept of the *milieu innovateur* as defined above does not seem to share many characteristics with the city: the only similarity, in theoretical terms, resides in the agglomeration and proximity element. But if one proceeds to a more accurate consideration, and in particular if one abstracts from the consideration of the physical element which is more easily attached to the common image of the 'city', presenting it as a built environment, more similarities emerge. In fact, taking up a theoretical perspective in terms of *relational capital, spatial interaction and learning processes*, one could easily find that the genetic elements of the city and the *milieu* are not so distant: they are in fact at least commensurable, comparable, though bearing a different level of complexity.

Under the generic conceptual umbrella of the agglomeration principle, which we consider as a common 'genetic' principle of both phenomena, lies a wide spectrum of different elements/processes/effects, which span from the development of a common identity and sense of belonging to the 'socialised' production of human capital and know-how; we argue that these elements and processes – which are not deterministically, but only probabilistically linked to the pure agglomeration fact – once empirically established are at the heart of both the innovative nature of the milieu and the 'progressive' role of the city.

A word of caution and prudence is necessary from the very beginning when dealing with such a multifaceted realm as the city. In fact:

- the city is a complex phenomenon, probably the most complex product of mankind. It is "un territoire particulier,..., le dispositif topographique et social qui donne leur meilleure efficacité à la rencontre et à l'échange entre les hommes" (Roncayolo, 1990). Therefore, it can be analysed under different perspectives: "comme structure materielle, comme système d'organisation sociale, comme ensemble d'attitudes et d'idées, comme costellation de personnes s'impliquant dans des formes types de comportement collectif" (Wirth, 1938);
- cities have evolved in history, performing different functions, and even today they are undergoing fast structural change. In particular, the form of the city is rapidly evolving, and its boundaries with respect to the non-city are blurring (Remy and

Voye, 1992): forms of low density peri-urbanisation, processes of 'metropolisation', edge-city developments on one side; evolution of the countryside in terms of infrastructure, social equipment, life-styles on the other (Camagni and Gibelli, 1996);
- there exist different kinds of cities: of different size (therefore performing different functions within the spatial division of labour), different specialisation, different location (ports, etc.);
- cities are differently linked together within wider regional spaces (urban systems, hierarchies, city-networks) and therefore their role and functions cannot be fully interpreted through the consideration of the isolated, standalone city;
- cities are indicated by great historians (Braudel, Pirenne) and sociologists (Weber, Sombart) as the birthplace of innovation (economic, political, cultural); but other functions are characteristically performed by the city, giving rise to an economic advantage: defense (once), control and power, cultural interchange and knowledge transfer.

As a consequence of the theoretical complexity and the empirical diversity of the object of this reflection, the limits and the characteristics of the approach have to be made clear. Firstly, we limit ourselves in a first approximation to economic aspects: the city as a particular and efficient form of organisation of economic relationships (though by the term 'economic relationship' we mean a much wider set of factors and interactions than the mainstream economic textbooks do). The interpretation given of the city's role and performance is therefore partial, though not trivial.

Secondly, the main dimensions under which the city is analysed are a relational one (the city as a set of territorial and social relationships) and a dynamic one (the city as a learning system). Thirdly, we assume, at least initially, an abstract and archetypal approach to the city (the City with a capital c), abstracting from geographical or historical differentiation and considering those characteristics of the urban environment which

- have an impact on economic phenomena and economic performance,
- explain the genesis of the city as an efficient form of organisation of economic relationships. As already said, these economic

functions are not the sole functions performed efficiently by the city, but are nevertheless (very) important;
- explain its innovative character, a character that historians and economists usually assign to it.

Consequently, we do not consider different, non-economic aspects, which have strong feedback effects on the economic performance of the city: city size, form, environmental quality, etc.

5.4.2. The economic role of the city and a taxonomy of urban agglomeration advantages

An economist looks at the city as a self-organising system (Camagni, 1996), whose competitive advantage resides in agglomeration (the city as a 'place'), accessibility (the city as a 'node' in global networks) and interaction (the city as 'relational capital'); this system is addressed to the achievement of collective goals such as economic efficiency, welfare, territorial power and control (at least for ruling classes).

In history, the success of this form of social organisation was striking, and it allowed the achievement of further general goals like cultural development, quality of life, individual freedom, and more generally democracy, and progress, modernisation of the society and innovation in the economy.

In a sense, we can affirm that the IM realises a short circuit between the general characteristics it shares with the city (agglomeration, accessibility, interaction) and the specific outcome, namely innovation, reducing the complexity of the full process of urban development and its high degree of roundaboutness, and forgetting about the other possible outcomes.

It is important to note that the characteristics of innovativeness in an abstract scheme is directly attributed to the city or the IM may well be absent in many (or most) empirical circumstances witnessing spatial agglomeration. In fact the existence of a city or of a *milieu* is only a relevant precondition for innovativeness, but its actual manifestation depends on finer local specificities and, on the aggregate level, is subject to stochastic processes.

Starting with the agglomeration element which characterises the urban environment, and which in some respects may also encompass the other two elements – external accessibility and networking goes hand in hand with urban size, and the same happens to internal interaction potential, a direct function of size and differentiation of urban activities – we can devise a taxonomy of the single sub-elements on which agglomeration advantages reside.

On one hand, a distinction can be made, in a quite traditional way, between 'hard' and 'soft' elements of agglomeration advantage, and, on the other hand, less traditionally, between the two main sources of the same advantage, namely *indivisibilities*, stemming from city size, and *synergy*, allowed by more subjective elements like interaction, cooperation and synergetic processes (Figure 5.3).

In the lower left side of the table, we find the advantages which derive from the provision and concentration of public goods such as infrastructure and overhead capital, public services, large urban functions like fairs, congress facilities, universities and the cultural heritage. On the other hand, in the lower right side we find advantages connected with the nature of big market of the city: market for products, human capital, private services on the demand side, and a diversified supply of intermediate inputs, on the supply side.

On the upper right side of the picture, we can find the elements which are more interesting in my view, which were pointed out in the recent past: elements connected with the synergetic action performed by the city. In fact we find

- accessibility to information – which is inherently a cooperative good – through informal, face-to-face and inter-personal contacts,
- explicit cooperation among actors, stemming from trust, common sense of belonging to a community sharing the same values, and
- implicit cooperation among actors, in the form of socialised production of skilled labour, human capital for high-level managerial functions, marketing (*image de marque*) and information transcoding.

Some of these functions may be embodied in the provision of physical or 'hard' elements like dedicated infrastructure or important urban projects realised through private/public partnership. Therefore, we find in the upper left part of the graph the socialised

Figure 5.3. Sources of urban agglomeration advantages

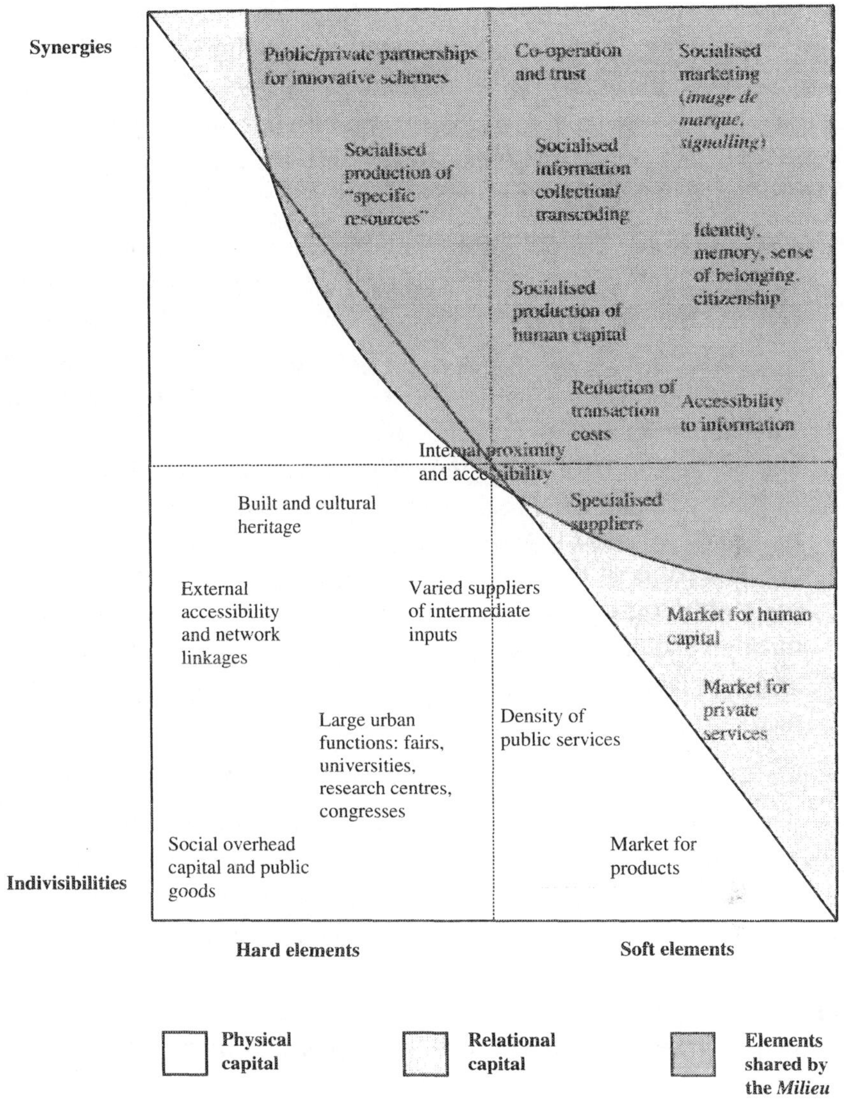

Source: Camagni, 1999.

provision of 'specific resources', to the use of typically urban productions or functions.

The lower triangle of Figure 5.3 encompasses what could be labelled as the 'functional capital' of the city, of a mainly physical nature; the upper right triangle, on the other hand, may be seen as representing the 'relational capital' of the city.

In our opinion, it is on the theorisation of the relevance of the relational capital of territorial systems that the contribution of this kind of reflection is likely to bring the most advanced results. And in fact the IM shares with the city many of the above-mentioned characteristics (the dark grey area in Figure 5.3) and may lend many theoretical and analytical tools to the interpretation of the city. In fact, urban relational capital resides in different elements:

(a) the synergy and cooperation element, embedded in the 'local milieu effect' and in territorial cooperation networks developed by the GREMI group (Aydalot, 1986; Maillat and Perrin, 1992) and subsequently theorised by the French proximity school[9] and by Storper (1995) with the concept of 'untraded interdependencies';
(b) the socialised nature of the production of specific resources, as skilled labour and human capital, or the socialised production of market signals (Gordon, 1989; Camagni, 1991a);
(c) the reduction of dynamic uncertainty, inherent in the processes of technological innovation and economic transformation, through socialised management/transcoding of information and ex ante coordination and control over competitors' moves, as shown before.

One important element that differentiates the IM from the city resides in the relevance of physical and economic size, which is crucial in the urban environment, as it was shown earlier through the indivisibility element. The nature of the city being a big market for products and for production factors, and particularly for labour, was stressed by Veltz (1993) as representing an important locational advantage of the city over the industrial district, another way of achieving the reduction of uncertainty (*ville-assurance tout risque*).

5.4.3. The theoretical relationships between the Milieu *and the City*

From arguments developed so far, the theoretical similarity between the City and the Milieu emerges with relative clarity. They share

---

[9] See, among others, Bellet *et al.* (1993), Rallet (1993), Dupuy and Gilly (1995), Rallet and Torre (1995) and Gilly and Torre (2000).

the elements of proximity, strong internal integration, synergy, and psychological and cultural identity. Furthermore, they share the functions of collective and socialised production of specific resources, human capital and market signalling and of supplying the substrate for collective learning processes.

Their special characteristics may be described as follows:

City	Milieu
Tendentially de-specialised	Tendentially specialised
Physical agglomeration important	Proximity important, even without agglomeration
General-purpose infrastructures	Oriented infrastructures
Private services with intersectoral market	Private services integrated in *filières*
Social heterogeneity	Social homogeneity
Identity defines economic 'vocation'	Economic 'vocation' defines identity

As said before, the City is a much more complex system, addressed towards major social goals which are not relevant for the Milieu; and they bear a physical dimension (built environment, size, built and cultural heritage) which is alien to the *milieu*.

Another logical path that can be traced in the case of both the concepts regards how to pass from the functional aspects of the territory to the IM aspects.

In the same way as the *milieu* represents the relational capital of local territorial systems, adding the elements of synergy, governance and identity, the *city as a milieu* represents the relational capital of the urban context (Figure 5.4). The innovative element of both the *milieu* and the city derives from the existence of collective learning processes and the development of a common 'vision' for the evolution of the local system.

But in the case of the city, another relevant situation may emerge (represented by the central column in Figure 5.4): the presence of an *urban milieu*, a network of informal or selected linkages developed around a specialisation sector or filière, developing inside the urban context or the urban production system. Empirical evidence suggests that many cases exist of such *milieux* or innovative *milieux* which characteristically exploit an urban atmosphere (and therefore an urban location), without implying that the entire city behaves like a *milieu*. The cases of the financial *milieu* in cities like Zurich, Geneva,

Figure 5.4. The urban milieu and the city as a milieu.

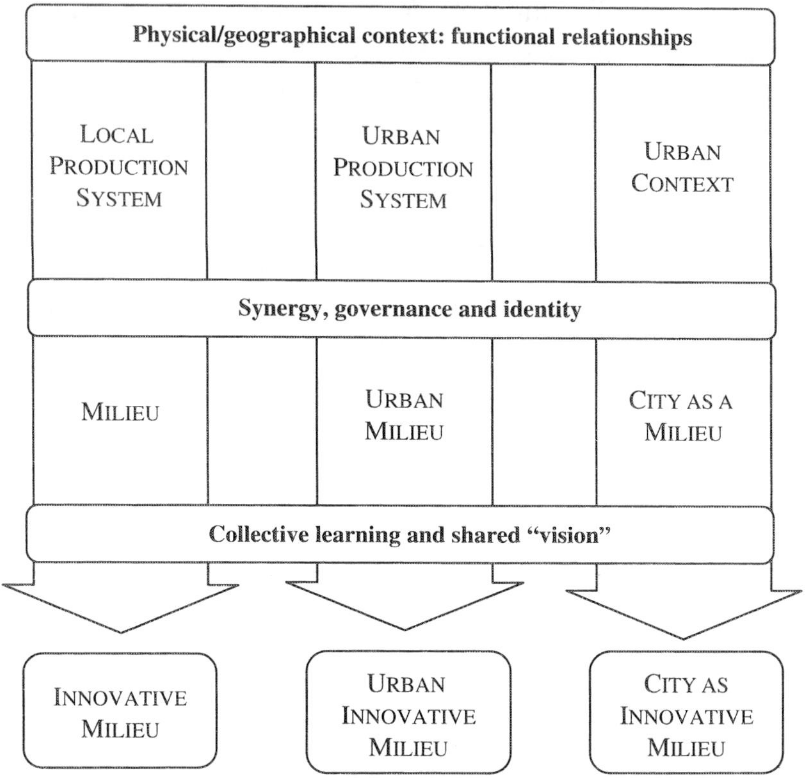

Source: Camagni, 1999.

Frankfurt; the innovative *milieux* developing around the fashion creation filière in Milan or Paris; the media or the communication *milieux* in Hamburg and Milan are important examples.

Still adopting a dynamic approach and the aim of interpreting innovation processes, existing literature attributes to the city some characteristics that may assign to it a dynamic comparative advantage. In fact, urban competitiveness and its continuous recreation in time may be linked to the following elements:

(a) the city is the natural location site of *production services* (in a degree which is proportional to their quality and rarity), a sector which is responsible for the level (and growth rate) of the efficiency of the local (urban, regional) industrial sector. According to Thompson (1968): "the economic base of the larger metropolitan area is the creativity of its universities and research parks, the sophistication of its engineering firms and financial institutions, the persuasiveness of its public relations and advertising agencies, the flexibility of its transportation networks and utility systems, and all the other dimensions of infrastructure that facilitate the quick and orderly transfer from old dying bases to new growing ones".[10] In the empirical analysis, we will call these kinds of advantages, which are typical of urban areas and which support innovative activity in cities, with the label 'dynamic urbanisation economies';

(b) the city is the natural location site of small and medium firms (incubator hypothesis) which are by definition the schumpeterian innovation agents;

(c) the city is the natural location site of industries and products in the early, pioneering phases of their life-cycle;[11]

(d) similar to the previous one is the hypothesis that metropolitan areas play a major role in the phases of radical reconvertion and rejuvenation of products, when a strict interaction is demanded among different functions of the firm, usually spatially dispersed: engineering (mastering of technologies), R&D (mastering of products), marketing (mastering of demand) (Camagni, 1988): a large city supplies a barycentric location for all these functions.

All these reflections were developed in the context of location theory; they may be easily utilised in an evolutionary context characterised by synergetics and learning processes, as we are trying to make here.

5.5. Towards a new urban governance: the tool of strategic planning

Most of the elements that we have listed as determinants of static or dynamic agglomeration advantages for the city refer more to process orienting decision making, at the individual and collective level,

---

[10] Please note the dynamic element constituted by the term 'transfer', meaning the continuous shift of local specialisation and the relaunching of the local competitiveness through it.

[11] This idea was first developed by Vernon (1957) with reference to a spatial setting, well before his well-known article (Vernon 1966) referring to industrial evolution.

rather than to pure objective facts or factors. But if the central issue in modern, innovation oriented economies, lies in decision making – how to reduce uncertainty on decision outcomes, how to limit transaction costs, how to realise ex ante co-ordination of complementary decision makers, how to enhance public consensus and avoid conflict – then the success factors in these economies have to be found in what are increasingly called *governance* structures.

The need for new governance tools derives directly from the increased complexity of economic systems and development processes, not just in the normal sense of plurality of actors, globalisation of forces and sophistication of processes and technologies, but, as we argued at the beginning, as a consequence of increased uncertainty of the context and limits of our cognitive capabilities.

All these elements have profound effects on the definition of policy styles, in particular in the urban context: in fact, if all we said is true, we have to shift attention from objects to subjects, from projects to actors, from resources to rules; in terms of the concepts utilised in this chapter, we have to shift attention from fixed capital to relational capital, from conception of single projects to the creation of a *milieu* effect – the best guarantee for building a long-term innovation and project capability throughout the entire urban territory.

Many local administrations are already engaged in this new challenge, and have found out the new tool, most suitable for the new task: urban *strategic planning*.

In the international literature, it is defined as the collective construction of a shared vision of the future for a given territorial area, through processes of participation, discussion, and listening. It is an agreement between administrators, actors, citizens, and various partners to implement this vision through a strategy and a series of projects, with varying interconnections with each other, which are justified, evaluated, and shared. Strategic planning can finally be defined as the coordination of responsibilities assumed by different actors for the implementation of such projects (Gibelli, 1996).

Strategic planning accordingly

- favours future-oriented analysis and scenario building,
- traces complexity and local specificity back to a single strategic path,

- operates in an openly pragmatic way, aware that it operates in a situation of limited rationality, and consequently behaves dynamically and flexibly when defining objectives and actions,
- relies on learning processes and iterative review processes,
- promotes wide consultation and the participation of interests and civil society,
- evaluates projects on the basis of their fit with the general strategy and principles of compatibility and sustainability in urban planning,
- assigns strategic relevance to the implementational phases of the plan, and
- entrusts a mainly persuasive and promotional function to the plan documents.

The strategic plan is not a plan *for* the city, implemented by the local administration, but it is a plan *of* the city, implemented through the widest possible participation of interests, groups and individual citizens, with the public administration being a facilitator, coordinator, auditor of compatibilities, and partial implementer.

Local public administration is responsible for checking the compatibility – not so much the financial as the planning and techno-functional aspects – of the various proposals emerging from the partnership process, as well as defining priorities. At the same time it has to also evaluate which of the different projects might possibly function as triggers and catalysts for self-sustaining processes, thus being crucial for implementing the overall strategy.[12]

---

[12] As can be seen, it is neither possible nor justified to attribute a deregulatory function to strategic planning, but rather strategic planning allows for true flexibility in plan choices and acceptance of private skills in project conception and implementation. It is possible to introduce deregulation into urban planning without using strategic planning, and, conversely, in its assessment and approval of projects, strategic planning relies on definite rules of land use which may be more or less restrictive and binding, depending on the 'style' of planning adopted. By the same token, although strategic planning is the most appropriate instrument for addressing issues of a city's strategic positioning in the context of a global economy and for checking if structures and infrastructure are adequate for the objectives set, it does not thereby automatically prioritise economic objectives relative to social improvement or quality of life. Priorities are defined though participative processes and can be very varied.

In the process of strategic planning, the city no longer appears as just a container and physical background for the processes of development or decline, triggered by external decisions, it is not the passive object of competitively determined decisions and outcomes. It is becoming – or is attempting to become – the driver of rational and shared decisions and an actor involved in global processes.

The main function of a city's traditional master plan was to reduce uncertainty for private operators about public choices. Strategic planning requires private operators to assume responsibility, it expects – and facilitates – ex ante coordination among different actors and synergies between the public and private sector. These are all factors which assist economic decision making and improve its effectiveness, whatever the substance of the decision itself.

Strategic planning has many attributes affecting all areas of social capital shown in Figure 5.2. It focuses on the interaction between actors and the public; it emphasises participative processes; it aims to construct a shared vision of the future; it attempts to stimulate all forms of synergy among actors and complementary projects as well as all possible ways of involving local stakeholders and investing them with responsibility; it is committed to defining clear rules of urban planning. While strategic planning assumes a certain minimum degree of local relationality, it has a fundamental role in creating relationality.

It is possible to contend that a new type of community governance is emerging – promoting and recognising social networks, focusing on shared values and identity and providing opportunities for discussing and coordinating issues – which can address evident cases of both 'market failure' and 'government failure'. It can do this by supplying public goods, for example in the self-organisation of neighbourhood amenities; by organising mechanisms for cooperatively sharing risks; and by evaluating public and private performance. In most cases the local community is able to effectively carry out collective actions where spontaneous mechanisms or actions by the public administration fail due to a lack of crucial information regarding the behaviour of various partners, their abilities or needs (Bowles and Gintis, 2000a).

As an innovative instrument of urban governance strategic planning allows:

- greater economic efficiency and lower entrepreneurial risk in urban projects,
- greater planning creativity,
- improved effectiveness in organising public assets, and jointly
- greater consistency for individual territorial projects, which are subject to both public and collective processes of evaluation, and
- greater participation of citizens in collective decisions, not only as a means for achieving the above aims but as a positive objective in itself.

5.6. Conclusions

The most important conclusion achieved in the chapter resides in pointing out the relevance of the *milieu* approach for a modern and renewed interpretation of the city as a spatial archetype. Cities and *milieux* share many characteristics, not really in their physical form but in their intrinsic economic role in the shaping of the spatial economy: thanks to geographical and cultural proximity, density of interaction and presence of social or relational capital, these 'organised territories' provide the governance structures that are needed to complement pure market forces in a context of increasing complexity and uncertainty. In particular, coordination of complementary and necessarily simultaneous decisions, collective action and local learning processes are highly enhanced – probabilistically – in an urban environment, and this explains the innovative or 'progressive' role of cities in history.

As a consequence, new governance and policy tools have to be devised, based on the strengthening of the *milieu* effect and the local relational capital.

Strategic planning supplies this tool, with interesting potential outcomes, already visible in the most advanced, theoretically sound international experiences.

References

Arrighetti, A. and G. Seravalli (eds.) (1999), *Istituzioni Intermedie e Sviluppo Locale*, Roma: Donzelli.

Arrighetti, A., A. Lasagni and G. Seravalli (2001), *Capitale Sociale, Associazionismo Economico e Istituzioni: Indicatori Statistici di Sintesi, Working Papers No. 4*, Dipartimento di Economia, Università di Parma.

Aydalot, Ph. (ed.) (1986), *Milieux Innovateurs en Europe*, Paris: GREMI.

Bagnasco, A. (2002), "Il capitale sociale nel capitalismo che cambia", *Stato e Mercato*, Vol. 2(65), pp. 271–303.

Becattini, G. (1990), "The Marshallian industrial district as a socio-economic notion", pp. 37–51, in: F. Pyke, G. Becattini and W. Sengenberger, editors, *Industrial Districts and Inter-firm Cooperation in Italy*, Geneva: ILO.

Bellet, M., G. Colletis and Y. Lung (eds.) (1993), "Economies de Proximités", *Revue d'Economie Régionale et Urbaine*, n. 3, Special Issue.

Bowles, S. and H. Gintis, (2000a). Social capital and community governance, Department of Economics, University of Massachusetts, December, mimeo.

Bowles, S. and H. Gintis, (2000b). Optimal parochialism: the dynamics of trust and exclusion in networks, Department of Economics, University of Massachusetts, February, mimeo.

Bruno, S. and A. De Lellis, (1994). The economics of ex-ante coordination, Università di Roma La Sapienza, April, mimeo.

Camagni, R. (1988), "Functional integration and locational shifts in the new technology industry", pp. 48–64, in: Ph. Aydalot and D. Keeble, editors, *High Technology Industry and Innovative Environment*, London: Routledge.

Camagni, R. (1991a), "Local milieu, uncertainty and innovation networks: towards a dynamic theory of economic space", pp. 121–144, in: R. Camagni, editor, *Innovation Networks: Spatial Perspectives*, London: Belhaven-Pinter.

Camagni, R. (ed.) (1991b), *Innovation Networks: Spatial Perspectives*, London: Belhaven-Pinter.

Camagni, R. (1996), *Principes et Modèles de l'Économie Urbaine*, Paris: Economica.

Camagni, R. (1999), "The city as a milieu: applying GREMI's approach to urban evolution", *Revue d'Economie Régionale et Urbaine*, Vol. 3, pp. 591–606.

Camagni, R. and R. Capello (eds.) (2002), *Apprendimento Collettivo e Competitività Territoriale*, Milan: Franco Angeli.

Camagni, R. and M.C. Gibelli (1996), *Cities in Europe: Globalisation, Sustainability and Cohesion, European Spatial Planning*, Rome: Presidenza del Consiglio dei Ministri, Dipartimento Politiche Comunitarie, Poligrafico dello Stato.

Camagni, R. and R. Rabellotti (1997), "Footwear production systems in Italy: a dynamic comparative analysis", pp. 139–164, in: R. Ratti, A. Bramanti and R. Gordon, editors, *The Dynamics of Innovative Regions*, Aldershot: Ashgate.

Capello, R. (2001), "Urban innovation and collective learning: theory and evidence from five metropolitan cities in Europe", pp. 181–208, in: M.M. Fischer and J. Froehlich, editors, *Knowledge, Complexity and Innovation Systems*, Berlin: Springer.

Capello, R. (2004), *Economia Regionale*, Bologna: Il Mulino.

Coase, R.H. (1991), "1991 Nobel Lecture: the institutional structure of production", in: O. Williamson and S. Winter, editors, *The Nature of the Firm*, New York: Oxford University Press.

Coleman, J.S. (1990), *Foundations of Social Theory*, Cambridge: Harvard University Press.

Cooke, P. and Clifton, N. (2002). Social capital and small and medium enterprise performance in the United Kingdom, presented at the workshop on Entrepreneurship in the modern space economy: evolutionary and policy perspectives, Amsterdam, June.

Crevoisier, O. and R. Camagni (eds.) (2000), *Les Milieux Urbains: Innovation, Systèmes de Production et Ancrage*, Neuchâtel: EDES.

Dasgupta, P. (2000), "Economic progress and the idea of social capital", in: P. Dasgupta and I. Serageldin, editors, *Social Capital: A Multifaced Perspective*, The International Bank of Reconstruction and Development, Washington: The World Bank.

Dupuy, C. and Gilly, J.-P. (1995). Dynamiques Industrielles, Dynamiques Territoriales, paper presented at the International Conference of ASRLF, held in Toulouse, 30–31 August, 1 September.

Gibelli, M.C. (1996), "Tre famiglie di piani strategici", pp. 15–53, in: F. Curti and M.C. Gibelli, editors, *Pianificazione Strategica e Gestione Urbana*, Firenze: Alinea.

Gilly, J.P. and A. Torre (eds.) (2000), *Dynamiques de Proximité*, Paris: L'Harmattan.

Gordon, R. (1989), "Entrepreneurs, firms and the social foundation of innovation", *Sociologie du travail*, n. 1.

Grootaert, C., Van Bastelaer, T., (2001). Understanding and measuring social capital: a synthesis of findings and recommendations from the social capital initiative, World Bank, *Social Capital Initiative Working Paper n. 24*, April, Washington, DC.

Hahn, F. (1991), "The next hundred years", *The Economics Journal*, Vol. 1(101), pp. 47–50.

Healey, P. (2004), "The treatment of space and place in the new strategic spatial planning in Europe", *International Journal of Urban and Regional Research,* March, pp. 45–67.

Maillat, D. and J.-C. Perrin (eds.) (1992), *Entreprises Innovatrices et Développement Territorial*, Neuchâtel, EDES: GREMI.

North, D. (1990), *Institutions, Institutional Change and Economic Performance*, Cambridge: Cambridge University Press.

Putnam, R.D. (1993), *Making Democracy Work*, Princeton: Princeton University Press.

Rallet, A. (1993), "Choix de proximité et processus d'innovation technologique", *Revue d'Economie Régionale et Urbaine*, Vol. 3, pp. 365–386.

Rallet, A. and A. Torre (eds.) (1995), *Economie industriale et économie spatiale*, Paris: Economica.

Ratti, R., Bramanti, A. and Gordon, R. (eds.) (1997), *The Dynamics of Innovative Regions*, Aldershot: Ashgate.

Remy, J. and L. Voye (1992), *La Ville:Vers Une Nouvelle Definition?*, Paris: L'Harmattan.

Richardson, G.B. (1960), *Information and Investment*, Oxford: Oxford University Press.

Romer, P.M. (1986), "Increasing returns and long-run growth", *Journal of Political Economy*, Vol. 94, pp. 500–521.

Roncayolo, M. (1990), *La Ville et ses Territoires*, Paris: Gallimard.

Solow, R. (1995), "But verify", *The New Republic*, Vol. 11, p. 36.

Stiglitz, J., (1999). Participation and development: perspective from the comprehensive development paradigm, International Conference on Democracy, Market economy and Development, Seoul.

Storper, M. (1995), "La géographie des conventions: proximité territoriale, interdépendences non-marchandes et développement économique", pp. 111–128, in: A. Rallet and A. Torre, editors, *Economie Industriale et Économie Spatiale*, Paris: Economica.

Thompson, W.R. (1968), "Internal and external factors in the development of urban economies", in: H.S. Perloff and L. Wingo, editors, *Issues and Urban Economics*, Baltimore: Johns Hopkins Press.

Veltz, P. (1993), "D'une géographie des coûts à une géographie de l'organisation: quelques thèses sur l'évolution des rapports entreprises/territoires", *Révue Economique*, n. 4.

Vernon, R. (1957), "Production and distribution in large metropolis", *The Annals of the American Academy of Political and Social Science*, pp. 15–29.

Vernon, R. (1966), "International investment and international trade in the product cycle", *Quarterly Journal of Economics*, May, pp. 190–207.

Von Hayeck, F.A. (1949), "The meaning of competition", *Individualism and Economic Order*, London/Henley: Routledge/Kegan Paul, pp. 92–106.

Williamson, O. (1985), *The Economic Institutions of Capitalism*, New York: Free Press.

Williamson, O. (2002), "The lens of contract: private ordering", *American Economic Review, Papers and Proceedings*, Vol. 92(2), pp. 438–453.

Wirth L. (1938) "Le phénomène urbain comme mode de vie" Grafmeyer Y. Joseph I. L'Ecole de Chicago, naissance de l'écologie urbaine 1938 Aubier-Montaigne Paris.

PART 2

Accessibility

Urban Dynamics and Growth, CEA, vol. 266
Roberta Capello and Peter Nijkamp (editors)

CHAPTER 6

Land-use, Transportation and Urban Development

K.J. Button[a], P. Nijkamp[b] and P. Rietveld[b]

[a]School of Public Policy, George Mason University, Fairfax, VA, USA
[b]Department of Regional Economics, Free University of Amsterdam, De Boelelaan 1105, 1081 HV, Amsterdam, The Netherlands

Abstract

Land-use, transportation and urban development are inevitable entwined. Over the years, research across a range of disciplines has sought to disentangle the directions and magnitudes of the causal links. The situation is still far from transparent. Economics has, however, shed some light on the matter, and this chapter provides a selective overview of how our understanding has moved forward during the past quarter century or so. This period has coincided with advances in relevant economic theory, notably in the area of demand analysis and congestion modeling, and applied techniques on the one hand, and with changes in urban form and spatial economic structures on the other. Edge cities, for example, have suddenly been 'discovered'. Given the extent of these changes, there is thus only a small amount of space devoted to the classic literature on the subject of land-use transportation interactions. The emphasis is much more on contemporary topics such as the impacts of traffic congestion on urban form and the role of changes in the nature of information systems on the way transportation affects urban development.

Keywords: urban land-use, traffic congestion, urban form, 'New Economic Geography', transportation modelling

JEL classifications: R11, R14, R41

6.1. Introduction

Urban land-use and transportation are closely interwoven. Urban density and sprawl are influenced by the volume and density of traffic as well as by the capacity and spatial structure of transport infrastructure. Many changes in urban form and scale are incremental but more significant shifts in form and structure can result from transportation infrastructure investments, innovative transportation management or new technologies.

This chapter deals with a vast topic and is thus inevitably selective in its coverage.[1] To begin with it deals almost exclusively with economics, as the theme of the book demands, but in doing so misses out very important literatures in fields such as sociology, geography and political science that have helped better our understanding of linkages between land-use, transportation and urban development. The coverage is also limited largely to recent advances, again following the theme of the book. 'Recent' in our context is from the mid-1980s or so.[2] But what it does try to do is to focus on some of the major developments in our economy of the ways urban form and location are influenced by transportation and, conversely how transportation is influenced by urban form.

The chapter gives relatively little emphasis to the way that simple transportation variables can be 'plugged' into land-use and urban form modeling.[3] The focus of much urban economics is not specifically on transportation, which is frequently treated as an exogenous factor and defined simply as an input variable (travel time

---

[1] One reason for the selectivity is to avoid excessive overlap with other volumes that have appeared recently dealing with similar themes. Early work linking transportation and urban form includes Mill and De Ferranti (1971), Solow and Vickrey (1971), Langley *et al.* (1973), Kraus (1974), Miyao (1978) and Kanemoto (1980). A survey of this literature can be found in Medda (2000).

[2] Many of the classic earlier papers that cover a similar theme to this chapter are to be found in Berechman *et al.* (1996) and Rietveld *et al.* (2003).

[3] The traditional way of treating transportation is to view it as a derived demand to meet the needs of other activities. The approach in urban and regional economics is often to take a difference stance and see transportation as an input variable in the longer term determination of the spatial distribution of production and consumption. This raises issues such as the role of traffic congestion in influencing urban sprawl and the growth of edge cities. It is really this latter type role that is the main focus of this chapter.

or distance). This is quite legitimate in many cases where the concern of the analysis is on other attributes of urban economies and systems. This type of analysis is covered in detail in other chapters in the volume. Here the attention is much more on the recent economic work that has been conducted seeking to get a better understanding of how transportation functions in an urban setting, and how the transportation variable can be refined should a more sophisticated variable be needed in land-use driven analysis. It is also important in instances where transportation matters are at the heart of the issue being addressed.

Some initial thoughts on previous trends are included by way of introduction and to establish threads, but as any serious student of the subject will soon realize, this is a very cursory account, and also a very selective one. The discussion is also largely limited to theoretical advances. There have been significant advances in the way data are collected, in econometric analysis and in the political economy of land-use and transportation policy, but space precludes any detailed assessment of this. There are passing comments on advances in these areas where there have been strong ties to theoretical developments but they are kept to a minimum. The chapter also looks at both the internal form of urban areas and, especially in the context of urban development, more general issues of urbanization and interactions between urban areas.

6.2. Linking land-use and transport

We start with a few words on the general state of the status of the economic work that is being done linking land-use, transportation and urban development. In many ways the perception of work in this area has changed little in recent years. Traditionally, and despite the early work of analysts such as Von Thünen, 'space' has not been a central subject for study by economists. The American Economics Association JEL Classification System places 'Urban, Rural and Regional Economics' only before 'Other Special Topics' at the end of its very comprehensive listing. The UK's Royal Economic Society in a 1991 issue of the *Economic Journal* celebrating it centenary, contained articles by eminent economists on 'The Next 100 Years'. There was no piece in the collection even remotely concerned with regional or transportation economics, let alone one linking them. Despite this there have been changes and one can perhaps trace them back to the work of Paul Krugman and, in a slightly different context, to that of Michael Porter.

Krugman has been credited with much of the recent work in what has become known as 'The New Economic Geography'.[4] Beginning in the early 1990s, Krugman (1991, 1998) began producing a series of papers that sought to bring more realism to analysis of international, and, subsequently, regional trade. There are numerous dimensions to his arguments, and in particular an emphasis on the importance of agglomeration economies, but they did bring about a greater focus on the role that such things as transportation costs play in trade and economic development at all levels of spatial aggregation. More specifically, he showed that when there are scale economies, and non-linear transportation costs are introduced into an interregional location model, the ultimate spatial distribution of activities critically depends on initial conditions including the starting distribution of activities and the nature of the non-linearity embodied in the activity transportation interaction.

Michael Porter's role has been somewhat different; it has much less to do with formal, mathematical modeling, and more to do with changing thinking about the competitive advantage of different locations. It has more to do with notions of industrial organization and economic development than with spatial economics in its more traditional sense. From the time of his seminal work on competitive strategy Porter (1980) has developed a series of ideas about management and competition that have influenced the way business strategy is viewed, and with this, the way location and production costs are implicitly treated in more formal modeling.

There have been many more generic developments in economics that have also influenced work on local spatial issues. In terms of style and approach, the economic analysis of land-use, transportation and urban development has followed that of all other fields of economics. If there was any agreement amongst the contributors to

---

[4] The *Spatial Economy* (Fujita *et al.*, 1999) offers perhaps the most complete work on new economic geography to date. More basic accounts are Ottaviano and Puga (1998) and Fujita and Thisse (2002). Krugman (1993) ties his trade theory ideas to those of location. For a critical assessment of new economic geography, see Neary (2001).

the centenary issue of the *Economic Journal* it was that, first, economics has become more fragmented, and second, that it has become more mathematics. Regarding the first, there are now certainly more specialized economics journals, and the electronic media has added to these. While the more general publications such as the *American Economic Review*, *Economic Journal*, *Journal of Political Economy*, etc. have traditionally carried only an occasional article on the interface of transport, land-use and urban economics there has gradually been a growth in more focused academic publications over the years. Although their names have sometimes changed, the *Journal of Regional Science* and the *Journal of Transport Economics and Policy* were launched in the 1960s, *Regional Science and Urban Economics* and the *Journal of Urban Economics* first appeared in the 1970s, and these were joined in the 21st century with the appearance of the *Journal of Economic Geography*.

Whether this trend towards greater specialization is good or not is a topic well beyond this narrow paper – so only a few observations. On the one hand there are arguments that it is to be expected in a maturing discipline, which one may argue economics is, and with the onset of greater professionalism (Friedman, 1991). It is, in this sense simply a manifestation of Adam Smith's arguments for divisions of labor and is just a reflection of trends seen earlier in the physical sciences. On the other hand it does probably mean that cross-fertilization of ideas across 'sub-areas' of economics is less easy than in the past and wheel reinventing may become an industry in its own right.

In terms of the use of mathematics, the interface between transportation and urban economics has been as absorbent as all other fields of economics. This has been combined with enhanced computer power and new estimation and programming techniques to produce an ever-increasing number of econometrics studies in the field (Anselin, 1992).[5] The availability of GIS data has provided a massive amount of ammunition for such work. With all this, however, has come the demise of the Marshallian method of

---

[5] There has also been an increasing interest in deploying cleometrics to look at transportation and urban including work by Gin and Sonstelie (1992).

economic analysis – the notion that you use mathematics to work through the logic of an argument but then move to express this in good prose. The tendency that many point to when refereeing academic papers, or examining PhD theses that much work seems to be based on 'a technique seeking a problem', is perhaps one outcome of this. Also, the divide between what was traditionally called political economy and economics would seem to be growing in spatial economics as in many other areas.

6.3. The distant past – up to the mid-1980s

In his classic work establishing the basis of neo-economics, Alfred Marshall, perhaps rather arrogantly, gave little credit to his predecessors, but von Thünen was one economist whose work he did acknowledge. Von Thünen's (1826) analysis highlighted ties between land values and transportation costs, and led directly to the seminal work of Alonso (1964) on urban form. But one should also go back to contemporaneous work with that of von Thünen, namely to the French engineering school of economists, Dupuit, Napier, Minard, etc. to find some of the earliest work tying transportation to spatial patterns (Ekelund and Hébert, 1999).

Later came the seminal work of Weber (1909) on location theory, where the ties between location and transportation were clearly articulated. This in turn led to the more detailed work of Isard (1951) and provides a clear lineage to modern regional economics. Predöhl's (1928) was more in line with much of modern general equilibrium economics, and provides a backdrop to Moses' (1958) contribution. Nevertheless, much of the work done in both the field of urban economics and transportation economics was largely what would today be called 'institutional economics'. It often contained a significant amount of normative discussion and frequently was focused on quasi-legal issues. There was also little that could be considered as a comprehensive theoretical approach and, for example, textbooks were much more individual topic based.

The 1950s saw significant advances in applied economics as computers began to appear, new data sources emerged, and the role of government expanded with a concomitant need for strategic analysis. Some of these techniques, although often not used immediately, had important implications for studying the interface between transportation and urban form. Path-breaking work appeared on both linear (Koopmans, 1951) and nonlinear (Kuhn and Tucker, 1951) programming that by the late 1950s was being regularly applied to optimization topics in location and transportation. Interest in micro-management of economic systems, often within a Keynesian framework, led to the development of regional and urban input–output analysis as a direct result of Leontief's (1953) pioneering work at the macroeconomic level. The techniques also provided an important component of equilibrium models that were beginning to be constructed.

From the transportation side, the gravity model was introduced in land-use-transportation planning exercises, initially more or less as a pragmatic tool with little intellectual underpinning, and adapted for use in spatial location analysis. The period also showed a renewed interest in the economic analysis of traffic congestion (Vickrey, 1959; Walters, 1961) and links between pricing of transportation infrastructure and investment in new capacity (Vickrey, 1969). This was reinforced by considerable intellectual efforts to incorporate non-market factors, such as the costs of travel time, into urban transportation modeling (Moses and Williamson, 1963; Beesley, 1965). Without this better description of how transportation systems function, and the implications of their market imperfections, any model of urban form would have been deficient.

The 1950s and the 1960s saw ad hoc technical advances, but fragmentation of methods and a problem-based ethos meant that a unified way of approaching location, transportation and urban form was lacking. The picture changed in the 1970s. The so-called New Urban Economics (Mills and MacKinnon, 1973; Richardson, 1977) emerged building much on neo-classical economics and focusing on broader questions of urban size and form. It involved some of the major economic figures of the latter part of the 20th century, including Nobel Prize winners James Mirlees (1972) William Vickrey (Solow and Vickrey, 1971) and Robert Solow (1972), and reintroduced more positive analysis to urban and transportation economics.

These analytical developments in urban economics also occurred at a time when the modern field of regional science began to attract attention. While the new urban economics largely helped define the study of intra-urban form, regional science offered additional insights into inter-relationships between cities. Regional science combined, amongst other academic disciplines, the skills of economists with those of human geographers, environmental scientists and applied mathematicians who sought to understand more completely location patterns and spatial dynamics. It has led not only to an area of study in its own right but also has stimulated, by bringing new tools and concepts to bear, significant new work linking land-use, transportation and urban form. In particular, it has provided powerful new ways of viewing urban areas as parts of larger spatial economic systems.

Although there has been a significant shift towards abstract theoretical modeling many of the recent advances at the interface of transportation and urban development have come about because of larger changes. Effectively, at least part of the research in recent years has reflected a changing urban policy and commercial environment.

On the transportation side, major changes have occurred in the way transportation services are delivered.[6] Advances in management science in the 1970s, particularly in costing and scheduling, have led to significantly different ways in which transportation logistics is now viewed. It has led not only to new demand patterns for urban transportation infrastructure but also to changes in related activities, such as warehousing, that have primary and secondary effects on urban forms. Equally, major developments in regulatory economics[7] and changes in regulatory regimes have impacted on modeling of institutional structures and their influence on transportation supply. There have been technology changes as the 'Information Society' has gradually emerged. It has affected the lives of people in general but also the demands they place on the transportation systems of cities, on the way suppliers of transportation services can respond and on urban geography (Gasper and Glaeser, 1996). There have also been significant changes since the 1980s in urban form. Urbanization has

---

[6] See papers in Brewer *et al.* (2001).

[7] In the case of transportation it is not simply that the new ideas of regulation were applied but in addition they were tailored in many cases to the particular characteristics of a network industry (Economides, 1996).

grown, suburbanization has grown but there has also been the emergence of new urban spatial structures, of which edge cities are perhaps the most discussed (Garreau, 1991).

The edge city was first described as a new form of urbanization. It refers to a spatial constellation where large companies form a new nucleus outside a city's central business district and induces new suburban development in the area. Various approaches have been advocated to explain this development. One explanation can be found in the equilibrium forces between agglomeration and dispersion forces that determine the spatial-economy of the city (Fujita and Mori, 1997). A second embraces the microlocational choices of business in an urban area (Krugman, 1996). Other approaches are reviewed in Medda (2000).

Henderson and Mitra (1996) and Glaeser and Kahn (2003) have developed economic models with explicit attention to agglomeration economies for firms in central locations. These central locations are hindered in their development by the high generalized transport costs of workers living in remote suburbia. Subcenters emerge as a result of the trade-off between lower transport costs of workers and lower agglomeration economies resulting from the spread of activities.

In recent years an interest has developed in the relationship between urban dynamics and modal choice (Newman and Kenworthy, 1999). In particular, the use of public transport and of other more environmentally benign modes of transport have attracted attention (Pucher and Lefèvre, 1996). The implications of urban public transport (or transit-oriented developments) have been studied not only from a transportation perspective, but also from an urban land-use viewpoint, for example, the emergence of subcenters along a transit connection (Calthorpe, 1993).

There are a number of different threads to what is emerging at the interface of economic work between land-use, transportation and urban development. These are not easily delineated but some effort is made here for reasons of exposition. These trends also reflect in many cases continuations of previous trends, advancing, refining and testing fairly well established basic concepts. There is, for example, an expansive contemporary literature that effectively refines Alonso's work. Others are breaks in trends, or possibly new trends. No attempt is made here to try to separate these various possibilities.

6.4. Congestion and pricing of transportation infrastructure

Traffic congestion is a major problem and is one of the main determinants of location decisions in cities. Car ownership and use have grown considerably since the 1970s, and the pattern of road use has altered in a way that often conflicts with the design aims of road networks. In particular, in the US in 1969 some 25% of aggregate urban car trips were for commuting reasons and there was the prospect of channeling this along a limited number of arteries. Now only 15% are exclusively for journey to work purposes, and trip chaining involving complex trips that embody such purposes as shopping, recreation and child movements as well as a movement to and from employment, is much more widespread making the traditional focus on radial urban transportation networks much less relevant.

The upsurge of interest amongst academics, as well as practitioners since the 1960s moving to better understanding the economics of congestion, and in making better use of urban transportation infrastructure, has led to refinements in assessing how the road price should be calculated. As Vickrey pointed out, there is in ideal circumstances, a need to vary the price according to traffic demand and costs. The time of day, the traffic mix, the physical features of the network and local road conditions (such as the weather and accidents) may influence these. While there has been a number of important contributions over the past 15 years, not all of these areas have been looked at in detail and, indeed, some very important issues such as pricing congestion caused by incidents have received limited attention. But certainly the economic understanding of urban traffic congestion has made considerable steps forward.

Congestion theory has also moved forward in a different sense over the past 40 years.[8] The initial model of traffic congestion is time independent – it is essentially a snapshot at a point in time. This approach can be refined by looking at how travel demand varies over time, and at how traffic flows evolve over time and space – the model becomes time dependent. It looks, for example, at the evolution of traffic congestion over a rush hour. The detailed

---

[8] The recent literature on traffic congestion and its economic pricing is large and is only skimmed here. For a survey of recent developments, see Lindsey and Verhoef (2001).

modeling takes a variety of forms that, however, often assume vehicles are precluded from overtaking, etc. and results emerge that are highly sensitive to factors such as demand elasticities and, related to this, the valuations placed on travel time. What they are essentially doing is treating congestion as arising from a bottleneck in the system. A second generic form of the dynamic congestion model is now also sometimes employed that is concerned with flow congestion. This approach does not completely eliminate travel delays at the social optimum. Both approaches involve the distribution of travel delays and scheduling of costs at the peak, and the duration of the peak in the untolled equilibrium and the social optimum, all being determined endogenously.

Road pricing is designed to produce a Pareto optimal use of a facility, but this is dependent upon the standard assumptions that surround first-best partial equilibrium analysis and especially that all other prices are equal to marginal costs. Moving into the realm of 'second-best', where the assumptions of the Pareto world are relaxed, is less tidy, and inevitably the efficient price becomes situation specific. The tendency is to relax one or two assumptions at most to seek an efficient outcome. In reality, a larger number of assumptions may not hold.

One issue that has long been of interest is the pricing of substitutes to road transport. Subsidies may be given to public transport to optimize the modal split of traffic. A subsidy for public transport, accurately reflecting the cross-elasticity of demand between the modes, could theoretically be used to attract sufficient motorists from the roads to reduce pressure on them. The difficulty, beside the standard concerns over X-inefficiency that accompany subsidies, is that while they can theoretically optimize mode split, they can lead to an excessive overall use of transport with implications for the local economy.

It is now realized that similar situations arise when only part of a road network is subjected to road pricing (Verhoef, 1996). This essentially leaves road users with a choice between a facility where access is unpriced but the service quality poor and one where there is an access price but congestion is optimized. The road price introduced on such routes must reflect the opportunity cost of not using the unpriced roads as well as the first-best toll on the priced road. Ignoring the cross-effects will lead to sub-optimally high use of the unpriced facility. Marchand (1968) was one of the first to look at this situation in terms of setting a charge that weighs the welfare benefits of reducing traffic on the unpriced facilities against the losses from deviating from first-best pricing on the road priced route. But more recently Verhoef *et al.* (1996) have shown that in some cases a Pigou optimal subsidy may involve a negative road price.

Second-best considerations may also be important in the context of complements and substitutes for a road priced facility. Parking is the most explored complementary good (Niskanen, 1992; Verhoef, 1996; Voith, 1998). In a first-best world parking fees would reflect the opportunity cost of taking up land to 'rest' a vehicle as well as the congestion costs associated with cruising around seeking a parking spot. In practice, parking is often provided free or perverse charging regimes are employed. Such regimes are frequently structured to limit certain categories of users, such as long stay users, irrespective of willingness to pay. But parking costs can affect land-use. They are a fixed cost of a trip, they are spread more over a longer trip than a shorter one and thus may affect congestion in suburban areas.

Much of the early work on road pricing treated all road users as identical. Variations in income and, linked to this, the valuations that they place on travel time saving as well as the size of vehicles driven, which affects road take, are now appreciated as being important consideration for both efficiency and equity reasons. At the very least, they raise questions about whether the road price should reflect these features. If the concern is purely with efficiency, and standard first-best conditions hold, then there is no need to consider income variations with time-varying road pricing. This is because all that matters in optimizing charges are the congestion costs imposed, and Arnott and Kraus (1998) provide rigorous proof of this.

Where there are alternative routes in the network (including different lanes on a road) and users have different utility functions, optimality is attained by offering various road price/congestion level combinations. This maximizes welfare by allowing those with a high travel time value to buy their way onto the faster routes, leaving those with lower time values on the cheaper, more congested ones. The situation can also be compared with that involving a variety of users with different trip-time preferences using a single route. Arnott *et al.* (1992) work on situations where there are temporally inflexible road prices that may involve either a single price for each of a pair of roads, or different prices, indicates the greater efficiency of the latter.

Another situation that has attracted attention includes that of drivers having different preferences for speed, that are unrelated to income. Here the models have often tended to exclude the possibility of overtaking making them particularly relevant to bridge and tunnel cases. The results suggest that the road price should be higher for slow vehicles to reflect their impact of slowing higher speed vehicles. Yet, because slow vehicles affect fewer fast vehicles on average and also because the speed of fast vehicles declines asymptotically to that of the slower vehicle, the price differential should *decrease* with the proportion of slow vehicles in the traffic stream. This may come as a surprise as it contradicts the standard result in transport economics that congestion charges increase with traffic demand (Rouwendal *et al.*, 2002). This approach would require differential approaches to road pricing depending on drivers speed preferences – in practical terms this may be done by having different pricing regimes for cars and trucks.

Urban motor traffic places considerable strains on the environment, and the costs involved are external to the market. These effects are both wide ranging, including the emission of a variety of local, regional and global atmospheric pollutants, noise, ground water contamination, and visual intrusion, and, in some cases, very large. The lack of adequate pricing of these inputs indicates that there should be adjustments to the road price to meet second best conditions. This is a large subject area. The issue is complex and the literature substantial, and is not dealt with here.[9]

There is another side to the second-best debate that should just be touched upon. There has been a mounting interest in how other elements of transportation should react in the absence of appropriate road pricing of motorists – what should be the second-best strategy for them to adopt? In this context Arnott and Yan (2000) have looked at pricing across multimodal systems, while Chia *et al.* (2001) examined the appropriate fuel taxation policy to pursue if appropriate congestion charges are absent.

---

[9] See papers in Hensher and Button (2003).

Added to these challenges of defining optimal charges are those of developing mechanisms for charging. These may be significant and may become entwined with other issues such as how much information the road user has on the 'product' that is being purchased (De Palma and Lindsey, 1998).

Incorporation of transactions costs inevitably leads to trade-offs, the outcomes of which are case specific. The simple cordon toll systems and area licensing present a clear picture to drivers of the price for entering a road. But charges under such regimes are not easily adjusted to reflect local traffic conditions, and will not accurately indicate the level of congestion that is likely to be encountered. Most electronic, real time charging regimes are sensitive to immediate congestion levels but do not tell the trip-maker in advance of the road price to be paid.

The idea of using electronic technology to collect a real time, variable road price was initially advanced by William Vickrey as early as the late 1950s. Despite the largely successful testing of electronic road pricing equipment in Hong Kong, transactions costs, issues of privacy and concern over technical reliability have favored area-based systems. This means that a road price does not vary directly with traffic conditions but rather location or time of day is used as a proxy for congestion levels. The Singapore and other schemes outlined in Small and Gomez-Ibanez (1998), together with the recent London system, use a combination of these surrogates largely for pragmatic reasons.

The efficiency of these discrete charges is less than of a fine tuned continuous road price, the loss of efficiency being a function of just how closely the proxies reflect changing congestion levels. This in turn often depends on the extent to which cordons coincide with the location of bottlenecks, the number of 'steps' involved and the prices charged at each step. Empirical estimates of the extent to which these systems lose efficiency are found in Arnott *et al.* (1993, 1998) and Chu (1999) and they seem to be significant.

6.5. Investment in transportation infrastructure

The notion that transportation infrastructure, or indeed virtually any form of public infrastructure provision, had an impact on economic development at any level of aggregation tended to lose favor in the

1970s and 1980s as supply side-economics and the associated emphasis on the benign effects of tax cuts dominated much economic thinking.[10] The appearance of a provocative empirical study by Aschauer (1989) rekindled old fires and led to a major reassessment of the inter-relationship between infrastructure investment and economic efficiency. It also coincided with emergent new theories of endogenous economic growth (the 'New Growth Theory') associated with Romer (1986) and Lucas (1988, 1990) and others. Basically, the new growth theory, redefining older theories of circular-and-cumulative causality, argues that, over time spatial economic growth will continually diverge because of dynamic economies of knowledge growth. Since this poses problems of equity, economists sought ways of efficiently bringing about spatial convergence.

Aschauer's analysis indicated that transportation infrastructure improvements can lead to decreases in transport costs, stimulate inter-area trade and potentially contain divergent growth. The intensity of competition increases because sectors in urban areas that were formerly sheltered are now confronted by relatively cheap imports. The result is that, while consumers in these cities may be able to buy goods and services at lower prices, employment in these sectors and in such regions declines. In exporting regions an increase in employment may be anticipated. The theory of trade, therefore, predicts that in each urban area employment in some sectors will expand, while in others it will contract as a result of the infrastructure improvement. The overall impact on any city will depend on, amongst other things, its sector structure. The flexibility of the labor force is also important, and a rigidity can mean that employment loss in one sector cannot be completely compensated for in others (Rietveld, 1995).

It should be noted that the majority of firms that relocate move a very short distance. Relocation of firms in response to infrastructure improvements mainly occurs on a local scale and is not a major

---

[10] While the focus here is more on the general urban development implication of transportation infrastructure there have been numerous studies which have refined our understanding of the implications of individual investments (e.g. Bollinger and Ihlanfeldt, 1997).

cause of differences in regional growth rates. A study of the location behavior of firms with more than 10 employees in the eastern part of the Netherlands, for example, found that 75% of those that relocated did so to places in the same municipality (Bruinsma *et al.*, 1997). A closer inspection reveals that in the relocation process, 42% of the firms remained at approximately the same distance from the nearest ramp of a highway, 41% moved to a closer location and 16% moved further away. Such findings underline that the rapid growth in the number of firms that is sometimes observed at particular places near newly improved highways is to a considerable extent the consequence of relocation within regions. These relocation processes are, of course, quite relevant at a local level, but from a broader regional or national perspective they are less important.

The implications of transport investments for the spatial behavior of firms and residents manifest themselves in new growth patterns of firms and populations. There is usually a mutual causality phenomenon involved – development creates the funds to finance infrastructure and infrastructure creates development potential. There is now a multiplicity of studies addressing such mutual relationships (Nijkamp, 1999).

There have also been advances in our understanding at a more microlevel of the links between the pricing of urban transportation infrastructure and optimal investment strategies. A market price serves not only to allocate current facilities optimally but also provides signals to where capacity should be expanded, and the revenues from the price provide resources for that expansion. In a perfect market situation, effectively Knight's (1924) position, there is no requirement for separate consideration of capacity expansion – the net revenue flows given the necessary guidance. Given the reality of public ownership of most road track, and the context of the recent emergence of interest in road pricing, studies quickly emerged looking at the link between this and road investment strategies (Mohring and Harwitz, 1962).

In the first-best situation with road pricing, where there are constant returns associated with roads, the investment decision is also straightforward in its generality. Road pricing reflects the opportunity cost of making use of a road. If this, the benefits of road use exceed, exceeds the opportunity costs of expanding the network then more road space, or investment in improved management systems (e.g. signaling or signage), is justified. Basically, if the road agency is earning a return above costs then investment should be considered by comparing the net present financial value of more road capacity with that of investing elsewhere.

The optimal amount of capacity with road pricing is arrived at with a constant cost of expansion when the road price reverts to covering the maintenance cost of the system. This is a standard finding for any economic activity when marginal cost pricing is ubiquitous. The situation becomes more complex if the long-run marginal cost is not constant or when capacity cannot be expanded incrementally (Hau, 1998). Nevertheless, in concept it still conforms to situations in other sectors involving capital capacity decisions under similar conditions.

Essentially if there are diseconomies of scale then, at the optimal capacity and with an optimal road price, economic rent will be earned. This reflects a scarcity rent on a fixed factor of production; in this case it will be land. Equally, if there are economies of scale, and in some instances when there are large indivisibilities in investments, then road pricing will not generate sufficient revenue to cover the full long-run costs of optimal road provision. Subsidies will be required and techniques such as cost–benefit analysis come into play in determining their level.

Much, therefore, depends on the view taken about the nature of the long-run cost curve. Whether the assumptions do hold is an empirical matter. Work by Small (1992), amongst others, indicates that in many cases the relevant cost conditions approximate to those conditions providing a direct link between an optimum road price revenue and optimum investment levels.

6.6. *Transportation, information and land-use*[11]

There has been considerable interest in the links between communications, transportation and urban land-use patterns, and in particular to investigating teleworking (telecommuting), as an alternative to

---

[11] This is a growing area of analysis and the rapidly expanding literature is barely touched upon here – a useful collection of papers is contained in Stough *et al.* (2003).

commuting to a non-home work site.[12] This is attractive politically because it avoids generally unpopular traffic constrained by measures such as road pricing.[13] There is also the spatial implication that if telework expands then conventional land-use patterns will change, and become less structured (Gasper and Glaeser, 1996).

There are, however, two broad possibilities about the links between transportation and telecommunication: they may be substitutions or complements (Salomon, 1985; Schuler, 1992).[14] Until recently, substitution, involving a physical trip to be replaced by electronic communications technology, has been the more popular of the hypotheses. Overall, attempts to identify a wide spectrum of substitution effects have generally been inconclusive (Salomon, 2000). A complicating factor is the oft-invoked assumption of this type of work that total interactions, whether by travel or communication, tend to be constant, may not be valid. If, for example, people have a constant travel time budget then the availability of telecommunications does not reduce the amount of travel undertaken.

Supporting the hypothesis that substitution is unlikely, at least in the short term, Nijkamp and Salomon (1989) point to the fact that over time the total amount of all forms of communication has increased. Incomes have risen, car ownership has risen and the socioeconomic life of Western economies has evolved to encourage greater levels of interpersonal contact. Further, high growth in the service sector and information industries has resulted in a greater inherent emphasis on face-to-face communication and the development of interpersonal relationships during work hours (Storper and Venables, 2002).

---

[12] Teleworking is defined here as any proportion of work done at home, which would normally be done at the workplace. This would include parts of days or weeks working at home, regional centers, satellite work centers, local work centers or neighborhood work centers. While the focus here is on the links between work travel, communications and land-use, work has also been undertaken on such things as teleshopping (Manski and Salomon, 1987) and on the impact of e-business (Borenstein and Saloner, 2001). See also Shapiro and Varian (1998) for a general discussion of the economics of a communications society and Button and Taylor (2001) for a wider review.

[13] One may also add that the improvements in telecommunication that are taking place can also add to the efficiency and effectiveness of policies such as road pricing (Verhoef *et al.*, 1996).

[14] At the theoretical level, Safirona (2002) offers a general equilibrium framework tying transportation, land-use and communications in a single model.

Increasingly, business and pleasure are merging, and networking is a valued skill in contemporary working life. However, at any point in time there are physical and logistical constraints on the capacity of individuals to meet face-to-face. This leads to location patterns with an emphasis on proximity and easy access to high-speed connections via airports and railways. Concerns about the environmental implications of further infrastructure expansion and the high costs of such investments, combined with changes in life-styles, suggest that this interaction curve is now beginning to flatten. Through telecommunications technology, a greater number of interactions are possible. The implication is that telecommunications has the scope to not only fill the potential interactions needed, but also to push up the potential for additional interactions.

The economic argument for the decline in spatial variation, and indeed the potential end of conventional cities, ties in closely with the degree to which electronic communications are substitutes for face-to-face contact (Gaspar and Glaeser, 1996). As we have seen the evidence on this is far from clear since face-to-face contacts are important for trust, which appears to be the basis of many types of social and commercial interactions (see, for example, Fukuyama, 1995). This implies that transaction costs are highly relevant for urbanization patterns, and that these costs cannot easily be substituted by ICT means. There has been a tendency historically for urbanization to be unevenly spread, but often to broadly follow a consistent hierarchical pattern (Krugman, 1996). Whether this pattern is sustainable in the context of the widespread use of the Internet is still uncertain and in need of further empirical work.

This lack of any confirmation with the neoclassical convergence theory for Internet supply is also tending to be replicated in terms of physical distribution. The developments in information systems and e-commerce more generally have led to a tendency to concentrate interchange and consolidation at a limited number of nodes. This fact, that largely refutes neoclassical economics, may not be that surprising, however, in light of the standard (Vernon, 1966) theory of product cycles. When new products emerge they tend to be located in regions with high quality labor and access to specialized information. Time erodes these needs as the product becomes more standardized. The Internet is too new to be able to ascertain whether it will follow this pattern or not. The situation at present is that economic theories abound linking the Internet and e-commerce to land-use developments, but the empirical support for any of them is still extremely tenuous.

6.7. Transportation demand modeling

Transportation modeling and forecasting, in part because of a certain intellectual arrogance on the part of economists, had until the mid-1970s been dominated by engineering models based around aggregate behavior and a rather mechanistic view of how travel choices are made.[15] The basic analysis involved a four-stage sequence that initially forecasts aggregate trip making in an urban area, then distributed this between origins and destinations, then to modes and finally assigned it to individual routes. The overall framework hardly corresponded to an economic model, although some of the sub-models used, most notably those used to distribute traffic between various origins and destinations, did have a basis in utility theory. But the aggregate trip generation model, in particular lacked a sound economic basis and explicitly assumed a zero cost of travel elasticity.[16]

This approach changed with the advent of discrete choice modeling for which Dan McFadden would later win the Nobel Prize for economic science (McFadden, 2001). This disaggregate to modeling approach, that initially emerged in the 1970s, has been considerably refined and developed since.[17] For example it now can embody more flexible forms (Bhatt, 2000).

There have been a significant number of developments to traditional modes of transportation demand that have refined our

---

[15] Exceptions to this include Beckmann *et al.* (1956), Meyer *et al.* (1965) and Mills (1972).

[16] While this was seen as an unrealistic situation as far back as the 1850s, it was largely dismissed until the sudden discovery by traffic engineers that there was 'latent' demand for road space when congestion exists, and that 'induced demand' emerges when capacity is added. Reducing the generalized cost of trip making stimulates more travel in the aggregate (Abelson and Hensher, 2001).

[17] There has also been important development in the nature of modeling. The widespread use of revealed preference models has been supplemented by the development of stated preference methodologies that examine how actors will act when confronted by hypothetical situations. This work is not covered in detail here but does allow, in particular economists, to explore, for example, the introduction of entirely new forms of transportation or unusual transport policy options in an urban area.

understanding of links between transportation and urban form. In the context of this review, Sasaki (1990) and DeSalvo (1996) developed a model linking income, residential location and transportation, and Mori and Nishikimi (2002) focus on transport density and industrial agglomeration. Sasaki (1989) examines the implications of using charging for two competing transportation modes in a city, and looks at the affects on outward physical expansion of the urban area.

An important policy discussion relates to the potential contribution of spatial planning to curb urban sprawl and the associated negative externalities of transport, such as congestion, noise and other emissions. Brueckner (2000) mentions three main reasons – market failures – to worry about urban sprawl. The first concerns the failure to take into account the social value of open space when it is converted to urban use. The second is the failure of individual commuters to recognize the social costs of congestion created by their use of the road network. The third failure is related to the real estate developers who do not take into account all of the public infrastructure costs generated by their projects. The first issue of negative externalities related to the use of open space may indeed be a reason for government intervention in the form of spatial planning in order to prevent excessive fragmentation of open land. However, to address the other two problems, pricing measures in transport would be natural tools to use, implying that spatial planning is typically a second best instrument, to be used when pricing measures are not feasible.

One of the problems is that from an empirical viewpoint the impact of spatial planning on transport seems to be smaller than is sometimes thought. For example, average daily travel distances in highly urbanized areas tend to be lower than in other areas, but the differences are limited (Rietveld, 2001). Also the effects of particular spatial planning doctrines such as compact cities, new towns or diffuse patterns have smaller effects than anticipated. One of the reasons is that most of the settlements in mature economies are already given, so that spatial planning has, by definition, a limited effect. Another point is that, although spatial planning may have substantial impacts on choice sets of people in their traveling behavior, the specific choice of a travel alternative will be driven by other factors, including prices. This underlines the second best character of spatial planning compared with pricing policies.

6.8. Conclusions

This survey has focused on some of the main areas where our understanding of the economic links between location, transportation and urban form have moved forward since the mid-1980s. It has been a time when economics itself has changed, and this has affected the feel of the work done and the methods used. There has been a considerable focus on the more efficient use of urban transportation networks, and especially the use of congestion charging to reduce traffic congestion levels. Other fields of focused research concern investment in urban public infrastructure and its effect on land-use and productivity. Also the theme of substitution vs. complementarity between physical transportation and interactions through telecommunications media use has been an important field of research. There is still much uncertainty about how this interaction affects urban growth and form. This theme also touches on the central question of agglomeration advantages and may therefore be expected to be a major focus for future theoretical and empirical research in urban economics. To undertake this, however, and also other areas of work of a similar vein, will require more and better hard data than have often been available in the past.

References

Abelson, P.W. and D.A. Hensher (2001), "Induced travel and user benefits: clarifying definitions and measurement for urban road infrastructure", in: K.J. Button and D.A. Hensher, editors, *Handbook of Transport Systems and Traffic Control*, Oxford: Pergamon.

Alonso, W. (1964), *Location and Land Use*, Cambridge, MA: Harvard University Press.

Anselin, L. (1992), "Space and applied econometrics", *Regional Science and Urban Economics*, Vol. 22, pp. 307–318.

Arnott, R. and M. Kraus (1998), "When are anonymous congestion charges consistent with marginal cost pricing?", *Journal of Public Economics*, Vol. 67, pp. 45–64.

Arnott, R. and A. Yan (2000), "The two-mode problem: second-best pricing and capacity", *Review of Urban and Regional Development Studies*, Vol. 12, pp. 170–199.

Arnott, R., A. de Palma and R. Lindsey (1992), "Route choice with heterogeneous drivers and group-specific congestion costs", *Regional Science and Urban Economics*, Vol. 22, pp. 71–102.

Arnott, R., A. de Palma and R. Lindsey (1993), "A structural model of peak-period congestion: a traffic bottleneck with elastic demand", *American Economic Review*, Vol. 83, pp. 161–179.

Arnott, R., A. de Palma and R. Lindsey (1998), "Recent developments in the bottleneck model", in: K.J. Button and E.T. Verhoef, editors, *Road Pricing, Traffic Congestion and the Environment: Issues of Efficiency and Social Feasibility*, Cheltenham: Edward Elgar.

Aschauer, D.A. (1989), "Is public expenditure productive?", *Journal of Monetary Economics*, Vol. 23, pp. 177–200.

Beckmann, M., C.B. McGuire and C.B. Winsten (1956), *Studies in the Economics of Transportation*, New Haven: Yale University Press.

Beesley, M.E. (1965), "The value of time spent in travelling: some new evidence", *Econometria*, Vol. 32, pp. 174–185.

Berechman, J., H. Kohno, K.J. Button and P. Nijkamp (eds.) (1996), *Transport and Land Use*, Cheltenham: Edward Elgar.

Bhatt, C.R. (2000), "Flexible models structures for discrete choice analysis", in: D.A. Hensher and K.J. Button, editors, *Handbook of Transport Modelling*, Oxford: Pergamon.

Bollinger, C.R. and K.R. Ihlanfeldt (1997), "The impact of rapid rail transit on economic development: the case of Atlanta's MARTA", *Regional Science and Urban Economics*, Vol. 42, pp. 179–204.

Borenstein, S. and G. Saloner (2001), "Economics and electronic commerce", *Journal of Economic Perspectives*, Vol. 15, pp. 3–12.

Brewer, A.M., K.J. Button and D.A. Hensher (eds.) (2001), *Handbook of Logistics and Supply-Chain Management*, Oxford: Pergamon.

Brueckner, J.K. (2000), "Urban sprawl: diagnosis and remedies", *International Regional Science Review*, Vol. 23(2), pp. 160–171.

Bruinsma, F.R., S.A. Rienstra and P. Rietveld (1997), "Economic impacts of the construction of a transport corridor: a multi-level and multi-approach case study for the construction of the A1 highway in the Netherlands", *Regional Studies*, Vol. 31, pp. 391–403.

Button, K.J. and S.Y. Taylor (2001), "Towards an economics of the internet and electronic commerce", in: S.D. Brunn and T.R. Leinbach, editors, *The Wired Worlds of Electronic Commerce*, London: Wiley.

Calthorpe, P. (1993), *The Next American Metropolis*, New York: Princeton Architectural Press.

Chia, N.C., A. Tsui and J. Whalley (2001), *Ownership and Use Taxes as Congestion Correcting Instruments*. NBER Working Paper 8278, Cambridge, MA.

Chu, X. (1999), "Alternative congestion pricing schedules", *Regional Science and Urban Economics*, Vol. 29, pp. 697–722.

de Palma, A. and R. Lindsey (1998), "Information and usage of congestible facilities under different pricing regimes", *Canadian Journal of Economics*, Vol. 31, pp. 666–692.

DeSalvo, J.S. (1996), "Income, residential location and mode choice", *International Journal of Industrial Organisation*, Vol. 40, pp. 84–99.

Economides, N. (1996), "The economics of networks", *International Journal of Industrial Organisation*, Vol. 14, pp. 673–699.

Ekelund, R.B. and R.F. Hébert (1999), *Secret Origins of Modern Microeconomics: Dupuit and the Engineers*, Chicago: Chicago University Press.

Friedman, M. (1991), "Old wine in new bottles", *Economic Journal*, Vol. 101, pp. 33–40.

Fujita, M. and T. Mori (1997), "Structural stability and evolution of the urban system", *Regional Science and Urban Economics*, Vol. 27, pp. 399–442.

Fujita, M. and J.-F. Thisse (2002), *Economics of Agglomeration; Cities, Industrial Location and Regional Growth*, Cambridge: Cambridge University Press.

Fujita, M., P.R. Krugman and A.J. Venables (1999), *The Spatial Economy: Cities, Regions and International Trade*, Cambridge, MA: MIT Press.

Fukuyama, F. (1995), *Trust*, London: Hamish Hamilton.

Garreau, J. (1991), *Edge City: Life at the New Frontier*, New York: Double Day.

Gasper, J. and E.L. Glaeser (1996), *Information Technology and the Future of Cities, Working Paper 5562*, Cambridge, MA: National Bureau of Economic Research.

Gin, A. and J. Sonstelie (1992), "The streetcar and residential location in nineteenth century Philadelphia", *Journal of Urban Economics*, Vol. 32, pp. 92–107.

Glaeser, E. and M.E. Kahn (2003), "Sprawl and urban growth", *Handbook of Urban and Regional Economics*, Amsterdam: North-Holland.

Glazer, A. and E. Niskanen (1992), "Parking fees and congestion", *Regional Science and Urban Economics*, Vol. 22, pp. 123–132.

Hau, T.D. (1998), "Congestion pricing and road investment", in: K.J. Button and E.T. Verhoef, editors, *Road Pricing, Traffic Congestion and the Environment: Issues of Efficiency and Social Feasibility*, Cheltenham: Edward Elgar.

Henderson, V. and A. Mitra (1996), "The new urban landscape, developers and edge cities", *Regional Science and Urban Economics*, Vol. 26, pp. 613–643.

Hensher, D.A. and K.J. Button (eds.) (2003), *Handbook of Transport and the Environment*, Oxford: Pergamon.

Isard, W. (1951), "Distance inputs and the space economy, the locational equilibrium of the form", *Quarterly Journal of Economics*, Vol. 65, pp. 373–397.

Kanemoto, Y. (1980), *Theories of Urban Externalities*, Amsterdam: North-Holland.

Knight, F.H. (1924), "Some fallacies in the interpretation of social cost", *Quarterly Journal of Economics*, Vol. 38, pp. 582–606.

Koopmans, T.C. (ed.) (1951), *Activity Analysis of Production and Allocation*, London: Wiley.

Kraus, M. (1974), "Land use in a circular city", *Journal of Economic Theory*, Vol. 8, pp. 440–457.

Krugman, P.R. (1991), "Increasing returns and economic geography", *Journal of Political Economy*, Vol. 99, pp. 483–499.

Krugman, P.R. (1993), "On the relationship between trade theory and location", *Review of International Economics*, Vol. 1, pp. 110–122.

Krugman, P.R. (1996), *The Self-organizing Economy*, Oxford: Blackwell.

Krugman, P.R. (1998), "Space: the final frontier", *Journal of Economic Perspectives*, Vol. 12, pp. 161–174.

Kuhn, H.W. and A.W. Tucker (1951), "Nonlinear programming", in: J. Neyman, editor, *Proceedings of the Second Berkeley Symposium on Mathematical Statistics and Probability*, Berkeley: University of California Press.

Langley, L., M. Ripper and P. Varaiya (1973), "Effects of congestion on the shape of a city", *Journal of Economic Theory*, Vol. 6, pp. 162–179.

Leontief, W.W. (1953), *The Structure of the American Economy*, Oxford: Oxford University Press.

Lindsey, R. and E. Verhoef (2001), "Traffic congestion and congestion pricing", in: K.J. Button and D. Hensher, editors, *Handbook of Transport Systems and Traffic Control*, Oxford: Pergamon.

Lucas, R.E. (1988), "On the mechanics of economic development", *Journal of Monetary Economics*, Vol. 22, pp. 3–42.

Lucas, R.E. (1990), "Why doesn't capital flow from the rich to poor countries?", *American Economic Review*, Vol. 80, pp. 92–96.

Manski, C.F. and I. Salomon (1987), "The demand for teleshopping: an application of discrete choice models", *Regional Science and Urban Economics*, Vol. 17, pp. 109–121.

Marchand, M. (1968), "A note of optimal tolls in an imperfect environment", *Econometrica*, Vol. 36, pp. 575–581.

MacFadden, D. (2001), "Economic choices", *American Economic Review*, Vol. 91, pp. 351–379.

Medda, F. (2000), *The Assembled City*, Amsterdam: Thesis Publishers.

Meyer, J.R., J.F. Kanin and M. Wohl (1965), *The Urban Transportation Problem*, Santa Monica: Rand Corporation.

Mill, E.S. and D.M. de Ferranti (1971), "Market choice and optimum city size", *American Economic Review*, Vol. 61, pp. 340–345.

Mills, E.S. (1972), *Urban Economics*, Glenview: Scott Foresman.

Mills, E.S. and J. MacKinnon (1973), "Notes on the new urban economics", *Bell Journal of Economics and Management Science*, Vol. 4, pp. 593–601.

Mirlees, J. (1972), "The optimum town", *Swedish Journal of Economics*, Vol. 74, pp. 114–135.

Miyao, T. (1978), "A note on land use in a square city", *Regional Science and Urban Economics*, Vol. 8, pp. 371–379.

Mohring, H. and M. Harwitz (1962), *Highway Benefits*, Evanston: Northwestern University Press.

Mori, T. and K. Nishikimi (2002), "Economies of transport density and industrial agglomeration", *Regional Science and Urban Economics*, Vol. 32, pp. 167–200.

Moses, L.N. (1958), "Location and the theory of production", *Quarterly Journal of Economics*, Vol. 78, pp. 259–272.

Moses, L.N. and H.F. Williamson (1963), "Value of time, choice of mode, and the subsidy issue in urban transportation", *Journal of Political Economy*, Vol. 57, pp. 211–221.

Neary, J.P. (2001), "Of hype and hyperbolas: introducing the new economic geography", *Journal of Economic Literature*, Vol. 39, pp. 536–561.

Newman, P. and J. Kenworthy (1999), *Sustainability and Cities*, Washington: Island Press.

Nijkamp, P. (1999), "Future European transport systems and sustainable spatial development", in: A. Reggiani and D. Fabbri, editors, *Network Developments in Economic Spatial Systems*, Aldershot: Ashgate.

Nijkamp, P. and I. Salomon (1989), "Future spatial impacts of telecommunication", *Transportation Planning and Technology*, Vol. 13, pp. 275–297.

Ottaviano, G. and D. Puga (1998), "Agglomeration in the global economy: a survey of the 'New Economic Geography'", *World Economy*, Vol. 21, pp. 707–731.

Porter, M. (1980), *Competitive Strategy: Techniques for Analyzing Industries and Competitors*, New York: Free Press.

Predöhl, A. (1928), "A theory of location in its relation to general economics", *Quarterly Journal of Economics*, Vol. 36, pp. 371–390.

Pucher, J. and C. Lefèvre (1996), *The Urban Transport Crisis in Europe and North America*, London: MacMillan.

Richardson, H.R. (1977), *The New Urban Economics*, London: Pion.

Rietveld, P. (1995), "Transport infrastructure and the economy", *Organisation for Economic Cooperation and Development, Investment Productivity and Employment*, Paris: OECD.

Rietveld, P. (2001), "Urban development and the demand for open space: experiences in The Netherlands", *Studies in Regional Science*, Vol. 31, pp. 347–360.

Rietveld, P., K.J. Button and P. Nijkamp (eds.) (2003), *Urban Transport*, Cheltenham: Edward Elgar.

Romer, P.M. (1986), "Increasing returns and long-run growth", *Journal of Political Economy*, Vol. 94, pp. 1002–1037.

Rouwendal, J., E. Verhoef, P. Rietveld and B. Zwart (2002), "A stochastic model on congestion caused by speed differences", *Journal of Transport Economics and Policy*, Vol. 36, pp. 407–445.

Safirona, E. (2002), "Telecommuting, traffic congestion, and agglomeration: a general equilibrium model", *Journal of Urban Economics*, Vol. 52, pp. 26–52.

Salomon, I. (1985), "Telecommunications and travel – substitution or modified mobility?", *Journal of Transportation Economics and Policy*, Vol. 19, pp. 219–235.

Salomon, I. (2000), "Can telecommunications help solve transportation problems?", in: K.J. Button and D. Hensher, editors, *Handbook of Transport Modelling*, Pergamon: Oxford.

Sasaki, K. (1989), "Transportation system change and urban structure in two-transport mode setting", *Journal of Urban Economics*, Vol. 25, pp. 346–367.

Sasaki, K. (1990), "Income class, modal choice, and urban spatial structure", *Journal of Urban Economics*, Vol. 27, pp. 322–343.

Schuler, R.E. (1992), "Transportation and telecommunications networks: planning urban infrastructure for the 21st century", *Urban Studies*, Vol. 29, pp. 297–310.

Shapiro, C. and H.R. Varian (1998), *Information Rules: A Strategic Guide to the Network Economy*, Cambridge, MA: Harvard Business School Press.

Small, K.A. (1992), *Urban Transportation Economics. Fundamentals of Pure and Applied Economics*, Chur: Harwood.

Small, K.A. and J.A. Gomez-Ibanez (1998), "Road pricing for congestion management: the transition from theory to policy", in: K.J. Button and E.T. Verhoef, editors, *Road Pricing, Traffic Congestion and the Environment: Issues of Efficiency and Social Feasibility*, Cheltenham: Edward Elgar.

Solow, R. (1972), "Congestion, density and the use of land in transportation", *Swedish Journal of Economics*, Vol. 74, pp. 161–173.

Solow, R.M. and W.S. Vickrey (1971), "Land use in a long narrow city", *Journal of Economic Theory*, Vol. 3, pp. 430–447.

Storper, M. and A.J. Venables (2002), Buzz, the economic force of the city, DRUID Summer Conference, Copenhagen.

Stough, R., Y. Hihano, K.J.I. Button and P. Nijkamp (eds.) (2003), *Transport and Information Systems*, Cheltenham: Edward Elgar.

Verhoef, E.T. (1996), *The Economics of Regulating Road Transport*, Cheltenham: Edward Elgar.

Verhoef, E.T., P. Nijkamp and P. Rietveld (1996), "Second-best congestion pricing: the case of an untolled alternative", *Journal of Urban Economics*, Vol. 40, pp. 279–302.

Vernon, R. (1966), "International investment and international trade in the product cycle", *Quarterly Journal of Economics*, Vol. 80, pp. 190–207.

Vickrey, W.S. (1959), *Statement on the Pricing of Urban Street Use*. Hearings, US Congress Joint Committee on Metropolitan Washington Problems, November 11, Washington.

Vickrey, W.S. (1969), "Congestion theory and transport investment", *American Economic Review (Papers and Proceedings)*, Vol. 59, pp. 251–260.

Voith, R. (1998), "Parking, transit, and employment in a central business district", *Journal of Urban Economics*, Vol. 44, pp. 43–58.

von Thünen, J.H. (1826), *Der isolierte Staat in Beziehung auf Nationalökonomie und Landwirtschaft*, Hamburg.

Walters, A.A. (1961), "The theory and measurement of private and social costs of highway congestion", *Econometrica*, Vol. 29, pp. 676–697.

Weber, A. (1909), *Ueber den Standort der Industrien*, Tübingen.

Urban Dynamics and Growth, CEA, vol. 266
Roberta Capello and Peter Nijkamp (editors)

CHAPTER 7

Transport Systems and Urban Equilibrium

Lars Lundqvist

Unit for Transport and Location Analysis, Department of Infrastructure, KTH-Royal Institute of Technology, SE-100 44 Stockholm, Sweden

Abstract

This chapter provides an overview of urban model development focusing on land-use/transport equilibrium. Since the 'new urban economics' tradition and a number of important contributions in urban modelling evolved in the period 1960–1980, there has been a rapid development of both theoretical and operational approaches to systems analyses of land-use and transportation. Most modelling approaches are aiming explicitly at analysing equilibrium on urban markets. However, we are also reviewing attempts to find urban land-use/transport equilibria that are optimal in some sense of welfare. Both theoretical (qualitative) and operational (quantitative) model developments are covered in the form of selected examples. The selection, although subjective, is hopefully representative for the major modelling traditions during recent decades. The chapter ends by summarising general conclusions and lessons learned from qualitative and quantitative analyses by state-of-the-art urban models. Some directions for future research are identified.

Keywords: transport systems, activity location, urban structure, urban equilibrium, optimum cities

JEL classifications: R12, R13, R14, R21, R41, R42, R48, R52

This research was supported by a publication grant from the Swedish Agency for Innovation Systems. Comments from two anonymous referees and from professor T. John Kim are gratefully acknowledged.

7.1. Introduction

Studies of interactions between transportation systems and urban development constitute one of the core areas of urban economics and regional science. There are strong linkages between changes in transportation and communication technologies and the historical development of urban form and land use patterns. Similar transitions have been witnessed in most countries: from monocentric urban structures linked to harbour or rail terminals to suburbanising metropolitan regions with a polycentric character and relying predominantly on automobile use.

Interactions between firms and households on urban markets rely on transportation opportunities and result in transportation demand. Flows of person transport and movements of goods and services give rise to externalities in terms of environmental pollution and congestion. Other externalities relate to the scale and density of localised urban activities. Internal or external economies of scale in production contribute to forces of agglomeration that are transforming the typical tendencies towards decentralisation experienced in most cities during recent decades.

The modern analysis of interactions between transportation and urban land use has developed along two major lines during the post World War II period.

On the one hand theoretical urban economic models have been developed based on microeconomic foundations and relying on equilibrium (or in some cases welfare economic) concepts. Following Alonso (1964), the 'new urban economics' tradition developed during the 1960s and 1970s. The typical model was formulated in continuous space with a single urban centre or Central Business District (CBD) surrounded by residential and transportation land uses. Qualitative results concerning how the land rent and residential density distribution depended on model parameters (e.g. transportation cost) could be derived. Many generalisations of this prototype model were studied leading to more complex models. In many cases, only numerical simulation results could be obtained. Recent models of this type and still based on simplified and symmetric geographies focus on pricing of congestion externalities and on impacts of economies of scale in production.

The second strand of model development started with the great American urban transport studies of the 1950s and 1960s. Operational urban models were designed for use in planning contexts with the aim of forecasting the impacts of transport investments on urban form and structure. Herbert and Stevens (1960) and Lowry (1964) are early examples of such operational urban models. Advancements in discrete choice theory, spatial interaction modelling and transport network equilibrium modelling during the 1970s greatly improved the theoretical and computational characteristics of these models and commercial model packages developed during subsequent decades. The basic questions are the same as in the theoretical urban economic modelling approach: how do transportation systems and urban land use interact through urban land, labour and commodity markets?

It is the aim of this chapter to provide an overview of the two model traditions outlined above for the analysis of transport systems and urban structure (Sections 7.3 and 7.4). The basic paradigm is partial or general equilibrium of the urban economy and welfare economic extensions of this paradigm. However, many operational urban models apply iterative techniques with delays between instant equilibrium in the transportation market and subsequent equilibrium allocations of urban activities. A few subjectively selected but hopefully representative examples will be briefly outlined in order to illustrate each model category. Before proceeding to the modelling case studies, we will introduce the notions of *equilibrium* and *optimum* in the context of urban systems analysis (Section 7.2). This chapter is concluded by a summary of lessons learned from the qualitative and quantitative modelling approaches (Section 7.5) and some prospects for future research (Section 7.6).

7.2. Equilibrium and optimum in urban systems analysis

The main hypothesis in this chapter is that interactions between transport systems and urban structure take place on urban markets characterised by strong equilibrating market forces: prices of commodities and services, wages, transportation costs and land rents. The basic assumption is that either a *competitive equilibrium* is established instantaneously or we are considering a reasonable time period for equilibrating mechanisms to approach convergence.

Utility maximising individuals (or households) and profit maximising firms constitute the main agents in qualitative urban economics models but also, directly or indirectly, in operational urban planning models. Individuals are typically assumed to maximise a utility function depending on at least consumption of goods and land. In many cases leisure constitutes a third argument of the utility function. Consumption is in some models related to shopping trips to various shopping centres. Firms are assumed to select profit maximising amounts of land and labour in each location within a constant returns to scale technology. Sometimes also intermediary deliveries are included in the cost function. In equilibrium households of a certain type are adjusting until they reach the same level of utility and firms are producing at zero profits. Prices, wages and land rents are determined so that the markets for commodities, labour and land are cleared.

It is obvious that many prerequisites for the competitive equilibrium to be Pareto efficient can be easily questioned: private goods, no indivisibilities, no externalities, non-increasing returns to scale, all agents are price takers, etc. If some of these restrictions are relaxed, e.g. by allowing congestion externalities in the transport system that are not fully internalised, the competitive equilibrium corresponds to a *second-best* situation. Introducing a toll system that implements social marginal cost pricing would increase the overall efficiency of the urban system and lead to a *first-best* solution. Such 'optimum' considerations will be discussed in subsequent sections of this chapter.

Transportation costs are either assumed to be exogenously given or related to link flows by appropriate volume-delay functions reflecting congestion. In many simplified geographical structures the route choice is unique or otherwise easily deduced. In operational urban models equilibrium route choices are usually assumed to fulfil the following conditions (Wardrop, 1952) for each origin–destination pair: (1) travel costs on all routes that are used are equal, and (2) travel costs on unused routes are no less than the travel costs on routes that are used. Travel costs by public transport modes are often based on cost minimising route choices. Travel demand in terms of the number of trips generated by a certain activity in a given location, the destination choice and the choice of travel mode is normally assumed to follow nested logit models, which can be derived from random utility maximisation on the part of households. In equilibrium the travel demand, when assigned to the transportation network, gives rise to travel costs that sustain exactly the same demand.

We have touched upon the relation between equilibrium and some notion of optimum above. This is easily illustrated in the analysis of equilibrium on a congested transportation market. If users are faced with private (generalised) costs, the resulting (*user optimal*) equilibrium is not efficient in terms of the total generalised cost occurring in the network. In order to minimise this total cost, social marginal transport costs need to be introduced (by appropriate link tolls), leading to a system optimal equilibrium. In many cases, it is relevant to search for *system optimal* policies (e.g. network investments minimising total transport costs) while at the same time acknowledging that consumers behave in a user-optimal way (according to user equilibrium). Such model formulations can often be stated as *bi-level* optimisation problems.

A further step towards *welfare maximisation* can be taken by explicitly stating some system-wide objective function with more or less emphasis on equity in terms of household utilities. Options for realising an optimum urban structure through decentralised decision-making on urban markets can then be subject to analysis.

7.3. Transport systems and urban land use in simplified urban geographies

The concepts of equilibrium and optimum will now be used in analysing interactions between transportation systems and urban structure. In this section simplified urban geographies in terms of transportation systems and symmetric land uses will be studied. A distinction is made between models based on a continuous treatment of space and models relying on a discrete subdivision of space into zones.

7.3.1. Symmetric continuous space: optimum and equilibrium

7.3.1.1. Equilibrium in a continuous space monocentric city

Wheaton (1974) studies the equilibrium structure of open and closed urban economies with all production in the city centre and with exogenously given commodity prices, wages and transport costs for commuting (as a function of distance from the CBD). Consumers are assumed to earn the same income and have the same preferences with respect to consumption of land and other commodities. The only requirement on the utility function is that both the composite commodity and land are 'normal' goods and have positive income effects. The rent and residential density distributions are endogenously determined. Land use for transport is not explicitly treated and the urban geography is circular and symmetric.

In the open urban economy the utility level of residents is adjusted to the utility level in the surrounding countryside. The land rent at the city boundary must equal the opportunity price for land (agricultural land rent). The size of the urban area (outer radius) and the size of the population are determined by the model. In contrast, the closed urban economy is characterised by a given total population. Instead, the utility level of residents is endogenous.

By using the equilibrium conditions requiring that all residents experience the same utility level and that the marginal rate of substitution equals the price ratio, the consumption of land and commodities can be solved as a function of income, distance to the centre and utility level. By inserting these results into the budget constraint, rent can be expressed as a function of the same variables. Finally, using the requirements that the value of the rent at the city boundary should equal the agricultural rent level and that the whole population should be accommodated within the city boundary, the remaining parameters can be solved for: city radius and total population in the open city, and city radius and utility level in the closed city.

After deriving some properties of the land consumption, commodity consumption and rent functions, Wheaton proceeds to conduct a comparative static analysis by varying exogenous parameters. Agricultural rent, income and transportation cost are varied in both the open and closed cities. In addition the utility level is varied in the open city and population size in the closed city.

Focussing on transportation costs and assuming that commuting cost is proportional to the distance to the centre, how does an increase in transportation cost affect the urban structure? The comparative static analysis shows that in the open city the increased commuting cost leads to lower rents, lower densities and a reduction of the total population and the city radius. Hence, the higher transportation costs lead to a smaller and less dense urban system at equilibrium. In the closed city, the corresponding impact is steeper rent and density functions, a reduced city radius and a reduced utility level. Also in the closed city, the urban area shrinks, but the requirement of a given total population leads to higher rent and density in the vicinity of the city centre.

Wheaton discusses how congestion costs would affect the results. He concludes that the qualitative results would still be valid: higher income would still lead to a larger urban area but to a lesser extent than in the uncongested case. This is because the more dispersed urban structure would lead to more traffic at any location and therefore higher commuting costs which would counteract the tendency towards dispersal.

Solow (1973) explicitly introduced congestion costs and allocation of land between residential and transport use outside the CBD. Also the redistribution of the total land rent as a lump sum social dividend is taken into account (in the case the city owns the land). It is shown by numerical simulations that the rent gradient of the competitive equilibrium in the case when congestion is not taken into account is too flat as compared to the case when congestion is included. It is also suggested that cost–benefit analysis based on market land values in the case when congestion is important but not internalised leads to an excessive use of land for transport, especially close to the CBD where the market land rent is too low. Hence, the distorted land values (because of non-internalised congestion costs) seem to guide land-use policy away from the optimum policy that maximises the common utility level. A number of further improvements of the model were listed: introduction of explicit housing consumption, both time and money costs of transport, a more sensitive congestion function and more than one income class.

Beckmann (1976) showed that contacts between households lead to higher densities and land values in the geographical centre of the city even without any predetermined CBD location. Substantial savings in the average transportation costs to all other households resulted from this compression of densities in the spatial equilibrium as compared to the equal density distribution.

7.3.1.2. Optimum in a continuous space monocentric city

Many classical analyses in the 'new urban economics' tradition rely on the assumption of a predetermined CBD where production takes place and income is generated and where radial commuting/shopping trips are required from the place of residence to the CBD. As noted above Beckmann (1976) derives higher residential densities at the centre of the region without treating production or shopping.

Dixit (1973) extends the analysis of Solow (1973) to include the scale of production and the area of the CBD within an optimising framework. In comparison with Solow (1973) congestion and land requirements for transportation are explicitly modelled in a more flexible way. The welfare criterion is an aggregation of household utilities where the attitude towards inequality can be varied in terms of a parameter. Other studies of optimum urban structures had shown that, in general, the utility levels of households are unequal unless very equity oriented welfare measures are applied. Dixit assumes economies of scale in production and diseconomies of scale in transportation (through congestion). The optimal city size is obtained from a trade-off between these externalities. Households derive utility from consumption of commodities and space. The time needed for commuting reduces the hours of labour input. Production uses intermediary input (assumed proportional to the land area of production) and labour.

The focus of the analysis is on the planned optimum urban structure. However, the optimum structure can be sustained by decentralised decision-making if labour is paid its marginal product, congestion tolls are levied and appropriate lump sum subsidies as a function of the distance from the CBD are applied (independent of distance in the special case of equal utilities). Some qualitative results can be derived as regards the density and rent functions, the utility distribution and the share of land used for transportation. More complete solutions to the model are achieved by numerical computation. For the case of the equal utility solution, the optimum utility level, wage rate, city radius and CBD radius could be determined for each prescribed population level and for each assumption concerning economies of scale. The welfare maximising population and city radius increase with economies of scale but with a decreasing rate. With the specific choice of parameters used in the computations, cities with more than about 200,000 inhabitants cannot be obtained. The optimum population turned out to be very sensitive to the importance of congestion in total commuting time. If the welfare function is changed to permit inequalities, a wide range of optimal utilities is obtained for parameter values assumed to approximate the case of the additively separable welfare function. Among further improvements Dixit mentions introduction of transport costs within the CBD, relaxation of the fixed land/composite input proportion, introduction of housing services, multiple products, and indivisibilities in transport.

In an earlier paper Mirrlees (1972) had treated the optimum town problem in the case of a completely separable welfare function with household utilities being a function of consumption of a composite commodity, consumption of residential space and distance to the CBD. He concluded that the optimum distribution of utilities of identical people is in general unequal and that the optimum urban structure can be sustained by a competitive equilibrium under an appropriate income distribution. A competitive realisation of the optimum is also possible when the use of land for transport is taken into account (the uncongested case is assumed) and when the local population density is introduced as an externality in the utility functions of households. In the latter case a system of subsidies (or taxes) related to commuting distance is needed. Another way of internalising this externality could be via a land market operating in terms of shares of housing estates with given neighbourhood densities. A simple rule characterising the optimum town size in a wider geography could be derived. Among interesting further developments, Mirrlees mentioned relaxing the assumptions of uniform tastes and a single commodity. He also emphasised the importance of studying second-best problems: what is the optimal policy when some land uses (e.g. transport) are not charged for, when the income distribution is not optimal, or when redistribution between town and country is not feasible?

Oron *et al.* (1973) also discuss relations between optimum and equilibrium urban structures in the case of a congested transport system. The concept of equilibrium pertains to five sectors: households, housing producers, commodity producers, land transactors and a transportation authority. The concept of optimum refers to the maximum utility level, which can be realised provided that equals are treated as equals. It was shown that the optimum allocation may be supported by a competitive price system if a warranted congestion toll is collected and redistributed as a lump sum subsidy. If less than the warranted congestion toll is collected, the competitive allocation is distorted and the competitive city tends to be more suburbanised than the optimum city. The result is obtained through numerical solution of a specific formulation of a general equilibrium model. The conclusion on suburbanisation departs from some earlier results based on partial equilibrium models, where household income and composite commodity price were assumed to be exogenously given.

7.3.2. Symmetric discrete space: optimum and equilibrium

7.3.2.1. Equilibrium in a symmetric and discrete space city

There is a limit to qualitative analyses of continuous space models. Often numerical solutions have been the only resort in order to illustrate properties of the models (see the summaries of Solow, Dixit and Oron *et al.* above). By changing from a continuous representation of space to a discrete setting, a greater freedom of model formulation is acquired.

Anas and Kim (1996) and Anas and Xu (1999) provide two recent examples of applications of computable general equilibrium models in symmetric and discrete urban space. Both studies focus on the role of the congestion externality in urban systems with endogenous job location. In addition, Anas and Kim (1996) treat scale economies in shopping as an example of agglomeration externalities leading to polycentric urban structures.

The urban geography of both studies is linear with equal width or wedge-shaped width. Production is assumed to use land and labour inputs in a constant returns to scale technology. Output is purchased by consumers, which are undertaking shopping trips to the sites of production. Consumers maximise utility derived from consumption of commodities, space and leisure and a stochastic component defined for each commuting arrangement (combination of residential location and job location). Consumers have strong preferences for the variety of commodities. Products of each location are considered as a different variety and consumers travel to all production locations for shopping with the frequency of shopping trips depending on prices and travel costs related to each shopping location. Income derives from labour income and a dividend representing equal distribution per head of total rent payments (and possibly excess or deficit transport financing). Finally, travel (commuting and shopping trips) takes place on roads produced by using land. Travel time is flow-dependent according to the usual type of congestion function. The land used for roads is financed either by social marginal congestion tolls or by a head tax. The general equilibrium conditions state that land, labour and product markets clear in each zone. An iterative procedure is used for finding equilibrium rents and wages, which are then used to calculate equilibrium prices and output levels.

From model simulations based on a stylised set of parameters, Anas and Xu (1999) showed that in the case of first-best allocation of land to roads (i.e. toll revenue just covers the land rent for roads in each zone) both residences and production are characterised by dispersed location. Production is more centralised than residential location. Land rents fall sharply from the geographical centre to the edge of the city while prices fall and wages rise moderately with distance from the centre. 62% of the land is used for transportation in the central zone while on average only 6.3% of total urban land is allocated to transport.

The main purpose of Anas and Xu (1999) is to show how the introduction of congestion tolls impacts on land use. This is done both for the case when land is efficiently allocated to transport and for the case when equal amounts of land is allocated to transport in each zone (but with the total amount of land used for transport being equal to that of the first-best base case solution). The two principles for land allocation are combined with two principles for financing of transport: through congestion tolls (possibly complemented by a head tax in the equal allocation case) or via a head tax only. It is shown that introduction of road tolls as the means of financing has a centralising effect on both employment and population and this effect is larger for employment. The rent gradient becomes steeper while the wage differences between zones are reduced. The same patterns are observed for both cases of land allocation to transport: efficient and equal amounts of land. The efficiency gains in terms of welfare from introducing tolls are small (0.03–0.04%), while the welfare gains from making road planning efficient are 3–5 times as big.

Translating the results to income, the introduction of tolls and efficient road planning lead to gains of about 3%. A comparison with corresponding monocentric cities shows that the dispersed equilibrium has 26% higher welfare. The gains from introducing tolls and efficient road planning are of the same order as in the dispersed equilibrium. The authors conclude that removing land use restrictions seems to be more important than congestion pricing.

Anas and Kim (1996) formulated a model permitting scale economies in shopping, i.e. consumers preferring larger shopping centres to smaller. The model also provided for more than one commodity, intermediate deliveries between industries and freight flows, but these options were not used in numerical examples. Instead the analyses concentrated on the possibilities of obtaining polycentric development patterns instead of the dispersed equilibrium when the scale economies in shopping are introduced and are sufficiently strong relative to the cost of congested travel. One peak, three peak and five peak symmetric polycentric urban structures and a completely mixed structure are potential multiple equilibria in an 11 zone equal width linear city. Sufficiently big perturbations would move the urban structure from any one of these equilibria to one of the others. High scale economies in shopping relative to the level of congestion make the monocentric urban structure socially preferable (according to welfare measured as expected utility). With lower scale economies equilibria with more centres give higher expected utility and eventually the completely mixed case is the socially preferred equilibrium. However, even the completely mixed case is more centralised in terms of production and has a steeper rent gradient than the corresponding dispersed urban equilibrium in the case of no scale economies (as in Anas and Xu (1999)).

Eliasson and Mattsson (2001) extend the analysis of land use impacts of congestion pricing to the case of a stylised symmetric city with an attractive transit system, with a slow traffic mode and with explicit treatment of shopping and service provision including deliveries from producers to these sectors. The increased flexibility in terms of modes and sectors is in contrast with more rigid land allocation principles and an absence of endogenous income formation (from labour market and social dividend). Exogenously given amounts of land are assumed for housing, shopping and service establishments in each zone while production is assumed to meet no land constraints and land use for the transport system is implicit. The location behaviour of producers, shops and service providers is described in terms of accessibilities rather than in terms of profit maximising behaviour. The location behaviour of shops and service providers introduces agglomeration economies.

The model was used to analyse the impacts of social marginal cost pricing and toll rings in a generic star-shaped city with radial transportation from suburbs to the geographical centre node. The results confirm the conclusion of Anas and Xu (1999) that first-best road pricing tends to centralise both population and employment to the inner suburb. A somewhat stronger centralising effect is obtained for shops and service providers. However, the relocation effects are very small as compared to traffic effects: total distance by car is reduced by 24% and total travel time by car is reduced by 57%. Introducing optimal congestion pricing reduces the car shares by 7–9% per link and increases the public transit share by 4–5% per link. Comparisons are made with the effects of an inner toll ring, located immediately outside the central zone, and an outer toll ring, located outside the inner suburbs. The traffic effects of the toll rings are substantial but obviously more local than those of congestion pricing. The location effects are small and depending on the location of the toll ring: the inner toll ring has a decentralising effect while the outer toll ring has a centralising effect.

A later extension of the model included a ring road between the inner suburbs, see Mattsson and Sjölin (2004). In the absence of road pricing, the ring road leads to centralisation of activities to the inner suburbs (and in the case of housing and employment also to the central zone). It also leads to more congestion on the radial links immediately outside the ring road but less congestion on other links. There is an increase of car use for all trip types, in particular for residents in the centre and in the inner suburbs. The total distance travelled by car increases by 9%. Introduction of optimal congestion pricing in the ring road scenario leads to very small changes in location but reduces the total car distance by 25% by redirecting traffic to public transport and to the slow mode. Toll rings instead of congestion pricing lead to smaller and more local traffic effects with similar location effects as reported above from Eliasson and Mattsson (2001). The total revenues from toll rings can never reach the same level as in the case of optimal congestion pricing and the reduction of total distance travelled by car by introduction of congestion pricing can only possibly be reached by enforcing very high toll levels for an outer toll ring.

7.3.2.2. Optimum in a symmetric and discrete space city

In this section, two early contributions to the analysis of transport systems and urban location using optimising approaches will be discussed. These can be seen as important linkages from the theoretical analyses in Section 7.3.1.2 to the operational planning models applied to real cities of Section 7.4.2.

Mills (1972) formulated a linear programming model for efficient resource allocation in an urban area. Predetermined volumes of a number of commodities are produced in the city and exported from the urban centre. Housing is produced for the workers of the city. The geography is a rectangular grid with transport in north, south, east and west directions. Hence the urban system is symmetric and the location pattern is only dependent on the distance to the centre. A certain number of technologies are available for production of commodities and housing, representing substitution between land, capital and labour. The land–capital substitution is of specific interest and is interpreted in terms of building height. Transportation requirements include transport of goods to the centre and commuting. Given a certain resource allocation to transport in terms of land and capital, the congestion level can be computed for any transport demand and the resulting social marginal costs. The model seeks an urban structure that minimises the total opportunity cost for land, labour and capital and the total social marginal transport cost. A land constraint requires that the total demand for land in each grid cell does not exceed the supply of land.

Given the assumptions of the model it can easily be shown that housing is at least as suburbanised as the production of commodities and there is no outward commuting in the efficient allocation. The model computes an optimal allocation of land between commodity production, housing and transportation in each zone, an optimal level of congestion and optimal capital–land substitution in production activities. The dual variables of the land constraints provide information on the land rents. Some preliminary computations with the model were reported in Mills

(1972), which were assumed to represent a typical US city of about one million inhabitants. Relatively small shifts in parameters led to transition from one extreme solution with equal amounts of employment and housing in each zone and no commuting to another extreme solution with all production concentrated around the city centre. The optimum degree of congestion was reflected in the fact that central transport costs were about five times as high as transport costs at the edge. The technologies used near the centre represented 10–15 story buildings with the capital intensity falling off rapidly with distance from the centre. The shadow prices of the land constraints revealed that land prices at the centre were 50–100 times the agricultural rent level.

The optimum allocation of the model can be realised through competitive markets provided that the public sector constructs an optimum transport system and sets optimum prices for its use. The price charged in each grid cell should be the marginal cost of an extra unit of transportation, including congestion cost at the optimal level of congestion. The capacity and land used for transportation should be according to the optimal allocation. If transportation is underpriced, this may lead to excessive congestion and distorted locations of housing and production (e.g. avoiding congestion by suburbanisation). By assuming that transport is priced at average costs rather than social marginal costs and that the optimal layout of the transportation system is retained, the welfare loss from inadequate pricing can be computed by the model. The model gives rise to an equilibrium, which corresponds to an efficient allocation given the transport pricing principle. Further misallocations and welfare losses may be the result of suboptimal planning of the capacities of the transport system, possibly due to distorted land rents. Among desirable further modifications of the model Mills mentions fixing activity amounts according to the historical developments of urban areas, relaxing the symmetry assumptions, introduction of additional transport modes, representation of urban policies, introduction of intermediary goods in production and permitting exports also from suburban locations. Many of these extensions are covered in models discussed in Section 7.4.

An extended model, based on very similar principles and geographical specifications as those of Mills (1972), was presented by Ripper and Varaiya (1974). The model permits exports and imports either through the city centre or through the periphery. A certain number of centralised goods can only be produced only in the city centre while other decentralised goods and housing can be produced at any location outside the city centre. In addition to land, labour and capital, production also requires intermediary inputs of goods. The specification of housing technology represents both the inputs for construction and maintenance of residences and the consumption of the households. Each household demands one unit of housing and provides one unit of labour. Transportation can occur both inwards and outwards in any grid cell (but not simultaneously for a certain commodity). Social marginal cost pricing is assumed and congestion is treated in the same way as in Mills (1972).

In a numerical example one centralised good and one decentralised good were assumed. The former was used as input in the latter and in consumption. Only the decentralised good was subject to an export requirement. A stable allocation of high-density housing and transportation close to the centre and a mix of low-density housing, decentralised production and transportation further out occurred. If all export is required to go through the city centre the urban structure becomes more compact with the inner grid cells completely used for transportation. The rent gradient is steeper in the case of export through the centre only. In both cases there is a considerable degree of suburbanisation of production. The conclusions of Mills (1972) concerning the optimal degree of congestion and the steep rent gradient are confirmed by Ripper and Varaiya. The authors also considered generalisations to two types of labour, possibly combined with a choice of consumption bundle, and to dynamic allocation of activities over two or three periods. In the dynamic model new variables were introduced reflecting the remaining part of the urban structure existing in the previous period and costs were included for destroying of existing structures. Among possible future extensions the authors mentioned technological change and accounting for the age and depreciated values of existing structures. Other extensions included relaxation of the geographical symmetry, introduction of other transport modes and studies of the implications for the city structure of the appearance of sub-centres.

7.4. Transport systems and urban land use in realistic urban geographies

In Section 7.3 theoretical analyses and some simulation results were reported for urban areas with simplified geographies: monocentric urban structures with a continuous representation of space and symmetric urban structures with a discrete representation of space. The applications referred to generic cities, which were assumed to be representative of certain types of urban systems. The purpose of this section is to provide examples of operational models applied to particular real cities. Our focus is still on relations between transportation systems and urban land-use derived from equilibrium or optimum principles. We distinguish between approaches based on a linked set of sub-models on the one hand and integrated approaches aiming for a simultaneous treatment of urban transport/land-use interactions.

7.4.1. Modular/iterative approaches: optimum and equilibrium

Most available operational land-use/transport models are using the iterative, modular approach. A typical model proceeds from a transportation equilibrium in one period affecting the land-use in the next period, which immediately affects the transportation equilibrium in that period. Hence, the transport behaviour is assumed to adapt quickly to changes in land use while the land-use response to changes in the transport market is lagged by one period (typically 2–10 years).

The International Study Group on Land-Use/Transport Interaction (ISGLUTI) conducted a comprehensive comparative analysis of this modular type of land-use/transport models (Webster *et al.*, 1988). Each model was applied to an extensive set of standard tests (40 in total) covering changes in both land-use and transport conditions of the particular reference city for that model. In a second phase some models were also applied to some of the other cities: one model (LILT) was applied to three cities and another model (MEPLAN) was applied to two cities. This resulted in three models analysing Dortmund and two models each analysing Leeds and Tokyo. Comparisons were reported in two issues of *Transport Reviews* (vol. 10–11, 1990–1991). Similar exercises have later been carried out within, e.g. EU-funded research projects like SPARTACUS and PROPOLIS.

Three of the models in the ISGLUTI study were applied to most of the policy tests: the LILT model (Leeds), the MEPLAN model (Bilbao) and the IRPUD model (Dortmund). Their land-use modules simulate urban development building on both equilibrium (LILT, MEPLAN) and disequilibrium principles (IRPUD). LILT establishes equilibrium without explicit market prices while MEPLAN derives equilibria based on market prices for land and floor space. Also IRPUD computes land and housing rents, which affect location behaviour with a 2-year time lag. The main results of transportation policy tests on urban development can be summarised in the following way. Employment location is more sensitive to changes in travel costs than is the distribution of population. Retail employment is more responsive than non-retail service employment, which in turn is more responsive than non-service employment. Individual social groups show more movement than the population overall. A few examples will be mentioned from the summary of policy impacts resulting from the study. Increasing costs for car use would hardly affect population while centralising retail because of modal shifts to public transport and walking. Average housing land prices and densities overall would be slightly reduced. Taxation of car ownership or central area parking spaces would both decentralise employment and increase average housing land price (especially increased parking fees). Higher costs for car ownership would increase the use of public transport and walking and slightly increase average densities.

In a later analysis of the Dortmund region using the IRPUD model system Wegener (1996) studied how various transport policies might contribute to achieving environmental objectives expressed as a reduction of CO_2 emissions. A base scenario between 1970 and 2015 is compared with the impact of policy scenarios implemented between 1994 and 2015. Three types of policy scenarios were considered: travel cost scenarios, travel speed scenarios and combination scenarios. The first two types are similar to the corresponding policy tests in ISGLUTI but the cost increases of cars are introduced faster and linked to increased fuel-efficiency and the speed improvements of public transport are linked to increased capacity in terms of trains and buses. Two combination scenarios were formulated: 'promotion of public transport' included increased car costs and parking charges, faster public transport and slower car traffic, while 'reduction of mobility' in addition to higher car costs and parking fees and slower car traffic also included higher public transit fares and slower public transport.

The prevailing trend in Dortmund 1970–1994 has been characterised by increasing mean trip length, increasing car km per capita, decreasing public transport share and increasing CO_2 emissions. In the base scenario the first two of these trends continue during 1994–2015 according to the model, while the decrease of the transit share and the increase of CO_2 emissions per capita are stopped and mildly reverted around 2000. Out of the 10 scenarios analysed by IRPUD, the promotion of public transport policy is most efficient in reducing car traffic (more than 50% compared to the base scenario in 2015) and CO_2 emissions (about 70%) without reducing the mean trip length by more than 10%. This is achieved by a more than doubling of the public transport share between 1994 and 2015. In the base scenario the Dortmund region is decentralising and characterised by increased travel distances. Only in transit-oriented scenarios is the trend towards decentralisation reverted and the location of housing and workplaces becomes slightly more compact. Workplaces respond more strongly to transport changes than housing. If travel speeds by car are reduced, workplaces move outward to be closer to residences (even when public transport speeds increase). The conclusion from the study is that present urban systems in Europe have huge potentials for reducing car trips and emissions without fundamentally changing the physical layout of the cities.

Johnston and de la Barra (2000) applied the market-based TRANUS land-use/transport model to Sacramento for the period 1990–2015. The model is of the spatial input–output equilibrium type. The approach is similar to that of MEPLAN (cf. ISGLUTI above). Three scenarios for investments in the transport systems between 2000 and 2010 were studied in addition to the trend scenario: high-occupancy vehicle (HOV) lanes, Beltways + HOV and light rail transit and pricing (parking charges). The LRT/pricing scenario reduced the total vehicle miles of travel by 24%. The results of the land-use projections were aggregated to eight super-zones. The number of households for any scenario in 2015, net of the trend scenario, varied less than 13% of the total growth in households for each super-zone 1990–2015. The corresponding variation of employment is smaller than 9% because employment locators are less sensitive to changes in land rents. The differences in land consumption are somewhat higher than the differences in households and employment (up to 20%). Based on knowledge of the region's land market the results seem broadly reasonable.

UrbanSim is an urban simulation system for applications within land use, transportation and environmental planning (see Waddell, 2002). It is not a single model, but combines models of different types (e.g. aggregate, disaggregate, top-down, bottom-up) in different modules. The core models are demographic and economic transition models, household and employment mobility models, household and employment location models, a real estate development model and a land price model. These models are processed sequentially year by year and do not enforce equilibrium at any point of time. Model generated events can be combined with user-specified events like planned major development projects. The behaviour of urban actors is influenced by regional accessibility, which is based on travel times and utilities from an external travel model. The location decisions in turn lead to changes in travel demand. The travel demand model is typically run only every 5 simulated years since its outputs tend to change more slowly than other values in the simulation.

The initial implementation of UrbanSim was in Eugene-Springfield, Oregon. The spatial analysis was carried out for about 15,000 grid cells of 150 by 150 m while transport analysis was done for 271 traffic analysis zones (TAZ). Households were classified by income (9 segments), age of head (7), persons in household (5), workers in household (3) and number of children (2). Employment was classified into 14 sectors. The model was calibrated on 1994 data and validated by application to the time period 1980–1994. The correlation between modelled and observed values of employment, population, non-residential floor space, housing units and land values was 0.80–0.83 on the cell level and 0.86–0.93 on the TAZ level. The model could predict the change of population and employment with an error less than 50 individuals for 57% (population) and 31% (employment) of the TAZ. The interpretation of the results should take into account that the metropolitan region is fairly small and the changes in terms of growth and policies during the simulated period were modest.

The IMREL model (Integrated Model of Residential and Employment Location) was developed in the Stockholm regional planning context by Anderstig and Mattsson (1991) and applied for scenario analyses and impact studies related to infrastructure investment policies (Anderstig and Mattsson, 1991, 1998; Johansson and Mattsson, 1995). The model combines a commuting behaviour model (nested logit models of destination and mode choices), stochastic equilibrium on the housing market, an empirical (logit) model of employment location and a normative welfare maximising model of housing location. Traffic assignment is handled in a separate model, which may be iterated with IMREL until some desired degree of consistency is achieved. The model is run in a comparative static way for the horizon year of a certain planning period taking into account constraints related to the existing settlement structure of the base year and various limits on the future development of new settlements. The residential location sub-model and the employment location sub-model are iterated until convergence. A particular feature is that households are assumed to value accessibility and outdoor space availability. The aversion against density introduces a spatial externality (cf. Mirrlees, 1972), which is handled in the housing location part of the model relying on welfare maximisation based on locational surplus (consumer surplus + producer surplus).

The impact on land use in the Stockholm region over a 30-year period of an infrastructure investment programme was estimated to a relocation of 1.5–2.0% of the total amounts of employment and population, both on the level of nine sub-regions and on the level of 92 zones (Anderstig and Mattsson, 1991). Employment showed to be slightly more responsive than population. The investment programme decentralised both employment and population and shifted development from north-eastern and eastern parts of the region to other directions. The impact on land use was linked to a slightly lower car share, longer average travel distances by car (about 6.5%) and higher VKT (about 5.5%) than in a reference scenario. A similar evaluation of the land-use/transport impacts 1990–2005 of a package proposal, which also included a cordon toll system, is reported in Johansson and Mattsson (1995).

The combined effect of infrastructure investments and the cordon toll is predicted to keep total VKT unchanged in comparison with a do-nothing scenario and reduce central city VKT by 17%. Population and employment is decentralised from the central zone (surrounded by the cordon toll) and from semi-central and eastern locations to peripheral zones. At the level of 99 zones, 1.3% of the total population and 2.3% of the total employment is relocated. Finally, Anderstig and Mattsson (1998) report land use impacts 2020 of the package proposal for the Stockholm region mentioned above and regional train investments in the Mälar Valley. At the level of 214 zones, around 4% of the total population and employment is relocated. The location impacts are mainly operating within counties and the relocation between counties is marginal (less than 1%).

7.4.2. Simultaneous/integrated approaches: optimum and equilibrium

An early approach to integrated modelling of residential and employment activities with regard to a given transport system was presented by Coelho and Williams (1978) and Wilson *et al.* (1981). The transport system is represented by exogenously specified generalised costs between zones (cf. IMREL above). Equivalent optimisation models are formulated for simultaneous location of basic employment, service employment and residents. The markets for labour, services and housing are balanced and modelled in the tradition of Lowry (1964). The objective function is expressed in terms of interaction benefits less establishment costs, where interaction benefits are formulated as locational surplus. A model for a long-run equilibrium design problem is stated, covering location of housing, basic employment and service employment (Lowry type model) within a spatial interaction context (see Wilson *et al.*, 1981). A land use constraint is included and other planning constraints may be added. Two short-run models for reallocation of housing and services, respectively, within an existing urban structure are also formulated.

A numerical example was provided illustrating the application of the long-run equilibrium design model to a 28-zone subdivision of Leeds. The existing population and service employment was taken as attractiveness weights in the locational surplus measure and establishment costs were assumed to be independent of location.

This base case corresponds to a fairly centralised pattern of service employment with basic employment predominantly in suburban locations. By equalising all attractiveness weights a 'homogenous' city is obtained where basic employment is allocated in such a way as to generate an even distribution of population and service employment. Finally, by letting the cost sensitivity parameter in the travel behaviour formulation tend to infinity, the transport cost minimising urban structure is simulated emphasising development along the south-east to north-west axis.

The Projective Optimisation Land-use Information System (POLIS) was developed in the San Francisco Bay Area in order to provide an operational model based on state-of-the-art urban modelling knowledge (Prastacos, 1986a). The model can be seen as a generalisation of the Coelho–Williams framework to several modes of transport and to the incorporation of agglomeration economies in industrial location. The model is quasi-dynamic or comparative static in the sense that it allocates the net increase of activity volumes between the base year and the planning horizon.

POLIS was calibrated on 1975 data for the Bay Area divided into 107 zones, with transport represented by two modes and with economic activities divided into four sectors (see Prastacos, 1986b). The calibration was done stepwise after initial efforts to define attractiveness weights for residences and employment sectors and zonal functions reflecting agglomeration economies by sector. The model was then used to simulate the development of the region during 1975–1980 and a careful analysis of various aspects of goodness-of-fit was reported. POLIS is more successful in predicting the zonal allocation of total employment than housing. The goodness-of-fits of the two basic sectors are on average less successful than those of the retail and service sectors. Therefore, high priority areas for further extensions include introducing more than one type of housing, housing supply equations and household formation as well as redefining the zonal agglomeration economies function and finding a more uniform sectoral disaggregation.

Operational urban models with origins in the tradition of Mills (1972) were developed by Kim since the late 1970s and summarised in Kim (1989). These models cover many of the extensions mentioned by Mills (1972): more than one transport mode, intermediary goods, export from several locations and relaxation of spatial symmetry assumptions. The non-linear model to be discussed here is based on a general zonal subdivision and on exogenously given transportation networks for different modes of transport. The road network is link based with given capacities on each link and with congestion represented by volume-delay functions. The other modes are assumed to be timetable based or distance based with given generalised costs between origin–destination pairs. Urban activities are interconnected by input–output deliveries of intermediate goods and final consumption. Households are treated as one sector delivering labour and consuming goods and services. Like the Mills model production and housing technologies are represented in terms of building height with resulting possibilities for land–capital substitution. The discrete technology levels were eventually replaced by cost minimising, continuous, land intensities (related to the price of land and capital in each zone), see Lundqvist (1998a). Transport patterns are assumed to be consistent with gravity-type spatial interaction behaviour (destination choices), logit-type modal choices and user-equilibrium route choices. The urban system is driven by exogenous export demands. Equilibrium between supply and demand for commodities is established in each zone. Equilibrium is also established on the land market in each zone and the dual variables of these constraints provide land rents by zone.

The model is formulated as a non-linear programming model with cost minimisation as the overriding objective. Transport costs are included in terms of average private costs. This is achieved by incorporating the objective of the equivalent optimisation formulation of user-equilibrium route choices. In this respect, the model differs from the basic model of Mills (1972), which was based on social marginal transportation costs. The model combines important elements from major modelling traditions in urban economics: spatial input–output, economic base theory, spatial interaction, network equilibrium, equilibrium on commodity and labour markets and competitive bidding for land.

The model was estimated, validated and applied to the Chicago region (see Rho, 1988; Kim, 1989; Lundqvist, 1998a). The region was divided into 74 zones and the urban economy was represented by four urban activities (manufacturing, trades, services, households). The model covered road and transit modes with the road network containing about 1000 nodes and 3000 links. Comparisons of the estimated and calibrated model results for Chicago 1980 with observed land-use and transport patterns reveal good correspondence in terms of major land-use patterns (relative concentration of employment), land intensities and land rents. The model overestimates the relation between central city land rents and peripheral land rents (central city land rents are overestimated by about 40%). The total numbers of person and freight trips are well estimated but the goodness-of-fit between observed and estimated origin–destination trip tables is less exact.

Two applications of the model in the Chicago context were reported: one related to the effects of closure of a major arterial street (Lake Shore Drive), and one related to the effects of changed sector composition. We will focus on the first of these. The transport network change leads to increased congestion and a decreased share of road traffic, increased attractiveness of central zones (located along an alternative expressway) and many peripheral zones, decentralisation of production and land rent changes reflecting these conditions. Due to suburbanisation the average travel distances increase.

As pointed out above the model of Kim (1989) represents the case of private average transport costs rather than the case of efficient urban structure relying on social marginal transport costs treated in Mills (1972). This difference can be eliminated by introducing social marginal cost pricing on each road link in the Kim model, which would transform the model from a second-best to a first-best formulation.

Although the models discussed in this section are formulated as non-linear optimisation models they are basically used to obtain market equilibrium urban patterns. They rely on exogenously specified transport systems. Among model approaches aiming for an optimal allocation of urban activities related to explicitly formulated urban planning objectives, the TOPAZ (Technique for the Optimum Placement of Activities in Zones) model can be used for finding net benefit maximising urban structures with embedded spatial interaction behaviour analogous to the Kim model. But TOPAZ can also be used for seeking urban structures that optimise more general community objectives (e.g. minimising energy use, minimising travel, maximising accessibility). Many applications on the urban level as well as to various urban subsystems are reported in Brotchie *et al.* (1980).

Other modelling approaches seeking optimal urban systems were developed in the Stockholm regional planning context (for an overview, see Lundqvist, 1979). Three distinguishing features of this suite of models are worth mentioning. Firstly, they allocate urban activities based on system wide accessibility (or contact cost) and space utilisation (or density) criteria without any predetermined city centre or export nodes, see Karlqvist and Lundqvist (1972) (cf. Beckmann (1976)). Secondly, one of the models provides for the possibility of a simultaneous (but aggregate) analysis of activity location and transport network design. Thirdly, the models apply an explicitly multi-criterion approach, allowing for trade-offs between planning objectives (such as accessibility, space/capita, infrastructure costs) and analysis of how the long-term freedom of action depends on short-term development alternatives (see Lundqvist, 1978).

7.5. Lessons

The overview of various approaches to modelling interactions between transport systems and urban structure suggests that a rich source of knowledge has accumulated. Most of these contributions analyse the urban system in a partial or general equilibrium sense. However, we have also discussed approaches that seek optimal urban systems relying on some concept of common welfare. Some notable achievements in the development of this knowledge accumulation are listed below:

- In partial equilibrium models of closed monocentric urban systems (given population, given income and prices, uncongested travel) an increased travel cost per distance unit leads to a shrinking urban area and higher rents and densities in the vicinity of the city centre.
- Internalising congestion costs makes the rent gradient steeper than the case when only average private transport costs are paid. Cost–benefit analysis based on market land values in the case when congestion costs are not internalised tend to lead to excessive use of land for transport, especially close to the CBD.

- Optimum city size can be seen as a trade-off between economies of scale in production (in the monocentric CBD) and diseconomies of scale in the transportation system through congestion. The optimal population may be very sensitive to the importance of congestion in total commuting time. Welfare maximising leads in general to unequal utility levels among households unless the welfare criterion is extremely equity oriented.
- In general equilibrium approaches, the competition for land between transport, households and commodity production constitutes an important element. Also, the links between congestion tolls and ways of financing the transport system on the one hand and household income on the other hand need to be incorporated. A first-best solution of the urban equilibrium relies on an optimal allocation of land to transport and appropriate congestion tolls. Generally both residences and production are dispersed and production is more centralised than residential location. Land rents fall sharply from the geographical centre while commodity prices fall and wages rise moderately with distance from the centre. A high proportion of the land is allocated to transport in the geographical centre.
- The efficiency gains in terms of welfare from introducing congestion tolls are small while the potential welfare gains from making road planning efficient are considerably higher. It is even more important for welfare to change from a monocentric urban structure to a dispersed equilibrium. Internalising congestion costs leads to centralisation of both population and employment. The rent gradient becomes steeper while the wage differences between zones are reduced.
- When scale economies in shopping are strong relative to the cost of traffic congestion, multiple polycentric urban equilibria emerge. Agglomeration economies (and diseconomies) and economies of scale are key ingredients in many urban system models.
- Linear programming models for efficient resource allocation in urban systems provide a link between theoretical urban economics and operational models of transport systems and urban structure.
- Operational urban models show that employment location is more sensitive to changes in travel costs than is the distribution

of population. Retail employment is more responsive than non-retail service employment, which in turn is more responsive than non-service employment. Individual social groups show more movement than the population overall.

- There is a large potential for changes in the use of the transportation system (e.g. reducing car trips) without fundamentally transforming the physical layout of the urban system.

7.6. Research directions

Most of the contributions discussed in this chapter suggest improvements and extensions in their respective models. Some of these suggestions have been indicated in earlier sections. Many of these ideas have also been taken up in subsequent research and are now part of the accumulation of knowledge.

Many of the models used to analyse monocentric and/or symmetric spatial settings in Section 7.3 combined the network design aspect (e.g. the allocation of land for road transport) with equilibrium or optimum location of urban activities. In contrast, almost all operational models discussed in Section 7.4 relied on an exogenously specified transport network.

Integration of normative urban design aspects with equilibrium analysis of urban markets might be achieved within the framework of bi-level modelling (see Lundqvist, 1996). On the top level, network improvements and developments of the existing settlement system could be analysed within a welfare maximising approach (cf. Lundqvist, 1998b). Given transport network structures and capacities as well as land release policies (by zone) from the top level, the bottom level would analyse location and travel behaviour on urban markets, e.g. by employing extended combined network equilibrium models of the type developed by Kim (1989). Other research directions along similar lines were suggested in Kim (1989, 1990).

Other research directions relate to the changing context of urban land-use/transport analysis: increasing labour participation rates, an aging population, shrinking household size, an emerging information society, intelligent transportation systems, focus on sustainable development and sustainable mobility, new forms of spatial planning, introduction of travel demand management, etc. These changes affect life styles, production and transportation patterns and will ultimately influence the relations between transportation systems and urban activity location. Some of these issues have been subject to analysis with existing models. Dynamic approaches, high resolution of time and space, activity based modelling approaches, micro-simulation and moving disequilibria are examples of developments and responses to the changing context that have been emphasised in recent surveys of operational urban models. The approaches covered in this chapter and the system-wide qualitative knowledge they have generated represent a platform of key concepts and accumulated experiences for further elaboration in future research.

References

Alonso, W. (1964), *Location and Land Use*, Cambridge, MA: Harvard University Press.

Anas, A. and I. Kim (1996), "General equilibrium models of polycentric urban land use with endogenous congestion and job agglomeration", *Journal of Urban Economics*, Vol. 40, pp. 232–256.

Anas, A. and R. Xu (1999), "Congestion, land use, and job dispersion: a general equilibrium model", *Journal of Urban Economics*, Vol. 45, pp. 451–473.

Anderstig, C. and L.-G. Mattsson (1991), "An integrated model of residential and employment location in a metropolitan region", *Papers in Regional Science*, Vol. 70, pp. 167–184.

Anderstig, C. and L.-G. Mattsson (1998), "Modelling land-use and transport interaction: policy analyses using the IMREL model", pp. 308–328, in: L. Lundqvist, L.-G. Mattsson and T.J. Kim, editors, *Network Infrastructure and the Urban Environment – Advances in Spatial Systems Modelling*, Berlin: Springer.

Beckmann, M.J. (1976), "Spatial equilibrium in the dispersed city", pp. 117–125, in: G.J. Papageorgiou, editor, *Mathematical Land Use Theory*, Lexington, MA: Lexington Books.

Brotchie, J.F., J.W. Dickey and R. Sharpe (1980), *TOPAZ – General Planning Technique and its Applications at the Regional, Urban, and Facility Planning Levels, Lecture Notes in Economics and Mathematical Systems 180*, Berlin: Springer.

Coelho, J.D. and H.C.W.L. Williams (1978), "On the design of land use plans through locational surplus maximisation", *Papers of the Regional Science Association*, Vol. 40, pp. 71–85.

Dixit, A. (1973), "The optimum factory town", *Bell Journal of Economics and Management Science*, Vol. 4, pp. 637–651.

Eliasson, J. and L.-G. Mattsson (2001), "Transport and location effects of road pricing: a simulation approach", *Journal of Transport Economics and Policy*, Vol. 35, pp. 417–456.

Herbert, D.J. and B.H. Stevens (1960), "A model for the distribution of residential activity in an urban area", *Journal of Regional Science*, Vol. 2, pp. 21–36.

Johansson, B. and L.-G. Mattsson (1995), "From theory and policy analysis to the implementation of road pricing: the Stockholm region in the 1990s", pp. 181–204, in: B. Johansson and L.-G. Mattsson, editors, *Road Pricing: Theory, Empirical Assessment and Policy*, Boston, MA: Kluwer Academic Publishers.

Johnston, R.A. and T. de la Barra (2000), "Comprehensive regional modeling for long-range planning: linking integrated urban models and geographic information systems", *Transportation Research Part A*, Vol. 34, pp. 125–136.

Karlqvist, A. and L. Lundqvist (1972), "A contact model for spatial allocation", *Regional Studies*, Vol. 6, pp. 401–419.

Kim, T.J. (1989), *Integrated Urban Systems Modeling: Theory and Applications*, Dordrecht: Kluwer Academic Publishers.

Kim, T.J. (1990), *Advanced Transport and Spatial Systems Models – Applications to Korea*, New York: Springer.

Lowry, I.S. (1964), *A Model of Metropolis*, Santa Monica, CA: Rand Corporation, RM-4035-RC.

Lundqvist, L. (1978), "Urban planning of locational structures with due regard to user behaviour", *Environment and Planning A*, Vol. 10, pp. 1413–1429.

Lundqvist, L. (1979), "Models of Stockholm: policy perspectives", *Sistemi Urbani*, Vol. 1, pp. 91–99.

Lundqvist, L. (1996), "Some bi-level problems in land-use/transportation modelling", *Middle East Forum*, Vol. 1, pp. 185–193.

Lundqvist, L. (1998a), "A combined model for analysing network infrastructure and land-use/transportation interactions", pp. 329–343, in: L. Lundqvist, L.-G. Mattsson and T.J. Kim, editors, *Network Infrastructure and the Urban Environment – Advances in Spatial Systems Modelling*, Berlin: Springer.

Lundqvist, L. (1998b), Strategic urban design and land-use/transportation markets – model formulations and applications to Stockholm. Paper presented at the 8th World Conference on Transport Research, Antwerp.

Mattsson, L.-G. and L. Sjölin (2004), "Transport and location effects of a ring road in a city with or without road pricing", in: D.-H. Lee, editor, *Urban and Regional Transportation Modeling: Essays in Honor of David Boyce*, Cheltenham: Edward Elgar.

Mills, E.S. (1972), "Markets and efficient resource allocation in urban areas", *Swedish Journal of Economics*, Vol. 74, pp. 101–113.

Mirrlees, J.A. (1972), "The optimum town", *Swedish Journal of Economics*, Vol. 74, pp. 114–135.

Oron, Y., D. Pines and E. Sheshinski (1973), "Optimum vs. equilibrium land use pattern and congestion toll", *Bell Journal of Economics and Management Science*, Vol. 4, pp. 619–636.

Prastacos, P. (1986a), "An integrated land-use – transportation model for the San Francisco region: 1. Design and mathematical structure", *Environment and Planning A*, Vol. 18, pp. 307–322.

Prastacos, P. (1986b), "An integrated land-use – transportation model for the San Francisco region: 1. Empirical estimation and results", *Environment and Planning A*, Vol. 18, pp. 511–528.

Rho, J.H. (1988), Implementation and Evaluation of a Non-linear Three Dimensional Urban Activity Model, PhD Thesis, Urbana: University of Illinois at Urbana-Champaign.

Ripper, M. and P. Varaiya (1974), "An optimizing model of urban development", *Environment and Planning A*, Vol. 6, pp. 149–168.

Solow, R.M. (1973), "Congestion cost and the use of land for streets", *Bell Journal of Economics and Management Science*, Vol. 4, pp. 602–618.

Waddell, P. (2002), "UrbanSim: modeling urban development for land use, transportation and environmental planning", *Journal of the American Planning Association*, Vol. 68, pp. 297–314.

Wardrop, J.G. (1952), "Some theoretical aspects of road traffic research", *Proceedings of the Institute of Civil Engineers, 1, Part II*, pp. 325–378.

Webster, F.V., P.H. Bly and N.J. Paulley (eds.) (1988), *Urban Land-Use and Transport Interaction – Policies and Models*, Avebury: Aldershot.

Wegener, M. (1996), "Reduction of CO_2 emissions of transport by reorganisation of urban activities", pp. 103–124, in: Y. Hayashi and J. Roy, editors, *Transport, Land-Use and the Environment*, Dordrecht: Kluwer Academic Publishers.

Wheaton, W.C. (1974), "A comparative static analysis of urban spatial structure", *Journal of Economic Theory*, Vol. 9, pp. 223–237.

Wilson, A.G., J.D. Coelho, S.M. MacGill and H.C.W.L. Williams (1981), *Optimization in Locational and Transport Analysis*, Chichester: Wiley.

Urban Dynamics and Growth, CEA, vol. 266
Roberta Capello and Peter Nijkamp (editors)

CHAPTER 8

Intra-metropolitan Agglomeration, Information Technology and Polycentric Urban Development

Jungyul Sohn[a,b], Geoffrey J.D. Hewings[b] and Tschangho John Kim[c]

[a]National Center for Smart Growth Research and Education, University of Maryland, College Park, MD 20742, USA
[b]Regional Economics Applications Laboratory, University of Illinois, Urbana, IL 61801, USA
[c]Department of Urban and Regional Planning, and Department of Civil and Environmental Engineering, University of Illinois, 210 Temple Buell Hall, 611 Taft Drive, Champaign, IL 61820, USA

Abstract

Defining the urban spatial structure of a city is closely related to identifying the spatial distribution of urban economic activities. By developing a spatial econometric framework, this chapter aims to investigate the urban spatial structure of Seoul in the 1990s, focusing on factors considered as the major driving forces of the location/distribution of economic activities: agglomeration economies, information technology and centrality. The result of the analysis of the Seoul Metropolitan Region shows that the spatial agglomeration pattern of establishments is sector specific while the impact of information technology on spatial distribution pattern has been similar (centripetal) across all economic sectors.

Keywords: spatial agglomeration, information technology, urban spatial structure, centrality, Seoul

JEL classifications: R12, R30

8.1. Introduction

Urban spatial structure has been defined in various ways using a variety of techniques. The majority of them focus on the spatial distribution of urban economic activities, not an unreasonable bias in that the driving forces of urban growth are mainly found in the activities of its economic sectors. Higher levels of economic activities in a city usually guarantee higher growth rates for the city as a whole at an aggregate level. Likewise, a certain area in a city with a higher proportion of economic activities can be viewed as the city center or core.

In many cities, the dominant economic force of growth has come from agglomeration economies. Geographical closeness of firms in urban areas has generated external economies at a metropolitan scale through direct industrial linkages and interactions (indirect) among them. While agglomeration economies are still dominant, more recent urban phenomena also need to be explained along with the role of information technology. While it seems to be agreed that information technology has had a substantial influence on the spatial behavior of individuals and economic agents, there has been an extensive debate on the direction of the effect, for example, if this technology leads to concentration or deconcentration of urban activities.

Recent theoretical and empirical investigations reviewed in the next two sections show such efforts to explain urban spatial structure with the impact of agglomeration economies and information technology. With a few exceptions, however, most studies focus on the impact of agglomeration economies at a citywide level (rather than examining intra-metropolitan distribution patterns) and a single industry agglomeration (but not necessarily inter-industry agglomeration). In addition, while a large number of studies on IT impact on urban form discuss the direction of its impact at a conceptual level, there are few empirical tests conducted especially in an intra-metropolitan context.

By developing a spatial econometric framework, this chapter aims to investigate the urban spatial structure in Seoul in the 1990s, focusing on a few factors considered as the major driving forces of the location/distribution of economic activities: agglomeration economies, information technology and centrality. The analysis will be completed by examining the spatial distribution of urban economic activities and also by exploring the major factors in explaining the different spatial distribution patterns.

A more detailed set of research questions are established as follows:

- Is industrial agglomeration prominent in shaping urban spatial structure?
  - Is the spatial association pattern among industries (urbanization economies) collaborative or competitive?
  - Does the spatial association pattern reveal a simultaneous location/allocation process for the sectors?
  - How does the spatial association pattern of sectoral location/relocation process reflect the previous economic environment; is it working as an incubator?
  - Does self-reinforcement force exist in space (localization economies)?
- What role does IT have on the spatial distribution pattern of firms: concentrated or dispersed?
  - Does it work as a locational incentive (attraction factor on activity level)?
  - Does it draw more economic activities around or repel such activities from it (spillover factor on distribution pattern)?
- Do firms have a tendency to be in or near the city centers?
  - Which sector is more location-dependent on the city center and subcenters?
  - Which type of center (the city center and subcenters) shows a stronger attraction force on which economic sector and which type of center reveals a stronger spillover effect around them on which sector?

Figure 8.1 shows three major factors in the research used in investigating the urban spatial structure. It also provides the major techniques to be used in linking those factors with the urban spatial structure along with the corresponding arrows.

The first part of the research (left part of Figure 8.1) will examine the relationship between the spatial agglomeration of firms and

Figure 8.1. Schematic research plan

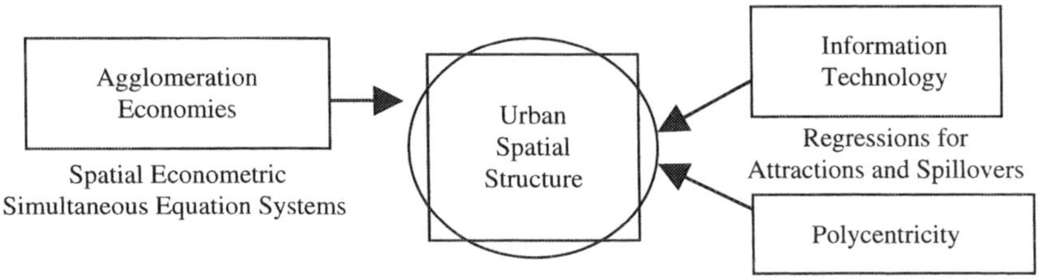

the urban spatial structure. Special focus will be on the role of localization economies and the urbanization economies in intra-metropolitan locations. The former is more related to *intra*industry linkages and the common infrastructures that are shared. The latter is more related to the *inter*industry linkage and the sectoral backward- and forward-linkages. There are two dimensions of the agglomeration economies to be considered in the analysis: scope and time frame. In essence, the first question probes whether the self-reinforcing forces of individual sectors are critical or whether the synergetic effects among different sectors are more significant. The second question explores whether the simultaneous interaction of firms within and between sectors is important or whether the previous economic setting is significant, for example, as an incubator. The major technique used in the analysis is a spatial econometric simultaneous equation systems, wherein a spatial weight matrix[1] is used to create the spatially lagged variables of the economic activities as well as the other relevant urban activities.

The second part of the research (top-right of Figure 8.1) will be performed on the relationship between information technology and the urban spatial structure. To provide empirical testing of the relationship, two aspects are identified related to the urban spatial structure: (1) the level of the economic activities or the number of firms in a certain sector in a zone (attraction effect) and (2) the distribution pattern of the economic activities in a certain sector surrounding a zone (spillover effect). As such, two sets of regression models are built for each sector. By examining these two aspects, the spatial effect of information technology will be uncovered. In addition to the information technology, centrality of each sector as another factor for determining the distribution pattern will also be examined (bottom-right of Figure 8.1).

8.2. Spatial clustering of economic activities

While there has been a great deal of attention focused on interrelations between economic activities and residences, it is hard to find research examining interactions among different industrial

---

[1] Spatial weight matrix is a matrix that represents the distance-decay effect among points or areas over space in various forms.

sectors within the urban economy.[2] As noted earlier, the distribution or location pattern of a single firm or a group of firms in a certain sector is affected by the pattern of residents. Similarly, one might expect that the distribution and location patterns of firms in other sectors would also have an influence on the distribution or location decision of an individual firm. Generally speaking, firms show some level of attraction to firms within the same industry or to firms providing important inputs or markets for products. The explanation for these phenomena is rooted in the concept of external economies or agglomeration economies.[3] Firms seem to be better off when they are spatially aggregated than when they are dispersed; this is especially the case when functional specialization between firms exists and geographic closeness matters.[4] Two perspectives can be adopted to explore the role of agglomeration economies, although they are not mutually exclusive. The intra-industry perspective, locating near firms in the same industry, provides the industrial firm with great access to more relevant infrastructure. The inter-industry perspective, locating near those firms in other economic sectors, offers the firm direct access to sources of backward and/or forward linkages. According to the more popular typology of agglomeration economies in the urban economics literatures, the former is termed as localization economies whereas the latter as urbanization economies (e.g. Pascal and McCall, 1980; Goldstein and Gronberg, 1984; Nakamura, 1985; Fogarty and Garofalo, 1988; McMillen and McDonald, 1998).[5] While the effect of those two economies may

---

[2] A notable exception would be the work on clustering; however, most of these applications focus on linkages within one spatial unit (a region or a metropolitan area). See a recent review by Feiser and Bergman (2000).

[3] The concept of increasing returns to scale provides a starting point of economic process of agglomeration economies. For the extensive review of the literatures on this topic, refer to Fujita and Thisse (1996).

[4] The terms, 'functional specialization' and 'geographic closeness' were two dimensions to explain the level of agglomeration economies in the paper of Bergsman *et al.* (1975).

[5] Some authors did not explicitly use those terms to present the characteristics of those two economies (e.g. Pascal and McCall, 1980; McMillen and McDonald, 1998) or had more than two types of economies classified (e.g. Pascal and McCall, 1980; Goldstein and Gronberg, 1984). But the fundamental idea behind those is still under the general classification scheme of localization and urbanization economies.

not be clearly divided, Fogarty and Garofalo (1988) adopted the function of the scale of the manufacturing sector (returns-to-scale) as the estimate of localization economy and the function of urban scale (efficiency parameter of the production function) as the estimate of urbanization economy in their model. They revealed that localization factors were significant in explaining productivity at the Standard Metropolitan Statistical Area level, but they suggested that urbanization economies might not be solely significant and rather urban spatial structure or arrangement of economic activity also needed to be considered in the model. With a more disaggregate classification of manufacturing sectors, Nakamura (1985) concluded that light manufacturing industries were more influenced by urbanization economies and heavy manufacturing industries more by localization economies in his research focusing on Japanese cities.

Much of the theoretical and empirical research on agglomeration economies has explored the relationship between agglomeration economies and other urban and economic features. For example, Abdel-Rahman (1990) developed a theoretical framework to examine the different size and type of cities associated with dominant agglomeration forces between cities. He argued that the industrial structure within the city or spatial proximity between related industries was an important factor in explaining the differences. Using an empirical model, Mitra (1999) also focused on the city size issue. Comparing city size and agglomeration economies measured as technical efficiency, he concluded that those two had a positive relationship even if diseconomies of scale would be realized over a certain threshold level of city size. Research with a focus on the relationship between agglomeration and other factors includes consideration of technical change (Calem and Carlino, 1991), the urban capital market (Helsley and Strange, 1991), and land rents and wages (Dekle and Eaton, 1999). While many authors linked agglomeration economies with economies of scale, Goldstein and Gronberg (1984) attempted to explain this relationship through appeal to economies of scope realized through a vertical integration process in production. Unlike most others who focused on manufacturing activity, Mun and Hutchinson (1995) extended the scope in a case study of Toronto that focused on intra-urban locations, in terms of the geographical scale, and on the office sector, as opposed to manufacturing. However, the existence of probable external economies between industries remains unexplored.

Another line of research has revealed a greater interest in the spatial context of these agglomeration economies. Lee (1981) adopted a standard distance measure from the economically weighted centroid to individual firms in each sector and contiguity measures as well to check the level of agglomeration economies or spatial concentration of individual industries. Since the measures focused on firms in the same sector, the scope of research was confined to exploring the level of intra-industry agglomeration economies. Maurel and Sedillot's (1999) paper was one of the few that dealt with intra- and inter-industry relations at the same time. Based on previous work, they derived several indices measuring the level of geographic concentration; these indices were interpreted as the correlation between the location decisions of two firms. In an application to French manufacturing industries, the authors sought to identify the spatial realization of agglomeration economies. While those indices are useful to check the overall pattern of spatial distribution, causality between industries with respect to the distribution may not be identified. Hanson (1996) proved in his empirical research that the spatial economic pattern experiences an iterative process of agglomeration, dispersion, and reagglomeration due to the changing role of external economies and diseconomies. Dekle and Eaton (1999) explored the distance decay effect of agglomeration economies and concluded that the financial sector was more sensitive to this effect than manufacturing and, as a result, had more limited geographical spillover effects.

Agglomeration economies are only part of the explanation for urban growth. DeCoster and Strange (1993) used the term 'spurious agglomeration' to represent the situation in which excessive concentration of economic activity occurs due to incentive programs. They noted that the system would be inefficient and could be relieved by certain tax measures. Both Moomaw (1985) and Hansen (1990) revealed that the contemporary economy was in transition from a stage that favored agglomeration economies to another; for example, Moomaw (1985) showed the productivity advantages of large cities have declined in eight 2-digit manufacturing sectors accounting for more than one-third of production worker employment. Hansen (1990) proposed that the productivity advantage of the center was offset by higher land and labor costs and the result suggested market forces might lead eventually to the decentralization of industries. Smith and Florida (1994) also mentioned that there also seemed to be conflicting effects of agglomeration economies with those on the negative side (diseconomies) including higher factor costs resulting from locations in brown-field sites, higher wages, higher levels of unionization, and greater social problems. In his probit model developed to test firms' location behavior, Cooke (1983) showed that traditional measures of agglomeration economies such as proximities to backward and forward linkage activities lacked explanatory power. Rather, as he noted, demand changes, initial plant size, or relative magnitude transport cost have gained more importance.

8.3. Information technology and urban spatial structure

Recent research on urban distribution patterns tends to focus more often on IT since information and communication technology has been seen as a way to overcome the costs of spatial separation. To some extent, the question whether IT has been important or not in explaining urban forms has been narrowed down to whether IT has an influence on the dispersal of urban activities.

Yen and Mahmassani (1997), among many others, have noted that the development of telecommunication technologies may affect land use patterns and play a role in the growth of economic activities and the spatial distribution of industry. Specifically, these authors suggested two aspects of office-location decisions by organizations in assessing the impact from the new technology. Those are (1) the need for certain organizations to locate where they can access telecommunication networks and (2) an increased opportunity for the organizations to locate their offices in the areas where infrastructure costs are generally lower than traditional office locations such as downtown areas. With respect to the residential distribution, they noted that insufficient evidence is available to confirm the impact of telecommuting on household residential location notwithstanding the expectation that increased mobility due to telecommunication technology makes it possible for households to move farther away from the urban core. It is not, however, completely unexpected considering that urban residential location is not based on economic criteria alone since access to urban amenities may keep people in cities.

Gaspar and Glaeser (1998) tried to uncover the relationship between IT and face-to-face interactions and/or the cities that facilitate those interactions. In an empirical analysis using telephone call data, the authors concluded that those two are complements rather than substitutes. As a result, the centralizing forces in cities did not seem to vanish. Instead of establishing a single framework to prove the relationship, they listed several different results from the regression analyses in different specifications. However, as the authors noted several times, it is very hard to separate the exclusive effect of IT in their regression models.

Gordon and Richardson (1997, p. 95) conjectured that such technology leads to a dispersion of economic activities and population, possibly up to the stage where 'geography is irrelevant'. They noted that high-rise or concentrated settlement has been dominant when transport or communication costs were high but the costs are likely to continue to fall in the future. It might be possible to summarize that, 'office work, rather than office workers, will do the traveling' (Drucker, 1989, p. 38). The critical issue here is whether transportation and communication are complementary or competitive: are they complementary, so that an increasing adaptation of communication technology induces higher transportation demands or are they competitive enough, so that communication technology is able to substitute for a portion of transportation demands? If the former is the case, geography matters even with the advent of the new communication technology.

Arguing against the optimistic view of technology, Salomon (1996) mentioned that there have been overly great expectations of the information age, for instance, that telecommunications can eliminate the effects of distance and as a result can have profound effects on the spatial organization of society. He also identified four assumptions underlying the proposition that cities will disperse due to an improved IT: (1) the substitutive relationship between transportation and telecommunications, (2) the substitution of information for material goods, (3) the ubiquity of telecommunications and (4) the recognition that dispersal has been constrained by congestion and travel costs. Even though he claimed that a complete change of urban form could not be expected in the information age, he agreed that there are some changes that may result from these technological changes. One example of the telecommunications' dispersal effect is an emergence of the back office activities located remotely from the core organization (Richardson and Gillespie, 1996; Salomon, 1996). Further, there exists a gap, according to Capello (1994), between an introduction of the new IT and the changes in the spatial pattern of firms. This is ascribed to an overestimation of technological potential and to an optimistic and superficial analysis on the relationship between the new technology and spatial restructuring. She also noted that in the long run, those technologies lead to a new production strategy such as the 'just-in-time' (JIT) system and it will require a physical proximity (either in an inter-urban or intra-urban context) between firms and eventually a spatial clustering of economic activities is expected. However, as Fujita and Hamaguchi (2001) noted, firms (specifically the buyers of intermediate goods in their research) can be more dispersed with a better-developed transportation/communication infrastructure as in the examples of many developed countries.

On a more conceptual level, the geography and/or distribution of economic activities can be redefined on the basis of information flows. Echeverri-Carroll (1996) noted that an effect of the geographical relationships between organizations cannot be conceptualized without understanding the intra-organizational and inter-organizational computer networks that bind particular locations together. Even though spatial decentralization continues to be relevant, the process is characterized by a much higher functional integration using the information network (Echeverri--Carroll, 1996). It is implied that network connectivity can be a more important factor in deciding the geographical relationships (i.e. concentrated or dispersed) than physical distance especially in the information age. It is obvious that the scope of geography here embraces not only territorial but virtual concepts. Echeverri-Carroll does not agree that such technology leads to the demise of the concept of 'distance'. Echeverri-Carroll concluded that since 'these technologies also impose higher investments on inter-firm linkages and more stringent restrictions on labor's skills and flexibility, both…restrain the location of industry in space (1996, p. 148)'.

Mokhtarian (1998) focused more on the spatial residential pattern of commuting. She noted that the effect of the new technology is not to reduce travel but to increase the flexibility of travel and, as a result, the total number of trips may be higher with a substantial portion of travel shifted to the off-peak periods. The ability to commute less often due to telecommuting often leads to a relocation of residence farther away from work sometimes enough for the total commuting VMT (vehicle miles traveled) on a smaller number of commuting days to exceed the previous levels (Mokhtarian, 1998). On a system-wide level, this trend may result in a decentralizing effect on urban form (for a more detailed discussion on the theoretical model of residential relocation due to telecommuting, refer to Stough and Paelinck, 1996 or Lund and Mokhtarian, 1994). Mokhtarian (1998) also suggested that this is not the case as far as part-time and short-term telecommuting is involved, in such cases, decentralization results from other reasons.

8.4. Spatial simultaneous equation systems

The model used is the spatial econometric simultaneous equation (SESE) systems, a spatial econometric version of three-way simultaneous equation model. It is in the form of a difference equation.[6] Each equation is designed to explain the distribution pattern of economic activity in each sector (manufacturing, retail, and service) and includes two endogenous variables (the other two sectors), one spatially lagged dependent variable, and six exogenous variables. The approach taken here is similar in spirit to a VAR model in that everything depends on everything else. The differences between the SESE model and the VAR model are twofold: a VAR model requires more than two time periods for a long run trend, and the SESE model attempts to capture the effect of neighboring zones.

The spatial distribution of business activities in the model is explained by several factors related to economic activities, labor force, and market situation. Factors related to economic activities are disaggregated into intra-sectoral and inter-sectoral effects. The former reflects what may be termed spatial autoregressive simultaneity while the latter focuses on feedback and spatial cross-regressive simultaneity.[7] Variables are selected to explain

[6] For the derivation process of the model, refer to Sohn and Hewings (2000).
[7] The terminologies used here were borrowed from Rey and Boarnet (2004).

the different aspects of urban agglomeration of employment within and between sectors; factors related to residence are population in general and labor force in a specific economic sector in terms of their residences. While the former is seen as a proxy for market potential for firms, the latter is considered as the labor pool of a specific industry. Generally, a larger population leads to higher level of demand for goods and services with the result that firms may prefer staying closer to areas with higher demands to maintain a larger market share by reducing transportation cost (and as a result, reducing the market price of goods and services). In a similar fashion, a larger labor pool leads to a higher level of cheap labor supply, thereby enabling firms to lower production costs and as a result the market price of those goods and services. In this sense, those two effects are expected to work as centripetal forces in certain zones. Variables used in the model are summarized in Table 8.1. It is noted that the numbers of establishments in manufacturing, retail, and service are aggregated based on the 1987 Standard Industrial Classification (SIC) used in the US Economic Census.[8]

Equations (8.1)–(8.3) show three-equation systems used in the analysis.

$$\begin{aligned}\Delta\mathrm{MAN}_i = {}& \varphi_0 + \varphi_1\Delta\mathrm{WMAN}_i + \varphi_2\Delta\mathrm{IWRET}_i + \varphi_3\Delta\mathrm{IWSER}_i \\ & + \varphi_4\mathrm{IWMAN}_i^{t-1} + \varphi_5\mathrm{IWRET}_i^{t-1} + \varphi_6\mathrm{IWSER}_i^{t-1} \\ & + \varphi_7\mathrm{IWMLF}_i^{t-1} + \varphi_8\mathrm{IWNMLF}_i^{t-1} \\ & + \varphi_9\mathrm{IWPOP}_i^{t-1} + v_{it}^{\mathrm{m}}\end{aligned} \tag{8.1}$$

$$\begin{aligned}\Delta\mathrm{RET}_i = {}& \gamma_0 + \gamma_1\Delta\mathrm{WRET}_i + \gamma_2\Delta\mathrm{IWMAN}_i + \gamma_3\Delta\mathrm{IWSER}_i \\ & + \gamma_4\mathrm{IWMAN}_i^{t-1} + \gamma_5\mathrm{IWRET}_i^{t-1} + \gamma_6\mathrm{IWSER}_i^{t-1} \\ & + \gamma_7\mathrm{IWRLF}_i^{t-1} + \gamma_8\mathrm{IWNRLF}_i^{t-1} \\ & + \gamma_9\mathrm{IWPOP}_i^{t-1} + v_{it}^{\mathrm{r}}\end{aligned} \tag{8.2}$$

[8] For more information, refer to http://www.census.gov/epcd/naics/nsic2ndx.htm.

Table 8.1. Variables included in the SESE model

Variable	Description	Source
ΔMAN_i, ΔRET_i, ΔSER_i	Change in the number of manufacturing, retail, and service establishment in zone *i* between 1991 and 1996	Korean report on establishment census
$\Delta WMAN_i$, $\Delta WRET_i$, $\Delta WSER_i$	Change in the number of manufacturing, retail, and service establishment around zone *i* between 1991 and 1996	
$\Delta IWMAN_i$, $\Delta IWRET_i$, $\Delta IWSER_i$	Change in the number of manufacturing, retail, and service establishment in and around zone *i* between 1991 and 1996	
$IWMAN_i^{t-1}$, $IWRET_i^{t-1}$, $IWSER_i^{t-1}$	Number of manufacturing, retail, and service establishment in and around zone *i* in 1991	
$IWMLF_i^{t-1}$, $IWRLF_i^{t-1}$, $IWSLF_i^{t-1}$	Manufacturing, retail, and service labor force in and around zone *i* in 1990	Korean population census
$IWNMLF_i^{t-1}$, $IWNRLF_i^{t-1}$, $IWNSLF_i^{t-1}$	Non-manufacturing, non-retail, and non-service labor force in and around zone *i* in 1990[a]	
$IWPOP_i^{t-1}$	Population in and around zone *i* in 1990	

[a] Total number of employed residents – number of employed residents in manufacturing, retail, and service.

$$\begin{aligned}\Delta SER_i = {} & \eta_0 + \eta_1 \Delta WSER_i + \eta_2 \Delta IWMAN_i + \eta_3 \Delta IWRET_i \\ & + \eta_4 IWMAN_i^{t-1} + \eta_5 IWRET_i^{t-1} + \eta_6 IWSER_i^{t-1} \\ & + \eta_7 IWSLF_i^{t-1} + \eta_8 IWNSLF_i^{t-1} \\ & + \eta_9 IWPOP_i^{t-1} + v_{it}^{s} \qquad (8.3)\end{aligned}$$

It should be noted that the independent variables are taken as values in the initial year, $t-1$. It is appropriate for the model to have a temporal gap between explanatory and dependent variables, considering causal relationships need some time to develop. In addition, as Anselin and Bera (1998) noted, when data are based on administratively determined units, there is no good reason to expect economic behavior to conform to these units. As a consequence, it might be reasonable to use variables with spatial weights (W)

to bridge the gap between the two as the model below specifies. In the estimation, a generalized method of moments estimator derived by Kelejian and Robinson (1993) and later revised by Kelejian and Prucha (1998) is used.

8.5. Regressions for attraction and spillover effects

In this section, two sets of regression models along with two dependent variables are defined to estimate the impact of IT on the activity level and distribution pattern of urban economic activities. For activity level, the number of establishments in 1996 in Seoul is used as the dependent variable. Urban activity is examined separately in three economic sectors (manufacturing, retail, and service). Different sectors are expected to show different patterns and explanations. By examining these variables, the analysis is able to determine whether IT-related variables or others are significant in explaining higher or lower levels of such activities. In this context, it is more related to an 'attraction' effect.

The second is the distribution pattern of economic activities around a certain zone of interest. This characteristic is measured by a local indicator of spatial association (LISA). This analysis adopts the G_i^* statistic for measuring the distribution pattern as shown in Equation (8.4).

$$G_i^* = \frac{\sum_j w_{ij} x_j - W_i^* \bar{x}}{s\{[(nS_{1i}^*) - W_i^{*2}]/(n-1)\}^{1/2}}, \qquad \text{all } j \tag{8.4}$$

where

x_j = observation in j

$W_i^* = \sum_j w_{ij}$

$\bar{x} = \frac{\sum_j x_j}{n}$

$s = \sqrt{\frac{\sum_j (x_j - \bar{x})^2}{(n-1)}}$

n = number of region

$S_{1i}^* = \sum_j w_{ij}^2$

It was first developed by Getis and Ord (1992) and later revised by Ord and Getis (1995). The uniqueness of this statistic is that a positive *z*-value for G_i^* statistic indicates spatial clustering of high values, whereas a negative *z*-value indicates spatial clustering of low values (Anselin, 1995, p. 23-2). By investigating these variables, the analysis is able to reveal whether the IT-related variables and others are significant in explaining the spatial distribution pattern of urban activities, with respect to the tendencies for concentration or dispersal. In this sense, it can be considered as a 'spillover' effect. Combining these two dependent variables enables the analysis to provide a comprehensive picture of the spatial distribution of urban activity patterns and the impact of information technology on it.

In sum, there are six dependent variables to be explained and as a result six regression models are built with a series of independent variables for Seoul. Equations (8.5)–(8.7) summarize this relationship. This model is expected to measure the IT impact on the location and distribution pattern of economic activities in a comprehensive way. On the other hand, however, it is not able to consider the influence driven by a certain type of agglomeration economies discussed in the previous section (localization economies).

$$\begin{aligned}\mathrm{MAN}_i \text{ or } G_i^{\mathrm{m}} = {} & \alpha_0 + \alpha_1\mathrm{IWITFIRM}_i + \alpha_2\mathrm{IWUTEL}_i \\ & + \alpha_3\mathrm{IWPTEL}_i + \alpha_4\mathrm{CBD}_i + \alpha_5\mathrm{IWJHR}_i \\ & + \alpha_6\mathrm{IWMLF}_i + u_i^{\mathrm{m}}\end{aligned} \tag{8.5}$$

$$\begin{aligned}\mathrm{RET}_i \text{ or } G_i^{\mathrm{r}} = {} & \beta_0 + \beta_1\mathrm{IWITFIRM}_i + \beta_2\mathrm{IWUTEL}_i \\ & + \beta_3\mathrm{IWPTEL}_i + \beta_4\mathrm{CBD}_i + \beta_5\mathrm{IWJHR}_i \\ & + \beta_6\mathrm{IWRLF}_i + u_i^{\mathrm{r}}\end{aligned} \tag{8.6}$$

$$\begin{aligned}\mathrm{SER}_i \text{ or } G_i^{\mathrm{s}} = {} & \chi_0 + \chi_1\mathrm{IWITFIRM}_i + \chi_2\mathrm{IWUTEL}_i \\ & + \chi_3\mathrm{IWPTEL}_i + \chi_4\mathrm{CBD}_i + \chi_5\mathrm{IWJHR}_i \\ & + \chi_6\mathrm{IWSLF}_i + u_i^{\mathrm{s}}\end{aligned} \tag{8.7}$$

The sets of independent variables are grouped in Table 8.2. The first group is about IT-related variables. While a direct measure of IT and/or usage level of those technologies are the most desirable data for this category, as noted earlier, it is very hard to obtain such information in a practical sense. Since there are few available data

Table 8.2. Classification of independent variables

Group	Variable	Description	Source
IT	$IWITFIRM_i$	Number of information intensive establishments in 1996	Korean report on establishment census
	$IWUTEL_i$	Estimated number of usual telecommuter in 1990	Korean population census
	$IWPTEL_i$	Potential number of telecommuter in 1990	
Centrality	CBD_i	Distance from the CBD	GIS data
	$IWJHR_i$	Job-housing ratio, during 1990–1991 period	Korean report on establishment census and Korean population census
People	$IWMLF_i$	Manufacturing labor force in 1990	Korean population census
	$IWRLF_i$	Retail labor force in 1990	
	$IWSLF_i$	Service labor force in 1990	

sources on the level of this function by disaggregate zones within an urban area, alternative indices will be used to represent the level of IT intensity. Those indices are (1) the number of information intensive firms, (2) the number of usual telecommuters, and (3) the number of potential telecommuters.

The first indicator is related to the level of information infrastructure that seems to be more closely linked to firm activities rather than individuals in households. Sinden (1995) used SIC 7902 (telecommunications) in research to examine the British economic restructuring process in the telecommunication services. More often than not, secondary data do not allow this level of detailed information to be adapted, especially when the research is conducted at a more disaggregate geographical level. As a result, several authors have tried to use surrogates. Moulaert and Djellal (1995) used NAE[9] 7703 (information technology and organization consulting sector) and 7704 (computer services) as alternatives in a research in France. Another study by Tofflemire (1992) focused on SIC 6000 (FIRE), 73 (business services), 81 (legal services), and 87 (engineering, accounting and management services) in the US. Both studies share common features in that they considered the level of producer services in a region as reflecting the level of IT.

[9] Nomenclature des activités économiques (Nomenclature of economic activities).

IWITFIRM in the present analysis is composed of those firms in KSCI 642 (telecommunication), K (financial institutions and insurance), L (real estate, renting and leasing), and M (business activities).

The 'usual' telecommuters (IWUTEL[10] in this analysis) are defined as those respondents that answered 'worked at home' to the census question asking workers by what mode they usually traveled to work, excluding self-employed workers and workers in particular occupations (Handy and Mokhtarian, 1996, p. 168).[11] While usual telecommuters are substantial participants in telecommuting, the number of potential telecommuters (IWPTEL) has to be estimated indirectly. Adopting Nilles' (1988) proposition that 50% of workers are 'information workers' and 80% of them are potential telecommuters, Handy and Mokhtarian (1996) defined potential telecommuters as those with certain occupations classified as telecommuting-conducive: executive, administrative, managerial; professional specialty; technicians and related; sales; and administrative support. Those two telecommuter variables are included in the model to measure the significance in explaining the urban spatial structure. They are used as a proxy for the level of IT infrastructure of a certain zone. In other words, a zone with a higher proportion of those information-related workers may have higher levels of information network infrastructure than others.

The second group of variables is advanced to examine the impact of centrality on the level of activity and distribution patterns. It is clear that the variable, CBD, focuses on the influence of the urban center. IWJHR is defined as the ratio between the number of employment opportunities and the number of residents in a zone. Usually, a center is characterized as having a higher number of economic activities and a relatively lower number of residents. Considering higher affordability of economic activities for urban

---

[10] As in the previous section, the prefix 'IW' implies that variables are spatially transformed.

[11] They include private household workers, protective services, farming, forestry and fishing, precision production, craft and repair, operators, assemblers, inspectors, transportation and material moving and handlers, helpers, laborers. At a more disaggregate areal level in the analysis, this number is estimated using the relative employment share of such occupations in corresponding areas.

Table 8.3. Possible outcomes of the model for IT impact

		Distribution Pattern (Spillover)	
		↑🡅	↑🡇
Activity level (attraction)	↑🡅	Higher activity level with concentrated pattern	Higher activity level with dispersed pattern
	↑🡇	Lower activity level with concentrated pattern	Lower activity level with dispersed pattern

Note: ↑ increase in IT variables (independent). 🡅🡇 increase or decrease in corresponding dependent variables.

rent than residential activities, it is not an unusual proposition. As a result, a higher rate of JHR is expected to represent stronger centrality trends; this feature is used for measuring the influence of urban sub-centers as well as the CBD.

The third group of variables is labor force as a measure for the labor market. All these three sectoral variables are used only in the equations for the corresponding economic activities.[12]

There are four possible outcomes for the relationship between the IT variables and the two dependent variables: activity level (attraction) and distribution pattern (spillover). Table 8.3 summarizes those outcomes. By applying this framework in interpretation, the analysis is able to determine whether the controversial propositions and hypotheses about the impact of IT are valid or not: whether the IT factor is crucial in explaining the location and/or distribution of economic activities and whether it leads to a concentration/dispersion of the activities or not.

There are also four ways to interpret the outcome of the centrality-related variables, especially with respect to the sign of the coefficients. Table 8.4 shows the conceptual classification of the impact of center-orientedness.

A positive sign for CBD implies that economic activities are more intense as we go farther from the CBD and a negative value indicates the opposite. In the case of positivity, we cannot expect the traditional monocentric shape of a city. On the other hand, a positive IWJHR reflects the higher level of economic activities in and around

[12] For example, the manufacturing equation only adopts $IWMLF_i$ as a measure of the labor market.

Table 8.4. Interpretation of the coefficients of center-orientedness related variables

		IWJHR	
		+	−
CBD	+	Subcenter	Non-center
	−	CBD and/or subcenter	CBD and/or Non-center

more generally defined city centers (the city center or subcenters). Each listing in the table is the anticipated dominant type(s) of centers within a city for each of the economic sectors with corresponding signs.

To remove the influence resulting from different areal size, some variables in both of the analyses that contain raw numbers (i.e. the number of establishments) are divided by the area to be transformed into standardized numbers – similar to the density measure.

8.6. Study area: Seoul metropolitan region

Seoul has experienced a dynamic development in a very short period of time. Even though it has been the capital for a long period of time, substantial changes occurred only after the 1960s following a late and rapid industrialization process. The Seoul Metropolitan Region (SMR) is composed of the City of Seoul, the City of Inchon and Kyungki Province. Related to the recent development in economic sectors, Seoul has experienced a more rapid population growth in the past 40 years. Up to 1980, the City of Seoul seems to have been a more attractive place for residents but in 1990, the growth rate of Seoul was lower than that of the SMR. During the 1990s, the City of Seoul has experienced a negative growth of population.

The City of Seoul has observed that the supply of housing fell far behind the demand of households between 1970 and 1990, partly due to the rapid growth of population and to the limitation of available land for residential development. In response to this development pressure, housing supply has rapidly increased in the rest of the SMR during the same period. An exodus from the city seems to be ongoing since the continuation of the suburbanization process that started several decades ago.

The labor force participation rate has increased while the unemployment rate has decreased during the same period in the City of Seoul. Employment in the City of Seoul has grown rapidly and many labor-intensive sectors have remained strong even though their share has decreased in the last few years. Similar trends and explanations are also applied to the rest of the SMR with varying degrees.

In terms of industrial structure, the tertiary industry dominates the SMR. More specifically, it shows a spatial division of industry between the City of Seoul and the rest of the SMR; the secondary industry dominates the rest of the SMR while the tertiary industry dominates the City of Seoul. As discussed later, strong law enforcement and limited land resources have pushed the secondary sectors (mainly manufacturing which usually requires more space than other sectors) outside the city boundary. Since 1980, the rest of the SMR has exceeded the City of Seoul in the share of manufacturing. The tertiary sectors, however, seem to enjoy the agglomeration economies in the central city.

The manufacturing sector has been in decline in the City of Seoul since 1968 in the midst of a rapid industrialization process that was under way in Korea. On the other hand, manufacturing employment in the suburban area, especially the near suburbs, has grown prominently in the same period. Limited land resources as well as strong government control on land development in the City of Seoul have accelerated this trend.

According to Lee (1996), the land price in Seoul has rapidly increased for the last 30 years. The national economy in the rapid growth process needed a growth pole for an efficient production and distribution system and, as a result, Seoul has been an exceptionally popular, but expensive place to live and run a business. The internal spatial structure of the land price reveals that the CBD area has maintained its dominance while there has also been a sign of polycentricity with a few distinct subcenters. In the 1990s, the rent has steadily increased in the city. While the northern side of the Han river (including the CBD) revealed higher rents during most of the 1990s, the difference between the rents in the northern and the southern sides has been very small.[13]

---

[13] The rent index between 1992 and 1997 is obtained from the Pacific Appraisal Company website (http://www.packor.com/pac/kr/m_info.html).

As mentioned before, the decentralization of the manufacturing sector in the SMR was largely influenced by strong government control. Most decentralization measures in the first half of the past 30-year period have been enacted through policies that provide incentives to entrepreneurs to move or establish their firms in a recommended area. A series of measures was introduced until the late 1970s. This policy has been changed into a more effective enforcement of deconcentration since the 1980s; while an incentive to those who follow government guideline continues to be provided, there are penalties for those who do not. While it is controversial whether or not the policy has been appropriate, it has been effective in pushing manufacturing firms out to suburban areas.

8.7. Intra-metropolitan agglomeration in Seoul

The empirical analysis on Seoul used the 61 administrative areas in the SMR. Estimated results are summarized in Table 8.5.

Table 8.5. Estimation results of the SESE model

Variable Name	Manufacturing	Retail	Service
Constant	2.04877 (0.673)	26.9594 (0.087)	−33.0335 (0.033)*
Exogenous variables			
$IWMAN_i^{t-1}$	−0.12815 (0.023)*	0.034530 (0.012)*	−0.02657 (0.798)
$IWRET_i^{t-1}$	0.06439 (0.167)	−0.43171 (0.000)**	0.23112 (0.002)**
$IWSER_i^{t-1}$	0.04952 (0.687)	0.58847 (0.031)*	−0.78706 (0.000)**
$IWMLF_i^{t-1}$	0.00438 (0.745)		
$IWNMLF_i^{t-1}$	−0.01961 (0.312)		
$IWRLF_i^{t-1}$		0.11911 (0.010)**	
$IWNRLF_i^{t-1}$		−0.01095 (0.711)	
$IWSLF_i^{t-1}$			0.05576 (0.304)
$IWNSLF_i^{t-1}$			−0.00760 (0.761)
$IWPOP_i^{t-1}$	0.00203 (0.747)	−0.00214 (0.854)	0.00171 (0.873)
Endogenous variables			
$\Delta IWMAN_i$		0.89440 (0.032)*	−0.09187 (0.840)
$\Delta IWRET_i$	0.17807 (0.044)*		0.54889 (0.001)**
$\Delta IWSER_i$	−0.00601 (0.964)	1.10948 (0.000)**	
$\Delta WMAN_i$	−0.59866 (0.766)		
$\Delta WRET_i$		−1.84009 (0.001)* *	
$\Delta WSER_i$			0.48444 (0.318)
Adj-R^2	0.5182	0.6322	0.5967

Note: * significant at 95% and ** significant at 99% (significance level in parenthesis).

In the manufacturing equation, $IWMAN_i^{t-1}$ was significant with a negative sign. This finding implies that new manufacturing firms do not like to establish their factories near other manufacturing firms. In a spatial context, this is more related to the outbound location/relocation process of such firms from the city of Seoul to surrounding areas. This movement is to a large extent the outcome of the industrial zoning policies taken for decades since the 1980s. The insignificance of $\Delta WMAN_i$ implies that such a dispersion of manufacturing factories are more or less randomly distributed over the region, so that no specific tendency can be traced related to the location/relocation of other firms in the sector. The negative or insignificant impact of the distribution of manufacturing sector on itself confirms Moomaw's (1985) argument at an intra-metropolitan scale; a large portion of manufacturing sector may have lost its productivity advantages within centers. In relation to the retail sector, $IWRET_i^{t-1}$ was not significant but $\Delta IWRET_i$ showed a positive sign. As can be seen in the retail sector equation, the manufacturing and retail sectors had a synergetic effect on each other. Both variables representing the service sector, $IWSER_i^{t-1}$ and $\Delta IWSER_i$, were not as significant. During the 1980s and 1990s, the business service sectors along with the FIRE (Finance, Insurance and Real Estate) industries became the top forward linkage sector in Korea[14] and especially in the SMR among many regions in that the SMR covers a large portion of the Korean economy. While such sectors might have substantial interaction with the manufacturing sector, they were still not large enough to have the service sector significantly connected with the manufacturing sector in a spatial context in the SMR. All the labor force related and market related factors, $IWMLF_i^{t-1}$, $IWNMLF_i^{t-1}$, and $IWPOP_i^{t-1}$, were not very significant, implying that the sector does not count much on the distribution of labor force and customers.

In the retail equation, both $IWMAN_i^{t-1}$ and $\Delta IWMAN_i$ were positive and significant. Combined with the result from the manufacturing equation estimation, both sectors were thought of as having a synergetic effect on each other in terms of the

[14] See Cho *et al.* (2000, p. 129) for more information on the forward linkage hierarchy in Korea.

location/relocation process. Both of the retail distribution factors, $IWRET_i^{t-1}$ and $\Delta IWRET_i$, showed a negative influence. Overall, this implied an expansion or dispersion of the sector over the region. In a sense, new retail firms tend to locate their stores far from the area where there were already a number of such functions ($IWRET_i^{t-1}$) and also far from the area where there were growing numbers of retail functions ($\Delta IWRET_i$). Avoiding competition is thought of as more critical than pursuing collaboration in terms of spatial distribution. The two service sector distributions, $IWSER_i^{t-1}$ and $\Delta IWSER_i$, showed a positive influence. Similar to the manufacturing sector, firms and people in the service sector are also a substantial market for retail firms. The relationship of the retail sector with manufacturing and service suggested the market-orientedness of the sector in the location/relocation process. It is also noted that the distribution of the labor force was considered important in the retail firm location decision. It may be related to the tendency that a larger number of people in the retail sector in the SMR are self-employees and their workplaces are not separated from the residences or at least are quite close to where they live. The two market factors, $IWNRLF_i^{t-1}$ and $IWPOP_i^{t-1}$, were not so significant in the analysis.

In the service equation, both $IWMAN_i^{t-1}$ and $\Delta IWMAN_i$ did not show any significance in explaining the distribution of the service sector, reflecting a weak spatial association between the two sectors. Again, as in the explanation for the manufacturing section, it is thought that both have not had a strong spatial bondage even though the two sectors (especially manufacturing and the producer services) have experienced an increasing level of linkage with each other. The weak spatial association between those two sectors might be ascribed to the fact that those business-related and/or producer service sectors have been a relatively smaller portion of all the service activities and the other non-business-related service sectors have not had such an intimate linkage with manufacturing. Both of the retail sector variables, $IWRET_i^{t-1}$ and $\Delta IWRET_i$, were positively significant. As before, a synergetic effect can be found between the two sectors. In a sense, it is partly related to the fact that a large number of retail and service firms have concentrated in places in and around the city of Seoul where manufacturing had prevailed in the previous stage. While $IWSER_i^{t-1}$ showed a negative impact, $\Delta IWSER_i$ provided a positive sign even though the latter was not very significant. In terms of the competition-collaboration tradeoff, service firms seemed to feel a more competitive environment in a saturated market (a negative sign of IWSER_i^{t-1}) and to feel a more collaborative environment in a growing market but with less significance (a positive sign of $\Delta\mathrm{IWSER}_i$). All the labor force and market oriented variables, IWSLF_i^{t-1}, IWNSLF_i^{t-1}, and IWPOP_i^{t-1}, were not significant in the model.

Explanatory power of the three equations represented by adjusted R^2 is satisfactory ranging from 0.52 to 0.63.[15] The result shows a higher level of R^2 for service and relatively lower level of R^2 for manufacturing. A lower level of the manufacturing sector may imply that something other than economic impetus, such as policy incentives, has played a critical role in shaping the distribution pattern of manufacturing firms.

Focusing on backward- and forward-linkage effects between industries, cross-reinforcing impacts between manufacturing and retail and retail and service were found, but there was no self-enforcing dynamics inside of each sector. In addition, there was no significant relationship between manufacturing and service. Based on this finding, it was not possible to state categorically whether spatial proximity between related industries is important (Abdel--Rahman, 1990) or if that proximity to backward and forward linkage activities lacks explanatory power (Cooke, 1983). The relationship in Seoul is summarized in Figure 8.2.

If the attention is directed to the previous economic settings or environment, *inter*industry relationships might be a little different from the one considered above. In Seoul, all three economic sectors showed a negative self-reinforcing force. The previous pattern of a certain sectoral activity worked against clustering of such economic activity and was not able to work as an incubator. Similar to the relationship in Figure 8.2, the retail sector lay in the middle of the spatial association, interacting with the manufacturing and service sectors. As before, it is hard to answer if the proximity to other sectors is important in general. It is also noted that a disaggregation

[15] It is also noted that the application of the SESE model on the Chicago region showed stability across different estimation methods. For more information, refer to Sohn and Hewings (2000).

Figure 8.2. Sectoral relationship based on simultaneous interaction

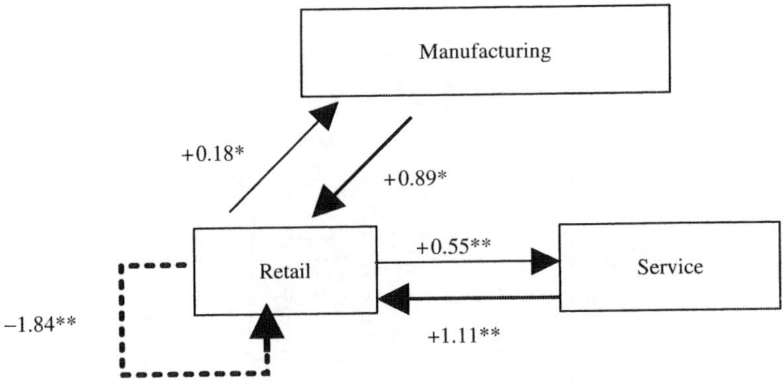

Note: Coefficients shown with significance level (** at 99% and * at 95%).

of the service sector may uncover a more accurate process of self-reinforcement. For example, some personal service firms similar to retail firms in their economic behaviors may also show a similar location/relocation pattern. On the other hand, such firms in business services may have different groups of customers and as a result, different location strategies. Figure 8.3 shows this relationship among the sectors in Seoul.

8.8. IT impact and polycentric urban development in Seoul

The estimation result of three sectors is listed in Table 8.6. In the activity equation of manufacturing, IWITFIRM was positively

Figure 8.3. Sectoral relationship based on previous setting

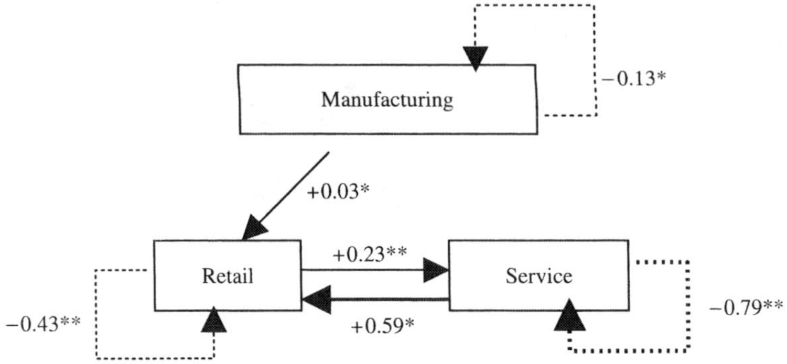

Note: Coefficients shown with significance level (** at 99% and * at 95%).

Table 8.6. ***Estimation results on regressions for attraction and spillover***

	Manufacturing		Retail		Service	
	Activity Level	Distribution Pattern	Activity Level	Distribution Pattern	Activity Level	Distribution Pattern
Constant	−183.285 (0.000)**	−0.4115 (0.531)	1.1153 (0.985)	1.5912 (0.045)*	22.01 (0.350)	1.4413 (0.026)*
$IWITFIRM_i$	1.2032 (0.000)**	−0.0147 (0.007)**	2.7760 (0.000)**	−0.0093 (0.079)	1.0396 (0.000)**	−0.0073 (0.153)
$IWUTEL_i$	0.2111 (0.038)*	−0.0020 (0.287)	0.2600 (0.086)	−0.0041 (0.032)*	0.2318 (0.000)**	−0.0011 (0.515)
$IWPTEL_i$	−0.0299 (0.001)**	0.0007 (0.000)**	−0.0916 (0.000)**	−0.0001 (0.716)	0.0386 (0.081)	−0.0002 (0.734)
CBD_i	0.0005 (0.368)	−0.00004 (0.000)**	−0.0001 (0.895)	−0.0001 (0.000)**	−0.00002 (0.947)	−0.0001 (0.000)**
$IWJHR_i$	1460.45 (0.000)**	20.1072 (0.001)**	515.126 (0.303)	10.4119 (0.106)	−0.8505 (0.997)	9.5869 (0.108)
Labor force	0.0529 (0.000)**	−0.0002 (0.367)	0.4518 (0.000)**	0.0023 (0.006)**	−0.0324 (0.554)	0.0016 (0.264)
Adj-R^2	0.9473	0.7122	0.9810	0.8028	0.9777	0.8289

Note: * significant at 95% and ** significant at 99% (significance level in parenthesis).
Labor force: Manufacturing: $IWMLF_i$, Retail: $IWRLF_i$, Service: $IWSLF_i$.

significant, implying that a higher level of information infrastructure attracted more businesses. Rather implicit IT indices compared to IWITFIRM, IWUTEL, and IWPTEL, showed conflicting signs. It seems that more manufacturing activities were attracted by a higher level of IT-related facilities, considering that IWITFIRM is more closely related to measuring IT infrastructure supporting business activities. CBD was not significant, reflecting a more dispersion of establishments from the city center; it also implied that the urban form might not be monocentric. IWJHR was positively significant. Combining these two variables, it can be concluded that while a dispersion of economic activities had occurred from the city center, the location/relocation process was mostly confined to the subcenters of the area. As mentioned earlier, a series of government-driven decentralization policies on the manufacturing sector seem to have been a major factor in explaining this pattern in Seoul. IWMLF was positive and significant, implying that the sector has sought the locations nearby the labor force. In the distribution equation, IWITFIRM was negative and significant and IWUTEL and IWPTEL had conflicting signs; CBD was negative and significant and IWJHR was positive and significant. More concentration can be seen near the CBD as well as the subcenters. IWMLF did not have a spillover effect in this case. Combining those two equations, IT-related variables worked as drawing more manufacturing activities, but less concentrated patterns. Center orientedness variables suggested less importance for the CBD and more significance for the subcenters in explaining the distribution pattern. It also confirmed a concentrated distribution pattern around the centers. The manufacturing sector in Seoul seemed to make use of the benefits of localized/urbanized economies (infrastructure or industrial linkage) by locating/relocating themselves close to such centers.

In the activity equation of retail sector, IT-related variables showed a similar pattern as the above case: positively significant IWITFIRM and conflicting IWUTEL and IWPTEL. CBD and IWJHR were not significant, reflecting that establishments in retail sector have not shown any preference on either centers or non-centers. IWRLF was positively significant and a market and a labor force were thought of as an important location factor. In the distribution equation, IWITFIRM was negative, but insignificant. IWUTEL and IWPTEL also had negative coefficients. CBD is negative and significant and IWJHR, the other center-oriented variable, was not significant, implying that spillover effect can only be observed near the city center. IWRLF was positively significant again, confirming the spillover effect of a market and a labor force in the retail sector on surrounding zones. As in manufacturing, IT related facilities worked as a facilitator to attract more intense retail activities, but to lead to a dispersed distribution pattern around. There was no explicit impact of centers in the city on retail locations; a more or less evenly and/or uniformly distributed pattern of the sector was expected.

In the activity equation of services, as in other sectors, IWITFIRM was positive and significant. Note, however, that the other two IT variables had positive coefficients in this case. Both of the center-orientedness variables did not show any significance in the model. A more evenly and randomly distributed pattern was expected to prevail in this sector as in retail. IWSLF as a market and a source of labor force was not considered important. In the distribution equation, all three IT related variables showed negative signs, resulting in dispersion patterns of the sectoral activities around. While the CBD variable is negative and significant, IWJHR was not significant and as a result, any explicit pattern around subcenters could not be traced. A more or less obscure pattern of distribution might be related, to some extent, to the complexity of the service sector as an aggregated entity. IWSLF was not significant in this case. Combining it with an insignificant coefficient of IWSLF in the activity equation, the location of this sector did not seem to be strongly influenced by markets or labor force. As in the other two sectors, IT related facilities worked as a facilitator to attract more service firms, leading to a less concentrated distribution pattern around. The estimation result of the two center-orientedness variables implied a more randomized or uniform spatial distribution of the sector, reflecting a mixed feature of service sector.

Overall, adjusted R^2 ranges from 0.71 to 0.98, showing that the regression models have sufficient explanatory power. In terms of attraction (activity level) forces of IT, retail showed the highest dependence while service showed the lowest. Manufacturing was somewhere between these two sectors. A relatively lower coefficient of services might be related to the mixed characteristics of the sector due to aggregation of heterogeneous service activities and/or

Table 8.7. ***Sectoral classification for IT impact***

		Distribution Pattern (Spillover)	
		↑ ↑	↑ ↓
Activity level (attraction)	↑ ↑		Manufacturing ($+1.2^{**}$ and -0.015^{**}), retail ($+2.8^{**}$ and -0.009), service ($+1.0^{**}$ and -0.007)
	↑ ↓		

Note: IWITFIRM coefficients in the attraction and spillover equations reported in parenthesis with significance level (** at 99%).

the dominance of traditional service firms in the composition of the sector. In terms of spillover (distribution pattern) effect, manufacturing had the highest, in absolute terms, followed by retail and service.

Information technology was very influential on the spatial agglomeration processes of firms in Seoul. The result on the attraction effect reveals that IT worked in a way to attract more activities. The spillover effect of IT, however, shows dispersion in Seoul, implying that the spatial extent of the positive IT influence might not be large. As Salomon (1996) and Yen and Mahmassani (1997) indicated, despite the dispersion-inducing factors of information technology on urban distribution, limited availability and accessibility of a well-equipped information network in certain areas seemed to restrict the favorable location of firms and as a result a more concentrated distribution pattern was observed. While it might turn out in the future that all the urban activities will be evenly dispersed with the complete coverage of IT facilities as Gordon and Richardson (1997) projected, at least for now, the IT infrastructure is working as a centripetal force with an uneven distribution of the technology.[16] Table 8.7 summarizes this finding based on the taxonomy in Table 8.3.

Among the center-orientedness variables in the equations for attraction, CBD was insignificant in all sectors. However, a positive coefficient in manufacturing and negative values in retail and service implied a dispersion of manufacturing firms from the CBD and a reconcentration of retail and service firms towards and around

[16] This tendency is also found in an inter-urban context. Fujita and Hamaguchi (2001) noted that customers of intermediate goods are more dispersed with the well-developed transportation/communication infrastructure than with the limited infrastructure.

the CBD. A more generalized variable to measure center-orientedness, IWJHR, showed a positive coefficient for manufacturing and insignificant coefficients for retail and service. In terms of the spillover effect, CBD was negatively significant in all sectors, implying more spillover effect near the CBD. IWJHR was insignificant in all sectors except manufacturing, in which spillover effect was expected around the city subcenters as well.

Based on the sign of the centrality coefficients, spillover effect can be found around the city center and subcenters in all sectors. The city center and subcenters were thought of as having great potential to attract various economic activities and working as an economic pole in a city to facilitate the development of surrounding zones. Attraction effects need to be explained by individual sectors; for example, manufacturing showed the most prominent pattern at subcenters. As explained earlier, government decentralization policies and firms' preference to seek localization/urbanization economies were among the major reasons for this pattern. On the other hand, retail showed center orientedness with larger department stores or discount outlets leading this trend especially in the 1990s. Finally, the service sector showed a contrasting distribution pattern: CBD and non-center orientedness. While the heterogeneous nature of the service sector complicates interpretation, the former process might be explained by sophisticated and professional service activities that have grown rapidly and the latter seems to be connected with the traditional

Figure 8.4. Sectoral map of attraction and spillover in Seoul.

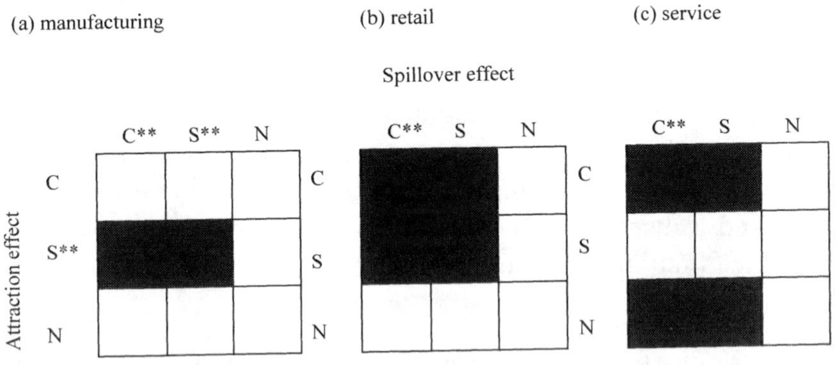

Note: C: CBD, S: suburban centers, N: non-centers,
** significant at 99%.

service sectors that have been somewhat ubiquitous over the region. Figure 8.4 summarizes this finding.

8.9. Conclusions

This chapter attempted to adapt the spatial econometric version of the simultaneous equation system to examine the spatial evidence of the agglomeration economies in Seoul in the 1990s. Thereafter, an attempt was made to explore the impact of information technology and center oriented tendency of economic activities on the distribution pattern of urban economic activities. For this purpose, a series of regression models were established for three sectors (manufacturing, retail, and service). Two aspects of the impact were examined: an attraction effect on a certain zone (level of activity) and a spillover effect on surrounding areas (distribution type). While detailed interpretations were provided in the previous sections, there are several interesting results to be noted.

First, in terms of backward- and forward-linkage effects between industries, cross-reinforcing impacts between manufacturing and retail and retail and service were found, but there were no self-enforcing dynamics within each sector. Secondly, if the attention is directed to the previous economic settings or environment, all three economic sectors showed a negative self-reinforcing force and the retail sector lay in the middle of the spatial association interacting with the manufacturing and service sectors. Thirdly, IT had a positive impact on the distribution pattern of urban economic activities with respect to attraction effect, but a negative impact with respect to spillover effect, implying that the IT influence may be limited to a very small area right next to IT facilities. Fourthly, in terms of center orientedness, the spillover effect could be found around the city center and subcenters in all sectors while the attraction effect revealed sector-dependent patterns.

One extension of the model would be to offer similar analysis at a more disaggregated sectoral level. Clearly, locational tendencies are going to be influenced dramatically by the variety of activities contained within any aggregate sector. In addition, the changing nature of production and service provision, generated in large part by lower transportation and communication costs, will contribute to changing the internal spatial dynamics of location within metropolitan areas. An industrial organizational approach might be another extension of the model especially related to office functions of service activities. Bailly (1995) noted that while the world cities have been centers for advanced producer services due to agglomeration economies (in an inter-metropolitan context), intra-urban distribution patterns revealed a dual structure: headquarter functions and high-level producer services were located near the urban cores and branch plant and small firms at the periphery. A more accurate picture of location variations can be obtained through the breakdown of organizations. Finally, as Hewings *et al.* (1998) noted, the hollowing out process, whereby intra-regional intermediation is exchanged for greater interregional dependencies, will generate different tendencies in intra-metropolitan location. In this context, the spatial scale of analysis may need to be broadened to a regional from a metropolitan level, which is able to handle intra-metropolitan as well as inter-metropolitan contexts at the same time.

References

Abdel-Rahman, H.M. (1990), "Agglomeration economies, types, and sizes of cities", *Journal of Urban Economics*, Vol. 27, pp. 25–45.

Anselin, L. (1995), *SpaceStat Version 1.80 User's Guide*, Morgantown: Regional Research Institute, West Virginia University.

Anselin, L. and A.K. Bera (1998), "Spatial dependence in linear regression models with an introduction to spatial econometrics", pp. 237–289, in: A. Ullah and D.E.A. Giles, editors, *Handbook of Applied Economic Statistics*, New York: Marcel Dekker.

Bailly, A.S. (1995), "Producer services research in Europe", *Professional Geographer*, Vol. 47, pp. 70–74.

Bergsman, J., P. Greenston and R. Healy (1975), "A classification of economic activities based on location patterns", *Journal of Urban Economics*, Vol. 2, pp. 1–28.

Calem, P.S. and G.A. Carlino (1991), "Urban agglomeration economies in the presence of technical change", *Journal of Urban Economics*, Vol. 29, pp. 82–95.

Capello, R. (1994), "Towards new industrial and spatial systems: the role of new technologies", *Papers in Regional Science*, Vol. 73, pp. 189–208.

Cho, B., J. Sohn and G.J.D. Hewings (2000), "Industrial structural change in the Korean economy between 1975 and 1995: input–output analysis", *Economic Papers*, Vol. 3, pp. 109–136.

Cooke, T.W. (1983), "Testing a model of intraurban firm relocation", *Journal of Urban Economics*, Vol. 13, pp. 257–282.

DeCoster, G.P. and W.C. Strange (1993), "Spurious agglomeration", *Journal of Urban Economics*, Vol. 33, pp. 273–304.

Dekle, R. and J. Eaton (1999), "Agglomeration and land rents: evidence from the prefectures", *Journal of Urban Economics*, Vol. 46, pp. 200–214.

Drucker, P.F. (1989), "Information and the future of the city", *Urban Land*, Vol. 48, pp. 38–39.

Echeverri-Carroll, E.L. (1996), "Flexible production, electronic linkages, and large firms: evidence from the automobile industry", *Annals of Regional Science*, Vol. 30, pp. 135–152.

Feiser, E.J. and E.M. Bergman (2000), "National industry cluster templates: a framework for applied regional cluster analysis", *Regional Studies*, Vol. 34, pp. 1–20.

Fogarty, M.S. and G.A. Garofalo (1988), "Urban spatial structure and productivity growth in the manufacturing sector of cities", *Journal of Urban Economics*, Vol. 23, pp. 60–70.

Fujita, M. and N. Hamaguchi (2001), "Intermediate goods and the spatial structure of an economy", *Regional Science and Urban Economics*, Vol. 31, pp. 79–109.

Fujita, M. and J.-F. Thisse (1996), "Economics of agglomeration", *Journal of the Japanese and International Economics*, Vol. 10, pp. 339–378.

Gaspar, J. and E.L. Glaeser (1998), "Information technology and the future of cities", *Journal of Urban Economics*, Vol. 43, pp. 136–156.

Getis, A. and K. Ord (1992), "The analysis of spatial association by use of distance statistics", *Geographical Analysis*, Vol. 24, pp. 189–206.

Goldstein, G.S. and T.J. Gronberg (1984), "Economies of scope and economies of agglomeration", *Journal of Urban Economics*, Vol. 16, pp. 91–104.

Gordon, P. and H.W. Richardson (1997), "Are compact cities a desirable planning goal?", *Journal of the American Planning Association*, Vol. 63, pp. 95–106.

Handy, S.L. and P.L. Mokhtarian (1996), "Forecasting telecommuting: an exploration of methodologies and research needs", *Transportation*, Vol. 23, pp. 163–190.

Hansen, E.R. (1990), "Agglomeration economies and industrial decentralization: the wage-productivity trades-offs", *Journal of Urban Economics*, Vol. 28, pp. 140–159.

Hanson, G.H. (1996), "Agglomeration, dispersion, and the pioneer firm", *Journal of Urban Economics*, Vol. 39, pp. 255–281.

Helsley, R.W. and W.C. Strange (1991), "Agglomeration economies and urban capital markets", *Journal of Urban Economics*, Vol. 29, pp. 96–112.

Hewings, G.J.D., M. Sonis, J. Guo, P.R. Israilevich and G.R. Schindler (1998), "The hollowing-out process in the Chicago economy, 1975–2011", *Geographical Analysis*, Vol. 30, pp. 217–233.

Kelejian, H.H. and I.R. Prucha (1998), "A generalized spatial two-stage least squares procedure for estimating a spatial autoregressive model with autoregressive disturbances", *Journal of Real Estate Finance and Economics*, Vol. 17, pp. 99–121.

Kelejian, H.H. and D.P. Robinson (1993), "A suggested method of estimation for spatial interdependent models with autocorrelated errors, and an application to a county expenditure model", *Papers in Regional Science*, Vol. 72, pp. 297–312.

Lee, K.S. (1981), "Intra-urban location of manufacturing employment in Colombia", *Journal of Urban Economics*, Vol. 9, pp. 222–241.

Lee, H.W. (1996), "Comparative researches on the fluctuations of land values and changes of internal structure in Seoul and Tokyo (written in Korean with English summary)", *Journal of the Korea Planners Association*, Vol. 31, pp. 121–138.

Lund, J.R. and P.L. Mokhtarian (1994), "Telecommuting and residential location: theory and implications for commute travel in the monocentric metropolis", *Transportation Research Record*, Vol. 1463, pp. 10–14.

Maurel, F. and B. Sedillot (1999), "A measure of the geographic concentration in French manufacturing industries", *Regional Science and Urban Economics*, Vol. 29, pp. 575–604.

McMillen, D.P. and J.F. McDonald (1998), "Suburban subcenters and employment density in metropolitan Chicago", *Journal of Urban Economics*, Vol. 43, pp. 157–180.

Mitra, A. (1999), "Agglomeration economies as manifested in technical efficiency at the firm level", *Journal of Urban Economics*, Vol. 45, pp. 490–500.

Mokhtarian, P.L. (1998), "A synthetic approach to estimating the impact of telecommuting on travel", *Urban Studies*, Vol. 35, pp. 215–241.

Moomaw, R.L. (1985), "Firm location and city size: reduced productivity advantages as a factor in the decline of manufacturing in urban areas", *Journal of Urban Economics*, Vol. 17, pp. 73–89.

Moulaert, F. and F. Djellal (1995), "Information technology consultancy firms: economies of agglomeration from a wide-area perspective", *Urban Studies*, Vol. 32, pp. 105–122.

Mun, S. and B.G. Hutchinson (1995), "Empirical analysis of office rent and agglomeration economies: a case study of Toronto", *Journal of Regional Science*, Vol. 35, pp. 437–455.

Nakamura, R. (1985), "Agglomeration economies in urban manufacturing industries: a case of Japanese cities", *Journal of Urban Economics*, Vol. 17, pp. 108–124.

Nilles, J.M. (1988), "Traffic reduction by telecommuting: a status review and selected bibliography", *Transportation Research A*, Vol. 22, pp. 301–317.

Ord, J.K. and A. Getis (1995), "Local spatial autocorrelation statistics: distributional issues and an application", *Geographical Analysis*, Vol. 27, pp. 286–306.

Pascal, A.H. and J.J. McCall (1980), "Agglomeration economies, search costs, and industrial location", *Journal of Urban Economics*, Vol. 8, pp. 383–388.

Rey, S.J. and M.G. Boarnet (2004), "A taxonomy of spatial econometric models for simultaneous equations systems", in: L. Anselin, R.J.G.M. Florax and

S.J. Rey, editors, *Advances in Spatial Econometrics*, Heidelberg: Springer, forthcoming.

Richardson, R. and A. Gillespie (1996), "Advanced communications and employment creation in rural and peripheral regions: a case study of the Highlands and Islands of Scotland", *Annals of Regional Science*, Vol. 30, pp. 91–110.

Salomon, I. (1996), "Telecommunications, cities and technological opportunism", *Annals of Regional Science*, Vol. 30, pp. 75–90.

Sinden, A. (1995), "Telecommunications services: job loss and spatial restructuring in Britain, 1989–1993", *Area*, Vol. 27, pp. 34–45.

Smith, D.F., Jr. and R. Florida (1994), "Agglomeration and industrial location: an econometric analysis of Japanese-affiliated manufacturing establishments in automotive-related industries", *Journal of Urban Economics*, Vol. 36, pp. 23–41.

Sohn, J. and G.J.D. Hewings (2000), *Spatial Evidence of Agglomeration Economies in Chicago, Regional Economics Applications Laboratory Discussion Paper (REAL 00-T-4)*, Urbana: University of Illinois.

Stough, R.R. and J. Paelinck (1996), *Substitution and Complementary Effects of Information on Regional Travel and Location Behavior, Proceedings, New International Perspectives on Telework Workshop*, London: Brunel University, pp. 380–397.

Tofflemire, J.M. (1992), "Telecommunication external economies, city size and optimal pricing for telecommunications", *Journal of Regional Science*, Vol. 32, pp. 77–90.

Yen, J. and H.S. Mahmassani (1997), "Telecommuting adoption: conceptual framework and model estimation", *Transportation Research Record*, Vol. 1606, pp. 95–102.

Urban Dynamics and Growth, CEA, vol. 266
Roberta Capello and Peter Nijkamp (editors)

CHAPTER 9

Dual Earners, Urban Labour Markets and Housing Demand

Jan Rouwendal[a,b] and J. Willemijn Van Der Straaten[a]

[a]Department of Spatial Economics, Free University, Amsterdam, The Netherlands
[b]Economics of Consumers and Households Group, Wageningen University, Wageningen, The Netherlands

Abstract

This chapter replicates Costa and Kahn's analysis of locational choices of highly educated couples for the Netherlands. We find increasing concentration of such power couples in the urbanised western part of the country. This trend occurs in spite of the absence of an urban wage premium and the concentration of traffic congestion there. We find that power couples locate more often in medium sized and larger cities than otherwise comparable households and that they are relatively often owner-occupiers and live in more expensive houses. Their commuting time is relatively short when it is taken into account that it is more difficult for those households to find suitable arrangements of employment and residence than it is for single earner households. A likely explanation for these findings is that power couples use their relatively large purchasing power to outbid other households from locations that are especially attractive to them, as is predicted by household location theory.

Keywords: dual earner households, power couples, urban wages, location choice, commuting distance, housing demand

JEL classifications: D1, R2

The authors are grateful to Henri de Groot and an anonymous referee for useful comments on an earlier version. The usual disclaimer applies.

9.1. Introduction

The share of dual earners in the household population increased considerably over the past decades. Dual earners used to be the exception, but nowadays they are the norm. Dual earner households are considerably different from single earners in many respects. There are obvious and substantial differences in the way they spend their time. Home production, housework and – when children are present – childcare have to be adapted to the working hours of two employed persons. There are also less self-evident ways in which the changing internal organisation of households influences our societies. One, which is the subject of this chapter, is the location behaviour of dual earners. Dual earner households are likely to differ in location behaviour from single earners because of the two commutes. This location problem has at least two important aspects. The most obvious one is that it may be more difficult to find a satisfactory employment–housing arrangement when two jobs and one residence are involved, instead of one job and one residence. As a consequence, commutes of dual earners are expected to be longer on average than those of single earners. A second aspect is that dual earners may not only differ in the location they choose within a region, but may also differ in the choices they make with respect to the region of residence. Costa and Kahn (2000) have recently emphasised the co-location problem experienced by 'power couples', that is dual earners that are both highly educated.[1] They argue that satisfactory career perspectives for both partners are only available in the dense local labour markets of metropolitan areas, and show that there is indeed an increasing concentration of 'power couples' in large metropolitan areas in the US in the second half of the 20th century.

In this chapter, we consider the location behaviour of dual earner households in the Netherlands, where the transition towards increased labour market participation of married and cohabiting women took place later than in the US. We find a similar concentration of highly educated dual earners in the Randstad

[1] Note that the definition of power couples refers only to the higher education of the two partners and does not require that both are employed. Even though most power couples are dual earners, this is not true for all of them.

(also called the Dutch Rimcity), the urbanised western part of the country. In contrast to Costa and Kahn, we find no evidence of increasing attractiveness of large urban areas for the control group of highly educated singles, as this group tends to become less concentrated in the Randstad. This leads us to a closer examination of the mechanism that explains the changing location patterns of both groups of households. The co-location hypothesis assumes that labour markets must be sufficiently dense in order to make a region attractive for power couples. However, a high density also offers better match opportunities for single earners (including singles). We examine the literature on urban labour market agglomeration in order to find examples of markets that function in such a way that they attract dual earners more than single earners.

An alternative (or additional) explanation for the observed location patterns in the Netherlands considers the effect of dual earners on the housing markets. Dual earners have a household income that is higher than that of single earners. As a consequence, they may be able to outbid the other household types from the most attractive parts of the housing markets in their preferred employment regions. Given the limited supply of newly constructed housing and the small geographical scale of the Netherlands, the other household types may choose lower-quality housing in these regions or choose a residence in the neighbouring region while accepting longer commutes.

The next section describes the geographical spread of power couples in the US and the methodology used by Costa and Kahn (2000). At the end of this section some labour market issues are investigated. Section 9.3 analyses the geographical spread of power couples in the Netherlands. Section 9.4 provides further empirical analyses of the urban wage premium in the Netherlands and housing market behaviour of dual earners. Section 9.5 concludes.

9.2. Locational choices of couples in the US

9.2.1. Summary Costa and Kahn

The article 'Power couples: Changes in the locational choice of the college educated' by Costa and Kahn (2000) analyses the trend of increasing concentration of college educated couples in large metropolitan areas. In contrast to these highly educated couples, there has been little change in the location pattern of couples in which neither spouse has a college education. As written in the introduction, the co-location problem experienced by power couples gives a good explanation of the trend. This chapter summarises the main results and discusses the methodology used by Costa and Kahn (2000). In Section 9.3 the results of the analysis of the location behaviour of couples in the Netherlands are described.

Costa and Kahn classify all households into five types: 'power' couples in which both spouses have a college education, 'part power' couples in which only one spouse has a college education, 'low power' couples in which neither spouse has a college education, and single households of the college educated and the noncollege educated. They distinguish three city size categories: large metropolitan areas (those with populations of at least 2 million), midsize metropolitan areas (those with populations of between 2 million and 250,000), and small and nonmetropolitan areas (metropolitan areas with populations of less than 250,000 and nonmetropolitan areas).

Their study attempts to provide an explanation for their finding that the probability of being located in a large metropolitan area rose by 0.174 between 1940 and 1990 and by 0.104 between 1970 and 1990 for power couples. In contrast, the increase in this probability for part-power couples was only 0.102 and 0.059, respectively (Costa and Kahn, 2000, p. 1293).

Costa and Kahn (2000) argue that the facilities that dense urban labour markets offer for both partners is the main explanatory factor for the rise of power couple concentration in large metropolitan areas. They arrive at this conclusion by comparing the changes in the location of power couples with those of other groups. According to these authors there are three main factors that determine the trends in the spatial distribution of people: the co-location problem, urban amenities of general interest (for example, higher returns for education or cultural activities) and urban amenities of specific interest for singles (for example, the existence of a marriage market as a result of an increase of young unmarried singles in large cities). They try to disentangle these three factors by means of two differencing procedures.

In the first procedure, changes in the spatial distribution of so-called 'coincidental couples' are subtracted from changes in the distribution of actual couples (power or low power couples). The former distribution is the one that could be expected on the basis of the distribution of singles (power or low power singles). It is determined on the basis of a simple theory of couple formation. The number of potential matches between males and females that could form within a particular area is the minimum of the number of males and females in a particular region.[2] The distribution of these potential matches is referred to by Costa and Kahn as the distribution of coincidental couples. It is interpreted as the location pattern that should be expected for power couples in the absence of the co-location problem.

Changes in the distribution of coincidental couples are driven by changes in urban amenities of general interest and changes in urban amenities of specific interest for singles. Changes in the actual distribution of couples are driven by co-location problems and urban amenities of general interest. An overview of the urban amenities for various types of couples is given in Table 9.1.

The effect of urban amenities of general interest can be removed by taking the difference between the changes in the spatial distribution of actual couples and coincidental couples. They compute this difference both for power couples and for low power couples.

The resulting double differences are determined by the co-location problem and by the urban amenities that are of specific interest for singles. In the second procedure, the difference of these double differences is computed. If it is assumed that the effects of urban amenities that are of specific interest to singles are equal for both power and low power singles, this leaves us with the effect of the co-location problem that is specific for power couples. Costa and Kahn conclude on the basis of the resulting triple difference that a substantial part of the total changes in the location pattern of power couples is caused by the co-location problem.

---

[2] Suppose that there are 100 single power men and women and that 40 of the men are in large cities and 60 are in small cities. The probability of a coincidental couple being in a large city is therefor 0.5. We therefore take our estimates of the probability of a single power man living in city size 's' ($p_s^{M,P}$) and our estimates of the probability of a single power woman living in city size 's' ($p_s^{F,P}$) and take the minimum of these probabilities ($\min(p_s^{M,P}, p_s^{F,P})$). We then estimate the probability that a coincidental couple will be living in a given city size ($min(p_s^{M,P}, p_s^{F,P})/\sum_s \min(p_s^{M,P}, p_s^{F,P})$) (Costa and Kahn, 2000, p. 1295).

Table 9.1. Urban amenities for various types of couples

Power Couples	Coincidental Power Couples	Low Power Couples	Coincidental Low Power Couples
Solution co-location problem for higher educated	Urban amenities of general interest	Solution co-location problem for lower educated	Urban amenities of general interest
Urban amenities of general interest	Urban amenities of specific interest for higher educated singles	Urban amenities of general interest	Urban amenities of specific interest for lower educated singles

9.2.2. Methodological issues

It should be clear that the procedure adopted by Costa and Kahn is based on strong assumptions with respect to the effect of the three factors that are distinguished upon the location patterns of the groups concerned. Changes in the share of a group of households in a particular region must be the sum of two effects. Moreover the coefficients referring to urban amenities in general must either be equal for the pairs (power couple, coincidental power couple) and (low power couple, coincidental low power couple) or for the pairs (power couple, low power couple) and (coincidental power couple, coincidental low power couple). Costa and Kahn assume the first possibility to be the relevant one. If one adopts the second possibility, the effect of the co-location problem can be found by subtracting the changes in the regional distribution of low power couples from those of high power couples and there would be no need to take triple differences. Finally, the coefficients referring to urban amenities of specific interest to singles should be equal for power and low power singles. These assumptions are not motivated by, for instance, a model of location choice of households. They should probably be interpreted as approximations to such a model, but it is far from clear that, for instance, a first-order approximation to such a model has the properties assumed by Costa and Kahn.

On the other hand, it is also clear that the differencing procedures used have attractive properties in that they correct potentially for effects of all types of omitted variables, even though it must be noticed that also here the assumption of linearity is crucial.

The analysis that has been carried out by Costa and Kahn results in the conclusion that the co-location problem is of considerable importance for the spread of power couples. However, it reaches this conclusion on the basis of a differencing procedure that intends to remove all other explanations. For this reason, it does not provide direct insight into the co-location problem itself. For that purpose we must turn to labour market analysis.

9.2.3. Urban labour market

Local labour markets play an important role in Costa and Kahn's analysis. First, they argue that only metropolitan labour markets are dense enough to offer both highly educated partners in power couples a reasonable career perspective. Second, they argue that an additional benefit is associated with an urban wage premium that appears to have increased over time. In this subsection we take a closer look at the characteristics that an urban labour market must have in order to facilitate solving the co-location problem. This will be done on the basis of theoretical literature. The existence of an urban wage premium in the United States is well established (see, for instance, Glaeser and Maré, 2001). For the Netherlands no prior studies are available and in Section 9.4 we investigate that issue.

9.2.3.1. The co-location problem

Before we start the review of labour market models that are relevant for the co-location problem, it is useful to return to the problem itself. Costa and Kahn (2000, p. 1288) introduce it as follows:

> All dual career households are more likely to be joint decision makers, and they face the difficulty of finding two jobs commensurate with the skills of each spouse within a reasonable commute from home.

This seemingly clear statement suggests that a dense labour market, with more jobs available within a reasonable commute from home, makes it easier to solve this difficulty. However, in a dense labour market there are not only more jobs, but also more workers. As a result, there will be more competition for jobs commensurate with the skills of each worker and in a labour market with imperfect information and other market imperfections, its effect may be similar to that of a lower density of jobs.

It should also be noticed that the co-location problem is defined as a problem that is specific to couples. In a given local labour market it may be difficult for both workers to find a job that is commensurate with their skills. However, if the difference between large and small labour markets is only in the number of jobs and not in the distribution of the skills required for carrying out these jobs, there seems no reason to talk about a specific co-location problem.

It is, of course, true that dual earners have to find a combination of two employment locations and one residential location with reasonable commutes, whereas single earners have a simpler problem to solve. However, this refers to location *within* an urban area (we discuss this aspect in the next section) and does not in itself make clear that denser labour markets are more attractive than others.

We may conclude that the co-location problem appears to be a bit more difficult to understand than one might think at first sight. For this reason it seems useful to take a closer look at labour market models that deal with the heterogeneity of workers in order to see how the co-location problem may emerge in such a context.

9.2.3.2. *The extent of the market, search and learning*

Frank (1978) was one of the first to provide a formal analysis of the co-location problem in his study of the effects of affirmative action. In his model jobs and workers are heterogeneous. The productivity of a particular match between a worker and a job is highest if the required skill level equals the offered skill level. If the two skill levels are unequal, productivity diminishes proportional to the absolute value of the difference. A single worker will locate in a large city if his personal match is better there than elsewhere. Dual earners will locate in the large city if the sum of the individual discrepancies is smaller in the large city than elsewhere. Frank discusses a particular simplified situation and shows that the probability that the latter event takes place is larger than the probability that the first event takes place. This suggests that dual earners will more often choose the large labour market than single earners.

Frank's analysis leads to a clear conclusion, but it is clearly partial in nature. This evokes the question whether his conclusion stands upright in a more general setting where less is assumed to be constant. Fortunately, there have been other analyses of the functioning of the type of labour market with heterogeneous labour that he studied, which are less restrictive. Kim (1989), referred to by Costa and Kahn, provided a model in which labour is heterogeneous in innate abilities of the workers. This setting is similar to that considered by Frank, but Kim also considers the effects of bargaining. One of his results is that in a sufficiently competitive market the wage is independent of the discrepancy between a worker's actual ability and the one that is optimal for the job in which he is employed. In such a market, the mechanism identified by Frank clearly does not work.

Kim also shows that in a larger market the wage level will be higher than in a smaller one. He does not discuss the peculiarities of dual earner households, but the relevant consequences of his model are not too difficult to grasp. In his model both wage earners can earn more in the large market than in the small one, independent of the discrepancy between their ability and the ideal one for the job in which they are employed. For both workers the advantage of the higher wage is equal and for each of them the benefit of locating in a large city is identical to that of a single worker. In this sense there is no real co-location problem in this model. Note also that this reasoning assumes that wage differences between large and small markets remain, whereas it must be expected that in general equilibrium context they will be removed.

When placed in a spatial setting, the analysis of Kim can be interpreted as an agglomeration effect. Helsley and Strange (1990) have indeed developed a model for a system of cities in which the better matching of skills and requirements is the reason why cities exist. A counteracting force is the consumption of land, which forces workers to commute over longer distances when city size increases. Workers move between cities in order to maximise expected utility. The equilibrium size of the city is determined by the agglomeration and disagglomeration effects.

Although Helsley and Strange only consider cities that are of equal size, it is possible to imagine that this model contains cities of different sizes, for instance, because of exogenous differences in the attractiveness of sites or the productivity of workers. In a general equilibrium setting all workers will experience the same utility, independent of their place of residence. There is no specific co-location problem.

Teulings and Gautier (2000) consider a model in which skills and job requirements are continuously distributed on the real line. Under conditions of perfect information, a Walrasian equilibrium exists in which each job type is matched to exactly one worker type. Imperfect information introduces 'noise' around the Walrasian equilibrium. In Teulings and Gautier (2002) they apply the model in a spatial general equilibrium setting. In equilibrium, all workers are indifferent between regions. The less dense regions are more specialised than the denser regions. In this setting a co-location problem may emerge. In the spatial general equilibrium, dense regions specialise in products that require a diversity of skills, whereas less dense regions specialise in products requiring specific skills. This means that the distribution of required skills depends on the density of economic activity in a region and has the largest variance in the densest regions.

If skills of individual workers are a random drawing from a given skill distribution (independent of their regions of origin) and if they match randomly into couples, it is obvious that some redistribution of the workers over the regions has to take place in order to establish this equilibrium. Workers with skills located at the tails of the distribution must move from rural to urban areas, or workers with skills located close to the median must move in the reverse direction (or both). Since there are relatively few workers with skills located at the tail of the distribution, the probability that they have a partner with skills located close to the median is relatively large. These partners may face conflicting interests with respect to their preferred location, and it is conceivable (even though we do not undertake a formal analysis) that the best way to solve this problem is to move to a dense urban region where job opportunities for both of them are available.

Teulings and Gautier stress that their results depend on their assumption that there are increasing returns to scale in the matching function. The survey of Petrongolo and Pissarides (2001), to which they refer, makes clear that empirical support for this assumption is weak. However, Teulings and Gautier argue that it may be difficult to observe these increasing returns when present, because of counteracting effects, such as an increase in reservation wages. A more recent paper of Petrongolo and Pissarides (2002) offers some support for this point of view, but also observes that it implies that the effect of increasing returns should show up in the wage rates paid in the denser labour markets. In Section 9.4 we will examine the existence of an urban wage premium for the Randstad area.

In conclusion it may be said that it is not easy to motivate the existence, let alone the significance, of the co-location problem on the basis of existing labour market models. We focused attention on models that explicitly incorporate the heterogeneity of labour, since this seems to be a main ingredient of the problem. A necessary condition for the co-location problem to emerge seems to be that the distribution of required skills differs over regions and has the largest variance in dense urban labour markets. This condition does not appear to be unrealistic. However, it is not obvious from Costa and Kahn's statement of the problem, and it is (no doubt for reasons of simplicity) not incorporated in most existing models.

9.2.3.3. Other labour market effects

The models discussed above stress the problems associated with matching heterogeneous workers to equally heterogeneous jobs and the advantages that large (urban) labour markets offer in this respect. However, this is not the only conceivable benefit of urban labour markets. Glaeser and Maré (2001) have studied the urban wage premium and find that on average it takes some time for immigrants to realise such a premium. If the main benefit of urban labour markets is a better match between required and actual capacities, it should be expected that this premium is realised immediately, or at least soon after arrival on the urban labour market. Since the data show that it usually takes some years for workers to realise this premium, it may be conjectured that something is missing from these matching models that is relevant in practice.

Glaeser (1999) argues that learning of workers from each other is the explanation. Experienced workers meet inexperienced colleagues and the exchange of ideas makes the latter more productive. The relevant contacts may occur on the job, but also outside working hours in informal contacts that are facilitated by the high density of the urban environment. Power couples may be in an especially advantageous position to realise such benefits. Research indicates that the diversity of activities is especially important for agglomeration effects and economic growth to realise. Iranzo (2003) presents evidence that wages are higher in cities with more dispersed human capital.

Peri (2002) developed a model in which learning externalities among educated workers are the reason for their concentration in urban areas. He argues that the development of ICT in the last quarter of the 20th century has increased the importance of such externalities and provides an alternative explanation for the increased concentration of power couples in urban areas, while it also explains a similar trend among highly educated singles in the US.

Although the co-location problem appears to be a self-evident problem, its relation with urban labour markets turns out to be subtler than one would expect at first sight. We will not pursue the issue further in this chapter, but note that the relation with diversity of required skills and increasing returns to scale are interesting areas for future research.

9.3. Location choices of couples in the Netherlands

In this section we examine the question whether the trend that was identified by Costa and Kahn for the US is also present in the Netherlands. In doing so, we should of course recognise the difference in size between both countries. The US has more than 200 million inhabitants, whereas the Netherlands has less than 20 million. The US has a number of metropolitan regions, whereas the Netherlands has essentially one – the Randstad. This means that we will actually consider the distribution of different household types over a small country consisting of one metropolitan area and its hinterland.

9.3.1. The data

The data that we use are those of a series of Housing Needs Surveys. The Housing Needs Survey is conducted every 4 or 5 years and the sequence started in 1973. We use the surveys of 1981, 1989, 1993 and 1998 in this chapter. Although the surveys are directed primarily towards the identification of housing needs, it contains a wealth of related relevant information about household characteristics and location.

The analysis of this section refers to respondents and their partners who were in the age interval |23–39| in order to facilitate the comparison with Costa and Kahn (2000). Power couples were identified as couples with both partners highly educated. In order to be qualified as highly educated an individual should have completed higher vocational training or have obtained a university degree. In part power couples one of the partners has received higher education, whereas the other has not. Neither partner in low power couples has received higher education.

Having received higher education became increasingly common among both males and females over the period under study, and the share of power couples increased from 7.0 in 1981 to 13.1 in 1998. Female labour force participation also increased considerably. In 1981, 43.7% of the low power couples and 63% of the power couples were dual earners. In 1998 these shares were 22% points higher for both groups.

9.3.2. *Regional division*

The Randstad has a number of special characteristics. First, it consists of a number of cities of different size that are separated by more or less rural areas. It is possible to consider the largest of these cities, Amsterdam, Rotterdam, Utrecht and The Hague as separate metropolitan areas. However, this approach has not been chosen because these cities are so close to each other that it is possible to commute from one to another. If we want to define a metropolitan area in such a way that (to a close approximation) all people who live within its boundaries are also employed there, any further decomposition of the Randstad becomes problematic. We have therefore adopted a commonly used regional division of the Netherlands into three areas: the Randstad (dark grey in Figure 9.1), a peripheral zone (white in Figure 9.1) and an area between these two that is referred to as an intermediate zone (grey in Figure 9.1). The Randstad comes close to a polycentric metropolitan area and if the location pattern of power couples in the Netherlands is close to that in the US, we expect to find a contrast between this region and the periphery, which is clearly nonmetropolitan. Inclusion of the intermediate zone in the analysis may provide additional useful information about location patterns. We will therefore consider the question how power couples have located over these three zones during the period 1981–1998.

It must be recognised that this regional division is conceptually different from that employed by Costa and Kahn. They were able to distinguish a large number of metropolitan areas of medium and large size that could be considered as separate labour markets because of spatial separations. We divide a small country into the

Figure 9.1. The Netherlands and the division in three zones

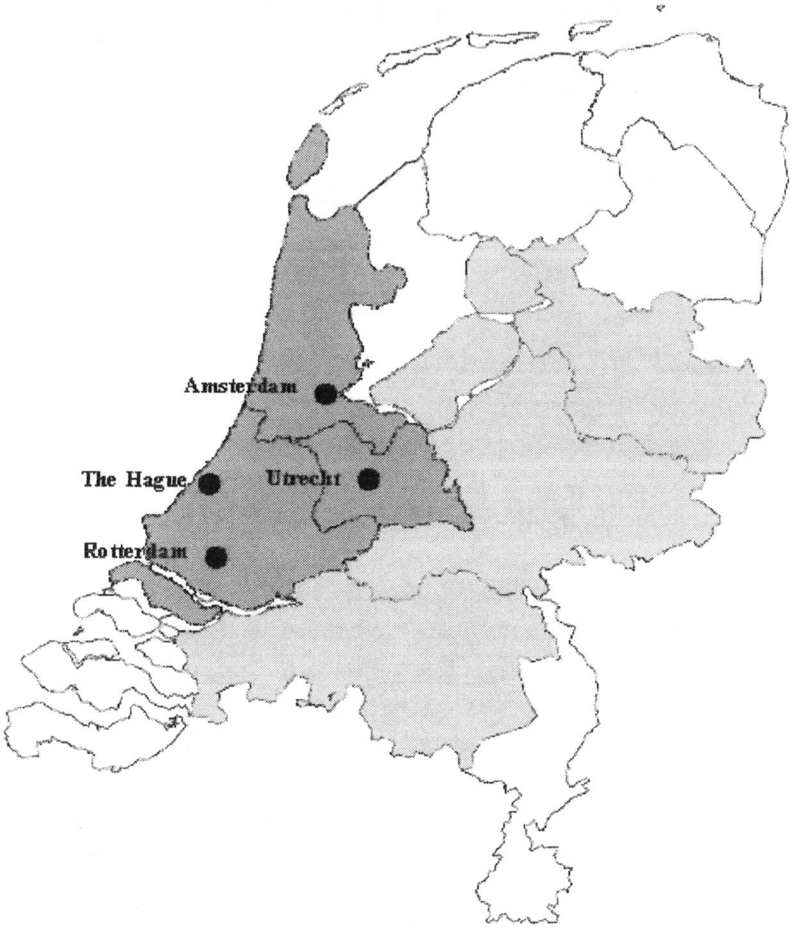

west, a heavily urbanised region, an intermediate zone, which surrounds the west geographically and the remainder of the country, referred to as the periphery. Commuting flows between these three regions are not negligible and they can therefore not be considered formally as completely independent regional labour markets. Similar remarks can be made with respect to their functioning as separate marriage markets for power singles.

9.3.3. Results

Table 9.2 shows the distribution of three types of households over the three regions discussed above. The table shows that power

Table 9.2. Spatial distribution and labour force participation of couples

	1981	1989	1993	1998
Power couples				
Randstad	50.3	49.0	52.0	55.0
Intermediate zone	30.1	33.7	32.4	32.0
Periphery	19.6	17.2	15.6	12.6
Part power couples				
Randstad	44.2	41.8	44.8	47.5
Intermediate zone	33.5	36.9	36.0	35.7
Periphery	22.3	21.4	19.2	16.8
Low power couples				
Randstad	41.5	39.8	40.3	39.6
Intermediate zone	37.5	38.5	38.3	39.6
Periphery	21.0	21.8	21.4	20.8

couples became increasingly concentrated in the Randstad, that the share of part power couples living in the Randstad also increased, but at a slower rate than the power couples. The share of low power couples decreased. This suggests that the Randstad became a more attractive residential area for power couples and one may wonder what causes this phenomenon.

Costa and Kahn conjectured that the facilities that dense urban labour markets offer for both partners constitute the main explanation of the rise of concentration of highly educated couples in the Randstad. We used their methodology, as described in Section 9.2.1, to compare the changes in the spatial distribution of coincidental couples with the changes in the spatial distribution of actual couples.

The results are reported in Table 9.3. The first four lines in this table summarise the changes in the location patterns of four couple types over the period concerned. The Randstad attracted a large share of power couples, but became less attractive for the other three types. The next two lines show the double differences.

The bottom line of each panel shows the computed effect of the specific co-location problem for power couples. One surprising aspect of these figures is that they suggest that locating in the intermediate zone is better for solving the co-location problem than locating in the Randstad. Another is that they are larger (in absolute value) than the changes reported in the first row of the table. This indicates that the increasing concentration of power couples

Table 9.3. Computation of co-location effects

	Randstad	Intermediate Zone	Periphery
All couples; differences 1998–1981			
Δ power couples	0.074	0.009	−0.083
Δ low power couples	−0.069	0.052	0.017
Δ coincidental power couples	−0.036	−0.036	0.071
Δ coincidental low power couples	−0.099	0.106	−0.008
Double differences 1998–1981			
Δ power couples − Δ coincidental power couples	0.109	0.044	−0.154
Δ low power couples − Δ coincidental low power couples	0.030	−0.054	0.024
Triple differences 1998–1981			
[Δ power couples − Δ coincidental power couples] − [Δ low power couples − Δ coincidental low power couples]	0.079	0.099	−0.178
Couples with both partners employed; differences 1998–1981			
Δ power couples	0.287	0.142	0.033
Δ low power couples	0.131	0.216	0.106
Δ coincidental power couples	−0.092	0.033	0.060
Δ coincidental low power couples	−0.106	0.091	0.015
Double differences 1998–1981			
Δ power couples − Δ coincidental power couples	0.380	0.110	−0.026
Δ low power couples − Δ coincidental low power couples	0.237	0.125	0.091
Triple differences 1998–1981			
[Δ power couples − Δ coincidental power couples] − [Δ low power couples − Δ coincidental low power couples]	0.142	−0.015	−0.118

in the Randstad and the intermediate zone took place despite a decreasing attractiveness of this region in terms of amenities of general interest.

Like Costa and Kahn we also carried out the procedure separately for power couples that are also dual earners, as a refinement of their initial analysis. The lower panel of Table 9.3 shows that the co-location effect for the Randstad is now 0.142, that for the intermediate zone −0.15 and for the periphery −0.118. These results suggest that the Randstad offers the best possibilities to solve the co-location problem.

9.3.4. Conclusions and comparison

The increasing concentration of power couples in the most urbanised part of the Netherlands is in line with the trend that has been observed by Costa and Kahn for the US. A replication of their analysis for Dutch data suggests that the co-location problem is an important determinant. In contrast to the United States, however, it seems to be the case that the effect of the co-location problem is counteracted by changes in the attractiveness of the Randstad in terms of urban amenities of general interest and amenities that are of special interest for singles. In other words, power couples would have become even more concentrated in the Randstad if urban amenities had remained as attractive as they were in the beginning of the 1980s. This leads to two further questions.

Why has the Randstad become less attractive apart from the possibilities it offers for solving the co-location problem? The data considered thus far cannot tell us the answer. Costa and Kahn relate the increased attractiveness of metropolitan areas in the US to changes in the urban wage premium. This suggests again that we should turn to labour market analysis for an explanation for what happens in the Netherlands. We do so in the next section.

9.4. Further empirical analysis

In this section we take a closer look at a number of labour and housing markets aspects that are relevant for understanding the location behaviour of dual earners in the Netherlands and, more specifically, power couples. We do this by means of an in-depth analysis of the data from the Housing Needs Survey of 1993. Concentrating on a single cross section offers us the opportunity to investigate a number of relevant topics in greater detail than would have been possible by examining a number of cross sections.

Before starting to discuss this issue, we should mention that the analyses that follow are not restricted to people aged 23–39, but concern in principle all respondents.

9.4.1. Urban wage premium

What do Dutch data tell us about the spatial aspect of the relation between education, household composition and wages? The Housing

Needs Survey of 1993 provides relevant information. We computed hourly wage rates for workers and regressed these on age, gender and education variables. The Housing Needs Survey does not indicate work experience, and the effects of this variable should therefore be expected to influence the estimates of other variables. When dealing with regional differences we have included dummies for the Randstad and the intermediate zone, treating the periphery as the reference region. Results are presented in Table 9.4.

Table 9.4. Hourly wage rates

Variable	(1)	(2)	(3)	(4)
Constant	*3.39*	*3.49*	*3.46*	*3.47*
Age	*0.0423*	*0.0400*	*0.0401*	*0.0401*
Age squared	*−0.000434*	*−0.000412*	*−0.000415*	*−0.000415*
Female	*−0.277*	*−0.219*	*−0.220*	*−0.218*
Only basic education	*−0.0965*	*−0.0995*	*−0.0973*	*−0.0973*
Higher vocational training	*0.203*	*0.209*	*0.207*	*0.208*
University	*0.308*	*0.329*	*0.376*	*0.375*
Female * higher vocational training	*0.0253*	−0.0180	−0.0185	−0.0179
Female * university	*0.0911*	0.0219	0.0201	0.0167
Single		*−0.207*	*−0.211*	*−0.210*
Dual earners		*−0.0474*	*−0.0495*	*−0.0560*
Power couple		−0.00570	−0.00609	0.00731
Female * single		*0.135*	*0.133*	*0.133*
Female * dual earners		−0.0560	−0.0557	−0.0568
Female * power couple		*0.0723*	*0.0719*	0.0411
Randstad			*0.0604*	*0.0499*
Intermediate zone			*0.0164*	0.0152
Randstad * university			*−0.0662*	*−0.0626*
Randstad * higher vocational training			−0.00852	−0.0103
Intermediate zone * university			*−0.0569*	*−0.0542*
Intermediate zone * higher vocational training			0.0155	0.0165
Randstad * dual earners				0.0153
Intermediate zone * dual earners				−0.0213
Randstad * power couple				−0.0213
Intermediate zone * power couple				−0.0119
Female * power Couples * Randstad				0.0509
Female * power couple * intermediate zone				0.0208
R^2	0.23	0.24	0.25	0.25

Dependent variable: ln of hourly wage rate. Italic figures indicate coefficients that are significant at $p = 0.05$. All equations are estimated on 47,403 observations using OLS.

The basic regression (1) relates natural logarithm of the hourly wage rate to age, gender and education. It does not yet include household characteristics or spatial variables. The natural logarithm of the wage rate increases with age, but at a decreasing rate. Female workers have a wage rate that is on average 28% lower than that of males. Those with only basic education earn 10% less, those with higher vocational or university training 20–30% more than the reference group with intermediate education. The positive effects of higher vocational or university training are somewhat higher for females than for males, although the difference is not sufficiently large to close the gender wage gap for the higher educated.

The next regression introduces household characteristics. Singles have lower wage rates than workers belonging to single earner households. If the single person is female, most of this effect disappears. Dual earners have on average a wage rate that is 4.7% lower than that of otherwise comparable single earners. Note that we have controlled for gender effects by including the product of the dummies for females and dual earners as a separate variable. The negative coefficient may indicate that both earners give up some of their career perspectives in order to solve the difficulties involved in finding a suitable arrangement of the other relevant aspects. The lower wage rate may therefore be an aspect of the co-location problem. Female earners in dual earner households are not significantly different from males in this respect. The indicator for power couples does not have a significant coefficient, but the female workers belonging to such a household have on average a higher wage rate than their otherwise comparable colleagues.

As a result of the introduction of household characteristics, the gender-specific effects of higher education become statistically insignificant. Female workers belonging to power couples earn 7% more than their otherwise comparable colleagues.[3] If we leave out the cross-effect of gender and power couple, the cross-effect of gender and university education becomes statistically significant again.

---

[3] See Bernasco (1994) for an analysis of the effects of resources of one partner on the career of the other.

In equation (3) we introduce dummies for workers located in the Randstad and in the intermediate zone. Estimation results show that otherwise equal workers who live in the Randstad earn on average a wage rate that is 6% higher than those living in the periphery, whereas workers who live in the intermediate zone earn 1.6% more. It is tempting to interpret these results as a confirmation of the existence of an urban wage premium caused by agglomeration effects of better matching or learning. However, the cross-effects of living in the Randstad or the intermediate zone and having a university education are negative, and make clear that these highly educated workers do not on average earn a higher wage outside the peripheral areas, and may indeed earn a somewhat lower wage rate in the intermediate zone. This surprising result does not show up for workers with a higher vocational training.

If the advantages of a large labour market that were highlighted in the models discussed above do indeed exist, one would expect that they are especially relevant for the university trained, who are often highly specialised. If the higher wage that is in general paid in the Randstad compensates for the disadvantages of working in an urban environment, such as congestion, traffic noise, high house prices, et cetera, one expects that university-trained workers have to be compensated as much as others. Whatever may be the explanation, the results cast doubt on the hypothesis that the urbanised part of the Netherlands is – apart from its potential to solve the co-location problem – an attractive place to live for university-trained people.

9.4.2. Dual earners and urban wage premium

If Frank (1978) and Costa and Kahn (2000) are right and there exists an important co-location problem, one would expect that dual earners in the peripheral parts of the country on average earn less than elsewhere. Even in a general equilibrium setting this effect should show up if a sufficiently large number of such couples preferred to stay in peripheral areas because of family ties that compensate for a somewhat lower wage, et cetera. In order to see whether such effects can be detected, we introduced a number of additional cross-effects into our estimating equation. Estimation results of equation (4) show that dual earners or power couples do not earn significantly more in the Randstad or in the intermediate zone than elsewhere in the country. Also, the wage rates of female workers in such households in the Randstad or the intermediate zone are not different from those living elsewhere. This shows that in the Netherlands workers belonging to power couples did not earn significantly more in the Randstad or the intermediate part of the country than in the periphery. Earlier regressions showed that female workers belonging to power couples have higher wage rates than their otherwise comparable colleagues, but this effect is no longer statistically significant in equation (4).

9.4.3. Conclusions so far

The results of the empirical analysis are at first sight somewhat puzzling. We could not identify an urban wage premium for university-educated workers, but workers with higher vocational training appear to earn somewhat higher wages in the Randstad and the intermediate zone.

However, this finding does not contradict the possibility that the denser labour market in these regions offers power couples possibilities to solve their co-location problem. Moreover, the absence of an urban wage premium for the Randstad, combined with the disadvantages of urban life such as congestion, noise, lack of recreational facilities, et cetera may well explain why the attractiveness of the Randstad in terms of urban amenities of general interest would in itself induce power couples to concentrate less in this region, as was found in the previous section.

The lack of direct evidence of a co-location problem in the form of lower wages for these workers in the periphery may be caused by the fact that many power couples avoid the realisation of such a situation by locating in the Randstad and intermediate zone. Moreover, power couples that succeed in finding a relatively bright career perspective for both partners will probably stay in the periphery, thereby introducing a selection effect that works in the opposite direction.

Our conclusion is therefore that the results of the empirical work reported in the second part of this section are consistent with those of the analysis carried out in Section 9.3. The effects of the co-location problem counteract those of other developments that would in themselves have induced power couples to locate outside the

Randstad and the intermediate zone, especially when they are educated at a university.

9.4.4. Dual earners and the housing market

9.4.4.1. General considerations

In this section we look at the location choice of power couples at a more detailed level by considering their housing consumption and location within urban areas.

The monocentric model provides a convenient starting point for studying the locational behaviour of households. We consider a basic version of the model in which all employment is located in the city centre (CBD) and all jobs are identical. Some workers belong to single earner households, others to dual earner households. The latter households have twice the income of single earner households, and two workers, instead of one, have to travel to the CBD each day. Unless the demand for residential land has a very high income elasticity, the model predicts that dual earners have steeper bid rent functions and will therefore concentrate around the city centre, leaving the suburban areas for the single earners. We will not elaborate this reasoning by means of a formal model. However, it would not be difficult, for instance, to introduce dual earners in this way into the model of Helsley and Strange (1990) discussed in Section 9.2 and show that in each city dual earners will locate close to the CBD, whereas single earners have larger home–work distances.[4]

In practice, things are not as simple as the highly stylised monocentric model suggests. A significant problem is that the predictions of the monocentric model with respect to the inner-city locations of various income groups are difficult to reconcile with empirical evidence about the location of high-income groups in urban areas in American cities. This was first shown by Wheaton (1977) and recently confirmed by Glaeser *et al.* (2000). Possible explanations for this anomaly are the durability of housing (older housing is concentrated in inner cities), and in the provision of special housing and (public) transport facilities in inner cities. Although European cities are considerably different from American

[4] See Freedman and Kern (1997) for a more extensive analysis and empirical work.

ones in a number of respects, the basic economic forces seem to be the same. In the Netherlands the largest cities have for long concentrated their housing policies on providing affordable housing for low-income groups, and this has resulted in a relatively large rental sector. At the same time, (national) tax facilities for owner-occupiers (including unlimited deductibility of mortgage rent from taxable income) are especially in favour of higher income groups. It is therefore not surprising that there is a close relationship between income and the probability that a household is owner-occupier. Moreover, the share of the rental sector in the dwelling stock is highest in the largest cities and, more generally, is increasing in city size. High-income households (to which most power couples belong), tend to be owner-occupiers and live outside the largest cities.

It is therefore possible that the empirical relationship between income and residential location within urban areas in the Netherlands is the opposite of what is predicted by the monocentric model, as is the case for the American cities. High-income households have to choose between two unattractive combinations: short commutes combined with old inner city housing and long commutes combined with modern housing.

Although Americans seem to have opted en masse for the latter possibility, it must not be forgotten that there is more than one difference between the monocentric model and reality. A second important discrepancy results from the fact that the monocentric model assumes that all employment is concentrated in the city centre, whereas in reality jobs are much more evenly spread over space. This results in better possibilities for combining a residential location outside the core urban area with a commute of limited length. The monocentric model is often thought of as fitting the American cities of some decades ago better than it ever fitted European cities. When the Randstad is regarded as a single urban area, it clearly has many employment centres and the large cities in this area (such as Amsterdam and Rotterdam) have more than one employment centre of their own. This means that it may have been possible for high-income households to realise the best of both worlds by realising a combination of modern owner occupied housing and limited home–work distances. In the remainder of this section we will investigate this issue in some detail.

On the other hand, it must also be recognised that dispersed unemployment increases the problems that dual earner households have in finding a satisfactory combination of employment and residential location. If the jobs of both workers are located at the same site, there is effectively only one job location, and the residential location problem is essentially the same as that of a single worker households. When the locations of the two jobs are different, as is usually the case in practice, the problem becomes more difficult to solve with limited commutes for both workers and this aspect of the co-location problem may well result in longer average commutes for both workers. This may especially be the case for workers belonging to power couples, who both face a thin labour market.

In what follows we have as a rule used the workers of single earner households as our reference group and included dummies for dual earners, singles and power couples and power singles. With respect to education we introduce dummies for higher vocational and university training and for basic education only, treating the group of workers with intermediate education as our reference group.

9.4.4.2. Commuting distances

We start with an examination of the commuting distance. On the basis of search theory it may be argued that this is a random variable and that the parameters of its distribution are dependent upon the spatial characteristics of relevant housing and employment opportunities in the area concerned.[5] These opportunities are in large part related to the worker's education, age and gender. Column (1) of Table 9.5 shows estimation results for a simple regression equation in which the natural logarithm of the commute, measured in kilometres is related to these basic worker characteristics. Males have longer commutes than females, older workers travel longer than younger workers and the higher educated travel longer than the low educated. Explanations for these well-known phenomena are that household responsibilities concentrated around the beginning and the end of the working day are often more pressing for females than for males, that older people are less mobile on the labour market

[5] See, for instance, Rouwendal and Rietveld (1994).

Table 9.5. Commuting distances

Variable	(1)	(2)	(3)	(4)
Constant	*0.182*	*0.258*	*0.213*	*0.424*
Age	*−0.00783*	*−0.00753*	*−0.00756*	*−0.00793*
Female	*−0.200*	*−0.270*	*−0.268*	*−0.241*
Only basic education	*−0.129*	*−0.130*	*−0.127*	*−0.136*
Higher vocational training	*0.137*	*0.134*	*0.0851*	*0.0873*
University	*0.0992*	*0.106*	−0.105	−0.0794
Female * higher vocational training	0.0472	−0.00137	−0.00803	−0.00489
Female * university	*0.309*	*0.238*	*0.234*	*0.245*
Ln(wage rate)	*0.383*	*0.371*	*0.362*	*0.350*
Working hours/40	*0.654*	*0.641*	*0.633*	*0.626*
Single		*−0.173*	*−0.189*	*−0.152*
Dual earners		0.0107	0.00529	0.0897
Power couple		0.0179	0.0140	0.0231
Female * single		0.118	0.111	0.0874
Female * dual earners		0.0680	0.0608	0.0276
Female * power couple		*0.106*	*0.169*	*0.171*
Number of persons 0–5		−0.00153	−0.00348	−0.00802
Number of persons 6–11		−0.0158	−0.0143	−0.0195
Number of persons 12–17		*−0.0341*	*−0.0328*	*−0.0380*
Female * number of persons 0–5		−0.0472	−0.0488	−0.0494
Female * number of persons 6–11		*−0.0855*	*−0.0886*	*−0.0888*
Female * number of persons 12–17		−0.0192	−0.0208	−0.0211
Randstad			*0.161*	*0.220*
Intermediate zone			*0.0947*	*0.121*
Randstad * university			*0.249*	*0.255*
Intermediate zone * university			*0.216*	*0.210*
Randstad * higher vocational training			0.0688	0.0656
Intermediate zone * higher vocational training			0.0604	0.0637
Female * Randstad * power couple			−0.129	−0.126
Female * intermediate zone * power couple			0.00422	0.000279
Size of municipality (0–5000 = reference)				
Inhabitants 5000–10,000				−0.000710
Inhabitants 10,000–20,000				−0.0809
Inhabitants 20,000–50,000				*−0.209*
Inhabitants 50,000–100,000				*−0.219*
Inhabitants 100,000–150,000				*−0.250*
Inhabitants 150,000–250,000				*−0.328*
At least 250,000				*−0.349*
R^2	0.089	0.092	0.096	0.101

Italic figures indicate coefficients that are significant at $p = 0.05$.

and therefore can easier adjust their residential location to their current job and that the relevant segments of the labour and housing markets for the higher educated are less thick than those relevant for other groups of workers. We added the hourly wage rate and the ratio between the actual number of working hours and those related to a fulltime job and found that part-time workers and low paid workers have relatively short commutes.

In equation (2) a number of household characteristics were added. Estimation results show that workers belonging to dual earner households do not on average have significantly longer commutes than other workers. Belonging to a power couple only makes a difference for females: they realise longer commutes. Workers have shorter commutes if they have children aged 12–17, and there is an additional effect of children between 6 and 11 on the commute of their mother.[6]

For the purposes of the present chapter, the most important aspect of equation (2) is that it provides only weak evidence of longer commutes for dual earners and, more specifically, power couples. These workers seem to be able to solve their co-location problem in a satisfactory way with commutes that are on average close to those of other workers. Before turning to the question how they are able to realise this outcome, we first incorporate the spatial dimension into the analysis.

Equation (3) indicates that the commutes of workers residing in the Randstad are in general longer than those in the intermediate zone, which are in turn longer than those in the peripheral parts of the Netherlands. The longer commutes of university-trained workers are restricted to the Randstad and the intermediate zone. The gender-specific effect of belonging to a power couple is now larger and does not appear to be spatially differentiated.

In order to investigate the effect of living in a particular urban or rural municipality on the commute, equation (4) adds a number of dummies for the size (measured as the number of inhabitants) of the municipality of residence. The shortest commutes are realised in the largest municipalities, and living in a rural area is clearly expensive

[6] It is, of course, likely that there will also be effects on labour force participation and number of hours worked.

in terms of commuting times. However, controlling for these effects does not change the estimates for the effects of dual earners or power couples substantially.

9.4.4.3. Location choice

If workers belonging to dual earner households do not realise longer commutes even though they have to solve a more difficult location problem, they will probably have used means that were not available to others in order to reach this goal. One possibility is that they have different preferences with respect to location. Another is that they use their higher income so as to outbid others from the locations that are especially attractive to them. Although the last regression discussed in the previous subsection shows that the former explanation cannot provide a complete explanation, it is nevertheless interesting to consider location choices of dual earners in some more detail. In the next subsection we discuss the other explanation.

Table 9.6 shows the results of a simple logit model for the distribution of households over the Randstad, the intermediate zone and the periphery. Households with a higher income are more likely to be found in the Randstad and intermediate zone. However, dual earners are in general underrepresented in the intermediate zone. Power couples show a different pattern: they are more often located in the Randstad. This is, of course, in line with the results of Section 9.3. One other variable in the model is 'power single' which indicates a single person household with higher vocational training or university education. These are more often living in the

Table 9.6. ***Distribution of households over Randstad, intermediate zone and periphery***

Variable	Randstad	Intermediate zone
Constant	*0.124*	*0.281*
Income (× 1000)	*0.000777*	*0.000436*
Age head household	*−0.00189*	*−0.00300*
Dual earners	−0.0701	*−0.0547*
Power couple	*0.198*	0.0755
Single	*0.357*	−0.0320
Power single	*0.547*	*0.296*
Loglikelihood	−58224	

Italic figures indicate coefficients that are significant at $p = 0.05$.

intermediate zone and the Randstad than other household types, after controlling for the other variables.

Table 9.7 reports results of a multinomial logit model referring to the choice of the size of the municipality. It shows that households with an older head and a higher income are less likely to be found in big cities. However, dual earners deviate from this trend and are over-represented in medium sized and large cities. Power couples are even more likely to be found in larger cities. Singles are also concentrated (at least in a relative sense) in the big cities, and this is even more true for power singles. This is consistent with Peri's argument that higher educated workers in general tend to concentrate in cities because of externalities associated with learning.

The results just discussed by and large confirm the analysis of Section 9.3, which showed that power couples tend to concentrate in metropolitan areas. On a national scale they are more likely to be found in the intermediate zone and the Randstad. Moreover, they tend to live often in medium and large cities and in this respect their location pattern resembles that of singles. However, as we have seen above, they do not realise the shorter commutes of singles, and the female workers in power couples realise longer commutes. This suggests that the dual earners are able to realise commutes that are on average approximately of the same length as those of single earners because their location pattern is different from that group. It is closer to that of singles. This location pattern facilitates shorter commutes than those realised by dual earners in general, but the dual

Table 9.7. Distribution of households over municipalities of different size

Variable	5000–10,000	10,000–20,000	20,000–50,000	50,000–100,000	100,000–150,000	150,000–250,000	At least 250,000
Constant	*1.87*	*2.85*	*3.01*	*2.77*	*2.28*	*2.10*	*2.31*
Income (× 1000)	−0.00464	−0.000145	−0.00117	*−0.00328*	*−0.00531*	*−0.0115*	*−0.0942*
Age head household	−0.00437	*−0.00596*	*−0.00828*	*−0.0135*	*−0.0159*	*−0.0213*	*−0.0156*
Dual earners	0.112	0.0899	0.144	*0.336*	*0.375*	*0.460*	*0.254*
Power couple	0.178	0.151	0.296	*0.320*	*0.780*	*1.04*	*0.854*
Single	−0.0408	0.0182	*0.230*	*0.526*	*0.736*	*1.04*	*1.01*
Power single	−0.108	−0.240	0.157	0.419	*0.857*	*1.03*	*0.957*

The table reports estimates of a multinomial logit model. Municipalities of the smallest size (0–5000 inhabitant) were taken as the reference class. Loglikelihood: − 100,056.26. Italic figures indicate coefficients that are significant at $p = 0.05$.

earners in cities actually do not realise these shorter commutes. The likely explanation is, of course, that the co-location problem prevents them from doing this.

9.4.4.4. Housing demand

We now turn to the second potential explanation mentioned above, according to which the workers belonging to dual earner households realise their relatively short commutes because of their higher income. This means, they outbid others in the locations that are especially attractive to them. This is only possible in the owner-occupied segment of the market, since rents are still by and large regulated in the Netherlands.[7] We would therefore expect that dual earners and power couples are over-represented in the owner-occupied part of the market and reside in relatively expensive housing (ceteris paribus).

Table 9.8 shows estimation results of a logit model for tenure choice. Households with a higher income are more likely to own. The effect of age is u-shaped, implying that old and young heads of households are less likely to be owner-occupiers. A set of spatial dummy variables is included in the model in an attempt to control for the peculiarities of the local housing market. They show that owner-occupation is less common in the intermediate zone and in larger municipalities. The results do not completely confirm the conjecture that dual earners are more likely to own. When controlling for income and age of the head of the household, dual earners are actually *less* likely to own a house than otherwise comparable single earners. Power couples are different, and the positive coefficient estimated for this group of households more than compensates for the negative coefficient for dual earners.

The negative coefficient for dual earners may be related to the fact that the income of the female worker was until recently not regarded as part of the household's permanent income. Until the 1980s many women in the Netherlands stopped working or switched to a part-time job when children were born. Also, the maximum possible mortgage

---

[7] Rent control in the Netherlands was in the early 1990s somewhat less restrictive than it used to be, but the market still used queuing as an important allocation mechanism. Only in a small part of the market (mostly containing expensive urban housing) there is a free market where prices adjust to equilibrate supply and demand.

Table 9.8. Logit model for the choice between rental and owner-occupied housing

Variable	Estimate
Constant	*−3.42*
Income	*0.0377*
Age head household	*0.122*
Age head household squared	*−0.00114*
Dual earners	*−0.263*
Power couple	*0.376*
Single	*−0.683*
Power single	*0.330*
Randstad	*0.0521*
Intermediate zone	*−0.170*
Size of municipality (0–5000 = reference)	
5000–10,000 inhabitants	*−0.175*
10,000–20,000	*−0.323*
20,000–50,000	*−0.733*
50,000–100,000	*−1.07*
100,000–150,000	*−1.26*
150,000–250,000	*−1.37*
At least 250,000	*−2.18*
Loglikelihood	−29,785

The dependent variable is a dummy for owner-occupied housing. Italic figures indicate coefficients that are significant at $p = 0.05$.

level was determined on the basis of the ratio between annual mortgage payments and income of the main breadwinner. The income of the female worker was not included in this computation, implying that dual earner households could borrow less than a single earner household with a comparable income. As a consequence, the effect of the (on average) higher income of dual on housing demand was small. This practice for determining a ceiling on the mortgage loan was abandoned in the beginning of the 1990s, and our estimation results can be interpreted as suggesting that power couples were the first to use the increased borrowing capacity that resulted.

Next we turn to the value of the owner-occupied house. Dual earner households have two incomes and per capita income in such households will therefore be higher than in otherwise comparable single earner households. Since housing is a normal good, this suggests that the demand for housing services of dual earners will be higher than that of other household types – that is, in the absence of credit constraints like the one discussed in the previous paragraph. The effect on housing demand may not be restricted to that of a higher income. We have already seen that dual earners are more often located in medium sized and large cities than single earners. House prices are typically higher in these areas, and this may have an additional positive impact on the value of houses occupied by dual earners. It has also been conjectured above that the relatively short commutes of dual earners are facilitated by their location choices. Urban economic models suggest that the mechanism underlying this process will be that dual earners outbid other potential buyers of their most preferred houses. If this happens, it may have an additional positive impact on the volume of housing services consumed. The credit constraint on dual earner households will probably have suppressed this effect, but perhaps less so if the dual earners are power couples.

Regression (1), reported in Table 9.9, shows that the value of owner-occupied housing is positively related to income and the age of the head of the household. Dual earners have, all else equal, somewhat less expensive housing, probably for the same reasons that made them less likely to be homeowners. Power couples differ

Table 9.9. The value of owner-occupied houses

Variable	(1)	(2)
Constant	*2.09*	*2.55*
Income (× 1000)	*0.479*	*0.0746*
Age head household	*0.00492*	*0.00435*
Dual earner	*−0.0660*	*−0.0297*
Power couple	*0.203*	*0.102*
Single	*−0.302*	*−0.0366*
Power single	*0.207*	*0.0627*
Randstad	*0.214*	*0.205*
Intermediate zone	*0.234*	*0.253*
Size of municipality (0–5000 = reference)		
5000–10,000	*−0.0465*	−0.0170
10,000–20,000	*−0.0895*	−0.0370
20,000–50,000	*−0.1114*	0.00996
50,000–100,000	*−0.2420*	*−0.0591*
100,000–150,000	*−0.1992*	0.0213
150,000–250,000	*−0.3413*	*−0.0808*
At least 250,000	*−0.5193*	*−0.0694*
λ		*−0.596*
R^2	0.1672	

Dependent variable is ln(self-reported value of the house). Italic figures indicate coefficients that are significant at $p = 0.05$.

here also and have more expensive housing. Singles have cheaper housing, although this effect is smaller if they are higher educated. Regression (2) refers to the same equation, but now with the Heckman correction for selectivity added.[8] Incorporation of this additional variable has a substantial impact on most of the included variables, but even though the power couple effect decreases in size, it remains large and highly statistically significant.

For our purposes, the main result is this large and positive coefficient for power couples, which outweighs the negative coefficient estimated for dual earners. The estimation results therefore provide evidence that power couples are paying even more for their owner occupied house than could be expected on the basis of their high incomes. This suggests that these households indeed outbid others from their most preferred houses, and that the implied commutes play an important role in their willingness to pay for a particular house.

9.5. Conclusions

This chapter focused attention on location choices of dual earners and of so-called power couples in particular. It appears from the analysis that the co-location problem is important for these households, both in choosing a particular region as their residence and in choosing a specific location within that region. Our replication of Costa and Kahn's analysis for the Netherlands indicates that power couples prefer the Randstad even though this area became less attractive for other reasons. Further analysis revealed that one of these other reasons is that wage rates for university-trained people are not higher in the Randstad or the intermediate zone than in the periphery, even though housing is considerably more expensive in the western part of the Netherlands, congestion problems are concentrated there, et cetera.

[8] The Heckman correction takes into account the fact that households who decide to become owner-occupiers may also be households that differ from the average household in their housing demand characteristics for reasons that are unobserved by the researcher. If this occurs, estimation of the housing demand equation with OLS gives biased results. The Heckman correction is discussed in any modern econometrics textbook.

Within their region of residence, power couples are more often located in the medium sized and large cities, which contributes to their on average relatively short commutes. However, even if we control for region and the size of the municipality, commuting distances of workers belonging to dual earner households and power couples are on average not different from other workers. There is one exception: females in power couples have a longer commute.

It seems therefore that dual earner households, and more specifically, power couples with both partners employed, are able to solve, or at least to mitigate, this aspect of their co-location problem by paying considerable attention to the choice of their residential location relative to the job locations of the two workers. For power couples, this conjecture was confirmed by our finding that they are more likely to be owner-occupiers and if they are, live in more expensive housing than could be expected on the basis of their other characteristics.

In summary, it may be said that we have found a consistent picture of the locational behaviour of power couples:

- they locate more often in the densely populated urbanised western part of the country than could be expected on the basis of their other characteristics;
- they do so, even though there is no urban wage premium for university educated people in this part of the country;
- they locate more often in medium and large sized cities than otherwise comparable households;
- they manage to realise housing–employment arrangements that imply commutes that are somewhat longer for the female workers, but not for the males;
- they are more often owner-occupiers than other dual earner households and seem to outbid others from the segments of the housing market they prefer most.

These findings suggest that the increasing importance of dual earner households and, more specifically, power couples with both partners employed has important consequences for the distribution of households over space, both at the national and at the regional scale in the Netherlands. At the national scale the significance of dense labour markets for solving the co-location problem is important for the Randstad because it counteracts other forces that make it less attractive as a residential location. At the regional scale the trend towards suburbanisation seems to be mitigated by relative attractiveness of larger municipalities for realising suitable employment–residential arrangements for dual earner households.

It may also be noticed that in several stages of this research we found power couples (with both partners employed) to behave substantially different from other dual earner households. It is unclear whether this effect is related only to the higher education, or perhaps also to their role as trendsetters. Cross section analysis is unable to differentiate between these two possible explanations. However, it is likely that the effect of the relaxed credit constraint on dual earner households will have spread over the whole group of these households in the years following 1993.

The increased concentration of power couples in the Randstad and the intermediate zone can be interpreted as part of a self-reinforcing process of spatial concentration. The brain drain makes the periphery a less attractive region for starting new business activities, and the inflow of highly educated workers makes the Randstad and the intermediate zone more attractive in this respect. As a result, local labour markets in the periphery become less dense, and those in the remainder of the country even more dense, accelerating the outflow of power couples. On the other hand, it must be observed that the results presented above also suggest that there exist important forces that make the Randstad less attractive as a region of residence. If these forces become stronger, the chain of cumulative causality may be broken. The probable result will be that the intermediate zone will gain, whereas the relative position of the periphery may also improve.

References

Bernasco, W. (1994), *Coupled Careers: The Effect of Spouse's Resources on Success at Work*. PhD Thesis, Utrecht University.

Costa, D.L. and M.E. Kahn (2000), "Power couples: changes in the locational choice of the college educated, 1940–1990", *Quarterly Journal of Economics*, Vol. 115, pp. 1287–1315.

Frank, R.H. (1978), "Family location constraints and the geographic distribution of female professionals", *Journal of Political Economy*, Vol. 86, pp. 117–130.

Freedman, O. and C.R. Kern (1997), "A model of workplace and residence choice in two-worker households", *Regional Science and Urban Economics*, Vol. 27, pp. 241–260.

Glaeser, E.L. (1999), "Learning in cities", *Journal of Urban Economics*, Vol. 46, pp. 254–277.

Glaeser, E.L. and D.C. Maré (2001), "Cities and skills", *Journal of Labor Economics*, Vol. 19, pp. 316–342.

Glaeser, E.L., M.E. Kahn and J. Rappaport (2000), *Why do the Poor Live in Cities?, Discussion paper 1891*, Harvard Institute of Economic Research.

Helsley, R.W. and W.C. Strange (1990), "Matching and agglomeration economies in a system of cities", *Regional Science and Urban Economics*, Vol. 20, pp. 189–212.

Iranzo, S. (2003), *Wages and Diversity of Skills in Cities, Working paper*, Davis: Department of Economics, University of California.

Kim, S. (1989), "Labor specialization and the extent of the market", *Journal of Political Economy*, Vol. 97, pp. 692–705.

Peri, G. (2002), "Young workers, learning and agglomerations", *Journal of Urban Economics*, Vol. 52, pp. 582–607.

Petrongolo, B. and C. Pissarides (2001), "Looking into the black box: a survey of the matching function", *Journal of Economic Literature*, Vol. 39, pp. 390–431.

Petrongolo, B. and C. Pissarides (2002), *Scale Effects in Markets with Search, Working paper*, Centre for Economic Performance, London School of Economics and CEPR.

Rouwendal, J. and P. Rietveld (1994), "Changes in the commuting distances of Dutch households", *Urban Studies*, Vol. 31, pp. 1545–1557.

Teulings, C.N. and P. Gautier (2000), *The Right Man for the Job, Discussion paper 038/3*, Tinbergen Institute.

Teulings, C.N. and P. Gautier (2002), *Search and the City, Discussion paper 061/3*, Tinbergen Institute.

Wheaton, W. (1977), "Income and urban residence: an analysis of consumer demand for location", *American Economic Review*, Vol. 67, pp. 620–631.

Urban Dynamics and Growth, CEA, vol. 266
Roberta Capello and Peter Nijkamp (editors)

CHAPTER 10

Land Use Regulation and Its Impact on Welfare

Stephen Sheppard

Department of Economics, Williams College, Williamstown, MA 01267, USA

Abstract

The long history and widespread application of land use regulation implies that these policies provide benefits for someone. It is not clear, however, that they represent a Pareto improvement or even generate positive net benefits in an economy. On the one hand, they may provide a natural response to externalities by reducing exposure to and generation of external costs. Alternatively, they may provide a mechanism for restricting the supply of space to extract surplus from potential residents. This chapter considers both theoretical and empirical approaches to the welfare analysis of land use regulation.

Keywords: land use regulation, planning, zoning, welfare economics

JEL classifications: R520, R130, R140

10.1. Introduction

The role of good neighbors in affecting welfare has been understood for as long as there have been cities (and probably before that). The classical Greek poet Hesiod noted that a "bad neighbor is as great a calamity as a good one is a great advantage", and no doubt there are earlier recorded examples of this fundamental insight. Efforts to arrange human settlements and cities so as to decrease the probability of being confronted with a bad neighbor are surely just as old. The regulation of who is allowed to build, in what fashion or in what location has been undertaken by most urban civilizations in one way or another. The citizens of Rome were required to leave a certain amount of space between their buildings and the street. London in the 16th century (as discussed by Evans (1999)) attempted to prohibit construction of any new buildings on unbuilt land further than three miles from the gates of the city. Thus while specific types of land use regulation, such as 'zoning', are clearly 20th century creations, the regulation of land use in some form is very old and probably intrinsic to human settlements.

Land use regulations are costly to enforce, and therefore must generate benefits for someone. It is less obvious that they generate benefits for all persons in the economy, or even a careful accounting of those who gain and those who lose would indicate that the regulations generate positive net benefits. In general, these policies are justified as a response to some market failure, and there may well be externalities that are controlled by the regulations so that they increase social welfare. Alternatively, they may be a way for existing owners of developed property to control supply and increase their wealth by increasing the price of these properties. The central goal of this chapter is to sort out these arguments, to present a theoretical model useful for evaluating them, and to consider the limited amount of empirical evidence that has been assembled regarding these important policies.

In order to understand these regulations and review the analyses that have been provided we must identify the variety of motivations that might lead to imposition of land use regulation. There are at least six reasons that can be identified for regulating land use. The first (and simplest) is the desire to give as many residents as possible 'good neighbors' by separating activities that are likely to impose external costs on one another. This makes particular sense when the external costs are such that they tend to dissipate over space so that spatial separation actually does reduce exposure to the externality. The second rationale is closely related, but not amenable to solution via spatial separation: to regulate land development and building when such development itself imposes external costs. This arises in the context of unsafe or unsound construction, building that casts shadows or limits a view, and building that reduces access to open space.

If the first two are the traditional reasons given by practitioners and planners for zoning, the third was the source of some early critiques of regulation. Land use may be regulated in an effort to exclude households whose income or preferences for public goods differ substantially from the existing community. This may be the result of a desire to exclude households who are likely to impose a net burden on the community (as in so-called *fiscal zoning*). Alternatively, it may result from the belief that the households generate a negative externality directly, or might alter the political composition of the community and thus the decisions about the types and quantities of public goods to provide for the community (*exclusionary and/or public choice zoning*).

A fourth (and frequently offered) explanation for land use regulation is that it provides an indirect response or mechanism for regulating other activities that are socially inefficient. Thus, land use regulations might be justified as a method of responding to the pollution or road congestion that results from commuting in a setting where charging congestion tolls or pollution taxes is for some reason not feasible. Congestion of other infrastructure such as schools or water systems might also fall within this category. This perspective suggests that land use regulation might constitute a second-best response to the inefficient activity that cannot be regulated directly.

A fifth explanation is that land use regulation provides a mechanism for compelling households to generate positive externalities or provide local public goods without raising revenues and purchasing these services directly. The regulations may compel residents to have large amounts of land (minimum lot size) the presence of which enhances the value of neighboring properties, or compel owners of land at the urban periphery to keep the land in agricultural use as a way of providing open space at the edge of the city. The residents could always tax themselves and purchase these arrangements via the market, but find land use regulation a more attractive option.

Sixth, and least attractive from a social welfare perspective, land use regulation might be used to create monopoly rents by restricting the supply of land for private use and thereby decreasing the availability of housing in order to increase values for home owners, or to restrict commercial space or activities that might compete with existing enterprises. Many critics (and much of the economics literature) focus attention on this explanation for the widespread use of land use regulation.

Thus, land use regulation may be adopted for reasons that are likely to enhance social welfare (such as reducing external diseconomies) and for reasons that might be expected to reduce social welfare (such as creating or defending market power). Because of very limited data on the extent and nature of land use regulations, it has been difficult to make an empirical determination of which motivation for land use regulation is dominant. As a result, much of the analysis has been within the context of theoretical models. In the sections below we first review some of these analyses that have sought to understand the impacts of land use regulation within the context of purely theoretical models. Following this, we present in some detail the core model of land use control that has been developed and adapted by Brueckner (1996) and others, along with a relatively simple extension of the model that helps us to understand the circumstances under which land use regulation might be welfare enhancing. After this, we discuss the limited efforts at empirical evaluation of land use regulation and its impacts. The final section offers some concluding remarks and directions for future work.

10.2. Evolution of the literature

While land use regulation is widespread, and while it has generated a reasonable literature devoted to its evaluation, it has generated fewer comprehensive studies than many other types of regulation. For example, regulation of telecommunications generates two or three times the number of published papers than regulation of land and housing markets, despite the fact that telecommunication services represent a far smaller share of total household expenditures than housing, and even (in many markets) less than the raw land on which housing is located. Why then are there relatively fewer analyses of the impact of land use regulation? One possible explanation is that while land use regulation can generate considerable monopoly rents, the rents and costs are dissipated over a much wider group of recipients than, for example, telecommunications regulation. As a result, there are fewer large agents supporting research in the area. An alternative suggestion is the lack of widely available data for testing and analysis of the impacts. While there are considerable data on property values, there are few national (and no international) data resources that describe the extent of land use regulation and enforcement. Lacking data, economists have begun with the construction of theoretical models of land use regulation in an attempt to identify circumstances in which the impacts of these constraints would be clearly welfare enhancing or welfare reducing.

10.2.1. Theoretical analysis of the efficiency of land use regulation

Theoretical insights concerning the expected impact of land use regulation on house prices have a relatively long history, and many of these have been derived as relatively straightforward applications of welfare theory to land markets. Bailey (1959) notes that the presence of externalities provides a central justification for the use of land use regulation. Most of these analyses have been based on incomplete and partial equilibrium models that provide insights but do not consider the impacts on both the residents of the community and the households potentially excluded by the regulations. One of the first comprehensive theoretical models devoted to this purpose was developed by Frankena and Scheffman (1981), whose analysis introduces several features that have been retained in many subsequent studies: identification of separate groups of consumers (current residents and potential residents) whose land consumption is fixed at one unit each without regard to the price of land. They consider urban areas that are 'small' in the sense that planning policies chosen within the city do not affect the welfare levels outside the city. The cities they study are open in that total population responds to economic conditions and policies within the city, but not the overall utility level of residents. They explore the central link between tax and local public expenditure policies and the planning policy that would be chosen by a majority of residents, providing a model for fiscal zoning. They reject the idea that overly restrictive zoning would be chosen in an environment with efficient taxation, and therefore argue that when inefficiently restrictive regulations arise they are attributable to the system of public finance rather than to land use regulation per se.

Cooley and LaCivita (1982) respond to this analysis with a critique of Frankena and Scheffman's (1981) model that excludes congestion and externalities that might be important justifications for land use regulation, and also assumes that land use regulations in an urban area do not impact the utility level of potential residents outside the city. They analyze a model with externalities and with congestion of local public goods, as well as tax-induced distortions. They identify conditions under which residents will find growth controls welfare enhancing. They do not present a comprehensive welfare analysis that includes both the residents and nonresidents. Like Frankena and Scheffman, they suggest that land use regulation may be justified in response to distortions arising from local public expenditures and taxation.

Epple *et al.* (1988) take the analysis of the tension between community 'insiders' and 'outsiders' one step further, presenting a model that explores the incentives of early residents in a community to regulate and control the conditions of entry for future residents. As with the other models reviewed, their analysis draws attention to the fiscal incentives for regulation of development. Land plays almost no role in the model. Despite this, the analysis is interesting in that it presents a context that draws attention to the difference in decisions that would be made by a profit-maximizing developer of a community, who will choose an efficient level of regulation (that will present some constraint on the development), and the decisions made by existing residents of a community, who will choose an inefficiently restrictive level of regulation.

Brueckner (1990) brings attention back to the central dilemma of the impact of land use regulations: are they welfare enhancing because they increase amenities (or reduce congestion) for residents or are they welfare reducing because they restrict supply in an attempt to generate monopoly rents? He works within the context of a simple model that abstracts from the impact of local public finance problems. The analysis focuses on an 'open city' model in which the welfare levels of land-consuming residents is exogenous, so that land use regulation cannot affect resident welfare. It does, however, affect aggregate land values and these are important to the utility of absentee landlords. His analysis shows that growth controls may raise the value of undeveloped land, and in this sense may be 'welfare enhancing'. The result is obtained under the assumption (similar to Frankena and Scheffman (1981)) that decisions made in the city have no impact on welfare levels in the urban system as a whole.

Engle *et al.* (1992) take up the same question regarding the dual nature of land use regulation in the context of an urban system with two cities. Only one of the cities can implement land use regulation. While they assume that the decisions made in the city that actively regulates the land market will not affect the welfare levels in the other city, this dual urban system sets the stage for future models that consider a more general range of impacts. The analysis retains the assumptions of constant land consumption, and introduces congestion and development externalities that provide a rationale for land use regulation. They show that in the absence of externalities, growth controls unambiguously reduce welfare. In the presence of externalities, they show that the results can go either way.

Sakashita (1995) responds to Engle *et al.* (1992) and the earlier Brueckner (1990) analysis with the concern that both analyses relied on aggregate land values to measure welfare effects. This is required because both were essentially open city models with the welfare level of residents exogenous. In some sense this was desirable because it allowed endogenous levels of population in the model. In other words, there was at least potential population growth to which land use regulations or growth controls might respond. On the other hand, it seems unsatisfactory that the only agents in the model whose welfare seemed to actually be affected by land use regulation were absentee landowners. Sakashita's answer to this problem was to generalize the Engle, Navarro and Carson two-city framework in an important way. His analysis takes the entire urban system (of two cities) as closed, with fixed total population.[1] This allows the utility level of residents to be determined endogenously in the model. One of the cities is active in that it can impose land use regulation, while the other is passive and does not regulate land use. Migration between the two cities in the system causes utility levels to equalize between them, so that decisions made within the active city can affect the welfare level in the passive city. He establishes that growth control policies

---

[1] It is worth noting here two other contributions that examine the comparative static impacts of land use regulation in the context of closed city models. Sheppard (1988) considers constraints in a single monocentric model with several population classes. Pasha (1992) considers the impact in a single monocentric city with suburban and central location, and also examines the impact of a 'fully closed' structure in which land rents contribute to the incomes of urban residents.

are justified when there are externalities resulting from population growth.

Two important papers made use of this closed system of cities to examine interactions between urban policy makers in different cities as they impose growth controls in reaction to policies chosen by other cities in the system. Brueckner (1995) and Helsley and Strange (1995) both consider strategic interactions between such cities, deriving Nash equilibria for the noncooperative game that results. Both obtain most of their results in a context where land consumption per household is constant. Growth controls are adopted only when they serve the interests of the particular adopting community. In the closed urban system, growth controls may preserve (or even enhance) amenities and still reduce welfare because of the impact on house prices. Helsley and Strange show that such welfare-reducing regulations can be adopted as equilibrium strategies even by profit-maximizing developers (who are essentially absentee landowners). Brueckner's analysis draws particular attention to the conflicting interests of 'renters' who reside in the communities but do not own land (nor benefit from its appreciation) and landowners who are not resident and directly benefit from land value appreciation. His analysis shows that growth controls help landowners and harm renters, and therefore will not be implemented unless landowners have control of policy. He further shows that the strategic interactions can result in change in growth control policies throughout the urban system in response to changes in characteristics of a single city.

Brueckner and Lai (1996) set out to extend this model of strategic interaction by bringing landowners into the community. The analyses discussed above show, in general, that landowners will want to impose socially inefficient levels of land use regulation. Whether this will remain true if landowners are resident in the community is a central concern to Brueckner and Lai (in addition to the greater realism of the model when landowners reside within the community). They show that landowners who are resident in the community have a reduced desire for land use regulation, although land use regulation (if chosen by landowners) remains inefficiently restrictive. Their analysis does not incorporate any externality or congestion that might justify the controls. Such a model is presented in Brueckner (1996). His model (reviewed in detail below) shows that even with an urban development externality growth controls chosen by landowners are, in general, inefficient.

Sasaki (1998) follows the Brueckner and Lai (1996) model and extends it to include congestion externalities and public goods production in a way that is more general but somewhat less transparent than the extension presented in Brueckner (1996). He obtains results on the changes in optimal city size and growth controls under a variety of alternative assumptions. A central result is that if congestion externalities affect only landowners, then their presence results in adoption of more restrictive land use regulation. If congestions affect all residents the impact is ambiguous. He also introduces production into the model to examine the impact that growth controls might have in restricting available labor supply. The labor market considerations would be expected to lead landowners to select less restrictive growth controls, and in Sasaki's model they do. Lai and Yang (2002) respond to Sasaki's analysis suggesting that the impact of commuting costs may result in growth controls being selected that may be either more or less restrictive than those chosen in the absence of production considerations.

10.2.2. Other possible effects

The logical development of models presented above has focused on the impacts of land use regulation on the price of land (or housing), the income of landowners, and in some cases on the availability of amenities and public goods, imposition of tax burdens, and reduction of congestion or other externalities. These combine to produce a variety of welfare effects that are the central interest of economists in evaluating land use controls.

There is a smaller literature that focuses on how land use regulation affects the risk of land ownership and property development. Land use regulations are the products of human institutions and are therefore subject to variability in both design and enforcement. This variability alters the risk of land ownership or risks associated with development and supply of housing. The change in risk may affect not only the value of land but also the timing of land development. As such, changes in risk resulting from land use regulation can be an additional source of welfare impacts. While we do not explore the models in detail here, regulation-induced changes in the risk of development may partially explain empirical results obtained linking land use regulation and land values.

The analysis presented in Mayo and Sheppard (1991, 2001) refers to the problem as 'stochastic development control' and shows that an increase in the variance of land use regulation can increase the value of vacant land (because it preserves the option of the developer to respond to future market conditions in the more uncertain environment). Riddiough (1997) focuses attention on a similar problem of 'taking risk' in which landowners are subject to the risk of complete or partial loss of use of land due to regulatory constraints. Mayer and Somerville (2000) present an empirical analysis of the impacts on new house construction attributable to regulatory constraints and the types of development delays that are the central variable element in the analysis of Mayo and Sheppard. Mayer and Somerville estimate that mean starts can be as much as 45% lower, and supply elasticity 20% lower in cities with extensive land use regulation.

The models that have been developed over the past two decades have implied that land use regulations are often inefficient, and this seems to hold even in models with congestion and urban development externalities. What is the exact source of this result? In Section 10.3 we carefully review a model of land use control that is sufficiently clear and simple as to help us isolate why the theoretical models have this implication. Following that we extend the model to consider a situation in which land use regulation is potentially beneficial.

10.3. A 'canonical' model of land use control

An approach that presents some central results of the impact of land use regulation is simply and elegantly presented in Brueckner (1996). In this model there are two classes of residents: owners and renters. The renters are mobile between the cities, and adjust (as in a traditional open city model) to equalize renter utility in all cities. In the simplest version of the model there are two cities: one 'passive' (indicated by subscript 0) in the sense that no attempt is made to control growth, and another that is 'active' (subscript 1) where land use or population is regulated by the owners to maximize their welfare.

Regulation of population and land use in this model are equivalent since land consumption does not vary with distance to the CBD, and only residential land use is allowed. The city is a 'long narrow city' that is one unit wide, so that if residents consume one unit of land each then P of them can live within P units of the CBD. Residents commute to the CBD of the city in which they live, and pay transportation costs of t per unit distance x. There is a potential disamenity due to congestion or loss of public good that rises with population P in the city, whose monetary value is βP, with $\beta \geq 0$. While preferences depend on both land consumption l and consumption of a composite good c, the constancy of land consumption over locations allows us to suppress the argument l and write utility as

$$u = c - \beta P \tag{10.1}$$

Equilibrium requires that renter utility be the same in both cities, so that $u_1 = u_0 = u$. Renter household income is exogenously given by y, and the value of land in agricultural use (which determines the edge of the passive city) is 0. Income, land consumption and transport costs per unit distance are identical in both cities, and all renters consume 1 unit of land, for which they pay land rent r. Let P_1 be the population of renters in the active city. Then the distance from the CBD to the edge of the city must also be P_1.

The budget constraint for renters in the active city is

$$c_1 + r = y - tx \Rightarrow \tag{10.2}$$

$$r = y - tx - c_1 \Rightarrow \tag{10.3}$$

$$r = y - tx - u - \beta P_1 \tag{10.4}$$

Let $\bar{x}_1$ be the distance to the edge of the active city in equilibrium. As noted above we must have $\bar{x}_1 = P_1$. In the absence of any land use regulation the edge of the city is determined by $r = 0$. Using this fact, we can solve for the distance $\bar{x}_1$ using Equation (10.4):

$$0 = y - t\bar{x}_1 - u - \beta P_1 = y - t\bar{x}_1 - u - \beta\bar{x}_1 \Rightarrow \tag{10.5}$$

$$\bar{x}_1 = \frac{y - u}{\beta + t} \tag{10.6}$$

Equation (10.6) gives the distance to the edge of the city in the absence of any land use regulation. The active city, however, is managed by the landowners to maximize their welfare. Suppose first that the active city controlled by landlords is small relative to the urban system so that they take the utility level u as fixed. It then seems reasonable for them to choose a constraint $\hat{x}_1$ (this is both the allowed population and the maximum extent of the built up area) that maximizes total land rents collected in the active city. Total rents are given by

$$V_1 = \int_0^{\hat{x}_1} (y - tx - u - \beta\hat{x}_1)\mathrm{d}x \tag{10.7}$$

Maximizing this expression and solving for $\hat{x}_1$ we obtain:

$$\hat{x}_1 = \frac{y - u}{2\beta + t} \leq \bar{x}_1 \tag{10.8}$$

with strict inequality as long as the externality associated with urban development is positive, i.e. $\beta > 0$.

The restriction in this case arises because the landlords internalize the urban development externality. Constraining urban expansion makes the city more pleasant to live in, and as a result land rents rise. The tradeoff is that fewer people live in the city to pay the rents. The landlords solve this tradeoff by choosing the constraint specified in Equation (10.8).

Now suppose the active city is large relative to the urban system so that landlords have the ability to influence the final utility level achieved by renters. To make things even simpler, suppose that there is no urban development externality ($\beta = 0$) so that in equilibrium we must have equal utility in each community which is only possible if $c_0 = c_1 = c$. The landlords simply use their market power and restrict the supply of land in the active community by choosing a maximum extent $\tilde{x}_1$ for the active city. In the passive city no constraint is imposed and land rents at the border are zero; so from Equation (10.3) we have:

$$0 = y - t\bar{x}_0 - c \Rightarrow c = y - t\bar{x}_0 \tag{10.9}$$

Let the total population in the system of two cities be $2P$. Since all renters must be accommodated in one of the two cities, and since $\bar{x}_0$

renters are accommodated in the passive city, we must have:

$$\bar{x}_0 + \tilde{x}_1 = 2P \Rightarrow \bar{x}_0 = 2P - \tilde{x}_1 \tag{10.10}$$

Combining this with Equations (10.3) and (10.9) gives an expression for land rent:

$$\begin{aligned} r &= y - tx - c \\ &= y - tx - y + t\bar{x}_0 \\ &= tx + t2P - t\tilde{x}_1 \end{aligned} \tag{10.11}$$

So that total land rent in the active city is given by

$$V_1 = \int_0^{\tilde{x}_1} (tx + t2P - t\tilde{x}_1)\mathrm{d}x \tag{10.12}$$

The rent maximizing constraint in this case is given by

$$\tilde{x}_1 = \frac{2P}{3} \tag{10.13}$$

This is the central 'supply-restriction' result of Brueckner and Lai (1996). With no urban development externality the two-city urban system would have

$$P = \bar{x}_0 = \bar{x}_1 = \hat{x}_1 > \tilde{x}_1 \tag{10.14}$$

The market power of landlords and the impact of an urban development externality can be combined into a more general urban system having $n - 1$ passive cities and one active one. Let total population in the urban system be nP so that in the absence of any constraints each city would have boundary equal to P. In this model the landlords will choose a boundary $\hat{\bar{x}}_1$ for the active city to maximize rents received. Land rents at the boundary $\bar{x}_0$ of each passive city must be zero, and this implies:

$$0 = y - t\bar{x}_0 - c_0 \tag{10.15}$$

Recalling that $\bar{x}_0$ will be the population of each of the passive cities, and $\hat{\bar{x}}_1$ will be the population of the active city, equal utility for renters in all cities implies that:

$$c_0 - \beta\bar{x}_0 = c_1 - \beta\hat{\bar{x}}_1 \tag{10.16}$$

Finally, all renters must be accommodated so that:

$$nP = \hat{\hat{x}}_1 + (n-1)\bar{x}_0 \tag{10.17}$$

Combining Equations (10.15–10.17) we can solve for the level of consumption in the active city as a function of the land use regulation:

$$c_1(\hat{\hat{x}}_1) = y + \frac{(t+\beta n)\hat{\hat{x}}_1 - (t+\beta)nP}{n-1} \tag{10.18}$$

Using the form of the land rent function represented in Equation (10.3) total land rents in the active city are

$$V_1 = \int_0^{\hat{\hat{x}}_1} (y - tx - c_1(\hat{\hat{x}}_1))\mathrm{d}x \tag{10.19}$$

The value of the constraint that maximizes total land rents is given by

$$\hat{\hat{x}}_1 = \frac{(t+\beta)nP}{(t+2\beta)n+t} \tag{10.20}$$

The landlord's choice for $\hat{\hat{x}}_1$ is clearly less than P. For $\beta = 0$ and $n = 2$ Equation (10.20) reduces to Equation (10.13) as expected. Increasing the urban development externality β causes the optimal constraint $\hat{\hat{x}}_1$ to fall, so that in a two-city system with $\beta > 0$ we have:

$$P = \bar{x}_0 = \bar{x}_1 > \tilde{x}_1 > \hat{\hat{x}}_1 \tag{10.21}$$

As the number of passive cities in the urban system grows, the rent-maximizing constraint level $\hat{\hat{x}}_1$ increases. In the limit

$$\hat{\hat{x}}_1 \longrightarrow \frac{P(t+\beta)}{(t+2\beta)} \qquad \text{as } n \longrightarrow \infty. \tag{10.22}$$

Thus the rent-maximizing land use regulation remains more restrictive than the free market solution even in the limit for a large urban system. This is to be expected since as n increases, the total population of the urban system increases as well so that the mean size of cities in the system does not shrink. Therefore, the urban development externality does not vanish and the landowners in the active city continue to respond to it.

The model now encompasses much of what seems to be intrinsic to all systems of land use regulation: there is a real urban development externality that affects the welfare of residents and provides an incentive to regulate development. There is also some measure of market power exerted by landowners, who might use land use regulation to increase their incomes. To complete the analysis we need to specify a social welfare function and determine the efficient level of land use regulation, then compare it with the outcome presented in Equation (10.20). The form of the utility function is such that utility is measured in units of composite good consumption. Therefore, a utilitarian social welfare function will be to combine aggregate consumption by renters, with aggregate land rents in the active and passive cities.

$$\mathrm{SWF}(x_1)=nP\underbrace{\left(y+\frac{(t+\beta)(x_1-nP)}{n-1}\right)}_{\text{Renter utility level}} + \underbrace{\int_0^{x_1}\left(y-tx-\overbrace{\left(y+\frac{(t+\beta n)x_1-(t+\beta)nP}{n-1}\right)}^{\text{Composite good consumption in active city}}\right)dx}_{\text{Total land rent in the active city}} + (n-1) \times \underbrace{\int_0^{\overbrace{\frac{nP-x_1}{n-1}}^{x_0}}\left(y-tx-\overbrace{\left(y+\frac{t(x_1-nP)}{n-1}\right)}^{\text{Composite good consumption in passive city}}\right)dx}_{\text{Total land rent in passive cities}} \quad (10.23)$$

Differentiating this expression with respect to the constraint level x_1, setting the result equal to zero and solving for the optimal constraint level yields the optimal level of constraint:

$$x_1^* = P \quad (10.24)$$

This implies that land use regulation in this context is not socially efficient.

It may seem surprising given that there is no scope for land use regulation to increase social welfare, even when $\beta > 0$ so that the urban development externality is present. The key observation is that this externality impinges both upon the active city and the passive cities. While it certainly might be in the interest of the active city (or its landowners) to restrict development, this cannot be justified more broadly on social welfare grounds since it simply imposes the external costs on the passive cities.

There are two other features of this model that might be regarded as contributing to this result. First is that the land consumption of renters does not vary so that there is no scope for using land use regulation to reduce total urban land consumption in the system of cities. Reducing total urban land consumption would not address the particular form of urban development externality present here, which depends on total urban population. One could introduce a different sort of urban development externality that depended on total urban land use, and this might offer scope for welfare-improving land use regulation although such possibilities would be sensitive to the elasticity of substitution between land and other consumption. With Leontief preferences (implicitly assumed in this model) there is no substitution possibility so that again we would expect land use regulation to be capable of increasing social welfare for landowners in the active city but to not be justified on broader welfare criteria. In urban systems with relatively flat density gradients it might be reasonable to suspect that land use regulation is socially inefficient.

A second feature of the analysis that might be relevant is the 'absentee' nature of landowners. In each of the scenarios considered so far, the landowners do not themselves experience the urban development externality. Therefore, the only way in which landowners themselves benefit from land use regulation is through the increase in land rents in the active city. This would seem to be an important feature of much land use regulation observed in actual cities: the regulations are demanded by local property owners who reside in the community. The question is whether these demands can be justified on social welfare grounds.

Brueckner (1996) presents a model that adds an urban development externality to the model of Brueckner and Lai (1996) in which all land is owned by a group of resident landowners. It is assumed that landowners consume one unit of land each and that renters consume $\alpha < 1$ units of land each. There are P_h immobile landowners in each city, and $2P_r$ mobile renters who can move between the active and passive cities in the system. Both landowners and renters commute to the CBD and pay identical transport costs t per unit distance. Both landowners and renters have preferences represented by a utility function of the form seen in Equation (10.1). In addition to income y^h from exogenous sources, the landowners receive a share of aggregate land rents R_i within the city $i = 1$ or 2.

With this structure it easy to derive the bid rents from the budget constraints of each group. The renter bid rent function is given by

$$r = \frac{y^r - tx - c_i^r}{\alpha} \tag{10.25}$$

The landlord bid rent is given by

$$r = y^h + R_i - tx - c_i^h \tag{10.26}$$

Since $\alpha < 1$ the bid rent of the renters is steeper than the landowners, they will occupy the more central locations. As above, the model makes use of the constant levels of land consumption to express total populations in terms of the ranges of locations occupied by each group. Let $\hat{x}_i$ represent the point at which the renter and landowner bid rents intersect in each city. Thus renters reside from 0 to $\hat{x}_i$ and $\hat{x}_i/\alpha$ can be accommodated within this area. Let $\bar{x}_i$ be the maximum extent of urban land use in each city, so that landowners reside from $\hat{x}_i$ out to $\bar{x}_i$, and $\bar{x}_i - \hat{x}_i$ will be accommodated within the city.

Equilibrium in the urban system is characterized by the following conditions:

Market clearing for landowners	$\bar{x}_i - \hat{x}_i = P_h$ for $i = 1$ or 2
Market clearing for renters	$\dfrac{\hat{x}_0 + \hat{x}_1}{\alpha} = 2P_r$
Renter utility equal in each city	$c_0^r - \beta(\bar{x}_0 - \hat{x}_0 + \frac{\hat{x}_0}{\alpha}) = c_1^r - \beta(\bar{x}_1 - \hat{x}_1 + \frac{\hat{x}_1}{\alpha})$
Boundary rent the in passive city	$y^h + R_0 - t\bar{x}_0 - c_0^h = 0$
Bid-rents intersect at $\hat{x}_i$	$\dfrac{y^r - t\hat{x}_i - c_i^r}{\alpha} = y^h + R_i - t\hat{x}_i - c_i^h$ for $i = 1$ or 2
Landowners income from land rents	$\int_0^{\hat{x}_i} \frac{y^r - tx - c_i^r}{\alpha} \mathrm{d}x + \int_{\hat{x}_i}^{\bar{x}_i} (y^h + R_i - tx - c_i^h) \mathrm{d}x = P_h R_i$ for $i = 1$ or 2

Assuming as above that the constraint level $\bar{x}_1$ would be set to maximize landowner utility yields landowner-utility maximizing constraint of

$$\bar{x}_1 = \frac{(3t\alpha + (4 - \alpha)\beta)P_h + 2(\alpha t + \beta)\alpha P_r}{3t\alpha + 4\beta} \tag{10.27}$$

Is this the social welfare maximizing constraint? Writing out the social welfare function in a manner analogous to Equation (10.23) and maximizing reveals that the optimal level of constraint is

$$x_1^* = P_h + \alpha P_r > \bar{x}_1 \tag{10.28}$$

This shows that even when landlords are resident in the community and hence direct beneficiaries of internalizing the urban development externality, it remains socially efficient to have no land use regulation. As above, if landowners are allowed to set the constraint level to maximize their utility the regulation will be inefficiently restrictive. The inefficiency of land use regulation is, therefore, not a product of landowners who do or do not benefit directly from the reduction in urban development. We turn next to consideration of the importance of constant individual land consumption.

10.4. Extending the model: potentially beneficial land use regulation

Now consider a simple adaptation of the Brueckner model presented in Section 10.3 that does admit of potentially welfare-improving land use regulation. As observed above, we need to consider a situation in which there is both an externality associated with urban development and the potential for consumers of land to respond to higher rents by economizing on their consumption.

As before, suppose there are two cities in a closed region. One is passive and cannot implement land use regulation and the other is active and can implement a containment policy to constrain the sprawl of the city. The first is indicated by a subscript 0 and the second by a subscript 1. There are two types of persons in the region: renters and landowners. There are $2P_r$ renters who all have income y and pay transport costs of t per unit distance from the center of the city in which they reside. Suppose all landowners are absentee and care only about the total value of their land. Renters have preferences that depend on consumption of a composite good C, private land consumption L and the extent of urban sprawl x_1. These preferences are represented by a utility function

$$U = C + 2\sqrt{L} - \beta x_i \tag{10.29}$$

The price of the composite good is 1, and the price of land is r. Assuming an interior solution, the demand for land is

$$L = \frac{1}{r^2} \tag{10.30}$$

Solving to obtain the demand for the composite good, substituting back into the utility function to obtain the indirect utility, and solving for land rent yields the land rent as a function of the utility level $\hat{u}$, transport costs, income, the urban externality parameter β, maximum extent of urban land use x_i, and distance from the CBD

$$r = \frac{1}{\hat{u} - y + \beta x_i + tx} \tag{10.31}$$

In this region the price of agricultural land is given by $r_a = 1$, and the boundary of the passive city is defined by the point where $r = r_a = 1$. Setting Equation (10.31) to 1 we can solve for the maximum extent of the passive city:

$$x_0 = \frac{1 - \hat{u} + y}{t + \beta} \tag{10.32}$$

With this condition defining the maximum extent of the passive city, the number of renters that will be accommodated is

$$\int_0^{x_0} \left(\frac{1}{\hat{u} - y + \beta x_0 + tx} \right)^2 \mathrm{d}x = \frac{1 - \hat{u} + y}{t\hat{u} - ty + \beta} \tag{10.33}$$

In the active city, the maximum extent of urban development x_1 is set as a policy variable, so we have:

$$\int_0^{x_1} \left(\frac{1}{\hat{u} - y + \beta x_1 + tx} \right)^2 \mathrm{d}x$$

$$= \frac{x_1}{(\hat{u} - y + \beta x_1)(\hat{u} + tx_1 - y + \beta x_1)} \tag{10.34}$$

Since all renters must be accommodated, equilibrium requires that

$$2P_r = \frac{1 - \hat{u} + y}{t\hat{u} - ty + \beta} + \frac{x_1}{(\hat{u} - y + \beta x_1)(\hat{u} + tx_1 - y + \beta x_1)} \tag{10.35}$$

While Equation (10.35) can be solved to obtain the utility level $\hat{u}$ as a function of x_1 and the parameters, the solution is complex (since Equation (10.35) is a cubic in $\hat{u}$) and not obviously informative. Greater insight into the nature of this economy can be obtained by considering an economy with no land use regulation ($x_1 = x_0$) and asking whether, on the margin, social welfare or the welfare of renters and landowners individually can be increased by introducing such regulation (reducing x_1).

In an unconstrained economy the population of renters would be divided equally between the passive and active city, so that the population of renters in each would be P_r. Combining this observation with Equation (10.33) allows us to solve for the renter utility level in an economy without land use regulation:

$$\hat{u} = \frac{1 + y + P_r(ty - \beta)}{1 + P_r t} \tag{10.36}$$

Equation (10.35) defines $\hat{u}$ implicitly as a function of x_1. Differentiating Equation (10.35) implicitly, and evaluating with x_1 set equal to the value of x_0 given in Equation (10.32), and with utility level given by Equation (10.36) we obtain:

$$\frac{\partial \hat{u}}{\partial x_1} = -\frac{(t + \beta)(P_r\beta(2 + tP_r) - 1)}{1 + P_r(2 + tP_r)(2t + \beta)} \tag{10.37}$$

Making the reasonable assumption that renter population is sufficiently large that $P_r\beta(2 + tP_r) - 1 > 0$, Equation (10.37) shows that starting from an unconstrained state, a marginal increase in the level of land use regulation (a marginal decrease in x_1) will increase renter utility. This happens even though the policy is only implemented in the active city. The process of equilibration in allocating population between the active and passive communities transmits the increase in welfare to all renters. Why this happens is clear: there is an external diseconomy caused by land consumption and this is partly internalized by a marginal increase in land use regulation.

If public policy were not concerned about the welfare of the landowners, then this would complete the analysis and it would be clear that, in general, some level of land use regulation would be beneficial for the economy. A comprehensive welfare analysis of the economy, however, must take into account the impact on the incomes of landowners as well. Using Equation (10.31) we can determine aggregate land rent in the two cities by

$$R = R_0 + R_1 = \underbrace{\int_0^{x_0} \frac{1}{\hat{u} - y + \beta x_0 + tx} \mathrm{d}x}_{R_0 = \text{Total rent in passive city}} + \underbrace{\int_0^{x_1} \frac{1}{\hat{u} - y + \beta x_1 + tx} \mathrm{d}x}_{R_1 = \text{Total rent in active city}} \tag{10.38}$$

If we use Equations (10.32) and (10.36) to evaluate aggregate rents in an unconstrained urban system, and evaluate the marginal impact of a change in land use regulation, we obtain:

$$\frac{\partial R}{\partial x_1} = \frac{(1 + tP_r)(2tP_r + \beta P_r(4 + tP_r) - 1)}{1 + P_r(2 + tP_r)(2t + \beta)} \tag{10.39}$$

The marginal impact on a utilitarian social welfare function from a change in x_1 is

$$\frac{\partial \mathrm{SWF}}{\partial x_1} = \frac{\partial R}{\partial x_1} + 2P_r \frac{\partial \hat{u}}{\partial x_1} \tag{10.40}$$

Land use regulation is socially desirable on the margin if $(\partial \mathrm{SWF}/\partial x_1) < 0$. Using the results from Equations (10.39) and (10.37), we can evaluate this marginal impact as

$$\frac{\partial \mathrm{SWF}}{\partial x_1} = \frac{P_r(3t+6\beta) + P_r^2(2t^2+3t\beta-2\beta(t-2\beta)) - t\beta P_r^3(t-2\beta) - 1}{1+P_r(2+tP_r)(2t+\beta)} \tag{10.41}$$

The sign of this impact clearly depends on the sign of the numerator. Assuming that $t > 2\beta$, we will always have $(\partial \mathrm{SWF}/\partial x_1) < 0$ for a sufficiently large population of renters, so that a marginal reduction in x_1, starting from an unconstrained economy, will improve social welfare.

The structure of this economy reveals that a marginal increase in land use regulation can improve social welfare in an unconstrained economy if the urban development externality β is neither too strong nor too weak. To see this, consider the numerator of Equation (10.41). Viewing this as a quadratic in β, we can solve for the roots to obtain:

$$\frac{\partial \mathrm{SWF}}{\partial x_1} = 0 \text{ when}$$

$$\beta = -\frac{6 + P_r t(1 - P_r t) \mp \sqrt{(52 - 28P_r t - 67P_r^2 t^2 - 18P_r^3 t^3 + P_r^4 t^4)}}{4P_r(2 + P_r t)} \tag{10.42}$$

This allows us to state the main conclusion of this section.

Proposition 1. *In an urban system with mobility between two cities and preferences represented by a utility function of the form given in Equation (10.29), land use regulation restricting development at the periphery of one city is potentially beneficial when the urban development externality β satisfies:*

$$\underbrace{-\frac{6 + P_r t(1 - P_r t) + \sqrt{(52 - 28P_r t - 67P_r^2 t^2 - 18P_r^3 t^3 + P_r^4 t^4)}}{4P_r(2 + P_r t)}}_{\beta_{\min}} < \beta < \overbrace{-\frac{6 + P_r t(1 - P_r t) - \sqrt{(52 - 28P_r t - 67P_r^2 t^2 - 18P_r^3 t^3 + P_r^4 t^4)}}{4P_r(2 + P_r t)}}^{\beta_{\max}}$$

When $\beta \leq \beta_{\min}$, the reduced external cost from urban development is not sufficient to compensate for the reduced land area earning rents higher than agricultural rents. When $\beta \geq \beta_{\max}$, then the increase in the welfare of residents from reducing urban development is large, but this increase in welfare reduces equilibrium land values in the passive city by an amount that exceeds the total benefit to renters in the active city.

As an example, in an urban area with 100,000 households and annual transport costs per unit distance $t = 20$, marginal land use regulation would be beneficial if $0.0002 < \beta < 10$. Note that β can be interpreted as the effective loss in consumption spending to a resident resulting from expanding the urban area by one unit distance.[2] If distance is measured so that adding one household at the periphery increases the size of the urban area by one unit distance, then β can also be interpreted as the external cost to each resident resulting from adding a new household to the community. Put this way restricting β to satisfy $0.0002 < \beta < 10$ does not seem unreasonable, but the question is an empirical one and perhaps most usefully considered in a model more complex than the one presented here.

Analysis of this model has in some sense confirmed the observations made above. For land use regulation to be beneficial (even potentially) there must be some externality associated with urban development, and there must be some scope for land consumption to respond to the regulatory constraint (each household cannot have fixed land consumption). These features have been incorporated here into the core model of land use regulation, and indeed we discover some scope for land use regulation to improve social welfare. Interestingly, even marginal land use regulation (a small restriction in peripheral development beginning from an unconstrained urban system) is not always beneficial. If the magnitude of the urban development externality is either too large or too small, then marginal land use regulation will reduce social welfare.

10.5. Empirical studies of land use regulation

While the theoretical models presented and reviewed above provide insights into the circumstances under which land use regulations might be inefficient or efficient, the first studies and investigations of the issue were essentially empirical. Hedonic analysis of housing

---

[2] In the model presented here, residential land consumption is, by definition, equal to 1 at the urban periphery. If this area were 1 acre per household it would make sense to measure distance in units whose square was equal to 1 acre. If annual transport costs were 500 per mile, this would give annual transport costs of approximately 20.

markets made it possible to examine the separate impact on house (or even land) prices of zoning and other types of land use regulation. While simply testing for impacts on prices does not provide a complete analysis of the welfare impacts of these regulations, it is surely a prerequisite for such analysis. In this section, we review a selection of the many studies that have documented impacts large or small on house prices attributed to land use regulation. We then give careful consideration to three studies that warrant close examination either because of the data they employ in examining these impacts or the comprehensive econometric techniques employed. Finally, we examine the distributional impacts that have been attributed to land use regulation.

10.5.1. Data and empirical evidence

We noted above that the lack of data on the nature and extent of land use regulation has posed a major obstacle to empirical testing and evaluation of these policies. In countries such as the US, land use regulation decisions are taken at the local government level. As a result there is no uniform national standard or enabling legislation that generates similar land regulation policies across many urban areas.[3] This limits the opportunity to observe variation in local economic conditions that is independent of the structure of land use regulation, and makes it difficult to identify the impact of these regulations. As a result, empirical studies of land use regulation tend to fall into one of the three categories: they may examine a cross-section of cities using an approximate measures of land use restrictions and use census data; they may focus on countries whose land use planning systems are based on a national approach with enabling institutions operative throughout the country and limited local options; or they may focus on evaluation of the impacts in a single urban area or local government level.

---

[3] For the US, this overstates the difficulty somewhat, since there are similarities between enabling statutes and regulatory approaches in a large number of states. There are, however, noted exceptions with some large states, like California and Texas, taking quite different approaches. In any event, there has been only very limited effort devoted to collecting comparable data across the US concerning levels of land use regulation.

One of the original studies of the impacts of land use regulation undertaken by Hamilton (1978) falls into the first category. He makes use of census data for 13 urban areas in the northeastern quadrant of the country. Lacking any direct measure of land use regulation or zoning laws, Hamilton argues that the ability to use zoning to extract monopoly rents for current landowners depends on the number of nearby competing jurisdictions. He constructs an index of local monopoly power to use zoning by taking the number of municipalities in the urban area or the number of municipalities relative to the population of the urban area. His results suggest that increasing the 'market power' of local governments in an urban area is associated with increasing house prices.

Cheshire and Sheppard (1989) provide an example of the second type of study. By focusing on the UK town and country planning system, they work within a national system of land use regulation, and have access to national data on the extent to which town planning constrains local provision of housing and other types of development. They use these data to select extreme cases of restrictive and permissive planning regimes, and use hedonic methods to determine the difference in house values that are attributable to the level of planning restrictions.

The analysis of Pollakowski and Wachter (1990) falls into the third group, as do the studies by Levine (1999) and Phillips and Goodstein (2000). Of these three, Pollakowski and Wachter employ a hedonic technique that carefully examines the impact of various levels of zoning restrictiveness on house prices within the community and within neighboring communities. They also are one of the first empirical studies to explicitly raise the concern of the welfare impact of zoning restrictions and the distributional impacts, although they are not able to reach definite conclusions concerning these issues. Their analysis focuses on a suburban area north of Washington, DC that has experienced rapid growth pressures and was an early adopter of comprehensive growth management regulations tied to local infrastructure capacity. Pollakowski and Wachter construct an index of zoning restrictiveness based on a weighted average of the amounts of land designated for various levels of development. They not only estimate clear impacts in which zoning restrictiveness increases house and land prices, but they also show that there is a clear 'spillover' impact in which restrictive land use regulation causes an increase in land prices in nearby communities. Spillover impacts are also implied by the theoretical models presented above, so such evaluation provides a possible mechanism for testing or even calibrating these models.

A further examination of impacts of land use regulations on constrained and on neighboring communities is undertaken by Levine (1999). This study combines information on 490 local governmental units in California. These governments have adopted a surprisingly wide variety and mixture of land use regulations, and Levine tests for impacts on creation of new housing, the total stock of housing, and the location choices of population subgroups dependent on certain types of housing. The study, therefore, provides both a test of the theoretical models examined above and also some insights into the distributional impacts of these regulations. Levine finds that land regulations seem to have had a clear impact of reducing the availability of both rental and owner-occupied housing. He further finds evidence that these reductions have been displacements of housing construction to jurisdictions that are more passive in their approach to land use regulation. He notes that the regulations seem to have had a disproportionate impact on low income and minority populations, who seem to have been displaced away from metropolitan areas. Finally, his analysis suggests that occasional attempts to implement growth encouragement policies have been mostly ineffective.

The evaluation of growth controls and the urban growth boundary in Portland, Oregon presented by Phillips and Goodstein (2000) is worthy of note because they focus attention on an urban area whose land use regulatory policies have generated considerable interest and commentary. They consider a cross-section of cities, and estimate a very simple model of house prices in a sample of 37 cities. They conclude that while Portland's land use policies have increased house prices, the actual magnitude of the increase that can be separately attributed to the urban growth boundary in Portland is very small – on the order of a 6% increase in house prices.

Before turning to some more recent or more complex empirical studies of land use regulation, we should note the study by Gin and Sandy (1994), who have examined the demand for growth controls by looking at the voting patterns for growth controls in a referendum conducted in San Diego, CA. Such empirical studies hold promise for testing models of endogenous land use regulation in which resident landowners are assumed to vote in their self-interest for land use regulations. They find that homeowners exhibit considerably higher demand for growth controls, as do residents in areas experiencing rapid growth.

10.5.2. Assessing the impact of land use regulation

Three empirical studies of the impacts of land use regulation warrant special attention, albeit for different reasons. The analysis of Pogodzinski and Sass (1994), examining house prices in Santa Clara, CA, presents one of the most thorough hedonic-based evaluations of the impact of zoning. They take special care to account for the extent to which zoning interacts with various local amenities and public goods to alter the rate of capitalization of these amenity values into house prices. They attempt to account for endogeneity in both the demographics of the communities and the types of regulatory policies adopted. This increases confidence in their final results, some of which are strikingly different from previous studies.

Pogodzinski and Sass find significant evidence of selection bias in choice of tax and zoning regime, and this raises doubts about the robustness of empirical studies that measure the impact of land use regulation without addressing this concern. They also find significant differences in the impacts of different types of land use regulation. Constraints on characteristics (such as height restrictions and lot size restrictions) appear to have significant impacts on the value of housing. In contrast to this, they find that land use zoning (such as allocating limited areas for construction of single family housing) essentially ‘follow the market’ so that more land is allocated in those areas where there is greater demand for such land use. This implies that the actual price impact of such regulations is minimal, and that any observed increases in property values are due not to the land use regulation itself, but rather to the unobserved characteristics of the property that is zoned for particular uses.

Evenson and Wheaton (2003) employ an important (and apparently unique) data resource that permits careful evaluation of actual land use and land use regulation across all cities and towns in the state of Massachusetts. These data have been collected from each local government with authority to regulate land use, and combined with satellite data in a Geographic Information System to facilitate comparison between the allocation of land for particular uses and the actual land use. This type of data collection should serve as a model for scholars and policy makers worldwide who are concerned about the impacts of land use regulation. The results obtained from this data are impressive. They are able to estimate the impact of income on the quantity of protected land, and verify that such land appears to be a normal good. They show that wealthier towns also devote less land to commercial and industrial uses, and find evidence that current and future buildings are highly correlated. They find no evidence for traditional fiscal zoning, in which minimum lot sizes are used to keep housing expensive and exclude the poor or others with large demands for local public goods. They find no link between income and the density of construction. They find general support for the conclusion reached by Pogodzinski and Sass that zoning tends to follow the market and provide future accommodation for current development patterns. Their analysis does not yet incorporate house prices, however, so that final conclusions regarding the impacts on household welfare are difficult to reach. It is clear, however, that these data and this type of analysis will be a rich source of insights into the impacts of land use regulation.

Glaeser and Gyourko (2003) combine Census, American Housing Survey and construction cost data for 42 MSAs to add further evidence concerning the impact of building restrictions on house prices. A key insight of their analysis is that for most urban areas in the US, house prices are very near or even below the costs of construction. They interpret this as showing that for most cities building restrictions are not acting to increase house prices. Those urban areas where construction costs are significantly below existing house prices are also characterized by other factors suggesting that building restrictions of some sort are a contributing factor. While this analysis does not directly bear on land use regulation, it is broadly consistent with the other recent analyses and suggests a new approach for undertaking cross-section evaluation of urban house prices and regulatory environments.

10.5.3. Distributional impacts

While several of the empirical studies reviewed above have directly addressed the impacts of land use regulation on house prices, few of them of tried to tie these impacts directly to welfare levels of residents. Even fewer have addressed the distributional impacts of land use regulations. This is not because there is a belief that such regulations affect all income groups equally. Concern about fiscal zoning and exclusion of low-income households underscore the expectation that some households receive benefits from land use regulation, while others bear costs. How these benefits and costs are distributed across society is clearly of central concern to any welfare evaluation of the policies. The paper by Cheshire and Sheppard (2002) is directly concerned about both the net welfare impacts in aggregate, and the distribution of these impacts over the population.

Undertaking such analysis is complex, and the paper makes a number of assumptions in order to carry it out. Using a hedonic model to obtain implicit prices for characteristics, the demand for land, housing characteristics, and local amenities is estimated. This estimated demand structure is used to calibrate a monocentric urban model to make inferences about the level of land use regulations being imposed by the planning system. This calibrated system is then used to evaluate alternative (counter-factual) planning regimes in which urban containment is relaxed, internal space availability is increased, etc. Each of these alternative planning regimes has implications for the availability of open space and amenities produced through land use regulation, and these are accounted for via the demand system.

As a result (and subject to the assumptions maintained) it is possible to estimate the value of the benefits being produced by the planning system, as well as the costs, and to make a net comparison. From a distributional perspective, it is also possible to impute these costs and benefits to the individual households in the sample and estimate the distributional impacts. These turn out to be very interesting. The estimates are obtained in the context of a single urban area that has previously been identified as one of the most restrictive in the UK. The estimates suggest that these restrictions have been set inefficiently, and that on the margin greater costs are being generated than benefits.

Furthermore, the policies have interesting distributional impacts. Calculating the income equivalent of the benefits and costs, calculations are made concerning the distribution of these over the sample income groups. In general, the benefits of the planning system are seen to be regressive in the sense that if the income equivalents of planning amenities are added to actual household income, the Gini index of income inequality increases. This should not be completely surprising in a system that concentrates great effort at preserving open space at the urban periphery through growth boundaries. The costs of these policies are also estimated to fall disproportionately on the wealthy, however, so that the net costs of the planning system are close to distributionally neutral. It would be of great interest to discover if these results were similar in other urban areas within the UK or in other cities subject to alternative types of land use regulation.

10.6. Conclusion

The potential importance of the spatial arrangement of land uses and different social classes has long been recognized, although the impacts on social welfare or the distribution of welfare that result from attempts to control this structure are unclear. Lack of data concerning the nature and extent of regulatory constraints on land use have hampered efforts to study these impacts, and only recently have the data started to present a somewhat consistent picture.

Lacking solid data upon which to base an estimate, and in an effort to determine and refine empirical strategies for making such estimates, analysts have developed a variety of theoretical models to help explain and predict the impacts. The models that are most successful have incorporated explicit consideration of both the impact on the community that is constrained by the regulation and other communities in the same urban system to which development and growth might be displaced. Many of the models incorporating these features suggest at most a very limited scope for land use regulations to improve overall social welfare. We have presented a simple extension of one of these models that suggests there is scope for beneficial land use regulation when there is a real urban development externality and when individual land use is responsive to changes in land rents.

There has been steady progress in empirical studies of the impacts of land use regulation. Most recently, new sources of data have begun to emerge that combine remote sensing, geographic information, and comprehensive recording of the amount of land area allocated to specific uses. Further developments in this direction hold promise for extending our understanding of these policies. Eventually, the welfare analysis of land use regulation must turn to the net effects on overall social welfare and the distribution of these costs and benefits across income and social groups. Some initial work has been done in this area. Like many areas of policy analysis, however, much work remains.

References

Bailey, M. (1959), "A note on the economics of residential zoning", *Land Economics*, p. 25.

Brueckner, J.K. (1990), "Growth controls and land values in an open city", *Land Economics*, Vol. 66, pp. 237–248.

Brueckner, J.K. (1995), "Strategic control of growth in a system of cities", *Journal of Public Economics*, Vol. 57, pp. 393–416.

Brueckner, J.K. (1996), "Modeling urban growth controls", *Office of Research Working Paper Number 96-0137.*

Brueckner, J.K. and F.C. Lai (1996), "Urban growth controls with resident landowners", *Regional Science and Urban Economics*, Vol. 26, pp. 125–143.

Cheshire, P. and S. Sheppard (1989), "British planning policy and access to housing: some empirical estimates", *Urban Studies*, Vol. 26, pp. 469–485.

Cheshire, P. and S. Sheppard (2002), "The welfare economics of land use planning", *Journal of Urban Economics*, Vol. 52, pp. 242–269.

Cooley, T.F. and C.J. LaCivita (1982), "A theory of growth controls", *Journal of Urban Economics*, Vol. 12, pp. 129–145.

Engle, R., P. Navarro and R. Carson (1992), "On the theory of growth controls", *Journal of Urban Economics*, Vol. 32, pp. 269–283.

Epple, D., T. Romer and R. Filimon (1988), "Community development with endogenous land use controls", *Journal of Public Economics*, Vol. 35, pp. 133–162.

Evans, A. (1999), "The land market and government intervention", pp. 1637–1669, in: P. Cheshire and E. Mills, editors, *Handbook of Regional and Urban Economics, Applied Urban Economics*, Vol. 3, Amsterdam: North-Holland.

Evenson, B. and W.C. Wheaton (2003), "Local variation in land use regulations", *Brookings Wharton Papers on Urban Affairs*, pp. 221–260.

Frankena, M.W. and D.T. Scheffman (1981), "A theory of development controls in a small city", *Journal of Public Economics*, Vol. 15, pp. 203–234.

Gin, A. and J. Sandy (1994), "Evaluating the demand for residential growth controls", *Journal of Housing Economics*, Vol. 3, pp. 109–120.

Glaeser, E.L. and J. Gyourko (2003), "The impact of building restrictions on housing affordability", *FRBNY Economic Policy Review*, Vol. 9, pp. 21–39.

Hamilton, B.W. (1978), "Zoning and the exercise of monopoly power", *Journal of Urban Economics*, Vol. 5, pp. 116–130.

Helsley, R.W. and W.C. Strange (1995), "Strategic growth controls", *Regional Science and Urban Economics*, Vol. 25, pp. 435–460.

Lai, F.C. and S.T. Yang (2002), "A view on optimal urban growth controls", *Annals of Regional Science*, Vol. 36, pp. 229–238.

Levine, N. (1999), "The effects of local growth controls on regional housing production and population redistribution in California", *Urban Studies*, Vol. 36, pp. 2047–2068.

Mayer, C.J. and C.T. Somerville (2000), "Land use regulation and new construction", *Regional Science and Urban Economics*, Vol. 30, pp. 639–662.

Mayo, S. and S. Sheppard (1991), "Housing supply and the effects of stochastic development control", *Oberlin College Discussion Papers in Economics*, Oberlin College Economics Department.

Mayo, S. and S. Sheppard (2001), "Housing supply and the effects of stochastic development control", *Journal of Housing Economics*, Vol. 10, pp. 109–128.

Pasha, H.A. (1992), "Comparatie statics analysis of density controls", *Journal of Urban Economics*, Vol. 32, pp. 284–298.

Phillips, J. and E. Goodstein (2000), "Growth management and housing prices: the case of Portland, Oregon", *Contemporary Economic Policy*, Vol. 18, pp. 334–344.

Pogodzinski, J.M. and T.R. Sass (1994), "The theory and estimation of endogenous zoning", *Regional Science and Urban Economics*, Vol. 24, pp. 601–630.

Pollakowski, H.O. and S.M. Wachter (1990), "The effects of land-use constraints on housing prices", *Land Economics*, Vol. 66, pp. 315–325.

Riddiough, T.J. (1997), "The economic consequences of regulatory taking risk on land value and development activity", *Journal of Urban Economics*, Vol. 41, pp. 56–77.

Sakashita, N. (1995), "An economic theory of growth control", *Regional Science and Urban Economics*, Vol. 25, pp. 427–434.

Sasaki, K. (1998), "Optimal urban growth controls", *Regional Science and Urban Economics*, Vol. 28, pp. 475–496.

Sheppard, S. (1988), "The qualitative economics of development control", *Journal of Urban Economics*, Vol. 24, pp. 310–330.

PART 3

Spatial Interaction, Migration and Commuting

Urban Dynamics and Growth, CEA, vol. 266
Roberta Capello and Peter Nijkamp (editors)

CHAPTER 11

Spatial Interaction Models: From the Gravity to the Neural Network Approach

Manfred M. Fischer[a,*] and Aura Reggiani[b]

[a]Department of Economic Geography and Geoinformatics, Vienna University of Economics and Business Administration, Vienna, Austria
[b]Department of Economics, Faculty of Statistics, University of Bologna, Bologna, Italy

Abstract

Spatial interaction models describe and predict spatial flows of people, commodities, capital and information. They are one of the oldest and most widely used of all social science models. This chapter provides a coherent state-of-the-art overview of the field that has witnessed the progression from gravity models to entropy maximising and random utility maximising models and finally to models based on neurocomputing principles that represent the most recent innovation in the design of spatial interaction models.

Keywords: spatial interaction, entropy maximisation, utility maximising choice behaviour, neural network models, the learning problem

JEL classifications: C31, C45, R19

11.1. Introduction

Spatial interaction models represent a class of methods which are appropriate for modelling data that are associated with a link or pair of locations (points, areas) in geographic space. They are used

* Corresponding author.

to describe, and hence help predict spatial flows of people, commodities, capital and information over geographic space. Such models are relevant in studies of transport planning, population migration, journeys-to-work, shopping behaviour, freight flows, and even the transmission of information at different scales of spatial resolution, including the intra-urban and the inter-regional. Almost all human relationships can be said to involve interactions that are impeded by distances of one form or another (Sen and Smith, 1995).

The subject of spatial interaction modelling has a long and distinguished history. Original formulations rested on analogies with the physical world of interacting particles and gravitational force. Contemporary models have led to the emergence of three major schools of analytical thought: the macroscopic one corresponding to a statistical equilibrium approach, the microscopic one corresponding to a choice-theoretic or utility maximising approach and the geocomputational school corresponding to a neural network approach to spatial interaction modelling. Each of these approaches will be briefly reviewed in this chapter, with a particular focus on the neural network approach that perceiving interaction models as universal function approximators represents the most recent innovation in the design of spatial interaction models. The section that follows sets forth the context and framework for the discussion.

11.2. Context and analytical framework

The phenomena of interest in this chapter may be described in their most general terms as interactions between populations of actors and opportunities distributed over some relevant geographic space. More specifically, interest is focused on those patterns of spatial interactions that may occur during some relevant period of time. Such interactions may involve movements of individuals from one location to another, such as daily traffic flows in which case the relevant actors are individual travellers (commuters, shoppers, etc.) and the relevant opportunities are their destinations (jobs, stores, etc.). Similarly, one may consider annual migration flows, in which the relevant actors are migrants (individuals, family units, firms, etc.) and in which case the relevant opportunities are their possible new locations. Interactions may also involve flows of information such as telephone calls or electronic messages. Here the callers or message senders may be the relevant actors, and the possible receivers of calls or electronic messages may be considered as the relevant opportunities (Sen and Smith, 1995). With this range of examples in mind, the purpose of this section is to outline a framework in which all such spatial interaction behaviour can be studied. To do so we introduce some notation first.

Notation. Suppose we have a spatial system consisting of I origins and J destinations and let t_{ij} denote the volume of interaction from spatial unit (location, zone, region) i to spatial unit (location, zone, region) j $(i = 1, \dots, I; j = 1, \dots, J)$. This information may be displayed in the form of an interaction matrix of the following kind

$$\boldsymbol{T}_{I \times J} = \begin{bmatrix} t_{11} & \cdots & t_{1j} & \cdots & t_{1J} \\ \vdots & & \vdots & & \vdots \\ t_{i1} & \cdots & t_{ij} & \cdots & t_{iJ} \\ \vdots & & \vdots & & \vdots \\ t_{I1} & \cdots & t_{Ij} & \cdots & t_{IJ} \end{bmatrix}. \qquad (11.1)$$

In some cases the sets of origins and destinations are the same and, thus, $\boldsymbol{T}_{I \times J}$ is a squared matrix. The interpretation of the main diagonal of the square $\boldsymbol{T}_{I \times I}$ depends on the specific application. For instance, it might represent internal telecommunication flows within region i $(i = 1, \dots, I)$. Often such values are not recorded. In other applications, for example shopping trips from residential areas to individual shopping malls, the number of origins and destinations may differ and $\boldsymbol{T}_{I \times J}$ will not be a square matrix (Fischer *et al*., 2001).

For all applications, the ith row of the matrix $\boldsymbol{T}_{I \times J}$ describes the outflows from region i to each of the J destinations, while inflows from each of the I origins into destination j are described by the jth column (see Equation (11.1)). From $\boldsymbol{T}_{I \times J}$ we can calculate the volume of interaction originating from region i or terminating in region j, that is

$$t_{i\bullet} = \sum_{j=1}^{J} t_{ij} \qquad i = 1, \dots, I \qquad (11.2a)$$

and

$$t_{\bullet j} = \sum_{i=1}^{I} t_{ij} \qquad j = 1, \ldots, J \tag{11.2b}$$

respectively. In turn, these marginal sums can be utilised to calculate the overall level of interaction that is defined as

$$t_{\bullet\bullet} = \sum_{i=1}^{I} \sum_{j=1}^{J} t_{ij}. \tag{11.2c}$$

11.2.1. The generic spatial interaction model of the gravity type

The primary objective in the analysis of interaction data is to understand or model the pattern of flows. The interest lies in analysing patterns of spatial interactions. Typically, we can recognise three important sets of factors:

- the spatial separation of origins i from destinations j,
- characteristics that determine the volume of flow, $t_{i\bullet}$, from each origin i (measures of origin propulsiveness), and
- characteristics of the destinations which are associated with their attractiveness (measures of destination attractiveness).

Let us briefly consider each of these sets of factors. The first relate to the way in which spatial separation constrains or impedes movement across geographic space. Spatial interaction may be influenced by various types of spatial separation between actors and opportunities. Spatial interaction data analysis is, in particular, interested in identifying those specific types of spatial separation that tend to impede or enhance the likelihood of interactions in a given spatial interaction context. The most obvious types of separation involve physical space. For example, the physical separation between the locations of shoppers and of retail stores will certainly influence the relative likelihoods among various possible shopping trips. But such separation relationships need not involve physical space. They can, for example, represent cultural and/or mental barriers as well. Overall, what we require is an appropriate measure of spatial separation (briefly termed distance between an origin and a destination) and we shall usually expect an inverse relationship between interaction and distance, known as distance decay or distance deterrence relationship.

The second set of factors refers to the ability of origins to generate flows. This might be measured in terms of a single variable. We shall expect to observe a positive relationship between the volume of flow generated from an origin and some appropriate measure of propulsiveness. A similar point may be made of the third set of factors in the spatial interaction problem, the representation of measures of destination attractiveness (Bailey and Gatrell, 1995).

The basic structure of the spatial interaction problem is, then, to express the volume of interactions in terms of origin, destination and spatial separation factors. Mathematically, the situation we are considering in this chapter is one of observations t_{ij} ($i = 1, \ldots, I$; $j = 1, \ldots, J$) on random variables, say N_{ij}, each of which corresponds to a movement of people (cars, commodities, telephone calls, etc.) between spatial locations i and j. In general, we are interested in models of the type

$$N_{ij} = \tau_{ij} + \varepsilon_{ij} \tag{11.3}$$

where observed flows t_{ij} are independent Poisson variates with $\tau_{ij} = E(t_{ij})$. The error ε_{ij} is noise, with the property $E(\varepsilon_{ij}|t_{ij}) = 0$ by construction. We aim to develop appropriate models for the systematic part, τ_{ij}, of the stochastic relationship with other random variables which are the forecasts.

Spatial interaction models of the gravity type (often termed gravity models)[1] simultaneously incorporate the effect of origin and destination characteristics as well as separation. Mathematically, they may be written as

$$\tau_{ij} = b_{(ij)}\, r_i\, s_j\, f_{ij}, \qquad i = 1, \ldots, I; \;\; j = 1, \ldots, J \tag{11.4}$$

where τ_{ij} denotes the estimated flow from i to j, and $b_{(ij)}$ a constant of proportionality, r_i represents a factor characterising the origin i of interaction, and s_j a factor characterising the destination j of

---

[1] The term arises because of the analogy of the models with Newton's law of gravity where the force of attraction is proportional to the product of the masses of the two bodies involved and inversely proportional to the square of the distance between them.

interaction, while f_{ij} is a separation factor (also termed deterrence function) that measures separation from i to j. The exact functional form of these three terms is subject to varying degrees of conjecture (see Fotheringham and O'Kelly, 1989).

11.2.2. Specification of the deterrence function

The separation factor reflects the attitudes of actors towards spatial separation, and, thus, constitutes the very core of spatial interaction models. Hence a number of alternative specifications of f_{ij} have been proposed. Two forms dominate the literature: the power specification and the exponential specification which may be combined into a more flexible two-parameter family of gamma deterrence functions for any positive scalar measure of separation or interaction costs, c_{ij}:

$$f_{ij} = f(c_{ij}) = (c_{ij})^{-\alpha} \exp(-\beta\, c_{ij}) \tag{11.5}$$

with corresponding cost sensitivity parameters α and β. This specification is especially advantageous from a statistical point of view when comparing the relative appropriateness of power specifications and exponential specifications (see, for example, Morrill and Pitts, 1967).

11.2.3. Four different cases

Unfortunately, model (11.4) does not always turn out to be adequate. This is because it may generate estimates of interaction which – when summed over the rows and columns of the interaction matrix – give results that are inconsistent with the a priori known number of flows leaving an origin or arriving at a destination, and we may wish the model to reproduce these given totals a priori (Bailey and Gatrell, 1995). Consequently, it is useful to distinguish various cases of the spatial interaction problem.

Depending on the kinds of a priori information given we may distinguish four obvious cases: (i) neither origin nor destination totals are a priori known (the *unconstrained case*), (ii) origin totals are a priori known (the *production constrained case*), (iii) destination totals are a priori known (the *attraction constrained case*), and finally (iv) both origin and destination totals are a priori known (the *production–attraction constrained case*). Cases (ii) and

(iii) are given the generic name of singly constrained and case (iv) is known as *doubly constrained*. These four cases lead to different specifications of the term $b_{(ij)}$ in Equation (11.4).

11.2.3.1. The unconstrained case

Here we do not need a conservation principle so that

$$b_{(ij)} = b \tag{11.6a}$$

where b is a factor independent of all origins and destinations. If $t_{\bullet\bullet}$ denotes the total number of interactions in the spatial system,

$$b = t_{\bullet\bullet}\Big(\sum_{\substack{i=1\\ i\neq j}}^{I} \sum_{j=1}^{J} r_i\, s_j\, f_{ij}\Big)^{-1}. \tag{11.6b}$$

11.2.3.2. The production constrained case

In this case the row totals $t_{i\bullet}$, are assumed to be a priori known. The conservation principle is enforced from the viewpoint of origins only. Thus

$$b_{(ij)} = b_{(i)} \tag{11.7a}$$

where $b_{(i)}$ is an origin specific constant given by

$$b_{(i)} = t_{i\bullet}\Big(r_i \sum_{\substack{j=1\\ j\neq i}}^{J} s_j\, f_{ij}\Big)^{-1} \qquad i = 1, \ldots, I \tag{11.7b}$$

that guarantees that

$$\sum_{j=1}^{J} \tau_{ij} = t_{i\bullet} \qquad i = 1, \ldots, I. \tag{11.7c}$$

11.2.3.3. The attraction constrained case

This case is the mirror image of the previous case. Now the column totals $t_{\bullet j}$ are given a priori. The conservation principle is enforced from the viewpoint of destinations. Thus

$$b_{(ij)} = b_{(j)} \tag{11.8a}$$

where $b_{(j)}$ is a destination specific constant given by

$$b_{(j)} = t_{\bullet j}(s_i \sum_{\substack{i=1 \\ i \neq j}}^{J} r_j f_{ij})^{-1} \qquad j = 1, \ldots, J \tag{11.8b}$$

that ensures

$$\sum_{i=1}^{J} \tau_{ij} = t_{\bullet j} \qquad j = 1, \ldots, J. \tag{11.8c}$$

11.2.3.4. The production–attraction constrained case

In this case $t_{i\bullet}$ and $t_{\bullet j}$ are given a priori. Thus, the conservation principle is enforced from the viewpoint of both of origins and destinations. Generally, it is assumed for simplicity (see Wilson, 1967, among others) that[2]

$$b_{(ij)} = b_{(i)}\, b_{(j)} \tag{11.9a}$$

where

$$b_{(i)} = t_{i\bullet}(r_i \sum_{\substack{j=1 \\ j \neq i}}^{J} b_{(j)}\, s_j f_{ij})^{-1} \qquad i = 1, \ldots, I \tag{11.9b}$$

$$b_{(j)} = t_{j\bullet}(s_j \sum_{\substack{i=1 \\ i \neq j}}^{J} b_{(i)}\, r_i f_{ij})^{-1} \qquad j = 1, \ldots, J \tag{11.9c}$$

---

[2] As an alternative it can be assumed (Tobler, 1983): $b_{(ij)} = b_{(i)} + b_{(j)}$. In this case the origin-specific and destination-specific constants, $b_{(i)}$ and $b_{(j)}$, respectively, appear as solutions of the system (Dorigo and Tobler, 1983):

$$b_{(i)} = (t_{i\bullet} - r_i \sum_{\substack{j=1 \\ j \neq i}}^{J} b_{(j)} s_j f_{ij})(r_i \sum_{\substack{j=1 \\ j \neq i}}^{J} s_j f_{ij})^{-1} \text{ and } b_{(j)} = (t_{\bullet j} - s_j \sum_{\substack{i=1 \\ i \neq j}}^{I} b_{(i)} r_i f_{ij})(s_j \sum_{\substack{i=1 \\ i \neq j}}^{I} r_i f_{ij})^{-1}$$

that guarantee that

$$\sum_{j=1}^{J} \tau_{ij} = t_{i\bullet} \qquad i = 1, \ldots, I \tag{11.9d}$$

$$\sum_{i=1}^{I} \tau_{ij} = t_{\bullet j} \qquad j = 1, \ldots, J. \tag{11.9e}$$

It should be noted that each $b_{(i)}$ is dependent on all the $b_{(j)}$ in this case, and vice versa. This means that Equations (11.9b) and (11.9c) have to be solved iteratively. The process that we have defined with the $b_{(i)}$ and $b_{(j)}$ factors can be considered to balance an existing interaction matrix – the $\boldsymbol{T}_{I\times J}$-matrix as given by Equation (11.1) – so that it conforms to the a priori given row and column totals (see Equations (11.9d) and (11.9e)). Finally, it is worthwhile to note that the $b_{(i)}$ and $b_{(j)}$ in the doubly constrained case are *not* the same as those appearing in the two singly constrained models. We do not distinguish them by a special notation because it is always clear in a particular context, where only one model is in use, which ones are involved.

11.3. The statistical equilibrium

In view of the interest generated by gravity models since the 1940s, it is not surprising that different theoretical approaches have been suggested to provide a more solid foundation for the wide spread use of gravity models. A probabilistic approach based on statistical equilibrium concepts from statistical mechanics was first proposed by Wilson (1967, 1970) and later extended by many others (see for example, Fisk and Brown, 1975; Snickars and Weibull, 1977; Roy and Lesse, 1981; Smith, 1988; Roy, 2004). Without loss of generality we will illustrate the idea behind the approach, using the production–attraction constrained case of spatial interaction as context. Since it is the expected or mean flow τ_{ij} that we attempt to model, we consider how we might theoretically expect a set of such mean flow to arise in an origin–destination-constrained case.

Recall that we are interested in bundles of individual flows – as described by t_{ij} – from a particular location i to a particular location j

and not in individual flows (so-called microinteractions). Then it can be shown that the probability occurring of a spatial interaction pattern – essentially a spatial interaction matrix – is proportional to the number of ways, say W, in which units being moved (person, commodity, etc.) could be arranged in these bundles to produce that pattern. If $\tau_{..}$ is the total of all such flows, combinatorial theory shows this number for a particular pattern to be

$$W = \frac{\tau_{..}!}{\prod_{i=1}^{I} \prod_{j=1}^{J} \tau_{ij}!} \tag{11.10}$$

We can, thus, derive a most probable spatial pattern by maximising this function which is called *entropy function* in physics, subject to constraints (11.9d) and (11.9e). But we have to add another constraint, which reflects the propensity of individuals to interact across space. If C denotes the total interaction cost, then

$$\sum_{i=1}^{I} \sum_{j=1}^{J} c_{ij} \, \tau_{ij} = C \tag{11.11}$$

where c_{ij} denotes the cost from location i to location j. Finding the most likely τ_{ij} out of all possible sets of flows that would satisfy the three constraints is essentially equivalent to choosing values for the τ_{ij} in such a way as to maximise W subject to the constraints.

In practice, it is more convenient to maximise $S = (1/t_{..}) \ln W$ rather than W. Since S is a monotonic function of W this does not alter the optimisation result, but is mathematically more tractable. If all the τ_{ij}-values are large, we can make use of Stirling's approximation[3] that

$$\ln \tau_{..}! = \tau_{..} \ln \tau_{..} - \tau_{..} \tag{11.12}$$

and

$$\ln \tau_{ij}! = \tau_{ij} \ln \tau_{ij} - \tau_{ij} \tag{11.13}$$

---

[3] Note, however, when the τ_{ij}-values are small, the approximation is rather poor.

to obtain

$$S = (1/\tau_{\cdot\cdot})(\tau_{\cdot\cdot} \ln \tau_{\cdot\cdot} - \tau_{\cdot\cdot} - \sum_{i=1}^{I} \sum_{j=1}^{J} \tau_{ij} \ln \tau_{ij} + \tau_{\cdot\cdot}) \tag{11.14}$$

Rearranging and ignoring constants this can be written as

$$S = -\sum_{i=1}^{I} \sum_{j=1}^{J} (\tau_{ij}/\tau_{\cdot\cdot}) \ln (\tau_{ij}/\tau_{\cdot\cdot}) \tag{11.15a}$$

or equivalently, defining p_{ij}, the proportion of all interactions that originate at location i and terminate at j, as $\tau_{ij}/\tau_{\cdot\cdot}$

$$S = -\sum_{i=1}^{I} \sum_{j=1}^{J} p_{ij} \ln p_{ij} \tag{11.15b}$$

which is the formula for the entropy of a distribution (Shannon and Weaver, 1949; Georgescu-Roegen, 1971) and can be interpreted as a measure of uncertainty about which microstate actually produces the observed macrostate.

The full maximisation problem for the production–attraction constrained case of spatial interaction can then be stated as

$$\max_{\boldsymbol{T}_{I\times J}} S = -\sum_{i=1}^{I} \sum_{j=1}^{J} p_{ij} \ln p_{ij} \tag{11.16a}$$

with $p_{ij} = t_{ij}/t_{\cdot\cdot}$, or equivalently

$$\max_{\boldsymbol{T}_{I\times J}} S = -\sum_{i=1}^{I} \sum_{j=1}^{J} (t_{ij}/t_{\cdot\cdot}) \ln (t_{ij}/t_{\cdot\cdot}) \tag{11.16b}$$

subject to the constraints (11.9d), (11.9e) and (11.11). This is a non-linear optimisation problem that can be solved using Lagrange multipliers to take account of the three constraints.

The result obtained is that any choice of τ_{ij} that will maximise the entropy function must satisfy the general equation

$$-\log (\tau_{ij}) - \lambda_i^{(1)} - \lambda_j^{(2)} - \beta c_{ij} = 0 \tag{11.17}$$

where $\lambda_i^{(1)}$ are the Lagrange multipliers used to take account of the origin constraint (11.9d); $\lambda_j^{(2)}$ those to take account of the destination constraint (11.9e); and β that to take account of the total cost constraint (11.11).

The doubly constrained entropy maximising model. Entropy maximisation, thus, leads to a model for τ_{ij} of the general form

$$\tau_{ij} = b_{(i)}\, b_{(j)}\, r_i\, s_j \exp(-\beta\, c_{ij}) \tag{11.18a}$$

with

$$b_{(i)} = \frac{1}{\sum_{j=1}^{J} b_{(j)}\, s_j \exp(-\beta\, c_{ij})} \tag{11.18b}$$

$$b_{(j)} = \frac{1}{\sum_{i=1}^{I} b_{(i)}\, r_i \exp(-\beta\, c_{ij})} \tag{11.18c}$$

where $r_i = \exp(-\lambda_i^{(1)})$ and $s_j = \exp(-\lambda_j^{(2)})$ simply re-express the Lagrange multipliers in a more convenient form. The r_i are interpreted as a set of parameters that characterise the propensity of each origin i to generate flows, the s_j a set of parameters that characterise the attractiveness of each destination j and β a distance deterrence effect. It is worth noting that the negative exponential separation function emerges directly from the derivation. Note, moreover, that if c_{ij} is taken as the logarithm of interaction cost then we arrive at a power-deterrence rather than an exponential-deterrence function.

11.4. The choice-theoretic approach

The entropy maximising approach was followed by a host of alternative derivations. The most important is the choice-theoretic approach that was first proposed by Niedercorn and Bechdolt (1969) and has generated a great deal of interest since then (see Golob and Beckman, 1971; Choukroun, 1975; Nijkamp, 1975; Smith, 1975, 1978; Batten and Boyce, 1986; Fotheringham and O'Kelly, 1989; Fotheringham *et al.*, 2000; Roy, 2004). The essential idea of this approach is to model spatial interaction behaviour within the microeconomic paradigm of random utility maximising choice behaviour (see Fischer and Nijkamp, 1985, among others). Without loss of generality we shall briefly review the theoretical derivation of the multinomial logit model (the most popular model) using the production constrained case for illustration purposes and individual travellers as relevant actors.

To base a spatial interaction model within the microeconomic framework, it is necessary to define a utility function that represents the net benefit of making a trip from i to j for a given purpose. If this was optimised then every individual in a given location i would choose the same destination what is unrealistic. One approach to circumvent this problem is to define an average utility[4] and then add a random component to reflect different perceptions in the population. For an (i,j)-combination, let v_{ij} be that average and let ε_{ij} be a random component. The utility u_{ij} accruing from an individual resident in i selecting destination j is assumed to be

$$u_{ij} = v_{ij}(z_{ij}, \alpha) + \varepsilon_{ij} \tag{11.19}$$

where z represents a vector characterising the choice-destination j and its separation from i. α is a vector of parameters in the utility function. A wide range of models can be derived by making different assumptions about the functional forms of v_{ij} and ε_{ij}.

Given that each alternative has a random component to its utility, we cannot say for certain which destination an individual will select. We can only evaluate each choice option based on the observable component and then derive a probabilistic statement based on the individual's discrete likelihood of choosing a particular destination (Ben-Akiva and Lerman, 1985). An individual will choose destination j if this alternative maximises his/her utility, that is if

$$p_{ij} = \text{Prob}[u_{ij} > u_{ik} \ \text{ for all } \ k \in \boldsymbol{J}, k \neq j] \tag{11.20}$$

that, on substituting Equation (11.19) and rearranging, can be

[4] Note that location [region] i may be conceived of as a collective decision unit.

written as

$$p_{ij} = \text{Prob}[\varepsilon_{ik} > u_{ij} - u_{ik} + \varepsilon_{ij} \text{ for all } k \in \boldsymbol{J}, k \neq j]. \tag{11.21}$$

Recognising that ε_{ij} and ε_{ik} are random terms drawn from a continuous distribution ranging from $-\infty$ to $+\infty$, Equation (11.21) can be written as

$$p_{ij} = \int_{x=-\infty}^{\infty} g(\varepsilon_{ij} = x) \prod_{k(k \neq j)} \int_{y=-\infty}^{v_{ij}-v_{ik}+x} g(\varepsilon_{ik} = y)\mathrm{d}y\,\mathrm{d}x \tag{11.22}$$

where $g(\cdot)$ represents some probability density function of the disturbance terms. A wide range of models can be derived by making different assumptions about the functional form of $g(\cdot)$. The most common assumption used is that the ε_{ij}-terms are distributed according to a Weibull distribution so that

$$g(\varepsilon_{ij}, \mu) = \mu \exp[-\mu(\varepsilon_{ij} - \eta)] \exp[-\exp[-\mu(\varepsilon_{ij} - \eta)]] \tag{11.23}$$

with μ denoting a positive scale parameter and η a parameter that determines the mode of the distribution. For obvious reasons, this function is sometimes known as the double-exponential distribution. It is not surprising that the Weibull distribution has a family resemblance to the entropy function (see Wilson, 2000).

Then it can be shown that the model which results is the multinomial logit model version developed by McFadden (1973) and others

$$p_{ij} = \frac{\exp(\mu\, v_{ij})}{\sum_k \exp(\mu\, v_{ij})} \tag{11.24}$$

that clearly forms the basis for the production constrained spatial interaction model. For example, if v_{ij} is defined as

$$v_{ij} = \alpha \ln z_{ij} \tag{11.25}$$

the resulting model form is equivalent to the production-constrained spatial interaction model described in the previous sections. Use of this functional specification is reinforced by its intuitively appealing representation of the behavioural response of individuals to changing levels of observable utility as illustrated for the case of retail choice in Fotheringham *et al.* (2000).

It is interesting to note that random utility maximising models are traditionally estimated from disaggregate data, while gravity and entropy maximising spatial interaction models are generally estimated from aggregate data. But it is perfectly reasonable and computationally feasible to recast the latter into a disaggregate form or the former into an aggregate one. The entropy-maximising and the choice-theoretic approaches can then be recognised as complementary views that may lead to identical parameter estimation equations and solutions for the same problem. The fact that spatial interaction models estimated utilising one approach have produced different results from similar models estimated via another approach is not necessarily due to any fundamental difference in principle, but may simply be associated to differences in the use of data, its aggregation and value judgements used in specifying explanatory variables[5] (Batten and Boyce, 1986).

The choice-theoretic approach to modelling spatial interactions provides an attractive statistical framework for testing spatial interaction hypotheses such as specific perceived distance hypotheses. In particular the multinomial logit model is directly applicable to spatial choice behaviour and – under appropriate utility hypotheses – guarantees spatial interaction models with exponential separation functions (see Batten and Boyce, 1986; Fotheringham and O'Kelly, 1989, and many others).

The approach was developed for aspatial contexts such as the choice of transportation mode. As such it is based on assumptions that – while appropriate in most aspatial contexts – however, are generally not tenable when applied to larger sized spatial interaction problems. In Equation (11.20) and subsequently, for example, the assumption is made that an individual is capable to evaluate all choice alternatives, i.e. destination choice j is compared with all the options in the full choice set. In principle, the individual is

---

[5] For further information concerning the reconciliation and unification of the entropy-maximising and the random utility maximising approaches, see Choukroun (1975); Anas (1981); Smith (1984); Batten and Boyce (1986).

assumed to be omniscient and capable to process vast amounts of information on all the alternatives. But it is increasingly recognised that individuals have a limited capacity for information processing, and Bettman (1979), for example, argues that the limit might be reached with as few as six or seven alternatives. Consequently, in most spatial interaction situations the number of alternatives is far too large to assume that individuals can evaluate all possible choice options.

This shortcoming of the approach has inspired Fotheringham (1983) to develop a new form of spatial interaction model, called a *competing destinations model*. The derivation of the model closely follows that of the logit model described in Equation (11.24) with the added flexibility to allow individuals to make choices from restricted choice sets rather than from the complete set of alternatives. This is achieved by replacing Equation (11.20) with

$$\begin{aligned} p_{ij} =& \text{Prob}[u_{ij} > u_{ik} + \ln p_i(k \in \boldsymbol{J}') \\ & \text{for all } \; k \in \boldsymbol{J}, k \neq j] p_i(j \in \boldsymbol{J}')] \end{aligned} \tag{11.26}$$

where $\boldsymbol{J}'$ is the restricted choice set in which an individual at i evaluates all alternatives and $\boldsymbol{J}$ is the full set of destination alternatives. Destinations that are not within $\boldsymbol{J}'$ are not evaluated and, thus, cannot be chosen by the individual. For more details see Fotheringham *et al.* (2000).

11.5. The neural network approach

The emergence of GeoComputation as a subject (see Longley *et al.*, 1998; Fischer and Leung, 2001) and the powerful and fast computing environment now upon us has inspired many scholars to apply neurocomputing principles and techniques to revisit old and to solve new spatial interaction problems (see Fischer and Getis, 1999; Fischer, 2001a). The interaction models derived are given a very general formulation represented in form of specific neural networks and viewed as universal function approximators. They tend to be especially useful in data rich spatial interaction environments where little is known about the true spatial interaction function (Fischer and Reismann, 2002b).

11.5.1. The unconstrained case of neural spatial interaction modelling

Suppose we are interested in approximating a spatial interaction function $\Phi : \Re^N \rightarrow \Re$ where $\Re^N$ is the N-dimensional input space and $\Re$ the one-dimensional output space. The function is not explicitly known, but given by a finite set of samples $M = \{(x_u, y_u), x_u \in \Re^N, y_u \in \Re, u = 1, \ldots, U\}$. The set M is the set of pairs of input and output vectors. The task is to find a continuous function that approximates set M. In real world contexts, U is a small number and the samples contain noise.

There is a growing literature that deals with alternative model specifications to approximate Φ. Examples include, among others, Gopal and Fischer (1993); Openshaw (1993); Fischer and Gopal (1994); Black (1995); Gopal and Fischer (1996); Nijkamp *et al.* (1996); Bergkvist and Westin (1997); Openshaw (1998); Reggiani and Tritapepe (1998); Bergkvist (2000); Thill and Mozolin (2000); Mozolin *et al.* (2000). It can be shown that all these models are members of the following general class of unconstrained neural spatial interaction models given by

$$\Omega^H(\boldsymbol{x}, \boldsymbol{w}^H) = \psi\Big(w_{00} + \sum_{h=1}^{H} w_{0h}\, \varphi\Big(\sum_{n=1}^{N} w_{1hn}\, x_n\Big)\Big) \tag{11.27}$$

where the N-dimensional Euclidean space (generally, $N = 3$) is the input space and the one-dimensional Euclidean space the output space. Vector $\boldsymbol{x} = (x_1, \ldots, x_N)$ is the input vector that represents measures characterising the origin and the destination of spatial interaction as well as their separation. $\boldsymbol{w}^H \equiv (\boldsymbol{w}_0, \boldsymbol{w}_1)$ is the $(HN + H + 1) \times 1$ vector of the network weights (parameters). There are H hidden units. The vector $\boldsymbol{w}_0$ contains the hidden to output unit weights, $\boldsymbol{w}_0 \equiv (w_{00}, w_{01}, \ldots, w_{0H})$, and the vector $\boldsymbol{w}_1$ contains the input to hidden unit weights, $\boldsymbol{w}_1 \equiv (w_{10}, \ldots, w_{1H})$ with $\boldsymbol{w}_{1h} \equiv (w_{1h1}, \ldots, w_{1hN})$. We allow a bias at the hidden layer by including w_{00}. A bias at the input array may be taken into consideration by setting $x_1 \equiv 1$. φ is a hidden layer transfer function, ψ an output unit transfer function, both continuously differentiable of order 2 on $\Re$. φ and ψ are often specified as logistic functions. Note that the model output function and the weight vector are explicitly indexed by the number, H, of hidden units in order to indicate the dependence (Fischer and Reismann, 2002b).

The novelty and fundamental contribution of the neural network approach to spatial interaction analysis derives not so much from the associated learning methods to determine the model parameters but from its focus on functions such as φ and ψ. Neural networks such as Ω^H can approximate any continuous function uniformly on compacta, by increasing the number of hidden units. There are also some results on the rate of approximation, i.e. how many hidden units are needed to approximate to a specified accuracy. As always, however, with such results, they are no guide to how many hidden units might be needed in any practical problem (Fischer, 2001b).

Neural network models of the form (11.27) may be of little practical value if a priori information is available on accounting constraints of the predicted flows. For this purpose Fischer *et al.* (2003) have developed a novel class of neural spatial interaction models that are able to deal efficiently with the singly constrained case of spatial interaction.

11.5.2. The class of singly constrained neural spatial interaction models

The models are based on a modular connectionist architecture operating under supervised learning algorithms. Modularity is seen here as decomposition on the computational level. The network is composed of two processing layers and two layers of network parameters. The first processing layer is involved with the extraction of features from the input data. This layer is implemented as a layer of J functionally independent modules with identical topologies. Each module is a feedforward network with two inputs x_{2j-1} and x_{2j} (representing measures of destination attractiveness and separation between origin and destination, in the origin-constrained case, for example), H hidden product units, and terminates with a single summation unit. The collective output of these modules constitutes the input to the second processing layer consisting of J output units that perform the flow prediction (Fischer and Reismann, 2002b).

This network architecture implements the *general class of neural models in the case of origin constrained (OC) spatial interaction* defined by

$$ {}^{\pi}\Omega^{\mathrm{OC}}(\boldsymbol{x}, \boldsymbol{w})_j = \psi_j\Big(\sum_{h=1}^{H} \gamma_h\, \varphi_h\Big(\prod_{n=2j-1}^{2j} x_n^{\beta_{hn}}\Big)\Big) \qquad j = 1, \ldots, J \tag{11.28} $$

with $\varphi_h : \Re \rightarrow \Re$, $\psi_j : \Re \rightarrow \Re$ and $\boldsymbol{x} \in \Re^{2J}$, i.e. $\boldsymbol{x} = (x_1, x_2, \ldots, x_{2j-1}, x_{2j}, \ldots, x_{2J-1}, x_{2J})$ where x_{2j-1} represents a variable pertaining to destination j ($j = 1, \ldots, J$) and x_{2j} a variable f_{ij} pertaining to the separation from region i to region j ($i = 1, \ldots, I$; $j = 1, \ldots, J$) of the spatial interaction system under scrutiny. β_{hm} ($h = 1, \ldots, H$; $n = 2j - 1, 2j$) are the input-to-hidden connection weights, and γ_h ($h = 1, \ldots, H$) the hidden-to-output weights in the jth module of the network model. The symbol $\boldsymbol{w}$ is a convenient shorthand notation of the $(3H)$-dimensional vector of all the model parameters. ψ_j ($j = 1, \ldots, J$) represents a non-linear summation unit transfer function and φ_h ($h = 1, \ldots, H$) a linear hidden product unit transfer function. For additional information, the reader is referred to the description in Fischer *et al.* (2003) and Fischer (2002b).

11.5.3. The modelling process

The neural network approach to modelling spatial interactions involves three major stages (Fischer and Gopal, 1994):

- The *first stage* consists of the identification of a model candidate from the general class of neural spatial interaction models of type (11.27) or (11.28). This involves both the specification of appropriate transfer functions ψ and φ, and the number, H, of hidden units.
- The *second stage* involves the network training (network learning, parameter estimation) problem, that is the determination of an optimal set of model parameters where optimality is defined in terms of an error (loss, performance) function.
- The *third stage* is concerned with testing and evaluating the out-of-sample (generalisation) performance of the chosen model.

Both the theoretical and practical side of the model selection problem have been intensively studied (see Fischer, 2000, 2001b among others). The standard approach for finding a good neural spatial interaction model is to split the available set of samples into three subsets: training, validation and test sets. The training set is used for parameter estimation. In order to avoid overfitting, a common procedure is to use a network model with sufficiently large H for the task, to monitor – during training – the out-of-sample performance on a separate validation set, and finally to choose the model that corresponds to the minimum on the validation set, and employ it for future purposes such as the evaluation on the test set.

It has been frequent practice to fix these sets. But recent experience has found this approach to be very sensitive to the specific splitting of the data. Fischer and Reismann (2002a) show that the bootstrapping pairs approach with replacement may be adopted to combine the purity of data splitting with the power of a resampling procedure to overcome the shortcomings of fixed data splitting. This approach, moreover, has the power to provide a better statistical picture of the prediction variability of the model under scrutiny.

11.5.4. The network learning problem and parameter estimation procedures

If we view models of type (11.27) or (11.28) as generating a family of approximations – as $\boldsymbol{w}$ ranges over $\boldsymbol{W}$, say – to the explicitly unknown spatial interaction function, then we need a way to pick the best approximation from this family. This is the function of network training (learning) or parameter estimation in neural network modelling.

The optimisation problem. It is convenient to consider network learning as unconstrained non-linear minimisation problem in which the objective function is defined by a loss (error, cost) function and the search space by the Q-dimensional parameter space where $Q = (MN + H + 1)$ in the unconstrained case and $Q = 3H$ in the singly constrained case of spatial interaction. Formally

$$\min_{w \in W} \lambda(\boldsymbol{x}, \boldsymbol{y}, \boldsymbol{w}) \tag{11.29}$$

where λ represents the loss function measuring the performance given the parameter $\boldsymbol{w}$ and observations $(\boldsymbol{x}, \boldsymbol{y})$. It is evident that the choice of the loss function plays a crucial role in the determination of the optimal parameter.

In a spatial interaction world it makes sense to assume observations generated as the realisation of a sequence of random vectors defined on a Poisson probability space and hence to choose $\lambda(\boldsymbol{x}, \boldsymbol{y}, \boldsymbol{w})$ to be the negative of the log-likelihood function (Fischer, 2002a). Thus

$$\min_{w \in W} \lambda(\boldsymbol{x}, \boldsymbol{y}, \boldsymbol{w}) = \min_{w \in W} \left[- \left(\sum_{u} y_u \ln \Omega(\boldsymbol{w}_u) - \Omega(\boldsymbol{w}_u) \right) \right] \tag{11.30}$$

$(\boldsymbol{x}, \boldsymbol{y}) = \{(\boldsymbol{x}_u, \boldsymbol{y}_u), u = 1, \ldots, U\}$ denotes a sequence of observations. The solutions of Equation (11.30) yield ML-estimates of the model parameters.

$\lambda(\boldsymbol{x}, \boldsymbol{y}, \boldsymbol{w})$ is non-negative, continuously differentiable on the Q-dimensional parameter space, which is a finite dimensional closed bounded domain and, thus, compact. The compactness of the parameter space is of great theoretical convenience. It can be shown that λ assumes its weight minimum under certain conditions, but characteristically there exist many minima in real world applications all of which satisfy

$$\nabla\lambda(\boldsymbol{x}, \boldsymbol{y}, \boldsymbol{w}) = 0 \tag{11.31}$$

where $\nabla\lambda$ denotes the gradient of λ. The minimum for which the value of λ is smallest is termed the global minimum and other minima are called local minima. Unfortunately there is no analytical solution to this learning problem. But computationally intensive procedures may be used to solve the problem, all of which yield ML-estimates of the model parameters. We may distinguish two classes of procedures to finding the minimum of the loss function: local search and global search procedures.

Local search procedures characteristically use derivative information of λ within a local iterative process in which an approximation to the function in a neighbourhood of the current point in parameter space is minimised. The approximation is often given by a first- or second-order Taylor expansion of the function.

The general scheme of the iteration process may be formulated as follows (Fischer, 2001b):

(i) choose an initial vector $\boldsymbol{w}$ in parameter space and set $\tau = 1$,

(ii) determine a search direction $\boldsymbol{d}(\tau)$ and a step size $\eta(\tau)$ so that

$$\lambda(\boldsymbol{w}(\tau) + \eta(\tau)\boldsymbol{d}(\tau)) < \lambda(\boldsymbol{w}(\tau)) \qquad \tau = 1, 2, \ldots \tag{11.32}$$

(iii) update the parameter vector

$$\boldsymbol{w}(\tau + 1) = \boldsymbol{w}(\tau) + \eta(\tau)\boldsymbol{d}(\tau)) \qquad \tau = 1, 2, \ldots \tag{11.33}$$

(iv) if $\mathrm{d}\lambda(\boldsymbol{w})/\mathrm{d}\boldsymbol{w} \neq 0$ then set $\tau = \tau + 1$ and go to (ii), else return $\boldsymbol{w}(\tau + 1)$ as the desired minimum.

Determining the next current point in the iteration process entails two problems. First, the search direction $\boldsymbol{d}(\tau)$ has to be determined, i.e. what direction in parameter space we want to go in the search for a new current point. Second, once the search direction has been found, we have to decide how far to go in the specified direction, i.e. step size $\eta(\tau)$ has to be determined. To solve these problems, normally two types of operation must be carried out: the computation or the evaluation of the derivatives of the loss function with respect to the model parameters, and the computation of the parameter $\eta(\tau)$ and the direction vector $\boldsymbol{d}(\tau)$ based upon these derivatives. The evaluation of the loss function is most commonly performed by the backpropagation technique which provides a computationally efficient procedure for doing this. Gradient descent, conjugate gradient and quasi-Newton procedures are characteristically used for the computation of the parameter $\eta(\tau)$ and the direction vector $\boldsymbol{d}(\tau)$. See Fischer (2001b) for more details.

Local search procedures find the local minima efficiently and work best in unimodal problems. But they have difficulties when the surface of the parameter space is flat (i.e. gradients close to zero), when there is a large range of gradients, and when the surface is very rugged. The search may progress too slowly when the gradient is small, and may overshoot where the gradient is large. When the error surface is rugged, a local search from a random starting point converges to a local minimum close to the initial point and worse solution than the global minimum (Fischer, 2001b).

Global search algorithms employ heuristics to be able to escape from local minima. These algorithms can be classified as probabilistic or deterministic. Of the few deterministic global minimisation methods developed, most apply deterministic heuristics to bring search out of a local minimum. Other methods, like covering methods, recursively partition the search space into subspaces before searching. None of these methods operates well or provides adequate coverage when the search space is large, as is usually the case in neural spatial interaction modelling.

Probabilistic global minimisation methods rely on probability to generate decisions. The simplest probabilistic algorithm uses restarts to bring search out of a local minimum when little improvement can be made locally. More advanced methods rely on probability to indicate whether a search should ascend from a local minimum: simulated annealing, for example, when it accepts uphill movements. Other probabilistic algorithms rely on probability to decide which intermediate points to interpolate as new trial parameter vectors: random re-combinations or mutations in evolutionary algorithms (see, for example, Fischer and Leung, 1998).

The success of global search procedures in finding a global minimum of a given function such as λ over $\boldsymbol{w} \in \boldsymbol{W}$ hinges on the balance between an exploration process, a guidance process and a convergence-inducing process. The *exploration process* gives the search a mechanism for sampling a sufficiently diverse set of parameters $\boldsymbol{w}$ in $\boldsymbol{W}$. This exploration process is generally stochastic in nature. The *guidance process* is an implicit process that evaluates the relative quality of search points and biases the exploration process to move toward regions of high-quality solutions in $\boldsymbol{W}$. The *convergence-inducing process* finally ensures the convergence of the search to find a fixed solution $\hat{\boldsymbol{w}}$. The dynamic interaction among these three processes is responsible for giving the search process its global optimising character (Hassoun, 1995). An example of a powerful global search procedure is Alopex, a correlation-based method for solving the maximum likelihood problem. The reader interested in details of the procedure is referred to Fischer and Reismann (2002b).

Global search procedures such as Alopex based search – as opposed to local search – have to be used in network training problems where reaching the global optimum is at premium.

The price one pays for using global search procedures is increased computational requirements. The intrinsic slowness of such procedures is mainly due to the slow but crucial exploration process. This may motivate the development of a hybrid procedure that uses global search to identify regions of the parameter space containing local minima and gradient information to actually find them (Fischer, 2002a).

11.6. Concluding remarks

In this chapter we have briefly and selectively reviewed some modelling efforts that have characterised the field of spatial interaction over the last few decades. We have witnessed the progression from gravity models to entropy maximising and random utility maximising models and finally to models based on neurocomputing principles. Despite the advances made in the theory of neural spatial interaction modelling there is no doubt in mind that parametric models based upon entropy maximising or random utility maximising principles will remain important tools for use in applied research. The appeal of these models can be attributed both to the simplicity of their mathematical form and the theoretical nature of their underlying assumptions. They tend to be less useful, however, in situations where little is known about the form of the spatial interaction function to be approximated.

The neural network approach to spatial interaction is attractive, particularly in data rich, but theory poor spatial interaction contexts. Neural network models can approximate virtually any spatial interaction function of interest to any desired degree of accuracy, provided sufficiently many hidden units are available. These results establish neural spatial interaction models as a class of universal approximators. As such, failures in applications can be attributed to inadequate learning, inadequate numbers of hidden units, or the presence of a stochastic rather than a deterministic relation between input and target. Consequently, the stages of model choice and parameter estimation are of crucial importance for the success of real world applications.

Several recent studies have illustrated that many aspects of the study of neural spatial interaction models lend themselves to rigorous mathematical analysis. This provides a firm foundation of neural spatial interaction modelling on which to base real world projects. Much progress has been made in the theory and methodology. But several important areas remain for further research. The design of a neural network approach suited to deal with the doubly constrained case is still missing. Finding good global optimisation procedures for solving the non-convex learning problems is still an important issue for further research even though some relevant work can be found in Fischer *et al.* (1999) and Fischer and Reismann (2002a,b). Also the model choice problem deserves further attention to come up with techniques that go beyond the current rules of thumb.

References

Anas, A. (1981), "The estimation of multinomial logit models of joint location and mode choice from aggregated data", *Journal of Regional Science*, Vol. 21, pp. 223–242.

Bailey, T.C. and A.C. Gatrell (1995), *Interactive Spatial Data Analysis*, Essex: Longman.

Batten, D.F. and D.E. Boyce (1986), "Spatial interaction, transportation, and interregional commodity flow models", pp. 357–406, in: P. Nijkamp, editor, *Handbook of Regional and Urban Economics*, Vol. I, Amsterdam: North-Holland.

Ben-Akiva, M. and S.R. Lerman (1985), *Discrete Choice Analysis. Theory and Application to Travel Demand*, Cambridge, MA: The MIT Press.

Bergkvist, E. (2000), "Forecasting interregional freight flows by gravity models", *Jahrbuch für Regionalwissenschaft*, Vol. 20, pp. 133–148.

Bergkvist, E. and L. Westin (1997), Estimation of gravity models by OLS estimation, NLS estimation, Poisson and neural network specifications. CERUM Regional Dimensions, Working Paper No. 6.

Bettman, J.R. (1979), *An Information Processing Theory of Consumer Choice*, Reading, MA: Addison-Wesley.

Black, W.R. (1995), "Spatial interaction modelling using artificial neural networks", *Journal of Transport Geography*, Vol. 3(3), pp. 159–166.

Choukroun, J.M. (1975), "A general framework for the development of gravity-type distribution models", *Regional Science and Urban Economics*, Vol. 5, pp. 177–202.

Dorigo, G. and W. Tobler (1983), "Push-pull migration laws", *Annals of the Association of American Geographers*, Vol. 73, pp. 1–17.

Fischer, M.M. (2000), "Methodological challenges in neural spatial interaction modelling: the issue of model selection", pp. 89–101, in: A. Reggiani, editor, *Spatial Economic Science: New Frontiers in Theory and Methodology*, Berlin: Springer.

Fischer, M.M. (2001a), "Spatial analysis in geography", pp. 14752–14758, in: N.J. Smelser and P.B. Baltes, editors, *International Encyclopedia of the Social and Behavioral Sciences*, Vol. 22, Oxford: Elsevier.

Fischer, M.M. (2001b), "Neural spatial interaction models", pp. 195–219, in: M.M. Fischer and Y. Leung, editors, *GeoComputational Modelling: Techniques and Applications*, Berlin: Springer.

Fischer, M.M. (2002a), "Learning in neural spatial interaction models: a statistical perspective", *Journal of Geographical Systems*, Vol. 4(3), pp. 287–299.

Fischer, M.M. (2002b), A novel modular product unit neural network for modelling constrained spatial interaction flows, *Proceedings of the IEEE 2002 World Congress on Computational Intelligence: 2002 Congress on Evolutionary Computation*, Piscataway, NJ: IEEE Press, pp. 1215–1220.

Fischer, M.M. and A. Getis (1999), "New advances in spatial interaction theory", *Papers in Regional Science*, Vol. 78, pp. 117–118.

Fischer, M.M. and S. Gopal (1994), "Artificial neural networks: a new approach to modelling interregional telecommunication flows", *Journal of Regional Science*, Vol. 34(4), pp. 503–527.

Fischer, M.M. and Y. Leung (1998), "A genetic-algorithm based evolutionary computational neural network for modelling spatial interaction data", *The Annals of Regional Science*, Vol. 32(3), pp. 437–458.

Fischer, M.M. and Y. Leung (eds.) (2001), *GeoComputational Modelling: Techniques and Applications*, Berlin: Springer.

Fischer, M.M. and P. Nijkamp (1985), "Developments in explanatory discrete spatial data and choice analysis", *Progress in Human Geography*, Vol. 9, pp. 515–551.

Fischer, M.M. and M. Reismann (2002a), "Evaluating neural spatial interaction modelling by bootstrapping", *Networks and Spatial Economics*, Vol. 2(3), pp. 255–268.

Fischer, M.M. and M. Reismann (2002b), "A methodology for neural spatial interaction modeling", *Geographical Analysis*, Vol. 34(2), pp. 207–228.

Fischer, M.M., K. Hlavackova-Schindler and M. Reismann (1999), "A global search procedure for parameter estimation in neural spatial interaction modelling", *Papers in Regional Science*, Vol. 78, pp. 119–134.

Fischer, M.M., M. Reismann and K. Hlavackova-Schindler (2003), "Neural network modelling of constrained spatial interaction flows: design, estimation and performance issues", *Journal of Regional Science*, Vol. 43(1), pp. 35–61.

Fisk, C.S. and G.R. Brown (1975), "A note on the entropy formulation of distribution models", *Operational Research Quarterly*, Vol. 26, pp. 755–758.

Fotheringham, A.S. (1983), "A new set of spatial interaction models: the theory of competing destinations", *Environment and Planning A*, Vol. 22, pp. 527–549.

Fotheringham, A.S. and M.E. O'Kelly (1989), *Spatial Interaction Models: Formulations and Applications*, Dordrecht: Kluwer.

Fotheringham, A.S., C. Brunsdon and M. Charlton (2000), *Quantitative Geography. Perspectives on Spatial Data Analysis*, London: Sage Publications.

Georgescu-Roegen, N. (1971), *The Entropy Law and the Economic Process*, Cambridge, MA: Harvard University Press.

Golob, T.F. and M.J. Beckman (1971), "A utility model for travel forecasting", *Transportation Science*, Vol. 5, pp. 79–90.

Gopal, S. and M.M. Fischer (1993), Neural net based interregional telephone traffic models. In *Proceedings of the International Joint Conference on Neural Networks IJCNN 93* Nagoya, Japan, October 25–29, pp. 2041–2044.

Gopal, S. and M.M. Fischer (1996), "Learning in single hidden-layer feedforward network models: backpropagation in a spatial interaction context", *Geographical Analysis*, Vol. 28(1), pp. 38–55.

Hassoun, M.H. (1995), *Fundamentals of Artificial Neural Networks*, Cambridge, MA: MIT Press.

Longley, P.A., S.M. Brocks, R. McDonnell and B. MacMillan (eds.) (1998), *Geocomputation: A Primer*, Chichester: Wiley.

McFadden, D. (1973), "Conditional logit analysis of qualitative choice behaviour", pp. 105–142, in: P. Zarembka, editor, *Frontiers in Econometrics*, New York: Academic Press.

Morrill, R.L. and F.R. Pitts (1967), "Marriage, migration, and the mean information field", *Annals of the Association of American Geographers*, Vol. 57, pp. 401–422.

Mozolin, M., J.-C. Thill and E.L. Usery (2000), "Trip distribution forecasting with multilayer perceptron neural networks: a critical evaluation", *Transportation Research B*, Vol. 34, pp. 53–73.

Niedercorn, J.H. and B.V. Bechdolt (1969), "An economic derivation of the 'gravity law' of spatial interaction", *Journal of Regional Science*, Vol. 9, pp. 273–281.

Nijkamp, P. (1975), "Reflections on gravity and entropy models", *Regional Science and Urban Economics*, Vol. 5, pp. 203–225.

Nijkamp, P., A. Reggiani and T. Tritapepe (1996), "Modelling inter-urban transport flows in Italy", *Transportation Research*, Vol. 4C(6), pp. 323–338.

Openshaw, S. (1993), "Modelling spatial interaction using a neural net", pp. 147–164, in: M.M. Fischer and P. Nijkamp, editors, *Geographic Information Systems, Spatial Modeling, and Policy Evaluation*, Berlin: Springer.

Openshaw, S. (1998), "Neural network, genetic, and fuzzy logic models of spatial interaction", *Environment and Planning A*, Vol. 30, pp. 1857–1872.

Reggiani, A. and T. Tritapepe (1998), "Neural networks and logit models applied to commuters' mobility in the metropolitan area of Milan", pp. 111–129, in: V. Himanen, P. Nijkamp and A. Reggiani, editors, *Neural Networks in Transport Applications*, Aldershot: Ashgate.

Roy, J.R. (2004), *Spatial Interaction Modelling. A Regional Science Context*, Berlin: Springer.

Roy, J.R. and P.F. Lesse (1981), "On appropriate microstate descriptions in entropy modelling", *Transportation Research*, Vol. 15B, pp. 85–96.

Sen, A. and T.E. Smith (1995), *Gravity Models of Spatial Interaction Behavior*, Berlin: Springer.

Shannon, C. and W. Weaver (1949), *The Mathematical Theory of Communication*, Urbana: University of Illinois Press.

Smith, T.E. (1975), "A choice theory of spatial interaction", *Regional Science and Urban Economics*, Vol. 5, pp. 137–176.

Smith, T.E. (1978), "A cost-efficiency principle of spatial interaction behavior", *Regional Science and Urban Economics*, Vol. 8, pp. 313–337.

Smith, T.E. (1984), "Testable characterizations of gravity models", *Geographical Analysis*, Vol. 16, pp. 74–94.

Smith, T.E. (1988), "A cost-efficiency theory of dispersed network equilibria", *Environment and Planning A*, Vol. 20, pp. 231–266.

Snickars, F. and J.W. Weibull (1977), "A minimum information principle", *Regional Science and Urban Economics*, Vol. 7, pp. 137–168.

Thill, J.-C. and M. Mozolin (2000), "Feedforward neural networks for spatial interaction: are they trustworthy forecasting tools?", pp. 355–381, in: A. Reggiani, editor, *Spatial Economic Science: New Frontiers in Theory and Methodology*, Berlin: Springer.

Tobler, W. (1983), "An alternative formulation for spatial interaction modelling", *Environment and Planning A*, Vol. 15, pp. 693–703.

Wilson, A.G. (1967), "A statistical theory of spatial distribution models", *Transportation Research*, Vol. 1, pp. 253–269.

Wilson, A.G. (1970), *Entropy in Urban and Regional Planning*, London: Pion.

Wilson, A.G. (2000), *Complex Spatial Systems: The Modelling Foundations of Urban and Regional Analysis*, Harlow: Pearson Education.

Urban Dynamics and Growth, CEA, vol. 266
Roberta Capello and Peter Nijkamp (editors)

CHAPTER 12

Commuting: The Contribution of Search Theory

Jos van Ommeren

Free University, FEWEB, De Boelelaan, 1081 HV Amsterdam, The Netherlands

Abstract

The main objective of this study is to demonstrate the relevance of search theory to analyse the consequences of the assumption that individuals are confronted with a distribution of commuting costs offers, which arises as a consequence of the absence of full information. Market imperfections (transaction costs, uncertainty) are at the heart of search theory.

Keywords: commuting, search, mobility

JEL classifications: R20, R64, J64

12.1. Introduction

Many studies have attempted to answer the following question: given the characteristics of an individual, where would this individual locate his/her residence given the workplace location? Economic theories have contributed a valuable answer to this question. Assuming the existence of a simplified static world with perfect markets, a utility maximising individual accepts the costs of the work trip, because the marginal commuting costs are compensated for by marginal benefits. In other words, the commuter accepts the disutility

---

Jos van Ommeren is affiliated as a Fellow to the Tinbergen Institute, Amsterdam. I would like to thank Marcel Hoogzaad for valuable assistance.

of commuting, because the commuter is compensated by either higher wages or cheaper housing. Based on this compensating principle, various approaches are plausible for analysing commuting behaviour. The main application is in the field of urban economics.

Models in urban economics are based on the assumption that firms and households compete for scarce land for production and housing activities. The underlying behaviour is thought of as bidding behaviour. Following the tradition of Alonso (1964), these urban models contain a typical structure of the urban area. Traditionally, the urban area contains only one centre in which employment is concentrated and land prices decrease gradually from the centre to the rural areas. Employees may choose to locate near the centre (enjoying low commuting costs, but expensive housing) or locate farther from the centre (enjoying higher commuting costs, but cheaper housing). The model has been extended towards more realistic applications, including decentralised employment (inclusion of wage gradients) and several business centres (Muth, 1969; White, 1988; Timothy and Wheaton, 2001).[1]

The standard urban economics model assumes that markets are perfect, and has therefore been criticised (Anas, 1982; Hamilton, 1982, 1989). In particular, imperfect information and moving costs are ignored.[2] Under these assumptions, workers would adjust their

---

[1] For example, Timothy and Wheaton (2001) recently demonstrate that the variation in commuting costs of individual workers employed at the same work location but living at different residence locations is capitalised into land rents. Variation in the average commuting costs between those employed at different work locations will be capitalised into wages. So, the spatial wage variation of individuals depends positively on the *average* commuting time of the individual's zone of employment. This result is essentially an extension of the studies by Muth (1969) and Mills (1972), which showed that given the same work location, differences in commuting costs capitalise into land rents, and the study by Moses (1962), which showed that given the same residence location, differences in commuting costs capitalise into wages. These models may be viewed as long run models if the local agglomeration effect offsets existing wage differentials so that spatial variation in wages and commuting costs is sustainable.

[2] As is nowadays well known Hamilton (1982) raised the question of whether, or not, commuting in US metropolitan areas is inefficient. He argued that 10 times more commuting actually occurs in metropolitan areas than is predicted by urban economic models. Cropper and Gordon (1991), Small and Song (1992) and Kim (1995) also provide evidence that more commuting occurs than the minimum amount required for workers to

residence or workplace location so that the costs due to commuting are fully compensated. In an economy with market imperfections, such as moving costs, however, this will rarely be the case (Weinberg *et al.*, 1981; Zax, 1991; Holzer, 1994). Market imperfections distort dwelling price adjustments and prevent households from moving. Therefore, some authors have expressed their concern about the pitfalls of modelling assuming perfect markets.

This suggests that a better understanding of commuting behaviour may be obtained by focusing on market imperfections, defined here as lack of perfect information and the presence of moving costs.[3] The aim of this chapter is therefore to analyse the consequences of the assumption that individuals are confronted with a *distribution of commuting costs offers*, which arises as a consequence of the absence of full information (and the presence of moving costs). The presence of a commuting costs offer distribution generates a search process. Conceptually, the search processes in the housing and labour market are to a large extent similar, including the decision to search and the decision to move. For simplicity, we focus first only on job mobility, but use the main insights also to understand residential mobility.[4]

---

commute between metropolitan area's existing houses and its existing jobs, but the best evidence suggests that the ratio of actual to minimum commuting is around 2.5 or even less (Kim, 1995). Overall, the 'wasteful' commuting controversy has shown that workers on average commute too much, but the amount of extra commuting may be lower than previously thought. Crane (1996) discusses how uncertainty concerning job locations would effect the ratio of actual-to-minimum commuting. We will explain later on that moving costs and job uncertainty are more relevant when employment is more dispersed.

[3] Factors such as imperfect information and moving costs are very difficult to observe. These factors are, however, directly related to job and residential moving behaviour, which one may observe.

[4] Similar to the labour market literature, search behaviour in the housing market has received extensive attention (Brown and Holmes, 1971; Speare *et al.*, 1975; Smith *et al.*, 1979; Smith and Mertz, 1980; Clark and Flowerdew, 1982; Clark and Smith, 1982; Smith and Clark, 1982; Huff, 1984; Rouwendal and Rietveld, 1988; Clark and Van Lierop, 1986; Wheaton, 1990; Pickles and Davies, 1991; Rouwendal, 1991, 1992; Kooreman and Rouwendal, 1992). Although the importance of commuting costs has been emphasised in the residential mobility literature, in the theoretical residential search literature, commuting costs have been largely ignored (exceptions include Huff, 1984; Rouwendal, 1992).

12.2. Search theory

12.2.1. The basic assumption

Our objective is to derive a theoretical framework that explains the on-the-job relocation behaviour of full-time employed individuals who explicitly take into account commuting costs. The relationship between job mobility and commuting costs will be described from a search-theoretical perspective. For more than two decades, job search theory has been one of the main theoretical and empirical tools for understanding the working of the labour market.[5] Search theory is appealing as it is based on the primary idea that individuals maximise utility by moving through different states and so it is explicitly dynamic. We make the assumption that job offers arrive exogenously at a specified rate and that these offers are instantly accepted or rejected. So job moving behaviour is due to a combination of chance – the arrival of an offer – and a decision-making process – the decision to accept an offer. Such a description reaches the heart of the matter.

In principle, search theory can readily be applied to determine the optimal decision rule of moving job: individuals are thought of as facing a set of alternative employment opportunities. Every job location uniquely determines the commuting costs. The costs and benefits of any job offer are examined, taking into account potential future job offers. These costs and benefits are a function of many characteristics, including personal and household characteristics, current job and dwelling characteristics and commuting distance. Macro factors like job availability and housing supply also play

---

[5] Job search theory has been originally developed to understand the search behaviour of unemployed persons. So job search theory has been chiefly used to explain the dynamics of unemployment. More recently, studies also aim to understand on-the-job search (Burdett, 1978; Hey and McKenna, 1979; Black, 1981; Hall, 1982; Kahn and Low, 1982, 1984; Holmlund and Lang, 1985; Hughes and McCormick, 1985; Mortensen, 1986; Hartog *et al.*, 1988; Van Ophem, 1991; Burgess, 1992; Lindeboom, 1992; Van den Berg, 1992, 1995; Hartog and Van Ophem, 1994; Pissarides and Wadsworth, 1994). In the labour market literature almost all emphasis has been on wages, while spatial aspects are of less importance. Similarly, in the job search literature there is scarcely any literature that addresses the location of the job; most study the search behaviour of job seekers. Notable exceptions are Simpson (1980), Sugden (1980), Holzer (1994), Rouwendal and Rietveld (1994), Van den Berg and Gorter (1997) and Rouwendal (1998, 1999).

an important role, as these factors determine whether better combinations of dwellings and jobs can be found.

A standard point of departure in the search literature is the distinction between the rate that a job or residence is offered, the so-called arrival rate, and the probability that the offer is accepted. In the job search literature, most results stem from the basic principle that jobs are characterised by wages and that workers prefer higher wages.

Whether an individual accepts a job offer not only depends on the direct gain in wage, but also on the *once-only costs associated with moving* and the *search costs*. In our search model, we have explicitly included the once-only costs associated with moving job. We assume that every worker is continuously engaged in search for a better job.[6]

It is worthwhile to emphasise that the model can easily be extended by incorporating search in the housing market, transitions to unemployment, varying search effort in the labour and housing market and by including more than one wage earner. For reasons of clarity, we do not discuss such an extension here.[7] In addition, we omit mathematical proofs.

The point of departure is that individuals are employed and search continuously for better jobs. Individuals derive utility from the wage w and commuting distance z (an alternative interpretation is to interpret z as commuting time). So, the instantaneous utility v

---

[6] The typical situation in which a worker searches continuously for a job given fixed turnover costs has been explored by Hey and McKenna (1979). They suppose that when workers evaluate a job offer they contemplate the notion of moving more than once in the future. So the difference between the new wage offered and the present wage needs to be *greater* than the costs of moving 'to guard against the possibility of getting another offer after moving that would have been preferred before moving, but which is not sufficiently high to induce a second change'. As a consequence, the more *moves* one expects, the larger this difference will be, because one does not want to pay too many times for the costs of moving.

[7] Simpson's (1980) model of workplace choice extends the theory of job search by explicitly considering the spatial dimension of job search (see also Maier, 1995 in this context). The searcher samples jobs of which the attributes are unknown, so he will sample more in areas close to his residence because of the higher time and commuting costs. So the standard job search assumption that jobs arrive completely at random and are exogenous to the job seeker is dropped. For an empirical investigation, see Rogers (1997).

experienced in a certain period of length Δt by an individual is a function of w and z and is equal to $v(w, z)\Delta t$.

We assume that $\partial v/\partial w > 0$ and $\partial v/\partial z < 0$. Hence, instantaneous utility is increasing in wage and is decreasing in commuting distance. The individual takes into account the once-only costs of changing jobs c. Furthermore, for convenience of notation, we suppose that the instantaneous utility v depends linearly on the moving costs c. So $v(c+\delta) - v(c) = -\delta$. We assume that individuals receive job offers, which arrive according to a Poisson process. Jobs arrive with arrival rate p. Moreover, we will assume that pooling of offers is not allowed: job and dwelling offers have to be refused or accepted before other offers arrive.

A job is entirely characterised by the wage, w, and the commuting distance, z. *Wage and commuting costs offers are random drawings* from a bivariate distribution F_{wz}. So, w and z may be dependent. The wage w is received until a new job is accepted. The commuting costs z are borne until the individual leaves the job. We suppose that w and z assume non-negative values. The maxima of w and z are denoted as $\bar{w}$ and $\bar{z}$, respectively.

12.2.2. The optimal strategy

For an employed person, we denote the (discounted) expected lifetime utility (indirect utility) received from the current wage and commuting costs as $V(w, z)$. V includes the possibility of better offers in the future. All benefits and costs are discounted at rate ρ. The individual is assumed to maximise lifetime utility $V(w, z)$. The basic decision the individual has to take is whether to accept a new job, taking into account the expected offers in the future. Consider a (short) interval of time length Δt. The lifetime utility for an employed person is then

$$
\begin{aligned}
V(w,z) = \frac{1}{1+\rho\Delta t}[&v(w,z)\Delta t \\
&+ p\Delta t E \max[V(w_{\mathrm{x}},z_{\mathrm{x}}) - c, V(w,z)] \\
&+ (1-p\Delta t)V(w,z) + o(\Delta t)]. \qquad (12.1)
\end{aligned}
$$

In this expression the expectation is taken with respect to the variables which got a subscript. So, the expectation is taken with respect to the distribution of the wage and commuting costs. The interpretation of Equation (12.1) is straightforward. The instantaneous utility $v(w,z)$ is proportional to the length Δt of the time interval. With probability $p\Delta t$ a job offer will be received, and that offer will be accepted if the value of the new position exceeds that of the current position. If the offer is rejected, the individual's position does not improve and he/she receives $v(w,z)$. If the job is accepted, one has to pay moving costs. With probability $1-p\Delta t$ the worker will not receive a job offer. The last term reflects the notion that as Δt approaches zero any non-proportionality of utility to the length of the time interval goes to zero at an even faster rate (Albrecht *et al.*, 1991). The latter term includes the probability of receiving more than one job offer. From now on, we will treat the workers decision problem in continuous time to simplify the analysis. So, we rewrite V, dividing by Δt, and let Δt approach zero. This gives

$$V(w,z) = \frac{1}{\rho}[v(w,z) + pE\max[V(w_X,z_X) - c, V(w,z)] - pV(w,z)] \tag{12.2}$$

Following Albrecht *et al.* (1991) we rewrite Equations (12.1) and (12.2) to prove that these formulae define the values of V. Let us choose a constant $M > p$. Multiplying the above formula by ρ, adding $MV(w,z)$ to both sides, and dividing through by $M+\rho$ gives the following expression:

$$V(w,z) = \frac{1}{M+\rho}[v(w,z) + pE\max[V(w_X,z_X) - c, V(w,z)] + (M-p)V(w,z)] \tag{12.3}$$

The above formula defines the value of V and the optimal acceptance rules. It is then straightforward to show that V is increasing in w and decreasing in z. Furthermore, the above formula implies that V is increasing in p and decreasing in c.

Individuals receive offers, which imply a change in the commuting costs. So, the optimal strategy is conditional on the commuting costs offered (Van den Berg and Gorter, 1997). The following arbitrary decision rule will be proposed which states which job offer induces a job acceptance, and which not. Given a job offer of wage w_y and commuting distance z_y, and given the current wage w, commuting distance z, and reservation wage $\text{res}_w(w, z|z_y)$:

$$\text{change job} \quad \text{if } w_y > \text{res}_w(w, z|z_y)$$
$$\text{otherwise do not change job.}$$

This decision rule defines the dynamic maximising problem because the individual maximises the lifetime utility by choosing the *optimal* value for res_w. The optimal values can be established by straightforward application of standard solution techniques to derive the optimal strategy of individuals (Hey and McKenna, 1979; Albrecht *et al.*, 1991; Burgess, 1992). First, one determines the value of the lifetime utility $V(w, z)$, given the current values of w and z and the arbitrary decision rules:

$$V(w,z) = \frac{1}{M+\rho}\left[v(w,z) + p\int_0^{\bar{z}}\int_{\text{res}_w(y)}^{\bar{W}} [V(x,y) - c - V(w,z)] \times \mathrm{d}F_{w,z}(x,y) + MV(w,z)\right]. \qquad (12.4)$$

The objective of the individual is to maximise lifetime utility. So, we derive the first and second-order conditions for the optimal decision rules. Setting the derivative of V with respect to res_w equal to zero gives the first-order condition, which determines the optimal strategy for the employed. Taking the derivative of V with respect to res_w, and setting the resultant equal to zero gives the following result:

$$V(w,z) - V(\text{res}_w(z_y), z_y) + c = 0, \qquad (12.5)$$

where res_w denote now the *optimal* reservation values. These expressions have a clear interpretation. The reservation wage is chosen, conditional on the commuting costs offer z_y, such that the lifetime utility associated with the optimal level of the reservation wage which induces a job movement is equal to the cost of moving to another job plus the lifetime utility associated with the current job.[8]

---

[8] It can be easily shown that the second-order condition for res_w is ensured.

12.2.3. Moving behaviour

According to the search model proposed above, job moving behaviour of an employed person is described by the transition rates. The transition rate of moving from the current job to another job $\theta_w(w, z)$ can be written as the product of the job offer arrival rate and the conditional probability of accepting a job offer.

$$\theta_w(w, z) = p \int_0^{\bar{z}} \int_{\text{res}_w(y)}^{\bar{w}} dF_{wz}(x, y). \tag{12.6}$$

Hence the transition rate of moving job depends negatively on the reservation wage. The optimal setting of the reservation wage in the current job depends on future job moving behaviour. Therefore, we define the job rates from a future position. We will focus on the transition rates in one of the minimum acceptable positions which may be obtained by a job move (the lifetime utility of this person after the move is equal to $V(\text{res}_w(z_y), z_y)$). The transition rate of moving job to another job can from this particular position be written as $\theta_w(\text{res}_w(z_y), z_y)$. So

$$\theta_w(\text{res}_w(z_y), z_y) = p \int_0^{\bar{z}} \int_{\text{res}_w^2(y)}^{\bar{w}} dF_{wz}(x, y), \tag{12.7}$$

where we use the heretofore undefined indices $\text{res}_w^2(z_y)$ which is defined as the minimum wage which induces a second job move at the same costs z_y. Thus, $\text{res}_w^2(z_y) = \text{res}_w(\text{res}_w(w, z|z_y), z_y|z_y)$. It can be easily shown that $\text{res}_w^2(z_y) > \text{res}_w(z_y)$ if $c > 0$.

12.2.4. The optimal reservation wage strategy

The optimal setting of the reservation wage res_w depends, among other things, on the functional form of the instantaneous utility function v. Interpretation of the optimal strategy is therefore facilitated by making assumptions about the functional form of the instantaneous utility function v. In this section, we will suppose that v is a linear function in w and z, so $v(w, z) = w - z$. So, $w - z$ may be interpreted as the net wage viz. the wage minus the commuting costs. The optimal reservation wage strategy can now be rewritten as (see van Ommeren *et al.*, 2000a):

$$\begin{aligned} \text{res}_w(z_y) = {} & w + (z_y - z) + c[\rho + \theta_w(\text{res}_w(z_y), z_y)] \\ & + p \int_0^{\bar{z}} \int_{\text{res}_w(y)}^{\text{res}_w^2(y)} [V(x, y) - V(\text{res}_w(z_y), z_y)] dF_{wz}(x, y). \end{aligned} \tag{12.8}$$

Interpretation of the right-hand side is facilitated by concentrating on the special case that the costs of moving job are zero ($c=0$, and thus, $\mathrm{res_w}=\mathrm{res_w^2}$). This explains the first two terms on the right side. The reservation wage can then be written as $\mathrm{res_w}(z_y)=w+z_y-z$. This result makes sense. Given a job offer at a distance z_y, the job seeker demands at least the current wage plus the change in the commuting costs due to the job move. Whether the reservation wage is set higher or lower than the current wage depends on whether the commuting costs will increase or decrease.[9] For this special case, the optimal reservation strategy can be rewritten as a function of the net wage: accept the job offer if $w_y-z_y>w-z$; otherwise, reject the job offer.[10]

Now suppose that the costs of moving job are positive ($c>0$).[11] Recall that persons take into account that after accepting a job offer, they may move job another time and that employed persons wish to be compensated for the moving costs. This explains the third term on the right side. The amount of compensation depends on the time spent in the new job. So, $c\rho$ can be interpreted as the long-run compensation and $c\theta_w(\mathrm{res_w}(z_y),z_y)$ as the compensation for leaving the new job voluntarily (see also Van den Berg, 1992). A well-known result in the literature is that the job seeker sets the reservation wage higher than would be necessary to be compensated for the job moving costs. This explains the fourth term. This term may be interpreted as a compensation to guard against the possibility of getting another wage offer after changing jobs that would have been preferred before changing, but which is not sufficiently high to induce a second change (Hey and McKenna, 1979; Burgess, 1992).

---

[9] Sugden (1980) was one of the first who recognised that search theory is well suited for handling problems of the spatial dimension of labour markets. He defined the net wage of a job as the wage earned minus commuting costs and showed that an increase in the cost of travel implies an increase in the reservation wage.

[10] Rouwendal and Rietveld (1994) examine the influence of commuting distance on the acceptance of a job offer for unemployed job seekers. The reservation wage (the minimum wage asked which induces job acceptance) can then be calculated as the sum of the minimum wage asked for where commuting costs absent plus the commuting costs. Van den Berg and Gorter (1997) derive the job acceptance strategy for unemployed persons who may move residence once. Also Rouwendal (1998) concentrates on the acceptance strategy of unemployed Dutch women.

[11] This case is an extension of a case thoroughly analysed by Hey and McKenna (1979) and Burgess (1992).

We have derived a search model that explains on-the-job relocation behaviour of employed individuals. An essential feature of the model is that workers who consider a job move take into account *future* job moving behaviour. The model includes commuting costs, which alter after job move. The optimal moving strategy depends on the current position of the individual in the labour market. In line with intuition, the reservation wage is increasing in the current wage and decreasing in the current commuting costs. One of the major consequences of market imperfections is then that those job offers at longer commuting distances are more likely to be accepted if they offer higher wages (see Manning, 2003). As a consequence, *wages should be a positive function of commuting costs*. Although in the current model, we emphasise the importance of compensation by means of wages, we should observe the same relationship for other forms of work compensation (e.g. pensions, company cars, etc). So, given a distribution of wage and commuting costs, wages and commuting costs are positively related, in line with empirical evidence.[12] Further, the reservation wage is increasing in the job arrival rate and decreasing in the moving costs (see Van den Berg, 1992).

12.2.4.1. Implications for commuting

It is a standard assumption in the economic literature that commuting costs are compensated either by lower housing prices or by higher wages (see Zax, 1991). It is straightforward to see that the current commuting costs may also be compensated by *future* wages or *future* place utilities, which are obtained via future job or residence relocations. As a result, the common result that commuters are compensated in the labour or housing market does not hold if market imperfections are prevalent. Due to market imperfections, individuals will temporarily accept commuting costs, which are compensated by future wages or

---

[12] The positive relationship between wages and commuting costs has many other explanations which are consistent with perfect markets (e.g. urban wage gradients, see White, 1999). Careful analysis by Manning (2003) indicates that the presence of market imperfections is one of the main reason that we observe this relationship.

future rents which may be obtained by future job or residence relocations.

The obtained result that mobile commuters do not demand full compensation for current commuting suggests intuitively that those workers who are more mobile in the housing market demand *less* compensation (in form of wages) for commuting than others, and are therefore more likely to accept a job offer. Moreover, workers with lower probabilities of receiving a residential offer, for example due to discrimination, will generally demand higher wages to compensate for the loss in commuting costs incurred. For competitive labour markets this may result in higher unemployment rates and a more frequent acceptance of wages which do not compensate to the same extent for the commuting costs as the workers with higher residential mobility rates (see Holzer, 1994). This may also explain the findings of Zax (1991) that blacks and females do receive less compensation than white males, as blacks and females are more restricted in the housing market.[13] This result also concurs with the notion that given a relocation of an employer, it is much more likely that those with low residential moving rates will leave the current employer (see Zax, 1991; Zax and Kain, 1991).

Individuals will differ in their commuting behaviour due to differences in moving costs and in the probability to receive an offer. Clearly, individuals who have lower moving costs and who are more likely to receive an offer reach better positions at lower (moving or waiting) costs, and are therefore more able to reach a more favourable situation *as time passes*. So, persons with a greater ability to adapt their housing situation to their work locations will have shorter commuting distances (see Rouwendal and Rietveld, 1994), or higher wages (see Van den Berg, 1992) and is also consistent with the view that persons with a greater ability to adapt their housing situation to their work locations (e.g. young persons) will *temporarily* accept longer commuting distances as they will be

[13] Females are thought to be more restricted if their spouse has a higher income, because it is less likely that it is beneficial to the household to move residence if it implies that the spouse with the highest income has to give up the job. It should be noted that discrimination would induce firms to pay lower wages to females and blacks, but it is difficult to see why they would offer less compensation for commuting.

compensated in the future. In the empirical investigation, we will come back to this issue.

Until now, we did not discuss the effect of exogenous differences in the disutility of commuting. However, there are many reasons to expect that individuals differ in this respect (see, for example, Van den Berg and Gorter, 1997). For example, married women may have a higher disutility of commuting due to higher household responsibilities (see Madden, 1981). Even small differences in the disutility of commuting are thought to have large consequences for the behaviour in the labour or housing market, as it reduces the opportunity to receive an acceptable offer substantially. However, in case a person is mobile in both the labour and housing market this effect will be weaker. As a consequence, differences in the disutility of commuting will particularly have an effect on the acceptance behaviour in one market as the moving in the other market is restricted.

According to a number of empirical studies, commuting distance is affected by the length of the time spent in the current job and residence (and thus by job and residential moving behaviour) (Madden, 1981; Singell and Lillydahl, 1986; White, 1986; Dubin, 1991). In these empirical studies, a variety of effects are reported. White (1986) finds that residence tenure has a significant negative effect on commuting. Madden (1981) finds negative effects for job tenure. Both results suggest that commutes are reduced by moves in the labour and housing market. Consistent with these findings, Rouwendal and Rietveld (1994) find that starters on the labour market have an above-average commuting distance. Singell and Lillydahl (1986) find that in two-earner households that recently have changed their residence, male commute times decline whereas female commute times increase. These effects may reverse when female earnings exceed males. Such a result is consistent with the observation that the high earning job is the most stable, so it is more worthwhile for the household to move closer to this job. Dubin (1991) reports that residential mobility has relative little importance on the extent to which workers use firm decentralisation to shorten their commutes, but job mobility is important. This suggests that if the commute is (exogenously) increased, workers tend to move job, and not residence, in line with high residential moving costs.

12.2.5. Adaptions and extensions

The above described job search model can be easily interpreted as a residential search model, not by modelling the distribution of wages, but by modelling the distribution of the indirect utility of a residence. The indirect utility, called the *place utility*, is defined as the utility experienced in a certain location (net of housing costs), see e.g. Wolpert, 1965 or Yapa *et al.*, 1971. The main result is now that the effects of commuting and residential moving costs on residential mobility are negative.[14]

The above job search model is a stylised model which can easily be extended in a large number of ways. The most fundamental extensions are the inclusion of residential mobility and two-earner households. Other extensions, although interesting in many applications, such as the inclusion of unemployment and endogenous on-the-job search effort do not change the results fundamentally (see van Ommeren, 2000a).

12.2.5.1. Job and residential mobility

In most search models, mobility in another market is ignored. The acceptance probabilities of jobs and dwellings however depend on each other, as a job or residence relocation implies a change of commuting distance which affect both acceptance probabilities. Application of the concept of search theory suggests that if individuals search on both markets, they will accept a dwelling or job offer only when the expected gains of an offer are higher than an acceptable minimum, taking into account future offers on *both* markets. The properties of the optimal strategy of the worker who faces job and residential offers may thus shed new light on the relationship between job and residential mobility and commuting. So, we are interested in a search model assuming that workers

---

[14] Kim (1992) views the hedonic price equation as the minimum asking rent by the landlords which determines the competition among landlords. A housing demand function takes into account the search process for a suitable housing unit. Since the search activities are not observable, the observed transaction data in a housing market are truncated. Hence the standard housing demand model would suffer from truncation bias. Joint estimation of the hedonic price and housing demand (reservation rent) equations shows that commuting time affects the reservation rent.

consider commuting distance and the once-only costs of changing jobs and residences in their search for jobs and dwellings.

The point of departure is that individuals are employed and search continuously for better jobs and dwellings. Individuals derive utility from the wage w, place utility r and commuting costs z. So, the instantaneous utility v experienced in a certain period of length Δt by an individual is a function of w, r and z and is equal to $v(w, r, z)\Delta t$. We assume again that $\partial v/\partial w > 0$, $\partial v/\partial z < 0$ and $\partial v/\partial r > 0$. Hence, instantaneous utility is increasing in wage and place utility, and is decreasing in commuting costs. The individual takes into account the once-only costs of changing jobs c_1, and residences c_2.

We are particularly interested in understanding the effect of changes in the housing market (job market) parameters on job (residential) mobility. One of the main conclusions drawn from a comparative static analysis is that the effect of the labour market parameters (c_1 and p_1) on job mobility is determined, but the effect of the housing market variables (c_2 and p_2) is in general ambiguous.[15] This somewhat disappointing result is, of course, the result of the generality of the model proposed, as we do not make any assumption on the ordering of the job and housing offers. Fortunately, unambiguous results can be obtained for some special cases which will be discussed in the next section.

Although the effect of the job moving costs on the residential moving rate, and the effect of the residential moving costs on the job moving rate are ambiguous, the effect of an increase in these costs on the *sum* of the job and residential moving rate is determined:

$$\partial(\theta_{\mathrm{w}} + \theta_{\mathrm{r}})/\partial c_1 \leq 0; \tag{12.9a}$$

$$\partial(\theta_{\mathrm{w}} + \theta_{\mathrm{r}})/\partial c_2 \leq 0, \tag{12.9b}$$

where θ_{r} denotes the residential moving rate. The effect of an increase in the job or residential moving costs on the *sum of* the job and residential moving rates can be shown to be negative. This result has a straightforward interpretation: the effect of an increase in the

[15] Similarly, the effects of the housing market parameters (c_2 and p_2) on residential mobility are all determined, but the effect of the labour market variables (c_1 and p_1) is ambiguous (see van Ommeren *et al.*, 2000a).

residential moving costs on the job moving rate is less than the (absolute value of the) effect on the residential moving rate (so, $\partial \theta_w / \partial c_2 < |\partial \theta_r / \partial c_2|$). This result is due to the fact that the job moving rate is affected by an increase in c_2, because the residence moving rate is affected by an increase in c_2. This result will be useful in the next subsection.

12.2.5.1.1. Special cases. The sign of the residential moving costs on the reservation wage of an employed person is in general ambiguous. In certain cases this sign can be determined. We will discuss some of these cases here.

(i) The wage distribution is degenerated; this is common in certain types of labour markets which are highly institutionalised and where wage differences are minimal (e.g. teaching at high schools). Suppose therefore that jobs are homogeneous such that the wage distribution is degenerated. Thus, jobs only differ with respect to their location. Then it can be shown that the job moving rate is increasing in the residential moving costs. The interpretation is straightforward. When jobs are homogeneous then job movements occur only to reduce the commuting distance. If residential moving costs are higher, then it will be relatively less costly to decrease the commuting distance by moving job than by moving residence.

(ii) The current commuting costs are zero (a large share of workers are close to this situation: in most countries, about 20% of the workers commute for less than 10 min). Any job move will then increase the commuting costs. Given high residential moving costs, it will be less beneficial to reduce the commuting costs by moving residence, so this discourages accepting a job offer. Hence, residential moving costs have a negative effect on job mobility.

(iii) The residence moving rate is zero, whereas it may become positive after a job move. This case is relevant for persons who do not plan to move residence (as they have found their 'ideal' house, so close to the situation of a perfect market), but who still expect to make a labour market move. Such a labour market move may increase the commuting costs, which gives an incentive to move residence after the job move. When the residential moving rate is currently zero, then an increase in the residential moving costs decreases the job moving rate. This result is due to the fact that after a certain job move, which may increase the commuting distance, the probability of moving residence may become positive. As a consequence, the search model predicts that in economies in which high residential moving costs are prevalent, it is likely that a (further) increase in the residential moving costs will induce persons to move job less often.

We conclude that the uncertainty on the sign of the residential moving costs on job mobility can be removed in specific cases: if the current commuting costs z are (close to) zero or if the residential moving costs are high then job mobility depends negatively on the residential moving costs; if the wage distribution is degenerated, then job mobility depends positively on the residential moving costs.

12.2.5.1.2. The order of job and residential moves. In the expressions derived above, the search model appears symmetric, which is counter-intuitive, because it is observed that in Europe, in *general*, workers will first accept a new job and then search for a new residence (see, for example, Verster, 1986; Camstra, 1994). This issue may be solved by focusing on the parameters of the search process on both markets. Consider the case when the ratio of the probability of receiving a residence offer to the probability of receiving a job offer is large. This case is relevant, because finding a job is, in general, far more difficult than finding another residence. Given the values of these parameters, the search model is consistent with the observation that after a job move which increases the commuting costs, an individual would almost immediately move residence as the probability of a residence offer is high. After a residence move, which increases the commuting costs, it may take considerable time before a worker will adjust the workplace location. This case seems to be the most common situation. However, under particular circumstances the opposite may be found for certain subgroups. For example, the typical low income employee in the Netherlands rents a subsidised residence in a market which is highly regulated, with a small probability that a residence offer will arrive.

Our model would then predict that this worker is inclined to refuse job offers which are farther from his current residence;[16] this typical low income employee would likely be more willing to move residence (given an offer) even if this implies that commuting costs are increased. In sum, the model indicates that workers generally accept first a new job and then move residence closer to the new workplace location, because of a differential rate of arrival of jobs and residences. This sequence may be reversed, depending on the conditions on the labour and housing market.

12.2.5.2. Empirical evidence

The empirical importance of search theory can be tested based on data of individual job and residential moving behaviour. We come up with the following testable hypothesis: *Residential, job mobility and search intensity should all be a positive function of commuting costs.*[17] I interpret this hypothesis as the main test of the importance of search theory, since it indicates that workers are not fully compensated for the costs associated with commuting by higher wages or lower rents. Several empirical applications of job search theory point to significant effects of commuting distance on *job mobility* (e.g. Van der Vlist, 2000; van Ommeren, 2000a).

---

[16] Note that empirical studies show that the unemployed workers tend to accept most job offers, but the job acceptance probability of the employed is much less than one.

[17] The positive relationship between work compensation (e.g. wages) and commuting costs implies that tests of the relationship between commuting costs and job/residential mobility (search) are biased towards not finding a relationship (the direction of the bias depends on the sign of the effect of the omitted variable on the dependent variable and the sign of the correlation between the omitted and included explanatory variable (see e.g. Stewart and Wallis, 1981, p. 164)). To the extent that one can never fully observe work compensation in all its facets, tests of the effect of commuting costs on job/residential mobility (search) are *conservative* tests. Furthermore, the measurement error in commuting costs are plausibly quite large, as the commuting costs are unobserved and normally measured by means of commuting distance or time. To the extent that employees will choose the optimal speed level, given the same environment, commuting distance is perfectly related to commuting costs. Nevertheless, employees face very distinct choices with respect to the optimal choice of the speed level (due to local congestion, variation in parking costs, variation in the supply of public transport), so commuting distance may only be weakly correlated to commuting costs. As a result, a test of a negative relationship between commuting costs and mobility (search) will be biased towards not finding a relationship due to measurement error (e.g. if the R^2 of a regression of commuting costs on commuting distance is 0.50, the estimated effect is about 50% of the real effect).

Van Ophem (1991), van Ommeren (1998) and Van Ham *et al.* (2001) examine the effect of commuting time on *on-the-job search* and show that the effect is positive and quite strong. Employees who commute more than 1 h one-way are about 50% more likely to be involved in on-the-job activities than those with short commuting times. The results by Van Ham *et al.* (2001) indicate that females are slightly more sensitive to commuting time than male employees. To the extent that males and females are equally sensitive to wages, this suggests that females' MWP for commuting time is slightly higher in line with evidence that females' commuting times are (slightly) less than those of males. van Ommeren (1998) demonstrates that on-the-job search activities are a convex function of commuting time. Such a finding is in line with the theoretical model, since the probability of finding a job closer to the residence is an increasing function of the current commuting distance. One exception is the study by Mekkelholt (1993), who does not find a positive significant relationship between commuting time and the probability of accepting a job, suggesting that those with higher commuting costs usually have higher wages which sufficiently compensate for the commuting costs. Van den Berg (1992) reports that individuals with longer commuting distances are more willing to change jobs.[18]

Henley (1998) analyses the effect of travel-to-work time of the household head for homeowners and reports a (statistically significant) positive effect on *residential mobility*. van Ommeren (2000a) also finds a positive effect of commuting distance on residential mobility, but Van der Vlist (2000) fails to find such an effect (plausibly due to the limited number of observations). The effect of changes in the commuting costs that are exogenous has also been studied (e.g. a workplace relocation by the employer, Van Engelsdorp Gastelaars and Maas-Drooglever Fortuijn, 1985; Holzer, 1994). Further, Verster (1986) finds that residential moves are still triggered by a workplace change after a considerable time lag. Summarising, there is a large empirical literature, which supports the hypothesis that workers with long commutes are more mobile in the labour and housing market.

---

[18] Van den Berg and Gorter (1997) find that the reservation wage of an unemployed seeker increases if a job is offered at a larger commuting distance from the current residence.

12.2.5.3. *Two-earner households*

The case of two-earner households deserves special attention because the two wage earners in the same household share a dwelling, but have separate working places. Having separate places of employment adds to the complexity of their spatial decision problem. The practical importance of this topic is evident from the currently high number of wage earners who form a two-earner household, and it is still growing, mainly because of increased participation of women in the labour market. Nevertheless, in general, not much is known about the moving and commuting behaviour for two-earner households both theoretically and empirically. The theoretical studies of commuting behaviour for two-earner households are mainly based on static urban equilibrium models (see, for example, White, 1977, 1986; Madden, 1981). Moreover, empirical studies of commuting behaviour for two-earner households are often interpreted by means of static models (see White, 1977, 1988; Madden, 1981; Curran *et al.*, 1982; Singell and Lillydahl, 1986; Dubin, 1991). Static models typically ignore the effect of future residential and job moves on the choice of the present commuting costs.

We suppose now that the household consists of two wage earners who currently earn wages w_1 and w_2, respectively. The present residence renders a place utility of value r. The commuting costs due to the travel between the current workplaces and the residence are denoted by z_1 and z_2 and are presumed to be a linear function of commuting distance. The instantaneous utility v experienced in a certain period by the household is a function of w_1, z_1, w_2, z_2 and r. So, $v = v(w_1, z_1, w_2, z_2, r)$. We assume that $\partial v/\partial w_i > 0$, $\partial v/\partial z_i < 0$, $i = 1, 2$, and $\partial v/\partial r > 0$. Hence, instantaneous utility increases in wages and place utility, and decreases in commuting costs. These assumptions allow for a wide range of particular forms of the instantaneous utility v.

The two wage earners of a two-earner household share a dwelling but have different working places so residential behaviour affects the job behaviour of both wage earners. This implies that the travel costs between the workplace locations of the wage earners, which is denoted by z_3, affect the residential and job moving behaviour of both wage earners. In Figure 12.1 we have drawn the relationship between z_1, z_2 and z_3.

Figure 12.1. The workplace and residential locations of two-earner household

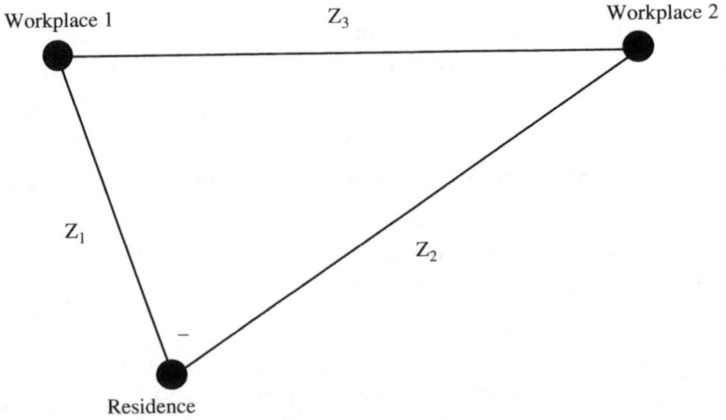

The main implication is that the travel costs between the two workplaces increase job mobility (and on-the-job search) of both wage earners and decrease residential mobility (and residence search), conditional on the commuting costs of both wage earners. This result makes intuitive sense. If the workplaces are close to each other, the commuting costs of both wage earners can be reduced by a residential move, so, as a result, there is less reason to reduce the commuting costs by a job move (see van Ommeren *et al.*, 1998, 2000b; van Ommeren, 2000b).

We are aware of two empirical tests based on job moving/search behaviour that only find weak evidence supporting the above implication using data from the Netherlands at the beginning of the 1990s (van Ommeren *et al.*, 1999, 2002b). This raises the question why the empirical evidence is not fully in line with theory. One explanation is certainly statistically, since the share of full-time employed two-earner households was still low in the beginning of the 1990s, so the number of observations of two-earner households is maybe limited. Another explanation is that in the Netherlands residential mobility rates are low and residential moving costs are high, so two-earner households typically ignore the possibility to adapt the residence location to improve the commuting costs of both wage earners.

12.2.6. Marginal willingness to pay

According to the labour market theory of equalising differences, an equilibrium locus of wage and job characteristics exists and can be estimated by an appropriately specified wage equation. The marginal willingness of workers to pay for job attributes is normally estimated using static hedonic wage methods which are based on the assumption of perfect mobility and information (see Rosen, 1974). Gronberg and Reed (1994) have proposed a method to estimate the willingness to pay for fixed job attributes based on job moving behaviour.[19] There has been in general interest in this topic (see, for example, Van den Berg and Gorter, 1997). Information on the willingness of workers to pay in order to avoid additional commuting might help to evaluate policy measures directed to the abatement of commuting. For example, the direct cost of an additional minute commuting due to increased congestion can be calculated.

The marginal willingness to pay for commuting can be defined as the wage the worker is willing to pay for an additional unit of commuting. So the workers' marginal willingness to pay for commuting costs z is defined by

$$\mathrm{MWP}(z) = \frac{\dfrac{\partial V(w,z)}{\partial z}}{\dfrac{\partial V(w,z)}{\partial w}}. \tag{12.10}$$

In words, the marginal willingness to pay for commuting costs z equals the ratio of the marginal change in lifetime utility V due to a marginal increase in z and the marginal change in lifetime utility V due to a marginal increase in the wage. It can then be easily shown that the following relationship holds:

$$\mathrm{MWP}(z) = \frac{\dfrac{\partial \theta_{\mathrm{w}}(w,z)}{\partial z}}{\dfrac{\partial \theta_{\mathrm{w}}(w,z)}{\partial w}}. \tag{12.11}$$

[19] Commuting can, however, hardly be regarded as a fixed non-wage job characteristic, as the worker may alter the commuting distance by moving residence. This method can however easily be extended to estimate workers' marginal willingness to pay for non-wage characteristics, which are not fixed to the job such as the commuting costs which may be reduced by moving residence closer to the job (e.g. van Ommeren *et al.*, 2000b).

Consequently, the workers' MWP for commuting costs is equal to the ratio of the marginal effect of commuting costs on the job-to-job moving rate and the marginal effect of the wage on the job-to-job moving rate. In a similar way, the MWP(z) can be derived given information on the marginal effect of commuting costs on job search intensity and the marginal effect of the wage on job search intensity (see van Ommeren and Hazans, 2003). It is common to write the moving rates θ_w as an exponential function of the wage and the commuting costs, so $\theta_w = \exp(\beta_w w + \beta_z z)$. In this case the MWP $= \beta_z/\beta_w$. Hence, the MWP can be calculated given the ratio of two (to be estimated) parameters. Given the specification $\theta_w = \exp(\beta_w \ln(w) + \beta_z z)$, MWP $= w\beta_z/\beta_w$.

We are aware of four empirical applications of estimates of MWP for commuting based on job mobility/search behaviour (Van Ophem, 1991; Van der Vlist, 2000; van Ommeren *et al.*, 2000b, 2002a. The latter is discussed in van Ommeren (2002)). These applications have been based on workers in the Netherlands. According to van Ommeren *et al.* (2000b), the MWP for an additional kilometre is -0.040 (s.e. is 0.020) and the implied MWP per working day of 8 h is estimated to be about -0.15 euro. The MWP for commuting distance implies a MWP for commuting time, which is about 1/3 of the hourly wage for the first half hour of commuting; for those who commute more than a half hour, the marginal value of time is about 2/3. These empirical findings are in line with other empirical studies (see Small, 1992). For slower means of transport (e.g. the bicycle) the implied marginal value of commuting time is lower. van Ommeren *et al.* (2002a) show that the MWP is higher for two-earner house holds, which is consistent with the notion that these households are more restricted in the housing market. The estimates of Van Ophem (1991) imply a MWP for commuting time which is substantially higher than the hourly wage, but due to the large standard errors, it is difficult to interpret the results. Another Dutch study by Van der Vlist (2000, p. 100) estimates the determinants of job mobility including commuting distance (in kilometres) and wages (in logarithm). Commuting distance has a positive effect ($\beta_z = 0.00048$) and the wage rate a negative effect ($\beta_w = -0.14$) on job mobility; the mean wage is 8.45 euro. So, the MWP $= 0.029$ euro (s.e. is about 0.015), which is close to the estimate by van Ommeren *et al.* (2000b).

12.2.7. Geographical structure

The relationship between residential moving, job moving and commuting behaviour depends on the geographical structure of the economy (see Zax, 1994; Crane, 1996). The geographical structure is captured in the model via the distribution of commuting costs. We will distinguish three stereotypes:

(i) An urban area with one Central Business District (CBD); the dwellings are distributed around the CBD. This stereotype traditionally used to characterise the geographical structure of large American cities.
(ii) A region where employment and residential locations are homogeneously distributed over space.
(iii) A region which consists of overlapping urban areas. Overlapping urban areas can be typified by many employment centres which are near to each other and where each employment centre is surrounded by residential areas. This type of geographical structure is more common in the Netherlands and Germany.

The essential characteristic of an urban area with one CBD is that all workplace locations are concentrated in one location; hence the commuting costs are fully determined by the residential location. This implies that the worker is merely interested in the difference between the wage and the commuting costs. In this type of geographical structure, individuals will ask full compensation for commuting based on *current* characteristics. In all other – more realistic – cases that employers are also located outside the CBD, individuals commuting behaviour will be affected by the location of the employer.

Whenever workers face a homogenous distribution of employers and dwellings (stereotype (ii)), *every* future job or residential offer implies a change in the commuting costs. In the case where employers and dwellings are geographically concentrated (type (iii)), future relocations of job and residence will be less influential compared to the situation when workers face a homogenous distribution of employers and dwellings, because it will be more likely that either the new residence or the new job is located near the current residence or job, respectively. Consequently, one of the implications is that if distributed over space, future job and residence relocations are more important as a factor, which determines commuting behaviour.[20] This elementary result might have consequences for commuting behaviour (and modelling of commuting behaviour): current trends in many countries show that employers suburbanise such that potential workplace locations become more scattered over space.[21] Similar trends are observable for the residential location of households. These trends indicate that the distribution may become more homogeneous. This implies that the possibility of future job and residence relocations will become more significant as decisive factors, which determine present day commuting behaviour.

Most models of urban form assume that influence of uncertainty job location mobility on both urban form and the choice of commuting length is absent (Crane, 1996). A number of studies have addressed this issue by analysing the effects of uncertainty job location on the urban form and choice of the commute length (Zax, 1994; Crane, 1996; Turnbull, 1998; van Ommeren *et al.*, 1999). Crane (1996) and Turnbull (1998) demonstrate that job site uncertainty in a multiple centre city is relevant, because the residential location is based not only on where the current job is located, but also on the expectation of where future jobs will be located. Both studies implicitly assume that residential moving costs constrain households to live in the same place.

These models essentially demonstrate that workers sort themselves spatially according to their job site stability. Workers also respond to the spatial variation in commuting cost risk which offsets

---

[20] In a careful analysis, Kan (2002) shows that individuals who are more likely to change job (and therefore workplace) are less likely to change residence.

[21] Holzer (1994) discusses how the urban geography of employers and residences may affect job search behaviour. An important outcome is that suburbanisation by firms seems particularly to affect those who are most restricted in the housing market. Gordon *et al.* (1989) investigate the influence of spatial structure on the mean commuting time. They demonstrate that the more polycentric and dispersed metropolitan areas facilitate shorter commuting times. The study does not separate out the two components of commuting time, the length of the trip and the speed of travel (reflecting congestion and choice of modality). They also show that smaller urban areas imply shorter commuting times. These results are consistent with the hypothesis that households choose residential locations around employment subcentres to minimise commuting costs. It rejects the notion that the more dispersed metropolitan areas induce congestion that increases the mean commuting time.

the tendency to sort by job site stability. So, allowing for job site uncertainty predicts less spatial sorting by commuting costs than the certainty models predict. House bid prices capitalise the spatial variations in expected travel cost (and its variance in the model of Turnbull (1998), which presumes risk averse workers), leading a flatter price surface than under certainty. Consequently, job uncertainty may generate a price peak between employment centres and lead to flatter rent gradients. The latter result is a consequence of the assumption of multiple employment centres and the presence of moving costs which prevents households from adapting towards the optimal residential location. One of the main consequences is that workers with the most stable jobs will have the shortest commutes, because these workers will choose the residential location which is optimally located relatively to the current employment location.[22] Another test of the hypothesis (which also avoids the endogeneity problem of past mobility measures) is to regress the commuting distance on the predicted value of the probability to move job (van Ommeren *et al.*, 1997). The main difficulty of such a test is that workers who are more mobile (for example, because of lower job moving costs) are more likely to reach preferred positions, indicating that mobile workers have shorter commutes. In line with the latter explanation, van Ommeren *et al.* (1997), find that workers who are more likely to move job have smaller commuting distances.

12.3. The observed commuting costs distribution

Here, we are interested to derive the *commuting costs distribution* G given the commuting costs *offer* distribution F as defined in

---

[22] In most studies of job mobility, it has been shown that the duration of employment has a strong negative effect on job-to-job mobility and job search. This suggests that workers with a long duration of employment should have shorter than average commutes, because these workers are less likely to change to another job. Levinson (1998) and van Ommeren (2000a,b) find support for this hypothesis. The result that workers commute more if they have recently moved job may be due to other factors as well. It may be that as the cost of commuting is decreasing over time, workers accept jobs, which are, on average, further from the home location. Rouwendal and Rietveld (1994) also concludes that the rapid increase in commuting distance may be in large part due to changes in the employment situation (while residential moves are more or less neutral in their effect on the average commuting distance).

Section 12.1.[23] We will assume here a random distribution of jobs and residences and presume the absence of variation in wages (see van Ommeren, 2004).

Given the presence of a commuting costs distribution and at random job search, but given infinitely high residential moving costs, workers will be able to decrease their commuting costs over time by finding workplaces closer to their residence, implying that the observed commuting costs distribution as used in the above job search models, becomes degenerate, since each worker will find the work location with minimal commuting costs after some time. A (sufficient) requirement for a non-degenerate equilibrium distribution is then that some workers leave the population of employed workers and are replaced by new workers, which, on average, have higher commuting costs. Let us therefore assume that employed workers are dismissed and thus become unemployed at rate λ (an element ignored in the search model of earlier sections).

The *observed* cumulative commuting distribution is denoted as $G(t)$, where $G(t)$ is the proportion of employed workers at a commuting costs no greater than t.[24] We will assume now that the unemployed and employed search randomly and the unemployed will only accept job offers when the commuting costs are *less* than the reservation commuting costs T, whereas the employed will only

---

[23] We will not deal with the (more difficult) question when a wage distribution will emerge (see, e.g. Burdett and Mortensen, 1998), and, for simplicity, assume a degenerate wage distribution, so the wage is identical for each job offer. We emphasise here that the reasons for the existence of a wage and commuting costs distribution are most likely distinct. The wage distribution is thought to be the result of an incentive by employers to offer a wage different from other employers (e.g. to increase the probability of job offer acceptance), whereas the observed commuting costs distribution is the result of the combination of spatial distribution of jobs and residences and the presence of moving costs and the job seekers' lack of information about where to search for better job opportunities. If moving residence (or workplace) would be costless or when job seekers have full information on the location of the nearest job vacancy, one would observe a degenerate commuting costs distribution. Further, the spatial distribution of jobs and residences is important. For example, the commuting costs distribution is degenerate for each individual in a monocentric urban model, where all workplaces are at the same location. This suggests that the commuting costs distribution is more useful as an analytical tool in regions with dispersed employment locations.

[24] This distribution is different from the commuting costs *offer* distribution $F(t)$ across firms, which is determined by the location distribution across employers.

accept jobs for which holds that the implied commuting costs are less than the current costs t. Given these assumptions, it can be shown that

$$G(t) = 1 - \frac{1 - F(t)/F(T)}{1 + kF(t)}, \tag{12.12}$$

for $t \leq T$, where $k = p/\lambda$, which denotes the ratio of the job arrival rate to the dismissal rate.

The above equation is insightful. In line with intuition, if $k = 0$ (so we presume the absence of on-the-job search), and if $F(T) = 1$ (so all job offers are accepted), then $G(t) = F(t)$. In case of the absence of on-the-job search $G(t) = F(t)/F(T) = F(t|t \leq T)$ for $t \leq T$. So, the observed commuting costs distribution is equal to the *conditional* commuting costs offer distribution, the condition being that the unemployed only accept offers within a certain range defined by T. We will show now that given the presence of a two-dimensional space, commuting distributions are obtained which are empirically relevant (see also Rouwendal and Rietveld, 1994). In contrast, the assumption of a one-dimensional space generates a distribution which is more difficult to bring in line with empirical evidence. One-dimensional space implies that $F(t)/F(T) = t/T$. We standardise T to 1, so

$$G(t) = \frac{(1+k)t}{1+kt}, \qquad t \leq 1 \text{ and } g(t) = \frac{1+k}{(1+kt)^2}. \tag{12.13}$$

Hence, it follows that the commuting costs density $g(t)$ is strictly decreasing in its argument. This is not in line with empirical evidence, which shows that the densities of commuting distance and time are both first increasing and then decreasing. Both distributions are normally approximated by a log-normal or a gamma distribution (see Rouwendal and Rietveld, 1994). Two-dimensional space implies that $F(t) = t^2/T^2$. Given $T = 1$, it follows:

$$G(t) = t^2 \frac{1+k}{1+k^2t^2}, \quad t \leq 1. \tag{12.14}$$

So $g(t) = [(1+k)2t]/(1+k^2t^2)^2$, which is not monotonic. Hence, it follows that the distinction between one- and two-dimensional space is crucially important. Given two-dimensional space, job search theory tends to generate empirically-relevant predictions of commuting costs distributions.

12.4. Conclusion

The objective of this study is to explain the contribution of search theory to the understanding of commuting behaviour. At the heart of search theory is the assumption of imperfect information. We have analysed the consequences of the assumption that individuals are confronted with a distribution of commuting costs offers, which arises as a consequence of the absence of full information on the availability of jobs (residences) at a certain location (and the presence of moving costs). One of the major consequences is that workers are not fully compensated for the commuting costs by either higher wages or lower rents (or both). This result is consistent with the finding of studies which find that mobility (and search intensity) in the housing and labour market depends positively on the length of the commute. Another consequence is that workers tend to commute over longer distances than they would given perfect information.

References

Albrecht, J.W., N. Holmlund and H. Lang (1991), "Comparative statics in dynamic programming models with an application to job search", *Journal of Economic Dynamics and Control*, Vol. 15, pp. 555–769.

Alonso, W. (1964), *Location and Land Use*, Cambridge, MA: Harvard University Press.

Anas, A. (1982), *Residential Location Markets and Urban Transportation: Economic Theory, Econometrics and Policy Analysis with Discrete Choice Models*, New York: Academic Press.

Black, M. (1981), "An empirical test of the theory of on-the-job search", *Journal of Human Resources*, Vol. 16, pp. 129–140.

Brown, L.A. and J. Holmes (1971), "Search behavior in an intraurban migration context: a spatial perspective", *Environment and Planning*, Vol. 3, pp. 307–326.

Burdett, K. (1978), "Employee search and quits", *American Economic Review*, Vol. 68, pp. 212–220.

Burdett, K. and D.T. Mortensen (1998), "Wage differentials, employer size and unemployment", *International Economic Review*, Vol. 2, pp. 257–273.

Burgess, S. (1992), "A search model with job changing costs: eurosclerosis and unemployment", *Oxford Economic Papers*, Vol. 44, pp. 75–88.

Camstra, R. (1994), *Household Relocation and Commuting Distance in a Gender Perspective, PDOD-paper no. 26*, Amsterdam: University of Amsterdam.

Clark, W.A.V. and R. Flowerdew (1982), "A review of search models and their application to search in the housing market", in: W.A.V. Clark, editor, *Modelling Housing Market Search*, London: Croom Helm.

Clark, W.A.V. and T.R. Smith (1982), "Housing market search behavior and expected mobility, theory 2: the process of search", *Environment and Planning A*, Vol. 14, pp. 717–733.

Clark, W.A.V. and W.F.J. Van Lierop (1986), "Residential mobility and household location modelling", pp. 97–132, in: P. Nijkamp, editor, *Handbook of Regional and Urban Economics*, Vol. I, Amsterdam: North-Holland.

Crane, R. (1996), "The influence of uncertain job location on urban form and the journey to work", *Journal of Urban Economics*, Vol. 37, pp. 342–356.

Cropper, M. and P. Gordon (1991), "Wasteful commuting: a re-examination", *Journal of Urban Economics*, Vol. 29, pp. 2–13.

Curran, C., L.A. Carlson and D.A. Ford (1982), "A theory of residential location decisions of two-worker households", *Journal of Urban Economics*, Vol. 12, pp. 102–114.

Dubin, R. (1991), "Commuting patterns and firm decentralization", *Land Economics*, Vol. 67, pp. 15–29.

Gordon, P., A. Kumar and H.W. Richardson (1989), "The influence of metropolitan structure on commuting time", *Journal of Urban Economics*, Vol. 26, pp. 138–151.

Gronberg, T.J. and W.R. Reed (1994), "Estimating workers' marginal willingness to pay for job attributes using duration data", *Journal of Human Resources*, Vol. 24, pp. 911–931.

Hall, R.E. (1982), "The importance of life time jobs in the U.S. economy", *American Economic Review*, Vol. 72, pp. 716–724.

Hamilton, B.W. (1982), "Wasteful commuting", *Journal of Political Economy*, Vol. 90, pp. 1035–1053.

Hamilton, B.W. (1989), "Wasteful commuting again", *Journal of Political Economy*, Vol. 97(1), pp. 497–1504.

Hartog, J. and H. Van Ophem (1994), "On-the-job search and the cyclical sensitivity of job mobility", *European Economic Review*, Vol. 38, pp. 802–808.

Hartog, J., E. Mekkelholt and H. Van Ophem (1988), "Testing the relevance of job search for job mobility", *Economics Letters*, Vol. 27, pp. 299–303.

Henley, A. (1998), "Residential mobility, housing equity and the labour market", *The Economic Journal*, Vol. 108, pp. 414–427.

Hey, J.D. and C.J. McKenna (1979), "To move or not to move", *Economica*, Vol. 46, pp. 175–185.

Holmlund, N. and H. Lang (1985), "Quit behavior under imperfect information: searching, moving, learning", *Economic Inquiry*, Vol. 23, pp. 383–393.

Holzer, H.J. (1994), "Work, search and travel among white and black youth", *Journal of Urban Economics*, Vol. 35, pp. 320–345.

Huff, J.O. (1984), "Distance-decay models of residential search", pp. 345–366, in: G.L. Gaille and C.J. Willmott, editors, *Spatial Statistics and Models*, Dordrecht: Reidel.

Hughes, G.A. and B. McCormick (1985), "An empirical analysis of on-the-job search and job mobility", *Manchester School*, Vol. 53, pp. 76–95.

Kahn, L. and S. Low (1982), "The relative effects of employed and unemployed search", *Review of Economics and Statistics*, Vol. 64, pp. 234–241.

Kahn, L. and S. Low (1984), "An empirical model of employed search, unemployment search and nonsearch", *Journal of Human Resources*, Vol. 19, pp. 104–117.

Kan, K. (2002), "Residential mobility with job location uncertainty", *Journal of Urban Economics*, Vol. 52, pp. 501–523.

Kim, S. (1992), "Search, hedonic prices and housing demand", *Review of Economics and Statistics*, Vol. 74, pp. 503–508.

Kim, S. (1995), "Excess commuting for two-worker households in the Los Angeles metropolitan area", *Journal of Urban Economics*, Vol. 38(2), pp. 166–182.

Kooreman, P. and J. Rouwendal (1992), Search behaviour and the Dutch housing market: a structural model. Wageningen.

Levinson, D.M. (1998), "Accessibility and the journey to work", *Journal of Transport Geography*, Vol. 6(1), pp. 11–21.

Lindeboom, M. (1992), *Empirical Duration Models for the Labour Market*, Amsterdam: Thesis publishers.

Madden, J.F. (1981), "Why women work closer to home?", *Urban Studies*, Vol. 18, pp. 181–194.

Maier, G. (1995), *Spatial Search, Structure, Complexity, and Implications*, *Studies in contemporary economics*, Heidelberg: Physica-Verlag.

Manning, A. (2003), "The real thin theory: monopsony in modern labour markets", *Labour Economics*, Vol. 10, pp. 105–131.

Mekkelholt, E.W. (1993), *Een sequentiele analyse van de baanmobiliteit in Nederland.* PhD thesis, Amsterdam.

Mills, E.S. (1972), *Studies in the Structure of the Urban Economy*, Baltimore: John Hopkins University Press.

Mortensen, D.T. (1986), "Job search and labor market analysis", in: O.C. Ashenfelter and R. Layard, editors, *Handbook of Labor Economics*, Amsterdam: North-Holland.

Moses, L.N. (1962), "Towards a theory of intra-urban wage differentials and their influence on travel patterns", *Papers and Proceedings of the Regional Science Association*, Vol. 9, pp. 53–63.

Muth, R.F. (1969), *Cities and Housing: The Spatial Pattern of Urban Residential Land Use*, Chicago, IL: The University of Chicago Press.

Pickles, A.R. and R.B. Davies (1991), "The empirical analysis of housing careers: a review and a general statistical modelling framework", *Environment and Planning A*, Vol. 23, pp. 465–484.

Pissarides, C.A. and J. Wadsworth (1994), "On-the-job search: some empirical evidence from Britain", *European Economic Review*, Vol. 38, pp. 385–401.

Rogers, C.L. (1997), "Job search and unemployment duration: implications for the spatial mismatch hypothesis", *Journal of Urban Economics*, Vol. 42, pp. 109–132.

Rosen, S. (1974), "Hedonic prices and implicit markets: product differentiation in pure competition", *Journal of Political Economy*, Vol. 82, pp. 34–55.

Rouwendal, J. (1991), "Housing choice and search behaviour in a disequilibrated market: an exploratory analysis", *Kwantitatieve Methoden*, Vol. 12.

Rouwendal, J. (1992), Ruimtelijke interactiemodellen en zoektheorie. Wageningen.

Rouwendal, J. (1998), "Search theory, spatial labor markets and commuting", *Journal of Urban Economics*, Vol. 43, pp. 1–22.

Rouwendal, J. (1999), "Spatial job search and commuting distances", *Regional Science and Urban Economics*, Vol. 29, pp. 491–517.

Rouwendal, J. and P. Rietveld (1988), "Search and mobility in a housing market with limited supply", *Annals of Regional Science*, Vol. 22(3), pp. 80–98.

Rouwendal, J. and P. Rietveld (1994), "Changes in commuting distances of Dutch households", *Urban Studies*, Vol. 31(9), pp. 1545–1557.

Simpson, W. (1980), "A simultaneous model of workplace and residential location incorporating job search", *Journal of Urban Economics*, Vol. 8, pp. 330–349.

Singell, L.D. and J.H. Lillydahl (1986), "An empirical analysis of the commute to work patterns of males and females in two-earner households", *Urban Studies*, Vol. 23, pp. 119–129.

Small, K.A. (1992), *Urban Transportation Economics, Fundamentals of Pure and Applied Economics*, Chur: Harwood.

Small, K. and S. Song (1992), "Wasteful commuting: a resolution", *Journal of Political Economy*, Vol. 100, pp. 888–898.

Smith, T.R. and W.A.V. Clark (1982), "Housing market search behavior and expected mobility, theory 1: measuring preferences for housing", *Environment and Planning A*, Vol. 14, pp. 681–698.

Smith, T.R. and F. Mertz (1980), "An analysis of the effects of information revision on the outcome of housing market search, with special reference to the influence of realty agents", *Environment and Planning A*, Vol. 14, pp. 681–698.

Smith, T.R., W.A.V. Clark, J. Huff and P. Shapiro (1979), "A decision-making and search model of intra-urban migration", *Geographical Analysis*, Vol. 11, pp. 1–22.

Speare, A., S. Goldstein and W.H. Frey (1975), *Residential Mobility, Migration and Metropolitan Change*, Cambridge, MA: Ballinger.

Stewart, M. and K.F. Wallis (1981), *Introductory Econometrics*, 2nd edition, Oxford: Basil Blackwell.

Sugden, R. (1980), "An application of search theory to the analysis of regional labour markets", *Regional Science and Urban Economics*, Vol. 10, pp. 43–51.

Timothy, D. and W. Wheaton (2001), "Intra-urban wage variation, employment location and commuting times", *Journal of Urban Economics*, Vol. 50, pp. 338–366.

Turnbull, G.K. (1998), "Housing prices and residential land use under job site uncertainty", *Journal of Housing Economics*, Vol. 7(1), pp. 1–20.

Van den Berg, G.J. (1992), "A structural dynamic analysis of job turnover and the costs associated with moving to another job", *Economic Journal*, Vol. 102, pp. 1116–1133.

Van den Berg, G.J. (1995), "Wage dispersion and mobility", *Economic Modelling*, Vol. 12(1), pp. 15–27.

Van den Berg, G.J. and C. Gorter (1997), "Job search and commuting time", *Journal of Business and Economic Statistics*, Vol. 15, pp. 269–281.

Van der Vlist, A. (2000), *Residential Mobility and Commuting*, The Netherlands: Tinbergen Institute Research Series.

Van Engelsdorp Gastelaars, R. and J.C. Maas-Drooglever Fortuijn (1985), "Personeel op drift: hoe reageren personeelsleden op een verplaatsing van hun bedrijf?", *Geografisch tijdschrift*, Vol. 19, pp. 181–191.

Van Ham, M., C.H. Mulder and P. Hooimeijer (2001), "Local underemployment and the discouraged worker effect", *Urban Studies*, Vol. 38(10), pp. 1733–1751.

van Ommeren, J.N. (1998), "On-the-job search behaviour: the importance of commuting time", *Land Economics*, Vol. 74(4), pp. 526–548.

van Ommeren, J.N. (2000a), *Commuting and Relocation of Jobs and Residences*, Aldershot: Ashgate.

van Ommeren, J.N. (2000b), "Job and residential search behaviour of two-earner households", *Papers in Regional Science*, Vol. 79, pp. 375–391.

van Ommeren, J.N. (2002), *Employed and Unemployed Search Activity: Estimating Individuals' Marginal Willingness to Pay for Attributes, Research Memorandum, 2002-8*, Amsterdam: Free University.

van Ommeren (2004), *The Commuting Costs Distribution.* Tinbergen Institute, Discussion paper, 45(3).

van Ommeren, J.N. and M. Hazans (2003), *Employed and Unemployed Search: the Marginal Willingness to Pay for Attributes in Lithuania, the US and the Netherlands*. ZEI, Working Paper, B15.

van Ommeren, J.N., P. Rietveld and P. Nijkamp (1997), "Commuting: in search of jobs and residences", *Journal of Urban Economics*, Vol. 42, pp. 402–421.

van Ommeren, J.N., P. Rietveld and P. Nijkamp (1998), "Spatial moving behaviour of two-earner households", *Journal of Regional Science*, Vol. 38(1), pp. 23–41.

van Ommeren, J.N., P. Rietveld and P. Nijkamp (1999), "Impacts of employed spouses on job moving behaviour", *International Regional Science Review*, Vol. 22(1), pp. 54–68.

van Ommeren, J.N., P. Rietveld and P. Nijkamp (2000a), "Job mobility, residential mobility, and commuting: a theoretical analysis using search theory", *Annals of Regional Science*, Vol. 34, pp. 213–232.

van Ommeren, J.N., G.J. Van den Berg and C. Gorter (2000b), "Estimating the marginal willingness to pay for commuting", *Journal of Regional Science*, Vol. 40(3), pp. 541–563.

van Ommeren, J.N., P. Rietveld and P. Nijkamp (2002a), "A bivariate duration model for job mobility of two-earner households", *European Journal of Operational Research*, Vol. 137(3), pp. 574–587.

van Ommeren, J.N., A. Van der Vlist and P. Nijkamp (2002b), *Transport-related Fringe Benefits: Implications for Commuting a Relocation*. Tinbergen Institute, Discussion Paper, 63(3).

Van Ophem, H. (1991), "Wages, nonwage job characteristics and the search behavior of employees", *The Review of Economics and Statistics*, Vol. 71, pp. 145–151.

Verster, A.C.P. (1986), *Locatiegedrag van beroepsbeoefenaars: de invloed van afstandsgebonden kosten*. PhD Thesis, Rotterdam.

Weinberg, D.H., J. Friedman and S.K. Mayo (1981), "Intraurban residential mobility: the role of transaction costs, market imperfections and household disequilibrium", *Journal of Urban Economics*, Vol. 9, pp. 332–348.

Wheaton, W. (1990), "Vacancy, search and prices in a housing market matching model", *Journal of Political Economy*, Vol. 98, pp. 1270–1292.

White, M.J. (1977), "A model of residential location choice and commuting by men and women workers", *Journal of Regional Science*, Vol. 17, pp. 41–52.

White, M.J. (1986), "Sex differences in urban commuting patterns", *American Economic Review*, Vol. 76, pp. 368–372.

White, M.J. (1988), "Location choice and commuting behavior in cities with decentralized employment", *Journal of Urban Economics*, Vol. 24, pp. 129–152.

White, M.J. (1999), "Urban areas with decentralised employment: theory and empirical work", in: P. Cheshire and E.S. Mills, editors, *Handbook of Regional and Urban Economics, Applied Urban Economics*, Vol. 3, Amsterdam: Elsevier.

Wolpert, G. (1965), "Behavioral aspects of the decision to migrate", *Papers and Proceedings of the Regional Science Association*, Vol. 15, pp. 159–169.

Yapa, L., M. Polese and J. Wolpert (1971), "Interdependencies of commuting, migration and job site relocations", *Economics Geography*, Vol. 47, pp. 59–72.

Zax, J.S. (1991), "Compensation for commutes in labor and housing markets", *Journal of Urban Economics*, Vol. 30, pp. 192–207.

Zax, J.S. (1994), "When is a move migration?", *Regional Science and Urban Economics*, Vol. 29, pp. 153–165.

Zax, J.S. and J.F. Kain (1991), "Commutes, quits, and moves", *Journal of Urban Economics*, Vol. 29, pp. 153–165.

Urban Dynamics and Growth, CEA, vol. 266
Roberta Capello and Peter Nijkamp (editors)

CHAPTER 13

Ethnic Concentration and Human Capital Formation

Thomas de Graaff[a] and Henri L.F. de Groot[a,b]

[a]Department of Spatial Economics, Free University Amsterdam, De Boelelaan 1105, 1081 HV, Amsterdam, The Netherlands
[b]Tinbergen Institute, Amsterdam, The Netherlands

Abstract

Concentration of immigrants and its associated externalities have become an important topic in contemporary international migration research, both from a methodological as well as an empirical perspective. The purpose of this chapter is twofold. First, it aims to provide an overview of that part of the migration literature that is concerned with the externalities created by the influx of immigrants. Second, it presents a stylized model in which human capital accumulation and ethnic cluster formation are explicitly incorporated. The model shows that lock-in effects can result from heterogeneous human capital and spillover effects on different spatial levels. Extensions of the model are discussed, together with their possible impacts on the spatial variation of the evolution of human capital stocks.

Keywords: ethnic concentration, human capital, migration, migration costs

JEL classifications: J61, R11

This chapter is a modified version of a chapter in the PhD thesis of Thomas de Graaff. The authors would like to thank Peter Nijkamp, Jos van Ommeren, Jan van Ours, Joop Hartog and an anonymous referee for useful comments. We are much indebted to Cees Gorter, who contributed greatly to this chapter and was an important source of inspiration.

13.1. Introduction

During the last decade, an increasing body of the empirical migration literature has dealt with the phenomenon of ethnic concentration, its causes and its economic consequences (Bartel, 1989; LaLonde and Topel, 1991; De Graaff, 2002). Traditional neoclassical migration models have been argued not to be very appropriate in explaining this concentration. They predict that workers migrate to those locations where their labor skills are scarce (see inter alia Greenwood and McDowell, 1986; Massey *et al.*, 1993; Borjas, 1994, for comprehensive overviews of the literature). Thus, one would expect that immigrants spread out over space and sectors. Empirical evidence clearly shows the opposite: immigrants are concentrated in specific cities, neighborhoods and sectors (see De Graaff, 2002, for a recent overview). Recently, economic models have been developed that are able to deal with the spatial concentration of foreign immigrants. A common characteristic of these theoretical models is the presence of externalities, which are created by migrant groups and that directly or indirectly affect the migration costs and human capital accumulation. The literature contains at least five approaches to model ethnic concentration. The most influential is that of Borjas (1992, 1995) who introduced the concept of ethnic capital, being a local, ethnic-specific spillover in human capital accumulation. In his view, immigrants cluster because of opportunities created by specific ethnic niches. According to Borjas, it is difficult to overestimate the importance of ethnic capital, or in his own words

> Ethnicity has an impact above and beyond both parental and neighborhood effects for persons who are frequently exposed to a particular ethnic environment.
> (Borjas, 1995, p. 389)

The advantage of his theory is that it can explain different degrees of clustering among ethnic groups. The second line of research is that of Stark (1991, 1994) in which low-skilled migrants are inclined to mix with high-skilled migrants in order to obscure their skill signals. Low-skilled migrants then take advantage of the imperfect information that (indigenous) employers have. Eventually, this will lead to lower wage offers to *all* immigrants caused by asymmetric information and thus depresses incentives to invest in human capital. Hendricks (2001) provides a third reason for migrants to cluster.

He argues that due to skill complementarity and imperfectly observable worker's skills, immigrants cluster. New immigrants choose to live near existing immigrants to take advantage of cross-matching. Employers then use ethnicity as a proxy for skills. This has two strong implications: (i) incentives for clustering vary between different ethnic groups and (ii) second-generation immigrants have less incentives to cluster than first-generation immigrants because usually the former group is higher skilled than the latter. Epstein and Hillman (1998) use another argument based on herd behavior to explain ethnic concentration. When a large group of previous migrants has gone to location *A*, new migrants will also be attracted to *A*, even though they have information that *B* would be the better choice. In this line of thinking, migrants take the decision of others into account, because other migrants may have access to information that they do not have. The fifth approach is that of Carrington *et al.* (1996), who introduce endogenous migration costs to explain ethnic clustering. The larger the stock of migrants in *A*, the lower are the migration costs for new migrants to go to *A*. Except for the approach of Borjas – who argues that ethnicity directly influences human capital accumulation – the other approaches acknowledge that some kind of network or information externalities are the cause for ethnic clustering, which in turn affects human capital accumulation.

Before focusing on the development of human capital of ethnic groups, it is insightful to look at the empirical evidence considering the relation between ethnic concentration and human capital formation. First, estimating the effect of ethnic concentration on economic outcomes is not straightforward. Ethnic composition is arguably an endogenous variable; migrants are selective in choosing for specific neighborhoods. Residential location of a migrant is therefore probably correlated with labor market outcomes due to unobserved attributes of the migrant.[1] There are some ways out to circumvent this endogeneity effect in the estimation procedure. Borjas (1992, 1995) uses parental choices of residential location to ensure the exogeneity of the economic outcomes of their offspring. Cutler and Glaeser (1997), Bertrand *et al.* (2000) and Dustmann

---

[1] See Manski (1993) for fundamental critique on estimating social effects without proper identification.

and Preston (2001) avoid the endogeneity problem by using fixed spatial effects (on a city level). These studies typically show that segregation in low-skilled areas is disadvantageous for ethnic minorities.[2] More recently, Edin *et al.* (2003) use a natural experiment to properly identify the concentration effect. They look at the economic performance of refugee immigrants who have been sorted by government authorities. Their results modify the previously described results. Living in an ethnically concentrated area improves the economic performance of immigrants ceteris paribus. However, the overall performance of the ethnic group is crucial. Immigrants who are clustered in a high-income area seem to benefit more than immigrants who are clustered in a low-income area.

To conclude, immigration tends to be associated with ethnic concentration and affects average levels of human capital within those clustered areas. This may have important consequences for human capital accumulation of not only the immigrants, but also for the indigenous population living close to the immigrant population. Since the seminal work of Lucas (1988), the notion that dispersion and creation of knowledge take largely place within the boundaries of the city or the neighborhood has become widespread and has been further developed by authors like Durlauf (1994), Borjas (1995) and Bénabou (1996a,b). The importance of human capital can be deduced from the fact that it is the main determinant of present income (Becker, 1975), and that it acts as the engine for future income growth (Romer, 1986; Lucas, 1988). So the externalities associated with immigration and its corresponding impacts on human capital accumulation are most likely the cause for lock-in effects of certain ethnic groups,[3] their inferior economic performance in most cases,[4] and their successful behavior in only few cases.[5] Therefore,

---

[2] Ethnic minorities are commonly defined as first and second generation immigrants.

[3] The Afro-American population group in the United States is a good example of an ethnic group that is characterized by lock-in effects, especially within certain neighborhoods in the larger cities.

[4] Most ethnic groups originating from the so-called guestworkers in the 1970s and 1980s in Western Europe perform economically consistently worse compared to the indigenous population. Later generations do not seem to be able to catch up quickly.

[5] Nowadays, successful migrant groups are usually formed by high-skilled immigrants (expats), with short migration spells.

the contribution of this chapter is twofold. First, we aim to give an overview of that part of the recent migration literature that deals with ethnic clustering and its consequences. Second, we present a basic model merging two main theories, namely, that of endogenous migration and that of human capital formation, which offers insight into the cause and dynamics of heterogeneity in human capital on the level of cities and neighborhoods.

The remainder of this chapter is organized as follows. Section 13.2 presents the basic model, which consists of a simple economy and a decision framework for potential migrants along the lines of Sjaastad (1962). This model mainly focuses on endogenous migration between two countries and defines individual human capital accumulation. Subsequently, Section 13.3 presents the analytical results in a framework with homogeneous human capital both in the source country and in the country of destination. In Section 13.4, we relax the assumption of homogeneity. First, we focus on the dynamics of human capital accumulation when migration is skill independent. Thereafter, we assume that migration costs depend both on skills and migrant networks and look at migration dynamics and human capital accumulation in a simulation framework. The last section concludes.

13.2. A model of migration and human capital accumulation

In this section, we propose a model in which we have two countries (source and destination) and where human capital and an exogenously given physical capital stock are the only production factors. Total production depends on total human capital, where human capital exhibits decreasing returns to scale. Individual wages are proportional to the individual's human capital. With equal human capital, there is an initial wage difference between the source and the destination country, due to country-specific technology.[6] Migration between the two economies is possible, but at a cost,

[6] In this framework, country-specific technology can also be interpreted as restrictions or barriers for production, such as climate, infertile land and geographical features (such as rivers, coastal waters and mountains). For historical examples of the effects of such restrictions on economic performance, see De Vries and Van der Wouden (1995) and Mokyr (1999). For empirical estimates of country-specific technologies, see Islam (1995).

which depends on the stock of migrants already in the destination country and on the amount of human capital that potential migrants possess. Here, we adopt the theory of Carrington *et al.* (1996) of endogenous migration costs. It states that costs of migration decrease in the size of the migrant network in the country of destination, where costs are not only monetary, but also involve psychological and information (search) costs. For example, the larger a migrant network, the easier it will be for a new migrant to find a job through the job referral system as suggested by Montgomery (1991). Munshi (2003) offers an empirical study of such a job referral system used by Mexican migrants in the United States and shows that an ethnic network for new migrants is indeed beneficial, at least in the short run.

Human capital develops endogenously over time. We assume that human capital accumulation depends on three factors (cf. Bénabou, 1996b). First, an individual's personal amount of human capital influences the process of accumulation. This reflects the influence of initial human capital endowment and transmission within the family. Second, the aggregate human capital in the individual's direct environment, i.e. the neighborhood, affects human capital formation. Apart from directly learning from neighbors, a person's direct environment plays a pivotal role in providing information, opportunities and role models. The third factor that plays a role in human capital accumulation is the aggregate amount of human capital in a country. This can be seen as reflecting the influence of average national human capital on the development of the individual's human capital, i.e. by the national education system. The influence of aggregate human capital is defined by constant elasticity to scale aggregator functions, where aggregate human capital is influenced by a human capital spillover parameter. If the spillover parameter is positive, then variety in human capital will lead to lower aggregate human capital. If it is negative, then it will lead to higher aggregate human capital. In the former case, human capital spillovers are hampered by lack of 'transactions' (Lazear, 1999). Basically, this means that there is insufficient communication due to a lack of trust, understanding or common language. The last two arguments build on the notion that when variety in human capital is too large, knowledge transfers are no longer possible. With a negative spillover parameter, there are returns to variety. In this case, individuals have a common basis (e.g. the same language) on which they are willing and able to communicate.

In addition to the development of human capital, the model can explain differences in population size between countries and regions. Due to immigration, richer countries have, ceteris paribus, a higher population growth. However, the distribution of the immigrant population in the country of destination is not even. Immigrants are attracted to those areas where former immigrants are situated, i.e. the city. Then, without intranational migration, these clustered areas will grow faster as compared to the rest of the country or the city, thus leading to an even more uneven distribution of the immigrant population. So, in this view, network externalities will lead to agglomeration externalities.

The next subsection describes the economic setting in more detail. Thereafter, we deal with the individual's decision to migrate.

13.2.1. The economy

We start with the construction of a simple economy. Suppose that total output $Y_{j,t}$ in a country j at time t is defined by the following production function:

$$Y_{j,t} = A_j \bar{K}_j^{1-\rho} \left(\sum_{i \in L_t} h_{i,j,t} \right)^{\rho}, \qquad \rho < 1, \tag{13.1}$$

where A_j is country-specific technology, $\bar{K}_j$ is the exogenously given capital stock in j, L_t is the labor force of country j at time t, and $h_{i,j}$ are individual-specific levels of human capital in country j. So, the total *effective* labor force in country j at time t equals $\sum_{i \in L_t} h_{i,j,t}$. Human capital is the only production factor for output. We assume that the accumulation of human capital is determined by the following relation (cf. Bénabou, 1996b):[7]

$$h_{i,t+1} = \Theta h_{i,t}^{\alpha} H_{k,t}^{\beta} H_t^{\gamma}, \qquad \alpha, \beta, \gamma < 1, \tag{13.2}$$

---

[7] Country indices j are omitted where it leads to no confusion.

where Θ is a constant, $H_{k,t}$ is composite human capital in neighborhood k at time t, and H_t is national human capital at time t. Furthermore, we assume that $\alpha + \beta + \gamma = 1$.[8] In contrast with Bénabou, we regard Equation (13.2) not only as the education that a child receives, but also as the knowledge that individuals gather during their working life. Thus, individual human capital accumulation depends on three factors: namely, the individual's inherited human capital, the human capital in the local neighborhood or direct environment of the individual, and the human capital accumulated in the entire society (all displaying diminishing returns to scale).

We assume local ($H_{k,t}$) and national (H_t) human capital to be defined by the following CES aggregators:

$$H_{k,t} = \left(\frac{1}{L_{k,t}} \sum_{i \in L_{k,t}} h_{i,t}^{(\varepsilon-1)/\varepsilon} \right)^{\varepsilon/(\varepsilon-1)} \tag{13.3}$$

and

$$H_t = \left(\frac{1}{L_t} \sum_{i \in L_t} h_{i,t}^{(\sigma-1)/\sigma} \right)^{\sigma/(\sigma-1)}, \tag{13.4}$$

with $L_{k,t}$ the labor force in neighborhood k at time t. In this case, if $\varepsilon > 0$ ($\sigma > 0$), then Equation (13.3) (Equation (13.4)) is convex in its argument. Heterogeneity is then a source of loss ($H_{k,t} < \bar{h}(H_t < \bar{h})$), where $\bar{h}$ denotes the average human capital, and individuals i are considered to be substitutes. The larger $1/\varepsilon(1/\sigma)$ is, the more $H_{k,t}(H_t)$ will converge to the minimum of $h_{i,t}$. On the other hand, if $1/\varepsilon < 0$ ($1/\sigma < 0$) then $H_{k,t}(H_t)$ is concave, heterogeneity is a source of gains, and individuals i are considered to be complements to each other. Therefore, the smaller $1/\varepsilon(1/\sigma)$, the more $H_{k,t}(H_t)$ will converge to the maximum of $h_{i,t}$.[9]

---

[8] This ensures endogenous human capital growth, because of non-diminishing returns in the human capital accumulation function (cf. Rebelo, 1991).

[9] Note that we consider individuals to be complements to each other when a heterogeneous labor force increases human capital accumulation, and they are substitutes for each other when a homogeneous labor force adds to human capital accumulation.

Let us assume that initial human capital is lognormally distributed across individuals, with $\ln h_{i,0} \sim N(\mu, \lambda^2)$. Now, we can directly link ε and σ with the human capital spillover functions, because $H_{k,t} = E[h_{i,t}]e^{-\lambda^2/2\varepsilon}$, and analogously $H_t = E[h_{i,t}]e^{-\lambda^2/2\sigma}$ (see Bénabou, 1996b).[10] This enables us to interpret ε and σ as parameters, that reflect a loss (gain) when ε and σ are positive (negative).

In order to gain insight into the behavior of the aggregator functions $H_{k,t}$ and H_t, Table 13.1 shows some values of $H_{k,t}(H_t)$ that are obtained for different values of $\varepsilon(\sigma)$ and different variances of $h_{i,t}$. Moreover, we vary the distribution of $h_{i,t}$, by using both a lognormal and a Bernoulli distribution.[11]

Table 13.1 reveals that in the presence of heterogeneity for positive $\varepsilon(\sigma)$, $H_{k,t}(H_t)$ will be lower than $E[h_{i,t}]$. So substitutability leads to losses from heterogeneity. In contrast, complementarity (negative $\varepsilon(\sigma)$) leads to gains from heterogeneity. This applies for both distribution functions, although the effects of increased heterogeneity are stronger in the case of the lognormal distribution than in the case of the Bernoulli distribution.[12]

Whether heterogeneity causes an increase or a decrease of the effective stock of human capital in a spatial area is an empirical question. Potentially, a larger diversification in human capital is beneficial, because of the different sets of knowledge available, indicating increasing returns to diversity. However, too large differences in human capital may prevent individuals from communicating effectively and therefore impede the transfer of knowledge, which leads to decreasing returns to diversity.[13]

---

[10] Because higher moments of the lognormal distribution are defined as $E[h_{i,t}^r] = e^{r\mu + r\lambda^2/2}$ (see, e.g. Mood *et al.*, 1974), we can write, for example, for the aggregator function

$$H_t = \left(\frac{1}{L_t} \sum_{i=1}^{L_t} h_{i,t}^{(\sigma-1)/\sigma} \right)^{\sigma/(\sigma-1)} = E[h_{i,t}]e^{-\lambda^2/2\sigma}.$$

[11] Note that, for a Bernoulli distribution we get $H_t = ((1-p)h_{1,t}^{(\sigma-1)/\sigma} + ph_{2,t}^{(\sigma-1)/\sigma})^{\sigma/(\sigma-1)}$.

[12] In general, one can prove that, if $h_{i,t}$ is a positive random variable, then $H_t(H_{k,t})$ will decrease in $1/\sigma(1/\varepsilon)$ (for more details, see Bénabou, 1996b).

[13] Lazear (1999) and De Graaff (2002) empirically deal with this trade-off between different sets of knowledge and the ability to communicate. Moreover, Chiswick (1991, 1998) finds a positive relation between earnings and indigenous language proficiency among immigrants.

Table 13.1. ***Simulation values of $H_{k,t}(H_t)$ with human capital ($h_{i,t}$) distributed over 1000 individuals***

$1/\varepsilon$, $1/\sigma$	Lognormal ($E[h_{i,t}]=1$); Var[$h_{i,t}$]			Bernoulli ($E[h_{i,t}]=1$); Var[$h_{i,t}$]		
	0	0.25	0.5	0	0.25	0.5
− 10	1.000	2.559	4.021	1.000	1.408	1.603
− 2	1.000	1.286	1.589	1.000	1.205	1.358
− 0.5	1.000	1.072	1.133	1.000	1.063	1.126
0.5	1.000	0.958	0.919	1.000	0.993	0.853
2	1.000	0.815	0.685	1.000	0.750	0.499
10	1.000	0.394	0.220	1.000	0.540	0.316

Having specified the composites of human capital, we now have to relate human capital to individual wages, in order to fully specify the dynamics of the model.

We assume that consumers have a linear utility function, so for each individual:

$$\max U_0 = \sum_{t=0}^{\infty} C_t \delta^t, \qquad \text{s.t.} \qquad C_t P_{C,t} \leq I_t, \tag{13.5}$$

where δ is the discount factor and $P_{C,t}$ the price of consumption goods. Since utility is linear, consumers have no incentive to smooth consumption. So they immediately spend all their income (I_t) acquired by working. Thus

$$\sum_i P_{C,t} C_{i,t} = P_{Y,t} Y_t. \tag{13.6}$$

For convenience, we normalize the output price $P_{Y,t}$ at 1. Total profits in the economy at time t are then equal to

$$\Pi_t = Y_t - \sum_{i \in L_t} w_{i,t}, \tag{13.7}$$

with w_i the individual wage. Profits accrue to the owners of capital, who have the same utility function as consumers (Equation (13.5)).

Rees, 1992), so

$$w_{i,t} = \rho A \bar{K}^{1-\rho} \left(\sum_{i \in L_t} h_{i,t} \right)^{\rho-1} h_{i,t}. \tag{13.8}$$

The relative wage between two (arbitrary) individuals, k and l, thus equals:

$$\frac{w_{k,t}}{w_{l,t}} = \frac{h_{k,t}}{h_{l,t}}, \tag{13.9}$$

from which we can derive, using Equations (13.1) and (13.8), that

$$\sum_{i \in L_t} w_{i,t} = \rho Y_t. \tag{13.10}$$

Basically, this defines the economy in a particular country. The next subsection presents the mechanisms driving migration between countries.

13.2.2. Endogenous migration

We suppose that there are two countries (each described by the basic economy, as laid out in the previous subsection) named the source (s) and the destination (d). Furthermore, the country-specific technology is strictly larger in the country of destination, due to historical or geographical factors, so $A_d > A_s$. Individuals migrate if their present discounted utility is larger in the country of destination than in the country of origin. The present discounted utility does not only depend on wages and the discount factor δ, but also on migration costs. The migration costs are assumed to depend on the stock of migrants and human capital according to

$$c_{i,t} = c(P_{M,t}, h_{i,t}), \tag{13.11}$$

with $P_{M,t}$ the stock of migrants in the receiving country. We assume that the migration costs decrease in the stock of migrants and individual human capital. Furthermore, $c_{i,t} > 0$ always holds for each potential migrant i. We assume that wages in d are initially higher than in s, due to better country-specific technologies A.[14] Then, migrants will always flow from s to d. Following the present value approach of Sjaastad (1962), we can now determine the maximum present discounted utility values for an individual i at time t in s, who migrates (M) and in the case in which he stays (S):

$$V_{i,t}^{S} = \frac{w_{i,t}^{s}}{1-\delta}, \tag{13.12}$$

$$V_{i,t}^{M} = \frac{w_{i,t}^{d}}{1-\delta} - c_{i,t}. \tag{13.13}$$

In this specification, individuals do not take future development of human capital, and thus wages, into account. So, we assume that individuals display myopic behavior instead of rational behavior.[15]

It is now useful to adopt from Carrington *et al.* (1996) the concept of the marginal migrant who has human capital $\tilde{h}$ viz. that migrant who is indifferent between migrating and staying. Because of the monotonic decrease of costs in the stock of migrants, we can also put $\tilde{h} = \phi(P_{M,t})$. Using Equation (13.12), the marginal migrant is indifferent between migrating or staying when the following equality holds (cf. Carrington *et al.*, 1996):

$$\frac{w_{i,t}^{d} - w_{i,t}^{s}}{1-\delta} = c(P_{M,t}, \phi(P_{M,t})). \tag{13.14}$$

Equation (13.14) basically presents a cost–benefit approach, where the costs exhibit a network externality.

Having specified the economies of the two countries and the mechanism driving international migration, we can now turn to the analysis of the relation between immigration and human capital dynamics. The next section analyzes migration and human capital

---

[14] If population sizes are highly unequal, then wages in a small country with an inferior technology can be higher than in a large country with a superior technology, because there are diminishing returns to human capital. So, due to differences in factor prices, people could migrate from d to s, even though country-specific technology might be higher in d. We avoid this situation by construction (i.e. the source country is assumed to be sufficiently large).

[15] Thus, migrants take into account the influence of the stock of migrants (M) on the payoff of migrating, but not the (marginal) impact of their and future migrants' migration on future wages.

accumulation in the case of within-country homogeneous human capital, so that the migration costs of Equation (13.11) only depend on the stock of migrants in d.

13.3. Within-country homogeneous human capital

In this section, we look at the equilibrium when both country s and country d have a homogeneous labor force. With homogeneous human capital in the source country, every individual would migrate if migration costs are below a certain threshold. Therefore, we have to calculate an instantaneous macro-migration flow, M_t, at time t. Because we only analyze the effects of one migration flow in this section, we omit the time-indices for the comparative static analysis for clarity reasons.

Table 13.2 provides the comparative statics for the size of the instantaneous migrant flow M. For more details and the specification of the macroeconomic equilibrium we refer to Appendix A13.

The comparative static results can be understood as follows (see Appendix A13 for technicalities). An increase in the human capital in the source country (h_s) has a positive impact, because migration costs will decrease and wages in the destination (d) will increase for the immigrant population relative to the source (s). Human capital in the country of destination (h_d) has a negative impact, because there are decreasing returns to total human capital. Therefore, if country d has a higher amount of average human capital than country s, then a smaller wage will accrue to the immigrant population due to the additive character of human capital in the production function that is characterized by decreasing returns to scale. If wages in s rise (due to a higher country specific technology (A_s)), then migration will decrease, because of a higher utility in s. The discount factor (δ) has a positive impact, because a higher δ means that future gains are more appreciated relative to the one-time migration costs that have to be incurred in the first period, so that more individuals will opt for the higher wages in d. An increase in ρ increases the returns to total human capital, and therefore increases individual wages. Because

Table 13.2. Comparative statics for M

h_s	h_d	A_s	A_d	δ	ρ	P_I	P_s	P_M
+	−	−	+	+	+	−	+	−/+

initially there is assumed to be a wage gap between s and d, wages in d rise relatively stronger, which will have a positive impact on the migration flow (M). If the indigenous population (P_I) in d increases, then the equilibrium value of M decreases, due to decreasing returns to total human capital, and thus lower wages in d. Finally, a larger size of the population in s will increase M, because of lower wages in s due to higher population pressure. The effect of P_M depends on two factors, namely, the endogenous migration costs and the effect of the size of the migrant population on the wages they earn in d. If P_M is sufficiently small, then the size of the migrant community contributes positively to the migrant inflow (which can be regarded as a positive network externality for the immigrants). If, however, it is smaller, then the stock of migrants has a negative influence on M (which is caused by the decreasing returns to scale in the production function).

We now turn to the dynamics of human capital accumulation. We still assume one (initial) migrant flow, in contrast to the next section where we also take the dynamics of migration into account. According to our model, migrants will only go to one neighborhood (say k) in d.[16] Figure 13.1 shows the geographical configuration.

On the neighborhood level there is now a mixture between a (growing) group of immigrants with an amount of human capital that deviates from that of the indigenous population and a group of indigenous individuals that gets smaller. For simplicity, we assume that there is no interregional migration from k to d. Although human capital within groups is homogeneous and we only have one migration flow, it is interesting to look at the dynamics of human capital accumulation in s, d and k. First, using Equation (13.2), it is easy to see that human capital in the source country accumulates as follows:

$$\ln h_{s,t+1} = \theta + \ln h_{s,t}, \tag{13.15}$$

[16] Regardless of where they live, immigrants earn the same wage in d. However, realistically, if immigrants take into account future human capital accumulation, then there is the possibility that they will pay higher migration costs to go to a less ethnically-clustered area, in order to gain a higher future wage for themselves or their children. Because we use myopic behavior of individuals, our model does not incorporate this possibility.

Figure 13.1. Geographical configuration

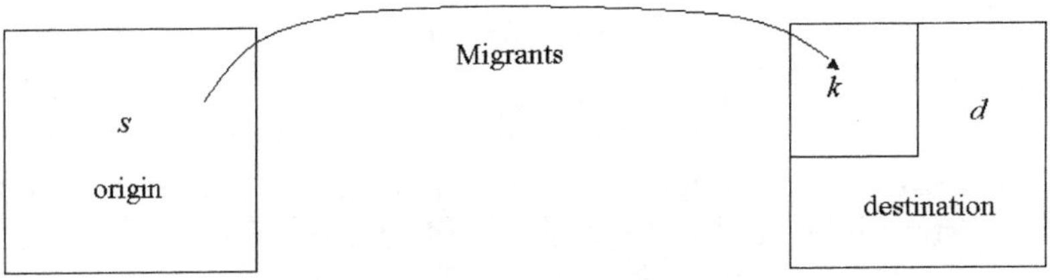

where the growth constant θ is defined as ln Θ. Immigrants in d living in k have the following human capital accumulation:

$$\begin{aligned}\ln h_{s,k,t+1} = \theta + \alpha \ln h_{s,k,t} + \frac{\beta\varepsilon}{\varepsilon - 1} \ln\left(m_{k,t} h_{s,k,t}^{(\varepsilon-1)/\varepsilon} + (1 - m_{k,t}) h_{d,k,t}^{(\varepsilon-1)/\varepsilon}\right) \\ + \frac{\gamma\sigma}{\sigma - 1} \ln\Big(m_t h_{s,k,t}^{(\sigma-1)/\sigma} + p_{k,t} h_{d,k,t}^{(\sigma-1)/\sigma} \\ + (1 - m_t - p_{k,t}) h_{d,t}^{(\sigma-1)/\sigma}\Big) \end{aligned} \quad (13.16)$$

Here, $m_{k,t}$ denotes the percentage of migrants in k at time t, m_t denotes the percentage of migrants in the whole country d at time t, $p_{k,t}$ the percentage of the indigenous population in k at time t relative to the whole country d, $h_{s,k,t}$ is the amount of human capital of migrants in k at time t, and $h_{d,k,t}$ is the human capital of the indigenous population in k at time t.[17] This specification follows from the fact that we consider the mixture between immigrants and the indigenous population, for example, in neighborhood k as a Bernoulli distribution. Human capital accumulation of the indigenous population in neighborhood k can be calculated analogously to the migrant population. Now, it is not difficult to see that the difference in human capital between migrants and the indigenous population in k is equal to (see also footnote 11):

$$\ln h_{s,k,t+1} - \ln h_{d,k,t+1} = \alpha(\ln h_{d,k,t} - \ln h_{s,k,t}). \quad (13.17)$$

So with homogeneous human capital in s and d, differences in the levels of human capital within a neighborhood are only caused by

[17] More specifically, we can define $m_{k,t} \equiv P_{M,t}/L_{k,t}$, $m_t \equiv P_{M,t}/L_t$ and $p_{k,t} \equiv P_{I,k,t}/L_t$, where $P_{I,k,t}$ is the indigenous population in k at time t, $L_{k,t} \equiv P_{M,k,t} + P_{I,k,t}$ and $L_t \equiv P_{M,t} + P_{I,t}$.

initial differences.[18] Moreover, in the long run, the difference in the levels of human capital will decrease. Eventually, all individuals in k will have equal amounts of human capital. So the values of human capital will converge.

Finally, after calculating human capital accumulation in the rest of the destination country and comparing it to neighborhood k, we can state that $\ln\ h_{d,t} > \ln h_{d,k,t}$ only if:

$$\frac{\varepsilon}{\varepsilon - 1}\ln(m_{k,t}h_{s,k,t}^{\varepsilon-1/\varepsilon} + (1 - m_{k,t})h_{d,k,t}^{(\varepsilon-1)/\varepsilon}) < \ln h_{d,t}, \qquad (13.18)$$

which occurs when $\varepsilon > 0$ (thus with decreasing returns to human capital diversity on a local level; see also Table 13.1), or when m_k is sufficiently large. However, the former mechanism may cause a persistent difference between the population in k and the population in the rest of the country d (as long as there is diversity in human capital), where the latter mechanism causes a difference that will eventually disappear. Thus, with initial homogeneous human capital, if diverse human capital within one neighborhood hampers human capital accumulation, then neighborhoods with more diverse human capital (due to immigration) lag in their human capital accumulation compared to neighborhoods with more equal human capital. This falling behind could be caused by less communication between people with different amounts of human capital, where benefits from heterogeneous human capital can be found in different sets of knowledge between individuals. However, communication between the indigenous and the migrant population should be high to make use of these different sets of knowledge. De Graaff (2002) looks more deeply into the trade-off between communication and information sets. In the next section, we drop the assumption of within-group homogeneity and analyze the more realistic case of heterogeneous human capital.

13.4. Heterogeneous human capital and migration

Because individuals differ widely in abilities, intelligence and their knowledge, homogeneity is a strong assumption. In this

[18] Note that we assume symmetry between the learning opportunities for the indigenous and the migrant population.

section, we therefore analyze human capital accumulation and migration patterns in the case where human capital is heterogeneous. Because of the analytical complexity, we resort to numerical methods.[19]

We assume (cf. Bénabou, 1996b) that the human capital of agents is initially lognormally distributed, just as in Section 13.2.1. Furthermore, we assume that individuals in s initially have a lower mean human capital than individuals in d, such that $\ln h_{s,0} \sim N(\mu_{s,0}, \lambda^2_{s,0}) < \ln h_{d,0} \sim N(\mu_{d,0}, \lambda^2_{d,0})$. In contrast to the static migration analysis in the previous section, we now analyze the complete transition path to a migration equilibrium allowing for multiple flows of immigrants.

In this subsection, we specify the individual moving costs of Equation (13.11) as

$$c_{i,t}(m_t, h_{i,t}) = \varrho m_t^{\nu} h_{i,t}^{\zeta}, \tag{13.19}$$

with ϱ being a scale parameter, $\zeta < 0$ and $\nu < 0$. For the special case of $\zeta = 0$ we refer to Appendix B13 for an analytical solution. Thus, individual migration costs decrease when human capital increases. In the more general case, the solution is no longer analytically tractable and we have to rely on numerical simulations to derive the dynamics of the full model. Table 13.3 displays the values of the parameters used.

We begin with a small initial stock of migrants compared to the total number of inhabitants in d. Furthermore, countries s and d have the same size. We assume that country-specific technology in d is higher than in s. Both countries exhibit equal (exogenous) growth in human capital. Initial human capital is assumed to be slightly larger in d than in s. Moreover, the development of human capital is mainly dependent on the size of initial human capital (α). The influence of the global environment (γ) is assumed to be more important than the influence of the local environment (β). Finally, the higher an individual's human capital, the lower are the migration costs, due to the negative human capital coefficient (ζ).

---

[19] Appendix B13 analytically deals with the accumulation of human capital, both in country s and in country d in the case in which the cost of migration does not depend on human capital.

Table 13.3. ***Simulation framework***

Initial stock of migrants: m_0	0.001
Initial population in the source (s)	10,000
Initial population in the destination (d)	10,000
Initial population in the neighborhood (k)	2500
Discount factor: δ	0.95
Country-specific technology in the source (s)	1
Country-specific technology in the destination (d)	2
Stock of given capital ($\bar{K}$) in s and d	1
Growth Θ (both in s and d)	1.03
Initial human capital in the source (s): μ_s	0
Initial human capital in the destination (d): μ_d	0.1
Elasticity of individual human capital: α	0.5
Elasticity of local human capital: β	0.2
Elasticity of national human capital: γ	0.3
Elasticity of total human capital: ρ	0.8
Human capital coefficient: ζ	-0.1
Scale parameter: ϱ	5×10^{-7}

In this chapter, we are mainly interested in the effects of the size of the network externality (ν), the local elasticity of substitution (ε) and the global elasticity of substitution (σ) on the patterns of immigration and the accumulation of human capital. First, we look at the effect of the network externality on the size and pattern of immigration. Figure 13.2 maps out the relation between the absolute stock of migrants in d and the size of the network externality (ν), using the configuration depicted in Table 13.3.

Figure 13.2. ***Migration patterns with different network externality sizes (with $\varepsilon = 0.5$ and $\sigma = -1$)***

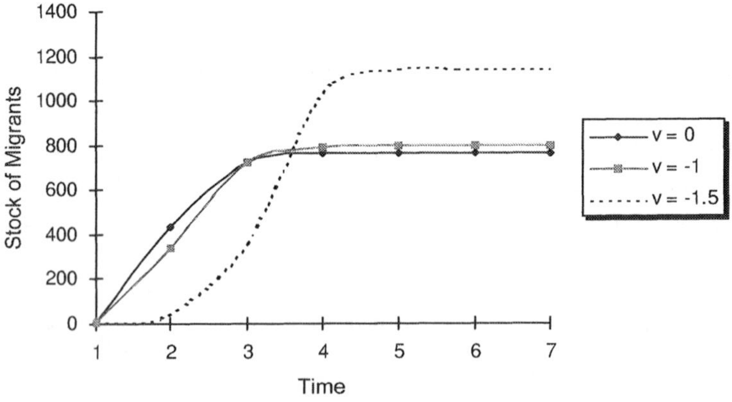

Figure 13.2 shows that a stronger network externality (v lower) causes a higher level of equilibrium in- and outmigration. This is caused by the low moving costs. Due to the specific functional form chosen in Equation (13.19), small stocks of migrants incur high moving costs when large network externalities are present. As a consequence, the size of migration is relatively small in early time periods compared to the size of migration when no network externalities are present.

The next question that arises is to what extent do the local (ε) and global (σ) elasticities of substitution have an impact on the human capital accumulation of individuals living in k, s and d? The answer to this question depends on the hypothesis chosen regarding these elasticities. We investigate here two possible cases.

13.4.1. Negative impact of the neighborhood

First, we assume that on a neighborhood level heterogeneity has a strong negative impact on the development of human capital, whereas it has a weakly positive impact on the development of human capital on a national level. So, ε must be slightly larger than 0 and σ must be negative and, in an absolute sense, larger than ε. We can interpret this case as a situation in which the migrant population and the indigenous population in k have difficulties in communicating, due to large cultural and language differences. They will not learn from each other, which hampers human capital accumulation. On the other hand, on a national level, the new influx of immigrants can be beneficial, because the enlarged population demands and supplies a larger variety of products, there are more niche markets, and there is a broadening of culture in general. Figure 13.3 shows the human capital accumulation for this particular case, when the neighborhood k is highly segregated.

In Figure 13.3, *k_d* and *k_s* denote, respectively, the indigenous and immigrant population in k. One can observe that, although the immigrants have the highest amount of human capital, it deteriorates quickly. The same holds for the indigenous population in k. Because heterogeneity slows down human capital accumulation, the human capital of individuals living in k deteriorates for a long time, until, in the end, it converges to that of the rest of the country (d).

Figure 13.3. Human capital accumulation (with $\varepsilon = 0.1$, $\sigma = -2$ and $\nu = -1.5$)

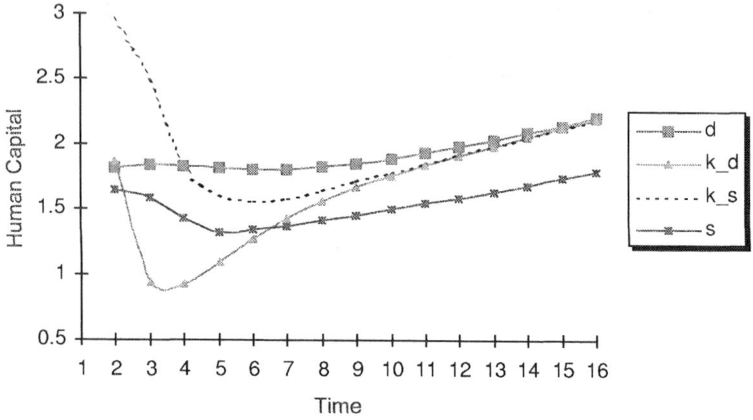

13.4.2. Positive impact of the neighborhood

In our second case, we assume that on a neighborhood level heterogeneity has a strongly positive impact on the development of human capital, whereas it has a weakly negative impact on the development of human capital on a national level. Then ε must be slightly smaller than 0 and σ must be positive and – in absolute value – be larger than ε. We may interpret this case as a situation where there are large learning effects when immigrants and the indigenous population live close together. Consider, for example, the case where the immigrant and native languages are rather similar. Then individuals have few barriers in communicating. These spillovers could then be beneficial for neighborhood k, but negative for the rest of the country d. The latter effect can be caused by a brain drain of immigrants with high human capital from the rest of the country to neighborhood k, because of the high human capital spillovers. Figure 13.4 shows the human capital accumulation for this configuration of parameters.

Figure 13.4 reveals that in the long run the human capital of immigrants again decreases. But because the human capital of the indigenous population increases, average human capital in k remains above that of the rest of d and will only slowly converge to the level of d. Average human capital in s will – in the long run – grow at the same rate as human capital in d, but its level will remain lower,

Figure 13.4. Human capital accumulation (with $\varepsilon = -0.1$, $\sigma = 2$ and $\nu = -1.5$)

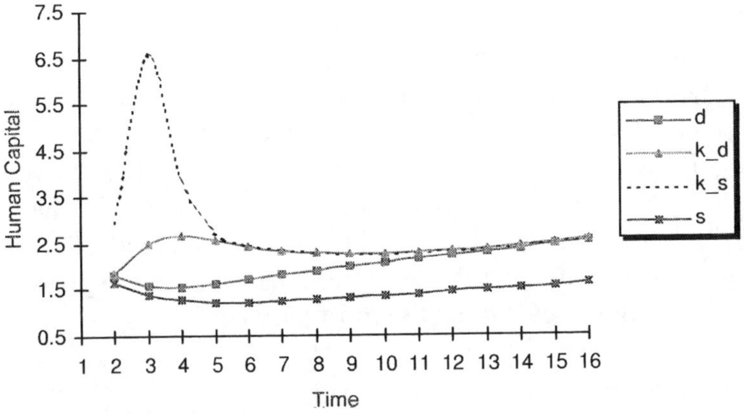

so that there is no absolute convergence, but only relative convergence. The main differences between Figure 13.3 and Figure 13.4 apply to the short run. Due to the composition effect, individuals living in *k* witness their human capital decrease in Figure 13.3 in the short run, while the human capital in *k* in Figure 13.4 receives a large boost in the short run. Furthermore, just as in Figure 13.3, initially the highly-skilled immigrants leave, causing a brain drain from *s* in the short run, from which average human capital in *s* only slowly recovers.[20]

Historical and contemporary evidence shows that both cases can occur. The processes shown in Figure 13.3 can be seen in most Western European countries and in the integration of former guest workers. Because these immigrants came from further away than before, communication between the immigrant and indigenous population was, and still is, difficult, and does not occur as often as within groups. Therefore, immigrants and the indigenous population act as substitutes on a local level in their human capital accumulation. On the other hand, migrants who act as complements for the indigenous population (cf. Figure 13.4) are nowadays typically high-skilled immigrants.

[20] Although the international literature often assumes that high-skilled immigrants have lower moving costs than low-skilled immigrants, there is actually not much empirical evidence for this assumption. Stalker (1994) suggests that it is not necessarily the most highly skilled on a national level who are the first to leave their source country, but the most highly skilled on a city or village level.

Obviously, scientists are among them, but also exchange students, expatriates, ICT-workers, and the like. Communication between these immigrants and the indigenous population is mostly in English and does not raise many problems. These immigrants – although they form a relatively small minority compared to the lower skilled migrants – usually live in upmarket neighborhoods. In addition to these two cases, other kinds of human capital accumulation processes can occur, depending on the values of ν, σ and ε and the relative levels of human capital. However, in most Western European countries a process with local substitution seems to be the most prevalent.

13.5. Conclusion

This chapter developed a model to analyze the relation between endogenous migration and human capital accumulation affected by knowledge spillovers among individuals. In the basic version of the model, we considered the simple case of homogeneous human capital within a country. Here, we concluded that the relation between the stock of migrants and the extent of immigration is ambiguous. If the stock of migrants is large, then network externalities ensure lower migration costs and thus a higher number of immigrants. However, if the stock of migrants becomes too large, wages will decrease because of decreasing returns to scale in aggregate human capital.[21]

For the more extended version of the model in which we allowed for heterogeneous human capital within a population, we had to resort to numerical methods. The results indicate that, within a country, the human capital of the migrants and that of the indigenous population will eventually converge, although only after a considerable time period. As expected, the results indicate the presence of a brain drain from the source country, because the more highly skilled are assumed to be the first to leave.[22] However, the average human capital of the

---

[21] Evidently, in the presence of downward wage-rigidity, these decreasing returns to scale assumptions will show up in relatively high unemployment rates as is consistent with stylized empirical facts.

[22] Recently, some authors (Stark *et al.*, 1997; Beine *et al.*, 2001) argued that brain drains could be potentially beneficial for source countries. The reason is that because of higher returns in potential destination countries more investments will be made by the population in the source country in human capital, so that in the long run growth and brain drains could be positively correlated.

high-skilled migrants will quickly decrease to that of the indigenous population in the same spatial area. Whether average human capital in this clustered area will be lower or higher than in the rest of the country, depends on the level of communication between the immigrant and indigenous population. If their individual human capital is complementary in human capital accumulation, so that individuals will learn from each other faster in the case of heterogeneous levels of human capital, then average human capital in the clustered area will be higher than in the rest of the country. On the other hand, if their human capital can be considered as substitutes, then average human capital in the clustered area remains lower than in the rest of the country for a long time. Because of the direct relationship between wages and human capital, this will have direct consequences for the distribution of earnings within a country.

We also considered the effect of stronger network externalities on the process of migration. First, these network externalities enhance migration. Second, they ensure that the size of the stock of migrants follows a sigmoid pattern over time. Initially, migration will start slowly, then it accelerates because of the network effect on the migration costs. Thereafter, migration will decelerate because of the diminishing returns to human capital, until, finally net migration is zero. All migration theories that take network externalities into account will have similar effects as in this chapter as long as low-skilled migrants will have incentives to cluster with high-skilled migrants. They can thus provide insight into the cause of spatial heterogeneity and its evolution over time. Moreover, they generate insights into the causes of lagging performance of particular ethnic groups in terms of skill acquisition.

Appendix A13. Comparative statics

Combining equilibrium Equation (13.11) with wage Equation (13.8) gives us the following relation:

$$\frac{(1-\delta)}{\rho}c(P_M,h_s)=\frac{(P_Ih_d+(P_M+M)h_s)^{\rho-1}}{P_I(\frac{h_d}{h_s})+(P_M+M)}A_d-\frac{((P_s-M)h_s)^{\rho-1}}{P_s-M}A_s. \tag{A13.1}$$

The cost of migration is equal for every migrant because of homogeneous human capital in s. Note, that in this specification everyone within one country with the same human capital receives the same wage, so, for example, migrants in d receive $(h_s/P_I h_d + (P_M + M)h_s)P_Y Y_d$. Moreover, because migrants leave s, population will decrease in s and marginal returns to human capital, and thus the wage, will increase in s.

To derive the comparative statics characteristics of the model, we can use the implicit function theorem (see, e.g. Chiang, 1984). First, we define the implicit function Φ as follows:

$$\Phi = \frac{1-\delta}{\rho} c(P_M, h_s) - h_s A_d (P_I h_d + (P_M + M)h_s)^{\rho-2} + A_s (P_s - M)^{\rho-2} h_s^{\rho}. \qquad \text{(A13.2)}$$

Then, using the implicit function theorem, we get

$$\frac{\mathrm{d}M}{\mathrm{d}x} = -\frac{\Phi_1'(x, M)}{\Phi_2'(x, M)},$$

if $\Phi_2'(x, M) \neq 0$ where x is a parameter of the model. The determination of the signs of the derivatives as presented in Table 13.2 is now straightforward, except for two. First, the sign of the partial derivative of Φ with respect to ρ is not immediately clear, but we can write:

$$\begin{aligned} \frac{\partial \Phi}{\partial \rho} = & -\frac{1}{\rho^2}\{(1-\delta)c(P_M, h_s) \\ & + h_s(P_I h_d + h_s(P_M + M))^{\rho-2} \\ & \times (\ln(P_I h_d + h_s)(P_M + M)) \\ & \times A_d \rho^2 - h_s((P_s - M)h_s)^{\rho-2} \\ & \times (\ln(P_s - M)h_s)A_s \rho^2\} < 0, \end{aligned} \qquad \text{(A13.3)}$$

where the sign of $\partial \Phi / \partial \rho$ is negative, provided that

$$P_I \frac{h_d}{h_s} + P_M + 2M + (1-\delta)c(P_M, h_s) > P_s \qquad \text{(A13.4)}$$

given that $A_s \leq A_d$.

Second, the partial derivative of Φ with respect to the stock of migrants is not immediately clear and equals:

$$\frac{\partial \Phi}{\partial P_M} = \frac{(1-\delta)}{\rho}\frac{\partial c(P_M,h_s)}{\partial P_M} + \frac{(2-\rho)h_s}{\rho}\frac{\rho(P_I h_d + h_s(P_M+M))^{\rho-1} h_s A_d}{(P_I h_d + h_s(P_M+M))^2}$$

$$= \underbrace{\frac{(1-\delta)}{\rho}\frac{\partial c(P_M,h_s)}{\partial P_M}}_{<0} + \underbrace{\frac{(2-\rho)h_s}{\rho(P_I h_d + h_s(P_M+M))} w^d_{h_s}}_{>0}, \tag{A13.5}$$

with $w^d_{h_s}$ the wage that someone with low human capital (h_s) will earn in d. It is now easy to see that eventually (for P_M sufficiently large) the second term will dominate the first term.

Appendix B13. Human capital accumulation when migration costs do not depend on human capital

We first assume that human capital in country s is equally distributed for each individual. Then growth in expected human capital in country s is straightforwardly denoted as (see also footnote 10):

$$\mu_{s,t+1} = \theta + \mu_{s,t} + \gamma\frac{\sigma-1}{\sigma}\frac{\lambda^2_{s,t}}{2}, \tag{B13.1}$$

with variance: $\lambda^2_{s,t+1} = (\alpha+\beta)^2\lambda^2_{s,t}$, which converges to 0. The human capital distribution in s remains lognormally distributed. However, this is not the case for the human capital distribution in d. In the country of destination, every group of migrants and indigenous population is lognormally distributed, and where total population in k or the rest of d is the sum of lognormal distributions. Therefore, analytical expressions for the growth in expected human capital in areas in d are – although feasible – not insightful. Instead, we may look at the differences in expected human capital accumulation. The difference in expected human capital growth between the indigenous and the migrant population in k can again be denoted as

$$\mu_{d,k,t+1} - \mu_{s,k,t+1} = \alpha(\mu_{d,k,t} - \mu_{s,k,t}). \tag{B13.2}$$

Because $\alpha < 1$, human capital within a neighborhood will converge by construct. If we, e.g. want to model the ethnic capital concept of Borjas (1992, 1995), then we have to allow for different human capital spillovers for the migrant and the indigenous population. Here, we are especially interested in the difference in human capital accumulation between the migration population inside and the indigenous population outside neighborhood k. Unfortunately, the expression is rather complex and equals:

$$\mu_{d,t+1} - \mu_{s,k,t+1} = \alpha(\mu_{d,t} - \mu_{s,k,t}) + \beta(\mu_{d,t} - E\ln[H_{k,t}]), \tag{B13.3}$$

with

$$E\ln[H_{k,t}] = \frac{\varepsilon}{\varepsilon - 1}\ln\left\{\sum_{\tau=0}^{t-1} m_{k,t,t-\tau}\left(E[h_{d,s,t,t-\tau}]\exp\left\{-\frac{\lambda^2_{d,s,t,t-\tau}}{2\varepsilon}\right\}\right)^{(\varepsilon-1)/\varepsilon} + \left(1 - \sum_{\tau=0}^{t-1} m_{k,t,t-\tau}\right)\left(E[h_{d,k,t}]\exp\left\{-\frac{\lambda^2_{d,k,t}}{2\varepsilon}\right\}\right)^{(\varepsilon-1)/\varepsilon}\right\}, \tag{B13.4}$$

where $\sum_{\tau=0}^{t-1} m_{k,t,t-\tau}$ denotes all migrant groups, who came to k in the past. All migrant and indigenous groups in Equation (B13.4) separate are still lognormally distributed. However, the expectation of $H_{k,t}$ is affected by some weighted average of expectations of the human capital of the various groups living in k. Basically, Equation (B13.4) has the same characteristics as those described in footnote 9. So again, whether human capital in k is larger than in d depends on the size of the group of migrants and the value of ε. If ε is larger than zero, then there are again decreasing returns to diversity. Otherwise there are increasing returns to diversity. Thus, with heterogenous human capital the analytical results do not change in nature. Due to the larger variation in human capital, it is likely that the dynamic evolution of human capital accumulation fluctuates more strongly when incorporating heterogenous human capital. In this setting, the parameter σ does not have any influence on differences in human growth within a country. If there are, e.g. national returns to diversity ($\sigma < 0$), then immigration will benefit a country as a whole.

Note that the variances in human capital will eventually converge to zero for all groups, due to the fact that we do not allow for idiosyncratic shocks in human capital accumulation. Therefore, human capital differences within a country will eventually disappear, because of the averaging influence of human capital in the neighborhood and the country as a whole. However, we know that idiosyncratic shocks do occur. For example, persons are born. Therefore, convergence does not necessarily occur in practice.

References

Bartel, A.P. (1989), "Where do the new U.S. immigrants live?", *Journal of Labor Economics*, Vol. 7, pp. 371–391.

Becker, G.S. (1975), *Human Capital: A Theoretical and Empirical Analysis with Special Reference to Education*, New York: Columbia University Press.

Beine, M., F. Docquier and H. Rapoport (2001), "Brain drain and economic growth: theory and evidence", *Journal of Development Economics*, Vol. 64, pp. 275–289.

Bénabou, R. (1996a), "Equity and efficiency in human capital investment: the local connection", *Review of Economic Studies*, Vol. 63, pp. 237–264.

Bénabou, R. (1996b), "Heterogeneity, stratification, and growth: macroeconomic implications of community structure and school finance", *American Economic Review*, Vol. 86, pp. 584–609.

Bertrand, M., E.F.P. Luttmer and S. Mullainathan (2000), "Network effects and welfare cultures", *Quarterly Journal of Economics*, Vol. 115, pp. 1019–1055.

Borjas, G.J. (1992), "Ethnic capital and intergenerational mobility", *Quarterly Journal of Economics*, Vol. 107, pp. 123–150.

Borjas, G.J. (1994), "The economics of immigration", *Journal of Economic Literature*, Vol. 32, pp. 1667–1717.

Borjas, G.J. (1995), "Ethnicity, neighborhoods, and human-capital externalities", *American Economic Review*, Vol. 85, pp. 365–390.

Carrington, W.J., E. Detragiache and T. Vishnawath (1996), "Migration with endogenous moving costs", *American Economic Review*, Vol. 86, pp. 911–930.

Chiang, A.C. (1984), *Fundamental Methods of Mathematical Economics*, Singapore: McGraw-Hill.

Chiswick, B.R. (1991), "Speaking, reading, and earnings among low-skilled immigrants", *Journal of Labor Economics*, Vol. 9, pp. 149–170.

Chiswick, B.R. (1998), "Hebrew language usage: determinants and effects on earnings among immigrants in Israel", *Journal of Population Economics*, Vol. 11, pp. 253–271.

Cutler, D.M. and E.L. Glaeser (1997), "Are ghettos good or bad?", *Quarterly Journal of Economics*, Vol. 112, pp. 827–872.

De Graaff, T. (2002), *Migration, Ethnic Minorities and Network Externalities*, Amsterdam: Thela Thesis.

De Vries, J. and A. Van der Wouden (1995), *The First Modern Economy: Success, Failure, and Perseverance of the Dutch Economy, 1500–1815*, Amsterdam: Balans.

Durlauf, S.N. (1994), "Spillovers, stratification, and inequality", *European Economic Review*, Vol. 38, pp. 836–845.

Dustmann, C. and I. Preston (2001), "Attitudes to ethnic minorities, ethnic context and location decisions", *Economic Journal*, Vol. 111, pp. 353–373.

Edin, P.-A., P. Fredriksson and O. Åslund (2003), "Ethnic enclaves and the economic success of immigrants – evidence from a natural experiment", *Quarterly Journal of Economics*, Vol. 118, pp. 329–357.

Epstein, G.S. and A. Hillman (1998), *Herd Effects and Migration*, London: CEPR.

Gravelle, H. and R. Rees (1992), *Microeconomics*, New York: Longman Publishing.

Greenwood, M.J. and J.M. McDowell (1986), "The factor market consequences of U.S. immigration", *Journal of Economic Literature*, Vol. 24, pp. 1738–1772.

Hendricks, L. (2001), "The economic performance of immigrants: a theory of assortive matching", *International Economic Review*, Vol. 42, pp. 417–449.

Islam, N. (1995), "Growth empirics: a panel data approach", *Quarterly Journal of Economics*, Vol. 110, pp. 1127–1170.

LaLonde, R. and R.H. Topel (1991), "Immigrants in the American labor market: quality, assimilation, and distributional effects", *AEA Papers and Proceedings*, Vol. 81, pp. 297–302.

Lazear, E.P. (1999), "Culture and language", *Journal of Political Economy*, Vol. 106, pp. S95–S126.

Lucas, E.P. (1988), "On the mechanics of economic development", *Journal of Monetary Economics*, Vol. 22, pp. 3–42.

Manski, C.F. (1993), "Identification of endogenous social effects: the reflection problem", *Review of Economic Studies*, Vol. 60, pp. 531–542.

Massey, D.S., J. Arango, G. Hugo, A. Kouaouci, A. Pellegrino and J.E. Taylor (1993), "Theories of international migration: a review and appraisal", *Population and Development Review*, Vol. 19, pp. 431–466.

Mokyr, J. (1999), *The British Industrial Revolution: An Economic Perspective*, Boulder: Westview Press.

Montgomery, J. (1991), "Social networks and labor market analysis", *American Economic Review*, Vol. 91, pp. 1407–1418.

Mood, A.M., F.A. Graybill and D.C. Boes (1974), *Introduction to the Theory of Statistics*, Singapore: McGraw-Hill.

Munshi, K. (2003), "Networks in the modern economy: Mexican migrants in the U.S. labor market", *Quarterly Journal of Economics*, Vol. 118, pp. 549–599.

Rebelo, S. (1991), "Long-run policy analysis and long-run growth", *Journal of Political Economy*, Vol. 99, pp. 500–521.

Romer, P.M. (1986), "Increasing returns and long-run growth", *Journal of Political Economy*, Vol. 94, pp. 1002–1037.

Sjaastad, L.A. (1962), "The costs and returns of human migration", *Journal of Political Economy*, Vol. 70, pp. 80–93.

Stalker, P. (1994), *The Work of Strangers: A Survey of International Labour Migration*, Geneva: International Labour Office.

Stark, O. (1991), *The Migration of Labor*, Cambridge: Basil Blackwell.

Stark, O. (1994), *Patterns of Labor Migration when Workers Differ in their Skills*, *Economic Aspects of International Immigration*, Berlin: Springer, pp. 57–74.

Stark, O., A. Prskawetz and C. Helmenstein (1997), "A brain gain with a brain drain", *Economics Letters*, Vol. 55, pp. 227–234.

PART 4

Urban Hierarchy

Urban Dynamics and Growth, CEA, vol. 266
Roberta Capello and Peter Nijkamp (editors)

CHAPTER 14

Advanced Insights in Central Place Theory

Shin-Kun Peng
Academia Sinica and National Taiwan University, Taipei 11529, Taiwan, ROC

Abstract

Theoretical models with the central place theory, emergence of the city, urban system, and rank-size rule are surveyed in a common framework. We also review some related econometric models and discuss their application, and also show some empirical evidences in a few regions or countries. The main focus is to selectively survey several ideas from location theory, regional science, and microfoundation of scale economics and production externality that directly address the following questions. First, we discuss the famous but problematic 'central-place theory' development to explain the pattern of city sizes and locations, where some empirical works are also studied. Second, we examine the crucial scale economics or external economics on the formation of a monocentric urban configuration and also review the econometric model. Third, we investigate the ad hoc but fruitful theoretical analysis on the emergence of an urban system and discuss a few related empirical works. Finally, we introduce the theoretical model of the 'rank-size rule', and employ this useful idea to examine the distribution of cities within a country in different periods with various empirical evidences.

I would like to thank Roberta Capello, Peter Nijkamp, and an anonymous referee for their valuable suggestions and helpful comments. I also thank research assistance support from the Institute of Economics of Academia Sinica by grant No. IEAS-03-06. The usual disclaimer applies.

Keywords: central place theory, monocentric configuration, urban system, rank-size rule

JEL classifications: R10, R12, R30

14.1. Introduction

Theoretical and empirical studies of trends in the spatial distribution of city size and a hierarchical urban system have engaged economists and geographers since the beginning of this century. Economists and geographers try to describe why firms and households concentrate in large metropolitan areas even though empirical evidence suggests that the cost of living in such areas is significantly higher than in the small areas. They must explain the reason why the formations of small and specialized clusters of firms and workers are not necessarily located within major cities, and they must specify why the hierarchical urban system has emerged within most countries during the last few decades. It has attracted renewed attention recently, in part for the following few reasons: First, the cumulative effects of important contributions to the literature on new urban economics and new economic geography have increased the need for new empirical works. Second, technological progress has brought about new types of innovative activities that benefit most from the development of urban areas and therefore tend to arise in developed urban areas. Third, the poverty of nations seems to be more and more related to the development of prosperous and competitive clusters of specific industries as well as to the emergence of large and diversified metropolitan areas (Glaeser, 1998; Thisse and van Ypersele, 1999). Finally, the increasing availability of the consideration of a production externality and the fast-growing development of new information and transportation technologies might suggest to us to re-examine the main cause for the formation of the various types and sizes of cities, as well as their distribution.

The main focus of this chapter is to selectively survey several ideas from location theory, regional science, and microfoundation of scale economics as well as production externality that bear directly on the question addressed above. First, we will introduce the famous but problematic 'central-place theory' development to explain the pattern of city sizes and locations, and provide some empirical works.

Second, we discuss the crucial scale economics or external economics on the formation of monocentric urban configuration and show some empirical studies. Third, we review the ad hoc but fruitful theoretical analysis on the emergence of an urban system and explore the related econometric models. Finally, we introduce the useful idea of the 'rank-size rule' model and employ it to examine the distribution of cities in a few countries in different periods with various empirical evidences.

In Section 14.2 we display the basic idea of central place theory which seems powerfully intuitive to explain why some activities that serve farmers (or consumers) cannot be evenly spread due to economies of scale. Therefore, it is obvious that the trade-off between scale economies and transportation costs will lead to the emergence of the concentration of economic activities that serve the surrounding consumers. Because economic activities differ in scale effects and transport costs, one can expect to find different types of clustered urban areas. Thus, we will discuss an econometric approach to estimate the empirical works.

In Section 14.3 we formulate a few models to generate a monocentric urban configuration. These models usually involve centrifugal and centripetal forces to endogenously determine the urban configuration in which all firms (or other economic activities) cluster in the central city. Because agricultural production requires both land and labor, agricultural workers (i.e. consumers on the demand side) must be spread out along the space; this creates an incentive to disperse manufacturing as well, so that they are close to the rural market and have access to cheaper agricultural products. On the other hand, the effect of scale economies due to many firms that produce different products being located in the same area or the Romer type of economic externality will induce all firms to agglomerate together. It has been shown that when the centripetal forces are strong enough to outweigh the centrifugal forces, the monocentric urban configuration arises. Also, we also review an econometric model to explore the formation and the pattern of agglomeration on the monocentric city.

In Section 14.4 we analyze what determines the number and size of cities of different types in an economy – that is, how to generate an urban system. These urban system models rely either on the trade-off between external economies and diseconomies in the Henderson

(1974, 1977) model, or on the consideration of the product variety with monopolistic competition as examined by Abdel-Rahman (1990, 1996), or on the trade-off between economies of scale and distance in central place theory as employed by Fujita and Mori (1997) and Fujita *et al.* (1999) in which the Christaller-type hierarchical urban system can emerge from a decentralized process. We will provide an empirical model to test the existence of a multi-centric city by using the density function for employment and population.

In Section 14.5 we introduce the theoretical model for the rank-size rule in the hierarchical urban system and make a comparison with some empirical observations for testing this rule within a country in various nations during different periods. Section 14.6 concludes.

14.2. Central place theory

The central place theory, which was initially developed by Christaller (1933), predicts ideal urban size distributions and functions. Christaller documented that each commodity has a given threshold of minimum demand as well as a fixed geographical domain beyond which people are unwilling to pay for it, suggesting that only a certain proportion of all settlements will offer higher order goods and services. For the similar formation, Lösch (1940) gave powerful intuition on the image of a hierarchy of an urban system in the history of spatial economic thought. Krugman (1996) made use of the 'heuristic model' to formulate the main concept of the central place theory. Imagine a featureless plain as shown in Figure 14.1, inhabited by an evenly-spread population of farmers of size $2S$, and assume that each farmer chooses one unit of land to live, thus making the line $2S$ long, and we measure it from the center, so that it extends from $-S$ to S. Suppose that some activities that serve the farmers cannot be evenly spread, because they are subject to economies of scale: manufacturing, administration, etc. For

Figure 14.1. The linear spatial economy

Source: Krugman (1996).

simplicity, assume that there only exists one type of manufacturing industry with identical firms of size M: all firms only have one plant, and thus each firm (or plant) provides a product with a fraction $1/M$ of the total demand for manufacturers. Assume also that each farmer demands enough manufacturers to directly support μ manufacturing workers, and also that each manufacturing worker directly supports μ manufacturing workers. Then the total manufacturing employment is then given by $2S/(1-\mu)$.

If the firm incurs a transportation cost t for each unit of product shipped one unit of distance and a fixed cost F_0 for each firm it chooses to operate, then the firm is simply assumed to choose the location to minimize the sum of fixed and transportation costs. We assume that all firms are concentrated at a single 'urban' location C. It is obvious that if C is at location 0, i.e. in dead center of the agricultural area, then there is no reason for individual firms to relocate away from C. What if C is off-center? Suppose that a firm considers the transportation costs and chooses a location D, possibility different from that of location C. The situation is illustrated in Figure 14.1. Notice that the firm ships its product to three different markets: the $S+D$ farmers to its left, at an average distance of $(S+D)/2$; the $S-D$ farmers to its right, at an average distance of $(S-D)/2$; and the $2S/(1-\mu)$ urban consumers, at the distance $|C-D|$. Thus, the firm's overall transportation cost is given by

$$T = \frac{t\mu}{M}\left[\frac{1}{2}(S+D)^2 + \frac{1}{2}(S-D)^2 + 2\frac{S}{1-\mu}|D-C|\right]. \quad (14.1)$$

A planner who could choose to locate all firms simultaneously in appropriate places would clearly set $D=C=0$. However, we want to see what happens as S increases, i.e. as the population grows, the agricultural frontier shifts out with it. Without a new firm, this entire market will be served from the existing city. Suppose a new firm is built at location E. As such, only farmers from 0 to $E/2$ will be served from the old location in center 0, as the new firm will serve farmers from $E/2$ to S. Thus, it is easy to show that the transport costs of serving these consumers would be

$$T = \frac{t\mu}{2M}\left[\frac{1}{2}(E)^2 + (S-E)^2\right]. \quad (14.2)$$

The optimal location for the new plant is to minimize the transport costs locating at $2S/3$, and the transport costs are $t\mu S^2/6M$. The cost of serving these same customers from the original plant is $t\mu S^2/2M$. It is hence worth introducing a new plant, at location $E = 2S/3$, only if the following condition is satisfied:

$$\frac{t\mu S^2}{3M} > F_0. \tag{14.3}$$

The saving in transport costs must therefore be larger than the fixed operating costs when a new plant emerges. This inequality implies that new cities will appear as soon as S reaches the critical value

$$S^* = \sqrt{3F_0 M/t\mu}. \tag{14.4}$$

Once the cities have come into being, however they will be locked in place by the same logic that locked the original city into place. We can then repeat the analysis as the population grows. When the agricultural frontier has extended S^* beyond the existing cities, a new pair will pop into existence, and so on. Thus, there will be a typical distance S^* between cities, depending positively on fixed costs and negatively on transport costs, and of course a typical size of cities as well.

In a numerical simulation model, Krugman (1993) documented that the process of city formation is one of cumulative causation, but the eventual locations of cities tend to have a roughly central place pattern. He developed a monopolistic general equilibrium model and showed something resembling the central place theory. In a few experiments the economy with a given set of parameters ended up with three cities, revealing that the economy produced only a single city in the case with either less differentiated products among manufacturing goods, a larger manufacturing share in the utility function, or lower transport costs to ship the manufacturing products. In the extension of the central place theory in terms of economic analysis with production cost and transportation cost, the case study of the Aichi urban system in Japan by Ishikawa and Toda (2000) also showed the validity of a central place system with market areas of different shapes.

In the econometric modeling approach, Mushinski and Weiler (2002) developed a simultaneous equation of Tobit model to estimate the geographic interdependence of a central place and its neighboring areas. In this model it is supposed that the number of establishments in an industry in the place ($\mathbf{X}_p$) depends on the number of establishments in an industry in the neighboring areas ($\mathbf{X}_n$) and other demand-related variables ($\mathbf{D}_p$). Randomness due to nature and unobserved variables is captured by a disturbance $\mathbf{u}_p$ with variance σ_p^2. In addition, the endogenous variables $\mathbf{X}_p$ are censored at zero, and also a similar relationship holds for $\mathbf{X}_n$. Thus, $\mathbf{X}_n$ is also a censored endogenous variable. Therefore, the model is developed as follows:

$$\mathbf{X}_p = \begin{Bmatrix} \beta_1 \mathbf{X}_n + \beta_p \mathbf{D}_p + \mathbf{u}_p \\ 0 \end{Bmatrix} \quad \text{if RHS} \quad \begin{Bmatrix} > \\ \leq \end{Bmatrix} 0 \qquad (14.5)$$

$$\mathbf{X}_n = \begin{Bmatrix} \beta_2 \mathbf{X}_p + \beta_n \mathbf{D}_n + \mathbf{u}_n \\ 0 \end{Bmatrix} \quad \text{if RHS} \quad \begin{Bmatrix} > \\ \leq \end{Bmatrix} 0. \qquad (14.6)$$

Equation (14.5) describes the 'place equation' and Equation (14.6) implies the 'neighbor equation'. It is assumed that the disturbances have a bivariate normal distribution with covariance ρ, they were estimated using maximum-likelihood techniques. Mushinski and Weiler (2002) employ the data from the United States of the 1992 Census of Retail Trade and 1992 City and County Data Book (both published by the Bureau of the Census) to estimate nine various industries. They obtained the interesting result that for six of the nine industries, the number of establishments in the neighboring areas ($\mathbf{X}_n$) has a statistically significant impact on the number of establishments in the place ($\mathbf{X}_p$) and has the expected sign, underlying the role of supply-side spatial competition. More importantly, the industries of gas stations, and both eating and drinking have both supply-side and demand-side interdependencies, both industries reveal a more frequent mobile clientele. More gas stations and restaurants or bars in outlying areas reduce the need for going 'into town' for these services, but more residents in the surrounding county also increase the likelihood of stopping by a place for precisely these same needs. In addition, they find that the industries of building supplies, food stores, and drug stores do not reveal any geographic interdependence. The absence of interdependence for these three industries implies that the presence and number of such stores in the place are unlikely to be affected by outlying residents or establishments.[1]

14.3. The existence of a monocentric configuration

The previous model of central place theory so far generates only one size of a city under the assumption of a homogenous firm in the manufacturing industry, and it has never been shown to emerge from a decentralized market process. It is still unclear whether one is looking at the supposed equilibria of a decentralized market process, the solution to a planning problem, or simply at a plausible but ad hoc classification scheme. It indeed falls far short of being what an economist would call a model. However, Ogawa and Fujita (1980) developed a model to consider business externalities among firms. They explained that agglomeration force is due to the existence of communication among firms permitting the exchange of information, and the information transmission often requires direct communication between firms that typically incur distance-sensitive costs. If we define $c(x, y)$ as the level of contact activity chosen by a firm at location x with the firm at location y, and $V[c(x, y)]$ denotes the total contribution of this contact level to the firm's revenue, and $g(y)$ is the density of firms at location $y \in X$ then the revenue of a firm at x is given by

$$pQ(x) = \int_X V[c(x, y)]g(y)\mathrm{d}y. \tag{14.7}$$

We in turn have the profit of firm at location x as

$$\pi(x) = \int_X \{V[c(x,y)] - [\eta_1(x,y) + \eta_2]c(x,y)\}g(y)\mathrm{d}y \\ - R(x)S_\mathrm{f} - W(x)L_\mathrm{f}, \tag{14.8}$$

where $\eta_1(x,y)$ represents the cost per unit of contact that the firm at location x must bear. We see that η_1 is supposed to be a function of the location of the two firms. However, during this contact activity, the firm at y also bears some cost η_2, which is typically independent of

---

[1] Please refer to Tables 1–3 in Mushinski and Weiler (2002) for the detailed regression results.

the firms' locations. Here, $R(x)$ and $W(x)$ stand for the unit land rent and wage rate at the location x, and S_f and L_f are the land and labor employed by the firm, respectively.

Given the firm's spatial distribution and contact level, each firm chooses its location x and its contact level $c^*(x, y)$ in order to maximize its profit. Since the optimal contact level of a firm at x with any firm at y can be determined independent from the whole distribution of firms, we can choose $c^*(x, y)$ to maximize $V[c(x, y)] - \eta_1(x, y)c(x, y)$. With the assumption of symmetricity between any pair of firm, we have $c(x, y) = c(y, x)$ and $\eta_1(x, y) = \eta_1(y, x)$. Let us define the local accessibility between each location pair (x, y) by

$$a(x, y) \equiv V[c^*(x, y)] - [\eta_1(x, y) + \eta_2]c^*(x, y). \tag{14.9}$$

It is convenient to formulate the aggregate accessibility of each firm at location x across the interaction space X as follows:

$$\begin{aligned} A(x) &\equiv \int_X a(x, y)g(y)\mathrm{d}y \\ &= \int_X \{V[c^*(x, y)] - [\eta_1(x, y) + \eta_2]c^*(x, y)\}g(y)\mathrm{d}y. \end{aligned} \tag{14.10}$$

Note that $A(x)$ implies the information exchange having the nature of a spatial externality. The amount of information received by a firm is exogenous; however, it still depends on its location relative to the others. Therefore, the profit function (14.8) can be simplified by

$$\pi(x) = A(x) - R(x)S_f - W(x)L_f. \tag{14.11}$$

In this specification, Ogawa and Fujita (1980) find that the monocentric configuration (i.e. firms cluster in the central area, and households reside the surrounded areas) is a spatial equilibrium if the commuting cost for households is relatively small in comparison with the contact cost for the firms.

A more important model has been developed by Fujita and Mori (1997). They made use of 'new economic geography' to specify a model to examine that Christaller-type hierarchies can indeed emerge from a decentralized market process. Specially, if we imagine that a dynamic process in which the population gradually increase will lead to a movement of the agricultural frontier and the occasional formation of new cities, then one can generate an emergent hierarchy of a central place.

In the beginning, consider a simple spatial model with two types of economic activities: the manufacturing sector and agricultural sector. Assume that all manufacturing firms are located within the city and agricultural land use is only in countryside. The agricultural sector is perfectively competitive and produces a single, homogenous good, whereas the manufacturing sector provides a large variety of differentiated manufacturing goods under the monopolistic competition framework by Dixit and Stiglitz (1977). For a city with a monocentric configuration that is a spatial equilibrium, the equilibrium must reach a point where no firm has an incentive to move away from the city and set up in the countryside.

Fujita *et al.* (1999) and Fujita and Thisse (2002) specified a potential function of a firm at location x in a city as

$$\Omega(x) = \left(\frac{w^{\mathrm{M}}(x)}{w^{\mathrm{A}}(x)} \right)^{\sigma}, \tag{14.12}$$

where σ represents the elasticity of substitution between any two manufacturing goods. Let $w^{\mathrm{M}}(x)$ be the zero-profit wage rate and the maximum wage rate that a firm located at x in a city is willing to pay, given that the rest of the economy remains unchanged. Moreover, $w^{\mathrm{A}}(x)$ is the real wage rate of an agricultural worker currently prevailing at each location x (which is also the real wage of manufacturing workers in the central city). Because $w^{\mathrm{A}}(x) = w^{\mathrm{M}}(0)$, the potential is unity in the city. Thus, a monocentric configuration is sustainable if and only if

$$\Omega(x) \leq 1 \qquad \text{for all } x \geq 0. \tag{14.13}$$

The above means that there is no alternative location where firms making zero profit could offer more than workers are currently making: the real wage that these locations offer is less than that in the agriculture sector or in existing cities. In other words, there is no place, other than the incumbent cities, at which firms can offer workers the prevailing equilibrium real wage while making nonnegative profits. In this case the equilibrium monocentric configuration is stable, and new cities cannot emerge. However, suppose that the population growth has just pushed the potential curve up to the point where it lies slightly above 1 in some locations. A small group of workers may then gain higher wages by moving to these locations. In short, we can expect new cities to emerge when and where the market potential curve keeps itself above 1.

Berliant *et al.* (2002) provided a framework that involves the production externality re-examining the emergence of urban configuration. They formally explore Jacobs' idea that uncompensated knowledge spillover is crucial for agglomeration by modeling location-dependent interfirm production externalities in a general-equilibrium linear-city framework. They formalized the knowledge spillover idea *a la* Shell (1966), Jacob (1969), and Romer (1986) in which knowledge spillover is regarded as uncompensated factor inputs in firms' production (rather than via transaction costs or firm profits independent of aggregate capital usage). That is, an individual firm, which is located at x in the city, employs capital (K), labor (L), and land $s(x)$ to produce goods using a constant returns to scale technology which exhibits a Cobb–Douglas form as the following:

$$Y(x) = AK^{\alpha}L^{\beta}[Q(x)\bar{K}]^{1-\alpha-\beta}D(x)^{1-\alpha-\beta}, \tag{14.14}$$

where $\alpha, \beta \in (0,1)$, $\alpha + \beta \in (0,1)$, $Y(x)$ is the output at location x, $\bar{K}$ is the aggregate captial usage for all firms, and $D(x)$ is the effective land input given by $D(x) = \min\{1, S(x)\}$ with $S(x) = 0, \forall\ s(x) < 1$.

The efficient use of land for each firm is at $s(x) = 1$, which simplifies the analysis greatly, as the model is not tractable otherwise. Here, $Q(x) = 2 - (x - \mu)^2 - \varepsilon\sigma_\tau^2 > 0$ measures the degree of effectiveness of interactions between a particular firm x and the others in the linear city given a configuration of type τ. Namely, σ_τ is the overall dispersion of firm sites by type τ urban configuration, where μ denotes the mean location of firm sites, and $\varepsilon \in (0,1)$ denotes the degree of penalty on the overall dispersion of firms and the second term specifies a quadratic cost function in terms of the distance between a particular firm site and the mean site.[2] Let z be the location index $z \in X \equiv [-1,1]$ and then denote the density of firms at location z under a particular urban configuration τ by $m_\tau(z)$, with a continuum of firms of mass M in

---

[2] A type τ urban configuration can be monocentric, duocentric, completely mixed, or incompletely mixed.

the interaction space X. This means that

$$\mu = \int_{z \in X} z m_T(z) \mathrm{d}z \tag{14.15}$$

$$\sigma_T = (2/M) \int_{z \in X} |z - \mu| m_T(z) \mathrm{d}z, \tag{14.16}$$

where the constant multiplier 2 is incorporated in the equation so that the maximum overall dispersion is normalized to unity.

One may regard the Q function as a proxy (with the first and second moments) for the (locational) distribution of firms. Apparently, this model is different from the Ogawa and Fujita (1980) model setting, whereby the externality is sourced from the aggregation of information exchange for each pair of firms. The model developed by Berliant *et al.* (2002) captures the Romer convention in which externalities enter the system based on an average or aggregation of individual measures.[3]

The simplicity of $Q(x)$ derives the analytical results, and explains that when the commuting cost is very low, firms are concentrated to take advantage of knowledge spillover while households have to commute and receive a high wage to offset the travel cost. Thus, the monocentric configuration emerges. That is, in this model a sufficiently large knowledge-spillover penalty on the overall dispersion of firms causes the formation of a monocentric urban configuration.

With the consideration of production externalities among firms in a circular city, Lucas (2001) and Lucas and Rossi-Hansberg (2002) formulated an explicit model of the interaction forces among firms and the interaction forces between the production sector and household sector. They specified a Cobb–Douglas form on both the production function and utility function. In this model setting the external effect for an individual firm plays a key role in the determination of the equilibrium configuration. They proposed a general algorithm for constructing equilibria and employed a calibration to document a situation where the commuting cost of

---

[3] In the examination of the effect of this Romer type of externalities on the formation of urban configuration, Berliant *et al.* (2002) proved that the multi-centric symmetric urban configuration cannot emerge in the competitive spatial equilibrium.

workers (i.e. consumers) is low enough and the production externality is high enough, such that the equilibrium of the urban configuration takes on a monocentric form, where a central business district is surrounded by a residential area. Therefore, the monocentric city emerges.

The econometric issue has undergone regression analysis by a few in the empirical literature. First, Ellison and Glaeser (1997) explored the levels of geographic concentration of the four-digit subindustries of each two-digit manufacturing industry based on data from the *1987 Census of Manufactures* in the US, and documented that there is no obvious single factor accounting for extreme concentration. They found that the most concentrated industry, furs, is probably explained both by the local transfer of knowledge from one generation to the next and as a response to buyers' search cost. Furs also have an unusually high ratio of value to weight that may make physical transportation costs less important. In addition, on the investigation of the effects of natural advantages on industry location, Ellison and Glaeser (1999) pointed out that most industry locations are related to resource and labor market natural advantages. There does remain a large number of highly concentrated industries where it seems that agglomeration must be explained by localized intra-industry spillover.

As an empirical work to estimate the effect of the production externality or spillover effect on the choice of industry location and using the MarketPlace database of the fourth quarter of 2000, Rosenthal and Strange (2001) regressed the Ellison and Glaeser (1997) index of spatial concentration on industry characteristics that proxy for the presence of knowledge spillover, labor market pooling, input sharing, product shipping costs, and natural advantage. The effect of agglomerative spillover on the spatial concentration indices is measured by

$$\lambda_{j,m} = \beta Z_m + \varepsilon_{j,m}, \tag{14.17}$$

where $\lambda_{j,m}$ is the localization statistic for the mth industry at the level of geography j, Z_m is the vector of industry characteristics with the associated coefficient vector, β, and $\varepsilon_{j,m}$ is assumed to be an independent and identically-distributed error.

Rosenthal and Strange (2001) estimated Equation (14.17) separately for the three geographic specifications, with λ_j measured at the state, county, and zip code levels. They found that agglomeration reduces the cost of innovation by enhancing knowledge spillover while also reducing the cost of labor and intermediate inputs through labor market pooling and input sharing. Thus, industries sensitive to innovation, labor, and intermediate input costs are more likely to agglomerate. That is, the evidence of a positive relationship between agglomeration and these other factors confirms that tendencies to innovate, pool labor, and share inputs all lead to an increase in agglomeration.[4] Specifically, knowledge spillover positively affects agglomeration only at the zip code level, possibly because such a spillover attenuates rapidly across space.

In addition, Rosenthal and Strange (2003) developed a model and examined the effect of industrial organization on the agglomeration economies. The pattern of agglomeration shows that if agglomeration economies are present, then the births of new firms will occur near concentrations of existing employment, all else being equal. If agglomeration economies are absent, then births will tend to disperse. Thus, they develop the econometric model as

$$B_{j,t} = \beta_{\mathrm{p}} Z_{\mathrm{p},j,t-1} + \gamma_{m,\mathrm{b}} + \varepsilon_{\mathrm{b},t} \tag{14.18}$$

$$N_{j,t} = \alpha_{\mathrm{p}} Z_{\mathrm{p},j,t-1} + \gamma_{m,\mathrm{n}} + \varepsilon_{n,t}, \tag{14.19}$$

where $Z_{p,j}$ denotes the local characteristics at zip code j, $\gamma_{m,\mathrm{b}}$ and $\gamma_{m,\mathrm{n}}$ control for all attributes common to a metropolitan area that affect productivity, and B_j and N_j are the number of births per square mile and total new-establishment employment in zip code j, respectively.

Rosenthal and Strange (2003) employed the data from the fourth quarter of 1997 in Dun & Bradstreet MarketPlace database and used the Tobit fixed-effects model. They showed the empirical result that own-industry competition encourages births and new-establishment employment in every industry except one. In contrast, other industry competition has a negative effect in both models for every industry.

---

[4] See Table 3 of Rosenthal and Strange (2001) for the detailed empirical results at the level of zip code, county, and state, respectively.

On the examination of the geographic variable on the agglomeration economics, they found the key geographic result that the localization economies attenuate rapidly in the first few miles, but slowly thereafter.[5] In addition, in the empirical work with an examination on the geographical location of patent citations, Jaffe *et al.* (1993) find that citations to domestic patents are more likely to be domestic, and more likely to come from the same state and metropolitan area as the cited patents. These results confirm the theoretical model by Berliant *et al.* (2002) and Lucas and Rossi-Hansberg (2002).

14.4. The emergence of urban system

An understanding of the analysis as in previous section only generates one type with an identical size of urban configuration. Now suppose that the population size is sufficiently large, or the manufacturing products are differentiated enough, or there are many different industries differing in their type of external economics and transport costs. First, in the original formulation by Fujita and Mori (1997), as the curve formulated by Equation (14.12) hits the value 1 at some location $\hat{x}$, the population size reaches the level $\hat{N}$ and thus makes this location as profitable as the city. If the population slightly rises above $\hat{N}$, then the potential curve exceeds 1 at $\hat{x}$. Therefore, $\hat{x}$ is now more profitable than the city, which brings firms an incentive to set up here. It implies that a new city will form at $\hat{x}$. Of course, under the assumption of symmetric space, the same arises at a location $-\hat{x}$. In turn, at the population size $\hat{N}$, the equilibrium monocentric configuration is transformed into a symmetric tricentric pattern. It appears to be very problematic to derive analytical results when the equilibrium involves more than three cities.

To examine this evolution of an urban system, Fujita *et al.* (1999) employed a numerical simulation to describe how the spatial urban system evolves over time as population (N) increases gradually. They documented that the monocentric configuration is a stable equilibrium when the population is relatively small. When population N increases up to a critical value, the monocentric

---

[5] Please refer to Tables 2 and 3 of Rosenthal and Strange (2003) for the detailed results by each industry.

configuration becomes structurally unstable. At this particular value of N, by the symmetric assumption in the urban configuration, the spatial economy experiences a catastrophic transition from a monocentric to a tricentric pattern. Therefore, this new type of tricentric urban configuration is a stable equilibrium. As population N keeps rising continuously, it will indicate that the tricentric pattern becomes structurally unstable. In turn, the tricentric pattern will transform into a pentacentric configuration. By a similar process, as N continues to rise, a pair of frontier cities emerges periodically as a result of a catastrophic transition of the spatial economy.[6]

Given the preliminary analysis above, Fujita *et al.* (1999) developed the model to investigate the endogenous formation of a hierarchical urban system and proposed an evolutionary approach which combines a general equilibrium model with adjustment dynamics. It is demonstrated that when the economy's population size increases gradually, the urban system self-organizes into a highly regular hierarchical system *a la* Christaller. Finally, they concluded that, as the number of cities increases, the urban system approaches into a highly regular network of cities, as conjectured in the central place theory.[7]

Important contributions to the literature on city size distributions are found in the work of Henderson (1974, 1977). He claimed that there are two forces to determine the size of a city. On the one side, a centripetal force is sourced from the external economies associated with a geographical concentration of an industry within a city, and a centrifugal force with diseconomies associated with large cities on the other. The net effect of this tension is that the relationship between the size of a city and the utility of a representative resident is an inverted U.

In the analysis of Henderson's model, he emphasized that cities differ from each other due to varying demands for their products, either as final goods or as intermediate goods. The Henderson model features a theory of city distribution that directly reflects preferences and explains that the types of goods produced in cities determine

---

[6] Please refer to Figure 7 in Fujita *et al.* (1999) for a detailed simulation.

[7] For a detailed evolutionary process of the hierarchical urban system, please refer to Figure 6 in Fujita *et al.* (1999).

the sizes of those cities. If the degree of external scale economies differs from one industry to another, then it will produce different city sizes, and if the types of goods currently in vogue change, then urban concentrations are expected to change. Therefore, for each type of city, there will be an optimum city size. In equilibrium, at the optimum size, each resident will yield the same utility level, but that size will vary depending on the type of city. This fundamental approach can also be used to examine the equilibrium as well as optimal city size associated with international trade theory (Henderson, 1987, 1988).

By observing the effect of externalities on the formation of an urban system, Marshall (1895) argued that intra-industry externality is a significant determination in the choice of industrial location and productivity. On the examination of an urban system associated with the Marshallian externality, Abdel-Rahman (1990, 1996) analyzed two types of cities. Based on the partial general equilibrium framework, the coexistence of both types of cities is pre-imposed in the model, showing that one city is specialized while the other is diversified. Peng (1991) re-examined the effect of inter-city transport cost associated with Marshallian externality on the formation of urban city structure. He showed that the city type (specialized or diversified) and city size will be determined by the trade-off effects between the inter-city transport costs of goods and the advantage (disadvantage) of Marshallian externality, and it will depend on which force is dominant.

It is important to point out that the coexistence of specialized and diversified cities does not have the result only from non-market forces, such as externalities in the literature. In a general equilibrium approach, Abdel-Rahman and Fujita (1990) and Abdel-Rahman (2000) presented that these models usually generate three types of urban configuration: (1) pure specialization, in which each city specializes in the production of one trade good, (2) pure diversification, in which all cities in the economy produce two goods, and (3) a mixed urban system, in which specialized and diversifications coexist in the economy. For each type of city, the city size is identical, and thus there is non-ranking for the same type of city, but the size may be different for different types of city.

Given the formation of an urban system above, it is important to emphasize first that even Fujita and Mori (1997) and Fujita *et al.* (1999) employed the 'new economic geography' type models to show that Christaller-type hierarchies can indeed emerge from a decentralized process. They usually ignored the intra-city transportation cost and only focused on the examination of inter-city transportation cost in the analysis of an urban system. That is, because they regarded a city as a point in the linear space, their production or consumption function does not include land employed. When the city size is small, this assumption could be acceptable. However, if the urban size is large enough, then the intra-city transportation cost could be higher than the inter-city transportation cost. In that case, a re-examination of the hierarchical urban system associated with both inter-city and intra-city transport costs would be more interesting in reality.

Although the Marshallian type of an urban system is involved in the intra-city commuting cost, in general, the diversified cities tend to be larger than specialized cities. Note that the Marshallian model has not presented the hierarchical urban system, as it specially displays that the city size is identical for the same type of city. Models of the Henderson genre typically involve two types of forces: a force that leads to a concentration of population and economic activities (centripetal force) and a force that leads to decentralization (centrifugal force). When these forces are balanced at the margin, the equilibrium city size is determined. These types of models also do not provide explanations for the spatial distribution of cities or hierarchical urban system. Thus, distance between cities is not taken into consideration (i.e. ignore the inter-city transport cost), which is in contrast to the pioneering work of Christaller (1933) and Fujita *et al.* (1999), in which distance between centers of cities was explicitly modeled.

It is also important to consider the variety in intermediate inputs and various types of labor to explore the formation of the urban area (or region). Abdel-Rahman (1994) examined the relationship between the economies of scope in intermediate goods and system of cities. Zhang and Sasaki (1997) developed a closed city model with a subcenter to examine the equilibrium urban shape and found that the establishment of a subcenter keeps most properties obtained in a monocentric city unchanged, and the utility level of residents necessarily increases as the subcenter location moves farther from the CBD. Furthermore, Peng *et al.* (2004) investigated the interactions between economic integration and population agglomeration in a middle product economy (i.e. consider the variety in intermediate inputs) displaying neoclassical growth. They showed that population agglomeration and output growth need not be positively related in the same urban or region.

Empirical studies of an urban system focus on estimating the monocentric and polycentric density function for employment and population, and investigate the relationship between the number of employment subcenters with population and commuting costs. Small and Song (1994) developed a polycentric urban system model by the negative-exponential functional form to examine the density function for employment and population as follows:

$$D_m = \sum_{n=1}^{N} A_n e^{-b_n r_{mn}} + \varepsilon_m, \qquad m = 1, 2, 3, \ldots, M, \tag{14.20}$$

where D_m is the observed density of population or employment in zone m; n is the number of employment centers; M is the total number of zones in a metropolitan area; r_{mn} is the distance between zone m and center n; ε_m is the error term of density associated with zone m; and A_n, b_n are parameters to be estimated for each subcenter n.

Equation (14.20) interprets that the polycentric form of an urban system collapses to the monocentric form with an additive error if the intercepts of all centers except one are zero. Therefore, we can test statistically whether the polycentric model explains the actual urban distribution better than the monocentric model. They obtained data on population and employment for 1135 zones of the Greater Los Angeles region from the California Department of Transportation and regressed Equation (14.20) by using 1970 and 1980 data. The results showed that the employment and population distributions were quite flat, with density declining by only 4–6% per mile. This indicates a high degree of dispersion for the urban region.

Small and Song (1994) test the validity of the monocentricity based on the statistic as

$$F = \frac{(\mathrm{SSR}^{\mathrm{M}} - \mathrm{SSR}^{\mathrm{P}})/2(N^{\mathrm{P}} - N^{\mathrm{M}})}{\mathrm{SSR}^{\mathrm{P}}/(M - 2N^{\mathrm{P}})}, \tag{14.21}$$

where $\mathrm{SSR}^{\mathrm{M}}$ and $\mathrm{SSR}^{\mathrm{P}}$ are monocentric and polycentric sums of square residuals and N^{M} and N^{P} denote the number of centers that are monocentric and polycentric, respectively. The testing results indicate that the monocentric model is soundly rejected based on the data in 1970 and 1980. Thus, it implies that the urban system of the Greater Los Angeles area is interpreted better by the polycentric than by the monocentric form.

A polycentric urban system in general combines many of the advantages of big and small urban areas. Whereas the conventional monocentric city offers firms the advantage of significant agglomeration economies, it also requires high wages to compensate for expensive and time-consuming commutes. However, what determines the number of subcenters within an urban area? The previous theory, especially that developed by Fujita and Mori (1997), suggests the most important determinants of subcenter counts are population and commuting costs. A large population allows firms to reproduce some of the agglomeration economies of the CBD in secondary employment centers. High commuting costs provide the incentive to form subcenters, because wages can be reduced at a suburban location that saves commuting time for workers.

McMillen and Smith (2003) employed cross-section data of 62 American urban areas in 1990 and use the McMillen (2001) subcenter identification procedure to determine the number of subcenters in these urban areas. They then regressed the number of subcenters on the population and commuting costs. The empirical study reveals that the number of subcenters rises with population and commuting costs. Specifically, these two variables alone account for nearly 80% of the variation in the number of subcenters across urban areas.[8] This finding confirms the theoretical predictions as in the model developed by Fujita and Mori (1997).

14.5. The rank-size distribution in an urban hierarchy

As noted by many economists as well as geographers and other social scientists, the rank-size rule (or power rule) has been found useful to describe the spatial distribution of cities. They allege that

[8] On the other hand, Gordon *et al.* (1989) examined the effects of metropolitan spatial structure on commuting behavior by using 1980 census data, finding that polycentric and dispersed metropolitan areas facilitate shorter commute times.

statistical regularity is an outgrowth of the application of the Pareto distribution to city size data. Stated in its various forms, the rank-size rule implies that city size multiplied by the rank in its system equals a constant, which has been debated, calculated, and dismissed several times over different periods across countries. We first describe the theoretical model of the rank-size rule, and then provide the various empirical evidences for testing the rank-size rule in different countries during various periods.

Many scholars in different disciplines have studied urban area sizes and concluded that they are highly skewed, and follow a Pareto distribution as $m_i = AN_i^{-a}$, where m_i is the rank of urban areas with N_i inhabitants, and A and a are constants to be determined from the data. In particular, Zipf (1949) claimed, on the basis of urban area size data in the United States and in other countries, that the distribution of urban area size follows a special case of the Pareto distribution in which $a = 1$, called the rank-size rule. It can be written as

$$m_i = AN_i^{-1}, \tag{14.22}$$

which says that the product of an urban area's size and rank is a constant. For example, if $m_i = 1$, Equation (14.22) results in $N_i = A$, i.e. A is the size of the largest urban area. It has an easy implication thereby the second largest urban area has half the population of the largest, the third largest has one-third the population of the largest, and generally an urban area of rank r has 1 rth the population of the largest urban area.

Why should urban area sizes follow this particular distribution? Beckmann (1958) developed a model to explain the Pareto distribution of city size. Suppose that each urban area performs certain functions for its own residents and for the residents of a set of smaller urban areas. If the functions performed by cities are the production of commodities, and the unit which measures output is chosen such that one unit of the commodity is consumed by each consumer, and if ρ is the labor input required per unit of output of the commodity, then the first assumption of Beckmann's model can be expressed as follows:

$$N_i = \rho M_i. \tag{14.23}$$

Let N_1 be the population of an urban area of the largest size, and M_1 is the population served by the largest urban area. And the population of and the population served by the urban area of the second largest urban size are N_2 and M_2, respectively. Generally, N_i and M_i are the population of and the population served by an urban area of ith largest urban size. In rank i, if there exists V number of cities, then this model implies that all urban areas of rank i ($i = 1, 2, 3, \ldots, s$) are the same size. If the population of the smallest urban size is N_s and the population served by the smallest urban size is M_s, respectively. A subsidiary assumption is that workers in the smallest urban areas serve themselves and some rural residents, N_r. This implies that

$$M_s = N_s + N_r. \tag{14.24}$$

The second basic assumption of Beckmann's model limits the number of urban areas of the next smallest size that can be served by an urban area of a given size. Specifically, it is assumed that an urban area of a given size can serve k^i urban areas of the next smallest size. Hence,

$$M_i = N_i + k^i M_{i-1}. \tag{14.25}$$

Using Equations (14.23)–(14.25) yields N_i as

$$N_i = N_r\left(\frac{\rho}{k^i}\right)\left(\frac{k}{1-\rho}\right)^i. \tag{14.26}$$

In addition, this model assumes that there is one largest urban area, k urban areas of rank 2, k^2 urban areas of rank 3, and generally k^{i-1} urban areas of rank i. By employing the rule of geometric series, and if i is fairly large, then the rank (m_i) of the middle urban area in the ith class can be approximated by

$$m_i = k^i\left(\frac{1}{k^i - 1} + \frac{1}{2}\right). \tag{14.27}$$

Combining Equations (14.26) and (14.27) implies that the product of rank and population for an urban area in the middle of the ith largest urban size class is

$$m_i N_i = A(1-\rho)^i, \tag{14.28}$$

which says that the product of rank and population size is a constant A times $(1-\rho)^i$. If, as seems reasonable, ρ is close to zero, then $(1-\rho)^i$ will be close to 1, and in turn, the rank times urban size will be nearly constant, which is the remarkable rank-size rule.

Beckmann's model shows that a very simple economic mechanism can generate a distribution of urban sizes that are similar to the Pareto distribution. However, it includes some extraordinarily assumptions in this simple model. First, it assumes a simple production in which labor is the only input, and input/output ratios are constant above the minimum output. Second, the entire model is built on the foundation of the number of rural residents that can be served by the cities of the smallest size. This determines the smallest urban size on the spatial distribution of urban areas, which in turn determines the spatial distribution of urban areas of the next smaller size, and so on. Third, the model ignores all geographical irregularities, natural resource availability, amenity resources, available of harbors and other discrete transportation modes, production externality sourced from the same industry and same city, and any climatic differences. Finally, the model treats urban areas as points, without regard for spatial phenomena within urban areas.

If we ignore the analysis of the specific location for each city and just focus on the urban system, then Beckmann and McPherson (1970) documented a discrete urban hierarchy with a given number, n, of levels. Each level j ($j = 1, 2, \ldots, l$; l is the upper level with the single largest city) is composed of v_j centers; and the jth level center has a population N_j; the total number of centers in the system is $V = \sum_{j=1}^{l} v_j$. Beguin (1985) suggested a city-size distribution which has a rather regular continuum, and empirical evidence often corresponds to the classical rank-size distribution, at least when the number V of centers is large; such a distribution is represented by

$$rN_{\mathrm{r}}^{\theta} = N_1^{\theta}, \tag{14.29}$$

where N_{r} is the population of the center of rank r ($r = 1, 2, 3, \ldots V$), where the centers are ranked in decreasing order of their population (N_1 is thus the population of the largest city) and θ is a constant.

One empirical study consists of 44 countries and this type of international data is usually problematic, in particular because it is difficult to assemble comparably-defined metropolitan areas. Rosen and Resnick (1980) showed that the relationship in Equation (14.29) is generally valid and that the average of θ is virtually one for metropolitan areas. That is, they suggested that most national metropolitan size distributions are well described by a power law with an exponent not too far from one. When the exponent gets closer to 1, the more carefully the metropolitan areas are defined.

In the other empirical observation by Carroll (1982), it was found that for at least 70 years that the distribution of larger cities in the United States was surprisingly well described by a power law, i.e. the number of cities with a population larger than N_i is approximately proportional to $N_i^{-\theta}$, with θ quite close to 1. Batten (1995) also provided some empirical evidence by using 1831–1982 French metropolitan data to examine the power law in the distribution of city size. Despite vast variations in the number and size of French cities during this period in France, he stressed that the rank-size rule seems to have been held over a 150-year period.

Eaton and Eckstein (1997) collected population data of 39 urban areas for 7 years (1876, 1911, 1936, 1954, 1962, 1982, and 1990) in France. The pooled rank-size regression is shown as

$$\ln(m_i) = 8.6541 - 1.0311\ln(N_i), \tag{14.30}$$

where m_i is the number of cities with population N_i or more, and N_i is the population of the ith largest city. The evidence for France supports the rank-size rule for all but the earliest period of observation, for which it is substantially below 1. The estimated value of the coefficient for 1876 is 0.87. In the context of the same paper, Eaton and Eckstein (1997) organized the 40 largest metropolitan data in Japan from 1925 to 1985, for every 5 years. The pooled rank-size regression for that is given by

$$\ln(m_i) = 15.7916 - 0.9649\ln(N_i). \tag{14.31}$$

This examination shows that the coefficient is slightly less than 1, and somewhat still lower than that in 1925, 1947, and 1950. From the comparison of Equations (14.30) and (14.31), it reflects the somewhat greater inequality of cities in Japan relative to France.

Krugman (1996) showed, by using 40 US metropolitan areas' data in 1991 to approach this idea and find a log-linear regression, that

$$\ln(m_i) = 10.549 - 1.004\ln(N_i). \qquad (14.32)$$

This implies that the distribution of city sizes in the United States has been well described by a power law with an exponent close to 1 at least for the past century. Krugman pointed out that it is only the upper tail of city size that obeys such a law and documented that the confirmation of the power law in the size of city distribution involves a sort of principle of self-similarity.

By using 665 cities of China in 1998, Song and Zhang (2002) similarly obtained the relationship as follows:

$$\ln(m_i) = 18.1953 - 1.0414\ln(N_i). \qquad (14.33)$$

This estimated coefficient (1.0414, with a standard deviation of 0.013) is statistically greater than 1, indicating that the city size distribution becomes more even than that predicted by the rank-size rule. However, if the examination is based on 1991 data, then it shows that the coefficient (0.9231) is less than that implied by the rank-size rule. This finding suggests that the city size distribution in China became more even during 1991–1998. They also described that the inclusion of new cities would make cities in the larger sample appear more evenly distributed.

Additional empirical evidence can be presented here for the case of Spanish cities. Lanaspa *et al.* (2003) took the 100 largest Spanish cities for each year during 1900–2000 and found that the estimation coefficient (b) in a log-linear regression, $\ln(m_i) = a - b\ln(N_i)$, displays a U-shape over time, with the minimum value being reached in 1970. As a result, the evolution through time of the estimations of the Pareto exponent can be deduced to two different patterns of behavior over this period. From the data during 1900–1970, they showed that the size distribution of the cities is increasingly divergent; while from 1970 to 2000, this distribution becomes more equal. In turn, they concluded that this change can be related to the decline of traditional industry, the progressive industrialization of the economy, and the phenomenon of counter-urbanization.

In summary of this section, based on the empirical testing that was selectively surveyed as above, we suggest that the city-size distribution within a country in many nations is approximately log-linear over a long time, thereby conforming to the rank-size rule. Despite vast variations in the number, period, and size of cities in different countries, the rank-size rule seems to hold over one decade.

14.6. Conclusions

We have surveyed theoretical models with the central place theory, emergence of the city, urban system, and rank-size rule with urban hierarchy in a common framework. We have also reviewed a few econometric approaches to study these theoretical models and discussed their application in the formation of city, the size and type of the city, and the city distribution in urban hierarchy with a few empirical works in various countries. First, we introduced the basic idea of central place theory. Second, we surveyed a few models to generate a monocentric urban configuration in the literature. Third, we discussed the determination of the number and size of cities with different types in a spatial economy and the generation of an urban system. Finally, we studied the rank-size rule in the hierarchical urban system and provided some empirical observations for testing this rule within a country in various nations during different periods.

References

Abdel-Rahman, H.M. (1990), "Agglomeration economies, types, and sizes of cities", *Journal of Urban Economics*, Vol. 27, pp. 25–45.

Abdel-Rahman, H.M. (1994), "Economies of scope in intermediate goods and a system of cities", *Regional Science and Urban Economics*, Vol. 24, pp. 497–524.

Abdel-Rahman, H.M. (1996), "When do cities specialize in production?", *Regional Science and Urban Economics*, Vol. 26, pp. 1–22.

Abdel-Rahman, H.M. (2000), "Cities systems: general equilibrium approaches", in: J. Huriot and J.-F. Thisse, editors, *Economics of Cities*, Cambridge: Cambridge University Press.

Abdel-Rahman, H.M. and M. Fujita (1990), "Product variety, Marshallian externalities and city size", *Journal of Urban Economics*, Vol. 33, pp. 189–222.

Batten, D.F. (1995), "Network cities: creative urban agglomerations for the 21st century", *Urban Studies*, Vol. 32, pp. 313–327.

Beckmann, M. (1958), "City hierarchies and the distribution of city size", *Economic Development and Cultural Change*, Vol. 6, pp. 243–248.

Beckmann, M. and J. McPherson (1970), "City size distribution in a central place hierarchy: an alternative approach", *Journal of Regional Science*, Vol. 10, pp. 25–33.

Beguin, H. (1985), "A property of the rank-size distribution and its use in an urban hierarchy context", *Journal of Regional Science*, Vol. 25, pp. 437–441.

Berliant, M., S.K. Peng and Wang Ping (2002), "Production externalities and urban configuration", *Journal of Economic Theory*, Vol. 104, pp. 275–303.

Carroll, G. (1982), "National city-size distributions: what do we know after 67 years of research?", *Progress in Human Geography*, Vol. 6, pp. 1–43.

Christaller, W. (1933), *Central Places in Southern Germany*. Fisher, Jena; English translation by C.W. Baskin, London: Prince Hall, 1966.

Dixit, A.K. and J.E. Stiglitz (1977), "Monopolistic competition and optimum product diversity", *American Economic Review*, Vol. 67, pp. 297–308.

Eaton, J. and Z. Eckstein (1997), "Cities and growth: theory and evidence from France and Japan", *Regional Science and Urban Economics*, Vol. 27, pp. 443–474.

Ellison, G. and E. Glaeser (1997), "Geographic concentration in U.S. manufacturing industries: a dartboard approach", *Journal of Political Economy*, Vol. 105, pp. 889–927.

Ellison, G. and E. Glaeser (1999), "The geographic concentration of an industry. Does natural advantage explain agglomeration", *American Economic Association Papers and Proceedings*, Vol. 89, pp. 311–316.

Fujita, M. and T. Mori (1997), "Structural stability and evolution of urban systems", *Regional Science and Urban Economics*, Vol. 27, pp. 399–422.

Fujita, M. and J.-F. Thisse (2002), *Economics of Agglomeration: Cities, Industrial Location, and Regional Growth*, Cambridge: Cambridge University Press.

Fujita, M., P. Krugman and A.J. Venables (1999a), *The Spatial Economy: Cities, Regions, and International Trade*, Cambridge, MA: The MIT press.

Fujita, M., P. Krugman and T. Mori (1999b), "On the evolution of hierarchical urban systems", *European Economic Review*, Vol. 43, pp. 209–251.

Glaseser, E.L. (1998), "Are cities dying?", *Journal of Economic Perspectives*, Vol. 12, pp. 139–160.

Gordon, P., A. Kumar and H. Richardson (1989), "The influence of metropolitan spatial structure on commuting time", *Journal of Urban Economics*, Vol. 26, pp. 138–151.

Henderson, J.V. (1974), "The size and types of cities", *American Economic Review*, Vol. 64, pp. 640–657.

Henderson, J.V. (1977), *Economic Theory and the City*, New York: Academic Press.

Henderson, J.V. (1987), "Systems of cities and inter-city trade", in: P. Hansen, M. Labbe, D. Peeters, J.-F. Thisse and J.V. Henderson, editors, *Systems of Cities and Facility Location*, Chur, Switzerland: Harwood Academic Publishers, pp. 71–119.

Henderson, J.V. (1988), *Urban Development: Theory, Fact and Illusion*, Oxford: Oxford University Press.

Ishikawa, T. and M. Toda (2000), "Some economic extensions of central-place theory involving profit maximization", *Urban Studies*, Vol. 37, pp. 481–495.

Jacob, J. (1969), *The Economy of Cities*, New York: Random House.

Jaffe, A.B., M. Trajtenberg and R. Henderson (1993), "Geographic localization of knowledge spillover as evidenced by patent citations", *The Quarterly Journal of Economics*, Vol. 108, pp. 577–598.

Krugman, P. (1993), "On the number and location of cities", *Economic Geography*, Vol. 37, pp. 293–298.

Krugman, P. (1996), "Confronting the mystery of urban hierarchy", *Journal of the Japanese and International Economics*, Vol. 10, pp. 399–418.

Lanaspa, L., F. Peuyo and F. Sanz (2003), "The evolution of Spanish urban structure during the twentieth century", *Urban Studies*, Vol. 40, pp. 567–580.

Lösch, A. (1940), *The Economic of Location*, Fisher, Jena; English translation, New Haven: Yale University Press, 1954.

Lucas, R.E. (2001), "Externalities and cities", *Review of Economic Dynamics*, Vol. 4, pp. 245–274.

Lucas, R.E. and E. Rossi-Hansberg (2002), "On the internal structure of cities", *Econometrica*, Vol. 70, pp. 1445–1476.

Marshall, A. (1895), *Principles of Economics*, London: Macmillan & Co.

McMillen, D.P. (2001), "Nonparametric employment subcenter identification", *Journal of Urban Economics*, Vol. 50, pp. 448–473.

McMillen, D.P. and S.C. Smith (2003), "The number of subcenters in large urban areas", *Journal of Urban Economics*, Vol. 53, pp. 321–338.

Mushinski, D. and S. Weiler (2002), "A note on the geographic interdependencies of retail market areas", *Journal of Regional Science*, Vol. 42, pp. 75–86.

Ogawa, M. and M. Fujita (1980), "Equilibrium land use patterns in a nonmonocentric city", *Journal of Regional Sciences*, Vol. 20, pp. 455–475.

Peng, S.K. (1991), "City type, city size, trade pattern, and interurban transportation costs in a spatial economy", *Environment and Planning A*, Vol. 23, pp. 1639–1652.

Peng, S.K., J.-F., Thisse and Ping Wang (2004), Economic integration and agglomeration in a middle product economy. Paper presented on the Midwest Economic Theory and International Trade Meeting, and the 50th North American Regional Science Association Meeting.

Romer, P. (1986), "Increasing returns and long-run growth", *Journal of Political Economy*, Vol. 94, pp. 1002–1037.

Rosen, K. and M. Resnick (1980), "The size distribution of cities: an examination of the Pareto law and primacy", *Journal of Urban Economics*, Vol. 8, pp. 165–186.

Rosenthal, S.S. and W.C. Strange (2001), "The determinants of agglomeration", *Journal of Urban Economics*, Vol. 50, pp. 191–229.

Rosenthal, S.S. and W.C. Strange (2003), "Geography, industrial organization, and agglomeration", *The Review of Economics and Statistics*, Vol. 85, pp. 377–393.

Shell, K. (1966), "Toward a theory of inventive activity and capital accumulation", *American Economic Review*, Vol. 61, pp. 62–68.

Small, K. and S. Song (1994), "Population and employment densities: structure and change", *Journal of Urban Economics*, Vol. 36, pp. 292–313.

Song, S. and K.H. Zhang (2002), "Urbanization and city size distribution in China", *Urban Studies*, Vol. 39, pp. 2317–2327.

Thisse, J.-F. and T. van Ypersele (1999), "The challenge raised by metropolitan and fiscal competition in economic development", *The World Economy*, Vol. 22, pp. 1201–1220.

Zhang, Y. and K. Sasaki (1997), "Effects of subcenter formation on urban spatial structure", *Regional Science and Urban Economics*, Vol. 27, pp. 297–324.

Zipf, G. (1949), *Human Behavior and the Principle of Least Effort*, New York: Addison-Wesley.

Urban Dynamics and Growth, CEA, vol. 266
Roberta Capello and Peter Nijkamp (editors)

CHAPTER 15

The City System Paradigm: New Frontiers[☆]

Hesham M. Abdel-Rahman

Department of Economics and Finance, University of New Orleans, Lakefront, New Orleans, LA 70148, USA

Abstract

This chapter provides a survey of recent developments of positive as well as normative theories of city systems. Static theory of city systems emphasizes the factors that result in the formation of cities through the interaction between two opposing forces: (i) agglomeration economy and (ii) agglomeration diseconomy. Furthermore, the theory examines the interaction between cities within the national economy through intercity trade and migration, which shape the internal population composition and the internal industrial structure of cities within the system. New development of this theory has been influenced by industrial organization and economic growth together with the new urban economic paradigm. This chapter focuses on the following questions: what are the factors that lead to the formation of cities? When do cities specialize in production, and when do they diversify? When do both specialized cities and diversified cities coexist? What determines the number and sizes of cities of different types in an economy? What are the factors that determine skill distribution and income disparities among different types of cities? What are the impacts of income inequalities on welfare? What are the tax and/or subsidy schemes that would result in a Pareto-efficient allocation of resources in a system of cities? Do we need the intervention of federal government in order to achieve a Pareto-efficient allocation of resources

[☆] The comments of anonymous referee on an earlier version of this chapter are appreciated.

in a system of cities? These questions are addressed in a spatial general equilibrium model of a closed economy consisting of a system of monocentric cities.

Keywords: system of cities, trade, industrial structure

JEL classifications: H41, R12, R13

15.1. Introduction

Why do population and economic activities concentrate in urban areas rather than disperse in space? Why do most urban systems tend to be dominated by a large metropolitan area such as London, Paris, Tokyo, and New York, which have a diversified industrial composition and a diverse labor force? Why does the size of diversified cities tend to be larger than specialized ones? Why do inter- and intra-city income disparities seem to be growing and that growth is more pronounced in developing countries? These are some questions that have attracted the attention of urban economists, regional scientists and economic geographers.

Addressing these questions are becoming more important over time for four main reasons: (1) the growing percentage of population living in urban areas documented by the United Nations Report, which indicated that the world urban population increased from 30% in 1950 to 45% in 1995 and is expected to be 50% by 2005. Furthermore, the percentage of increase was larger in industrialized countries where it increased from 61% in 1960 to 73% in 1993. On the other hand, the number of cities with population over 10 million in the world increased from only two cities in 1950 to 15 cities in 1995 and is expected to grow to 26 by 2025 (United Nations, 1996); (2) most of the non-agriculture GDP in industrialized countries is produced in urban areas; (3) labor productivities and industrial growth are positively related to the local size of the industry as well as the industrial composition in the city in which the industry locates.[1] It has also been documented that some industries grow faster in cities with a diversified

---

[1] See Rosenthal and Strange (2004) for a comprehensive survey of this literature.

industrial structure while other industries grow faster in specialized cities;[2] (4) trade liberalization and globalization, characterized by the emergence and the expansion of the European Union and with NAFTA, GATT, and some trade agreements between some European countries and some developing countries in North Africa. As a result of all these, the role the national government plays in influencing trade has been declining while the role of a city in a network of urban systems is becoming more important. Thus, it will be the comparative advantage of a city, not national trade policy that will shape the future of trade patterns. All of the above leads us to conclude that a comprehensive theory of a system of cities is not only an interesting theoretical exercise but also an essential part in understanding economic growth and international trade in a broader context.[3]

In an attempt to study city systems, urban economists, regional scientists, and economic geographers observed two striking features about most urban systems: the first is the regularity of the size distribution of cities, whether the country is a developed or developing country, characterized by a hierarchal structure in which each country's urban system consists of relatively small numbers of large cities and a large number of small cities, which is known as the rank-size rule;[4] the second striking feature about city systems is their industrial composition. Particularly, most city systems are characterized by a hierarchal structure. In the top level of the hierarchy we have the largest city in the system, for example New York, London, Paris, and Tokyo, which is characterized by a diversified industrial structure while in the bottom of the hierarchy we have a large number of small cities which are characterized by relatively specialized industrial structures.

---

[2] Glaeser *et al.* (1992) found that diversity of industrial composition in cities stimulates urban growth; Henderson *et al.* (1995) found that only new industries are attracted to diversified cities while mature industries grow faster in specialized cities.

[3] For a survey of growth and system of cities see Berliant and Wang (2004, this volume) and Abdel-Rahman and Anas (2004).

[4] The rank-size rule indicates that the population of each city in a system of cities, multiplied by its population rank equals the size of the largest city. For a survey of the rank-size rule and related work see the chapter by Gabaix and Ioannides (2004).

The objective of this chapter is to provide a survey of recent developments in the theory of systems of cities that attempts to explain the forces behind the existence of cities, the internal industrial structure of cities, the city labor composition, and the interaction between cities in a network of city systems. More precisely, the chapter addresses the following questions: *what are the main factors that lead to the formation of cities? When do cities specialize in production and when do they diversify? When do both specialized and diversified cities coexist? What is the role of trade in a system of cities? What determines the number and size of cities of different types in an economy? What are the factors that determine income disparities among different types of households?* These questions are addressed in a simple spatial general equilibrium model of a closed economy consisting of a system of cities. To simplify the basic model that will be used to review these issues we will impose some assumptions that facilitate such a task in a limited space. Thus, we will adopt the monocentric city model pioneered by Alonso (1964) and Muth (1969).[5] In this model the Central Business District (CBD) is the only employment center in the city. Models of a system of cities that will be discussed are in the spirit of Mills (1967) and Henderson (1974), who typically consider two types of forces: forces that lead to a concentration of population and economic activities (*agglomeration forces*) and the force that leads to deconcentration (*dispersion force*). When these forces are balanced at the margin, the equilibrium city size is determined which implies the existence of an optimal city size. These types of models do not provide an explanation of the spatial distribution of cities.

Since the pioneering work of Christaller (1933) and Losch (1954) urban theorists have been attempting to explain the spatial distribution, the hierarchal structure, and recently the growth of city systems. In this endeavor two prototypes of models have evolved in the literature: the first is the Alonso–Mills–Henderson (A–M–H) type model, labeled as the New Urban Economics (NUE), that explicitly considers the internal structure of the city and ignores the agricultural land use and the spatial distribution of cities

---

[5] All researchers in the area have imposed most of the assumptions that we adopt.

in the system over the landscape; the second is the Fujita–Krugman type model, that has been labeled the New Economic Geography (NEG), that explicitly considers the spatial distribution of cities and fixed agricultural land use but ignores the internal structure of the city.[6] This chapter will focus primarily on models of urban systems in the spirit of A–M–H. The NUE models of systems of cities has been influenced by: (i) conventional urban economics emphasizing the tension between economies due to the concentration of population and economic activities and diseconomy due to spatial concentration; (ii) the theory of industrial organization; and (iii) the theory of economic growth.

The first impact of the NUE-type model of systems of cities is due to the initial theory of city size which emphasized the interaction between indivisibility in production, at the firm level, as the source of concentration and diseconomy due to commuting costs, where the equilibrium city size is determined when these two opposing forces are balanced at the margin (Mills, 1967; Dixit, 1973).[7] The second impact on NUE-type model is due to the theory of industrial organization which occurred in a search for a more realistic model of city formation with many firms operating in the city. The Marshallian externality, also known as the black box, Marshall (1890) and Chipman (1970) provided an operational model that can provide a story of urban agglomeration economy (Henderson, 1974).[8] Later, after the seminal contribution of Dixit and Stiglitz (1977) to the Chamberlin (1933) model of the product differentiation and monopolistic competition to the industrial organization literature, the model provided a tractable framework as well as a microfoundation of the black-box externality in the NUE models of systems of cities (Abdel-Rahman and Fujita, 1990).

The Dixit and Stiglitz model also was the cornerstone upon which the NEG was built. The key features in the NEG model were increasing returns together with linear space, a fixed agriculture sector, and the Samuelson's iceberg transportation cost represented

---

[6] See Krugman (1991) and Fujita *et al.* (1999) and for a complete survey of the NEG see Ottaviano and Thisse (2004).

[7] See also Dixit (1973) for an optimal city size model.

[8] See also Kanemoto (1980) for a variable lot size model.

the essential ingredients of an operational spatial general equilibrium model. The third and relatively the most recent influential impact on the NUE type model is due to the theory of endogenous growth since most of the factors that have been identified as engines of economic growth such as knowledge spillover and human and physical capital accumulation (Romer, 1986, 1987; Lucas, 1988) are phenomena that take place in urban areas. However, this influential theory has only affected the NUE-type models but not the NEG.[9]

The organization of this chapter is as follows: Section 15.2 presents the internal structure of the city and household behavior as well as some important functions that will be used in characterizing the equilibrium city size. Section 15.3 discusses various agglomeration forces and the institutional mechanisms of city formation. Section 15.4 reviews models that result in identical cities. Section 15.5 presents models of specialized systems of cities with trade. Section 15.6 reviews models of specialized and diversified cities within the systems. Section 15.7 is devoted to the modeling of a system of specialized cities with heterogeneous households. Section 15.8 develops Pareto-efficient models of resource allocation in a system of cities and presents a comparison between equilibrium and the first best. Section 15.9 offers directions of future research.

15.2. The internal structure of the city

In this section, we will describe a monocentic city model to familiarize the reader with the internal structure of each city within the system. In this model it is assumed that each household resides at one location and has a single job that requires commuting to the CBD where all firms are located.[10] For simplicity, we postulate that each household consumes one unit of land. In addition, all households in the economy are assumed to have an identical utility function of the following Cobb–Douglas form:

$$u = x_1^{\alpha_1} x_2^{\alpha_2} \qquad \alpha_1, \alpha_2 \in (0, 1) \;\; \alpha_1 + \alpha_2 = 1 \tag{15.1}$$

---

[9] For a survey of models of economic growth and system of cities, see the chapter by Berliant and Wang (2004, this volume) and Abdel-Rahman and Anas (2004).

[10] This model has been adopted by most of the contributors to this literature.

where x_1 and x_2 are quantities of private goods consumed. We now specify the budget constraint facing a household residing at distance r from the CBD as

$$P_1x_1(r) + P_2x_2(r) + R(r) + Wtr = Y \qquad (15.2)$$

where P_i is the price of x_i $i = 1, 2$, $R(r)$ is the unit land rent at distance r, W the wage rate, t the amount of time required to commute one unit distance, and Y denotes the household income.[11] Given that each household is endowed with one unit of time, then the net labor supply of a household residing at distance r is $1 - tr$. On the other hand, one can interpret this as a household working unit of time and paying out of pocket commuting cost of Wt per unit distance. Both interpretations are consistent with the above budget constraint.

From the first-order conditions of the maximization of Equation (15.1) subject to Equation (15.2), we obtain the demand for x_i as

$$x_i(r) = \alpha_i P_i^{-1}[Y - Wtr - R(r)] \qquad i = 1, 2 \qquad (15.3)$$

By substituting Equation (15.3) into Equation (15.1), we derive the indirect utility function for a representative household in a given city as

$$V(P, W) = AP_1^{-\alpha_1}P_2^{-\alpha_2}[Y - Wtr - R(r)] \qquad (15.4)$$

where $A = \alpha_1^{\alpha_1}\alpha_2^{\alpha_2}$. In terms of land ownership, there are two types of land ownerships that can be used in models of cities. The first type is the case of absentee ownership of land, where owners of land in the city live outside the city. The second type is public ownership in which each household owns an equal share of the land rent generated in the city in which he resides.[12] The absentee landlord case is not considered here because we are interested in a fully closed model where all money generated in the system is

[11] The assumption that commuting cost is in terms of time will be relaxed later where we will consider a monetary commuting cost.

[12] Another type of land ownership is if each household owns an equal share of the land rent generated in the economy. This will be equivalent to our approach if the economy has identical cities. But if the economy has more than one type of city the efficiency will be distorted.

spent in the system. The reason for this is when we consider the first best problem in Section 15.8, we adopt a global optimal for the economy and we want the equilibrium to be comparable to the optimal.

Given the case of public ownership of land, the household income is given by

$$Y = \frac{\text{ALR}}{N} + W \tag{15.5}$$

where ALR is the aggregate land rent in a given city, and N is the size of the city. Recall that each household consumes precisely one unit of land. Thus, the total population of the city is given by

$$N = \int_0^b 2\pi r \, \mathrm{d}r = \pi b^2 \tag{15.6}$$

where b represents the urban fringe distance for the city. The total commuting cost (TCC) as a function of city size is

$$\text{TCC} = \int_0^b Wt2\pi r \, \mathrm{d}r = 2\mu W N^{3/2} \tag{15.7}$$

where $\mu = t/3\pi^{1/2}$.

Equilibrium requires that all workers in a given city achieve the same utility level. Hence, from Equations (15.4) and (15.5), at each r we derive that $W - Wtr - R(r) = W - Wtb - R(b)$. Recall that for convenience, we normalize the opportunity cost of land to be zero so that $R(b) = 0$.[13] In equilibrium the land rent schedule is $R(r) = Wt[b - r]$. Using Equation (15.6) and integrating the above land rent schedule, we can derive the ALR in each city as

$$\text{ALR} = \int_0^b R(r)2\pi r \, \mathrm{d}r = \mu W N^{3/2} \tag{15.8}$$

Equilibrium require that residential location cost must be the same for all r, thus the aggregate location cost (ALC) of N households in

[13] This assumption implies that there is no agricultural land use in the economy.

the city is

$$\text{ALC} = \int_0^b [(R(r) + Wtr]2\pi r \, \mathrm{d}r = 2W\mu N^{3/2} \tag{15.9}$$

Observe that the ALC increases twice as much as the ALR. Substituting Equations (15.5)–(15.9) into Equation (15.4), we derive the indirect utility function for a representative household in a given city as

$$V(N, P, W) = AP_1^{-\alpha_1} P_2^{-\alpha_2} W[1 - 2\mu N^{1/2}] \tag{15.10}$$

It can be seen from the above equation that $V(\cdot)$ is increasing in household net labor supply, given by the term between brackets, and wage but decreasing in prices. Furthermore, it can be seen that the indirect utility is decreasing in city size, which represents the dispersion force resulting from the higher commuting costs, which result from the physical expansion of the city. The above equation is important for two reasons: first, as will be seen later, it is essential in solving for city size; second, because it illuminates the fact that *if there exists no benefit to city residents due to concentration, i.e., no agglomeration economy, the city will not exist.*

By inverting Equation (15.10) with respect to W, we can derive the inverse population supply function

$$Y(N, U) = \frac{AP_1^{\alpha_1} P_2^{\alpha_2} U}{[1 - 2\mu N^{1/2}]}$$

This function represents the income necessary to attract an additional household to a city of size N so that all city residents would achieve the national utility level U. It can be seen that it is increasing in N, P, and U. By multiplying $Y(\cdot)$ by N, we can derive the total cost,

$$C(N, U) = \frac{AP_1^{\alpha_1} P_2^{\alpha_2} UN}{[1 - 2\mu N^{1/2}]}$$

which represents the total cost, residential and consumption costs, of maintaining U for a city with N residents.[14]

---

[14] For the derivation of the population supply function and the total cost function with variable lot size see Fujita (1989), Chapter 5.

15.3. Agglomeration economies and city systems

Agglomeration forces in the city system literature can be classified into three main categories depending on whether they influence the producer, the consumer or both.[15] Factors affecting the producer include the various forces that lead to the geographical concentration of production activities such as (1) economies of scale at the firm level, leading to the formation of company towns; (2) localization economies, which is economy of scale at the level of the industry in a given city, leading to the formation of a specialized city; and (3) urbanization economies, which is scale economy at the level of the urban area, leading to the formation of a multi-product, diversified city. Localization economies, as indicated by Marshall (1890, p. 271), describe the advantage of this externality:

> If one man starts a new idea, it is taken up by another and combined with suggestions of their own; and thus it becomes the source of further new ideas... Again, the economic use of expensive machinery can sometimes be attained in a very high degree in a district in which there is a large aggregate production of the same kind

Most of the advantages mentioned by Marshall, whether in the labor market, the intermediate inputs, or the information spillover, are intra-industry. In other words, they are due to concentration of firms in the same industry in a given urban area defined by Weber (1929) and followed by Hoover (1948) as localization economies. Romer (1986) and Lucas (1988) identified these types of externalities as the engine of economic growth. We can identify the sources of these externalities as (1) the *information externalities* stemming from interaction among agents and face-to-face communication, which enhance productivity and foster innovation; (2) the access to the wide range of *specialized intermediate input*; (3) *the matching of labor with heterogeneous skill in the labor market* which reduces the search cost; (4) the *matching of used assets in the capital market* which enhances the salvage values of assets from failed projects in large cities; and (5) the *acquisition of task-specific skill*, human capital, which enhances productivity.[16]

---

[15] This classification is borrowed from Abdel-Rahman (2000b).

[16] See Duranton and Puga (2004) for a survey of micro-foundation models of these externalities.

On the other hand, urbanization economies result from (1) cross-product technological externalities (Jacobs, 1969);[17] (2) the use of shared specialized intermediate input or local public good in production; (3) economies of scope in production; and (4) research and development of prototype. On the consumer side, we find the forces leading to the concentration of population through improvement in utility such as (1) the existence of natural amenity and (2) the provision of various local public goods.

Finally, models of agglomeration economies on the consumer and the producer side in the literature are due to product differentiation and monopolistic competition in which *the desire of consumers to consume a large variety* of goods and services together with economy of scale result in the formation of a city (Dixit and Stiglitz, 1977).

15.3.1. Equilibrium system of cities

Models of city systems can be classified, from the point of view of industrial composition of cities, into three main groups: (1) models which result in a specialized system of cities, (2) models which result in a diversified system of cities, and (3) models which result in coexistence of specialized and diversified cities. In the case of specialized system of cities, the main issues to be addressed are the factors that determine the size and the number of cities within the system. For a diversified system of cities, the main issues to be addressed are the factors that determine the industrial mix in the city as well as the size and the number of cities within the system. Finally, in the case of a system of cities where specialized as well as diversified cities coexist, the issues to be addressed are the factors that determine the structure of city systems, the factors that determine the cities' industrial compositions, as well as the sizes and the numbers of each type of city within the system.

Furthermore, models of systems of cities can be classified with respect to market structure within each city in the system into three groups: (1) models in which the market structure is perfectly competitive, (2) models in which the market structure is

---

[17] See Jacobs (1969) for an argument on the impact of interaction on creativity in production.

monopolistically competitive, and (3) models in which both perfectly competitive as well as monopolistically competitive markets exist.

Moreover, models of system of cities can be classified from the point of view of intercity trade into: (1) models in which there is no intercity trade, and (2) models in which there is intercity trade with positive or zero transportation costs.

In addition, models of city systems can be further classified, from the point of view of the types of workers/households, into two types: (1) models of identical households, in which all households are assumed to have identical income; and (2) models with a multi-type household, in which households are assumed to have different incomes.

Finally, one can classify models of systems of cities based on the type of institutional mechanism underlying the formation of cities in the system as: (1) city formation by local government or community, (2) cities formed by profit maximizing developers, and (3) cities formed by atomistic agents.

15.3.2. Institutional city formation mechanisms

The first institutional mechanism of city formation is local government (Henderson, 1974). The role of the local government is to set up the city by providing a public good or the basic infrastructure needed for the development of the city and collect any tax or provide any subsidy. Local government acquires the land needed for the city development and rents the land to households in a competitive market. Given that households are identical, they must achieve the same utility level in equilibrium. Thus, the objective of the local government is to maximize the utility of the representative household in the city by choosing the level of provision of a public good (if needed) and the city size. Furthermore, local government must balance the budget.

The second mechanism is when a profit maximizing developer forms and manages the city. It is assumed that the developer owns all the land required for city development and then sublets the land for city residents in a competitive land market. There are large numbers of potential sites for city development in the economy. Each developer controls one site that he uses for city development.

Furthermore, it is assumed that the number of cities in the economy is large and that each city is small relative to the national market. Thus, developers are competitive in the output and the labor markets. As a result the developer behaves as a utility and a price-taker in the national market. The objective of the developer is to maximize profit from the city development, defined as the difference between the revenue from the city and the cost of maintaining a national utility level U. The developer chooses the number of residents in such a way that each resident will not do worse than achieve the national reservation utility level. If a city developer makes profit, then new developers will enter the city development market and set up new cities. Competition among city developers would result in zero profit in equilibrium. The above two mechanisms will be equivalent under the assumptions that the city development market is contestable and that there is no limit to the number of cities.[18] Thus, both city formation mechanisms would result in the same city size.[19]

The third institutional city formation framework is self-organization.[20] In this framework the city size is determined through the defection of atomistic agents. Each agent is assumed to be small and behave so as to maximize his own profit or utility. It is assumed that households/firms can move freely and costlessly between cities to maximize utility/profit. If the utility of a representative household is strictly concave in city size, as it will be shown later, as the size of the city increases the utility will increase and households will continue to move into the city.[21] In this framework, all city sizes from the peak of the utility and beyond are stable equilibrium. Thus, this city formation mechanism results in

---

[18] Helsley and Strange (1994) showed that in a game-theoretic framework with a fixed finite number of cities the system of cities would be inefficient. Henderson and Thisse (2001) examined a strategic community development model of differentiated communities in which the number of communities is endogenous.

[19] For a game-theoretic approach to city formation mechanism see Helsley and Strange (1994). See also Henderson and Thisse (2001) for strategic community development with households differentiated by income.

[20] See Becker and Henderson (2000) for a comparison between formation of cities by large agent and by atomistic defection.

[21] One can specify a model in which the indirect utility function is strictly concave, as we will see later and as in Fujita (1989).

multiple equilibrium city sizes. This is unlike the previous two cases of city formation mechanisms where the city size that corresponds to the peak of the utility represents a unique equilibrium city size.

In most of Henderson's work he emphasized the role of developers or large agents in setting up cities. On the other hand, NEG models take the other extreme in which atomistic agents form cities. However, realistically city formation is a mix of large agents first starting up cities followed by independent agent or small developers.

15.4. Identical cities without trade

In this section we review models of a city system in which we have identical households in the economy and in which the industrial composition of cities within the system is predetermined. Furthermore, within this framework no interactions occur between cities except through costless migration.[22] The problem that will be considered is of a closed economy consisting of a system of identical cities, where the number of cities is endogenous. We postulate an economy sufficiently large so that we have a large number of cities. Thus, we treat the number of cities as a continuous variable. We will present four models of city formations. The first model is based on the provision of a local public good (Flatters *et al.*, 1974). The second model is based on external economy of scale (Chipman, 1970). The third model is based on the desirability of a final good industry to use a differentiated intermediate input in production (Ethier, 1982). Finally, we consider a model based on consumer preference for variety (Dixit and Stiglitz, 1977) in the consumption of non-traded goods.[23] The main issues that will be discussed are the factors that determine city size and the number of cities in the economy.

Let us first describe the common framework that we use to review this literature. We consider a closed economy consisting of a system of circular cities spreading over a flat, featureless plane. The economy is populated with $\bar{N}$ identical households where

---

[22] All models of a system of cities considered in this survey assume costless migration between cities.

[23] See Stahl (1983) for an initial work on the impact product variety in consumption on migration.

each one is endowed with one unit of labor (this assumption will be relaxed later in Section 15.7). Each household is free to choose the city in which to reside and work. Households migrate between cities, at zero migration cost, in search for the highest utility possible. Two final goods, X_1 and X_2, are produced within the system. Cities are formed in this economy by local governments. A local government of each city rents the land in the city at the opportunity cost, which is assumed to be zero. Then the local government sublets the land to households at the market rent. Each city government chooses the population size of the city, N, and the level of a public good if provided in the city, such that the utility of a representative household is maximized.

15.4.1. Public good

Suppose that the city is formed due to the provision of a local public good financed collectively by city residence (Flatters *et al.*, 1974; Stiglitz, 1977; Arnott, 1979; Arnott and Stiglitz, 1979; Kanemoto, 1980). We will assume that the city produces two private goods, X_1 and X_2 and a local public good Z (city good). The consumer has a utility function given by $u = x_1^{\alpha_1} x_2^{\alpha_2} Z^{\alpha}$. The private good is produced competitively under constant returns, with labor as the only input in production, $X_i = \eta_i L_i$ $i = 1, 2$ and $\eta_1 = 1$. Given that good X_1 is the numeraire, $P_1 = 1$. The public good is produced with the use of the private good X_1.

Residents pay a lump-sum tax, T, to finance the public good. $Z = TN$. Thus, given the disposable income $I = 1 - 2\mu N^{1/2} - N^{-1}Z$ and Equation (15.10), the indirect utility is

$$V = AP_2^{-\alpha_2}[1 - 2\mu N^{1/2} - N^{-1}Z]Z^{\alpha} \tag{15.11}$$

Now let us solve the problem in two steps. In the first step we choose Z for a given population size N. This will result in the following first-order condition

$$\frac{\partial V}{\partial Z} = \alpha[1 - 2\mu N^{1/2} - N^{-1}Z]Z^{1-\alpha} - N^{-1} = 0 \tag{15.12}$$

The above condition is the Samuelson condition for the optimal provision of a public good. This condition requires equality between the marginal cost of the public good, 1, and the marginal benefit, given by the sum of the marginal rates of substitution between the public good and the after tax disposable income

$$N\frac{\partial V/\partial Z}{\partial V/\partial I}$$

Solving Equation (15.12) for Z and substituting into Equation (15.11), we have

$$V = AP_2^{-\alpha_2}\left(\frac{1}{1+\alpha}\right)^{1+\alpha}\alpha^{\alpha}N^{\alpha}[1 - 2\mu N^{1/2}]^{1+\alpha}$$

Then, maximizing with respect to N, we have

$$\frac{\partial V}{\partial N} = -\mu N^{1/2} + \frac{\alpha}{1+\alpha}(1 - 2\mu N^{1/2}) = 0 \qquad (15.13)$$

After multiplying though by N we have the condition for the Henry George Theorem, which requires that the cost of provision of the public good be financed by confiscating the aggregate land rent in the city. Thus, the Pigouvian tax, T, will be imposed on each resident such that the total tax revenue is equal to the aggregate land rent. Conditions (15.12) and (15.13) can be solved for the equilibrium city size, N^*, and the equilibrium provision of the public good, Z^*, as

$$N^* = \left[\frac{\alpha}{[1+3\alpha]\mu}\right]^2 \qquad Z^* = \left[\frac{\alpha}{[1+3\alpha]\mu^{2/3}}\right]^3 \qquad (15.14)$$

Proposition 1. *The equilibrium city size and the equilibrium level of provision of the public good are increasing in the share parameter of the public good and decreasing in commuting cost.*

It can be seen from the above proposition that as $\alpha \rightarrow 0$ then $Z^* \rightarrow 0$ and $N^* \rightarrow 0$. In other words, if the public good is not provided the city will not exist.

Assuming a very large economy, the stable equilibrium number of cities in the economy, M^*, is given by $M^* = \bar{N}/N^*$. Observe that this equilibrium is stable since all households in the economy will achieve the highest possible utility level, and thus no movement of population between cities would improve welfare. To see this, suppose that some local governments would form cities larger than N^*.

In this case, atomistic defection of households from these cities to other smaller cities would improve their utilities, and thus larger cities cannot be sustained. On the other hand, suppose that some local governments form cities smaller than N^*. In this case, households would improve their utilities by defecting from these cities to other larger cities in the system, and thus smaller cities cannot be sustained. Thus, the only stable city system is when the population of all cities in the system is N^*.

Proposition 2. *A city system exhibits constant returns to aggregate population growth.*

The above result emphasizes that as the total population of the economy grows, the number of cities grows at the same rate. Thus, the economy would accommodate the new population by spawning new cities of size N^*.

15.4.2. Marshallian externality

Now we will consider a model in which cities are formed due to an external economy of scale (Henderson, 1974; Kanemoto, 1980; Upton, 1981).[24] Suppose that both X_1 and X_2 are private goods where industry 1 is produced by a production function subject to external economy of scale while industry 2 is produced with CRS as in Section 15.4.1. Furthermore suppose that both goods are non-traded between cities. Thus, this economy will result in a system of identical cities where each city produces both goods X_1 and X_2 (we will call this a system of diversified cities). The aggregate production function for the first good is $X_1 = L_1^{1+\varepsilon_1}$ while the aggregate production function for X_2 is the same as in Section 15.4.1. Given that workers are paid their private marginal product, the wage of workers in industry 1 and 2 are $W_1 = L_1^{\varepsilon_1}$ and $W_2 = \eta_2 P_2$. The objective of the city developer is to maximize the utility of a representative household in the city. Assuming that full employment will prevail in every city, then the labor supply for both

---

[24] Unlike the other authors, Kanemoto (1980) considered an agricultural sector in his model.

industries is given by

$$L_1 + L_2 = N - 2\mu N^{3/2} \tag{15.15}$$

All cities in the economy are identical. Therefore, the aggregate demand of good X_2 in each city is given by integrating Equation (15.3) as

$$AD_2 = P_2^{-1}\alpha_2 \int_0^b [W - Wtr + (\text{ALR})N^{-1} - R(r)]\mathrm{d}r \tag{15.16}$$

From Equations (15.5)–(15.8) and by equating the aggregate demand to the aggregate supply, we have

$$L_2 = \alpha_2[N - \mu N^{3/2}] \tag{15.17}$$

Thus, the net labor supply is allocated between both the industries proportional to the share parameter in the utility function. Substituting $W_1 = L_1^{\varepsilon_1}$, Equations (15.15) and (15.17) into Equation (15.10) we have

$$V = A\alpha_1^{\varepsilon_1} P_2^{-\alpha_2} N^{\varepsilon_1}[1 - 2\mu N^{1/2}]^{1+\varepsilon_1} \tag{15.18}$$

Given that local government is the price-taker in the output market, it will maximize the above indirect utility with respect to N. Solving the first-order conditions, we have

$$N^* = \left[\frac{\varepsilon_1}{(1 + 3\varepsilon_1)\mu^{3/2}}\right]^2 \tag{15.19}$$

Proposition 3. *The equilibrium city size is increasing in the external economy of scale parameter.*

Observe that if $\varepsilon_1 \to 0$ the city will not exist. In other words, if economy of scale does not exist in industry 1, the city will not exist. Furthermore, by comparing Equations (15.14) and (15.19) we can conclude that both the external economy of scale model and the public good model would result in the same city size if the parameters ε_1 and α were properly chosen.[25]

---

[25] This is only true under the above specification for public good. As it will be discussed in Section 15.8 the normative properties of both models are different.

15.4.3. Differentiated intermediate input

Consider an economy producing a homogenous good X_1 with the use of labor and a differentiated intermediate input (Ethier, 1982). This model presents a micro-foundation of the black box approach to external economy of scale that has been used in the city size model (Abdel-Rahman and Fujita, 1990). Here we explicitly model the availability of specialized services, such as repair and maintenance services, engineering and legal support, transportation and communication services, and financial and advertising services that have been used as the major cause of external economies of scale. Furthermore, these services constitute a significant share of employment in almost all industrialized countries (see Hansen, 1990 for the US case). Thus, to understand city formation and industrial composition, we have to take into consideration the service sector.

The central idea behind the model is that increasing returns to scale and the desire for a variety of intermediate inputs provide the basic source of industrial agglomeration and city formation (see Abdel-Rahman and Fujita, 1990). The production function of the final good is given by

$$X_1 = L_1^{\beta_1}\left[\left(\sum_i^n q_i^{\rho_1}\right)^{1/\rho_1}\right]^{(1-\beta_1)} \qquad \beta_1, \rho_1 \in (0,1) \tag{15.20}$$

where q_i is the quantity of differentiated intermediate input i and L_1 is the quantity of labor used in the production of X_1. As $\rho_1 \rightarrow 1$, intermediate inputs become close to perfect substitutes and firms in industry X_1 derive less productivity from variety. Alternatively, as $\rho_1 \rightarrow 0$, intermediate inputs become highly differentiated and firms in industry X_1 derive high productivity from variety. The production function of X_2 is as in Section 15.4.2.

The problem of the firm is to choose inputs L_1 and $\{q_i\}$ so as to maximize its profit given by $\pi_1 = X_1 - WL_1 - \sum_{i=1}^m P_i q_i$. From the first-order conditions, we have

$$L_1 = \beta_1 X_1 W \tag{15.21}$$

$$q_i = \left[(1-\beta_1)X_1 \sum_{l=1}^n q_i^{\rho_1} p_i^{-1}\right]^{1/(1-\rho_1)} \tag{15.22}$$

Following Dixit and Stiglitz, we assume that each of the n_1 intermediate inputs is produced by a single firm and by an increasing returns technology

$$L_{qi} = f_1 + c_1 q_i \tag{15.23}$$

where f_1 is the fixed labor requirement and c_1 is the marginal labor requirement. All firms in the city pay the same wage, W, and are perfectly competitive in the labor market. In the output market, the firms are monopolistically competitive and achieve Chamberlin equilibrium (Chamberlin, 1933).[26] The demand for all intermediate inputs is symmetrical and the Cournot-Nash markup condition is approximately

$$p_{1q} = Wc_1\rho_1^{-1} \tag{15.24}$$

where ρ_1 is related to the price elasticity of demand for each product. From the zero profit condition, we can derive the equilibrium quantity of each intermediate input as

$$q_1 = \frac{f_1\rho_1}{c_1(1-\rho_1)} \tag{15.25}$$

Substituting Equation (15.24) into Equation (15.22) we can derive the equilibrium employment in each firm in the intermediate input sector as

$$L_q = f(1-\rho)^{-1} \tag{15.26}$$

Full employment in a given city is given by

$$N - 2\mu N^{3/2} = L_1 + L_2 + n_1 L_{1q} \tag{15.27}$$

The RHS of Equation (15.27) is the aggregate labor demand and the LHS is the aggregate labor supply in each city. Substituting Equations (15.17) and (15.26) into Equation (15.27) we have

$$L_1 + n_1(1-\rho_1)^{-1}f = \alpha_1[N - \mu N^{3/2}] \tag{15.28}$$

---

[26] This imperfect competition in the output market is the basis of the market failure in this model. This will be discussed in Section 15.8.

From Equations (15.21), (15.25) and (15.28) and the zero profit condition in the intermediate inputs we have

$$L_1 = \alpha_1 \beta_1 (N - \mu N^{3/2}) \tag{15.29}$$

$$n_1 = \alpha_2 f^{-1}(1 - \rho_1)(1 - \beta_1)(N - \mu N^{3/2}) \tag{15.30}$$

The reason for the formation of cities in this economy is the existence of economies of scale in the production of each intermediate good. On the other hand, the reason for the concentration of firms is the desire of industry X_1 to employ a large variety of intermediate services given that such services are non-traded within the urban system. It can be seen from Equation (15.30) that the number of intermediate services is strictly concave function of city size. It can then be shown that the wage prevailing in the city is given as

$$W(n_1) = \beta_1^{\beta_1}(1 - \beta_1)^{(1-\beta_1)} c_1^{-(1-\beta_1)} \rho_1^{(1-\beta_1)} n_1^{(1-\rho_1)(1-\beta_1)/\rho_1}$$

This implies that larger the variety of intermediate goods the higher the productivity of labor in industry X_1. Substituting Equation (15.30) into $W(n_1)$ we have

$$W(N) = B_1(N - \mu N^{3/2})^{(1-\rho_1)(1-\beta_1)/\rho_1} \tag{15.31}$$

where $B_1 = \beta_1^{\beta_1}(1-\beta_1)^{(1-\beta_1)/\rho_1} c_1^{-(1-\beta)} \alpha_1^{(1-\rho_1)(1-\beta_1)/\rho_1} f_1^{-(1-\rho_1)(1-\beta_1)/\rho_1} \times \rho_1^{(1-\beta_1)}(1-\rho_1)^{(1-\rho_1)(1-\beta_1)/\rho_1}$.

Proposition 4. *The wage equation as a function of city size is structurally the same as the wage equation in the case of external scale economy.*[27]

This relationship is supported by empirical evidence (Hansen, 1990). Furthermore, the wage is strictly concave in city size, which represents the reason behind city formation in the presence of product variety in intermediate goods. *It is interesting to note that this model generates a wage equation which is structurally the same as the one assumed in the Marshallian externalities model* (see Abdel-Rahman and Fujita, 1990). In other words,

---

[27] This equation provides a micro-foundation for the existence of localization economies.

the product differentiation model is used as a micro-foundation for Marshallian externalities. Substituting Equation (15.31) into Equation (15.10), we derive the indirect utility function as a function of city size as

$$V(N) = AB_1 N^{(1-\rho_1)(1-\beta_1)/\rho_1}(1 - 2\mu N^{1/2})^{[(1-\rho_1)(1-\beta_1)/\rho_1]+1} \quad (15.32)$$

As in the previous model the local government maximizes the utility of a representative household by choosing city size. It can be shown that the above utility function is strictly concave in N with a unique maximum, which represents the equilibrium size of a given city in the economy like in the case of *Marshallian externality* (Abdel-Rahman and Fujita, 1990). The increasing segment of the indirect utility is due to the productivity gain that results from increasing the range of product variety, i.e. the agglomeration force dominates. On the other hand, the decreasing segment is due to high commuting costs that result from the physical expansion of the city, i.e. the dispersion force dominates. The equilibrium city size is given by

$$N^* = \left[\frac{(1-\beta_1)(1-\rho_1)}{[\rho_1 + 3(1-\beta_1)(1-\rho_1)]\mu} \right]^2 \quad (15.33)$$

Proposition 5. *The equilibrium city size increases with the degree of preference for variety, i.e. the smaller the parameter* ρ_1 *and the share parameter for the differentiated input in the production function,* $(1-\beta)$.

The intuition behind this result is that the higher the desire for variety in the final good industry, the larger the number of firms and consequently the larger the city size. Furthermore, the larger the share of the intermediate input in production, the more intensive is its use in production and hence the larger the city size. Here also, the total number of cities would be given as $\bar{N}/N^*$.[28]

---

[28] It can be seen that the equilibrium number of cities is decreasing in ρ_1 and $(1-\beta_1)$.

Now suppose that the production function of X_2 is given by

$$X_2 = L_2^{\beta_2}\left[\left(\sum_i^{n_2} q_i^{\rho_2}\right)^{1/\rho_2}\right]^{(1-\beta_2)} \qquad \beta_2, \rho_2 \in (0,1) \qquad (15.34)$$

If the intermediate inputs are non-traded, then both the industries share the use of the same differentiated services, and we have the following result.

Proposition 6. *This economy will result in a system of diversified cities where all cities in the economy will produce both consumption goods as well as the corresponding differentiated intermediate services.*[29]

Thus, sharable inputs represent a reason for the formation of diversified cities in the economy.

15.4.4. Differentiated consumption good

Next we will discuss the case of a differentiated, non-traded consumption good (Dixit and Stiglitz, 1977). Abdel-Rahman (1988) and Hobson (1987) among others adapted the Dixit and Stiglitz model into a special context where the desire of households to consume a variety of non-traded differentiated services represents one of the main reasons for the existence of large cities. Consider the case in which x_1 in Equation (15.1) is a sub-utility of the CES form.[30] Then Equation (15.1) would be

$$u = \left[\left(\sum_i^{n_2} q_i^{\sigma}\right)^{1/\sigma}\right]^{\alpha_1} x_2^{\alpha_2} \qquad \alpha_1, \alpha_2, \sigma \in (0,1) \qquad (15.35)$$

where q_i is the quantity of the variant i of a non-traded differentiated good or service such as restaurants or movie theaters. The budget constraint for a household is now given by $\sum_{i=1}^{n} P_i q_i(r) + x_2(r) + R(r) + Wtr = Y$. From the maximization of Equation (15.35) subject to the above constraint we derive the household demand for each

---

29 For this model see Abdel-Rahman (1990b).

30 See also Rivera-Batiz (1988) for a Dixit and Stiglitz model in a regional context.

differentiated service as

$$q_i(r) = \left\{ \alpha_1 P_i^{-1} \left(\sum_{i=1}^{n} q(r)^{\sigma} \right)^{-\sigma} [Y - Wtr - R(r)] \right\}^{1/(1-\sigma)} \qquad (15.36)$$

The demand for x_2 is as in Equation (15.3). Substituting Equations (15.3) and (15.36) into Equation (15.35), we derive the indirect utility function, $V = A\xi^{-(1-\alpha_1)}W[1 - 2\mu N^{1/2}]$, where $\xi = [\sum_i^n P_i^{-\sigma/(1-\sigma)}]^{-(1-\sigma)/\sigma}$. Each of the differentiated services, q_i, is assumed to be produced with an identical production process as in Section 15.4.3. Good X_2 is produced as before by labor only with a simple constant-returns-to-scale technology. Assuming that full employment prevails in each city, the total population of a representative city is given as $L_2 + nL_1 = N - \mu N^{3/2}$. All cities in the economy will be identical. The employment in industry 2 will be given as Equation (15.17) while the employment in each differentiated firm will be given as Equation (15.26); thus from the full employment condition we can derive the number of firms as

$$n = f^{-1}(1 - \sigma)\alpha_1[N - \mu N^{3/2}] \qquad (15.37)$$

Observe that the number of variety is strictly concave in city size. This suggests that the formation of large cities involves a substantial amount of product diversity. In other words, increasing product diversity is a sufficient condition for the formation of larger cities. This result is supported by casual observation that all large cities offer a large variety of goods and services. Local government form cities by choosing the city size that maximizes the utility of a representative household

$$V = CN^{(1-\sigma)\alpha_1/\sigma}[1 - \mu N^{1/2}]^{1+((1-\sigma)\alpha_1/\sigma)} \qquad (15.38)$$

where $C = Ac^{-\alpha_1}\sigma^{\alpha_1}((1 - \sigma)f^{-1})^{(1-\sigma)\alpha_1/\rho}$. Observe that this function is strictly concave in N, which shows that product variety of the non-traded, differentiated service is sufficient for the formation of multiform city. The increasing segment of the indirect utility is due to the increase in product variety, i.e. the agglomeration force dominates. On the other hand, the decreasing segment is due to high commuting costs resulting from the physical expansion of the city, i.e. the dispersion force dominates. Maximizing Equation (15.30) with respect to N we can derive the equilibrium size of a representative city as follows:

$$N^* = \left[\frac{\alpha_1(1-\sigma)}{\{\sigma + 3\alpha_1(1-\sigma)\}\mu} \right]^2 \tag{15.39}$$

Proposition 7. *The equilibrium city size is increasing in σ and α_1.*

Thus, the greater the desire of the consumer for variety and the greater the share of the differentiated goods in the utility function, the greater the number of varieties ultimately resulting in a larger city and thus the larger is the city size. The equilibrium number of cities is given by $\bar{N}/N^*$, since all cities in the economy are identical. From the above models we can conclude the following:[31]

Proposition 8. *All of the above four models result in structurally the same indirect utility function of the form $V(N) = AN^a(1 - 2\mu N^{1/2})^{1+a}$ and the same city size of the form*[32]

$$N^* = \left[\frac{a}{(1+3a)\mu} \right]^2$$

However, unlike cities formed due to a public good or Marshallian externality, for models of product variety on the consumption side or on the production side, if $a \to 0$, N^* will not be zero. This result occurs because the city can still be formed due to internal return to scale in a single firm in sector q.[33]

Helsley and Strange (1991) presented a non-spatial model that would result in agglomeration. They adapted a matching and search model for the capital market to a non-spatial model. Their idea hinges upon the assumption that salvage values of assets in large cities are higher than in small ones. The reason is that a bank allocates credit to projects that have addresses in a characteristic space, and if the project fails, then the bank repossesses the asset. The second best use of

---

[31] It can be seen that the equilibrium number of cities is decreasing in α_1, σ, and t.

[32] There exists a set of parameters, a and t, such that N^* represents a maxima of $V(N)$.

[33] In this case the city size will not be determined by the indirect utility function but by the full employment condition for a company town.

an immobile and specialized asset is more valuable where the density of possible uses is greater, as it is in large cities.

15.5. Specialization and trade

In all of the models presented in Section 15.4 each city in the economy produces all of the goods consumed. Now we will analyze problems of equilibrium resource allocation in a closed economy consisting of a system of different types of cities, where the number of cities is endogenous and in which cities interact with other cities in the system though trade. Two types of models will be considered. The first type of model considers trade with costless transportation, while the second type of model considers trade with a positive iceberg transportation cost. In these models, cities in the system will specialize in the production of a particular good and trade with the rest of the cities in the system as long as there is no benefit in producing more than one good in the same city. This will happen because if cities can trade and no benefit exists due to diversity, locating more than one industry in the city would increase the population and commuting costs and thus result in lower utility for city residents.

Henderson (1974) presented the first model in which trade with costless transportation cost and specialization occurs in a system. We will discuss a version of this model in which cities are formed due to an external economy of scale. Suppose that both X_1 and X_2 are private goods where $X_i = L_i^{1+\varepsilon_i}$ $i = 1, 2$. This economy will result in a system of specialized cities where one type of city will specialize in the production of good X_1 while the other will specialize in the production of good X_2. The reasons for specialization are external economies of scale and costless transportation cost. Given that workers are paid their private marginal products, the wage is $W_i = P_i L_i^{\varepsilon_i}$, where $P_1 = 1$. The developer of each type of city i will maximize the utility of a representative household in the city, as in Section 15.4.2. Assuming a very large economy, the stable equilibrium relative number of cities is given as

$$\frac{M_1^*}{M_2^*} = \frac{\alpha_1 N_2^*}{\alpha_2 N_1^*} \tag{15.40}$$

Proposition 9. *The relative number of cities in the economy is increasing in the demand parameter and decreasing in city size, but independent of the total population.*

Given that households are identical, all households in the economy will achieve the same common utility level.[34] Observe that this is a stable equilibrium since all households in the economy will achieve the highest possible utility level, and thus no movement of population between cities would improve welfare.[35] Furthermore, as the total population of the economy increases the economy will accommodate the new population by spawning new cities of both types while the relative number of city will remain unchanged.

Now we will consider another type of costless trade model. Specialization in this context is a result of the desire of each final good industry to use different group of differentiated services in production. Thus, one type of city specializes in the production of good X_1 and a group of differentiated services as in Section 15.4.3, while other type of cities specialize in the production of good X_2 and a different group of differentiated services. The production function of the final good is given by Equations (15.20) and (15.34). The behavior of the firms in the differentiated intermediate input/service is as presented in Section 15.4.3. Again following Dixit and Stiglitz, we assume that each of the n_i products is manufactured by a different firm and by an increasing returns technology as given in Equation (15.23). The above model can generate cities of different sizes. For example, *if the economy produces two final goods where each good uses a different group of non-traded, differentiated services, the economy will generate two types of cities, each specializing in the production of one final good and the group of differentiated services used in its production.* This will be the case as long as no benefit can be obtained from locating both the industries within the same city.

---

[34] The equal utility condition is used to solve for the equilibrium price of good X_1 in terms of good X_2.

[35] Wilson (1987) considers another reason for specialization in a non-spatial model where communities are formed due to the provision of public good. The economy produces two tradable private goods produced under constant returns with land and labor. In this framework, communities will specialize in production of one private good due to the difference between private goods in production.

Henderson and Abdel-Rahman (1991) adapted the Dixit and Stiglitz model to examine the formation of a system of cities where each city specialized in the production of a single differentiated good traded at zero transportation cost.[36] The objective was to explain national economies of scope arising from national product diversity. In other words, large nations are able to provide a larger variety of goods than small nations. In this framework, the reason for the formation of cities in the economy is the presence of economies of scale in the production of each variant, as given by Equation (15.29). Given that a differentiated product is traded at zero transportation cost, each city will specialize in the production of one variant and the production of the non-traded good. This result occurs because producing more than one differentiated good in a given city will increase commuting costs without generating any productivity gain. Production sectors are the same as in Section 15.4. However, full employment in each city requires that

$$N - 2\mu N^{3/2} = f + cq + L_x \tag{15.41}$$

Solving for q and L as in Section 15.4 and substituting into Equation (15.41), we have

$$f(1-\rho)^{-1} = \alpha_2[N - \mu N^{3/2}] \tag{15.42}$$

The above equation determines the equilibrium city size. Thus, we can state the following:

Proposition 10. *Equilibrium will result in a system of identical cities where each city produces one differentiated good and* x_2.

Observe that if the utility were a function of only the differentiated product then the model would result in a system of company towns as in Henderson and Abdel-Rahman (1991).

Proposition 11. *National economy of scope will increase with population growth.*

---

[36] Hochman (1997) examined the same model with variable lot size.

Population growth in the above system would result in more cities and thus more differentiated products. These additional differentiated products will result in higher utility since the consumer derives more utility with an increase in variety as implied by the utility function.

15.6. Specialization vs. diversification

This section surveys models that generate cities of different types, sizes, and industrial composition. The fundamental question addressed in these types of models is to explain the reasons behind the formation as well as the coexistence of specialized and diversified cities. Four types of models are reviewed in this section. In the first model, the formation of diversified cities is due to cross-product technological externalities. In the second model, the reason for the formation of diversified cities is economy of scope in production. In the third type of model, the reason for the formation of diversified cities is intercity transportation costs. Finally, in the fourth type of model, the reason for the formation of diversified cities is product cycles. These models can be classified into two groups: (1) models that result in either specialized or diversified cities, and (2) models that result in specialized, diversified, and mixed cities in the system. Since almost all systems of the cities that we observe in developing and developed countries have both diversified and specialized cities, models that result in mixed city systems are more realistic.

15.6.1. Cross-product externality

The first model that addressed the formation of diversified cities and the coexistence of diversified and specialized cities was Abdel-Rahman (1990a). The model is of two sectors and two cities embedded in an open economy. Both goods X_1 and X_2 are traded at zero transportation costs. Both good X_1 and X_2 are produced with the following production functions

$$X_{1j} = f(L_1, L_2)L_{1j} \qquad f'_1, f'_2 > 0 \tag{15.43}$$

$$X_{2j} = g(L_2)L_{2j} \qquad g'_2 > 0 \tag{15.44}$$

where L_1 and L_2 are the quantity of labor in industry X_1 and X_2 in a given city while L_{1i} and L_{2i} are quantity of labor in firm j in industry 1 and 2, respectively. Thus, industry 1 has *urbanization economy* (Jacobs, 1969), because it reflects external economies of scale that operate across industries in the same city. Industry 2 has *localization economies* (Marshall, 1890). The author examined a system consisting of a diversified city and a city specializing in industry 2.[37] The household side of the model is as in Section 15.2.[38] This framework enables us to analyze a system with two types of cities. The formation of a diversified city producing both goods X_1 and X_2 is due to the cross-product externality, while the reason for the formation of a specialized city producing only X_2 is Marshallian externality. Some of the main results of the model are: *first, the diversified city is larger than the specialized one if at least one of the two industries exhibits external diseconomy of scale*; *second, the specialized city will have more employment in the industry in which the city specialized when compared to the diversified city*. However, the model assumed the coexistence of the diversified and the specialized cities rather than derive the condition for that endogenously. Furthermore, the model is a partial equilibrium where the number of cities is given endogenously.

15.6.2. Transportation costs

Next we discuss models in which the number of cities as well as the industrial composition is determined endogenously. In other words, these models will lead to different equilibrium configurations, i.e. a different city system for each set of parameter values. The first model by Abdel-Rahman (1996) uses product variety in the intermediate good sector, which was introduced in Section 15.4. The model generates two equilibrium configurations: (i) pure specialization, in which each city specializes in the production of

---

[37] Jacobs (1969) stressed the importance of this inter-industry spillover as the force underlining the process of innovation in which the automobile industry emerged from a diversified economy.

[38] The author used a variable lot size model and a general functional form utility instead of the fixed lot size and a Cobb–Douglas utility function as in Section 15.2.

one group of intermediate producer services and one final good; and (ii) pure diversification, in which all cities in the economy produce two groups of intermediate producer services and two final goods. In the context of this model, cities may specialize in the production of a particular final good. The reason for that is *the desire of the final good industry to take advantage of productivity gains offered by using a wider variety of non-traded differentiated services* (Abdel-Rahman and Fujita, 1990). On the other hand, the reason for diversification is *the high intercity transportation costs* of the final goods.

The production function X_1 and X_2 are given as in Equations (15.20) and (15.34) where X_1 uses a group of differentiated services q_1, while X_2 uses a group of differentiated services q_2. Thus, the production sectors are the same as presented in Section 15.4.3. We assume Samuelson's iceberg transportation costs for both the traded goods. In this context, when a unit of a good is transported from one city to another, regardless of the distance between them, only a given fraction, $\tau_{1i} < 1$ if $i \neq 1$ and $\tau_{2i} < 1$ if $i \neq 2$, of this good will arrive. Hence, transportation costs are an inverse function of τ_i. The budget constraint for a given household at distance r_d from the CBD in a given diversified city of type d is given as in Equation (15.2). While the budget constraint for a household at distance r_i from the CBD in specialized city of type $i = 1,2$ is given by

$$\tau_{1i}^{-1} x_1 + P_2 \tau_{2i}^{-1} x_2(r_i) + R(r_i) = Y_i - W_i t r_i \qquad i = 1,2 \qquad (15.45)$$

where $\tau_{22}, \tau_{11} = 1$. It has been shown that equilibrium utility level in a purely specialized equilibrium is increasing in τ_{i1} and τ_{i2} (see Abdel-Rahman, 1996). In other words, *higher transportation costs lead to lower equilibrium utility*. Thus, there exist different combinations of τ_{i1} and τ_{i2} that can sustain a given equilibrium utility. This is because higher transportation costs imply that fewer resources will be used for consumption goods. In a purely diversified equilibrium configuration all cities in the economy will produce both final goods and both groups of intermediate goods. Thus, the equilibrium utility level is independent of τ_{i1} and τ_{i2} since no trade will occur in a diversified system. Thus, we can conclude the following:

Proposition 12. *There exists a set of values* $\{\tau_{i1}, \tau_{i2}\}$ *that will result in a purely specialized system of cities and a set of values* $\{\tau_{i1}, \tau_{i2}\}$ *that will result in a purely diversified system of cities.*

The intuition behind this result is that a specialized system will be the equilibrium outcome if the benefit from product differentiation dominates the loss due to transportation costs. On the other hand, the system will be diversified if the reverse is true. This result can be tested empirically by comparing the industrial compositions of cities in developed countries that possess developed inter-city transportation systems with that of cities in developing countries. An alternative approach in testing the result is to compare cities' industrial compositions over time in a given country.[39]

Anas (2004) extended Henderson and Abdel-Rahman (1991) by introducing positive transportation cost into the model. He assumed iceberg transportation costs, as in the previous model and considered a symmetric CES utility function, $u = [nq_i^{\sigma} + (m-1)nq_{-i}^{\sigma}]$, where q_i is the quantity of each variety produced in the city where the consumer resides, and q_{-i} is the quantity of each variety produced in other cities. It has been shown that the indirect utility is increasing in the aggregate population as long as transportation cost is finite. The above model will result in either a system of identical cities each of which produce the same variety with no trade as in Abdel-Rahman (1988) if $\tau = 1$, or a system of company towns as in Henderson and Abdel-Rahman (1991) if $\tau = 0$.

15.6.3. Economy of scope

Abdel-Rahman and Fujita (1993) introduced the first model that can explain the coexistence of a diversified and specialized city based on the concept of economies of scope (Panzar and Willig, 1981).[40] In this framework, the reason for the formation of a diversified city is the benefit of joint production. These benefits are cost savings resulting from the existence of economies of scope in

[39] See Anas and Xiong (2003) for the same model but with traded intermediate services as well as final goods with positive transportation costs.

[40] See also Goldstein and Gronberg (1984) for a discussion of economies of scope in an urban context.

which the total cost of producing more than one good jointly in one firm/city is less than that of producing them separately. On the other hand, the reason for the formation of specialized cities is due to scale economies at the firm level resulting from the existence of fixed costs in production. Unlike the models presented in Section 15.6.2, cities are formed here by surplus maximizing developers. The model generates three equilibrium configurations: (1) pure specialization, in which each city specializes in the production of one traded good; (2) pure diversification, in which all cities in the economy produce two goods; and (3) a mixed system, in which specialized and diversified cities coexist in the economy. The total labor requirement, $L(X_1, X_2)$, for the production of outputs X_1 and X_2 in a given city is given as

$$\begin{array}{ll} F_1 + X_1^2 & \text{if } X_1 > 0 \text{ and } X_2 = 0 \\ F_2 + X_2^2 & \text{if } X_2 > 0 \text{ and } X_1 = 0 \\ F_d + X_1^2 + X_2^2 & \text{if } X_1 > 0 \text{ and } X_2 > 0 \end{array} \tag{15.46}$$

where F represents fixed labor requirements. Hence the average labor requirement is U-shaped, which represents the reason for the formation of a city. Thus, given the national utility level, U, and a price vector, $\mathbf{P}$, the developer must choose the optimal household consumption bundle (x_1, x_2) by solving the following cost minimization problem:

$$\min_{x_1, x_2} = \sum_{i=1}^{2} P_i x_i, \qquad \text{s.t. } x_1^{\alpha_1} x_2^{\alpha_2} = U \tag{15.47}$$

From the above problem we derive the total consumption cost for a city having population, N, such that each household achieves the national utility level, U, as

$$C(U, N) = A^{-1} P_1^{\alpha_1} P_2^{\alpha_2} UN \tag{15.48}$$

Therefore, the surplus, S, from the development of a specialized city of type i, where $i = 1, 2$, is given by

$$\begin{aligned} S_i &= P_i(N - 2\mu N^{3/2} - F_i)^{1/2} - A^{-1} P_1^{\alpha_1} P_2^{\alpha_2} UN \\ i &= 1, 2 \end{aligned} \tag{15.49}$$

where the first term on the LHS of Equation (15.49) represents the total revenue from production, which is obtained from Equations (15.7) and (15.46), and the full employment condition in each city. The surplus from the development of a diversified city is given by

$$S_d = [P_1^2 + P_2^2]^{1/2}(N - 2\mu N^{3/2} - F_d)^{1/2} \\ - A^{-1}P_1^{\alpha_2}P_2^{\alpha_2}UN \tag{15.50}$$

The developer's problem is to maximize the surplus from the city development by choosing the city size. Given the above behavior, we can conclude the following:

Proposition 13. *There exists a set of parameters $\{F_1, F_2, F_d\}$ such that:*

(i) there exists a unique equilibrium purely specialized city system;
(ii) there exists a unique purely diversified city system;
(iii) there exists a unique mixed city system.

Pure specialization, *S*, where each city specializes in the production of one good, exists if the fixed costs of the specialized cities, F_i, are relatively small compared to the fixed cost of the diversified city, F_d. Conversely, in order to have a pure diversification, fixed cost, F_d, must be relatively small compared to F_i. In other words, production in diversified cities has strong economies of scope. Thus, diversified cities are more economical to form than specialized cities. Finally, in order to have equilibrium configurations of mixed-1 type (mixed-2 type), fixed cost F_1 (F_2) must be relatively small compared with F_d, while F_d must be relatively small compared with F_2 (F_1).

Proposition 14. *Whenever a specialized city and a diversified city coexist in an economy, the diversified city is larger than the specialized city.*

Although the above model generates a mixed system of cities, all cities in the economy are company towns. But company towns are not the most observed urban settlement. However, Abdel-Rahman (2000a) introduced a model that generates company towns (Dixit, 1973) or multi-firm cities (Henderson, 1974) based on the parameters of the model. The model is of a specialized system of cities in a closed economy producing one final good. The good is produced with the cost function, $C(X, G) = wX^2 + F(G)$, where G is the public infrastructure (such as water, electricity, sewage, and training facilities) and where F is a fixed set up cost. It is assumed that F is a decreasing function of G. Profit maximizing developers form cities in the economy, as discussed in Section 15.3.2. A developer can form a company town or a multi-firm city. The developer can produce good X by a single firm with fixed setup cost in a company town, or he can produce it in a multi-firm city. If the developer invests in public infrastructure, he will reduce the fixed set up cost for each firm in the city. This fixed cost reduction represents an agglomeration force that may result in the formation of a multi-firm city. *As a result of this specification, one can have a system of identical company towns or a system of identical multi-firm cities based on whether or not the developer will find it profitable to invest in public infrastructure.* The type of system that will emerge depends on the impact of the investment in public infrastructure on the reduction of the fixed set-up cost for each firm. This idea can be used for joint production if the public infrastructure is used by two industries. In this case, the joint use of public infrastructure can result in the formation of a diversified city. Thus, we can generate the same equilibrium configuration in which each city in the system is a multi-firm city.

Another extension of Abdel-Rahman and Fujita in the direction of multi-firm cities is Abdel-Rahman (1994). He extended the above model in two respects: first, by introducing a market for intermediate goods (or services) which leads to a multi-firm city; second, by examining the impacts of economies of scope due to lower variable costs resulting from the interaction and coordination between two production processes. As a result, it has been shown that *if the economies of scope are in the form of lower fixed costs* (Abdel-Rahman and Fujita, 1993), *the only possible equilibrium configurations are pure specialization and pure diversification* (unlike Abdel-Rahman and Fujita, 1993). *However, if economies of scope are in the form of lower variable costs, mixed systems are also possible.* Therefore, the model generates different equilibrium parameter spaces depending on the form of the economies of scope, i.e. whether and to what degree economies of scope affect fixed or variable costs.

The model was also extended by Tsukahara (1995) to explain the joint provision of an essential public good, such as police and fire protection, and on optional public good, such as a museum or a stadium. It was assumed that the costs of providing public goods are only fixed costs. In this context, *it was shown that if the fixed cost required for the provision of the essential public good were large, the result would be a large city size. Therefore, the per capita cost of providing the optional public good is low, and joint provision will lead to higher utility*. On the other hand, if the cost of providing the essential public good were small, the result would be a small city size. Therefore, the per capita cost of providing the optional public good would be high and joint provision would not be the equilibrium outcome.

15.6.4. Product cycle

Most recently, specialization and diversification has been analyzed in a dynamic model that resulted in a system of cities in which diversified and specialized cities coexist. Duranton and Puga (2001) developed the first product cycle model in the systems of cities literature. In this model, a metropolitan area plays the role of a nursery for new products. The authors employed a model of product development where firms experiment with prototypes in a diversified city until they find the ideal production process. After the firm identifies the ideal production process, the firm moves to a specialized city to start mass production. Henderson *et al.* (1995) provided empirical support for this product cycle model in a system of cities. The main result of the paper was to identify the conditions that result in a unique steady state in which specialized and diversified cities coexist. However, the size of the diversified city was the same as that of the specialized one, which is neither consistent with the empirical observation nor with other theoretical models that result in the coexistence of diversified and specialized cities.[41]

---

[41] For models in which diversified and specialized cities coexist see Abdel-Rahman and Fujita (1990) and Abdel-Rahman (1994).

15.7. Heterogeneous household and income disparities

Models with heterogeneous types of workers enable us to analyze the factors that determine skill distribution and income inequality and their impacts on welfare. These issues are becoming more important given the rise in income disparities nationally during the post-World War II period. This rise has been materialized by a dramatic decrease in the real wage of low-skilled labor as well as an increase in the wage of highly skilled labor.[42] Furthermore, large cities tend to be populated with a labor force characterized by a wide variety of skills, while small cities tend to be populated by a labor force with relatively specific skills. As a result, income disparity is relatively large in large cities compared to small and medium-sized cities. Thus, there is a need for models that examine the relationship between skill distribution and income disparity between cities within a hierarchical structure as well as within different types of cities. This section is devoted to survey the initial research in this direction. Two types of models are reviewed. The first type of model analyzes the case of exogenous types of households/workers (like Helsley and Strange, 1990; Kim, 1991; Abdel-Rahman and Wang, 1995, 1997; Becker and Henderson, 2000), while the second type of model analyzes the case where the types of households/workers are determined endogenously through a self-selection mechanism (Abdel-Rahman, 2002). The first group of models can be further classified into models that result in the same equilibrium wage and utility level for households with different skill characteristics (like Helsley and Strange, 1990; Kim, 1991) and model that result in different equilibrium wages and utility level for households with different skill (like Abdel-Rahman and Wang, 1995, 1997; Abdel-Rahman, 1998, 2002; Becker and Henderson, 2000).[43]

Becker and Henderson (2000) introduced a model of intra-industry specialization (Becker and Murphy, 1992), which resulted

---

[42] See Juhn *et al.* (1993) for these trends in the US and Machin (1996) for the UK among others.

[43] Henderson and Thisse (2001) in a non-spatial model examined the distribution of households differentiated by income between communities where community developers provide public good and behave strategically. In that model, the income distribution is given exogenously and public goods are differentiated by quality.

in a system of identical cities. The economy is populated with two types of households: workers and entrepreneurs. The entrepreneurs allocate one unit of time between mastering task-specific skills, distributed on a unit circle, and managing workers in performing each task. *The paper presented a micro-foundation model of externality in which increasing the number of entrepreneur in the city results in higher level of specialization.* Furthermore, the paper examined the formation of cities by large developers, self-organization, and a combination of both. In addition, the paper characterized the conditions required in achieving efficacy under each city formation mechanism.

Abdel-Rahman (1998) presented another approach in modeling income inequalities and analyzed its impacts on social welfare. The model is of a two-sector economy consisting of two types of cities where the number of cities of each type is determined endogenously. Cities are formed as a result of investment in public infrastructures. This investment leads to a reduction in commuting costs and consequently to an increase in the time that households can utilize for work and leisure. In the context of the model, the level of investment in public infrastructures is chosen optimally. The model analyzed a sorting equilibrium in which skilled labor locates in one type of city, while the unskilled labor locates in the other type. Wages in both labor markets are determined competitively. *This model explains the variation in city sizes as a result of differences in households' value of time.*

Abdel-Rahman and Wang (1997) considered an economy populated by a continuum of unskilled and skilled workers.[44] The economy produces two homogeneous goods, X_i, $i = 1, 2$. Skilled or unskilled workers produce good X_1 with CRS production function as in Section 15.3, but only skilled workers can produce good X_2. Unskilled workers are homogeneous. On the other hand, skilled workers are heterogeneous in their skill characteristics. Their types are uniformly distributed on a circle with unit circumference.

---

[44] This model is an extension of Helsley and Strange (1990) and Kim (1991) where they considered only heterogeneous skilled labor force. The basic difference between Helsley and Strange, and Kim is that in the former, skilled workers and firms have incomplete information, while in the latter, both skilled workers and firms have complete information about the job requirements and worker skill characteristics.

The wage for the unskilled workers is determined competitively, whereas the wage for skilled workers is determined via a Nash bargaining rule (Diamond, 1982) with the high-tech firms. Firms are also distributed on the unit circumference where each firm of a given type in the high-tech industry has a different skill requirement. Furthermore, firms pay a fixed entry cost. If a firm of a given type employs a worker who poses the same skill requirement, a perfect match, the maximal output per worker will be generated. On the other hand, if a firm employs skilled workers with skill characteristics different from the firm skill requirement there will be a productivity loss from the mismatch proportional to the distance between the firm and the worker on the unit circumference. *The model presented a micro-foundation for a localization economy, in which an increase in the size of the labor force enhances matching in the labor market and increases productivity and wage.*[45] The formation of local cities and the metropolis in this economy requires a public infrastructure (such as an intercity transportation system) produced using the numeraire good. City residents share the total cost of this public infrastructure. Thus, this represents the incentive for the agglomeration behavior in both types of cities.

Proposition 15. *The model generates a core–periphery system of cities in which the economy would have a single metropolis, populated by skilled workers, and a number of local cities, populated by unskilled workers.*[46]

It has been shown that the main determinants of geographical income inequalities include: (i) traditional economic factors, such as labor productivity and the mass of unskilled labor; (ii) spatial factors, such as commuting costs; and (iii) search and matching factors, such as entry and mismatch costs, relative bargaining power, and the mass of skilled workers. Furthermore, they showed that some factors that *decrease interregional income inequality reduces social welfare.*

---

[45] The above model does not structurally generate the same indirect utility or city size as in Proposition 7.

[46] The above result is unlike Helsley and Strange (1990) where they generated a system of identical cities where each city is populated with a continuum of skilled workers with differentiated skills.

Therefore, public policies designed to reverse the observed urbanization trend of widening geographical income inequality may not necessarily improve welfare.

One important assumption in the model is that selective migration is not allowed. For example, a matched high-tech firm and skilled workers cannot form a new city. This assumption results in single metropolis. However, violation of this assumption would result in a system of company towns where each company town would have a single firm and it would be populated with the workers that best match the skill requirement of the firm. The reason for this is there would be no benefit of having more than one firm located in the same city. As a result, the growth of the population of skilled workers in the economy would lead to higher productivity. Since in this case, the population growth leads to a better match between firms and workers, it also leads to more specialization and higher utility in the economy. Thus, we conjecture that this model can result in national economy of scope as in Henderson and Abdel-Rahman (1991).

Abdel-Rahman (2002) extended the previous models by incorporating endogenous determination of skill distribution in the model as well as the determination of the structure of the system of cities. Workers in this model are identical in terms of productivity. However, they are heterogeneous in their potential ability and are uniformly distributed on a unit interval. Potential ability can be materialized as productivity if workers acquire specialized training and produce X with the use of specialized technology. Workers in a given city can acquire specialized training by paying a training cost, where the average training cost in a given city is decreasing in city size. This represents one reason for the formation of a high-tech city. In other words, the (per capita) cost share represents the second agglomeration force, which also results in city formation. Workers utilizing the specialized training will concentrate in the city to get the advantage of lower per capita cost shares. Semi-skilled labor can also produce good X if they acquire basic training and use general technology. Furthermore, it is assumed that productivity is an increasing function in the level of basic training in a given city.

Proposition 16. *There exists a set of parameters such that some workers will acquire specialized training and some will acquire basic training.*

The above result represents one of the equilibrium that has been analyzed in the paper, which is the most observed outcome. Under this equilibrium, *the model will generate two types of cities: high-tech cities populated by skilled workers and low-tech cities populated with semi-skilled workers. The number of high-tech cities as well as the number of low-tech cities is determined by the parameters of the model.* Furthermore, the resulting city system could be a single large high-tech metropolis (core) and a system of small local cities (periphery) as in Abdel-Rahman and Wang (1995, 1997). However, unlike in Abdel-Rahman and Wang, the system of cities and the distribution of skill are outcomes of the model and not imposed exogenously.

15.8. Efficient system of cities

The concept that we will adopt in this section is the Pareto efficiency of resource allocation in a closed economy where the number of cities is endogenous. In the process of calculating the efficient allocation, we assume the economy is sufficiently large so that we have a large number of each type of city within the economy.[47] This is to be consistent with the equilibrium that was discussed in Section 15.4. Our objective is to address three fundamental questions concerning efficiency: (1) what is the source of inefficiency or market failure in the model? (2) Does the first best require federal or only local government intervention? (3) What are the instruments, a tax or a subsidy or both, that must be used to achieve efficiency?

Consider the case in which the only reason for city formation in the economy is the provision of a public good as in Section 15.4.1. Suppose that the economy is populated with $\bar{N}$ identical households, each endowed with one unit of labor. The objective of a central planning authority is to maximize a social welfare function defined as the aggregate utility of all households in the economy. This problem is equivalent to maximizing the utility of a representative household given that all cities are identical. This utility is maximized subject to two types of constraints: technology

[47] This assumption is imposed so that we can ignore the lumpiness problem (Henderson, 1985 Chapter 11, 240–242).

constraints, given in Section 15.4.1, and resource constraint, $N - kN^{3/2} = L_2 + L_1$. The problem of the central planning authority is obtained by substituting the constraints into the utility function. From this we have

$$\max_{L_1,Z,N} u = \eta^{\alpha_2} N^{-1}(L_1 - Z))^{\alpha_1}(N - 2\mu N^{3/2} - Z)Z^{\alpha} \tag{15.51}$$

The first-order conditions (after rearranging terms) are

$$\frac{\partial u}{\partial L_1} = -\frac{\alpha_2}{[N - 2\mu N^{1/2} - L_1]} + \frac{\alpha_1}{L_1 - Z} = 0 \tag{15.52}$$

$$\frac{\partial u}{\partial Z} = \frac{\alpha_2}{Z} - \frac{\alpha_1}{L_1 - Z} = 0 \tag{15.53}$$

$$\frac{\partial u}{\partial N} = -\frac{\alpha_2(1 - 3\mu N^{1/2})}{[N - 2\mu N^{1/2} - L_1]} + \frac{1}{N} = 0 \tag{15.54}$$

Condition (15.52) requires that the marginal cost of a worker to industry X_1 be equal to the marginal benefit. Condition (15.53) is the Samuelson condition for the optimal provision of a public good. Condition (15.54) requires the equality of the marginal benefit of a household to the city, i.e. the reduction in per capita cost of a public good, given by the first term, and the marginal cost, i.e. the rise in land rent, given by the second term. If we multiply through by N we get that the aggregate land rent is equal to the cost of the provision of a public good, which is the Henry George condition. Observe that the first best optimal result is consistent with the decentralized equilibrium presented in Section 15.4.1.

Proposition 17. *To achieve the social optimal city system, the developer of each city must tax the aggregate land rent to finance the provision of a public good.*

Now consider the case of a system of cities with externality in production. Suppose that all households in the economy have identical utility function given by Equation (15.1). Both X_i are produced with external economies of scale $X_i = L_i^{1+\varepsilon_i}$ $i = 1, 2$. Thus, this economy will result in a system of specialized cities. One type of city will specialize in the production of good X_1, while the other will specialize in the production of good X_2.

The objective of a central planning authority is to maximize the utility of a representative household given equal utilities of household living in city type 1 and 2. This utility is maximized subject to two technology constraints: the full employment constraint, $N_i - kN_i^{3/2} = L_i$, $i = 1, 2$, the population constraint for the economy, $\bar{N} = \sum_i M_i N_i$, where M is the number of cities, and the resource constraints, $M_i X_i = x_{i1} M_i N_1 + x_{i2} M_2 N_2$, $i = 1, 2$. Thus, the problem of the central planning authority is given by

$$\max_{x_{11}, x_{12}, x_{21}, x_{22}, N_1, N_2, M_1, M_2, u} \quad u \tag{15.55}$$

subject to

$$\lambda_i : u = x_{1i}^{\alpha_1} x_{2i}^{\alpha_2}, \qquad i = 1, 2$$

$$\gamma_i : M_i(N_i - 2\mu N_i^{3/2})^{1+\varepsilon_i} = x_{i1} M_i N_1 + x_{i2} M_2 N_2, \qquad i = 1, 2$$

The first order conditions (after rearranging terms) are

$$\frac{\partial L}{\partial x_{1i}} = \frac{\alpha_1 \lambda_i u}{x_{1i}} - \gamma_1 M_i N_i = 0, \qquad i = 1, 2 \tag{15.56}$$

$$\frac{\partial L}{\partial x_{2i}} = \frac{\alpha_2 \lambda_i u}{x_{2i}} - \gamma_2 M_i N_i = 0, \qquad i = 1, 2 \tag{15.57}$$

$$\frac{\partial L}{\partial N_i} = \gamma_i [(1+\varepsilon_i)(1 - 3\mu N_i^{1/2})(N_i - 2\mu N_i^{3/2})^{\varepsilon_i} - x_{1i} M_i] - \gamma_2 x_{21} M_i = 0, \quad i = 1, 2 \tag{15.58}$$

$$\frac{\partial L}{\partial M_i} = \gamma_i [(N_i - 2\mu N_i^{3/2})^{1+\varepsilon_i} - x_{1i} N_i] - \gamma_2 x_{21} N_i = 0, \quad i = 1, 2 \tag{15.59}$$

$$\frac{\partial L}{\partial u} = 1 - \lambda_1 - \lambda_2 = 0 \tag{15.60}$$

The interpretations of the Lagrange multipliers are: λ_i represents weights attached to different households to obtain equal utility; γ_i are shadow prices of goods 1 and 2. The first two conditions require equality between the marginal rate of substitution and the relative shadow prices. The third condition requires the equality of the marginal benefit of a household to the city, i.e. the external economy of scale, given by the first term, and the marginal cost, i.e. the rise in land rent, given by the second term. If we multiply through by N, we get that the aggregate land rent is equal to the gap between the social and the average marginal product, which is the Henry George condition. The fourth condition equates the cost of adding one city of type i, given by the consumption costs, to its benefit, the value of output produced in the city. Solving the first-order conditions for N and M, we have

$$N_i^{**} = \left[\left(\frac{\varepsilon_i}{1+3\varepsilon_i}\right)\mu^{-1}\right]^2, \quad i=1,2 \tag{15.61}$$

$$M_i^{**} = \alpha_i \bar{N}\left[\left(\frac{1+3\varepsilon_i}{\varepsilon_i}\right)\mu\right]^2, \quad i=1,2 \tag{15.62}$$

Proposition 18. *The first best optimal solution is consistent with the decentralized equilibrium presented in Section 15.4.2.*

So all that we need to achieve the social optimal city system is for the developer of each city to tax the aggregate land rent and use it to finance a lump-sum subsidy to workers in the city. The amount of this tax will be the difference between the private marginal product of the worker and the social marginal product. There are two observations about the first best optimal solution. The first is that the city size is only dependent on the parameters for commuting costs and the parameter for the scale economy. The second observation is that the demand parameters do not affect the city size, but they do affect the number of cities of each type in the system. Finally, an increase in the population of the economy will leave the relative number of cities unaffected. In other words, the city size distribution will remain the same.

Turning now to the social optimal system of cities with differentiated intermediate inputs. Suppose that households have the same utility function as in Section 15.4.3, but X_1 and X_2 are produced with two groups of differentiated inputs. The final goods are traded between cities with zero transportation costs, while the differentiated inputs are not traded. Furthermore, suppose that each of the differentiated inputs is produced by $L_{ij} = c_i q_{ij} + f_{ij}$; $i = 1, 2$; $j = 1..., n$. In a symmetric case, given equal utility of households in different type of cities, the central planning authority's problem is

$$\max_{x_{11}, x_{12}, x_{21}, x_{22}, N_1, N_2, n_1, n_2, u} u \tag{15.63}$$

subject to

$$\lambda_i : u = x_{1i}^{\alpha_1} x_{2i}^{\alpha_2}, \qquad i = 1, 2$$

$$\gamma_i : M_i n_i^{(1-\beta_i)/\rho_i} L_i^{\beta_i} q_i^{1-\beta_i} = x_{i1} M_i N_1 + x_{i2} M_2 N_2, \qquad i = 1, 2$$

$$\delta_i : n_i(f_i - c_i q_i) + L_i = N_i - 2\mu N_i^{3/2}, \qquad i = 1, 2$$

The first-order conditions after rearranging terms are

$$\frac{\partial L}{\partial x_{1i}} = \frac{\alpha_1 \lambda_i u}{x_{1i}} - \gamma_1 M_i N_i = 0, \qquad i = 1, 2 \tag{15.64}$$

$$\frac{\partial L}{\partial x_{2i}} = \frac{\alpha_2 \lambda_i u}{x_{2i}} - \gamma_2 M_i N_i = 0, \qquad i = 1, 2 \tag{15.65}$$

$$\frac{\partial L}{\partial q_i} = \gamma_i \left[\frac{(1 - \beta_i) X_i M_i}{q_i} \right] - \delta_i c_i n_i = 0, \qquad i = 1, 2 \tag{15.66}$$

$$\frac{\partial L}{\partial L_i} = \gamma_i \left[\frac{\beta_i X_i M_i}{L_i} \right] - \delta_i = 0, \qquad i = 1, 2 \tag{15.67}$$

$$\frac{\partial L}{\partial N_i} = \gamma_i [(1 - 3\mu N_i^{1/2}) - x_{1i} M_i] - \gamma_2 x_{21} M_i = 0, \qquad i = 1, 2 \tag{15.68}$$

$$\frac{\partial L}{\partial M_i} = \gamma_i [X_i - x_{1i} N_i] - \gamma_2 x_{21} N_i = 0, \qquad i = 1, 2 \tag{15.69}$$

$$\frac{\partial L}{\partial u} = 1 - \lambda_1 - \lambda_2 = 0 \tag{15.70}$$

The interpretations of the first and second conditions are as in the previous case. The third condition requires the equality of the marginal benefit of an additional differentiated input to the city, given by the second term, and the marginal cost, i.e. the rise in fixed cost, given by the first term. Solving the first order conditions for N, we have

$$N_i^{**} = \left[\frac{1}{f_i \mu^2} \left(\frac{(1-\rho_i)(1-\beta_i)}{[\rho_i + 3(1-\rho_i)(1-\beta_i)]} \right)^3 \right] \qquad i = 1,2 \quad (15.71)$$

Proposition 19. *The equilibrium city size, given in Section 15.4.3, is not first best optimum.*

First, observe as in the previous case that the city size is independent of the demand parameters. Second, the number of cities of each type is the same as in the previous model. Third, observe that the first best optimal result is not the same as the decentralized equilibrium presented in Section 15.4. However, if local government regulates the price of the intermediate input to be equal the marginal cost and offers lump-sum subsidies to the suppliers of the differentiated input, the equilibrium city size and the equilibrium number of firms will be a first best optimum. Furthermore, the subsidies can be financed by the aggregate land rent generated by the city. Thus, all that we need to achieve the social optimal city system is for the local government of each city to tax the aggregate land rent to finance the fixed cost in each city. A similar model that would lead to the same result is the model of differentiated products that are consumed by households (Dixit and Stiglitz, 1977).[48]

Now consider the case of national economy of scope arising from product diversity in the production of consumption goods. This argument supports the idea that large countries can support a greater range of differentiated product, which improves the utility of households in the nation. In this framework, national diversity involves diversity in the type of cities and in the traded good in which the cities specialize. Consider now the case in which the differentiated good is traded at zero transportation cost, and those households have the utility function $u = n^{1/\sigma} q$. In this framework, each city in the economy will specialize in the production of one q. Thus, the number of cities in the economy will be the same

[48] See Hobson (1987) and Abdel-Rahman (1988) for this result.

as the number of varieties. The planner problem is

$$\max_N u = \{c^{-1}\bar{N}^{1/\sigma}n^{1/\sigma}(N - 2\mu N^{3/2} - f)\} \tag{15.72}$$

The first-order condition for the above problem (after rearranging terms) is

$$\frac{\partial L}{\partial N} = (N - 3\mu N^{3/2}) - \frac{1}{\sigma}(N - 2\mu N^{3/2} - f) = 0 \tag{15.73}$$

From the above first-order condition, the city size is given by

$$\left(N - \left[\frac{2 - 3\sigma}{1 - \sigma}\right]\mu N^{3/2}\right) = f(1 - \sigma)^{-1} \tag{15.74}$$

The above city size equation will result in a larger size than equilibrium, which is given by $(N - 2\mu N^{3/2}) = f(1 - \sigma)^{-1}$. This equilibrium city size is determined by the full employment condition for a representative city.

Proposition 20. *Equilibrium will underprovide the number of variety, and therefore cities in the economy.*

Since this will be a system of company towns, in this case a local developer can correct this market failure by maximizing the profit from the production of the traded good and the tax collected from city residents. This must be maximized subject to full employment in the city, national demand for the traded good produced in the city, and a utility constant that indicate that the developer has to maintain the national utility level for city residents. Given this behavior of local developers the social optimal city size will be achieved. As can be seen in all of the above models, even though the equilibrium does not correspond to the social optimum, all that is required to correct this market failure is local government intervention, see Henderson and Abdel-Rahman (1991).[49]

This result of achieving first best with a decentralized equilibrium only holds if the utility function does not have an outside good. However, if there is an outside good, as in Equation (15.35), a social

[49] Also see Anas (2004) for this model with positive transportation costs.

optimum cannot be achieved through a decentralization mechanism. Thus, in this case, we need a central planner in order to reach a social optimum. This is also required in Anas and Xiong (2003), in which two homogenous goods as well as differentiated intermediate inputs are traded between cities at positive transportation cost. The reason for that is the city developer or local government cannot internalize the externalities associated with the formation of new cities in the economy. Thus, in these cases, federal government intervention is required to internalize these externalities.

15.9. Conclusion

In spite of the major developments in the theory of systems of cities over the past two decades or so, there are still major challenges ahead toward the formation of more complete theory. In most of the city systems that were discussed in this survey, the system of cities generated is either a specialized or diversified system. However, in most systems of cities that we observe, in developed or developing countries, specialized and diversified cities coexist. Thus, there is a need for more intensive research that will explain this phenomenon.

On the other hand, most observed city systems are characterized by a dominant city, which, in turn, is characterized by a diverse industrial structure and a diverse labor force. In other words, there is a need for more work on the core–periphery model that can explain the hierarchal structure of an urban system. Some attempts have begun in this direction, such as Abdel-Rahman and Wang (1995, 1997) and Abdel-Rahman (2002). However, there is a need for models that generate the core–periphery structure endogenously. Now with trade liberalization and free trade agreements the role of national governments in facilitating trade has been awakened, and the role of cities competing internationally has been expanding. Thus, we need models that can explain the impact of free trade agreements on the structure of the system of cities. In other words, we need models that can merge international trade theory with the theory of systems of cities. Finally, a technically challenging model is the one that will integrate spatial dimension to the NUE model of systems of cities. This will integrate the NEG model, in which the distance between cities is taken into consideration, and NUE model, in which the internal spatial structure of the city is taken into consideration. The simplest approach of integrating these models is by considering a linear city model with the explicit introduction of an agricultural sector.

References

Abdel-Rahman, H.M. (1988), "Product differentiation, monopolistic competition and city size", *Regional Science and Urban Economics*, Vol. 18, pp. 69–86.

Abdel-Rahman, H.M. (1990a), "Agglomeration economies, types, and sizes of cities", *Journal of Urban Economics*, Vol. 27, pp. 25–45.

Abdel-Rahman, H.M. (1990b), "Shareable inputs, product variety, and city sizes", *Journal of Regional Science*, Vol. 30, pp. 359–374.

Abdel-Rahman, H.M. (1994), "Economies of scope in intermediate goods and a system of cities", *Regional Science and Urban Economics*, Vol. 24, pp. 497–524.

Abdel-Rahman, H.M. (1996), "When do cities specialize in production?", *Regional Science and Urban Economics*, Vol. 26, pp. 1–22.

Abdel-Rahman, H.M. (1998), "Income disparity, time allocation and social welfare in a system of cities", *Journal of Regional Science*, Vol. 38, pp. 137–154.

Abdel-Rahman, H.M. (2000a), "Multi-firm city versus company town: a micro foundation model of localization economy", *Journal of Regional Science*, Vol. 40, pp. 755–769.

Abdel-Rahman, H.M. (2000b), "City system: general equilibrium approaches", in: J.-M. Huriot and J.-F. Thisse, editors, *Economics of Cities: Theoretical Perspectives*, Cambridge, MA: Cambridge University Press.

Abdel-Rahman, H.M. (2002), "Does the structure of an urban system affect income disparities?", *Journal of Regional Science*, Vol. 42, pp. 389–410.

Abdel-Rahman, H.M. and A. Anas (2004), "Theories of system of cities", in: J.V. Henderson and J.-F. Thisse, editors, *Handbook of Regional and Urban Economics*, vol. 4, Amsterdam: Elsevier.

Abdel-Rahman, H.M. and M. Fujita (1990), "Product variety, Marshallian externalities and city sizes", *Journal of Regional Science*, Vol. 30, pp. 165–183.

Abdel-Rahman, H.M. and M. Fujita (1993), "Specialization and diversification in a system of cities", *Journal of Urban Economics*, Vol. 33, pp. 189–222.

Abdel-Rahman, H.M. and P. Wang (1995), "Toward a general-equilibrium theory of a core-periphery system of cities", *Regional Science and Urban Economics*, Vol. 25, pp. 529–546.

Abdel-Rahman, H.M. and P. Wang (1997), "Social welfare and income inequality in a system of cities", *Journal of Urban Economics*, Vol. 41, pp. 462–483.

Alonso, W. (1964), *Location and Land Use*, Cambridge, MA: Harvard University Press.

Anas, A. (2004), "Vanishing cities: what does the New Economic Geography imply about the efficiency of urbanization?", *Journal of Economic Geography*, Vol. 4, pp. 181–199.

Anas, A. and K. Xiong (2003), "Intercity trade and the industrial diversification of cities", *Journal of Urban Economics*, Vol. 54, pp. 258–276.

Arnott, R.J. (1979), "Optimal city size in a spatial economy", *Journal of Urban Economics*, Vol. 6, pp. 65–89.

Arnott, R.J. and J.E. Stiglitz (1979), "Aggregate land rents, expenditure on local public goods and optimal city size", *Quarterly Journal of Economics*, Vol. 93, pp. 471–500.

Becker, G. and K. Murphy (1992), "The division of labor, coordination costs, and knowledge", *Quarterly Journal of Economics*, Vol. 107, pp. 1137–1160.

Becker, R. and J.V. Henderson (2000), "Intra-industry specialization and urban development", in: J.-M. Huriot and J.-F. Thisse, editors, *Economics of Cities: Theoretical Perspectives*, Cambridge, MA: Cambridge University Press.

Berliant, M. and P. Wang (2004), "Dynamic urban models: agglomeration and growth", in: R. Capello and P. Nijkamp, editors, *Urban Dynamics and Growth*, Amsterdam: Elsevier.

Chamberlin, E.H. (1933), *The Theory of Monopolistic Competition*, Cambridge, MA: Harvard University Press.

Chipman, J.S. (1970), "External economies of scale and competitive equilibrium", *Quarterly Journal of Economics*, Vol. 84, pp. 347–385.

Christaller, W. (1933), *Die Zentralen Orte in Süddeutschland*, Darmstadt: Wissenschaftlische Buchgesellschaft.

Diamond, P. (1982), "Aggregate demand management in search equilibrium", *Review of Economic Studies*, Vol. 49, pp. 217–227.

Dixit, A. (1973), "The optimum factory town", *Bell Journal of Economics and Management Science*, Vol. 4, pp. 637–654.

Dixit, A. and J.E. Stiglitz (1977), "Monopolistic competition and optimal product diversity", *American Economic Review*, Vol. 67, pp. 297–308.

Duranton, G. and D. Puga (2001), "Nursery cities: urban diversity, process innovation, and the life cycle of products", *American Economic Review*, Vol. 91, pp. 1454–1477.

Duranton, G. and D. Puga (2004), "Micro-foundation of urban agglomeration economies", in: J.V. Henderson and J.-F. Thisse, editors, *Handbook of Regional and Urban Economics*, Vol. 4, Amsterdam: Elsevier.

Ethier, W. (1982), "National and international returns to scale in the modern theory of international trade", *American Economic Review*, Vol. 72, pp. 389–405.

Flatters, F., J.V. Henderson and P. Mieszkowski (1974), "Public goods, efficiency and regional fiscal equalization", *Journal of Public Economics*, Vol. 3, pp. 99–112.

Fujita, M. (1989), *Urban Economic Theory: Land Use and City Size*, London: Cambridge University Press.

Fujita, M., P. Krugman and A.J. Venables (1999), *The Spatial Economy: Cities, Regions, and International Trade*, Cambridge, MA: MIT Press.

Gabaix, X. and Y.M. Ioannides (2004), "The evolution of city size distribution", in: J.V. Henderson and J.-F. Thisse, editors, *Handbook of Regional and Urban Economics*, vol. 4, Amsterdam: Elsevier.

Glaeser, E., H.D. Kallal, J.A. Scheinkman and A. Schleifer (1992), "Growth in cities", *Journal of Political Economy*, Vol. 100, pp. 1126–1152.

Goldstein, G.S. and T.J. Gronberg (1984), "Economies of scope and economies of agglomeration", *Journal of Urban Economics*, Vol. 16, pp. 63–84.

Hansen, N. (1990), "Do producer services induce regional economic development?", *Journal of Regional Science*, Vol. 30, pp. 465–476.

Helsley, R.W. and W.C. Strange (1990), "Matching and agglomeration economies in a system of cities", *Regional Science and Urban Economics*, Vol. 20, pp. 189–212.

Helsley, R.W. and W.C. Strange (1991), "Agglomeration economies and urban capital market", *Journal of Urban Economics*, Vol. 29, pp. 96–112.

Helsley, R.W. and W.C. Strange (1994), "City formation with commitment", *Regional Science and Urban Economics*, Vol. 24, pp. 373–390.

Henderson, J.V. (1974), "The sizes and types of cities", *American Economic Review*, Vol. 64, pp. 640–656.

Henderson, J.V. (1985), *Economic Theory and the Cities*, New York: Academic Press.

Henderson, J.V. and H.M. Abdel-Rahman (1991), "Urban diversity and fiscal decentralization", *Regional Science and Urban Economics*, Vol. 21, pp. 491–510.

Henderson, J.V. and J.F. Thisse (2001), "On strategic community development", *Journal of Political Economy*, Vol. 109, pp. 546–569.

Henderson, J.V., A. Kuncoro and M. Turner (1995), "Industrial development in cities", *Journal of Political Economy*, Vol. 103(5), pp. 1067–1090.

Hobson, P. (1987), "Optimal product variety in urban areas", *Journal of Urban Economics*, Vol. 22, pp. 190–197.

Hochman, O. (1997), "More on scale economies and cities", *Regional Science and Urban Economics*, Vol. 27, pp. 373–397.

Hoover, E.M. (1948), *Location Theory and the Shoe and Leather Industries*, Cambridge, MA: Harvard University Press.

Jacobs, J. (1969), *The Economy of Cities*, New York: Vintage.

Juhn, C., K. Murphy and B. Pierce (1993), "Wage inequality and rise in returns to skill", *Journal of Political Economy*, Vol. 101, pp. 410–442.

Kanemoto, Y. (1980), *Theories of Urban Externalities*, Amsterdam: North-Holland.

Kim, S. (1991), "Heterogeneity of labor markets and city size in an open spatial economy", *Regional Science and Urban Economics*, Vol. 21, pp. 109–126.

Krugman, P. (1991), "Increasing returns and economic geography", *Journal of Political Economy*, Vol. 99, pp. 483–499.

Losch, A. (1954), *Die Raumliche Ordnung der Wirtschaft*, New Haven, CT: Yale University Press, Translated as *The Economics of Location* by W.H. Woglom, and W.F. Stopler.

Lucas, R. (1988), "On the mechanics of economic development", *Journal of Monetary Economics*, Vol. 22, pp. 3–42.

Machin, S. (1996), "Wage inequality in the UK", *Oxford Review of Economic Policy*, Vol. 12, pp. 47–64.

Marshall, A. (1890), *Principles of Economics*, London: MacMillan.

Mills, E.S. (1967), "An aggregative model of resource allocation in a metropolitan area", *American Economic Review*, Vol. 61, pp. 197–210.

Muth, R. (1969), *Cities and Housing*, Chicago, IL: University of Chicago Press.

Ottaviano, G. and J.-F. Thisse (2004), "Agglomeration and economic geography", in: J.V. Henderson and J.-F. Thisse, editors, *Handbook of Regional and Urban Economics*, vol. 4, Amsterdam: Elsevier.

Panzar, J.C. and R.D. Willig (1981), "Economies of scope", *American Economic Association Papers and Proceedings*, Vol. 71, pp. 268–272.

Rivera-Batiz, F.L. (1988), "Increasing returns, monopolistic competition and agglomeration economies in consumption and production", *Regional Science and Urban Economics*, Vol. 18, pp. 25–153.

Romer, P. (1986), "Increasing returns and long run growth", *Journal of Political Economy*, Vol. 94, pp. 1002–1037.

Romer, P. (1987), "Growth based on increasing returns due to specialization", *American Economic Review*, Vol. 77, pp. 56–62.

Rosenthal, S.S. and W. Strange (2004), "Evidence on the nature and sources of agglomeration economies", in: J.V. Henderson and J.-F. Thisse, editors, *Handbook of Regional and Urban Economics*, vol. 4, Amsterdam: Elsevier.

Stahl, K. (1983), "A note on the microeconomics of migration", *Journal of Urban Economics*, Vol. 14, pp. 318–326.

Stiglitz, J.E. (1977), "The theory of local public goods", in: M.S. Feldstein and R.P. Inman, editors, *The Economics of Public Services*, London: MacMillan.

Tsukahara, K. (1995), "Independent and joint provision of optional public services", *Regional Science and Urban Economics*, Vol. 25, pp. 411–425.

United Nations (1996), *United Nations Human Development Report*, Washington, DC: United Nations Development Program.

Upton, C. (1981), "An equilibrium model of city sizes", *Journal of Urban Economics*, Vol. 10, pp. 15–36.

Weber, A. (1929), *Theory of Location of Industries*, Chicago, IL: University of Chicago Press.

Wilson, J.D. (1987), "Trade in a Tiebout equilibrium", *American Economic Review*, Vol. 77, pp. 431–441.

Urban Dynamics and Growth, CEA, vol. 266
Roberta Capello and Peter Nijkamp (editors)

CHAPTER 16

The City Network Paradigm: Theory and Empirical Evidence

Roberto Camagni and Roberta Capello
Politecnico di Milano, Milan, Italy

Abstract

Real city systems in advanced countries have significantly departed from the abstract Christallerian pattern of a nested hierarchy of centers and markets. The reduction in transport costs and the demand for 'variety' by the consumer have broken the theoretical hypothesis of separated, gravity-type, non-overlapping market areas. 'Location economies' as described by Hoover, and synergy elements operating through horizontal and vertical linkages among firms, have led to the emergence of specialized centers, in contrast to the typical despecialisation pattern deriving from the theoretical model. High-order functions sometimes locate in small (but specialized) centers where the model' expectations are only for lower-order functions. In recent years, network behaviour has extensively been analysed as the emerging model for economic growth. By network behaviour, a metaphor for cooperative behaviour among individuals, corporate or territorial partners is intended - which is increasingly becoming the reference paradigm in an era of continuing innovation and fast technological change, in the presence of "market failure" when dynamic and innovative behaviours are concerned and of the high costs of a growth strategy based on the sole internal know-how. The chapter provides insights into the economic rational of this behaviour, by approaching the issue treating cities and territories both as individuals and as collective actors. Moreover, the chapter provides empirical evidence on the existence of "city networks" and on the advantages cities gain from being part of a network.

Keywords: urban systems, urban networks, urban hierarchies

JEL classifications: D70, H77, R00

16.1. Introduction

During the last decade, networks and networking have become very fashionable concepts and terms in regional science, particularly in regional and urban geography (Castells, 1996; Malecki, 1997; Nijkamp, 2003). We speak about network firms, network society and network economy; in an explicitly spatial context, we speak about network cities, city networks, *reti urbane* and *réseaux de villes* to define new organisational forms of the urban structure and new tools for urban policies. For some people they are only catch-words, for others a true new scientific paradigm.

In the opinion of the authors, we are in fact confronted with a new paradigm in spatial sciences, providing that some precise conditions are met:

- its exact meaning is thoroughly defined,
- its theoretical economic rationale is justified,
- the novel features of its empirical content are clearly identified and distinguished from more traditional spatial facts and processes that can be easily interpreted through existing spatial paradigms.

In fact, if we examine the third condition, the concept of spatial networks is sometimes merely used as a substitute for 'interaction': an exchange of goods, services, information and contacts among places and nodes. In this case, the traditional paradigm (and related models) of spatial interaction can be easily utilised, unless one could demonstrate that the probability of such exchanges is mainly independent of distance and the size of nodes.

By the same token, the term 'network' is sometimes used to interpret relations and flows that take place within an urban hierarchy among centres of different hierarchical level. Also in this case, we do not need a new concept to identify well-known phenomena, unless interactions take place among centres belonging to the same hierarchical level – which are not supposed to maintain relationships with each other according to the standard central place model.

While interaction is supported by research work interested in the 'space of flows', other studies focusing on a description of the 'space of places' often use the network concept as a synonym (or an explanation) for polycentrism, a merely geographical and descriptive concept.

Finally, the same term is sometimes used to identify social linkages that characterise local communities or associations kept together by ethnic, linguistic, civic or even criminal goals. While in this case the micro-foundations of these phenomena are studied by other disciplines, such as sociology or political science, their spatial effects may well be interpreted through other existing concepts, like that of social capital (Coleman, 1990; Putnam, 1993; Bowles and Gintis, 2002; Durlauf, 2002a,b; Glaeser *et al.*, 2002) or local *milieu* (Camagni, 1991; Maillat *et al.*, 1993; Aydalot, 1986).

The issue is mainly terminological, but nevertheless central to our discourse: we prefer to use the term 'network' for selective and formalised linkages among well-defined economic actors and spatial units, trascending proximity relationships and taking place at a trans-territorial level. On the other hand, we use the term *milieu* or social capital to encompass informal and 'atmosphere' relationships, taking place inside local territories thanks to cultural proximity and social cohesion.

Once agreed on what precedes, we have to admit that there are spatial phenomena that cannot be interpreted through the usual tools – spatial interaction, urban hierarchy, social capital – for which the new concept, namely city networks, could be of use. We refer here to spatial interaction taking place for selected and targeted goals, irrespective of distance; relationships between centres of the same size and hierarchical level, performing the same tasks and functions on the territory; linkages among local actors giving rise to a network surplus as a consequence of synergies and cooperation;[1] to spontaneous or organised division of labour among centres in a regional context.

In a scientific context where the economy is increasingly seen as a system or web of links between individuals, firms and institutions, with links depending on experience and evolving through learning processes (Malecki and Oinas, 1999), the relevant theoretical building blocks on which the network concept or paradigm may be constructed are the following:

- recognition of cooperation as a new organisational and behavioural form for companies, intermediate between hierarchy

---

[1] For a review on the concept of 'network surplus' see Capello (1994, Chapter 2).

(internal development and merging of external activities through direct control) and resort to the market, following the well-known works of Williamson (1985) and Coase (1988). Cooperation networks among firms collaborating with each other on technological advances and innovation projects were earlier phenomena that were abundantly explored in the past;

- recognition of cities as economic actors, competing between each other and developing their own development strategy. This statement finds its economic rationale in various recent theoretical reflections, referring first of all to the possibility that individual economic actors inside the city could find a common interest in cooperation and collective action; a similar interest, common to the entire local community, might be defined through new urban governance tools and interpreted by the urban administration vis-à-vis other external interests, namely those of foreign multi-national and multi-localised firms. Secondly, cities and territories compete on the basis of an 'absolute' or 'competitive' advantage, and not on the basis of the well-known Ricardian principle of 'comparative advantage'. This fact requires them to pursue an intentional development strategy if the above-mentioned common interest is expressed in terms of the majority of local people wanting to remain in their traditional birthplace (Camagni, 2002).

As a consequence, it is possible to say that networking – intended in a micro-economic sense as cooperation among individuals, firms and institutions concerning collective action, public/private partnerships and the supply of public goods – may become a scientific paradigm for interpreting the macroscopic spatial behaviour of collective agents like cities, competing and cooperating in the global arena where locations of internationally mobile factors (professionals, corporations, institutions) are negotiated and large territorial projects are decided.

The paradigm of city networks, complementary to the traditional one of urban hierarchy and initially proposed by the Southern European tradition of spatial analysis (Dematteis, 1985, 1990; Camagni, 1993a), has gained interest and support in other scientific and policy contexts. Recently it was supported by the EU spatial strategy document, the European Spatial Development Perspective (ESDP) (particularly in its Glasgow and Potsdam drafts, 1997 and

1998, but also in its final draft approved in Potsdam by the Ministers responsible for spatial planning from European member states) (European Union, 1999) in the specification given by one of the present authors, namely that of 'complementarity' and 'synergy' in city networks (Camagni, 1993a).

The chapter is organised as follows. Section 16.2 is devoted to the micro-economic analysis of firm cooperation networks, in terms of institutional economics and game theory. Section 16.3 deals with the interpretation of cities as economic actors, concerning the way cities engage in inter-regional trade and competition and in the consequent necessity for them to define shared development strategies. Section 16.4 interprets competition and cooperation among cities ('city networks') in theoretical terms, while Sections 16.5 and 16.6 are devoted to two empirical exercises: the first concerning the very existence of city networks as selective interaction among pairs of centres, more intense than predicted by a pure gravity model; the second defining the advantages of networking for cities engaged in a supra-national cooperation network, the Healthy Cities sponsored by the World Health Organisation of the United Nations.

16.2. Cooperation networks among firms: the emerging economic paradigm

In searching for the rationale of the new spatial paradigm, some recent theoretical reflections on firm behaviour may be used and analogies with the approaches used in other disciplines or branches of economic theory explored. In particular the concept of 'firm networks', utilised in the theory of the firm to encompass all those new organisational and contractual forms that imply 'cooperation' among firms – strategic alliances, technological and commercial cooperation, joint ventures, consortia, and so on – looks crucial in this respect (OECD, 1986; Foresti, 1986; Johannisson, 1987; Camagni and Gambarotto, 1988; Maillat *et al.*, 1993). In fact, it may help us to understand not only the economic, but also the spatial consequences of those firm behaviours which are intermediate between competition and internal development, between 'market' and 'hierarchy' in the terminology of the institutional and transaction-cost approach to the firm (Williamson, 1985).

Cooperation – technological, commercial and financial – appears as a new economic paradigm in an era of continuing innovation and rapid technological change, where 'market failure' occurs as far as dynamic and innovative behaviour is concerned. Here the market does not deliver correct and timely signals and a growth strategy solely based on internal know-how has high costs (Camagni, 1991). The objectives of the new behavioural model may be summarised in the following:

- reaching sufficient *scale economies*, through the merging of R&D facilities, production or marketing structures,
- controlling the market of *complementary assets*, necessary for assuring fast reaction capability, and
- controlling the *development trajectories* of crucial complementary assets, in order to assure continuous innovation capability.

This new behavioural form has two major advantages. It avoids the high transaction costs which are inevitable when crucial inputs are demanded through the market and it reduces the high costs involved in adopting a strategy of internal development of a new technology or competence.

The new cooperation strategy is typical of firms operating in high-tech sectors, but more traditional sectors are also increasingly adopting the same strategy in their search for renewal and restructuring processes. This strategy involves a different attitude to spatial relationships: it requires not just the simple control of product markets or input markets, but also direct linkages with other innovative *milieux* where specific know-how or technology is developed, or with firms which were previously either competitors or simple providers of production inputs.

Cooperation also has some definite costs, however, which should be taken into account by the firm, and by theories attempting to interpret the new behavioural pattern. Some of these costs, or risks, refer to the relationships among partners – possible opportunistic behaviour and coordination costs, asymmetry in information and bargaining power, which affect the allocation of advantages among partners (Camagni, 1993b); others refer to the limited appropriability of the outcome of the cooperative game and the possibility of free riding, all elements that impinge on the probability of a cooperative game being initiated.

In general, the choice whether to engage in a cooperative or collective action will depend on whether the benefits of cooperative action (B_c), reduced by regulation and coordination costs (C_r, C_c) exceed the benefits of individual action (B_i) reduced by transaction costs (C_t):

$$B_c - (C_r + C_c) > (B_i - C_t)$$

Regulation costs depend on the possibility of free riding, the principal barrier to cooperative behaviour, and the consequent existence of incomplete markets: some goods will never be produced, in spite of there being a demand, because it is impossible to exclude people from their free use ('public goods'), and this will lower collective welfare. Rational individual behaviour will involve a defection from cooperative behaviour, but if generalised, results in an irrational collective outcome, namely a lower utility even at the individual level. This is a well-known case of market failure and social dilemma.

Game theory helps us to determine the conditions for increasing the probability of cooperative behaviour. Repeated games supported with proper strategies by the players – such as tit-for-tat with a cooperative opening, or a 'nice' and more generous strategy, definitely superior when there is imperfect knowledge about the choices of partners – present higher probability of a cooperative outcome with respect to one-shot games; while external constraints to opportunistic behaviour, or disincentives to defection, may modify the structure of the payoff matrix with similar effects (Arrighetti, 2003).

Interestingly enough, particular elements of the territorial context, like the presence of reciprocal trust, reputation and social or 'relational' capital reduce the regulation costs and the risks of defection; the same elements at the same time reduce the temptation to defect, as they provide a signal of huge and certain social sanctions to opportunistic behaviour.

The second type of cooperation cost regards coordination costs that arise due to the fact that the outcome (*payoff*) of the cooperative action is uncertain, depending widely on the characteristics and the content of the action itself. Only in textbook examples may they be assumed to be perfectly known beforehand. The costs refer to the elaboration of a common scheme/strategy, in the presence of information asymmetries, and to uncertainty about the cooperative attitudes of the different partners.

The same territorial elements that were indicated above, and in particular the presence of relational capital, reduce coordination costs and may be seen as favourable preconditions for the building of collective actions.

16.3. Cities as collective actors

The second theoretical building block we referred to in the introduction is the hypothesis that cities and territories behave like collective actors. This hypothesis is based on the following reflections.

First of all, it is possible to affirm that regions, local territories and cities, due to their intrinsic openness both to the movement of goods and movement of factors, operate in the context of inter-regional trade within a regime of 'absolute advantage' and not within a regime of 'comparative advantage' (Camagni, 2002).[2] As a consequence, if their absolute competitiveness is inadequate or declining with respect to the other regions, the spontaneous adjustment mechanisms which in the latter regime and in the case of nations always assure them a role in the international division of labour – namely devaluation of the currency and wage-price flexibility – either do not exist or are inadequate to re-establish equilibrium. Conditions of weakness, due to inadequacies in production factors, adverse geographic circumstances or poor accessibility, may well result in mass unemployment and, if public transfers of income are not sufficient, emigration and possible abandonment.

Secondly, it is widely accepted that local firms, in their search for competitiveness, rely not only on public goods, human capital and social overhead capital, but increasingly on selected external assets and 'specific resources' that cannot be easily obtained via spontaneous market developments. Therefore, firms are

---

[2] This is contrary to conventional wisdom. As Armstrong and Taylor affirm: "That trade is based on comparative advantage and not absolute advantage is universally accepted and rarely tested" (Armstrong and Taylor, 2000, p. 123). In our opinion, this statement, when referred to regions, should not be accepted.

increasingly engaged in a cooperative process with other local firms, (collective) actors and the public administration for the conception and provision of these resources (Colletis and Pecqueur, 1995; Cooke and Morgan, 1998).

This cooperation among firms and social actors is facilitated by particular territorial conditions, determined by a particular richness of inter-firm interactions or 'untraded interdependencies' (to use Michael Storper's expression), (Storper, 1995). These conditions generate cumulative learning processes, enhancing the innovativeness and the competitiveness of the local territorial system.

A good way of depicting this process is through the concept of an innovative *milieu* – consisting of shared values, common representations and codes, a strong sense of belonging, trust, common professional background and economic specialisation.[3] In a turbulent environment characterised by difficulty in information collection, processing and assessment, strong interdependence between the decisions of different actors and great complexity in the external environment, economic actors find in the local *milieu* the necessary support for coping with uncertainty (Aydalot, 1986; Camagni, 1991).

The third reflection concerns the fact that firms increasingly use locations as competitive tools, exploiting their global mobility in order to optimise production and distribution costs. Location territories, on the other hand, are not just the passive objects of location decisions by firms, but communities made up of economic subjects which act in their own interest by trying to keep or attract firms. Local workers, subcontracting firms, suppliers of intermediate inputs, services and factors, are all agents which can achieve their goal not just by competing on prices and wages with other communities (sites), but also by upgrading the quality of their service through direct or indirect tools which involve the community and the local public administration. Locations are in a sense bought and sold on a global market, where demand and supply confront each other.

---

[3] The 'innovative *milieu*' is defined as the set of relations uniting a local production system, a set of actors and their representations, and an industrial culture, which together generate a localised dynamic process of collective learning (Aydalot, 1986; Camagni, 1991). See also the chapter by Camagni in this volume.

As a consequence of the above, local territories, cities and *milieux* compete (and, as we shall see further on, cooperate) with each other, building their own 'competitive' or 'absolute' advantage. In fact, beyond the specific competitive advantages strategically *created* by single firms, we place an increasing and crucial importance on territorial synergies, on cooperation capabilities enhanced by an imaginative and pro-active public administration, on externalities provided by local and national governments, on the specificities historically built by a territorial culture.[4] As is clear, they are all artificial or created advantages, open to the proactive, voluntary action of local communities and their governments.

But if individual firms and individual actors undertake collective activities, facilitated by (and creators of) trust and local social capital; and if significant cognitive synergies, readily apparent in the local *milieu*, result from their various interactions; and finally if these actions and these processes draw additional vitality from cooperation with local public administrations, especially in the creation of attractive territorial projects and innovative schemes; then it appears justifiable to go beyond methodological individualism – which regards only single firms and individuals as economic actors – and to argue in favour of the logical validity of a 'collective' concept such as that of *territory* or city, and to affirm that territories and cities compete among themselves, using the creation of collective strategies as their instrument.

16.4. The structure of the urban system: from city hierarchy to city networks

16.4.1. The need for a new paradigm

According to the textbooks of theoretical geography and urban economics, the analytical model which still best describes the structure of the city system in strictly economic and locational terms is Christaller's and Lösch's central-place model developed in the 1930s and 1940s. After the basic refinements introduced by Isard, Beckmann and McPherson, a huge literature has grown upon

---

[4] As Porter puts it: "There is growing recognition that company success also has much to do with things that are outside the company", such as "supplier relationships and the benefits of partnering" (Porter, 2001, p. 140; Scott, 2001).

the same logical foundations and simplifying assumptions with the works of Parr, Beguin, Mulligan and others, but the basic economic characteristics of the initial model have not changed: it still remains the most elegant, abstract but consistent representation of the hierarchy of urban centres.[5]

Nevertheless, real city systems in advanced countries have significantly departed from the abstract Christallerian pattern of a nested hierarchy of centres and markets. The reduction in transport costs and the demand for 'variety' by the consumer have broken the theoretical hypothesis of separated, gravity-type, non-overlapping market areas. 'Location economies' as described by Hoover, and synergy elements operating through horizontal and vertical linkages among firms, have led to the emergence of specialised centres, in contrast to the typical despecialisation pattern deriving from the theoretical model. High-order functions sometimes locate in small (but specialised) centres where the model's expectations are only for lower order functions.

This evidence is not at all new, and the deficiencies of the model are often highlighted, but to change the underlying assumptions would mean changing the model itself, and no other set of clearly defined hypotheses has ever replaced the original ones.

Yet, as mentioned, other evidence is at variance with the logic of the model. Urban policies are increasingly addressed towards economic goals: towards enhancing the efficiency of the local production fabric, attracting new sectors and functions, expanding the markets of local firms through better external transport and communication linkages. According to the logic of the model, these kinds of goals lack any economic rationale: the location of sectors and the roles of single centres are defined on the sole basis of city size.

Our hypothesis, or theoretical conjecture, is that a new form of interurban interaction and consequently a new organisational structure of the entire urban system may be conceived, based on the new paradigm of firm behaviour illustrated above: something that we can call *city networks*. The theorisation that follows builds mainly on the contribution of one of the present authors (Camagni, 1993a).

---

[5] See, among others, Isard (1956), Beckmann and McPherson (1970), Mulligan (1979, 1984), Parr (1978, 1981) and Beguin (1984, 1988).

16.4.2. The three logics of spatial behaviour of the firm

From a theoretical and abstract point of view, it is possible to identify three logics of spatial behaviour of the firm: we may call them the territorial, the competitive and the networking logic (Table 16.1).

According to the first logic, the territorial one, a firm sells (and buys) from the geographical space it controls gravitationally. Space is, therefore, organised into the well-known Löschian honeycomb of market areas, where the friction of space, embodied in transport costs, at the same time differentiates the products of competing firms and represents the strongest entry barrier into the market. The crucial function of the firm is production and its strategy consists in the control of the market area defined around its geographical location.

According to the second logic, the competitive one, the market of a firm is not restricted to the local territory, as transport costs do not play a relevant role; the firm may sell anywhere, trying to control the widest *share* of the global market. Competitiveness, differently achieved and interpreted by the different firms, becomes the crucial element in the economic arena, and marketing the crucial function of the firm; the market of each production unit is limited by both its relative economic strength and by consumers' demand for 'variety'. 'Two-way' trade, or the geographical interchange of the same products in two directions, becomes the rule as, for example, with Turin people no longer being obliged to buy only Fiat cars.

In its search for effectiveness and economies of scale, the firm is increasingly organised into specialised units, performing only one of the functions of the production cycle: manufacturing, R&D, marketing or general management. This specialisation pattern, which takes advantage of both the scale and location economies (as each functional unit may be localised in the most appropriate spot, given the characteristics of its production inputs) replaces the integrated organisational model of the previous case.

Space and spatial lack of homogeneity are no longer a simple constraint to the output market, but are directly exploited by the firm in a global optimisation process which not only takes into consideration accessibility to geographical markets, but also accessibility to labour, skills and other production inputs. The location of the firm is, therefore, determined by geographical and historical specificities, and no longer by a single logic, as happens

Table 16.1. The three logics of spatial organisation

Levels and aspects	Organisational Logics		
	Territorial	Competitive	Network
Firm			
Nature	Local market firm	Export firm	Network firm
Crucial function	Production	Marketing	Innovation
Strategy	Control of market areas	Control of market shares	Control of innovation assets and their trajectories
Internal structure	Single unit	Specialised functional units	Punctually integrated units
Entry barriers	Spatial friction	Competitiveness	Continuing innovation
City systems			
Principles	Domination	Competitiveness	Cooperation
Structure	Nested Christallerian hierarchy	Specialisation	City networks
Sectors	Agriculture, government, traditional tertiary activities	Industry: industrial districts and filière of specialisation	Advanced tertiary activities
Efficiency	Scale economies	Vertical/horizontal integration	Network externalities
Policy strategy	None: size determines function	Traditionally: none, as export base determines growth. Nowadays: strengthening of competitive advantage of each centre	Intercity cooperation; intercity transport and communication network provision
Intercity cooperation goals	None (except military or diplomatic goals)	Intercity division of labour	Economic, technological and infrastructure collaboration
Networks of cities	Hierarchical, vertical networks	Complementary networks	Synergy networks, Innovation networks
Single city			
Nature	Traditional city	Fordist city	Information city
Form	Relative internal homogeneity	Monofunctional zoning	Multifunctional zoning, polycentric city
Policy goals	Power and image	Internal efficiency (clockwork city)	External effectiveness and attractiveness
Symbols	Palace, cathedral, market	Chimney, skyscraper	Airport, trade fair

in simplified general equilibrium models. With the latter, location becomes completely random. Only in the case where a production input may be realistically assumed as perfectly immobile (though scattered in geographical space) and accessibility to it as costly, a regular pattern of locations may be rebuilt on the basis of spatial *input* markets (Parr, 1989).

According to the third logic, the network logic, innovation becomes the crucial function of the firm and the control of innovation assets and their time trajectories its main goal. The firm, wherever located, may overcome crucial know-how weaknesses in its internal structure and surrounding 'milieu' by linking-up with other firms and by establishing cooperation agreements.

These trans-territorial linkages would appear to eliminate the spatial or geographical dimension; but in fact they do not:

- emphasise the need for the firm to be present in the information and communication nodes of the worldwide technological, commercial and financial networks (the 'world cities'), and
- show the crucial need for the firm to present itself as an efficient partner, this attribute being reached either through a strong internal culture or through its location in a 'district', rich in Marshallian 'industrial atmosphere' (Camagni, 1991).

As a consequence of the new organisational logic, the geography of locations shows a centripetal bias, created both by the demand for accessibility to the nodes of the international information network and by the search for new synergies within the firm. In this second respect, the pattern of dispersed, monofunctional and specialised units is replaced by a pattern of functional reintegration in centrally located 'mission units', where the maximum degree of innovativeness may be achieved through the physical proximity of engineering, production, marketing and research functions (Camagni, 1988).

16.4.3. The structure of the urban system

How is it possible to pass from the locational logic of the single firm to the general spatial allocation of activities and functions?

It is well known that in what we called the 'territorial logic', agglomeration economies may explain the coexistence of lower order functions in centres where higher order functions are already located, and that gravity-type considerations may attract different firms towards the centre of their market areas, where demand density is higher.

In the 'competitive' logic, on the other hand, agglomeration may derive from supply rather than demand considerations: the agglomeration of firms belonging to the same sectors ('district economies') or the same industrial complex (control of component suppliers, 'filières' of local specialisation) allows higher levels of static and dynamic efficiency to be achieved, giving rise to specialised industrial areas and 'innovative milieux' (Aydalot, 1986), made up of vertically or horizontally integrated firms (the long-standing concept of 'localisation economies').

The third logic is more complicated. In spatial terms it involves the presence of:

- nodes of localised and specialised know-how (in 'poles', 'districts', 'parks', 'valleys', 'corridors', etc.) interlinked through cooperation agreements and financial/technological/marketing alliances, or
- multi-functional nodes interconnecting different economic and spatial networks. In this respect, the old concept of 'urbanisation economies' is revitalised here in terms of interaction and synergy of network functions: the city gains a role as a node of interchange and interconnection among a set of worldwide networks of physical and information interactions.

It is widely known that scale economies and generic agglomeration economies are the main efficiency elements that shape the spatial structure of location centres under the first logic. On the other hand, economies of vertical and horizontal integration are the main efficiency elements in the second logic, and 'network externalities' in the third one. In this last respect, the network operates as a 'club good' delivering advantages only to the members of the club, an intermediate structure between private and public goods (Capello, 1994).

The three logics of spatial organisation presented here are, of course, to be considered as theoretical archetypes, and not directly as historical behavioural patterns. In some respects, they have always coexisted, as they apply specifically to different sectoral specificities (to the primary, secondary and tertiary or information sectors, respectively). Nevertheless, as these sectors or functions have prevailed in different and successive periods in recent history, the three logics may be assumed, albeit very cautiously, as the leading paradigms of different 'accumulation régimes'.

16.4.4. The city network paradigm

The territorial logic is the basic theoretical underpinning of the Christallerian hierarchy of centres. This logic applies well, even if in abstract and simplified terms, to the spatial behaviour of the following activities:

- agricultural production and markets (except for 'industrialised' agriculture producing mass 'commodities' and 'specialised' agriculture producing diversified products, like special wines, etc.),
- public administration and government functions, and
- private and public service activities; in particular, 'traditional' ones (retail and wholesale trade, health and education, etc.) but also modern ones (private consultants, banking and insurance, advertising, etc.) and, in general, activities where the customer bears the transport cost.

Therefore, the Christaller model applies well to those societies where these sectors account for the overwhelming share of economic activity. However, the model presents many drawbacks which significantly limit its empirical relevance in modem societies. More specifically:

(i) it overemphasises the role of transport costs, a fact that reduces its usefulness for the interpretation of industrial location and markets;

(ii) it neglects input–output relationships, and in particular *horizontal linkages* among specialised firms and, in spatial terms, horizontal linkages among specialised centres of *similar size*, performing different but complementary functions (in the model, only *vertical*, hierarchical linkages among centres of different size and rank are considered);

(iii) it neglects 'network externalities', or the 'synergetic surplus' that may come to the partners (firms or cities) of a cooperation network. These externalities may be utilised to explore the concept of 'city networks', as will be explained below.

These are important limitations to the theoretical assumptions of the model, and in fact empirical observations provide conflicting evidence regarding its outcomes. In particular, we may observe:

(i) processes of city specialisation, especially in industry but also in services, which contradict the prediction of Christaller's model about the hierarchical despecialisation of each centre (Cappellin and Grillenzoni, 1983);
(ii) an incomplete presence of the whole range of functions in each city (all the bundles of goods and services of equal or inferior rank) (Emanuel and Dematteis, 1990);
(iii) the presence of high-order functions in centres of lower order (Dematteis, 1985, 1990); and
(iv) horizontal linkages between similar functions (and cities), e.g. the financial network among top cities in the worldwide hierarchy.

Under these circumstances, our hypothesis is that a new paradigm of spatial organisation should be considered, the *network paradigm*, which links with the new logics of spatial behaviour we have labelled as the 'competitive' and the cooperative, 'network' logic.

As far as the 'competitive' logic is concerned, it lies at the basis of the well-known phenomenon of industrial districts, specialised by sectors or by 'filière', and, as a result, a host of territorial relationships among centres based on privileged complementarity relations in both production and marketing. These relationships mainly occur at the intra-regional level, as is the case, for example, with the specialised centres of the car industry filière (Turin area, Toyota City) or of the textile-fashion creation filière in the Lombardy region, with a spatial division of labour between headquarter, fair and design functions, mainly located in Milan and Como (for silk), and manufacturing and equipment-producing functions, located in suburban centres.

The third logic, the 'network' logic, in turn determines a set of privileged synergetic relationships between centres that cooperate or interact in the same fields or functions, through information, communication or transport networks. In parallel to a previous statement concerning network relationships between firms, a city network may be considered as a 'club good', providing externalities to the partners which cooperate on the basis of horizontal linkages and perform the same functions. Also in this case, networks might be seen as a way of generating (urban) scale economies in a cooperative way, without implying a growth of the single centres, and of distributing the consequent advantage among the partners.

Therefore, in the organisation of the city-system two kinds of city networks appear:

(A) *Complementarity networks*, made up of specialised and complementary centres, interlinked through a set of input–output and market relationships. Interurban division of labour at the same time ensures that there is a sufficiently large market area for each centre and that scale and agglomeration economies are achieved. Good examples of these networks are provided by the specialised cities in Randstad Holland or in the Veneto region in Italy;

(B) *Synergy networks*, made up of similar, cooperating centres. In this case the necessary economies of scale are provided by the network itself, which integrates the market of each single centre. Examples of these networks are the already mentioned financial cities, whose markets are virtually integrated through advanced telecommunication infrastructures, or tourist cities connected through cultural or historical 'itineraries'.

A third category, or better a sub-category of the second one, might also be found, namely:

(C) *Innovation networks*, made up of centres cooperating on *specific* projects in order to reach a sufficient critical mass, both with respect to demand or to supply considerations. Examples of these networks are the recent cooperation agreements between cities in France, Germany, Belgium, and Holland in the fields of infrastructure provision (airports), technological services, etc.

It might be important to note that these three types of city networks refer to the three main goals (and categories) of the new network behaviour of firms which we have mentioned before: provision of complementary assets, scale economies through cooperation and innovation, respectively.

These considerations may be synthesised in the following definition: *city networks are systems of relationships and flows, of a mainly horizontal and non-hierarchical nature, among complementary or similar centres, providing externalities or economies, respectively, of specialisation/complementarity/spatial division of labour and of synergy/cooperation/innovation.*

Immaterial relationships and physical flows of people and goods are highly complementary, as widely known. In similar ways, city networks rely on both physical integration – through transport and communication networks – and virtual, economic integration – information exchange, financial transactions, cooperation in multiple fields, including provision of high-order public services in culture and education. Thanks to the existence of cooperation networks among centres, high-order functions may be supplied by medium-sized centres, provided that the entire regional market is assured by a spontaneous or agreed division of labour among similar centres.[6]

This is mainly a deductive 'conjecture', in search of a corroboration of the underlying theory through proper empirical validation. Many aspects still require further, detailed examination, such as the economic effectiveness and the laws of motion of the new organisational logic and the way in which the new hypothesised network linkages may be observed and measured.[7]

The main difficulty in this field is that the nature of the problem requires 'flow indicators' between centres, while at this detailed territorial level mainly 'stock indicators' exist.

16.5. Do city networks really exist? An econometric experiment

Some years ago, an empirical experiment was run in order to detect and identify possible 'city networks' in the real world of spatial

---

[6] On this idea, the French and German strategies for *réseaux de villes* or *städtenetze* are built.

[7] In this last respect, the Dematteis geographical school in Italy has attempted, for almost a decade, to empirically reveal the network linkages among lower rank centres in the Po valley (from provincial capitals downwards). The linkages inspected refer to our first category – complementarity linkages between specialised centres – but the results are not yet conclusive, in our opinion. After appropriately measuring the shifts between the actual and the theoretical sectoral mix in each centre, the existence of direct complementarity relationships is mainly inferred deductively in the case of couples of neighbouring centres of similar size presenting, respectively, a very high and a very low employment share in some sectors (Emanuel and Dematteis, 1990).

interactions, starting from the definition given above (Camagni *et al.*, 1994). Flow data were measured by using telephone calls between 'telephone districts' in Northern Italy in 1990 (before the advent of cellular phones).[8]

Reliable data were available and they referred to a totally ubiquitous network, eliminating potentially misleading supply effects linked to the territorial architecture of physical networks – that exist for example using transportation data. A 'network' relationship between couples of centres was hypothesised when actual communication flows significantly exceeded the interaction expected on the basis of a traditional doubly constrained entropy model. By using the latter, it is in fact possible to capture all relevant but generic interaction that occurs within a regional or national urban system as a consequence of size and distance, while we are interested in capturing the extra interaction occurring as a consequence of special and selective economic relationships between the centres.

A rectangular matrix (36 × 119) was used for flows between the Lombardy telephone districts and the districts of the entire Northern Italian macro-region. The model was calibrated at different spatial scales, as it is far from clear to what extent the new principle works and changes as territorial scale is gradually reduced. Interaction was therefore studied separately, and the model was calibrated for different types of interactions, in order to take into account possible non-linearities in the parameters:

- between the Lombardy districts and the North-Italian districts (36 × 119),
- among districts inside the Lombardy region (36 × 36),
- among sub-districts (called telephone 'sectors') in the Lombardy region (178 × 178),
- in all cases, including and excluding the capital city of Milan.[9]

---

[8] Data refer to morning traffic (in order to capture mainly business interaction) and its intensity is measured in Erlang, during one year, 1990.

[9] Model fitting improves substantially when Milan is considered: R-square increases from 0.45 to 0.88 in the first case, encompassing the whole North Italian territory, and from 0.46 to 0.73 in the second case (only Lombardy region). The spatial impedence factor β also increases passing from 0.89 to 1.61 in the first case and from 0.75 to 1.35 in the second case. This seems to point out that a host of weak interactions takes place at the local level, obscured by the gravitational power of Milan, and following a logic which is only partially gravitational in nature.

The results of the empirical exercise were encouraging and in many respects counter-intuitive and different from expectations (Camagni *et al.*, 1994). In particular they showed that city networks, as we define them, do in fact exist, and that:

- they are not ubiquitous but very selective in space,
- they do not substitute more traditional, hierarchical forms of spatial organisation; rather, the two organisational forms of the city system appear to be complementary,
- a super-gravitational effect of the capital city, Milan, is apparent on the major Northern Italian cities (Figure 16.1: see the white empty bars, indicating flows underestimated by the model, or a wider real interaction), indicating a strong hierarchical structure;
- a similar super-gravitational effect of Milan is exerted on the major cities in Lombardy (Figure 16.2),
- in the context of the Lombardy region, city networks show up as statistically significant in three spatial conditions, characterised by a high synergetic potential: (a) within the metropolitan area of Milan, linking its major sub-centres; (b) within some industrial districts featuring distinct sectoral specialisation and economic

Figure 16.1. The super-gravitational effect of Milan in Northern Italy

Source: Camagni *et al.*, 1994.

Figure 16.2. The super-gravitational effect of Milan in the Lombardy region

Note: Super gravitation is expressed by white empty bars, indicating an underestimate of real interaction by the model.
Source: Camagni *et al.*, 1994.

'vocation' – Como (silk production), Lecco (mechanical engineering), Seregno (furniture and furniture design), Busto-Gallarate-Legnano (textiles and light mechanical engineering), Vimercate (high-tech industries) (Figure 16.3); and (c) in the eastern part of the region, linking major provincial capital cities (Bergamo, Brescia, Mantova, Cremona, and Verona) (Figure 16.4). These latter centres are in a sense structuring a strong alternative to the dependency to Milan via a city network, with a prominent role of the largest ones, namely Brescia;

– however, city networks did not show up over longer distances, for example, along the main Po valley central axis, where they were

Figure 16.3. City networks in the Milan metropolitan area

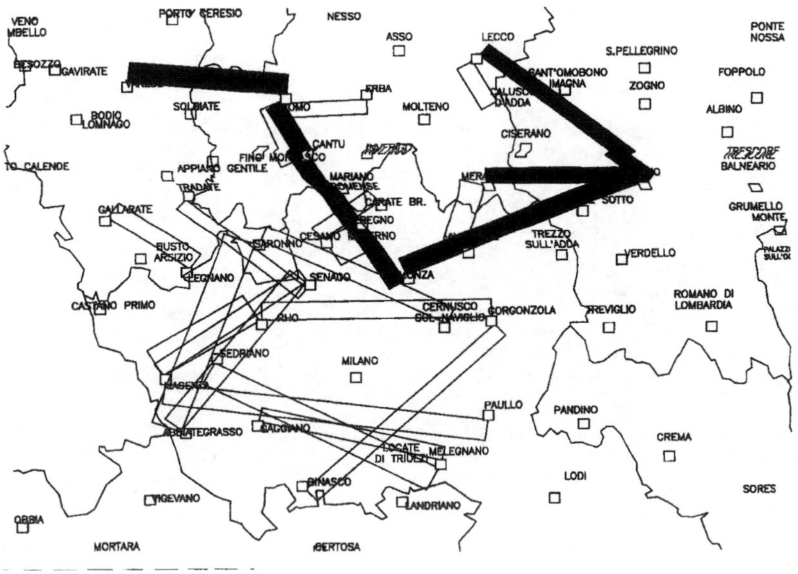

Note: Analysis is carried out at the lower territorial level, the 'telephone sector'; empty bars indicate an under-estimate of real interaction by the model, black bars an over-estimate.
Source: Camagni *et al.*, 1994.

expected, except for some bilateral links (Turin-Novara, Bergamo-Brescia) and the already mentioned gravitation on Milano.

As said before, these results were encouraging, but further empirical analysis is still needed in order to corroborate the theoretical conjecture regarding city networks.

16.6. Do city networks really generate advantages for city partners? Some empirical evidence

16.6.1. A measurement of 'network surplus'

Although it is essential to provide empirical evidence for a rather abstract theory, the empirical analyses presented above leave unmeasured some of the main features relating to a network theory. Does network behaviour really generate advantages for partner cities? Can these advantages be measured in terms of more efficient and effective urban policies implemented at the local scale?

Figure 16.4. City networks in the Lombardy region

Note: Analysis is carried out at the wider territorial level, the 'telephone district'; empty bars indicate an under-estimate of real interaction by the model, black bars an over-estimate.
Source: Camagni *et al.*, 1994.

From what was said in the theoretical framework, the main economic rationale for network behaviour is no longer to minimise transport costs and maximise control over non-overlapping market areas. It is rather to exploit scale economies in complementary relationships and synergic effects (external economies) in cooperative activities, achieved through participation in a network; i.e. the *network externality* (or network surplus) element is the main economic advantage explaining network behaviour.

Although crucial in explaining the rationale of network behaviour, the network externality element is difficult to translate into a measurable concept. A network externality is intended to mean a situation where the profits of one actor (a city) are affected by the actions of other actors (cities); network participation is expected to provide partner cities with greater performance and efficiency in terms of successful urban policies implemented. Moreover, intense use of the network is expected to strengthen network externalities positively; a high number of best practice and success stories provided by city partners and replicated in a city are expected to increase the success of local policies in that particular city.

Empirical evidence measuring the advantages cities achieve from networking behaviour has recently been collected, based on a direct survey developed within the 'Healthy City Network' (now on HC network) of the World Health Organisation (WHO) (Capello, 2000).[10] This network links together cities of different sizes and countries in a joint programme dealing with quality of life and health in urban areas: common (interurban) projects are developed in the field of health and related sectors (e.g. tobacco use, urban traffic control, AIDS, social equity) under the management of the World Health Organization (1994, 1995). The existence of this network provides a concrete example of an institutional network of different cities which aims to (a) widen information exchange among public managers on policy strategies in specific health-related areas; (b) provide an opportunity for cities to launch and run joint policies in specific health-related areas.

Through a direct survey, three main indicators were constructed.[11] The first was a connectivity indicator, a weighted sum of the number of business meetings cities have participated in and the number of Multi-City Action Plans (i.e. interurban policy programmes) partner cities are involved in. The second indicator,

---

[10] The authors of this chapter personally ran the interviews during a meeting of the network, held in Gothenburg in April 1997, where all member city representatives replied to a questionnaire, requiring direct, fair, complete and concise answers. From the questionnaire, complex indicators were constructed, based on multiple responses, and providing information on the structural characteristics of the cities, so as to examine different city behaviours according to their different characteristics; on goals and commitments, so as to check the hierarchy of goals, the seriousness of commitment, and learning processes in participating in the network; on initiatives, actions and projects launched thanks to the network; on specific projects, so as to examine individual capabilities of using the network for implementing projects conceived and suggested by each city.

[11] Member cities were asked to formulate intersectoral health promotion plans with a strong environmental component and to secure the resources for implementing them; these should include an intersectoral policy committee, mechanisms for public participation and a project office with full-time staff. More recently multi-action city plans (now on MCAPs) have been established to bring groups of cities together to address key issues such as equity, AIDS, tobacco use and traffic control. Dissemination of Healthy City strategies has been accelerated by the growth of national and subnational networks, which run in parallel with the international ones (WHO, 1995).

showing intensity of use, was formulated as the ratio between the number of cities from which each city copied an urban policy and the weighted sum of the number of Multi-City Action Plans each city is involved in and of the number of meetings cities have participated in.[12] This index provides a measure of the intensity of use of the HC network for each city. The higher the indicator, the greater the network exploitation. The last indicator is an indicator of urban performance, measured in terms of local policies implemented thanks to the existence of the network. It is built as the ratio between the number of successful policies achieved by each city thanks to cooperation within the network and the total number of local successful policies developed in each city.

The city network framework suggests that network participation should result in better performance in terms of successful urban policies implemented. This means that we expect a positive relationship between the degree of connectivity and urban performance in terms of urban policies. Imposing a direction of causality between the two indicators, we expect that:

$$y = f(c)$$

where y is the performance of the city in terms of successful local policies implemented thanks to the network and c the degree of connectivity.

Although our impression is that passive participation in a network may lead to specific local advantages, as far as achieving locally unavailable information is concerned, we expect that active participative behaviour should positively strengthen network externalities. If this is true, we should find that the explanatory power of the previous relationship increases when the intensity of use of the network is inserted into the previous model, which becomes:

$$y = f(c, i)$$

where y is performance in terms of local policies implemented, c the degree of connectivity and i the indicator of intensity of use of the network.

---

[12] The weights are the same as the ones used for the connectivity indicator.

Table 16.2. Estimates of urban network externality measurement

	First Regression[a]	Second Regression[a]
Constant	0.08 (−1.09)	−0.05 (−0.55)
Network connectivity	0.005 (2.21)	0.006 (2.51)
Intensity of use of the network		0.47 (2.23)
R-square	0.12	0.25

T-student values in parenthesis.
[a]Dependent variable: urban performance in terms of local policies implemented thanks to the network.

The econometric analyses gave the following results (Table 16.2):

- the higher a city's degree of connectivity to a network, the higher is the city's performance in terms of local policies implemented thanks to the existence of the network (and vice versa). This confirms that cities gain an advantage through the network, despite the intensity of use of the network itself;
- if a city plays an active role in the network, the resulting advantages increase (the interpretative power of the model increases with an R-square increase from 0.12 to 0.25).[13]

16.6.2. Preconditions for the exploitation of network surplus

The exploitation of network surplus (or network advantages) calls for specific preconditions in network participation, namely:

- *Commitment of the city to participating in the network.* Competence goals require serious investment and participation in the network, e.g. attendance at business meetings by qualified representative administrators and an interest in organising meetings of the network.
- *Flexibility regarding organisational change.* A sustainable long-term urban programme requires organisational changes in the public administration's procedures; the transition towards inter-sectoral approaches to urban problems (among different responsible bodies) and concertative procedures between institutions

[13] The aim of our exercise was not to explain the performance of the city in terms of local policies implemented via the network, but to estimate its relationship with network connectivity and intensity of use. Therefore, the fact that the statistical significance of the whole model (R-square) remains low does not affect our results.

and citizens, are difficult changes for administrations to make; concertative procedures and intersectoral approaches require functional integration, the elimination of existing positions and the creation of new professional profiles.

- *Open mentality to network behaviour.* Active participation in the network requires city managers to adopt attitudes accepting the organisational changes necessary to achieve network externalities. This attitude is frequently the result of prior experience with network strategies.

A strong interrelation is expected between these elements and the achievement of the main strategic network externalities. A greater commitment to participation, a more open mentality to network behaviours and increased organisational change all lead to higher network externalities. Therefore, within a city network, different attitudes in the way cities approach the network should emerge according to the different goals they have. Among them, the most strategic network behaviour is achieved when the above preconditions are met.

Based on a cluster analysis, some distinct behaviours among cities emerge (Table 16.3).

- A group of cities show *opportunistic behaviour.* In this cluster the main advantage achieved through the network is political legitimation for local policies, and the cities do not join the network for 'humanitarian' reasons. These are, in general, the largest cities, characterised by a low unemployment rate. These cities are aware of the advantages achieved, but seem to play a 'hit-and-run' strategy; they in fact seem to use the network for their short-term purposes (such as political legitimation of a specific local project), and at the same time do not manifest a willingness to develop this behaviour further. These cities do not exploit strategic advantages, such as know-how acquisition, and have local policies enjoying a high degree of success. Opportunistic behaviour, using the network only to achieve political legitimation, seems to be accompanied by a very low commitment to participating in the network.
- A group of cities manifests *explorative behaviour.* These cities participate in the network with serious commitment and make

Table 16.3. Behaviours of cities within the network: results from the cluster analysis

	Opportunistic behaviour (Cluster 1)	Explorative behaviour (Cluster 2)	Efficiency behaviour (Cluster 3)	Strategic behaviour (Cluster 4)	Sample mean
Commitment to participating in the network	*−0.96*	**1.06**	−0.5	−0.29	0
Degree of satisfaction	**0.13**	*−0.11*	**0.11**	−0.07	0
Success policy actions due to the existence of the network	**0.16**	−0.12	−0.19	**0.35**	0
Humanitarian reasons in joining the network	*−0.2*	0.04	0.07	−0.04	0
Awareness of the results achieved through the network	**0.45**	0.28	−0.19	*−0.43*	0
Advantages of scale economies	−0.31	−0.22	**0.89**	*−0.68*	0
Willingness to participate in the future	*−0.39*	−0.008	−0.13	**0.45**	0
Open mentality to network behaviour	**0.6**	−0.02	*−0.52*	**0.37**	0
Advantages of information gathering	−0.08	0.03	**0.09**	*−0.13*	0
Advantages of political legitimation	**0.99**	−0.14	*−0.18*	−0.15	0
Reasons of information gathering for joining the network	**0.12**	−0.05	**0.27**	*−0.38*	0
Learning in the use of the network	**0.42**	**0.32**	0.04	*−0.8*	0
High degree of success of local policy actions	*−0.14*	−0.14	*−0.25*	**0.65**	0
Advantages of choosing partners	**0.52**	*−0.15*	−0.07	0.007	0
Advantages of know-how acquisition	*−0.75*	*−0.23*	0.14	**0.61**	0

(*continued*)

Table 16.3 Continued

	Opportunistic behaviour (Cluster 1)	Explorative behaviour (Cluster 2)	Efficiency behaviour (Cluster 3)	Strategic behaviour (Cluster 4)	Sample mean
Urban size	**696,800**	*296,166*	324,144	382,051	379,444
European cities	1	1	*0*	1	1
Western cities	1	1	*0*	1	1
Unemployment rates	**6.2**	*13.4*	10.3	*13.8*	11.5

Values characterising clusters in a positive sense, in bold. Values characterising cluster in a negative sense, in italics.

significant investment in the experience. However, they are not characterised by a particular network advantage and they have created a learning process through which the network can be exploited.

– A group of cities is characterised by pure *economic efficiency behaviour*. Cities belonging to this group enter the network with simple information-gathering goals, and achieve typical advantages which increase economic efficiency, such as increased information and economies of scale. Eastern and non-European cities belong to this cluster. The profile of this cluster is of a group of cities whose primary goal in joining the network is very simply to gather information, thereby increasing their economic efficiency thanks to the network. They express greater satisfaction compared to the other clusters. These are also cities with no tradition of success stories in local projects.
– A group of cities shows *strategic behaviour*. In this cluster, cities gain the greatest and most strategic advantage from the network – the acquisition of know-how – and are the ones achieving a greater number of local success stories from the network. This result once again highlights the relationship between the intensity of use of the network and a high extent of locally successful policies achieved through the network: cities able to obtain new knowledge from the network are also those cities achieving the greatest advantages in terms of local policies. Moreover, these cities are also those characterised by an open mentality to network behaviour and do not join the network for simple reasons like information gathering. Interestingly, these cities are also

those with a high rate of unemployment, once again indicating that the network does not provide greater advantages for the richest cities.

In summary, the descriptive statistical analysis provides prima facie evidence that different behavioural patterns exist in the network, with the most cooperative behaviour leading to better performance in terms of a city's success in implementing local policies. However, the achievement of important advantages from the network also requires commitment to participation and an open mentality to network behaviour.

16.7. Conclusions

In this chapter, it is shown how the logics that shape the city system are more complicated than the simple 'territorial' and hierarchical logic of the traditional central-place model. A firm's control in the market of outputs, inputs and innovative assets is attained not only managing a gravitational area, but also and increasingly through cooperative, trans-territorial network relationships.

The new behavioural logic of the firm parallels and partly determines the new organisational logic of the city system, where phenomena of specialisation and networking also appear. Processes of urban complementary specialisation in a regional context allow the single centres to take advantage of the entire regional market in the specialisation sector, reaching relevant economies of scale and hosting functions that, in the traditional hierarchical model of urban system, were reserved to higher order centres. The spontaneous or partly planned division of labour among the Dutch cities of the Randstad or the one among the cities of the polycentric Veneto region are good examples. By the same token, centres of similar rank and size, specialised in similar sectors, may cooperate in order to reach a superior critical mass, benefiting from network externalities. Examples range from the case of world financial cities, operating on a unique and integrated market through communication networks and institutional arrangements, to second-order tourist cities, integrated in 'itineraries' through information and organisational networks, reaching in this way a sufficient attractiveness in order to compete with the champion cities.

In this sense, we distinguish between complementarity and synergy networks. Economies of specialisation and division of labour, on one hand, and scale economies reached through the network on the other represent the economic rationale of the new spatial paradigm in the two respective cases. Similar goals may lead cities to cooperate in order to establish innovative territorial schemes and projects, like a common airport serving the entire network.

The new concept, and the new logic, of city networks do not replace but complement the traditional hierarchical logic of the city system, which remains the fundamental spatial logic of many sectors (such as consumer services, traditional agriculture and public administration) and is still visible in the territory as the historical organisational form and driving force of spatial structure from times when these sectors were the leading ones. From a terminologic point of view, our proposal is to use the term ‘city network’ only for the new theoretical and empirical realm of selective horizontal linkages among centres of similar size and rank, leaving the usual terms of hierarchy and spatial interaction, respectively, for vertical, hierarchical linkages and for generic, gravity-type relationships.

The paradigm of city networks has recently been enriched by empirical evidence able to measure two important aspects of the city network paradigm. The first concerns the possibility of detecting and empirically identifying ‘city networks’ themselves in the real world of spatial interactions. In the Lombardy region, not only were Christallerian gravity-type relationships envisaged but also cooperative relationships among cities of similar rank and size. The second important aspect regards the possibility of empirically determining whether the network logic, which is cooperative in nature, brings specific advantages to the partners (firms, institutions or cities). A ‘network surplus’ was in fact evident and partly measured in the case of an international city network, the network of Healthy Cities.

The new pattern of territorial relationships opens up new opportunities for the planning activity, as a city is confronted with expanded alternatives regarding its development path. In particular, through cooperation inside a city network, each centre could be able to develop high-order functions – and benefit from the consequent income levels and ‘surplus’ – without increasing its own size, as it would be implied by the constraints of scale economies and indicated by the central place model. There is scope, therefore, for intentional city strategies, both at the level of the single centre and at the level of the entire city-system.

References

Armstrong, H. and J. Taylor (2000), *Regional Economics and Policy*, Oxford: Blackwell.

Arrighetti, A. (2003), *Economia dell'Azione Collettiva, Dispense*, Facoltà di Economia, Parma: Università di Parma.

Aydalot, Ph. (ed.) (1986), *Milieux Innovateurs en Europe*, Paris: GREMI.

Beckmann, M.J. and J. McPherson (1970), "City size distribution in a central place hierarchy: an alternative approach", *Journal of Regional Science*, Vol. 10, pp. 25–33.

Beguin, H. (1984), "The shape of city-size distribution in a central place system", *Environment and Planning A*, Vol. 16, pp. 749–758.

Beguin, H. (1988), "La région et les lieux centraux", pp. 231–275, in: C. Ponsard, editor, *Analyse Économique Spatiale*, Paris: Presse Universitaire de France.

Bowles, S. and H. Gintis (2002), "Social capital and community governance", *The Economic Journal*, Vol. 112, pp. F419–F436.

Camagni, R. (1988), "Functional integration and locational shifts in the new technology industry", pp. 48–64, in: Ph. Aydalot and D. Keeble, editors, *High Technology Industry and Innovative Environments: the European Experience*, London: Routledge.

Camagni, R. (1991), "Local milieu, uncertainty and innovation networks: towards a dynamic theory of economic space", pp. 121–144, in: R. Camagni, editor, *Innovation Networks: Spatial Perspectives* London: Belhaven-Pinter.

Camagni, R. (1993a), "From city hierarchy to city network: reflections about an emerging paradigm", pp. 66–87, in: T.R. Lakshmanan and P. Nijkamp, editors, *Structure and Change in the Space Economy, Festschrift in Honour of Martin Beckmann*, Berlin: Springer.

Camagni, R. (1993b), "Interfirm industrial networks: the costs and benefits of cooperative behaviour", *Journal of Industry Studies*, Vol. 1, pp. 1–15.

Camagni, R. (2002), "On the concept of territorial competitiveness: sound or misleading?", *Urban Studies*, Vol. 13, pp. 2395–2412.

Camagni, R. and F. Gambarotto (1988), "Gli accordi di cooperazione come nuove forme di sviluppo esterno delle imprese", *Economia e Politica Industriale*, Vol. 58, pp. 93–140.

Camagni, R., L. Diappi and S. Stabilini (1994), "City networks in the Lombardy Region: an analysis in terms of communication flows", *Flux*, Vol. 15, pp. 37–50.

Capello, R. (1994), *Spatial Economic Analysis of Telecommunications Network Externalities*, Aldershot: Ashgate.

Capello, R. (2000), "The city network paradigm: measuring urban network externalities", *Urban Studies*, Vol. 37(11), pp. 1925–1945.

Cappellin, R. and R. Grillenzoni (1983), "Diffusion and specialisation in the location of service activities in Italy", *Sistemi Urbani*, Vol. 1, pp. 249–282.

Castells, M. (1996), *The Rise of the Network Society*, London: Blackwell.

Coase, R. (1988), *The Firm, the Market, and the Law*, Chicago: University of Chicago Press.

Coleman, J.S. (1990), *Foundations of Social Theory*, Cambridge, MA: Harvard University Press.

Colletis, G. and B. Pecqueur (1995), "Politiques technologiques locales et création des ressources spécifiques", pp. 445–463, in: A. Rallet and A. Torre, editors, *Economie Industrielle et Économie Spatiale*, Paris: Economica.

Cooke, P. and K. Morgan (1998), *The Associational Economy, Firms, Regions and Innovation*, Oxford: Oxford University Press.

Dematteis, G. (1985), "Verso strutture urbane reticolari", pp. 121–132, in: G. Bianchi and I. Magnani, editors, *Sviluppo Multiregionale: Teorie, Metodi, Problemi*, Milano: Franco Angeli.

Dematteis, G. (1990), "Modelli urbani a rete: considerazioni preliminari", pp. 27–48, in: F. Curti and L. Diappi, editors, *Gerarchie e Reti di Città: Tendenze e Politiche*, Milano: Franco Angeli.

Durlauf, S. (2002a), "Symposium on social capital: introduction", *The Economic Journal*, Vol. 112, pp. F417–F418.

Durlauf, S. (2002b), "On the empirics of social capital", *The Economic Journal*, Vol. 112, pp. F459–F479.

Emanuel, C. and G. Dematteis (1990), "Reti urbane minori e deconcentrazione metropolitana nella Padania centro-occidentale", pp. 233–261, in: D. Martellato and F. Sforzi, editors, *Studi sui Sistemi Urbani*, Milano: Franco Angeli.

EU (1999), *Spatial Development Perspective*, Bruxelles: Committee for Spatial Development.

Foresti, J. (1986), *Une Toile d'Araignée se Referme Autour de l'Industrie Européenne: les Accords de Coopération Industrielle*, Bruxelles: Commission des Communautés Européennes.

Glaeser, E., D. Laibson and B. Sacerdote (2002), "An economic approach to social capital", *The Economic Journal*, Vol. 112, pp. F437–F458.

Isard, W. (1956), *Location and Space-Economy*, Cambridge, MA: MIT Press.

Johannisson, B. (1987), "Organising: the network metaphor", *International Studies of Management and Organisation (Special volume)*, Vol. 17(1).

Maillat, D., M. Quévit and L. Senn (eds.) (1993), *Réseaux d'innovation et Milieux Innovateurs: un Pari pour le Développement Régional*, Neuchatel: EDES.

Malecki, E.J. and P. Oinas (eds.) (1999), *Making Connections: Technological Learning and Regional Economic Change*, Aldershot: Ashgate.

Malecki, E.J. (1997), "Entrepreneurs, networks and economic development", *Advances in Entrepreneurship, Firm Emergence and Growth*, Vol. 3, pp. 57–118.

Mulligan, G. (1979), "Additional properties of a hierarchical city-size model", *Journal of Regional Science*, Vol. 1, pp. 1–42.

Mulligan, G. (1984), "Agglomeration and central place theory: a review of the literature", *International Regional Science Review*, Vol. 1, pp. 1–42.

Nijkamp, P. (2003), "Entrepreneurship in a modern network economy", *Regional Studies*, Vol. 37(4), pp. 395–405.

OECD (1986), *Technical Cooperation Agreements Between Firms: Some Initial Data and Analysis*, May.

Parr, J.B. (1978), "Models of the central place system: a more general approach", *Urban Studies*, Vol. 15, pp. 35–49.

Parr, J.B. (1981), "Temporal change in a central-place system", *Environment and Planning A*, Vol. 13, pp. 97–118.

Parr, J.B. (1989), Competition, supply areas and industrial location, paper presented at the 111 World Congress of the Regional Science Association, Jerusalem, April, Mimeo.

Porter, M. (2001), Regions and the new economics of competition, pp. 139–157, in: A. Scott, editor, *Global City-Regions: Trends, Theories, Policies*, New York: Oxford University Press.

Putnam, R.D. (1993), *Making Democracy Work*, Princeton, NJ: Princeton University Press.

Scott, A. (ed.) (2001), *Global City-Regions: Trends, Theory, Policies*, New York: Oxford University Press.

Storper, M. (1995), "La géographie des conventions: proximité territoriale, interdépendences non-marchandes et développement économique", pp. 111–127, in: A. Rallet and A. Torre, editors, *Economie Industrielle et Économie Spatiale*, Paris: Economica.

WHO (1994), *Action for Health in Cities*, Copenhagen: Regional Office for Europe, World Health Organisation.

WHO (1995), *Twenty Steps for Developing a Healthy Cities Project*, Copenhagen: Regional Office for Europe, World Health Organization.

Williamson, O. (1985), *The Economic Institutions of Capitalism*, New York: The Free Press.

PART 5

Urban Competitiveness

Urban Dynamics and Growth, CEA, vol. 266
Roberta Capello and Peter Nijkamp (editors)

CHAPTER 17

Dynamic Urban Models: Agglomeration and Growth

Marcus Berliant[a] and Ping Wang[b,c]

[a]Department of Economics, Washington University, St Louis, MO 63130-4899, USA
[b]Vanderbilt University, Nashville, TN, USA
[c]NBER, Cambridge, MA, USA

Abstract

Theoretical models of urban growth are surveyed in a common framework. Exogenous growth models, where growth in some capital stock as a function of investment is assumed, are examined first. Then endogenous growth models, where use of some factor by a firm increases the productivity of other firms, are studied. These are all models with perfect competition among agents. Next, models with imperfect competition are discussed. There are two varieties: those employing a monopolistic competition approach to product differentiation, and those employing explicit externalities but lacking some markets. Finally, avenues for future research are explored. Correlations between agglomeration and growth in the various models and data are compared.

We thank Gilles Duranton, Masa Fujita, Bob Helsley, Yoshi Kanemoto, Shin-Kun Peng, David Pines, Will Strange, Jacques Thisse, Mark Wright, an anonymous referee for this volume, and participants in seminars at the Chinese University of Hong Kong, Tamkang University and the Regional Science Association International Meetings in Philadelphia for valuable comments and suggestions. The first author gratefully acknowledges financial support from the American Philosophical Society. Parts of this chapter were completed while the first author was visiting the Division of Humanities and Social Sciences at the California Institute of Technology, the second author was visiting Academia Sinica, and both were visiting the Kyoto Institute of Economic Research at Kyoto University. We are grateful to all these institutions for their kind hospitality. Needless to say, the usual disclaimer applies.

Keywords: agglomerative activity, marshallian externalities, matching, urban growth

JEL classifications: C78, D51, R12

17.1. Introduction

Over the past two centuries, long-term trends of urbanization of population and sustained economic growth across both developed and developing countries are clear. This urbanization trend features an on-going increase in the number as well as the size of cities.[1] The observation is robust despite some sharp differences in the patterns of agglomeration, particularly between US and Asian/European cities in the past three or four decades.[2] What are the determinants of the rates of city growth and the speed of spatial agglomeration? Why do some cities rise and some fall in the process of economic development?[3] In spite of the lack of a complete microeconomic structure mimicking the real world economy, various dynamic urban models have attempted to address these important issues. The primary purpose of this critical survey is to synthesize the existing literature on agglomeration and growth so as to promote better understanding of the underlying driving forces of spatial agglomeration and the channels through which agglomerative activity fosters urban growth.

---

[1] For example, based on a consistent, old census definition, the number of American cities with population of 2500 or more increased from 30 in 1800 to more than 4000 in 1950. While urban population as a percentage of total population rose from about 6 to 60% over the same period, the population of the largest city, New York, increased from approximately 0.1 million to 10 million.

[2] Among many others, a crucial difference has been the suburbanization trend (within metropolitan areas) in the US since 1960, accompanied by the decay of central cities. This phenomenon is not prevalent in European countries, for instance.

[3] For instance, there have been dramatic changes in the national rankings of US cities over the past century. Notable cities rising in the rankings include Lexington, Los Angeles, Nashville, New Orleans and San Antonio, whereas cities falling in the rankings include Albany, Baltimore, Louisville, Pittsburgh, and St Louis. In contrast, Eaton and Eckstein (1997) find few new agglomerations added to the stock of 35 cities in France (1876–1990) and 40 in Japan (1925–1985), with *no* cities vanishing (ranked below 50) over the respective time periods.

There have been several recent developments uncovering key determinants of spatial agglomeration as well as establishing empirical regularities concerning cities and growth. As stressed by Marshall (1895), Kuznets (1962), Pred (1966), and Jacobs (1969), knowledge spillovers are one primary force for agglomerative activities. Jaffe *et al.* (1993) show that knowledge spillovers are localized in the sense that patents are more likely to cite previous patents from the same area and spillovers can cross national borders only with delay. Glaeser *et al.* (1992) and Henderson *et al.* (1995) find that spillovers occur both within and between industries and characteristics of urban areas play a role in the location decisions of industries. While Rauch (1993), Saxenian (1994) and Glaeser (1999) stress the role of cities in transmitting knowledge, Helsley and Strange (2002) describe how agglomeration facilitates innovation by lowering the costs of innovation through a higher density of factor suppliers.

Just what is the relationship between agglomeration and growth in cities? To address this issue, one must go beyond a simple measure of urban population to examine more thoroughly employment and income measures of a given city[4] (Table 17.1).

There is a clear positive correlation between city employment and median income, though no clear relation with city population.[5]

To study further the relationship between agglomeration and growth, see Table 17.2.[6]

If the change in city share of employment is a proxy for the change in agglomeration, and the change in real median income is a proxy for growth, then the link between agglomeration and growth is

---

[4] Source: Authors' computations from US Census Bureau data http://www.census.gov/Press-Release/www/2002/dp.comptables.html where median income is in 1999 dollars per household. As this is a survey of theory, we will not perform formal econometric tests.

[5] Obviously, one could use metropolitan area statistics, but this tends to mask migration of population and employment from the city to the suburbs. For example, 2000 census data provided in http://www.census.gov/population/cen2000/phc-t3/tab02.pdf shows us that the population of St Louis increased by 4.5% in the last decade, while we know that population has exited St Louis city at a rapid rate. Table 17.2 addresses some of these issues in more detail.

[6] Source: Authors' computations from US Census Bureau data http://www.census.gov/Press-Release/www/2002/dp.comptables.html where median income is in 1999 dollars per household.

Table 17.1. **Population, employment and median income in selected cities**

City/year	1990	2000	Percent Change (%)
Los Angeles			
Population	3,485,398	3,694,834	6.0
Employment	1,670,488	1,532,074	−8.3
Median income	40,346	36,687	−9.0
Chicago			
Population	2,783,726	2,895,964	4.0
Employment	1,207,108	1,220,040	1.1
Median income	34,314	38,625	12.6
Akron			
Population	223,019	217,088	−2.7
Employment	94,103	99,310	5.5
Median income	29,066	31,835	9.5
San Antonio			
Population	935,927	1,144,554	22.3
Employment	389,727	488,747	25.4
Median income	30,769	36,214	17.7

clearly more complicated than what we see from Table 17.1. In Table 17.2, there is no apparent link between agglomeration of jobs and growth.[7]

Consider next the general trend of urbanization since 1950 in Table 17.3.[8]

So we can see that population has been moving from rural to urban areas, resulting in agglomeration at least in a macro sense. Between the 1980 census and the 1990 census, 137 cities with population greater than 100,000 experienced population growth, while 56 experienced population loss.[9]

Conceptually, we regard a city as a 'settlement that consistently generates its economic growth from its own local economy' (cf. Jacobs, 1969, p. 262). As both the economic growth of cities and the agglomeration of economic agents are apparently endogenous, a formal model containing these endogenous variables, as well as exogenous variables, is required to understand how they are related.

[7] Y. Kanemoto has provided the authors with analogous tables for cities in Japan. The same conclusion can be drawn from this data.

[8] Source: http://www.census.gov/population/censusdata/table4.pdf

[9] Source: http://www.census.gov/population/censusdata/c1008090pc.txt

Table 17.2. Agglomeration versus growth: a first look

City/Year	1990	2000	Percent Change (%)
Los Angeles			
City share of county employment	0.397	0.388	−2.3
Median county income	45,617	42,189	−7.5
Chicago			
City share of county employment	0.500	0.500	0
Median county income	42,627	45,922	7.7
Akron			
City share of county employment	0.4	0.38	−0.05
Median county income	37,830	42,304	11.8
San Antonio			
City share of county employment	0.78	0.82	5.1
Median county income	33,824	38,328	13.3

Of course, differences in exogenous variables across cities can help explain different patterns of agglomeration and growth. Our purpose here is to provide an integrative and simple framework for studying the agglomeration of economic agents and the growth of city economies.

We limit the scope of our survey to models of real dynamics and growth. By this, we mean that there is a variable, usually the change in a capital stock, controlled by an agent in the current period that affects feasible allocations in the future. Models not falling into this description include those that examine disequilibrium migration dynamics, for instance stability of a static migration equilibrium. Another class not falling into this description is the class of models that are essentially static in nature, but some exogenous parameter such as population is presumed to grow at a rate related to time. The static equilibrium is examined in each period. Such models are

Table 17.3. Urbanization Trend in the US, 1950–1990

Year	Percent of Population that is Urban
1950	64.0
1960	69.9
1970	73.6
1980	73.7
1990	75.2

essentially an extended comparative statics exercise, as the change in the parameter is not the result of the choice of an agent.

In Section 17.2, we develop an integrated canonical Walrasian framework, synthesizing a variety of conventional models of exogenous urban growth studied by Fujita (1976, 1982), Anas (1978, 1992), Kanemoto (1980), Henderson and Ioannides (1981) and Miyao (1987). We characterize the steady-state equilibrium using three approaches: (i) the Solow–Swan aggregate production approach, (ii) the Allais–Phelps golden rule solution, and (iii) the Ramsey–Cass–Koopmans optimal exogenous growth framework. By utilizing an integrated setup, we allow easy cross-model comparisons.

In Section 17.3, we generalize the integrated framework constructed in Section 17.2 to allow for endogenous growth so that we can examine thoroughly two-way dynamic interactions between spatial agglomeration and urban development. In particular, we focus on models with perfect competition, including Palivos and Wang (1996), Eaton and Eckstein (1997), Black and Henderson (1999), Lin *et al.* (2004), and Rossi-Hansberg and Wright (2003). We fully explore a simple one-sector framework that highlights the trade-off between increasing returns from production externalities (a centripetal force of city formation) and transportation costs (a centrifugal force), followed by an illustration of a more general two-sector setup.

In Section 17.4, we consider dynamic urban growth models without a perfectly competitive labor, intermediate good, or consumption good market. The first part of this section introduces imperfect market structures, concentrating in particular on the case of monopolistic competition. Different from the perfect competition setup, these models consider increasing returns from differentiated products as the centripetal force of city formation. The second part of this section shifts attention to non-Walrasian setups that incorporate market frictions into the urban growth framework. This is motivated by its useful implications for urban employment, as illustrated in Helsley and Strange (1990), Abdel-Rahman and Wang (1995), Coulson *et al.* (2001) and Brueckner and Zenou (2003). Specifically, we focus on a matching model of agglomeration and growth based on a recent study by Berliant *et al.* (2003), where the role of horizontal knowledge exchange in spatial interactions is explored. We illustrate why improvements in the effectiveness of knowledge exchange may lead to higher growth and examine the channels through which an increase in the rate of economic growth may lead to spatial agglomeration. The conclusions obtained herein will lend theoretical support to the empirical facts presented in Section 17.1.

In Section 17.5, we provide critiques of the literature and avenues for future research. In particular, we elaborate on the size and number of black boxes in the models mentioned in Sections 17.2–17.4. Also, we would like to point out a problem with the definition of social optimum compared to that of Pareto optimum, and how this might be fixed. Finally, we discuss ways to formulate testable hypotheses that can distinguish between the different models. Such empirical tests may help delineate the role of urban growth models in explaining the dynamic process of city formation and development.

17.2. From Solow–Swan to Ramsey urban growth models

Modern growth theory was spurred by two seminal contributions by von Neumann (1937) and Harrod (1939) that provide a mathematical treatment of long-run economic growth in a Ricardian economy. Harrod's steady-state equilibrium is inherently unstable, as it requires a knife-edge condition specifying a fixed relationship between three exogenous constants (the fixed capital–output ratio, the constant population growth rate and the exogenous savings rate) that can hold true only for a set of parameters of measure zero. This undesirable property gave rise to the birth of neoclassical growth theory, led by Solow (1956) and Swan (1956), allowing a variable capital–output ratio using a neoclassical aggregate production function to generate a well-defined non-degenerate steady-state equilibrium. Over the next three decades, this aggregate production function was the heart of exogenous growth theory.

In this section, we develop an integrated framework, synthesizing a variety of urban growth models. We characterize the steady-state equilibrium using the Solow–Swan aggregate production approach and the golden rule solution as well as the optimal exogenous growth framework. The primary purpose of utilizing an integrated setup is to enable cross-model comparison in a parsimonious manner.

17.2.1. The aggregate production approach to urban growth

Consider a stylized neoclassical production function with final goods output (Y) produced using reproducible physical capital (K) and raw labor (N): $Y = F(K, L)$, where effective labor ($L = AN$) is raw labor augmented by a Harrod-neutral (labor-augmenting) technology and $A > 0$ is a scaling factor reflecting the current state of technology. The production function F is strictly increasing and strictly concave in each argument; it is continuously differentiable with derivatives denoted by subscripts; and satisfies constant-returns-to-scale ($F(aK, aL) = aF(K, L)$ for all $a > 0$), a boundary condition ($F(0, 0) = 0$), and Inada conditions ($\lim_{K\to 0} F_K(K, L) = \infty$, $\lim_{L\to 0} F_L(K, L) = \infty$, $\lim_{K\to\infty} F_K(K, L) = 0$, and $\lim_{L\to\infty} F_K(K, L) = 0$). Using constant returns, we can write the output per worker in efficiency units ($y = Y/L$) as a well-defined function of the capital–effective labor ratio $(k = K/L) : y = f(k)$, where $f(k) = F(k, 1)$.

Let S denote aggregate savings, I gross investment, γ the (constant) rate of technical progress, and n the (constant) rate of population growth. The Solow–Swan model can be summarized by the following three fundamental relationships (all in per worker forms):

(i) (fixed savings rate, s) $S/L = sf(k)$;
(ii) (full employment) $\dot{L}/L = n + \gamma$;
(iii) (capital accumulation) $\dot{k}/k = (I/L)/k - n - \gamma$.

A loanable funds market equilibrium requires that savings per worker equals investment per worker, i.e. $S/L = I/L$, or,

$$\dot{k} = sf(k) - (n + \gamma)k \tag{17.1}$$

In the steady state, capital per effective unit of labor reaches a constant (i.e. $\dot{k} = 0$). It follows immediately that

$$sf(k) = (n + \gamma)k \tag{17.2}$$

determining the steady-state equilibrium level of capital accumulation.

In the 1960s and 1970s, urban economists who emphasize the supply-side often adopted the Solow–Swan framework (see a long list of papers cited by Miyao (1987)). It is not the purpose of this chapter to present individual contributions of this vintaged literature. Rather, we would like to highlight the spirit of this conventional urban growth framework by simply modifying two fundamental relationships. Consider an area of interest – a city, a community or a region. Let w and r denote the (real) wage rate (in effective units) and the (real) interest rate, respectively, and w^A and r^A, the (exogenous) corresponding rates outside the area of interest (sometimes referred to as the national average or the agricultural/rural rates). In an open city setup, we can postulate that labor migration depends on the wage gap $((w - w^A)/w^A)$ whereas capital flows are determined by the interest rate differential $(r - r^A)$. Then we have

$$\dot{N} = nN + b^N(w/w^A - 1)N \tag{17.3}$$

$$\dot{K} = sF(K, AN) + b^K(r - r^A)K \tag{17.4}$$

Thus, Equation (17.1) becomes

$$\dot{k} = sf(k) - [(n + \gamma) + b^N(w/w^A - 1) - b^K(r - r^A)]k$$

$$= sf(k) - (\nu + \gamma)k \tag{17.5}$$

where $\nu = I/K = n + b^N(w/w^A - 1) - b^K(r - r^A)$ measures the investment ratio. Equation (17.5) can be used together with the envelope conditions ($w = f(k) - kf_k(k)$ and $r = f_k(k)$) to yield the following:

$$\left(s - \frac{b^N}{w^A}k\right)f(k) + \left(b^K + \frac{b^N}{w^A}k\right)f_k(k)k = [(n + \gamma) + b^K r^A - b^N]k \tag{17.6}$$

It is convenient to call the LHS of Equation (17.6) 'adjusted gross savings' (AGS) and the RHS 'adjusted gross investment' (AGI), which can be plotted in Figure 17.1 to pin down the steady-state equilibrium value of capital per effective labor unit (k^S). A useful example is the Cobb–Douglas production function with $f(k) = k^\alpha$, where the capital income share, $\alpha \in (0, 1)$, is constant over time. In this case, the steady-state equilibrium solution can be conveniently derived as $k^S = [s/(\nu + \gamma)]^{1/(1-\alpha)}$. Next we discuss the comparative

Figure 17.1. Steady-state equilibrium in the Solow–Swan model of urban growth

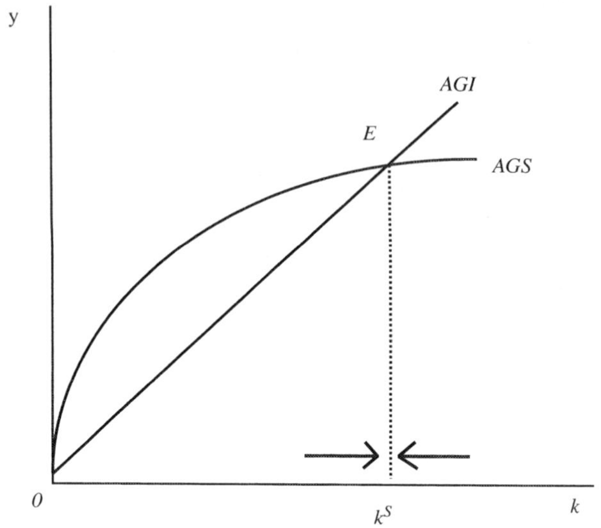

statics of this model. We can see that the AGS locus shifts out when the savings rate (s) increases, the outside alternative wage (w^A) is higher, or the state of the city is more sensitive to capital flows but less to labor migration (higher b^K, lower b^N). On the other hand, the AGI locus rotates clockwise in response to (1) a decrease in the population growth rate (n), or (2) a decrease in the outside alternative interest rate (r^A), or (3) a decrease in the sensitivity of the city to capital flows (lower b^K), or (4) to an increase in the sensitivity to labor migration (higher b^N). In all these cases, equilibrium achieves greater accumulation of capital and per capita output (Ay). When the rate of technical progress (γ) increases, however, the steady-state value of output per effective unit of labor (y) drops, but per capita output rises.

Since per capita output depends solely on the rate of technical progress, any growth effects discussed above are more precisely short-run (transitional) effects. Thus, it is important to examine the stability properties of the model. Conventionally, it is widely claimed that the steady-state equilibrium of urban growth is a saddle in (K, N) space based on Equations (17.3) and (17.4) (cf. Miyao, 1987). We would like to point out such a claim is unfortunately incorrect. To see this, differentiate Equation (17.5) with respect to k and manipulate it by utilizing the steady-state equilibrium relationship (17.6) to obtain:

$$\frac{d\dot{k}}{dk} = \left(b^k + \frac{b^N}{w^A}k\right)f_{kk}k - b^N - \frac{s}{k}w < 0 \tag{17.7}$$

This implies the steady-state equilibrium is globally stable. In other words, capital per unit of effective labor will always converge monotonically to its steady-state value regardless of the initial state.

In summary, we conclude this subsection with the following comparative statics:

(i) a higher savings rate or a lower population growth rate promotes a city's capital accumulation, thereby fostering short-run urban growth in per capita output;
(ii) while an increase in outside wages reduces net population growth and raises city growth, an increase in outside interest rate lowers the supply of local funds and suppresses short-run growth;
(iii) when the city's wages fall below the national average or its interest rate exceeds the national level, less barriers to interregional labor migration or capital flows tend to increase per capita output growth in transition;
(iv) in contrast with the above-mentioned factors, local technical progress advances the city's output growth not only in transition, but also in the long run;
(v) all regions adopting the same technology converge to the same steady-state in per capita output.

It is worthwhile to discuss the consequences of incorporating agglomeration economies or urban congestion next. First, should there be scale economies in the production function, all but the result concerning interregional migration remain qualitatively unchanged. While the presence of local scale economies favors a more centralized interregional migration policy, the presence of global scale economies tends to discourage such a policy. Second, what urban congestion (in the form of traffic or pollution) adds to the basic framework is the social cost associated with capital accumulation and population growth. Thus, the consideration of congestion provides a justification for a more centralized interregional migration policy.

17.2.2. The golden rule solution

This supply-oriented neoclassical growth model relies entirely on the mechanics of the aggregate production technology without an explicit account of a representative agent's optimizing behavior. To remedy this problem, one may apply the golden rule solution (or a 'maximum maximorum') proposed by von Neumann (1937), Allais (1947) and Phelps (1966). Specifically, the golden rule is reached when the steady-state level of consumption per worker (c) is maximized. Utilizing the steady-state equilibrium relationship derived from the supply-oriented problem (Equation (17.5) with $\dot{k} = 0$), we have

$$\max_k c = A[f(k) - (\nu + \gamma)] = A\{f(k) - [(n + \gamma) + b^N(w/w^A - 1) - b^K(r - r^A)]k\} \tag{17.8}$$

The first-order condition is

$$f_k(k) = \nu + \gamma = (n + \gamma) + b^N(w/w^A - 1) - b^K(r - r^A) \tag{17.9}$$

It is easily seen from Equation (17.9) that the golden rule accumulation of capital per effective worker (k^{G}) may be higher or lower than the Solow–Swan steady state (k^{S}), depending on whether the city's net interest income exceeds local savings or not. Under the Cobb–Douglas production function, the golden rule solution for capital accumulation is $k^{\mathrm{G}} = [\alpha/(\nu + \gamma)]^{1/(1-\alpha)}$. Thus, the Solow–Swan steady state over-accumulates capital relative to the golden rule if the exogenous savings rate is higher than the capital income share. How would this open city compare to a closed city in the absence of labor migration or capital flows? The latter case is equivalent to assuming that $w = w^A$ and $r = r^A$. From Equation (17.9), we can see that an open city reaches a higher level of output per capita if the local interest rate is higher than the national average and the local wage rate is lower than the national average. Finally, we substitute the envelope conditions ($w = f(k) - kf_k(k)$ and $r = f_k(k)$) into Equation (17.9) to obtain:

$$-\frac{b^N}{w^A}f(k) + \left(1 + b^K + \frac{b^N}{w^A}\right)f_k(k)k = (n + \gamma) + b^K r^A - b^N \tag{17.10}$$

By comparing Equation (17.10) with Equation (17.6), one can see that except for the result concerning changes in the exogenous savings rate, all the comparative statics obtained in Section 17.2.1 remain qualitatively valid for the golden rule solution.

17.2.3. The optimal exogenous growth framework

An obvious problem associated with the golden rule solution is that individuals ignore the time path of consumption, focusing on nothing but its steady-state level. When the subjective rate of time preference is positive, the solution is misleading, as it ignores an intertemporal cost of physical capital investment. This shortcoming has been noted by Cass (1965) and Koopmans (1965), who suggest a return to a pivotal but largely overlooked work by Ramsey (1928). Since then, this optimal exogenous growth model has become the predominant framework in urban growth theory (e.g. see Fujita, 1976, 1982; Anas, 1978, 1992; Kanemoto, 1980; Henderson and Ioannides 1981; Miyao, 1987).

Specifically, denote by $\rho > 0$ the subjective rate of time preference and $u(c)$ the felicity function. Taking factor prices as parametrically given, the representative agent in the city of interest faces the following optimization problem:

$$\max_{c} U = \int_0^{\infty} u(c)\, \mathrm{e}^{-\rho t}\, \mathrm{d}t$$

$$\text{s.t.} \qquad \dot{k} = f(k) - (\nu + \gamma)k - \frac{c}{A} \tag{17.11}$$

where $A(0) = 1$, $K(0)/N(0) = k_0 > 0$ is given, and recall that $\nu = n + b^N(w/w^A - 1) - b^K(r - r^A)$. Thus, what concerns the individual is the lifetime utility (U) that is associated with the properly discounted valuation of the path of consumption over their entire life span.

Denoting λ as the co-state variable associated with the capital evolution equation of the city (17.11), we can set up the current-value Hamiltonian as

$$\mathcal{H} = u(c) + \lambda[f(k) - (\nu + \gamma)k - c/A] \tag{17.12}$$

Straightforward application of the Pontryagin Maximum Principle yields

$$Au_c(c) = \lambda \tag{17.13}$$

$$\frac{\dot{\lambda}}{\lambda} = \rho + \nu + \gamma - f_k(k) \tag{17.14}$$

While the first-order condition (17.13) ensures intertemporal consumption efficiency, the Euler equation (17.14) governs the evolution of the shadow price of capital. The transversality condition is

$$\lim_{t \to \infty} \lambda(t)k(t)\, \mathrm{e}^{-\rho t} = 0 \tag{17.15}$$

Totally differentiating Equation (17.13) and utilizing Equation (17.14) to eliminate the co-state variable, we obtain the 'Keynes–Ramsey equation':

$$\dot{c} = \sigma(c)c[f_k(k) - (\rho + \nu)] \tag{17.16}$$

where $\sigma(c) = -(u_c/u_{cc}c) > 0$ measures the Fisherian intertemporal elasticity of substitution. In other words, consumption per worker grows over time if the marginal product of capital exceeds its user cost $(\rho + \nu)$.

Equations (17.11) and (17.16) constitute the dynamical system of this optimal urban growth model in (c, k) space. In the steady state $(\dot{k} = 0$ and $\dot{c}/c = \gamma)$, we get

$$\begin{aligned} f_k(k) &= [\rho - (1 - \sigma^{-1})\gamma] + \nu \\ &= (\rho + n + \sigma^{-1}) + b^N(w/w^A - 1) - b^K(r - r^A) \end{aligned} \tag{17.17}$$

$$\begin{aligned} c = A[f(k) - \nu k] = Af(k) - A[(n + \gamma) + b^N(w/w^A - 1) \\ - b^K(r - r^A)]k \end{aligned} \tag{17.18}$$

Equation (17.17) is the modified golden rule that equates the marginal product of capital with its user cost, whereas Equation (17.18) equates net output per worker with consumption per worker. The dynamics of the system can be depicted in Figure 17.2.

Equation (17.16) clearly indicates that the dynamics of consumption per worker evaluated at the steady state is independent of c, so the $\dot{c} = \gamma c$ locus is vertical. Totally differentiating Equation (17.11) and evaluating it at the steady state yields a hump-shaped $\dot{k} = 0$ locus, where the peak corresponds to the conventional golden rule solution of capital per effective worker (k^{G}). The intersection of these two loci (point E) determines the steady-state optimal growth

Figure 17.2. Equilibrium dynamics in the optimal urban growth model

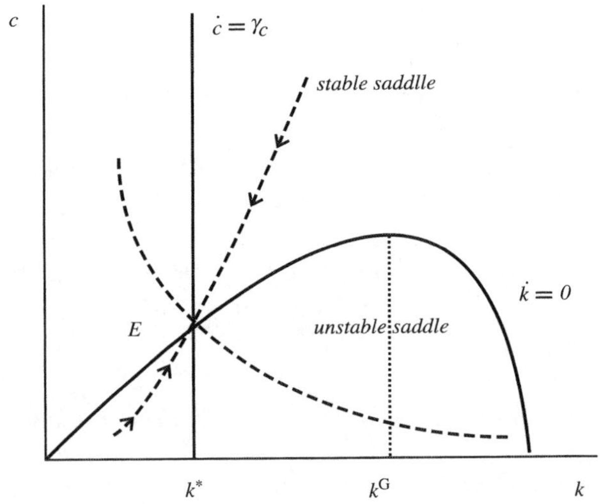

equilibrium value of capital per effective worker (k^*). Notice that along the unstable saddle, either the non-negativity constraint on consumption per worker ($c > 0$) or the transversality condition (17.15) will be violated. The dynamic equilibrium is thus uniquely determined by the stable saddle along which c (the control variable) must jump to the corresponding value for each given value of k (the state variable).

Comparing Equation (17.17) with Equation (17.9), one can see that the former has an addition of the term $\rho - (1 - \sigma^{-1})\gamma$, which is positive under the transversality condition. By diminishing marginal product, it follows immediately that relative to the optimal growth solution, the golden rule solution must over-accumulate physical capital in the steady state and is hence dynamically inefficient. In addition to previous comparative-static results (which continue to hold qualitatively), we also find that either an increase in the time preference rate or a decrease in the intertemporal elasticity of substitution discourages steady-state capital accumulation and lowers per capita output. As before, the rate of output growth is solely driven by the exogenous technical progress rate. Furthermore, we can characterize the optimal saving rate in the city, which falls (rises) with capital per worker if the intertemporal elasticity of substitution is sufficiently high (low) relative to the capital income share, $kf_k(k)/f(k)$.

In spite of their substantial influence on the theory of urban dynamics, exogenous growth models have several shortcomings. For brevity, we only discuss three. First, since the rate of exogenous technical progress is the lone driving force of the advancement of the city economy, little has been added to the understanding of the determinants of city growth. Second, along the stable saddle, cities with the same technology converge monotonically to the steady state, thus failing to explain why some cities rise but others fall. Finally, with regard to urban policy issues, all the instruments can only affect city growth in transition unless they can affect the rate of technological change.

17.3. From exogenous to endogenous urban growth models

Since the seminal work by Romer (1986), growth theorists have revived interest in long-run development by devoting effort toward understanding the underlying forces for economic advancement under the so-called endogenous growth framework. A central feature of the endogenous growth model is to consider the marginal product of all reproducible factors (including human, knowledge, physical and research capitals) to be bounded below by a positive constant. This is ensured when the aggregate production function exhibits:

(i) constant returns (cf. Rebelo, 1991; Bond *et al.*, 1996; Benhabib *et al.*, 2000),
(ii) asymptotically constant returns (cf. Pitchford, 1960; Jones and Manuelli, 1990),
(iii) or increasing returns (cf. Romer, 1986; Lucas, 1988; Boldrin and Rustichini, 1994),

with respect to all reproducible factors simultaneously.

The development of endogenous growth theory based on the one-sector Ramsey–Cass–Koopmans setup consists of important contributions by Romer (1986), Jones and Manuelli (1990) and Rebelo (1991). In these papers, general capital is interpreted as either knowledge capital or a combination of both physical and human capital. When the model exhibits constant returns (cf. Rebelo, 1991), the economy jumps instantaneously onto the balanced growth path (BGP) along which consumption, capital and output all grow at a common rate. Yet, even under constant returns, one may obtain very rich growth dynamics in a two- or multi-sectoral setting, following the classic work by Uzawa (1965). Typical examples can be found in the two-sector, constant-returns endogenous growth model of Bond *et al.* (1996) or in the two-sector, social constant-returns (private diminishing-returns) endogenous growth model of Benhabib *et al.* (2000). In these models, there is a capital good that is a perfect substitute for the final consumption good (normally referred to as physical capital) and a non-consumable pure capital good (usually referred to as human capital, knowledge capital or research capital). As far as research capital (or inventive activity) is concerned, it is natural to permit some degree of monopoly power, as suggested by Shell (1966). Accordingly, one may consider R&D and growth in a monopolistically competitive framework (cf. Romer, 1990; Grossman and Helpman, 1991) or in a monopoly setup (cf. Aghion and Howitt, 1992). We summarize this literature in Table 17.4.

In this section, we will discuss an array of endogenous growth models of cities, organized by the number of sectors in the model and the way the population size of the city is determined.

17.3.1. A basic one-sector endogenous urban growth model

There is no doubt that the simplest form of the production function satisfying the required property for endogenous growth is the so-called AK-model constructed by Rebelo (1991). Specifically, output is assumed to be linear in a general capital input: $Y = AK$ with $A > 0$. In terms of urban economics where endogenous population is a critical issue, however, this production function fails to capture the implications of an endogenous labor force. Thus, a more appropriate

Table 17.4. Summary of endogenous growth models

Returns to Scale	One-Sector	Multi-Sector
Constant or Asymptotically Constant	Jones and Manuelli (1990) and Rebelo (1991)	Bond *et al.* (1996) and Benhabib *et al.* (2000)
Increasing	Romer (1986)	Lucas (1988), Romer (1990), Grossman and Helpman (1991), Stokey (1991), Aghion and Howitt (1992) and Boldrin and Rustichini (1994)

form that follows the spirit of Romer (1986) is $Y = AK^{\alpha}N^{1-\alpha}\bar{K}^{1-\alpha}$, where A is assumed to be a positive constant (i.e. the exogenous technical progress rate γ is set to zero), $\bar{K} = K$ is the aggregate level of capital in the city and $\alpha \in (0, 1)$. Notice that this production function exhibits constant returns in all reproducible factors (K and $\bar{K}$) and constant returns in all private factors (K and N). While the former ensures a positive marginal product of reproducible capital, the latter guarantees 100% distribution of gross revenues to private factors and zero profit in equilibrium. The term $\bar{K}$ is designed to capture the Marshallian externality in a given city, created by uncompensated positive spillovers of general capital as suggested by Jacobs (1969). Such spillovers may be consequences of knowledge overlaps, peer-group learning effects, neighborhood externalities and/or invention applications. In the context of urban economics, the general capital stock can be seen as consisting of physical capital (structures) or human capital (knowledge), depending on the application.

To accept balanced growth, consider a felicity function with constant elasticity of intertemporal substitution (σ): $u(C/N) = [(C/N)^{1-\sigma^{-1}} - 1]/(1-\sigma^{-1})$, where $\sigma > 1$ and C is total consumption. Since per capita output is equal to $Y/N = A(K/N)^{\alpha}(\bar{K}/N)^{1-\alpha}N^{1-\alpha}$, we can specify the optimization problem of a representative agent in the urban area as

$$\max_{C/N} U = \int_0^{\infty} \frac{(C/N)^{1-\sigma^{-1}} - 1}{1-\sigma^{-1}} \mathrm{e}^{-\rho t}\mathrm{d}t$$

$$\text{s.t.} \left(\frac{\dot{K}}{N}\right) = A\left(\frac{K}{N}\right)^{\alpha}\left(\frac{\bar{K}}{N}\right)^{1-\alpha} N^{1-\alpha} - \nu\frac{K}{N} - \frac{C}{N} \tag{17.19}$$

where $K(0)/N(0) = k_0 > 0$ is given, and recall that $\nu = n + b^N(w/w^A - 1) - b^K(r - r^A)$ measures the investment ratio. Moreover, since wages are growing and unbounded with general capital in this setup, it is crucial to assume that the alternative wage will also grow at the same rate as the city's aggregate capital stock so that a BGP is attainable. More explicitly, we let $w^A = w_0\bar{K}$, which is taken as parametrically given by each individual agent, with the coefficient w_0 to be determined in the balanced growth equilibrium.

Following similar techniques, it is not difficult to show that all per capita quantities (consumption, capital, and output) exhibit common growth at rate θ, and the modified golden rule can be written as

$$\left(\frac{\dot{C}}{N}\right) = \sigma\left(\frac{C}{N}\right)\left[\alpha A N^{1-\alpha} - (\rho + \nu)\right] \tag{17.20}$$

In other words, the consideration of city-wide positive spillovers creates a 'scale effect' in the sense that a larger population size spurs city growth.

In equilibrium, factors must be paid at their marginal products: $w = (1-\alpha)AN^{-\alpha}K$ and $r = \alpha A N^{1-\alpha}$. Thus, we can rewrite the investment ratio as a function of N alone:

$$\nu(N) = (n + b^K r^A - b^N) + [b^N(1-\alpha)A/w_0]N^{-\alpha} - b^K \alpha A N^{1-\alpha} \tag{17.21}$$

which is strictly decreasing and strictly convex in N. It is useful to note that in the exogenous growth framework, the investment ratio is simply a constant in the steady state. Thus, the dependency of the investment ratio on the size of city population is again a consequence of the scale effect created by the Marshallian externality.

In the conventional exogenous growth model, the system becomes stationary once all growing variables are divided by the technology scaling factor. In contrast, the technology-scaling factor is fixed in the present setup where the source of growth arises endogenously from the accumulation of general capital in the absence of diminishing returns. As a consequence, it is convenient to transform the system by using the 'great ratios' – in particular, the consumption–capital ratio, $\chi \equiv C/K$. Combining Equations (17.19) and (17.20), one gets:

$$\dot{\chi} = \chi[\chi - (1 - \alpha\sigma)\alpha A N^{1-\alpha} + (1-\sigma)\nu(N) - \sigma\rho] \tag{17.22}$$

which depends on χ and N. Next, substituting the equilibrium values of wages into Equation (17.3) yield:

$$\dot{N} = \{(n - b^N) + [b^N(1-\alpha)A/w_0]N^{-\alpha}\}N \tag{17.23}$$

which is driven by N alone.

Along a BGP, consumption and capital grow at a common rate and there must neither be net migration nor net capital flows interregionally, i.e. we must have $\dot{\chi} = 0$, $\dot{N} = nN$, and $w_0 = [\alpha^{\alpha}(1-\alpha)^{1-\alpha}A(r^A)^{-\alpha}]^{1/(1-\alpha)}$. We can substitute w_0 into Equation (17.23) to obtain:

$$\dot{N} = \{(n - b^N) + b^N[r^A/(\alpha A)]^{\alpha/(1-\alpha)}N^{-\alpha}\}N \tag{17.24}$$

which implies the rate of city population growth is decreasing in the size of the city, eventually converging to a maximum bound. By straightforward manipulation, the BGP equilibrium values of (χ^*, N^*) can be solved:

$$N^* = [r^A/(\alpha A)]^{1/(1-\alpha)} \tag{17.25}$$

$$\chi^* = (1 - \alpha\sigma)r^A + (1 - \sigma)n - \sigma\rho \tag{17.26}$$

Equations (17.22) and (17.24) constitute the dynamic system of (χ, N). The local dynamic properties of the BGP equilibrium can be characterized by evaluating the trace and the determinant of the corresponding Jacobian matrix at the BGP values (χ^*, N^*), which are, respectively,

$$\text{trace} = (n - \alpha b^N) + \chi \qquad \text{and} \qquad \text{determinant} = (n - \alpha b^N)\chi \tag{17.27}$$

For a given natural rate of population growth n, the BGP is a saddle if and only if $n < \alpha b^N$. In other words, saddle path stability requires that (i) the city is responsive to migration (so b^N is high enough) and (ii) the agglomerative externality from uncompensated spillovers is not too large (so α is high enough). Otherwise, when these conditions do not hold, namely when $n > \alpha b^N$, the city must either shrink continually or grow unboundedly over time, implying that the BGP equilibrium is a source (locally unstable). The BGP equilibrium and the transition path can be depicted in Figure 17.3 for the case of $\sigma < 1$.

17.3.2. A modified one-sector model of endogenous urban growth

The model delineated above is essentially a straightforward extension of the conventional exogenous urban growth model to allow general capital that is not subject to diminishing returns in the

Figure 17.3. Equilibrium dynamics in the endogenous urban growth model

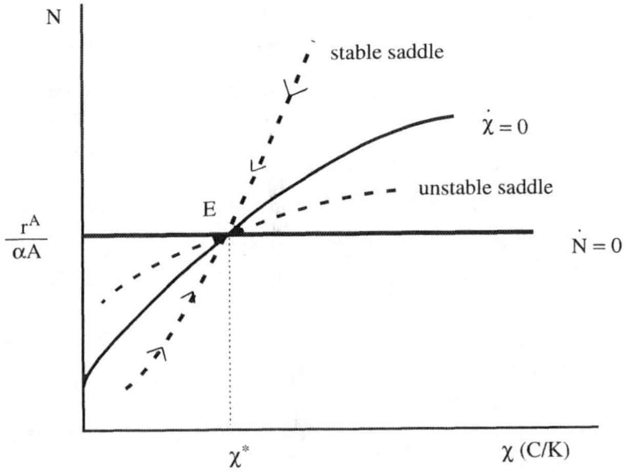

long run. While the evolution of city population depends on an ad hoc adjustment process based on interregional wage differentials, the model itself also lacks an explicit spatial structure. These shortcomings will be remedied in this section, which follows closely the work by Palivos and Wang (1996).

Consider a Lösch–Alonso circular monocentric city framework with completely inelastic demand for land and an exogenous, time-varying city border of $b(t)$ at time t. For simplicity, it is assumed that there is no exogenous population growth ($n = 0$), i.e. any evolution of city population must be a result of migration. The fixed density of land used by a consumer at distance $z \in [0, b(t)]$ from the central business district (CBD) $m(z) = m$ (for all z) and hence the population density is $2\pi z/m$. The population identity is $N(t) = \int_0^{b(t)} (2\pi z/m)\mathrm{d}z$, or, equivalently,

$$b(t) = [mN(t)/\pi]^{1/2} \tag{17.28}$$

Let all production and transaction activities take place in the CBD. Let $C(t)$ be the total amount of composite good produced by the economy at time t. Per capita land use is assumed to be constant across locations, and the only other good consumed by agents is composite good. Hence the gross per capita consumption of composite good, prior to the subtraction of transport cost from the

CBD to the consumer's house, for all agents living in all locations is $C(t)/N(t)$. Assume that the transportation cost facing each agent takes a modified iceberg form: $T(z,t) = \tau z C(t)/N(t)$, where $\tau > 0$ measures the transportation cost of one unit of consumption good per unit distance. Further assume that the city government sublets the publicly owned land to city residents at the competitively determined rent $R(z,t)$. Since all households are identical and the lot size is fixed, locational equilibrium implies $T(z,t) + R(z,t)$ must be constant for all $z \in [0, b(t)]$. By normalizing the rural land rent to zero, we have:

$$R(z,t) = T(b,t) - T(z,t) = \tau[b(t) - z][C(t)/N(t)] \tag{17.29}$$

which is, as one would expect, decreasing in the distance from the CBD.

Straightforward integration, together with the expression (17.28), yields the total transportation cost (TTC) and the total land rent (TLR):

$$\text{TTC}(t) = (2/3)\tau(m/\pi)^{1/2}[N(t)]^{3/2}[C(t)/N(t)] \tag{17.30}$$

$$\text{TLR}(t) = (1/3)\tau(m/\pi)^{1/2}[N(t)]^{3/2}[C(t)/N(t)] \tag{17.31}$$

Therefore, the resource constraint for the city economy in per capita form can be specified as

$$\left(\frac{\dot{K}}{N}\right) = A\left(\frac{K}{N}\right)^{\alpha}\left(\frac{\bar{K}}{N}\right)^{1-\alpha} N^{1-\alpha} - \left[1 + \tau\left(\frac{m}{\pi}\right)^{1/2} N^{1/2}\right]\frac{C}{N} \tag{17.32}$$

where the last term is the sum of per capita spending, including consumption, transportation cost, and land rent for each agent.

Thus, the representative agent's optimization problem is

$$\max_{C/N} U = \int_0^{\infty} \frac{(C/N)^{1-\sigma^{-1}} - 1}{1 - \sigma^{-1}} \mathrm{e}^{-\rho t} \mathrm{d}t \tag{17.33}$$

subject to Equation (17.32). The modified golden rule now becomes

$$\left(\frac{\dot{C}}{N}\right) = \left(\frac{C}{N}\right)\sigma(\alpha A N^{1-\alpha} - \rho) \tag{17.34}$$

Along a BGP, one can utilize Equations (17.32) and (17.33) to compute the lifetime utility of the representative agent, for a given value of N, as

$$U(N) = -\frac{(1/\sigma)^{1/\sigma}[\sigma/(1-\sigma)]}{(k_0)^{(1-\sigma)/\sigma}\{\rho + [(1-\sigma)/\sigma]A\alpha N^{1-\alpha}\}}$$

$$\times \left\{ \left[\frac{1+\tau(m/\pi)^{1/2}N^{1/2}}{\rho + (\sigma^{-1}-\alpha)AN^{1-\alpha}} \right]^{(1-\sigma)/\sigma} \right\}$$

$$\equiv -U^D(N)\{U^C(N)\} \tag{17.35}$$

Thus, population size affects lifetime utility through two channels: (i) it raises individual welfare via a positive scale effect on city growth (embedded in U^D via the effective discount rate $\rho + [(1-\sigma)/\sigma]A\alpha N^{1-\alpha}$) and (ii) it creates an ambiguous effect on welfare via its impact on the initial level of per capita consumption (captured by the term in curly brackets, $U^C(N)$). If the initial consumption effect is locally negative around the BGP, there exists an individual welfare maximizing size of city population.

Let G denote government spending. One may construct a social optimum where the benevolent (individual welfare maximizing) city government chooses consumption, capital accumulation and population subject to the resource constraint and the (period by period) government budget constraint:

$$\left(\frac{\dot{K}}{N}\right) = A\left(\frac{K}{N}\right)^{\alpha}\left(\frac{\bar{K}}{N}\right)^{1-\alpha} N^{1-\alpha} - \left[1 + \frac{2}{3}\tau\left(\frac{m}{\pi}\right)^{1/2} N^{1/2}\right]\frac{C}{N} - \frac{G}{N} \tag{17.36}$$

$$G = \frac{1}{3}\tau\left(\frac{m}{\pi}\right)^{1/2} N^{3/2} \tag{17.37}$$

where the government spending is entirely financed by the total city land rent (the Henry George approach). The socially optimal consumption growth is governed by the following equation:

$$\left(\frac{\dot{C}}{N}\right) = \left(\frac{C}{N}\right)\sigma(AN^{1-\alpha} - \rho) \tag{17.38}$$

Comparing Equation (17.38) with Equation (17.34), one can easily see that the social rate of return on capital is higher than the private rate of return, thus yielding a higher city growth under the social optimum. Intuitively, this is due to the presence of the free-rider problem in a decentralized equilibrium where an individual fails to account for the positive spillover effect of their investment in capital.

Moreover, the lifetime utility along a BGP is given by

$$U^{S}(N) = U(N)\left[\frac{\rho + (\sigma^{-1} - \alpha)AN^{1-\alpha}}{\rho + (\sigma^{-1} - 1)AN^{1-\alpha}}\right]^{(1-\sigma)/\sigma} \equiv U(N)Y(N) \tag{17.39}$$

Note that provided $\sigma < 1$, we have $U(N) < 0$, $Y(N) < 1$ and $\mathrm{d}Y(N)/\mathrm{d}N < 0$. Manipulation of Equations (17.35) and (17.39) thus implies a lower net marginal benefit of city size in the private sense (NMB) than that in the social sense (NMB^{S}), as in Figure 17.4. From Equations (17.34) and (17.38), we can also plot the corresponding city growth rates (θ and θ^{S}) as (strictly increasing and strictly concave) functions of city population.

In summary, we have: (i) the social welfare and city growth achieved under a decentralized equilibrium is lower than those under a social optimum, and (ii) in a decentralized equilibrium, the city is under-populated relative to the social optimum. This latter result is in sharp contrast with the conventional textbook proposition where as a result of a negative traffic congestion externality, equilibrium cities are over-populated (cf. Kanemoto, 1980; Fujita, 1989). For an entirely different reason, Abdel-Rahman (1990) finds under-populated cities in a decentralized equilibrium. He argues that individuals fail to account for the benefit from the reduction of the city infrastructure fixed cost spread over residents from an incremental increase in population. The results have useful empirical implications. Conventionally, whether a city is over or under-populated is tested based on the Henry George Theorem using the land value to Pigouvian subsidy

Figure 17.4. Decentralized equilibrium vs. social optimum

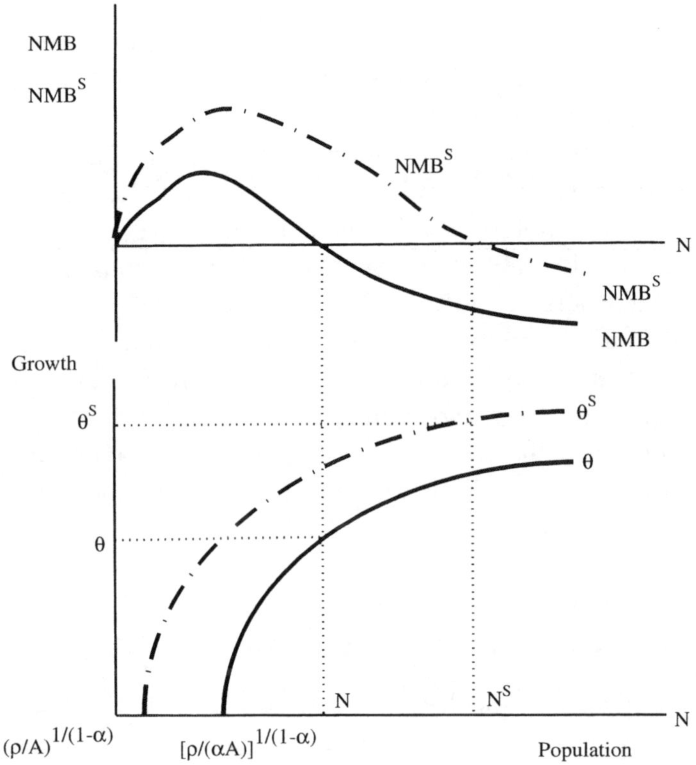

ratio (e.g. see Kanemoto *et al.*, 2003). From the arguments above, however, a more rigorous test may include the imputed value of uncompensated knowledge spillovers at the city level and urban-specific public infrastructure that are not fully capitalized in land value. See Palmon and Smith (1998) for a careful empirical study documenting partial, but not full, capitalization.

17.3.3. Housing dynamics and zoning

In the previous subsections, the housing stock was regarded as part of general capital with investment in housing a perfect substitute for the final consumption good. By construction, the relative price of housing (in units of the final consumption good) is unity. In this subsection, we follow the model developed by Lin *et al.* (2002) to allow for a determination and full characterization of intertemporal housing prices within the generalized *AK*-framework.

In particular, we modify the model in Section 17.3.2 by

(i) interpreting the consumption good as household consumption, the capital stock as housing capital and the production activity as household production;
(ii) decomposing housing capital as the product of housing quality (denoted Q, which can grow unboundedly) and housing quantity (denoted q, which is bounded above), i.e. $K/N = Qq$;
(iii) allowing for the non-productive use of housing capital to augment leisure and enhance utility;[10]
(iv) considering variable heights of buildings under a zoning policy such that the population density is uniform across locations within the monocentric city;
(v) eliminating the scale effect by using the 'average' housing stock to measure positive spillovers (especially, in the presence of congestion, the use of aggregate housing stock is difficult to justify);
(vi) focusing on a closed city with migration only within the city and with the city population growing at an exogenous rate n.

Letting ζ denote the fraction of housing capital devoted to the productive activity, the fraction $1 - \zeta$ of housing capital is then allocated to non-productive home entertainment. Consider the following felicity function of the Cobb–Douglas form satisfying the constant elasticity of intertemporal substitution property: $u(C/N, (1-\zeta)(K/N)) = \{[(C/N)^{\xi}((1-\zeta)(K/N))^{1-\xi}]^{1-\sigma^{-1}} - 1\}/(1-\sigma^{-1})$, where $\xi \in (0,1)$. In other words, housing capital augments leisure time, contributing to greater utility. Next, the household production technology is given by $Y = A(\zeta K/N)^{\alpha} \times (\bar{K}/N)^{1-\alpha}$. Moreover, we specify a plausible floor area ratio (FAR) schedule: $\Psi = \bar{\Psi} - \psi_0 \psi(z)$, where $\psi_z > 0$, $\psi_{zz} < 0$, $\psi_0 > 0$, and $\bar{\Psi} > \psi(b)$, for all $z \in [0, b]$. As observed in practice, this zoning restriction permits higher buildings closer to the CBD. Under this FAR schedule, the population density at distance z from the CBD is $2\pi z/[q(z,t)/\Psi(z)]$, and, by construction, the population identity requires $2\pi z/[q(z,t)/\Psi(z)] = N(t)/b(t)$. Furthermore, since a

---

[10] E.g. gardens.

greater FAR can cause heavier traffic congestion, we postulate that the unit transportation cost is an increasing function in the FAR: $\tau = \tau(\Psi)$.

Denoting by $p(z,t)$ the relative price of housing investment (in units of the final consumption good) at location z in time t, the representative household's intertemporal optimization problem can then be specified as

$$\max_{C/N} U = \int_0^{\infty} \frac{\{(C/N)^{\xi}[(1-\zeta)(K/N)]^{1-\xi}]^{1-\sigma^{-1}} - 1\}}{1-\sigma^{-1}} \mathrm{e}^{-\rho t} \mathrm{d}t \qquad (17.40)$$

$$\text{s.t.} \left(\frac{\dot{K}}{N}\right) = \frac{1}{p(z,t)} \left\{ A\left[\frac{K}{N}(z,t)\right]^{\alpha} \left[\frac{\bar{K}}{N}(z,t)\right]^{1-\alpha} - [1 + \tau(\Psi(z))z] \frac{C}{N}(z,t) \right\} \qquad (17.41)$$

The tradeoff between consumption and leisure is governed by

$$\frac{1-\xi}{\xi} \frac{C/N}{(1-\zeta)K/N} = \frac{\alpha A \zeta^{\alpha-1}}{1+\tau(\Psi)z} \qquad (17.42)$$

which yields

$$\chi = \frac{\xi}{1-\xi} \frac{\alpha A(1-\zeta)\zeta^{\alpha-1}}{1+\tau(\Psi)z} \qquad (17.43)$$

The modified golden rule for consumption and productive time allocation are

$$\left(\frac{\dot{C}}{N}\right) = \left(\frac{C}{N}\right) \sigma \left(\frac{1}{p} \alpha A \zeta^{\alpha-1} - \rho\right) \equiv \left(\frac{C}{N}\right) \sigma \Theta(\zeta, p) \qquad (17.44)$$

$$\begin{aligned} \dot{\zeta} &= -\frac{\zeta}{1-\alpha+\zeta/(1-\zeta)} \left[\Theta(\zeta,p) - A\zeta^{\alpha} + \frac{\xi}{1-\xi} \alpha A(1-\zeta)\zeta^{\alpha-1} \right] \\ &\equiv -\frac{\zeta}{1-\alpha+\zeta/(1-\zeta)} X(\zeta,p) \end{aligned} \qquad (17.45)$$

where it is clearly seen that $\Theta_{\zeta} < 0$, $\Theta_p < 0$, $X_{\zeta} < 0$, and $X_p < 0$ and that, utilizing Equations (17.41), (17.43) and (17.44),

$$\dot{\chi} = \chi X(\zeta, p) \qquad (17.46)$$

The evolution of the relative price of housing satisfies:

$$\dot{p} = \frac{(1-\alpha)(1-\xi+\sigma^{-1}\xi)+\sigma^{-1}\zeta/(1-\zeta)}{1-\alpha+\zeta/(1-\zeta)} X(\zeta,p) - \sigma^{-1}\Theta(\zeta,p) \tag{17.47}$$

which is a function of ζ and p alone.

Since households are free to relocate, locational equilibrium implies that at any point in time,

$$u\left[\frac{C}{N}(z,t),(1-\zeta(z,t))\frac{K}{N}(z,t)\right] = u\left[\frac{C}{N}(b,t),(1-\zeta(b,t))\frac{K}{N}(b,t)\right]$$
$$\forall z \in [0,b(t)] \tag{17.48}$$

Along a BGP, consumption, housing quality, housing capital, and output all grow at a common rate $\theta = \Theta(\zeta,p)$, which can be written as (recalling Equation (17.44)):

$$\theta = \frac{1}{p}\alpha A\zeta^{\alpha-1} - \rho \tag{17.49}$$

Moreover, along a BGP, $\dot{\chi} = 0$ and from Equations (17.45) and (17.46),

$$\theta = A\zeta^{\alpha} - \frac{\xi}{1-\xi}\alpha A(1-\zeta)\zeta^{\alpha-1} \tag{17.50}$$

Normalizing the housing price on the urban fringe to $p(b) = \bar{p}$ and utilizing the assumption of a uniform distribution of population, the population identity, the definition of housing capital as well as Equations (17.43) and (17.44), we can derive the BGP value of housing prices at each location:

$$p(z) = \bar{p} + \frac{\xi}{1-\xi}\frac{\alpha A}{\theta}[B(b,z)-1]\zeta^{\alpha-1}(z)[1-\zeta(z)] \tag{17.51}$$

where

$$B(b,z) = \frac{K_0(b)}{K_0(z)} = \frac{\Psi(b)b}{\Psi(z)z}\frac{q_0(b)}{q_0(z)},$$

measuring the ratio of the initial capital stocks between b and z, and satisfying $B(z;b) > B(b;b) = 1$. It can be easily verified that B is decreasing in ψ_0 and increasing in $\bar{\Psi}$. By the standard 'land abundance' argument, we expect that for a given population, an increase in urban fringe drives down housing prices, which is guaranteed by postulating $\mathrm{d}B(z;b)/\mathrm{d}b < 0$.

Equations (17.49)–(17.51) jointly determine the BGP values of (θ, ζ, p). Straightforward comparative-static analysis suggests:

(i) an increase in urban fringe (b), under a given population, encourages the productive use of housing capital, lowers housing prices at all locations and leads to a permanently higher city growth rate;
(ii) an increase in the unit transportation cost (τ) enlarges the ratio of housing stock at the border to that at any arbitrary location and raises housing prices at all locations, thus reducing the productive use of housing capital and the rate of growth of the city;
(iii) a loose zoning restriction on FAR, relaxing it more than proportionately for locations near the city center (higher ψ_0), encourages the productive use of housing capital, lowers equilibrium housing prices and fosters economic growth; however, a uniformly loose zoning restriction on the FAR (higher $\bar{\Psi}$) generates reverse outcomes.

Finally, we turn to examining the stability of the dynamical system summarized by Equations (17.45) and (17.47) in (ζ, p) space, because of the recursive nature of the model. By totally differentiating the system and evaluating the Jacobian at the BGP values, we can compute its determinant and the trace:

$$\text{determinant} = -\frac{\alpha A p \Theta_p \sigma^{-1} \zeta^{\alpha}}{1-\alpha+\zeta/(1-\zeta)} \times \left[1 + \frac{\xi}{1-\xi}\frac{1-\zeta}{\zeta}\left(1-\alpha+\frac{\zeta}{1-\zeta}\right)\right] > 0 \tag{17.52}$$

$$\text{trace} = -\frac{1}{1-\alpha+\zeta/(1-\zeta)} \times \left[p\Theta_p(1-\alpha)(1-\xi)(\sigma^{-1}-1)+\zeta X_{\zeta}\right] > 0 \tag{17.53}$$

Notice that in this 2×2 system, ζ and p are jump variables that can be adjusted instantaneously. Since the determinant is positive, the two characteristic roots must have the same sign. This together with positive trace implies both the roots are positive. Similar to the ad hoc one-sector urban growth model presented in Section 17.3.1, the BGP equilibrium is locally determinate in this one-sector economy, contrasting with findings in Romer (1986) and Rebelo (1991) where the economy must jump instantaneously to the BGP. Therefore, household relocation and locational equilibrium in an optimizing setting act as stabilizing forces, enabling saddle-path stability of the BGP equilibrium in the city economy. One may wonder whether this stability property continues to hold in a fully specified two-sector framework with heterogeneous non-housing and housing capitals, to which we now turn.

17.3.4. Two-sector endogenous urban growth and stability

The materials presented in this subsection extends the original work by Anas *et al.* (1995) and Chang (1999). Other contributors in this area include Turnovsky and Okuyama (1994), Black and Henderson (1999), Li (2002), and Rossi-Hansberg and Wright (2003), to name but a few. Consider a city economy featuring a spatial structure as delineated in Section 17.3.2. However, there are now two productive sectors: a final (composite) consumption – (non-housing) investment good sector, and a housing investment sector. The focus herein is on the interactions between the two sectors and their consequences for the growth dynamics of the underlying city economy.

Assume that labor, with endowment normalized to one, is inelastically supplied to production of the final good (at a fraction ζ) and housing investment (at a fraction $1-\zeta$) and with housing capital yielding no direct consumption value. Housing capital (H) can be combined with non-housing capital (K) to produce the final good. Both reproducible capitals generate sector-specific positive spillovers and both production technologies exhibit constant social returns, taking the Cobb–Douglas form. Specifically, denoting x_{ij} as the fraction of factor i allocated to sector j, the output of sector j is then given by

$$Y_j = A_j(x_{Kj}K/N)^{\alpha_{Kj}-\varepsilon_{Kj}}(x_{Hj}H/N)^{\alpha_{Hj}-\varepsilon_{Hj}}(\bar{x}_{Kj}\bar{K}/N)^{\varepsilon_{Kj}}(\bar{x}_{Hj}\bar{K}/N)^{\varepsilon_{Hj}} \tag{17.54}$$

where α_{ij}, $\varepsilon_{ij} \in (0,1)$, $i = K, H$ and $j = 1, 2$. While $\alpha_{ij} - \varepsilon_{ij}$ measures the private output elasticity of factor i in sector j, α_{ij} represents the corresponding social output elasticity with ε_{ij} indicating the degree of sector-specific externality. By constant social returns, we have $\alpha_{Kj} + \alpha_{Hj} = 1$. Further, we assume that the two technologies are different in the sense of both private and social returns, i.e. the matrices $[\alpha_{ij} - \varepsilon_{ij}]$ and $[\alpha_{ij}]$ are non-singular.

The representative agent's optimization is, therefore, given by:

$$\max_{C/N} U = \int_0^\infty \frac{(C/N)^{1-\sigma^{-1}} - 1}{1-\sigma^{-1}} \mathrm{e}^{-\rho t}\mathrm{d}t$$

subject to the evolution equations for per capita capital and housing:

$$\left(\frac{\dot{K}}{N}\right) = A_1\left(x_{K1}\frac{K}{N}\right)^{\alpha_{K1}-\varepsilon_{K1}}\left(x_{H1}\frac{H}{N}\right)^{\alpha_{H1}-\varepsilon_{H1}}\left(\bar{x}_{K1}\frac{\bar{K}}{N}\right)^{\varepsilon_{K1}} \times\left(\bar{x}_{H1}\frac{\bar{K}}{N}\right)^{\varepsilon_{H1}} - n\frac{K}{N} - \left[1+\tau\left(\frac{m}{\pi}\right)^{1/2}N^{1/2}\right]\frac{C}{N} \tag{17.55}$$

$$\left(\frac{\dot{H}}{N}\right) = A_2\left(x_2\frac{K}{N}\right)^{\alpha_{K2}-\varepsilon_{K2}}\left(x_{H2}\frac{H}{N}\right)^{\alpha_{H2}-\varepsilon_{H2}}\left(\bar{x}_{K2}\frac{\bar{K}}{N}\right)^{\varepsilon_{K2}}\left(\bar{x}_{H2}\frac{\bar{K}}{N}\right)^{\varepsilon_{H2}} - n\frac{H}{N} \tag{17.56}$$

the factor reallocation constraints, $x_{i1} + x_{i2} = 1$ $(i = K, H)$, and the given values of the initial stocks, $K(0)/N(0)$ and $H(0)/N(0)$. To ease the complexity of notation, let $x_j \equiv x_{Kj}$ and hence $x_{Hj} = 1 - x_j$.

Denote the nominal shadow price of factor i as W_i and the nominal shadow price of sector j output as P_j. Following the techniques developed by Bond *et al.* (1996), we can combine the first-order conditions (with respect to C/N and x_j) and the Euler equations (with respect to K/N and H/N) to yield:

$$c^{-\sigma} = P_1 \tag{17.57}$$

$$W_i = P_j(\alpha_{ij} - \varepsilon_{ij})A_j(x_{Kj}K/N)^{\alpha_{Kj}}(x_{Hj}H/N)^{\alpha_{Hj}}/(x_{ij}i/N) \tag{17.58}$$

$$\frac{\dot{P}_1}{P_1} = (\rho + n) - \frac{W_K(P)}{P_1} \quad \text{and}$$
$$\frac{\dot{P}_2}{P_2} = (\rho + n) - \frac{W_H(P)}{P_2} \tag{17.59}$$

for $i = K, H$ and $j = 1, 2$, where $P = (P_1, P_2)$.

We now transform the endogenous variables into stationary forms by dividing each of them by their growth factors: $k \equiv (K/N)/e^{\theta t}$, $h \equiv (H/N)/e^{\theta t}$, $y_j \equiv Y_j/e^{\theta t}$, $w_i \equiv W_i\, e^{\sigma\theta t}$, and $p_j \equiv P_j\, e^{\sigma\theta t}$ (see Mulligan and Sala-i-Martin (1993) and Benhabib *et al.* (2000), for a discussion of this time-elimination method for solving recursive optimal growth models). We can therefore obtain the 4×4 transformed system in (k, h, p_1, p_2) as follows:

$$\dot{k} = y_1 - (n + \theta)k - [1 + \tau(m/\pi)^{1/2}N^{1/2}]p_1^{-1/\sigma} \tag{17.60}$$

$$\dot{h} = y_2 - (n + \theta)h \tag{17.61}$$

$$\dot{p}_1 = (\rho + n + \sigma\theta)p_1 - w_K(p) \tag{17.62}$$

$$\dot{p}_2 = (\rho + n + \sigma\theta)p_2 - w_H(p) \tag{17.63}$$

where $p = (p_1, p_2)$.

To characterize this dynamical system, we follow Bond *et al.* (1996) and utilize the duality properties of two-sector dynamic general equilibrium models. In equilibrium, competitive profit yields the following Samuelsonian relationship:

$$p_j = \left(\frac{w_K}{\alpha_{Kj} - \varepsilon_{Kj}}\right)^{\alpha_{Kj}} \left(\frac{w_H}{\alpha_{Hj} - \varepsilon_{Hj}}\right)^{\alpha_{Hj}}, \qquad j = 1, 2 \tag{17.64}$$

The unit factor input coefficients are

$$a_{ij} = \frac{(\alpha_{ij} - \varepsilon_{ij})p_j}{w_i}, \qquad i = K, H, \;\; j = 1, 2 \tag{17.65}$$

For convenience, let $[a_{ij}]$ denote the input coefficient matrix. Using full employment, we have

$$[a_{ij}]\begin{bmatrix} y_1 \\ y_2 \end{bmatrix} = \begin{bmatrix} k \\ h \end{bmatrix} \tag{17.66}$$

Further, based on the duality relationships, the cost share coefficients can be written as

$$\hat{a}_{ij} = a_{ij}\left(\frac{\alpha_{ij}}{\alpha_{ij} - \varepsilon_{ij}}\right), \qquad i = K, H, \ j = 1, 2 \tag{17.67}$$

Accordingly, we can denote by $[\hat{a}_{ij}]$ the cost share coefficient matrix. The Samuelsonian relationship (17.64) therefore implies:

$$\begin{bmatrix} p_1 \\ p_2 \end{bmatrix} = [\hat{a}_{ij}]'\begin{bmatrix} w_K \\ w_H \end{bmatrix} \tag{17.68}$$

We are now prepared to analyze the dynamics of the system (17.60)–(17.63). Denote output, factor input, and factor price vectors as $y = (y_1, y_2)$, $\ell = (k, h)$ where $w = (w_K, w_H)$. Further denote ι as the 2×2 identity matrix. Utilizing Equations (17.65) and (17.67), we can derive the Jacobian of this 4×4 dynamical system in (k, h, p_1, p_2) evaluated at their BGP values as:

$$J = \begin{bmatrix} \left[\frac{\partial y}{\partial \ell}\right] - (n+\theta)\iota & \left[\frac{\partial y}{\partial p}\right] - [\wp_{ij}] \\ \underset{\sim}{0} & (\rho + n + \sigma\theta)\iota - \left[\frac{\partial w}{\partial p}\right] \end{bmatrix}$$

$$= \begin{bmatrix} [([a_{ij}])^{-1} - (n+\theta)\iota] & \left[\frac{\partial y}{\partial p}\right] - [\wp_{ij}] \\ \underset{\sim}{0} & [(\rho + n + \sigma\theta)\iota - ([\hat{a}_{ij}]')^{-1}] \end{bmatrix} \tag{17.69}$$

where

$$[\wp_{ij}] = \begin{bmatrix} -(1/\sigma)[1 + \tau(m/\pi)^{1/2}N^{1/2}]p_1^{-(1/\sigma)-1} & 0 \\ 0 & 0 \end{bmatrix}$$

Since p_1 and p_2 are both jump variables whose values may be adjusted instantaneously at any point in time, the 4×4 dynamical system is locally indeterminate (i.e. a sink) if there are less than two roots with positive real parts.

From Equations (17.60), (17.61) and (17.66), we have

$$\underset{\sim}{0} = \ell - [a_{ij}]y = \ell - [a_{ij}] \begin{bmatrix} (n+\theta)k + [1+\tau(m/\pi)^{1/2}N^{1/2}]p_1^{-1/\sigma} \\ (n+\theta)h \end{bmatrix}$$

or, by rearranging,

$$[\iota - [a_{ij}](n+\theta)]\ell = [a_{ij}] \begin{bmatrix} [1+\tau(m/\pi)^{1/2}N^{1/2}]p_1^{-1/\sigma} \\ 0 \end{bmatrix}$$

$$= [1+\tau(m/\pi)^{1/2}N^{1/2}]p_1^{-1/\sigma} \begin{bmatrix} a_{11} \\ a_{21} \end{bmatrix} \qquad (17.70)$$

On the other hand, Equations (17.62), (17.63) and (17.68) together yield:

$$\underset{\sim}{0} = [\hat{a}_{ij}]'w - p = [\hat{a}_{ij}]'p(\rho + n + \sigma\theta) - p$$

or, equivalently,

$$[[\hat{a}_{ij}]'p(\rho + n + \sigma\theta) - \iota]p = \underset{\sim}{0} \qquad (17.71)$$

Thus, the 2×2 matrix $[(\rho + n + \sigma\theta)\iota - ([\hat{a}_{ij}]')^{-1}]$ in the Jacobian expression (17.69) must be singular. This implies the Jacobian itself must also be singular and hence the four transformed endogenous variables are linearly dependent along the BGP. Because of this singularity property, a quick examination of Equations (17.69) and (17.70) suggests that the unit transportation cost (τ) and the spatial density parameter (m), both entering Equations (17.69) and (17.70) symmetrically with respect to (k,h), will not affect the sign of the roots of J.

We can now conclude that, regardless of the transportation and spatial density parameters

(i) price dynamics are recursive; they have a zero root and a negative root if the final goods sector (sector 1) is more capital intensive in the social sense, i.e. $\alpha_{11}/\alpha_{12} > \alpha_{21}/\alpha_{22}$;

(ii) quantity dynamics feature at least one negative root if the final goods sector (sector 1) is more housing intensive in the private sense, i.e. $(\alpha_{11} - \varepsilon_{11})/\ (\alpha_{12} - \varepsilon_{12}) < (\alpha_{21} - \varepsilon_{21})/(\alpha_{22} - \varepsilon_{22})$;

(iii) when the final goods sector (sector 1) is more capital intensive in the social sense and more housing intensive in the private sense, the Jacobian J has one zero root and at least two negative roots – in this case, dynamic indeterminacy arises where the BGP equilibrium is a sink.

These corroborate the findings in Bond *et al.* (1996) once we (i) reinterpret Marshallian externalities in sector-specific factors as distortionary factor taxes/subsidies and (ii) establish a formal link between private (social) factor intensity rankings and physical (value) factor intensity rankings. The inconsistency in the two-factor intensity rankings upsets the polarization theorem established in Bond *et al.* that is necessary for saddle-path stability in this class of two-sector endogenous growth models.

Summarizing, our results point out the possibility of dynamic indeterminacy in this generalized two-sector endogenous urban growth model, regardless of the transportation and spatial density parameters. An immediate economic implication is that urban growth and housing price dynamics in transition to a long-run balanced growth equilibrium may be very different across cities. Two crucial ingredients generating dynamic indeterminacy are (i) sector-specific externalities and (ii) the presence of two sectors for manufacturing the two reproducible factors. These allow for the dynamically reinforcing interactions under which self-fulfilling prophecies can lead to different transition paths toward the long-run BGP.

Along these lines, one may consider an alternative model by replacing housing and non-housing capitals with physical and human capitals as in conventional two-sector endogenous growth models. If the productivity of aggregate final good production at the city level depends on a city's aggregate employment and human capital, as specified in Rossi-Hansberg and Wright (2003), one may then expect similar dynamic properties to arise. It may also be interesting to allow unbalanced development between the final good sector and the housing sector in a way similar to Kongsamut *et al.* (2001). Such an extension is interesting as it may enable a thorough examination of housing dynamics with long-run trends in relative housing prices in units of the final composite consumption good.

17.3.5. Endogenous growth in a perfectly competitive economy with a system of cities

The urban growth models with perfect competition elaborated in the previous subsections can easily be extended to feature a system of cities. A natural way to do this is to consider a fixed, finite number of cities where there are knowledge spillovers across cities. For example, one may follow Eaton and Eckstein (1997) by assuming that the engine of growth is human capital and that the existing stocks of average human capital in all cities contribute to the rate of accumulation of human capital in the representative city of interest.[11] Thus, an exogenous spatial correlation matrix gives the influence of the stock of human capital in one city on the effectiveness of time spent learning in other cities, where the diagonal elements are thus the effect of aggregate human capital in a city on the effectiveness of time spent learning on its own residents. In balanced growth equilibrium, individual human capital accumulation must converge to the city average, whereas net migration flows between each pair of cities must all be zero. The latter cross-city locational equilibrium condition is reached when the positive effects of human capital accumulation less the negative congestion effects of population growth are equal in all cities.

Controlling for other factors, the city contributing more to knowledge spillovers has a relatively high stock of human capital, a relatively high level of productivity, and a relatively large population. This, therefore, yields a prediction that in the long run, larger cities feature higher wages per worker, though all cities grow at a common, balanced rate. In other words, there is no positive correlation between population agglomeration and economic growth across cities.

In principle, we may allow the degree of knowledge spillovers between each pair of cities (i.e. the off-diagonal elements of the spatial correlation matrix) to depend on the distance between cities in a fashion similar to Ogawa and Fujita (1980), Fujita and Ogawa (1982) and Berliant *et al.* (2002). This extension may allow a

---

[11] E.g. Rochester, NY.

complete characterization of the dynamic patterns of city formation in the process of economic development. Alternatively, one may rely on vertical integration rather than knowledge spillovers to study dynamic patterns of agglomeration in a system of cities. Peng *et al.* (2003) make an attempt at such an endeavor in a simple two-region setup.

17.4. Urban growth models with an imperfectly competitive market

In Sections 17.2 and 17.3, our spatial economy features perfect competition, despite the presence of Marshallian externalities in the form of uncompensated knowledge spillovers. In this section, we turn to considering dynamic urban growth models without a perfectly competitive labor, intermediate good, or consumption good market. Section 17.4.1 introduces imperfect market structures, focusing in particular on the case of monopolistic competition. Section 17.4.2 introduces non-Walrasian setups that incorporate market frictions into the urban growth framework.

17.4.1. The role of Marshallian externalities and imperfect competition

The New Economic Geography provides the type of circular causation that is amenable to dynamic modeling. There is a large literature studying the static version of the model, including Abdel-Rahman (1988), Abdel-Rahman and Fujita (1990), Krugman (1991, 1993), and Berliant and Kung (2002), to name but a few. Comparative statics exercises where exogenous variables, such as population, are presumed to grow at an exogenous rate tied to time can be found in Fujita and Krugman (1995) and Fujita and Mori (1997). For instance, in a stripped down form, growth in Henderson and Ioannides (1981) is tied exclusively to an increase in the number of cities as a response to an exogenous increase in population. Our focus is instead on the dynamic growth version of the model, as described in the introduction of this survey.

To our knowledge, the most comprehensive treatment of the New Economic Geography with dynamics is found in Fujita and Thisse (2002, Chapter 11); see also a more recent survey by Ottaviano and Thisse (2003). The basic, static model with two regions, monopolistic competition, iceberg transportation cost, and a preference for consuming a variety of commodities is modified by introducing a perfectly competitive R&D sector that operates under an endogenous growth externality: the production of patents in a region per unit time is a function of knowledge capital in the economy and the number of skilled workers in that region, while knowledge capital is a function of both individual knowledge in the region and the entire economy. Individual knowledge is a function of the total stock of patents produced up to any given time. There is a cost to migration. If a firm wants to produce a particular variety, then it must pay a fixed cost, i.e. the cost of a patent.

The basic result from this model is reinforcement of agglomeration in equilibrium. Agglomeration and growth are perfectly correlated, in part because having an agglomerated R&D sector causes higher growth due to the endogenous growth externality. This reinforces circular causation.[12]

While the behavior of such a model is very interesting and complex, the fundamentals do not connect very well with the theoretical structures we have surveyed so far. In particular, the R&D sector does not look like part of an endogenous growth model – see Equation (17.19). The allocation of skilled workers across regions and the stock of total patents in the entire economy (that affects individual knowledge) completely determine patent production rather than having *consumption good output* determined by an individual agent's investment in physical or human capital with an externality component based on the total investment in the city.

The predecessors of Fujita and Thisse are related as follows (see also a comprehensive survey by Duranton and Puga, 2003). Ioannides (1994) employs a multi-monocentric city framework with monopolistic competition in the output market. Overlapping generations yield life-cycle effects. In contrast with the balance of the literature, each city is assumed to produce only one commodity, and these cities are presumed to be symmetric (though the commodity that each produces is different). Each city has public capital, and transportation cost is decreasing in this capital. There is

---

[12] Both the workings of the model and the notation are very complicated. It seems important for future work in this literature to emphasize simplicity in order to keep the models tractable.

free mobility. In a steady-state equilibrium, city size is constant and the number of cities grows exponentially in population. Product variety is the engine of endogenous growth.

Walz (1996) employs a framework with two regions, free migration, and monopolistic competition in the intermediate goods sector. The fixed cost for production of intermediate goods is, like Fujita and Thisse, the cost of a patent. Various patterns can emerge in equilibrium, depending on initial parameters (to some degree exhibiting a bang-bang pattern as a function of endowments, due to free migration).

Baldwin and Forslid (2000) employ a two-region model with monopolistic competition in the consumption good sector and migration proportional to the wage differential between regions.[13] Increasing returns in the consumption good sector are generated by a one time fixed cost of physical capital that must be sunk by every firm that enters. A capital or investment sector subject to an endogenous growth externality, in that production is a function of a weighted sum of lagged total capital production in the regions, is employed under perfect competition. The results in the paper are purely numerical. Aside from studying the effect of integration of the two regions on equilibrium, the usual correlation between agglomeration and growth is found. This framework generates a model that is simple relative to the others studied in this subsection. We think that it would be interesting to explore subgame perfect equilibria that are not steady states in this model.

Finally, Duranton (2004) embeds a product-variety model of endogenous growth into a typical urban framework. Individual agents choose their locations to optimize utility. Investment in innovative activities is the main driving force of city growth. In locational equilibrium, urban population is proportional to the number of differentiated products produced in the local economy.

---

[13] With the dynamic general-equilibrium framework, it is desirable to use an explicit migration cost, as in Fujita and Thisse (2002), rather than an ad hoc adjustment process (in a way analogous to the static framework of Krugman (1991)). This allows explicit calculations on the part of consumers of the benefits and costs of migration and migration dynamics. Although Baldwin and Forslid (2000, Footnote 4) attempt to justify such an adjustment process, we find the assumption that migration costs are quadratic in the rate of migration to be strange.

Under proper conditions, the steady-state population distribution of cities can generate Zipf's law.

17.4.2. The non-Walrasian approach to agglomeration and growth

The previous sections of this survey have examined the dynamics of urban growth where the accumulation of capital has occurred in markets, either perfectly or imperfectly competitive. There have been externalities that have not been priced, for example endogenous growth externalities; so the equilibrium allocations were not efficient. But the markets for private goods were complete.

Recently, further abstractions have been attempted in models to provide microfoundations for the localization and urbanization externalities that are simply assumed to take a particular reduced form in standard dynamic regional models. In order to simplify matters, it is assumed that externalities take a pure form, in that there are no markets at all, so all interaction takes place in a non-market setting. This is what we call the non-Walrasian approach. Of course, it is hoped that these pure externality models will eventually provide microfoundations for reduced forms used in other models, and in this way the non-Walrasian models will be integrated with market or Walrasian models.

The non-Walrasian approach to urban economics begins with Helsley and Strange (1990), where matching of heterogeneous labor leads to the formation of a system of identical cities where workers with different skills achieve the same utility level in equilibrium. Abdel-Rahman and Wang (1995, 1997) go beyond this early urban labor matching framework to permit ex post heterogeneity and as a consequence, the formation of a core–periphery urban structure and the dispersion of wage incomes. In two recent studies, Coulson *et al.* (2001) and Brueckner and Zenou (2003) examine in an urban labor market with search and matching frictions why spatial mismatch occurs in the sense that unemployment in central cities and job vacancies in suburbs coexist.

Since none of these previous papers consider urbanization dynamics and city growth, our discussion is based on unpublished work by Berliant *et al.* (2003). First we shall describe the basic model that does not involve growth, and then give its extensions to various growth contexts.

This model has a focus on horizontal knowledge differentiation, where knowledge is represented by a circle of unit circumference.[14] The basic model has a continuum of consumers, each of whom has a knowledge type on the circle. Agents in a city meet each other pairwise and randomly, according to a Poisson process where the rate at which meetings occur for each agent is proportional to the number of unmatched agents in a city. When agents meet, they learn each other's type, and decide whether or not to produce together. Production uses the pair's time and results in the output of a homogeneous consumption commodity. If a pair decides to produce, they match, and detach after a time determined by another Poisson process. If they decide not to match, they detach immediately and search for other partners.

Consider any pair that meets. The model attempts to capture the idea that if the pair has knowledge that is too close together, there are few complementarities between the pair and production is relatively low. Similarly, if the pair has knowledge that is very distant, communication problems dominate, and production is also very low. Thus, it is assumed that production as a function of the two types who meet is linearly increasing beginning with zero distance between the knowledge types of the pair as the pair becomes more diverse up to some optimal distance. Then it is linearly decreasing beyond that optimal distance. In other words, production as a function of knowledge type distance on the circle is piecewise linear and single peaked.

Felicity is simply the quantity of consumption good produced and consumed in a time period, assumed to be zero if production is not taking place. Time is continuous and infinite. Utility is the (expected) present discounted value of consumption, where the discount rate is the same for all agents (so the only difference between agents is their knowledge types). Agents choose a range of others with whom they will accept matches and production; this range will typically be the union of two intervals around the optimal match type on opposite sides of the agent. The length of these intervals is called the 'knowledge spread'. In other words, the choice of agents with whom to match will be symmetric. Since only

---

[14] Thus, the setup contrasts sharply with the vertical knowledge differentiation model of Jovanovic and Rob (1989).

symmetric equilibria in pure strategies are considered, if one agent in a pair wants to match, so does the other. In general, there is a tradeoff in the choice of knowledge spread between the quality and quantity of matches.

In a closed city model with exogenously given population, a steady-state equilibrium exists and is unique. Moreover, higher population implies a lower knowledge spread, as agents are more selective about their partners. Finally, the per capita rate of matches (which one might interpret as patents for the purposes of empirical implementation) is higher in cities with larger population.

In an open city model, population is endogenous and determined by a per capita cost of city infrastructure that is linear in population. Entry into a city occurs until the value of an unmatched agent (or potential entrant) is equal to the infrastructure cost. Even though both the population and knowledge spread are endogenous, the same correlations between population and knowledge spread on the one hand, and population and matches on the other, hold when exogenous variables such as the technology are changed.

In this context, there are two external effects that interact. First, there is the classical urban economic congestion externality, that agents entering the city consider only the average congestion cost and not the marginal congestion cost imposed on others in the city. This leads to cities that are overpopulated in equilibrium relative to the social optimum. Second, there is an externality in the search process, since more unmatched agents mean a higher frequency of meetings and thus higher utility for unmatched agents. An agent's decision concerning knowledge spread has an external effect on other agents, since it affects the number of unmatched agents and thus the utility of others in the city. This leads to knowledge spreads that are too large in equilibrium relative to social optimum, and cities that are too small. Overall, equilibrium knowledge spread and population can be anything relative to the social optimum.

This basic model leaves agents unchanged after they meet, match, and produce. It is extended to the growth context, where knowledge accumulates in various ways, as follows. First, it is assumed that there is an endogenous growth externality in the production function when two agents match and produce, following in the spirit of Laing *et al.* (1995). This externality lasts over the infinite horizon, and matches are thus permanent. Second, it is assumed that production of every match is multiplied by a decreasing function of the city's average knowledge spread, assuming that more specialization leads to higher growth. Third, the circumference of the knowledge circle is allowed to expand with the city's average knowledge spread, representing a larger diversity of possible knowledge when people are matching more broadly. Fourth, agents' knowledge types are allowed to change when they match. They are assumed to move closer to their partners after a match. These extensions can be combined in any way, and despite these variations, all of the basic results outlined above continue to hold.

17.5. Avenues for future research

We have surveyed dynamic models of the urban economy that feature various forms of capital accumulation and thus growth. There are several features of these models that require elaboration and improvement in order to be useful, especially to empiricists.

First, it is apparent from our survey that an almost universal feature of the models in the literature is that agglomeration and growth are perfectly correlated.[15] Of course, in the real world this is not true; see, for example, Tables 17.1 and 17.2. Eventually, in cities with very large populations, gains from agglomeration diminish and congestion costs increasing in population will overtake the benefits of agglomeration, yielding cities that do not grow forever. It seems very important to include this in the models.

Second, it is important to develop testable hypotheses that can distinguish among the various models. This is especially important for detecting market failures, so appropriate public policies can be formulated if there are large welfare losses associated with equilibrium. Which models are best capable of predicting the dynamic process of city development and decline? The comparative statics we have derived in Sections 17.2 and 17.3 of this essay should be useful for this purpose. However, there are few comparative statics available for the models in Section 17.4.1; most are

---

[15] Notable exceptions can be found in Eaton and Eckstein (1997), Berliant *et al.* (2003), and Rossi-Hansberg and Wright (2003). The last paper features returns to scale that eventually diminish within each city, but returns to scale in the economy can be constant due to the addition of new cities.

rudimentary descriptions of either symmetric or core–periphery equilibria. Priority should be given to developing potentially testable comparative statics.

Third, in all of the models we have discussed, there are 'black boxes' – pieces of the model that are not based on microeconomic optimizing behavior. In other words, a reduced form is assumed. Often this is used to simplify the inner workings and notation, but it is often not clear that the black box can be opened and its contents found to be consistent with a simple micro model. Examples drawn from earlier sections include the aggregate production function and fixed savings rate used in the exogenous growth literature, the aggregate production function used in the endogenous growth literature, the representative consumer assumption (implying no heterogeneity in preferences), and assumptions about population growth or migration dynamics. In the non-Walrasian literature, there are assumptions made about the reduced forms of population arrival (both of rates and types), and the average cost of population in a city is taken as exogenous. For instance, an alternative to assuming that types enter uniformly into a city would be that the entry of types of agents depends on the unmatched types already there.

For each of these black boxes, one can ask whether they are consistent with a standard microeconomic model of location within a city, say a monocentric model, where all choices, including location, are made by consumers. The exercise in much of the literature is opening a black box to find more but smaller black boxes inside, occupying the same volume as the original black box. It would be better, of course, to simply provide the micro foundations for assumed behavior and dispose of the boxes. Berliant and Fujita (2004) attempt to make some progress along these lines.

Fourth, one of the underlying, strong forces causing economic growth is the migration of skilled workers, who embody knowledge, between cities. The consequence is the transmission of ideas across space and time. Construction of models of this phenomenon and derivation of their empirical consequences should be a priority.

The fifth and final issue that we wish to discuss is the concept of 'social optimum' in open city models. These models generally have the feature that population in the city is endogenous. Only one city is considered. Social optimum is often defined to be the solution to an optimization problem that has a utilitarian objective for agents in the city and standard feasibility constraints for resources in the city. Population is assumed to flow between the city and the rest of the world. How is this concept related to the standard notion of efficiency, Pareto optimality? Taken at face value, Pareto optimality is not well defined in an open city model, since the agents present in the city and model are not well defined. Whose welfare is accounted for? The concept of social optimum is really trying to capture efficiency in a system of (identical) cities. The number of such cities would have to be infinite, as the concept of social optimum assumes that there is no shadow cost for removing population from another city. The assumptions under which a social optimum is the same (or results in the same allocations) as a Pareto optimum in such an extended but closed model should be clarified and examined, and a theorem proved. Evidently, the theorem would involve two parts: conditions under which any Pareto optimum is a social optimum, and conditions under which any social optimum is a Pareto optimum. Models that do not satisfy the conditions of the theorem should avoid use of the concept of social optimum.

References

Abdel-Rahman, H.M. (1988), "Product differentiation, monopolistic competition and city size", *Regional Science and Urban Economics*, Vol. 18, pp. 68–86.

Abdel-Rahman, H.M. (1990), "Agglomeration economies, types and sizes of cities", *Journal of Urban Economics*, Vol. 27, pp. 25–45.

Abdel-Rahman, H.M. and M. Fujita (1990), "Product variety, Marshallian externalities and city size", *Journal of Regional Science*, Vol. 30, pp. 165–183.

Abdel-Rahman, H. and P. Wang (1995), "Toward a general-equilibrium theory of a core-periphery system of cities", *Regional Science and Urban Economics*, Vol. 25, pp. 529–546.

Abdel-Rahman, H. and P. Wang (1997), "Social welfare and income inequality in a system of cities", *Journal of Urban Economics*, Vol. 41, pp. 462–483.

Aghion, P. and P. Howitt (1992), "A model of growth through creative destruction", *Econometrica*, Vol. 60, pp. 323–351.

Allais, M. (1947), *Economie et Intéret* (in French).

Anas, A. (1978), "Dynamics of urban residential growth", *Journal of Urban Economics*, Vol. 5, pp. 66–87.

Anas, A. (1992), "On the birth and growth of cities: laissez-faire and planning compared", *Regional Science and Urban Economics*, Vol. 22, pp. 243–258.

Anas, A., T. Palivos and P. Wang (1995), *Durable Housing in an Endogenous Growth Model with Dynamic Indeterminacy, Working Paper*, University Park, PA: Penn State University.

Baldwin, R. and R. Forslid (2000), "The core–periphery model and endogenous growth: stabilising and destabilising integration", *Economica*, Vol. 67, pp. 307–324.

Benhabib, J., Q. Meng and K. Nishimura (2000), "Indeterminacy under constant returns to scale in multisector economies", *Econometrica*, Vol. 68, pp. 1541–1549.

Berliant, M. and M. Fujita (2004), *Knowledge Creation as a Square Dance on the Hubert Cube*. Mimeo.

Berliant, M. and F.-C. Kung (2002), *The Indeterminacy of Equilibrium City Formation Under Monopolistic Competition and Increasing Returns, Working Paper*, St Louis, MO: Washington University.

Berliant, M., S. Peng and P. Wang (2002), "Production externalities and urban configuration", *Journal of Economic Theory*, Vol. 104, pp. 275–303.

Berliant, M., R. Reed and P. Wang (2003), *Knowledge Exchange, Matching, and Agglomeration*. Mimeo.

Black, D. and J.V. Henderson (1999), "A theory of urban growth", *Journal of Political Economy*, Vol. 107, pp. 252–284.

Boldrin, M. and A. Rustichini (1994), "Indeterminacy of equilibria in models with infinitely-lived agents and external effects", *Econometrica*, Vol. 62, pp. 323–342.

Bond, E., P. Wang and C. Yip (1996), "A general two-sector model of endogenous growth with human and physical capital: Balanced growth and transitional dynamics", *Journal of Economic Theory*, Vol. 68, pp. 149–173.

Brueckner, J.K. and Y. Zenou (2003), "Space and unemployment: the labor-market effects of spatial mismatch", *Journal of Labor Economics*, Vol. 21, pp. 242–266.

Cass, D. (1965), "Optimal growth in an aggregative model of capital accumulation", *Review of Economic Studies*, Vol. 32, pp. 233–240.

Chang, S. (1999), *Three Essays on Housing and Urban Dynamics*. PhD Dissertation, Penn State University, University Park, PA.

Coulson, N.E., D. Laing and P. Wang (2001), "Spatial mismatching in search equilibrium", *Journal of Labor Economics*, Vol. 19, pp. 949–972.

Duranton, G. (2004), *Some Economics for Zipf's Law: Romer and Simon Unified*. Mimeo.

Duranton, G. and D. Puga (2003), "Micro-foundations of urban agglomeration economies", in: J.V. Henderson and J.-F. Thisse, editors, *Handbook of Urban and Regional Economics*, Vol. 4, Amsterdam: North-Holland, Forthcoming.

Eaton, J. and Z. Eckstein (1997), "Cities and growth: theory and evidence from France and Japan", *Regional Science and Urban Economics*, Vol. 27, pp. 443–474.

Fujita, M. (1976), "Spatial patterns of urban growth: optimum and market",

Fujita, M. (1982), "Spatial patterns of residential development", *Journal of Urban Economics*, Vol. 12, pp. 22–52.

Fujita, M. (1989), *Urban Economic Theory: Land Use and City Size*, New York, NY: Cambridge University Press.

Fujita, M. and P. Krugman (1995), "When is the economy monocentric?: von Thünen and Chamberlin unified", *Regional Science and Urban Economics*, Vol. 25, pp. 505–528.

Fujita, M. and T. Mori (1997), "Structural stability and evolution of urban systems", *Regional Science and Urban Economics*, Vol. 27, pp. 399–442.

Fujita, M. and F. Ogawa (1982), "Multiple equilibria and structural transition of non-monocentric urban configurations", *Regional Science and Urban Economics*, Vol. 12, pp. 161–196.

Fujita, M. and J.-F. Thisse (2002), *Economics of Agglomeration*, Cambridge: Cambridge University Press.

Glaeser, E.L. (1999), "Learning in cities", *Journal of Urban Economics*, Vol. 46, pp. 254–277.

Glaeser, E.L., H.D. Kallal, J.A. Scheinkman and A. Shleifer (1992), "Growth in cities", *Journal of Political Economy*, Vol. 100, pp. 1126–1152.

Grossman, G. and E. Helpman (1991), *Innovation and Growth in the Global Economy*, Cambridge, MA: MIT Press.

Helsley, R.W. and W.C. Strange (1990), "Matching and agglomeration economies in a system of cities", *Regional Science and Urban Economics*, Vol. 20, pp. 189–212.

Helsley, R.W. and W.C. Strange (2002), "Innovation and input sharing", *Journal of Urban Economics*, Vol. 51, pp. 25–45.

Henderson, J.V. and Y.M. Ioannodes (1981), "Aspects of growth in a system of cities", *Journal of Urban Economics*, Vol. 10, pp. 117–139.

Henderson, J.V., A. Kuncoro and M. Turner (1995), "Industrial development in cities", *Journal of Political Economy.*, Vol. 103, pp. 1067–1090.

Ioannides, Y.M. (1994), "Product differentiation and economic growth in a system of cities", *Regional Science and Urban Economics*, Vol. 24, pp. 461–484.

Jacobs, J. (1969), *The Economy of Cities*, New York, NY: Random House.

Jaffe, A.B., M. Trajtenberg and R. Henderson (1993), "Geographic localization of knowledge spillovers as evidenced by patent citations", *Quarterly Journal of Economics*, Vol. 108, pp. 577–598.

Jones, L. and R. Manuelli (1990), "A convex model of equilibrium growth: theory and policy implications", *Journal of Political Economy*, Vol. 98, pp. 1008–1038.

Jovanovic, B. and R. Rob (1989), "The growth and diffusion of knowledge", *Review of Economic Studies*, Vol. 56, pp. 569–582.

Harrod, R.F. (1939), "An essay in dynamic theory", *Economic Journal*, Vol. 49, pp. 14–33.

Kanemoto, Y. (1980), *Theories of Urban Externalities*, Amsterdam: North-Holland.

Kanemoto, Y., T. Kitagawa, R. Kurima, H. Saito, and E. Shioji (2003), *Is Tokyo Too Big? An Application of the Henry George Theorem*. Mimeo.

Kongsamut, P., S. Rebelo and D. Xie (2001), "Beyond balanced growth", *Review of Economic Studies*, Vol. 68, pp. 869–882.

Koopmans, T.C. (1965), "On the concept of optimal economic growth", in: S. Pietro, editor, *The Econometric Approach to Development Planning*, Amsterdam: North-Holland.

Krugman, P. (1991), "Increasing returns and economic geography", *Journal of Political Economy*, Vol. 99, pp. 483–499.

Krugman, P. (1993), "On the number and location of cities", *European Economic Review*, Vol. 37, pp. 293–298.

Kuznets, S. (1962), "Population change and aggregate output", in: *Population Change and Aggregate Output in Developed Countries*, Princeton, NJ: Princeton University Press (for NBER), pp. 324–340.

Laing, D., T. Palivos and P. Wang (1995), "Learning, matching and growth", *Review of Economic Studies*, Vol. 62, pp. 115–129.

Li, Y. (2002), *Three Essays on Public Finance and Urban Endogenous Growth Theory*. PhD Dissertation, State University of New York, Buffalo, NY.

Lin, C., C.C. Mai and P. Wang (2004), "Urban Land use and housing policy in an endogenously growing monocentric model", *Regional Science and Urban Economics*, Vol. 34, pp. 241–261.

Lucas, R.E., Jr. (1988), "On the mechanics of economic development", *Journal of Monetary Economics*, Vol. 22, pp. 2–42.

Marshall, A. (1895), *Principles of Economics*, London: Macmillan.

Miyao, T. (1987), "Dynamic urban models", in: E. Mills and P. Nijkamp, editors, *Handbook of Regional and Urban Economics, Urban Economics*, Vol. 2, Amsterdam: North-Holland.

Mulligan, C. and X. Sala-i-Martin (1993), "Transitional dynamics in two-sector models of endogenous growth", *Quarterly Journal of Economics*, Vol. 108, pp. 737–773.

von Neumann, J. (1937), "A model of general economic equilibrium", *Review of Economic Studies*, Vol. 13, pp. 1–9.

Ogawa, M. and M. Fujita (1980), "Equilibrium land use patterns in a nonmonocentric city", *Journal of Regional Science*, Vol. 20, pp. 455–475.

Ottaviano, G. and J.-F. Thisse (2003), "Agglomeration and economic geography", in: J.V. Henderson and J.-F. Thisse, editors, *Handbook of Urban and Regional Economics*, Vol. 4, Amsterdam: North-Holland, Forthcoming.

Palivos, T. and P. Wang (1996), "Spatial agglomeration and endogenous growth", *Regional Science and Urban Economics*, Vol. 26, pp. 645–669.

Palmon, O. and B.A. Smith (1998), "New evidence on property tax capitalization", *Journal of Political Economy*, Vol. 106, pp. 1099–1111.

Peng, S., J. Thisse and P. Wang (2003), *Economic Integration and Agglomeration in a Middle Product Economy*. Mimeo.

Phelps, E.S. (1966), *Golden Rules of Economic Growth*, New York: Norton.

Pitchford, J.D. (1960), "Growth and the elasticity of substitution", *Economic Record*, Vol. 26, pp. 491–504.

Pred, A. (1966), *The Spatial Dynamics of U.S. Urban–Industrial Growth, 1800–1914*, Cambridge, MA: MIT Press.

Ramsey, F. (1928), "A mathematical theory of savings", *Economic Journal*, Vol. 38, pp. 543–559.

Rauch, J. (1993), "Productivity gains from geographic concentration of human capital: evidence from cities", *Journal of Urban Economics*, Vol. 34, pp. 380–400.

Rebelo, S. (1991), "Long-run policy analysis and long-run growth", *Journal of Political Economy*, Vol. 99, pp. 500–521.

Romer, P. (1986), "Increasing returns and long-run growth", *Journal of Political Economy*, Vol. 94, pp. 1002–1037.

Romer, P. (1990), "Endogenous technological change", *Journal of Political Economy*, Vol. 98, pp. S71–S102.

Rossi-Hansberg, E. and M. Wright (2003), *Urban Structure and Growth*. Mimeo.

Saxenian, A. (1994), *Regional Advantage: Culture and Competition in Silicon Valley and Route 128*, Cambridge, MA: Harvard University Press.

Shell, K. (1966), "Toward a theory of inventive activity and capital accumulation", *American Economic Review*, Vol. 61, pp. 62–68.

Solow, R. (1956), "A contribution to the theory of economic growth", *Quarterly Journal of Economics*, Vol. 70, pp. 65–94.

Stokey, N. (1991), "Human capital, product quality, and growth", *Quarterly Journal of Economics*, Vol. 106, pp. 587–616.

Swan, T.W. (1956), "Economic growth and capital accumulation", *Economic Record*, Vol. 32, pp. 334–361.

Turnovsky, S. and T. Okuyama (1994), "Taxes, housing, and capital accumulation in a two-sector growing economy", *Journal of Public Economics*, Vol. 53, pp. 245–267.

Uzawa, H. (1965), "Optimal technical change in an aggregative model of economic growth", *International Economic Review*, Vol. 6, pp. 18–31.

Walz, U. (1996), "Transport costs, intermediate goods, and localized growth", *Regional Science and Urban Economics*, Vol. 26, pp. 671–695.

Urban Dynamics and Growth, CEA, vol. 266
Roberta Capello and Peter Nijkamp (editors)

CHAPTER 18

New Economic Geography Explanations of Urban and Regional Agglomeration

Kieran P. Donaghy

Department of Urban and Regional Planning, University of Illinois at Urbana-Champaign, 111 Temple Hoyne Buell Hall, 611 East Lorado Taft Drive, Champaign, IL 61820, USA

Abstract

The purpose of this chapter is to examine the 'new economic geography' (NEG) in terms of how well it explains urban and regional agglomeration. The chapter reviews Krugman's core–periphery model of regional agglomeration with an eye toward what motivates the analysis and how the model accomplishes its ends. It then proceeds to an examination of how several other types of agglomeration, both urban and regional, can be modeled by modifying key assumptions. The chapter concludes with an assessment of what the NEG has accomplished and where its challenges lie, particularly in regard to empirical application.

Keyword: agglomeration economies

JEL classifications: R11, R12, R30

18.1. Introduction

Patterns of urban and regional agglomeration have received much attention in urban economics over the last several decades. Casual

---

Research assistance by Nazmiye Balta and suggestions and constructive criticisms by Lewis Hopkins, the editors, and two referees are gratefully acknowledged.

observation would suggest that this is attributable in no small way to a number of dramatic contemporaneous developments that have called for study and explanation. These include (1) the emergence of new urban forms, such as edge cities; (2) the increasing spread of the urban footprint through suburbanization, sprawl, and leapfrogging development; (3) the formation and reinforcement of industrial belts and megalopolises; and (4) the change in economic relationships between cities that have come about through globalization and the attendant fragmentation of production systems.

In view of the number of factors involved, explaining patterns of agglomeration convincingly is not easy; giving formal expression to intuitions of what is involved is less so; and determining which of possibly many factors identified by theorists have been causally effective in the cases of particular cities and regions is perhaps most difficult. The 'new urban economics', which Button (2000) identifies with contributions by Beckmann, Mills, and Muth in the 1960s and 1970s, and subsequent work, was intended to explain the emergence of simple urban forms: monocentric cities formed around central business districts, surrounded by residential suburbs. Important as these contributions were in getting analytical urban economics untracked, the evolution of simple urban forms into the complex conurbations noted above has limited the relevance of these contributions and placed greater explanatory demands upon urban economists. Considerable progress has been made in both formal modeling and empirical analysis. Anas *et al.* (1998) provide a comprehensive survey of many of the developments in the analysis of urban form, whereas Fujita (1996) and Berliant and Wang (2004) have reviewed contributions to formal modeling of the growth of urban economies as it relates to agglomeration. More recently Black and Henderson (1999) have contributed a general model of endogenous urban growth in a city system in which they explore not only how urbanization affects the efficiency of the growth process but also how growth affects patterns of urbanization.

Turning to specific agglomeration phenomena, we note that extensions of the 'new urban economics' have produced formal models that explain how polycentric cities can emerge from monocentric ones (Fujita and Ogawa, 1982), why leapfrogging or discontinuous development might occur (Ohls and Pines, 1975; Fujita, 1982; Wheaton, 1982), and why in view of current transport pricing and governmental subsidies urban (suburban and ex-urban) sprawl is a reasonable outcome (Brueckner, 2000, 2001, 2003; Brueckner and Kim, 2003). Henderson and Mitra (1996) have modeled how edge cities form through the intermediation of developers and government policies.

Stylized explanations of the formation of industrial belts and megalopolises, the entrenchment of primary cities, the formation of suburban concentrations of industry, and the formation of new cities in central place systems have also been advanced along other lines of inquiry by the so-called 'new economic geographers', whose work is informed largely by international trade theory. The new economic geography (NEG) is perhaps best described as an approach to modeling the spatial distribution of agglomerations of economic activities. This approach explicitly incorporates increasing-returns-to-scale production technologies, consumer preferences for greater variety of goods, monopolistic competition between firms, costly trading of goods between locations (hence a role for geography or distance), and processes of cumulative causation, which give rise to pecuniary externalities and the self-organizing formation of various patterns of agglomeration.[1] Intuitive accounts of the roles played by the above-mentioned factors were expounded by Marshall (1936) and Isard (1956) and more recently by Jacobs (1969, 1984), but formal incorporation of these factors into models of agglomeration has been a relatively recent accomplishment (Abdel-Rahman and Fujita, 1990). The principal aim of the NEG thus far seems to be the development of 'clarifying examples' of agglomeration (Krugman, 1995; Ottaviano and Thisse, 2001), which calls for (if not entails) simplification wherever possible to increase the transparency of the work done by each assumption in a model. In fact, one

---

[1] In their review of the NEG, Papageorgiou and Pines (1999) identify another basic assumption, or an implication of the previously mentioned assumptions, on which models in this tradition are based, even if not explicitly acknowledged in the literature: "each individual interacts not only with one firm as an employee or, perhaps, with several firms as a consumer, but with every manufacturing firm in the economy wherever it is located." Fujita and Thisse (2002) identify the intellectual lineage of all assumptions at work in NEG models.

of the principal exponents, Paul Krugman, openly acknowledges the highly unrealistic nature of the assumptions employed and the heavy reliance upon 'modeling tricks', an issue we will return to below.[2]

The purpose of this chapter is to examine a part of this growing body of literature in terms of how well it helps us to make sense of some agglomeration phenomena which other approaches have been less successful in explaining. The plan of the chapter is as follows. In Section 18.2 we will review the most basic of the NEG models, Krugman's (1991a) core–periphery model, and its assumptions and implications, with an eye toward what motivates the analysis and how the model accomplishes its ends. While this model was intended to account for *regional* agglomeration, variations of it have been used to explain different types of urban agglomeration as well. So we proceed in Section 18.3 to examine how several different agglomeration phenomena can be modeled by modifying key assumptions. In Section 18.4 we take stock of what the NEG has accomplished and where its challenges lie. In particular, we consider issues of empirical application, other research programs with which it has affinities, and promising directions it might take.

18.2. Krugman's core–periphery model

What is perhaps the most well-known model of the NEG is that of Krugman (1991a).[3] The question motivating its development is "Why and when does manufacturing become concentrated in a few regions, leaving others relatively undeveloped" (Krugman, 1991a, p. 484)? To answer this question Krugman strips the problem down to its ultimate simples. In the economy modeled there are two

---

[2] Krugman characterizes his modeling approach as a combination of: "Dixit-Stiglitz (consumer preferences), iceberg (transportation technologies), the computer and evolution (Krugman, 1998)".

[3] The model is recapitulated in appendices of Krugman (1991b, 1995) and in Chapters 4 and 5 of Fujita *et al.* (1999b), with varying notation. It is also discussed in Chapter 9 of Fujita and Thisse (2002). Many of the ideas exposited in Krugman (1991a) were previously introduced in Abdel-Rahman and Fujita (1990). In reviewing the model, I present a logically consistent hybrid of its various published forms and supply intermediate steps in its derivation for readers not already familiar with the moves of the monopolistic competition/IRS dance. See Gandolfo (1998) for a helpful discussion of the literature on monopolistic competition in international trade.

industries, agriculture and manufacturing, and R regions.[4] The agricultural good is produced with a constant-returns-to-scale (CRS) technology, whereas production in manufacturing enjoys increasing-returns-to scale (IRS). The agricultural good can be shipped costlessly, whereas manufactured goods cannot.

18.2.1. Consumer behavior

All individuals in this economy are assumed to have the same utility function

$$U = C_{\mathrm{M}}^{\mu} C_{\mathrm{A}}^{1-\mu} \qquad \text{where } 1 \geq \mu \geq 0, \tag{18.1}$$

in which C_{A} denotes the consumption level of the agricultural good and C_{M} the consumption level of an aggregate (or composite) of manufactured goods. From Equation (18.1) one may infer that manufacturers' share of expenditures will always be μ and farmers' share will be $1 - \mu$. C_{M} is a constant-elasticity-of-substitution (CES) aggregate of the consumption levels of N differentiated varieties or 'brands' of manufactured goods, c_i, $i = 1, \ldots, N$,

$$C_{\mathrm{M}} = \left[\sum_{i=1}^{N} c_i^{(\sigma-1)/\sigma} \right]^{\sigma/(\sigma-1)}, \tag{18.2}$$

where $\sigma = 1/(1 - \rho) > 1$ is the elasticity of substitution between the varieties, and ρ, $1 \geq \rho \geq 0$, is the substitution or preference intensity parameter.[5] Given income, Y, the price of the agricultural good, P_{A}, and the price for each manufactured good, p_i, the consumer's problem is to maximize utility subject to the budget

---

[4] Fujita *et al.* (1999) remark that 'agriculture' may be viewed as the residual sector encompassing industries other than agriculture that are location bound. In Krugman (1991a) the number of regions considered is two. The following gloss of the model follows the more general multi-regional presentation of Krugman (1995) and Fujita *et al.* (1999b).

[5] The closer is ρ to 0, the more are manufactured goods perfect substitutes for each other; the closer is ρ to 1 the greater is the consumer's demand for variety. This characterization of consumer preferences was introduced by Spence (1976) and Dixit and Stiglitz (1977).

constraint

$$P_A C_A + \sum_{i=1}^{N} p_i c_i = Y. \tag{18.3}$$

Because the functional forms of Equations (1) and (2) are homogeneous of degree one, this optimization problem can be solved in two parts. For a chosen consumption level of aggregate manufactures, C_M, the compensated demand for the ith variety is

$$c_i = \frac{p_i^{-\sigma}}{\sum_{i=1}^{N} p_i^{1-\sigma}} C_M, \tag{18.4}$$

and the minimum cost of purchasing a unit of the composite good, P_M, is

$$P_M = \left[\sum_{i=1}^{N} p_i^{\rho/(\rho-1)} \right]^{(\rho-1)/\rho}. \tag{18.5}$$

The compensated demand for the ith variety can be written in terms of P_M as

$$c_i = \left(\frac{p_i}{P_M} \right)^{-\sigma} C_M. \tag{18.6}$$

Rewriting the budget constraint as

$$P_A C_A + P_M C_M = Y, \tag{18.7}$$

and solving for the utility maximizing levels of C_A and C_M one obtains the uncompensated consumer demand functions

$$C_A = (1 - \mu) Y / P_A, \qquad C_M = \mu Y / P_M. \tag{18.8}$$

From Equations (18.6) and (18.8) the amounts demanded of the manufactured good varieties are given by

$$c_i = \mu Y \frac{p_i^{-\sigma}}{P_M^{-\sigma-1}}. \tag{18.9}$$

Substitution from Equation (18.8) into Equation (18.1) yields the indirect utility function

$$U = \mu^{\mu}(1 - \mu)^{\mu} Y P_{\mathrm{M}}^{-\mu} P_{\mathrm{A}}^{-(1-\mu)}, \tag{18.10}$$

which is useful for demonstrating consumer preference for variety. Assuming that the price of each variety of manufactured good is the same, $\bar{p}$, then from Equation (18.5), the price index of the aggregate of manufactured goods can be written as

$$P_{\mathrm{M}} = N^{(\rho-1)/\rho}\bar{p}. \tag{18.11}$$

From Equation (18.11) it is clear that this price index is decreasing in the number of manufactured good varieties. Moreover, the substitution of Equation (18.11) into Equation (18.10) yields a restatement of the indirect utility function that, in view of the conditions placed upon μ and ρ, is clearly increasing in the number of manufactured good varieties, N,

$$U = \mu^{\mu}(1 - \mu)^{\mu} Y N^{\mu(1-\rho)/\rho} \bar{p}^{-\mu} P_{\mathrm{A}}^{-(1-\mu)}. \tag{18.12}$$

18.2.2. Producer behavior

In this economy there are two factors of production: agricultural and manufacturing labor. Since agriculture is a CRS activity, agricultural labor used in producing any quantity of $Q_{\mathrm{A}j}$, the agricultural good in region j, $L_{\mathrm{A}j}$, can be set equal to the production level by suitable choice of units, i.e.

$$L_{\mathrm{a}j} = Q_{\mathrm{A}j}. \tag{18.13}$$

Because of IRS in manufacturing, the amount of labor required to produce the ith manufactured good in region j, $L_{\mathrm{M}ij}$, includes a 'fixed-cost' amount, α and an amount that varies with the production level of the good, $Q_{\mathrm{M}ij}$,

$$L_{\mathrm{M}ij} = \alpha + \beta Q_{\mathrm{M}ij}. \tag{18.14}$$

Assuming that the prevailing wage rate paid to workers in region j is w_j, the cost of producing a unit of manufactured good i is

$$w_j L_{\mathrm{M}ij} = w_j(\alpha + \beta Q_{\mathrm{M}ij}). \tag{18.15}$$

Hence, the marginal cost is $w_j\beta$ and the average cost is $w_j(\alpha/Q_{Mij} + \beta)$.

The total amount of manufacturing labor available in region j at any given time, L_{Mj}, is then

$$L_{Mj} = \sum_{i=1}^{N} L_{Mij}, \qquad j = 1, 2. \tag{18.16}$$

Most generally, let L_A and L_M denote the economy-wide supplies of the two types of labor, which are allocated in fixed amounts across the regions. Regional shares of agricultural labor, ϕ_j, are exogenously given (and in Krugman (1991a) are equal). Regional shares of manufacturing labor, λ_j, evolve over time. At any given point of time, the regional full-employment conditions will be

$$L_{Aj} = \phi_j L_A,$$

$$\sum_{i=1}^{N} L_{Mij} = \lambda_j L_M. \tag{18.17}$$

Agricultural workers are assumed to be immobile, but manufacturing workers are assumed to migrate to locations where higher real wages are paid. Defining the average real wage, $\bar{\omega}$, to be

$$\bar{\omega} = \sum_{j=1}^{R} \lambda_j \omega_j,$$

where ω_j denotes the wage paid to manufacturing workers in region j, Krugman (1995) assumes that migration occurs according to the following law of motion,

$$\frac{d\lambda_j}{dt} = \gamma\lambda_j(\omega_j - \bar{\omega}), \tag{18.18}$$

where γ is the disequilibrium adjustment parameter. The interpretation of Equation (18.18) is that "workers move away from locations with below average real wages and toward sites with above average real wages (Krugman, 1995, p. 96)."[6]

[6] While this disequilibrium–adjustment or error–correction formulation of dynamics is admitted to be ad hoc, even if commonly used in evolutionary game theory (Fujita *et al.*, 1999, p. 77), it can be motivated as the solution to a multilevel optimization problem. (See, e.g., Salmon, 1982.)

The firms in Krugman's model are assumed to be non-strategic monopolistic competitors of the Chamberlinian type.[7] Each produces a single good and considers itself unable to influence either the aggregate demand for manufactures, C_{M}, or the price index, P_{M}. Hence, in view of the compensated demand functions (18.6), the firm will perceive itself to be facing a downward sloping demand curve. This will be the case in a monopolistically competitive market in which economies of scale lead each firm to produce a differentiated good for which it has no competitors and the number of firms will be identical to the number of products, N. From Equation (18.6), the inverse demand function for manufactured good i can be written as

$$p_i = (P_{\mathrm{M}} C_{\mathrm{M}}^{1/\sigma}) c_i^{-1/\sigma}, \tag{18.19}$$

and the marginal revenue as

$$\mathrm{d}[(P_{\mathrm{M}} C_{\mathrm{M}}^{1/\sigma}) c_i^{(\sigma-1)/\sigma}]/\mathrm{d}c_i = [(\sigma - 1)/\sigma] p_i. \tag{18.20}$$

If the firm follows the policy of equating marginal costs with marginal revenues, i.e.

$$\beta w_j = [(\sigma - 1)/\sigma] p_i, \tag{18.21}$$

and w_j is the wage rate of workers in region j, its profit-maximizing strategy will be to set its price as a fixed mark-up over marginal costs, i.e.

$$p_i = [\sigma/(\sigma - 1)] \beta w_j. \tag{18.22}$$

Noting that Equation (18.22) implies that all manufactured goods i produced in region j will be priced equivalently, we can write $p_i = p_j$. If firms are free to enter or exit the market until profits are zero, the output of any manufactured variety i will be

$$Q_{\mathrm{M}i} = \frac{\alpha}{\beta}(\sigma - 1). \tag{18.23}$$

This remarkable condition implies that *all* manufactured goods will be produced at the *same* level and scale, and this output level,

---

[7] See Anas and Li (2001) for a reworking of the model with strategic competition among firms.

$Q_{Mi} = Q^*$, depends only on the parameters characterizing the production technology and the consumers' sub-utility function. Moreover, the proportion of total manufactured goods (or varieties) produced in any region j, N_j/N, will be equivalent to that region's share of the manufacturing labor force, λ_j, or

$$N_j/N = \lambda_j. \tag{18.24}$$

18.2.3. Transportation costs

While, by assumption, it is costless to transport agricultural goods between regions, a fixed proportion of a unit of a manufactured good, τ, is used up per unit of distance the good is shipped.[8] If the amount of good i, x_{ijk}, is shipped over a distance of D_{jk}, from region j to region k, the amount that arrives, z_{ijk}, is given by

$$z_{ijk} = e^{-\tau Djk} x_{ijk}. \tag{18.25}$$

This transport technology implies that if a manufactured good produced at location j is sold at the producer's, mill, or f.o.b. (free on board) price, p_j, then the user's, delivered, or c.i.f. (carriage, insurance, and freight) price at any location k, P_{jk}, is

$$P_{jk} = p_j \, e^{\tau Djk}. \tag{18.26}$$

In view of Equation (18.26), the price index of the aggregate of manufactured goods (Equation (18.5)) may take on a different value in each region. Taking into account iceberg transport costs of goods shipments received from regions $j = 1, \ldots, R$, and the implication of Equation (18.22) that all varieties produced at any location j have the same price, the *true price index* of the manufactured goods aggregate faced by consumers at location k, T_k, is given by

$$T_k = \left[\sum_{j=1}^{R} N_j (p_j \, e^{\tau Djk})^{1-\sigma} \right]^{1/(1-\sigma)} \qquad j = 1, \ldots, R. \tag{18.27}$$

From Equation (18.9), the amount of a manufactured good i

---

[8] This is Samuelson's (1952) 'iceberg' transport technology.

produced in j demanded in k will be

$$\mu Y_k(p_j\, e^{\tau Djk})^{-\sigma} T_k^{(\sigma-1)}, \tag{18.28}$$

where Y_k is aggregate income at location k. For this amount to arrive at k, $e^{\tau Djk}$ times this amount must be shipped. The total sales of a variety of manufactured good produced at j to all locations, $k = 1, \ldots, R$, will then be[9]

$$Q_{Mj} = \mu \sum_{k=1}^{R} Y_k(p_j\, e^{\tau Djk})^{-\sigma} T^{\sigma-1}\, e^{\tau Djk}. \tag{18.29}$$

And from Equation (18.23), we know $Q_{Mj} = Q^*$. From the pricing rule (18.22) the nominal wage paid at j can also be expressed as

$$w_j = \left(\frac{\sigma-1}{\sigma}\right)\left[\frac{\mu}{Q^*}\sum_{k=1}^{R} Y_k(e^{\tau Djk})^{1-\sigma} T^{\sigma-1}\right]^{1/\sigma}. \tag{18.30}$$

Since the cost of living index in each region j will be $T_j^{\mu} P_{Aj}^{1-\mu}$, the real wage of manufacturing workers in j, ω_j, will be

$$\omega_j = w_j T_j^{\mu} P_{Aj}^{1-\mu}. \tag{18.31}$$

18.2.4. Normalizations and short-run equilibrium

To simplify notation and the analysis, Krugman normalizes several quantities. Since, by assumption, it is costless to transport agricultural goods, the wage rate of farmers will be the same in all regions. The size of the labor force – farmers and workers – in the economy can be normalized to 1. Then, letting manufactured goods' expenditure share μ also denote the number of workers, the number of farmers is $1 - \mu$. This normalization has the added effect of setting economy-wide income to 1. If all prices and wages are expressed in terms of the agricultural good, nominal income in

---

[9] From Equation (18.29) we may infer that, no matter how far a good manufactured in region j is shipped, nor how proportional the resulting increase in its delivered price, the mill price elasticity of aggregate demand for the good will be each consumer's constant price elasticity of demand, σ.

region j is given by

$$Y_j = (1 - \mu)\phi_j + \mu\lambda_j w_j. \tag{18.32}$$

Setting $\beta = (\sigma - 1)/\sigma$, the f.o.b. price of manufactured goods in j will equal the wage rate

$$p_j = w_j. \tag{18.33}$$

and the equilibrium output level, Q^*, will equal the size of the workforce, L^*, of any manufacturing firm in the economy

$$Q^* = L^* = L_j. \tag{18.34}$$

Setting $\alpha = \mu/\sigma$, the number of firms related to the size of the manufacturing labor force becomes $N_j = L_{\mathrm{M}j}/\mu$ and the zero-profit output level $Q^* = L^* = \mu$. These normalizations permit the true-price index and wage Equations (18.27) and (18.30) to be rewritten as

$$T_j = \left[\frac{1}{\mu}\sum_{k=1}^{R} L_j(w_k\,\mathrm{e}^{\tau Djk})^{1-\sigma}\right]^{1/(1-\sigma)}$$

$$= \left[\sum_{k=1}^{R} \lambda_k(w_k\,\mathrm{e}^{\tau Djk})^{1-\sigma}\right]^{1/(1-\sigma)}, \tag{18.35}$$

and

$$w_j = \left[\sum_{k=1}^{R} Y_k(\mathrm{e}^{\tau Djk})^{1-\sigma} T^{\sigma-1}\right]^{1/\sigma}. \tag{18.36}$$

Finally, with the agricultural good as numeraire, and $P_{\mathrm{A}} = 1$, the real wage equation reduces to

$$\omega_j = w_j T_j^{-\mu}. \tag{18.37}$$

The nominal-income, true-price-index, nominal-wage, and real-wage Equations (18.32) and (18.35)–(18.37) together, when solved for all R regions, characterize a short-run equilibrium in this economy. In a stable equilibrium the real wage paid in each region will be the same, hence there will be no incentive for workers to move.

Except in special cases, these 4 R non-linear equations do not admit of an analytical solution. They are solved numerically in conducting simulations. Krugman (1995) and Fujita *et al.* (1999b) conduct comparative-static exercises with a two-region version of the model for which the true-price and wage equations have been linearized around a symmetric equilibrium – i.e. where $L_1 = L_2$, $T_1 = T_2$, and $w_1 = w_2$ — and demonstrate several important properties of the short-run solution. Other things being equal, if a region has a larger share of manufacturing firms, it will have a lower true-price index because a smaller share of its region's consumption of manufactured goods will bear transport costs (the *price-index effect*). A region with the larger home market will have a disproportionately larger manufacturing sector, hence will also export manufactured goods (the *home-market effect*). A region with a higher demand for manufactured goods (and more varieties of goods) will offer a higher real wage to manufacturing workers (because the price index is decreasing in the number of varieties). And a region with large concentrations of manufacturing will tend to have a large demand for manufactured goods, reflecting the demand for goods by its workers. All of these short-run effects reinforce any initial advantage a region might enjoy through a long-run process of cumulative causation.

An important feature of the model is the presence of both centrifugal (or dispersive) and centripetal (or aggregative) forces. The former appear in the form of immobile farmers demanding manufactured goods, while the latter are manifested in linkages and scale economies. In numerical simulations conducted with the model for two regions and different parameter settings, reported in Krugman (1991a), there are only two possible long-run equilibrium outcomes: a symmetric distribution between the regions of manufacturing and aggregation of all manufacturing in one region. In simulations with the multiple-region version of the model reported in Krugman (1995) symmetric equi-spaced distributions of manufacturing centers result. Generally speaking, with lower transportation costs, there is greater concentration, and with higher transportation costs there is less.

18.3. Developments in the new economic geography

Since the appearance of Krugman (1991a) there has been a proliferation of studies in which assumptions of the original model have been modified.[10] In this section we take up a subset of these studies which examine what difference modifications in assumptions about consumer utility, transportation costs, production technology, labor and firm mobility, and adjustment dynamics make for explanations of various urban and regional agglomeration phenomena.[11] (It should be borne in mind that because the labor force and output levels of firms are constant in the economy of Krugman's (1991a) model, there is no growth as such, only regional aggregations and reaggregations of economic activities.)

Tabuchi (1998) examines possible causes for concentration and dispersion of firms and workers between regions by considering urban agglomeration economies due to product variety and agglomeration dis-economies due to intra-city congestion. Hence, he attempts a synthesis of Krugman (1991a) with Alonso (1964) and Henderson (1974) that involves several modifications. First, he incorporates the land-consumption (consumption-of-space) assumption of Alonso's model, which introduces the impacts of the price mechanism of the land rent market, and second, the cost of commuting. These changes are introduced through the utility function and the budget constraint. What Tabuchi finds, *contra* Krugman (1991a), is that for certain parameter settings there is "a U-shaped relationship between the decrease in transportation costs and spatial agglomeration" (Tabuchi, 1998, p. 334).[12]

Lanaspa and Sanz (2001) consider a series of modifications to the assumption that transport costs vary proportionately with distance and independently of other elements of the model. In particular, they examine how congestion costs, which increase with the size of a region, and infrastructure, which requires a threshold population level to be put in place, affect the distribution of manufacturing activity among regions. They find that, for certain parameter value settings, stable *asymmetric* equilibria may be obtained, providing

---

[10] Many have been incorporated into the chapters of Fujita *et al.* (1999b) and Fujita and Thisse (2002). More complete surveys of studies in this line of inquiry concerning other agglomeration effects are provided by Ottaviano and Puga (1998), Papageorgiou and Pines (2000) and Fujita and Thisse (2002).

[11] While several of the studies considered below deal with regions and not cities, the analytical approaches they take are amenable to the study of cities and urban agglomeration phenomena.

[12] This same result is obtained by Helpman (1998), inter alia, although for different reasons.

justification for the existence of "economic landscapes in which large industrial belts coexist with smaller ones" (Lanaspa and Sanz, 2001, p. 437).

Both of the two previously discussed studies have examined two-region scenarios in which the distance between the centroids of the regions (or CBDs of the urban centers of the regions) are given. Alternatively, Mori (1997) considers an economy distributed over a *continuous interval* of space. In this economy there are costs associated with the transport of agricultural goods as well as manufactured goods. Both firms and workers can move. Mori makes use of a 'market potential' function, defined as the ratio of economy-wide demand for a firm's output to its equilibrium output level, to indicate when and where it is profitable for firms to locate and hence cities to form. He identifies conditions under which falling transportation prices in the presence of increasing returns to scale imply the formation of a megalopolis consisting of large core cities connected by an industrial belt, instead of a greater concentration of manufacturing activity at point locations. The result of both firms and workers seeking the arrangement that is most advantageous to them is that no land that is exclusively agricultural in use is left between any two cities as firms aggregate toward an interval. Mori suggests that to increase realism, the 'agricultural' area in his model might be replaced by a residential area of workers who commute to urban jobs.

Also relaxing the assumption that shipment of agricultural goods is costless, Fujita and Krugman (1995) seek to derive conditions under which all manufacturing activity will concentrate in a central city and the conditions under which this pattern will be sustainable.[13] The economy they consider is spread across a one-dimensional continuous interval of space. In this model, the normalization of labor employed in Krugman (1991a) is relaxed and the population is allowed to grow gradually. The forces of aggregation and dispersion remain as in Krugman (1991a). Fujita and Krugman find for a range of values of the preference intensity parameter ρ that "when the population increases from a low level, the benefits of a larger manufacturing sector dominate, but as

---

[13] Note that even concentration represents a balance of agglomerative and dispersive forces. A condition that parameters must satisfy to prevent agglomerative forces from pulling all manufacturing activity into a 'black hole' is that $(\sigma - 1)/\sigma > \mu$.

the population continues to increase the disadvantages of an ever more distant agricultural frontier prevail" (Fujita *et al.*, 1999b, p. 140).[14] They determine that the population size that maximizes real wages is greater when the variety of manufactured goods is greater, when manufactured goods' expenditure share is larger, and when the costs of transporting both agricultural and manufactured goods are lower. Using a market potential function, in which potential is defined as the ratio of the maximum real wage that zero-profit manufacturing firms could pay workers at a given distance from the city center to the real wage paid to farmers at the same distance, Fujita and Krugman identify the critical population level at which it is profitable for a manufacturing firm to exit the central city, hence the population level at which the monocentric structure of the spatial economy is destroyed. They also identify conditions that combinations of key parameters must satisfy if the monocentric equilibrium is to be sustained.

In an extension of the previously discussed paper, involving multiple cities and multiple groups of industries, Fujita *et al.* (1999a) develop a general spatial equilibrium model whose equilibrium solution yields a hierarchical central place system of cities. Agglomeration forces continue to derive from consumer preference for variety in manufactured goods and scale economies in their production, while dispersion forces derive from the demand for consumer goods by the agricultural populations. The setup of the analysis assumes that the economy already has an established structure of multiple cities, in which new cities can nonetheless emerge. 'Higher order' cities have more industries than 'lower order' cities and achieve their status by being upgraded from lower order cities. Adjustment dynamics come from labor migration in response to real-wage differentials and in equilibrium the real wage is equalized across locations. A given system of cities is sustainable if the market potential (as defined in terms of the ratio of real wages in manufacturing and agriculture) at any location other than those already occupied is less than 1. To study the self-organization of the city system, the authors conduct dynamic simulations in which they introduce hierarchical groups of manufactured goods. As in

---

[14] Fujita and Krugman (1995) reappears as Chapter 9 of Fujita *et al.* (1999b).

the previously discussed study, an exogenously growing population drives the economy's growth. As the economy's population increases, existing cities bifurcate and urban populations disperse until, in the long run, there are, for a given order of city, equal-sized cities with equal market areas.[15] And as in Christaller's central place system, higher order cities produce all the manufactured goods produced by lower order cities and more.

In the next two studies considered intermediate goods play a prominent role. Fujita and Hamaguchi (2001) develop a monopolistic competition model of a spatial economy in which manufacturing firms require a large variety of intermediate goods (instead of consumers demanding a large variety of final manufactured goods).[16] Agglomeration forces "arise from the vertical linkages between the manufacturing and intermediate-good sector", whereas "dispersion forces arise from the demand for the manufactured goods by the agricultural workers who are spatially dispersed due to the necessity of land input" (Fujita and Hamaguchi, 2001, p. 79–80). The authors argue for the relevance of this model to the explanation of resurgence of some declining cities in developed countries because of the recent increase in demand for producer services as firms seek to outsource and reduce overhead costs. Solutions of the model yield two types of monocentric configurations. In the first, an 'integrated city equilibrium', transaction costs of intermediate goods are high and both manufacturing and intermediate sectors agglomerate in a single city. In the second, an 'I-specialized city equilibrium', the city specializes in the provision of intermediate goods. Fujita and Hamaguchi find that the former case represents a 'primacy trap' in which population growth alone will never lead to the formation of new cities.

Walz (1996) considers a two-region economy in which growth emanates from product innovation in the intermediate-goods sector,

---

[15] A significant part of the analysis in the NEG not discussed in this chapter has to do with bifurcations of solutions that occur at critical parameter values. While actual cities do not bifurcate, the bifurcating solutions of models indicate the possibility of multiple equilibria and the historical contingency of actual city-system formations. Fujita and Thisse (2002) take such model solutions to imply a 'putty-clay' geography.

[16] The model is in a sense dual to Fujita and Krugman (1995), in which the agglomeration forces arise from consumers' love for variety.

whose increase in varieties leads to higher productivity in the final-goods sector. While final goods are freely traded, intermediate differentiated goods, which are supplied by monopolists, are costly to trade. The production of intermediate goods requires specific R&D investments and the interregional trade in R&D products is associated with control and information costs. Unlike most other models in the NEG, explicit intertemporal optimizations by firms and consumers yield micro-foundations for dynamic adjustments in the model. Solution of the model determines the locations of goods production, R&D activities, and mobile workers. Under various conditions, the solution of the model results in a core–periphery pattern, equal-sized regions, or diverse production patterns and growth rates. Interestingly, and not unlike historical experience with various regional policies, policies designed to bring regions into convergence may actually result in unintended consequences and increase the gap between their production and growth structures. To the extent that intermediate good producers are located in cities, lessons from this model may be extended to city systems as well.

18.4. Accomplishments and challenges

Several broad assessments of what the NEG has accomplished have already been published.[17] Our concern lies more narrowly with the ability of this body of work to help us to understand certain aspects of urban and regional agglomeration. The authors of the papers we have considered have succeeded in developing models whose solutions are consistent with different patterns of agglomeration, giving formal expression to powerful intuitions and endogenizing factors previously taken to be exogenous. In so doing they have provided materials for possible explanations of these phenomena. These achievements represent a significant contribution to 'positive' urban economics. But several of the most central contributors to this work, Fujita *et al.* (1999b), have also called for empirical investigation of the models they have developed and

---

[17] See inter alia, Ottaviano and Puga (1998), Papageorgiou and Pines (1999), Martin (1999), and the critical forum in the Journal of Economic Geography, 2001, 1:131–152.

normative applications thereof. So we consider a set of issues concerning how well these models might fit into a larger, empirically oriented, research program of explaining agglomeration phenomena.

Certainly all modeling and theorizing entails making some simplifying (and occasionally over-simplifying) assumptions, which, we might agree with Krugman (1995), are essential for conducting instructive thought experiments. Problems with simplification arise when we move from thought experiments with models to causal explanations of real-world phenomena. We need to be concerned about the extent to which the worlds of NEG models *do not* correspond to the world we are trying to explain. For, *pace* Friedman (1953), to test and confirm causal explanations empirically, the models we use to articulate those explanations must pick out *actual* causal mechanisms, even if highly stylized (Miller, 1987; Runde, 1998). Hence, in applied work we need at least roughly accurate characterizations of the behavior we hope to explain and, no less importantly, measurable variables. To meet these requirements, *operationalized* NEG models may need to incorporate more of the actual 'furniture of the world' and take on board considerably more complexity than they now carry.[18] We consider several aspects of NEG models for which lack of realism presents problems for applied work.

That the Dixit-Stiglitz setup may not correspond to any actual state of affairs, as Krugman (1995) readily acknowledges, is a worry, in view of the work it is called upon to do in most NEG models. And as Strange (2001) points out, while the Dixit-Stiglitz specification is a convenient way to get aggregation economies into a model, it *does* matter to the analysis whether aggregation economies derive from consumer preferences for variety, from demand for variety of intermediate goods by firms producing final goods (Ethier, 1982), or from labor–market pooling, input sharing, or educational

[18] Note that demand for parsimony of Occam's razor only applies to models and explanations that are equally successful in accounting for the phenomena in question – i.e. in identifying effective causes at sufficient causal depth. A separate concern lies with the extent to which the results of NEG models are robust to choice of functional forms and transport technologies. See, e.g., Fujita and Hamaguchi (2001) on this acknowledged weakness.

improvement and skill acquisition (Quigley, 1998). We should also be mindful that, *wherever* they derive from, agglomeration economies will affect patterns of agglomeration differently at different distances (Anas *et al.*, 1998).

The role given to farmers' demand for final goods in attenuating centripetal forces in these models is, in the view of Pines (2001), certainly unrealistic in developed economies, even if we read 'agriculture' as a residual sector, as Fujita *et al.* (1999b) suggest. Not only is the characterization of the role of agricultural production and land use problematic, but also are the characterizations of consumers and firms. There are at least two issues here. Contrary to the implication of the basic assumptions of NEG models (see footnote 1), consumers do not interact even remotely with every firm producing final goods; alternatively, neither does every such firm interact with every firm producing intermediate goods. Moreover, in the 'information' or 'e-commerce' economy consumers and firms do not purchase goods and services in patterns that would suggest the operation of a central place system. (See, e.g. Pred, (1977) for a somewhat dated but still realistically relevant depiction of the interactions of firms in city systems of advanced economies.)

The characterization of firm mobility also needs refinement. Jones and Kierzkowski (1990, 2001), whose work is also based on models of monopolistic competition in international trade, have focused on one of the most fundamental empirical regularities of worldwide production and distribution: the vertical and horizontal fragmentation of firms and industries. The reality that they depict is not one of firms moving 'lock, stock, and barrel' from one region or point of aggregation to another. In this world, where production processes are being broken out more and more finely into blocs, the production of intermediate goods and producer services is dispersing increasingly across many regions (of many countries). And the largest share of goods in transit consists of semi-finished goods shuttling between firms' *own* far-flung establishments. It is *this* dynamics of globalization, induced by information and communication technology innovations, underpriced transport, and the availability of low-wage labor, that is radically altering the economic geographies of North America, Europe, and Asia and needs to be formally endogenized in models purporting to explain the ongoing evolution in the spatial distribution of economic activities.[19]

Other important actors on the urban scene who are conspicuous by their absence in the NEG stories about agglomeration phenomena are developers and government. Certainly these actors strongly influence decisions of firms and workers through the provision of infrastructure, the entrepreneurial undertaking of risky investments, and the jump-starting of aggregation. Indeed, it is arguable that some growth phenomena, such as edge cities, cannot be adequately explained without making reference to the behavior of these actors.

As discussed above, one of the important contributions of the NEG, and Fujita *et al.* (1999a) in particular, has been to demonstrate how a central place system can emerge from the basic assumptions of NEG models. One concern, however, is that (to the best of my knowledge) these models have not been able to replicate the evolution of uneven development patterns as have other models of self-organizing complex systems in the tradition of, say, Allen (1998). Admittedly, models in the latter tradition lack economic micro-foundations or a role for price mechanisms and so are not fully in the 'urban economics tent'. But as Anas *et al.* (1998) have noted, attempts to supply micro-foundations for Allen-type models along the lines of Chen (1993, 1996) may hold promise.

Finally, we turn to empirical investigations of the NEG. Most economists who have tried to test propositions implied by the NEG have carried out studies with data collected at the regional level (see, e.g., Hanson (1997) and Fingleton (2001), and papers discussed by Ottaviano and Puga (1998)). It is encouraging for the NEG project that evidence of IRS economies of production have been found at this level.[20] A limited number of studies have also been carried out at urban and sub-regional scales. Analyzing data at the urban scale, Glaeser *et al.* (1992) have found evidence of agglomeration economies arising from a demand for variety of goods and Fingleton

---

[19] Fujita *et al.* (1999a) acknowledge the need to incorporate multi-locational firms in NEG models. Krugman and Venables (1995) have shown how NEG models can illuminate other aspects of globalization that contribute to regional and international inequality.

[20] In addition to the papers cited, Donaghy and Dall'Erba (2003), using a growth model based on a generalized CES production technology, have obtained direct estimates of returns-to-scale parameters in the regional economies of Spain that indicate IRS in all regions.

(2003) has found evidence of increasing returns to the density of economic activities. LaFountain (2003) has found that pecuniary externalities are a source of agglomeration for some industries, but not others, at the county level. We discuss the latter two studies in greater detail.

Fingleton (2003) has found that as the density of workers in local areas of Great Britain increases there is a more than proportionate increase in wage levels. He infers from this that there is also a more than proportionate increase in the level of output of final goods and services. In the greater London area, in particular, he has found that enhanced worker efficiency is relative to in-commuting, which in turn is contingent upon transportation infrastructure.

Working with 2-digit SIC code industry data at the county level, LaFountain (2003) has examined empirical support for three alternative models of firms' location decisions and agglomeration. These include 'market access' models of the NEG, in which pecuniary externalities are the source of agglomeration, production 'externality' models, in which firms' production possibilities depend on the action of firms at the same location, and 'natural advantage' models, in which the endowments of different locations render them more or less attractive to different types of producers.[21] LaFountain has found that for six of the 18 industries she has analyzed – food, printing and publishing, stone, clay and glass, fabricated metals, industrial machinery and equipment, and miscellaneous manufacturing industries – the data are consistent with predictions of the market access model.[22]

Issues raised by the NEG might also be investigated empirically at *sub*-urban scales and with more explicit spatial content, in view of the fact that we now possess methodologies in applied spatial

---

[21] In a predecessor to LaFountain (2003), Kim (1995) distinguished between the three location-decision models for the US manufacturing sector as a whole but did not consider the role that urbanization externalities may play or identify which model best explains the location decisions of firms in individual industries.

[22] LaFountain (2003) has also found that data on seven industries – paper, chemicals, petroleum, and coal products, primary metals, electronic and other electric equipment, transportation, equipment, instruments, and furniture and fixture industries – are consistent with predictions of the 'natural advantage' model, and that data on the textiles and apparel industries are consistent with the predictions of the 'externality' model. Data on the lumber and Wood product and the rubber and plastics industries are not consistent with predictions of any of the three classes of models.

econometrics to investigate spatial dynamic models in one or two dimensions of space (Donaghy, 2001; Donaghy and Plotnikova, 2004). While questions about the availability of spatial time-series with sufficient observations to test such models may exist, it would appear that the NEG is at a stage of development where applied empirical work may help to push it along and supply important feedback to the theorizing that has been conducted to date through abstract modeling and numerical simulations. As it assumes greater realism, the NEG may in turn be able to contribute insights that will inform concrete normative assessments and suggest appropriate policy interventions.

References

Abdel-Rahman, H. and M. Fujita (1990), "Product variety, Marshallian externalities, and city size", *Journal of Regional Science*, Vol. 30, pp. 165–183.

Allen, P. (1998), *Cities and Regions as Self-Organizing Systems: Models of Complexity*, Amsterdam: Gordon and Beach Science Publishers.

Alonso, W. (1964), *Location and Land Use*, Cambridge: Harvard University Press.

Anas, A. and Y. Li (2001), "Strategic competition and the new economic geography", Paper prepared for presentation at the *North American meetings of the Regional Science Association International, Charleston, South Carolina*, November, 2001.

Anas, A., R. Arnott and K.A. Small (1998), "Urban spatial structure", *Journal of Economic Literature*, Vol. 36, pp. 1426–1464.

Berliant, M. and P. Wang (2004), "Dynamic urban models: agglomeration and growth", in: R. Capello and P. Nijkamp, editors, *Urban Dynamics and Growth: Advances in Urban Economics*, Amsterdam: Elsevier.

Black, D. and V. Henderson (1999), "A theory of urban growth", *Journal of Political Economy*, Vol. 107, pp. 252–284.

Brueckner, J.K. (2000), "Urban sprawl: diagnosis and remedies", *International Regional Science Review*, Vol. 23, pp. 160–171.

Brueckner, J.K. (2001), "Urban sprawl: lessons from urban economics", pp. 65–89, in: W.G. Gale and J.R. Pack, editors, *Brookings-Wharton Papers on Urban Affairs*, Washington, DC: Brookings Institution.

Brueckner, J.K. (2003), "Transport subsidies, system choice, and urban sprawl", *Working Paper*.

Brueckner, J.K. and H.-A. Kim (2003), "Urban sprawl and the property tax", *International Tax and Public Finance*, Vol. 10, pp. 5–23.

Button, K. (2000), "Where did the 'new urban economics' go after 25 years?", pp. 30–50, in: A. Reggiani, editor, *Spatial Economic Science: New Frontiers in Theory and Methodology*, Berlin: Springer.

Chen, H.-P. (1993), "Theoretical derivation and simulation of a nonlinear dynamic urban growth model". *PhD Dissertation*, Department of Economics, University of California Irvine.

Chen, H.-P. (1996), "The simulation of a proposed nonlinear dynamic urban growth model", *Annals of Regional Science*, Vol. 30, pp. 305–319.

Dixit, A.K. and J.E. Stiglitz (1977), "Monopolistic competition and optimum produce diversity", *American Economic Review*, Vol. 67, pp. 297–308.

Donaghy, K.P. (2001), "Solution and estimation of spatial dynamic models in continuous space and continuous time", *Journal of Geographical Systems*, Vol. 3, pp. 257–270.

Donaghy, K.P. and S. Dall'Erba (2003), "Structural and spatial aspects of regional inequality in Spain", revised draft of paper presented at the *United Nations University World Institute for Development Economics Research (WIDER) Project Meeting on Spatial Inequality in Development*, 29–30 May, Helsinki.

Donaghy, K.P. and M. Plotnikova (2004), "Econometric estimation of a spatial dynamic model in continuous space and continuous time: an empirical demonstration", in: A. Getis, J. Mur and H.G. Zoller, editors, *Frontiers of Spatial Econometrics*, London: Palgrave.

Ethier, W. (1982), "National and international returns to scale in the modern theory of international trade", *American Economic Review*, Vol. 72, pp. 389–405.

Fingleton, B. (2001), "Theoretical economic geography and spatial econometrics: dynamic perspectives", *Journal of Economic Geography*, Vol. 1, pp. 201–226.

Fingleton, B. (2003), "Increasing returns: evidence from local wage rates in Great Britain", *Oxford Economic Papers*, Vol. 55, pp. 716–739.

Friedman, M. (1953), *Essays in Positive Economics*, Chicago: University of Chicago Press.

Fujita, M. (1982), "Spatial patterns of residential development", *Journal of Urban Economics*, Vol. 12, pp. 22–52.

Fujita, M. (1996), "Urban land use theory", pp. 111–188, in: R. Arnott, editor, *Regional and Urban Economics, Part I*, Amsterdam: Harwood Academic Publisher.

Fujita, M. and N. Hamaguchi (2001), "Intermediate goods and the structure of an economy", *Regional Science and Urban Economics*, Vol. 31, pp. 79–109.

Fujita, M. and P. Krugman (1995), "When is the city monocentric?: von Thünen and Chamberlin unified", *Regional Science and Urban Economics*, Vol. 25, pp. 505–528.

Fujita, M. and H. Ogawa (1982), "Multiple equilibria and structural transition of non-monocentric urban configurations", *Regional Science and Urban Economics*, Vol. 12, pp. 161–196.

Fujita, M. and J.-F. Thisse (2002), *Economics of Agglomeration: Cities, Industrial Location, and Regional Growth*, Cambridge: Cambridge University Press.

Fujita, M., P. Krugman and T. Mori (1999a), "On the evolution of hierarchical urban systems", *European Economic Review*, Vol. 43, pp. 209–251.

Fujita, M., P. Krugman and A.K. Venables (1999b), *The Spatial Economy: Cities, Regions, and International Trade*, Cambridge: The MIT Press.

Gandolfo, G. (1998), *International Trade Theory and Policy*, Berlin: Springer.

Glaeser, E.L., H.D. Kallal, J.A. Scheinkman and A. Shleifer (1992), "Growth in cities", *Journal of Political Economy*, Vol. 100, pp. 1126–1152.

Hanson, G.H. (1997), "Increasing returns, trade, and the regional structure of wages", *Economic Journal*, Vol. 107, pp. 113–133.

Helpman, E. (1998), "The size of regions", in: D. Pines, E. Sadka and I. Zilcha, editors, *Topics in Public Economics. Theoretical and Applied Analysis*, pp. 33–54, Cambridge: Cambridge University Press.

Henderson, V. (1974), "The sizes and types of cities", *American Economic Review*, Vol. 64, pp. 640–656.

Henderson, V. and A. Mitra (1996), "The new urban landscape: developers and edge cities", *Regional Science and Urban Economics*, Vol. 26, pp. 613–643.

Isard, W. (1956), *Location and Space Economy*, Cambridge: The MIT Press.

Jacobs, J. (1969), *The Economy of Cities*, New York: Vintage.

Jacobs, J. (1984), *Cities and the Wealth of Nations: Principles of Economic Life*, New York: Vintage.

Jones, R.W. and H. Kierzkowski (1990), "The role of services in production and international trade: a theoretical framework", in: R.W. Jones and A.O. Krueger, editors, *The Political Economy of International Trade*, Oxford: Blackwell.

Jones, R.W. and H. Kierzkowski (2001), "A framework for fragmentation", in: S.W. Arndt and H. Kierzkowski, editors, *Fragmentation: New Production Patterns in the World Economy*, New York: Oxford University Press.

Kim, S. (1995), "Expansion of markets and geographic distribution of economic activities: the trends in U.S. regional manufacturing structure 1860–1987", *Quarterly Journal of Economics*, Vol. 110, pp. 881–908.

Krugman, P. (1991a), "Increasing returns and economic geography", *Journal of Political Economy*, Vol. 99, pp. 483–499.

Krugman, P. (1991b), *Geography and Trade*, Cambridge: The MIT Press.

Krugman, P. (1995), *Development, Geography, and Economic Theory*, Cambridge: The MIT Press.

Krugman, P. (1998), "Space: the final frontier", *Journal of Economic Perspectives*, Vol. 12, pp. 161–174.

Krugman, P. and A. Venables (1995), "Globalization and the inequality of nations", *Quarterly Journal of Economics*, Vol. 110, pp. 857–880.

LaFountain, C. (2003), Where do firms locate? Testing competing models of agglomeration, *Working Paper*, Arlington, TX: Department of Economics, The University of Texas at Arlington.

Lanaspa, L.F. and F. Sanz (2001), "Multiple equilibria, stability, and asymmetries in Krugman's core–periphery model", *Papers in Regional Science*, Vol. 80, pp. 425–438.

Marshall, A. (1936), *Principles of Economics*, 8th edition, London: Macmillan.

Martin, R.L. (1999), "The new 'geographical turn' in economics: some critical reflections", *Cambridge Journal of Economics*, Vol. 23, pp. 65–91.

Miller, R. (1987), *Fact and Method: Explanation, Confirmation, and Reality in the Natural and Social Sciences*, Princeton: Princeton University Press.

Mori, T. (1997), "A modeling of megalopolis formation: the maturing of city systems", *Journal of Urban Economics*, Vol. 42, pp. 133–157.

Ohls, J.C. and D. Pines (1975), "Discontinuous urban development and economic efficiency", *Land Economics*, Vol. 51, pp. 224–234.

Ottaviano, G.I.P. and D. Puga (1998), "Agglomeration in the global economy: a survey of the 'new economic geography'", *World Economy*, Vol. 21, pp. 707–731.

Ottaviano, G.I.P. and J.-F. Thisse (2001), "On economic geography in economic theory: increasing returns and pecuniary externalities", *Journal of Economic Geography*, Vol. 1, pp. 153–179.

Papageorgiou, Y. and D. Pines (1999), *An Essay on Urban Economic Theory*, Boston: Kluwer Academic Publishers.

Pines, D. (2001), "'New economic geography': revolution or counter-revolution?", *Journal of Economic Geography*, Vol. 1, pp. 139–146.

Pred, A. (1977), *City Systems in Advanced Economies*, New York: Halsted.

Quigley, J.M. (1998), "Urban diversity and economic growth", *Journal of Economic Perspectives*, Vol. 12, pp. 127–138.

Runde, J. (1998), "Assessing causal explanations", *Oxford Economic Papers*, Vol. 50, pp. 151–172.

Salmon, M. (1982), "Error correction mechanisms", *Economic Journal*, Vol. 92, pp. 615–629.

Samuelson, P.A. (1952), "The transfer problem and transport costs: the terms of trade when impediments are absent", *Economic Journal*, Vol. 62, pp. 278–304.

Spence, A.M. (1976), "Product selection, fixed costs, and monopolistic competition", *Review of Economic Studies*, Vol. 43, pp. 217–235.

Strange, W.C. (2001), "Review of Y.Y. Papageorgiou and D. Pines, An Essay on Urban Economics Theory", *Journal of Economic Geography*, Vol. 1, pp. 252–253.

Tabuchi, T. (1998), "Urban agglomeration and dispersion: a synthesis of Alonso and Krugman", *Journal of Urban Economics*, Vol. 44, pp. 333–351.

Walz, U. (1996), "Transport costs, intermediate goods, and localized growth", *Regional Science and Urban Economics*, Vol. 26, pp. 671–695.

Wheaton, W.C. (1982), "Urban residential growth under perfect foresight", *Journal of Urban Economics*, Vol. 12, pp. 1–21.

Urban Dynamics and Growth, CEA, vol. 266
Roberta Capello and Peter Nijkamp (editors)

CHAPTER 19

Agglomeration and Knowledge Diffusion

Johannes Bröcker
Institute of Regional Research, University of Kiel, D–24098, Kiel, Germany

Abstract

According to New Growth Theory one cannot rely on the convergence mechanisms inherent in traditional neoclassical constant returns to scale models. Convergence as well as divergence is possible, in general, depending on the assumptions about technology, factor mobility, and ease of knowledge diffusion. The paper shows by a two-regions endogenous growth model under what conditions divergence, convergence, or a stable center–periphery structure emerge. The model allows for different degrees of knowledge diffusion as well as for different degrees of labor and capital mobility. The paper also evaluates dynamic market equilibria with respect to allocative efficiency. It is shown that the market solution tends to be under-agglomerated, except for parameter constellations generating particularly low agglomeration forces. If agglomeration forces are low enough, no concentration emerges, and this is also socially desirable. For higher agglomeration forces, however, concentration becomes desirable though the market may not bring it about or brings it about to an insufficient degree only.

Keywords: convergence, divergence, agglomeration, endogenous growth, knowledge diffusion

JEL classifications: R110, R130, O330, O410, F430

Thanks to three anonymous referees for helpful suggestions to extend the paper in several dimensions. I could take account of some but not all of them due to time and space limitations.

19.1. Introduction

Convergence vs. divergence in a dynamic spatial economy is a classical issue in economics, that gained a lot of attention after the early contributions of Myrdal (1957) and Hirschman (1958), and later by Kaldor (1970). These authors tried to show that diverging tendencies would necessarily dominate converging ones in a growing market economy. Mobility of goods, people, and capital would enhance this divergence. No explicit attention was given to mobility of knowledge and ideas, so that these authors left open whether the easing of communication over space would strengthen diverging or converging forces. Solovian growth theory took the counterpart by proving that under constant returns to scale incomes converge, and that convergence is even accelerated by any kind of mobility. Information mobility, though, was not really an issue in these theories either, because equal access to technology in all countries or regions was assumed from the very beginning.

There was little progress in theory until the upswing of New Economic Geography and New Growth Theory, both taking increasing returns and imperfect markets as natural starting points for explaining endogenous growth as well as endogenous agglomeration. Both branches have recently been joined in the work of Walz (1996, 1999), in a series of papers by Baldwin and Forslid (1997, 2000), Baldwin (2001) and by Martin and Ottaviano (2001) and Fujita and Thisse (2002, Chapter 11), in their recent monograph. In these models growth results from a steadily increasing diversity of goods. Investment comes in the form of ideas allowing to introduce new goods. Ideas are generated by a costly research process, as in the pioneering work of Romer (1990). Different spatial patterns emerge depending on the mobility of goods, capital, and knowledge spillovers.

Due to the fact that monopolistic markets with transport costs as well as consumption and investment decisions of forward looking agents have to be handled in these models, they are highly complex. Hence, we try a simpler way by deriving long-term growth from Marshallian externalities of capital. Due to these externalities capital has non-decreasing returns on the global level. Knowledge spillovers are hampered by distance, however.

Our main question is how this distance deterrence of information flows affects the spatial distribution of economic activity as well as global efficiency. Furthermore, we try to figure out the interplay of factor mobility with the mobility of ideas. Due to the simplicity of the Marshallian externality approach we are able to draw clear-cut conclusions, both positive and normative ones. This simplicity comes at a cost, of course: a microfoundation of the innovation process is lacking in our approach, and we are unable to study also the interplay between factor mobility and mobility of ideas on the one hand, and freeness of trade on the other. This is a major issue of the cited literature.

19.2. A growth model with two regions

19.2.1. Firms

We set up a simple growth model with a standard neoclassical production function with three factors of production:

- an immobile factor stock that cannot be accumulated; one may regard it as immobile labor or a combination of immobile labor, land and other natural resources;
- mobile labor;
- capital, which is understood as a combination of real capital and knowledge; knowledge has a local impact on output as well as a global one, that is an extra unit of capital in a region increases output in the region where it is invested, as well as in the other region, though to a lesser extent due to the intervening impediments of communication.

Let us consider two regions (i or $j = 1, 2$) that are completely identical except that one region may cover a larger stock of the immobile factor. One may think of two regions with the same area, but one having a larger density of the immobile part of the population. This asymmetry is exogenously given. We will also consider the completely symmetrical case, where both regions are a priori identical in every respect. The asymmetric case helps to understand the impact of an exogenous starting advantage of a region, everything else being the same in both regions.

Both regions produce the same homogeneous output Y_i by the same Cobb–Douglas technology,

$$Y_i = aL_i^{\lambda} M_i^{\mu} K_i^{\alpha} G_i^{\gamma}, \tag{19.1}$$

with immobile labor L_i, mobile labor M_i, and capital K_i. Furthermore, output depends on regional access to global knowledge measured by an index G_i. Global knowledge is generated as an externality of capital and knowledge accumulation in both regions. Regional access to this knowledge is assumed to decrease with increasing distance, as convincingly demonstrated empirically by Jaffe *et al.* (1993). Hence we assume

$$G_1 = K_1 + \beta K_2 \text{ and } G_2 = K_2 + \beta K_1,$$

with parameter $0 < \beta < 1$ measuring the intensity of interregional communication. $\beta = 0$ means infinite impediments to communication; there is no flow of knowledge. $\beta = 1$ means no impediments to communication, access to knowledge externalities is everywhere the same, irrespective of the location where the respective knowledge has been accumulated. The literature has extensively dealt with the limiting cases $\beta = 0$ (pure local knowledge) and $\beta = 1$ (pure global knowledge), which both characterize an extreme and unrealistic world.

The homogeneous good is freely traded between regions without transportation cost. It is either consumed or invested. One unit of the good is transformed one-to-one into one unit of installed capital. The increase of the capital stock per unit of time $\dot{K}_i$ in region i is gross investment I_i minus depreciation δK_i, i.e.

$$\dot{K}_i = I_i - \delta K_i.$$

Once nailed to the ground in one region, installed capital cannot be relocated (or can be relocated only at a cost that never makes a relocation worthwhile).

The parameters λ, μ, α, and γ are the partial production elasticities of the respective factors. They are all assumed to be positive and less than one. Furthermore, we assume constant global returns to capital, which means $\alpha + \gamma = 1$. This is the 'knife-edge' assumption usually made in growth modeling. It guarantees convergence to a constant global steady state rate of growth.

A deviation from this assumption generates either explosive growth ($\alpha + \gamma > 1$), or stagnation ($\alpha + \gamma < 1$), contrary to observation. It is not nice that one has to build the theory on the assumption of a certain parameter to equal exactly unity without a deeper theoretical justification why the technology should tend to attain this parameter value. This is, however, a common drawback of the current state of endogenous growth theory, and we must take it as a still unresolved puzzle.

The effect of global knowledge represented by the index G_i is assumed to be entirely external. That means it is available as a public good, firms do not pay for making use of it. To be sure, there are also private knowledge flows in the economy; users in one location pay for knowledge accumulated elsewhere. But we disregard this kind of knowledge flows for the sake of simplicity. The elasticities λ, μ, and α may represent internal as well as external effects of the respective factors of production, such that $\lambda = \lambda_p + \lambda_e$, $\mu = \mu_p + \mu_e$, and $\alpha = \alpha_p + \alpha_e$. λ_p and λ_e stand for the elasticities representing the internal and external effect, respectively, similarly for μ and α. We can apply perfect competition theory of pricing if we assume $\lambda_p + \mu_p + \alpha_p = 1$. A special case is where the only externality is that of G_i. In order to allow for perfect competition pricing for this case we must assume $\lambda + \mu + \alpha = 1$.

19.2.2. Households

Households are either mobile or immobile workers. Mobile workers are perfectly mobile, not facing any relocation costs. Hence, they always choose the location offering the highest wage for mobile labor, that is the location with the highest private marginal productivity of mobile labor. All households are assumed to have identical linear-homogeneous preferences, such that they may be represented by a single representative household maximizing the isoelastic utility

$$U = \int_0^{\infty} \frac{\sigma}{\sigma - 1} C(t)^{1-1/\sigma} \exp(-\rho t) \mathrm{d}t, \tag{19.2}$$

subject to the budget constraint requiring the present value of consumption not to exceed the present value of labor income. $C(t)$ is consumption at time t, ρ is the subjective discount rate, and σ is the elasticity of intertemporal substitution. Households maximize over an infinite horizon; their utility is regarded as the utility of an immortal family.

There is a perfect asset market. Households save by accumulating a risk free asset. As the economy is closed, the asset value equals the value of the stock of capital at any point in time. That means the asset value equals $q_1K_1 + q_2K_2$, where q_i denotes the stock price of capital installed in region i.

19.2.3. Dynamic equilibrium

We begin with deriving the migration equilibrium for mobile workers. Due to the Cobb–Douglas form of the production function the marginal return of mobile labor in region i is $\mu_p Y_i/M_i$, such that $M_1/M_2 = Y_1/Y_2$ at any time due to perfect mobility. Hence, by Equation (19.1) we obtain

$$Y_1/Y_2 = \ell^\lambda (Y_1/Y_2)^\mu k^\alpha \left(\frac{k+\beta}{1+\beta k}\right)^\gamma$$

with $\ell := L_1/L_2$ and $k := K_1/K_2$. Solving for Y_1/Y_2 yields

$$m := M_1/M_2 = Y_1/Y_2 = \left[\ell^\lambda k^\alpha \left(\frac{k+\beta}{1+\beta k}\right)^\gamma\right]^{1/(1-\mu)}. \tag{19.3}$$

Given ℓ and the elasticities, the distribution of mobile labor across regions only depends on the distribution of capital across regions, not on the total capital stock.

We normalize the total stock of mobile labor to unity, $M_1 + M_2 = 1$, such that $M_1 = m/(m+1)$ and $M_2 = 1/(m+1)$. Inserting this into Equation (19.1) and dividing by K_i yields the average capital productivity

$$y_1 := Y_1/K_1 = aL_1^\lambda \left(\frac{m}{m+1}\right)^\mu K_1^{\alpha-1} G_1^\gamma,$$

and noting that $\alpha - 1 = -\gamma$ by assumption this is

$$y_1 = aL_1^\lambda \left(\frac{m}{m+1}\right)^\mu (1 + \beta/k)^\gamma, \tag{19.4}$$

and similarly

$$y_2 = aL_2^{\lambda}\left(\frac{1}{m+1}\right)^{\mu}(1+\beta k)^{\gamma}. \tag{19.5}$$

Note that the private marginal productivity of capital in region i is $\alpha_p y_i$. Hence, these marginal productivities also only depend on the distribution of capital across regions, not on the total capital stock. This is of course the implication of the assumed constant global returns to capital.

Next, we derive households' consumption demand. Maximizing Equation (19.2) subject to the budget constraint yields the well-known Keynes–Ramsey rule

$$\hat{C} = \sigma(r - \rho)$$

with interest rate r and consumption growth rate $\hat{C} = \dot{C}/C$. We generally denote the time derivative of a time dependent variable X as $\dot{X}$ and its growth rate as $\hat{X} := \dot{X}/X$. Furthermore, the transversality condition must hold stating that the present value of the household's assets must tend to zero as t tends to infinity (Barro and Sala-i-Martin, 1995, Section 19.1).

Next we have to deal with investments and the asset market. As the homogeneous good can be transformed into installed capital one-to-one without adjustment cost, investment demand would be infinite if q_i was larger than one in one of the regions. Private optimizing investors are willing to invest in region i, as long as $q_i = 1$. Hence, the complementarity

$$I_i \geq 0, \quad q_i \leq 1, \quad I_i(1 - q_i) = 0$$

must hold in both regions.

There are two kinds of assets that can be held, capital in region 1 and capital in region 2. Both must yield identical private rates of return,

$$rq_i = \dot{q}_i + \alpha_p y_i - \delta q_i.$$

There are two possible cases to be distinguished:

(1) $q_i = 1$ in a time span of positive duration. Then $\dot{q}_i = 0$, and hence $r = \alpha_p y_i - \delta$. In this case $I_i \geq 0$.
(2) $q_i < 1$. Then $I_i = 0$.

Positive investment in both regions is only possible if $r = \alpha_p y_i - \delta$ holds in both regions, that is if $y_1 = y_2$. Note, however, that $q_1 = 1$ and $q_2 < 1$ does not necessarily imply that capital is more productive in region 1 than in region 2. Even if productivity is bigger in region 2, stock prices $q_1 = 1$ and $q_2 < 1$ can occur, provided that q_2 declines sufficiently fast. We will see below that this situation in fact is observed near unstable steady states.

Finally, the goods market equilibrium condition

$$K_1 y_1 + K_2 y_2 = I_1 + I_2 + C$$

closes the system.

It turns out that there are three types of dynamic equilibria:

(1) *Balanced steady state*. In this case investment in both regions is positive. This implies $q_1 = q_2 = 1$, and hence $y_1 = y_2$, which in turn implies $\hat{K}_1 = \hat{K}_2$, because k must stay constant. Therefore, the interest rate $r = \alpha_p y_1 - \delta = \alpha_p y_2 - \delta$ is constant and $\hat{C} = \sigma(r - \rho)$ is also constant. Defining $Y := Y_1 + Y_2$ and $K := K_1 + K_2$ we can write

$$Y = AK$$

with $A := y_1 = y_2$. Hence, the entire economy becomes a so-called AK-economy (Barro and Sala-i-Martin, 1995, Section 19.3.1). Goods market equilibrium and transversality can only hold if $\hat{Y} = \hat{C}$. Hence, we have

$$\hat{Y} = \hat{C} = \hat{K} = \sigma(\alpha_p A - \delta - \rho). \tag{19.6}$$

The saving rate is

$$s := \frac{Y - C}{Y} = \frac{\dot{K} + \delta K}{K} \frac{K}{Y} = (\hat{K} + \delta)/A.$$

The higher capital productivity (which is equalized across regions) and the lower δ, the higher is the growth rate. Furthermore, the smaller ρ and the higher σ, the higher is the growth rate. If consumers are more patient or less keen on intertemporal smoothing of consumption, the economy grows faster. These are of course standard results of endogenous growth theory.

(2) *Concentrated steady state*. This is the limiting case for k tending to infinity or to zero, provided that $\mu > \gamma$. If k tends to infinity in the course of time, output and mobile labor are eventually completely concentrated in region 1. Hence, m tends

to infinity and eventually we have $y_1 = aL_1^\lambda$ by Equation (19.4). We will see below that y_1/y_2 tends to infinity; hence y_2 tends to zero. Therefore in the end we again have an AK-economy

$$Y_1 = AK_1$$

with $A := y_1 = aL_1^\lambda$ constant, and with steady state growth rate as in Equation (19.6). Similarly for k tending to zero.

(3) *Transition with k increasing or decreasing*. It is sufficient to deal with 'transitions from the left', i.e. with k increasing over time. For this case one can show that $q_1 = 1$ and almost always $q_2 < 1$. That q_1 cannot be less than one is obvious: $q_1 < 1$ would imply $\hat{K}_1 = -\delta$ and, as $\hat{K}_2 \geq -\delta$, it would imply $\hat{K}_1 - \hat{K}_2 \leq 0$ contradicting $\hat{k} > 0$. For proving that almost always $q_2 < 1$, assume to the contrary that $q_2 = 1$ in a proper time interval. Then $\dot{q}_1 = \dot{q}_2 = 0$, and hence $y_1 = y_2$ during that interval, contradicting non-constancy of k. Therefore in a transition with increasing k there is only investment in region 1 and no investment in region 2, that is K_2 declines with rate δ.

The transition is described by the differential equations

$$\dot{K}_1 = Y_1 + Y_2 - \delta K_1 - C, \qquad (19.7)$$

$$\dot{K}_2 = -\delta K_2, \qquad (19.8)$$

$$\dot{C} = C\sigma(\alpha_p y_1 - \delta - \rho), \qquad (19.9)$$

$$\dot{q}_2 = \alpha_p y_1 q_2 - \alpha_p y_2, \qquad (19.10)$$

which hold as long as $q_2 < 1$. Equation (19.7) is the goods market equilibrium condition, Equation (19.8) is due to non-investment in region 2, Equation (19.9) is the Keynes–Ramsey rule, and Equation (19.10) is the asset market equilibrium condition.

Equations (19.7)–(19.10) can be transformed into a system of two coupled autonomous differential equations in k and $c := C/K_2$. Substituting $\dot{K}_2 k + \dot{k}K_2$ for $\dot{K}_1$ in Equation (19.7), using $\dot{K}_2$ from Equation (19.8) and dividing through by K_2 yields

$$\dot{k} = ky_1 + y_2 - c. \qquad (19.11)$$

Furthermore, substituting $\dot{K}_2 c + \dot{c}K_2$ for $\dot{C}$ in Equation (19.9), using $\dot{K}_2$ from Equation (19.8) and dividing through by K_2 yields

$$\dot{c} = c[\sigma(\alpha_p y_1 - \delta - \rho) + \delta]. \qquad (19.12)$$

In the following we study transitions to balanced steady states 'from the left', i.e. transitions with increasing k, by numerically integrating the systems (19.11), (19.12) and (19.10) backwards in time, starting from known steady state values for k, c, and q_2. These steady state values are

- the $\bar{k}$ that solve the equation $y_1 = y_2$ at points where $f(K)$ defined in Equation (19.14) below has a negative slope (there may be one or two of them),
- $\bar{c} = (1+\bar{k})(y_1 - \hat{K} - \delta)$, with $\hat{K}$ from Equation (19.6), and
- $\bar{q}_2 = 1$.

q_1 stays constant on a path with k increasing over time. Transitions from the right are obtained by just exchanging indices of regions. As the two-dimensional system defined by Equations (19.11) and (19.12) is autonomous, the saddlepath stability of the dynamic equilibria could be shown by the standard two-dimensional phase space analysis, but this is omitted here.

For simulating a transition to a steady state concentrated in region 1 we integrate a similar system in the variables $h := 1/k$ and $\tilde{c} := 1/K_1$ backwards in time starting at $\bar{h} = 0$, $\bar{q}_2 = 1$, and $\bar{\tilde{c}} = aL_1^\lambda - \hat{K} - \delta$. $\hat{K}$ is given by Equation (19.6) with $A = aL_1^\lambda$. Note that, unlike balanced steady states, concentrated steady states are only attained in infinite time. (It is therefore necessary to substitute another variable, for example path length, for time in the integration.)

19.3. Dynamics: convergence and divergence

For studying the dynamics of this economy, one has to see how capital productivities in both regions depend on the distribution of capital across the regions, given that mobile labor is distributed such that wages are equalized across regions at any time. Let R denote the ratio of marginal productivities of capital (region 1 over region 2). It is a function of the ratio of capital stocks $k = K_1/K_2$ and the ratio of stocks of immobile labor $\ell = L_1/L_2$:

$$R = \left[\ell^\lambda k^{\mu-\gamma}\left(\frac{k+\beta}{1+\beta k}\right)^\gamma\right]^{1/(1-\mu)}. \qquad (19.13)$$

To see this, divide Equation (19.3) by k and note that $\alpha - (1 - \mu) = \mu - \gamma$. R is larger than one, equal to one, or less than one, if and only if the expression

$$f(k) = \frac{\mu - \gamma}{\gamma} \log k + \log \frac{k + \beta}{1 + \beta k}. \tag{19.14}$$

is larger than, equal to, or less than $f^* := -(\lambda/\gamma)\log \ell$. In particular, in the symmetric case $R > 1$ ($R = 1$, $R < 1$), if and only if $f(k) > 0$ ($f(k) = 0, f(k) < 0$). A balanced steady state is attained at point(s) k^* solving $f(k^*) = f^*$.

Intuitively, if f is positively sloped at k^*, an increase of k generates a capital productivity advantage of region 1 making region 1 even more attractive for capital investment. Hence, we call anything making f positively (negatively) sloped a divergence (convergence) force. (Note though, that with perfect foresight the answer as to whether k, starting at some k_0 close to k^*, will increase or decrease is a bit more involved, as will be shown in a moment.)

The first term in Equation (19.14) shows a divergence (convergence) force, if $\mu > \gamma$ ($\mu < \gamma$). Noting that $\alpha + \gamma = 1$, the condition $\mu > \gamma$ is equivalent to $\mu + \alpha > 1$, that is to increasing local returns of the mobile factors K and M. In other words, the first term shows a divergence (convergence) force, if mobile factors exhibit increasing (decreasing) local returns. The second term in Equation (19.14) is a divergence force, except for $\beta = 1$ (perfect knowledge mobility), where it vanishes. For $\beta < 1$ the term is strictly increasing in k with lower bound $\log \beta$ and upper bound $-\log \beta$. Obviously, the smaller β, the stronger is the divergence force.

Hence, if mobile factors exhibit increasing local returns, there are only divergence forces, while there is a convergence and a divergence force if mobile factors exhibit decreasing local returns. The convergence force will eventually always dominate, if k gets sufficiently large or sufficiently small, while the dominance at the balance point k^* at or near zero depends on parameters. This gives rise to three possible scenarios to be explained in the following subsections.

19.3.1. Divergence

If $\mu > \gamma$ (or equivalently $\mu + \alpha > 1$), $f(k)$ is strictly increasing for all k (see Figure 19.1).

Figure 19.1. Divergence

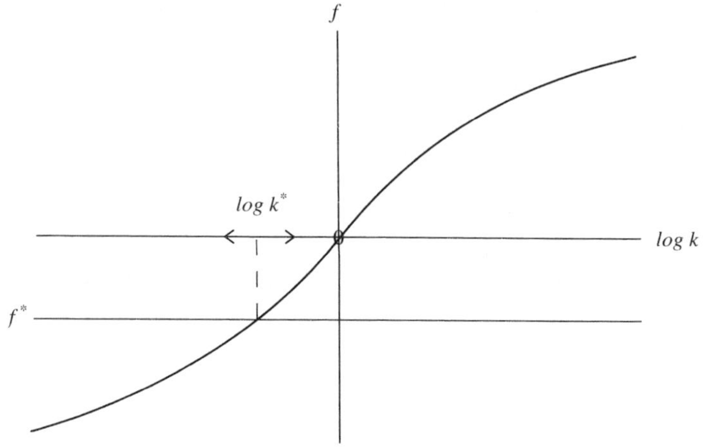

Note that this case would be impossible with $\lambda + \mu + \alpha = 1$. As explained before, we can however allow for $\lambda + \mu + \alpha > 1$ and still maintain marginal productivity factor payments if we assume these elasticities to have internal and external components, with the former ones adding up to unity.

A naive model with myopic agents investing exclusively in the region with the currently higher marginal productivity of capital would reveal k^* in Figure 19.1 to be the watershed between two possible time paths. If k starts from $k_0 > k^*$, capital moves to region 1 with higher capital returns, and mobile labor follows. This even increases the incentive to move, such that eventually capital and mobile labor are completely concentrated in region 1. Similarly one ends up with perfect concentration in region 2 if one starts from $k_0 < k^*$.

With perfect foresight the intuition that k^* is unstable still turns out to be true. If k slightly deviates from k^*, there is no equilibrium path leading back to k^*. For a transition from the left, q_2 would have to approach unity from above; otherwise shares in K_2 would yield higher returns than those in K_1, because left of k^* marginal productivity in region 2 exceeds that in region 1. A similar argument holds for transition from the right.

The dynamics leading the economy to a concentrated equilibrium either in region 1 or in region 2 with perfect foresight are more complicated than in a myopic world. Figure 19.2 illustrates the symmetrical case by plotting asset prices over k in

Figure 19.2. Asset prices in transition to concentrated equilibria

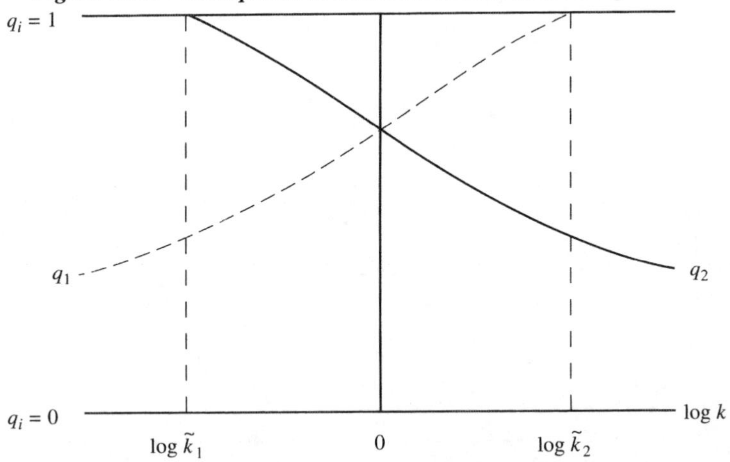

phase space. The solid (dashed) curve shows $q_2(q_1)$ for a transition to a concentrated steady state in region 1 (region 2). Let us call the sets $\{k|k \geq \tilde{k}_1\}$ and $\{k|k \leq \tilde{k}_2\}$ the attraction domains of regions 1 and 2, respectively. For any k in the attraction domain of region 1 (region 2) there is an equilibrium path leading to a concentrated steady state in region 1 (region 2). Note that the two domains overlap. Within the overlap the path is non-unique. Which one of both possible paths is chosen depends on self-fulfilling prophecies.

Consider for example a start at $k_0 = 1$. If asset owners are convinced that the economy moves towards a concentrated equilibrium in region 1, then q_2 drops to some price less than one and all investment goes to region 1. In a myopic world agents holding shares in the capital stock of region 2 would want to be compensated by an asset price below one such that $q_2(\alpha_p y_1 - \delta) = \alpha_p y_2 - \delta$. For $k = k^* = 1$ this would just mean $q_1 = q_2 = 1$, because $y_1 = y_2$ at k^*. Hence, in a myopic world attraction domains would not overlap, as already shown above. With perfect foresight, q_2 must be less than one at k^*, because agents predict q_2 to fall.

Figure 19.3 illustrates the asymmetric case with region 1 having more immobile factors (i.e. $\ell > 1$). The attraction domains shift leftwards, but still overlap, and the unstable balanced steady state is still in the interior of the overlap.

Figure 19.3. Asset prices in transition to concentrated equilibria

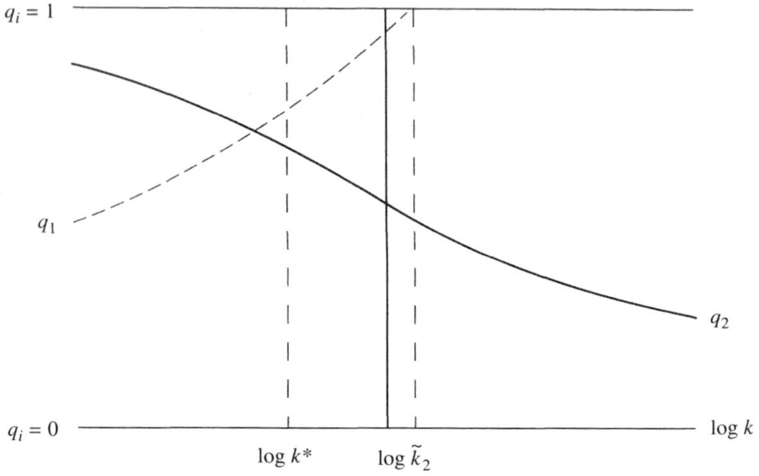

19.3.2. Agglomeration

If $1 > \mu/\gamma > 2\beta/(1+\beta)$ (area AA in Figure 19.6), mobile factors exhibit decreasing local returns. But if capital is equally distributed across regions ($k = 1$, i.e. $\log k = 0$) the divergence effect of the second term in Equation (19.14) dominates, that is the divergence

Figure 19.4. Agglomeration

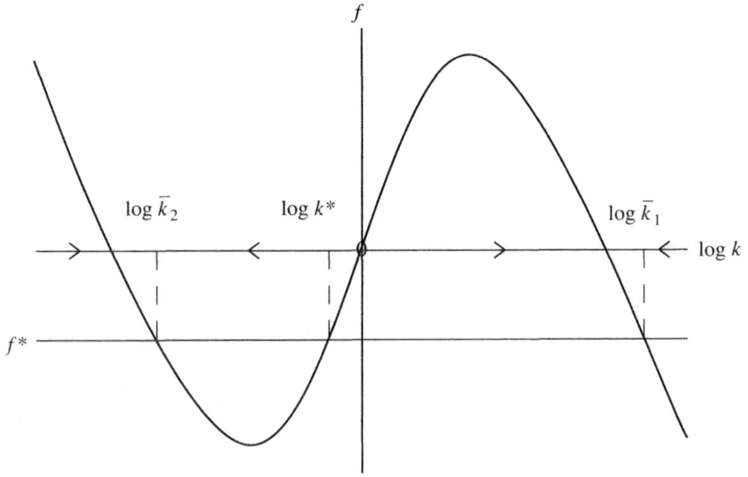

effect caused by the global external effect of capital. If capital is concentrated sufficiently in one or the other region, however, the first term eventually dominates, which works into the direction of convergence due to decreasing local returns of mobile factors.

Depending on ℓ (or f^*), there are two possible scenarios:

(1) If regions are sufficiently similar regarding their stocks of immobile factors, i.e. if ℓ is sufficiently close to unity, then there are three balanced steady states, an unstable one in the middle and two stable ones (called agglomerations) with high concentration of mobile factors either in region 1 or in region 2 (see Figure 19.4).
(2) If the regions sufficiently differ with regard to their stocks of immobile factors, then there is only one balanced steady state. It is stable and located in region 1 (region 2), if $\ell > 1$ ($\ell < 1$); see Figure 19.5.

Let us discuss the more difficult first case with three equilibria. In a myopic world, starting from $k_0 > k^*$ would always lead to an agglomeration in region 1, because an increasing concentration of capital in region 1 makes it even more attractive to invest in region 1, as long as capital is not too concentrated in region 1 (Figure 19.4).

Figure 19.5. Region 2 is too small for an agglomeration

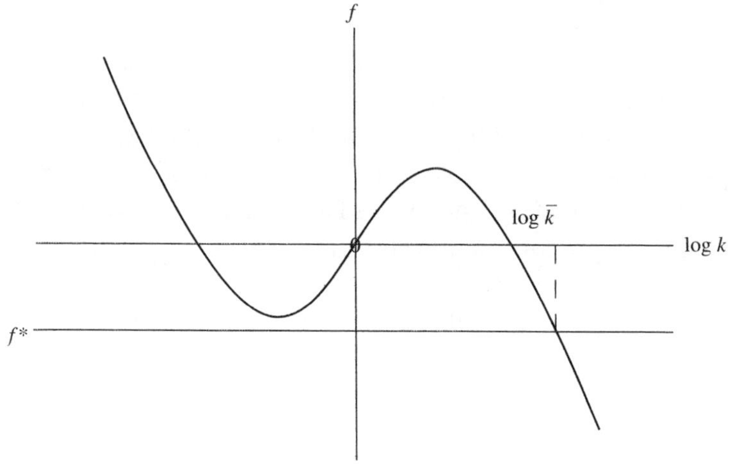

Figure 19.6. ***AA – the market generates agglomeration, efficiency requires a higher degree of agglomeration; CA – the market generates convergence, while efficiency requires agglomeration; CC – the market generates convergence, which is efficient***

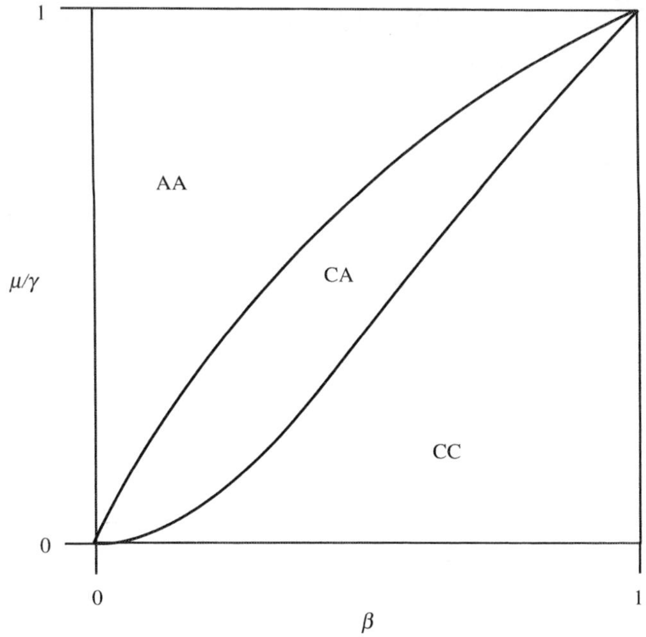

Later, decreasing local returns dominate and bring the agglomeration process to a halt, once the balanced steady state distribution $\bar{k}_1$ is attained. This happens after a finite transition period. Similarly, the economy converges to an agglomeration in region 2, if it starts from $k_0 < k^*$.

With perfect foresight we again observe overlapping attraction domains of the two balanced steady states, and the unstable balanced steady state k^* is in the interior of the overlap. Figure 19.7 illustrates the symmetrical case. The solid (dashed) curves show transitions to an agglomerated steady state in region 1 (region 2). Curve segments left of the steady states (transitions from the left) show q_2, while q_1 is unity. Curve segments right of the steady states (transitions from the right) show q_1, while q_2 is unity.

Figure 19.8 illustrates the asymmetric case. Increasing ℓ shrinks the attraction domain of region 2 right of $\bar{k}_2$ and extends the attraction domain of region 1 left of $\bar{k}_1$. It actually may extend to infinity, if ℓ is large enough (as it is the case in the simulation results shown in the figure).

***Figure 19.7.* Asset prices in transition to agglomerated steady states**

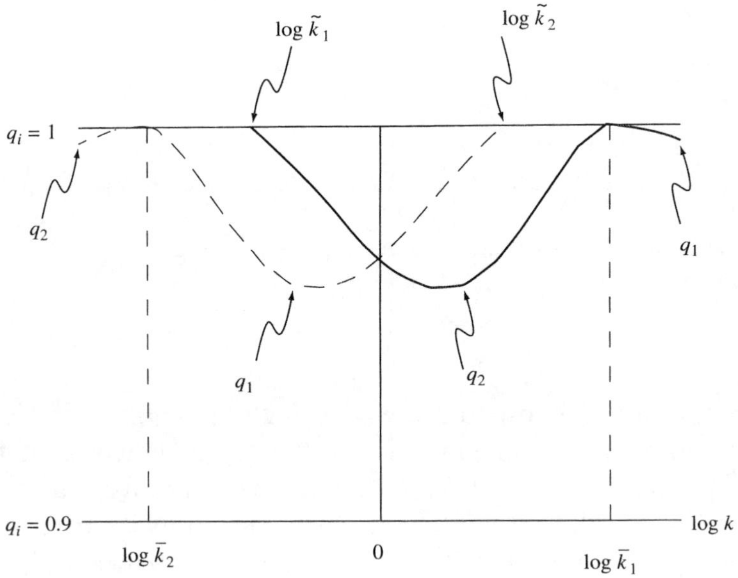

***Figure 19.8.* Asset prices in transition to agglomeration steady states**

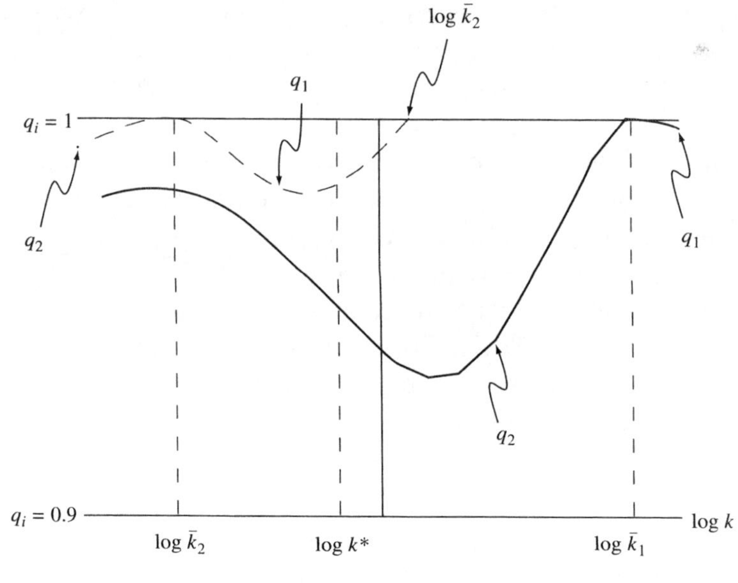

Figure 19.9. Catastrophic dissolution of an agglomeration in region 2

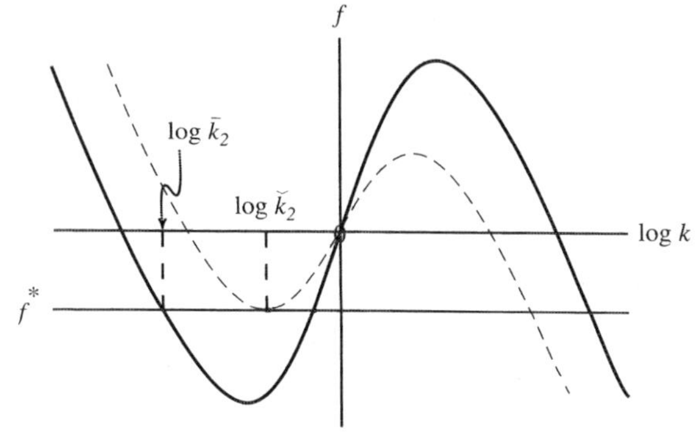

The dynamic response to decreasing agglomeration forces in case of an agglomeration locked in the smaller region is worth studying ($\bar{k}_2$ in Figure 19.4 or 19.9). Assume that improved means of communication allow for a better access to knowledge in distant locations (β increases), then the curve in Figure 19.9 becomes flatter. k increases from $\bar{k}_2$ to $\breve{k}_2$ and the attraction domain vanishes. Now the agglomeration in region 2 can no longer survive. At this point in time at the latest, one will observe a catastrophic transition; the agglomeration in region 2 vanishes and a new agglomeration will emerge in region 1. All investments go to region 1. Mobile labor will follow capital to region 1. Empirically we would first observe gradual convergence and then a leapfrog: the former poor region would rapidly overtake. Eventually the distribution converges to a new steady state with the larger region being the agglomerated center and the smaller one the poorer periphery.

In a myopic world the dissolution of the agglomeration in region 2 cannot occur before this bifurcation point is attained. Under perfect foresight, however, it can. Take the balanced steady state $\bar{k}_2$ with an agglomeration in region 2 in Figure 19.8 as a case in point. If agents collectively do not believe in an agglomeration in region 2 anymore and notice the transition path to an agglomeration in region 1, q_2 drops down, investment in region 2 stops, all investment goes to region 1 and k goes up until $\bar{k}_1$ is attained.

19.3.3. Convergence

If $2\beta/(1+\beta) > \mu/\gamma$ (areas CA and CC in Figure 19.6) the economy converges to the unique balanced steady state, which is

Figure 19.10. Convergence

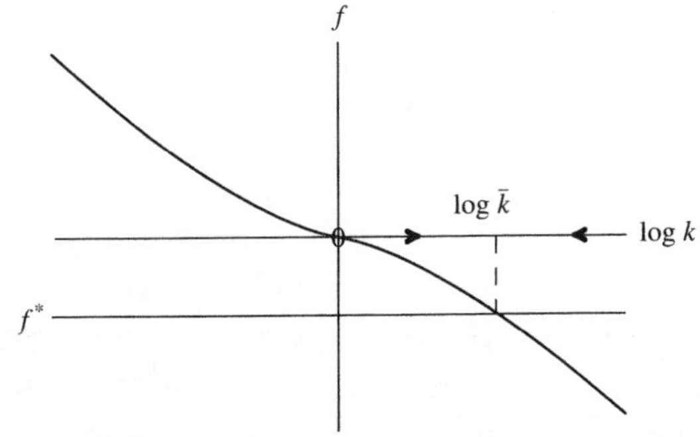

now stable (see Figure 19.10). Locally decreasing returns dominate even for a balanced distribution of capital. In the symmetrical case ($\ell = 1$) there will be unconditional convergence. From any starting point factor prices are equalized in the course of time, and eventually the economy grows in a balanced steady state with a constant rate. There is conditional convergence, if $\ell \neq 1$. Then the economy converges to a unique steady state as well ($\bar{k}$ in Figure 19.10). But income per capita and the wage rate of immobile labor are higher in the larger region than in the smaller one in the steady state.[1] We dispense with plotting asset prices over k. The pattern is obvious for this case: q_2 (q_1) approaches unity from below, if k approaches $\bar{k}$ from the left (right).

19.4. Efficiency

What about allocative efficiency of the outcome of market forces in this economy? Imagine an omniscient planner aiming at maximizing the household's utility subject to the technological constraints of the economy. He solves the problem

$$\max_{C,K_1,K_2} \int_0^\infty \frac{\sigma}{\sigma - 1} C(t)^{1-1/\sigma} \exp(-\rho t) \mathrm{d}t$$

[1] Let $\ell > 1$, as assumed in the figure. Then $\bar{k} > 1$, hence $G_1 > G_2$ in the steady state. This capital productivity advantage of region 1 must be compensated by a higher capital intensity; this implies a higher per capita income in region 1.

subject to

$$\dot{K}_i = I_i - \delta K_i, \qquad I_i \geq 0, \tag{19.15}$$

$$\begin{aligned} I_1 + I_2 + C &= K_1 y_1 + K_2 y_2, \\ K_i(0) &= K_i^0. \end{aligned} \tag{19.16}$$

y_1 and y_2 are as given by Equations (19.4) and (19.5), with m from Equation (19.3). Strictly speaking, the planner also has to choose m. He would however make the same choice as the market: he would distribute mobile labor across regions such that $m = Y_1/Y_2$. Though we allowed for a positive externality of mobile labor, it leaves the migration decision undistorted. This is why we can eliminate the choice dimension m, taking for granted that at any point in time mobile labor is distributed optimally across regions.

Letting π_i denote the co-state variables associated with Equation (19.15), and η denote the Lagrangian multiplier associated with Equation (19.16), we obtain the first-order conditions derived from the present value Hamiltonian:

$$C^{-1/\sigma}\exp(-\rho t) = \eta, \tag{19.17}$$

$$\pi_i - \eta \leq 0, \qquad I_i(\pi_i - \eta) = 0, \tag{19.18}$$

$$-\dot{\pi}_i = -\pi_i + \eta F_{K_i}, \tag{19.19}$$

$$\lim_{t \to \infty} \{\pi_i K_i\} = 0. \tag{19.20}$$

F_{K_i} denotes marginal social productivity of capital, i.e.

$$F_{K_i} := \frac{\partial F(K_1, K_2)}{\partial K_i},$$

with $F(K_1, K_2) := K_1 y_1 + K_2 y_2$.

Now define the 'social interest rate' $v := -\hat{\eta}$ and $\phi_i := \pi_i/\eta$. Then taking logs of Equation (19.17) and derivating with respect to time yields

$$\hat{C} = \sigma(v - \rho).$$

Equation (19.18) becomes

$$\phi_i \leq 1, \;\; I_i \geq 0, \;\; I_i(1 - \phi_i) = 0.$$

Inserting $\dot{\pi}_i = \phi_i\dot{\eta} + \dot{\phi}_i\eta$ into Equation (19.19) and dividing by η yields

$$v\phi_i = \dot{\phi}_i + F_{K_i} - \delta\phi_i.$$

Finally, Equation (19.20) becomes

$$\lim_{t\to\infty} \{\eta\phi_i K_i\} = 0,$$

saying that the present value of both types of assets has to vanish as time goes to infinity. 'Present value' here means to discount with the social interest rate v rather than the private interest rate r. Putting it differently, the transversality condition now reads

$$\lim_{t\to\infty} \left\{\phi_i(t)K_i(t)\exp\left(-\int_0^t v(\tau)\mathrm{d}\tau\right)\right\} = 0.$$

These equations coincide with those describing the decentralized market equilibrium, if we substitute r and q_i for v and ϕ_i, except that in the planner's problem the social marginal capital productivity F_{K_i} takes over the role of the private marginal capital productivity $\alpha_{\mathrm{p}}y_i$ in the market solution.

To be sure, the long run steady state of the market economy is inefficient in that the saving rate and, as a consequence, the growth rate are too small as compared to the optimal path. This is due to the fact that the households' saving decision is based on the private rate of return to capital, while the planner bases his decision on the higher social rate of return, which includes the positive capital externality. Just as the market equilibrium path, the optimal path converges to either a balanced or a concentrated steady state. The optimal steady state growth rate is (similar to Equation (19.6))

$$\hat{Y} = \hat{C} = \hat{K} = \sigma\,(F_K - \delta - \rho).$$

F_K is the social marginal productivity of capital, which is larger than the private one because of the two externalities, possibly a local one stemming from $\alpha > \alpha_{\mathrm{p}}$, and a global one stemming from the effect of G_i in the production function. Hence, the optimal steady state rate of growth is always larger than the steady state rate of growth in a decentralized market equilibrium. If the steady state is balanced, then $F_K = F_{K_i}$ with k such that $F_{K_1} = F_{K_2}$.

We will see in a minute, that this k is always larger than the $\bar{k}$ equalizing private rates of capital return.

Our focus is now on the spatial distribution in the steady state: how does the capital distribution in the optimal steady state compare with the market solution? Does the market generate over-agglomeration or under-agglomeration? I confine the analysis to the symmetrical case ($\ell = 1$).

The optimal k is found by derivating $Y = Y_1 + Y_2$ with respect to k, holding $K = K_1 + K_2$ constant. The resulting derivative has the same sign as the function $h(k) = f(k) + g(k)$ with

$$g(k) = \log\left(\alpha + (1-\beta)\gamma \frac{k}{k+\beta}\right) - \log\left(\alpha + (1-\beta)\gamma \frac{1}{1+\beta k}\right).$$

$g(k)$ is strictly monotone increasing, goes through zero for $k = 1$ (i.e. $\log k = 0$) and approaches the lower bound $-\bar{g}$ to the left and the upper bound $\bar{g}$ to the right, with

$$\bar{g} = \log(1 + (1-\beta)\gamma/\alpha) > 0$$

(see Figure 19.11). $g(k)$ is the positive agglomeration externality, which is neglected by private investors when choosing the location of investment. The smaller β (that is the less easily innovation diffuses over space) and the larger γ/α (that is the more important global effects of capital as compared to local effects), the larger is this externality.

Figure 19.11. Function g(k) representing the agglomeration externality

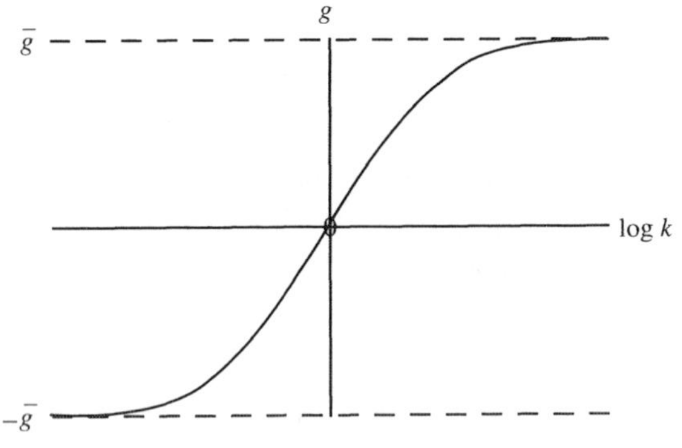

Comparing $h(k)$ with $f(k)$ allows to distinguish the following cases:

(1) $\mu/\gamma > 1$: The market leads to a complete concentration of capital in one region, and this is optimal.
(2) $1 > \mu/\gamma > 2\beta/(1+\beta)$ (Area AA in Figure 19.6): The market leads to an incomplete concentration of capital in one region; a higher concentration would be optimal.
(3) $2\beta/(1+\beta) > \mu/\gamma > c(\beta,\alpha)2\beta/(1+\beta)$ with a positive factor $c(\beta,\alpha) < 1$ depending on β and α (area CA in Figure 19.6):[2] Capital converges against a symmetrical distribution, but an incomplete concentration would be optimal.
(4) $c(\beta,\alpha)2\beta/(1+\beta) > \mu/\gamma$ (area CC in Figure 19.6): Capital converges against a symmetrical distribution, and this is optimal.

To summarize: the market allocation is either optimal – case 1 with complete concentration, case 4 with equal distribution –, or leads to an insufficient concentration (cases 2 and 3). In case 2 the market induces agglomeration, but not to a sufficient degree. In case 3 the market forces of agglomeration are not strong enough to bring the desirable agglomeration about.

19.5. Conclusion

This chapter showed how factor mobility and knowledge diffusion influence the spatial distribution of economic activity in a growing economy. Long run growth is due to a positive external effect of knowledge, such that capital, which is understood to consist of real capital and knowledge capital, has globally constant returns to scale. Access to knowledge is, however, not everywhere the same. Each region only partly participates in the knowledge generated elsewhere.

If mobile factors have jointly increasing local returns to scale, then we always observe a diverging growth path such that in the long run all activity is completely concentrated in one region.

---

[2] In Figure 19.6 the boundary between areas CA und CC is drawn for realistic parameter values $\gamma = \alpha = 1/2$.

The concentration as such is efficient, while the steady state rate of growth is too small as compared to an efficient path, because saving decision neglect the positive capital externality. The latter result is standard in endogenous growth models with Marshallian capital externalities, but does not necessarily hold in models where creative destruction of innovation is admitted.

If mobile factors have jointly decreasing local returns, the economy either attains an agglomerated steady state with most but not all activity concentrated in one location, or it converges to a steady state with total equality of the regions or only moderate differences (in case of exogenous size differences of the regions). Agglomeration is the more likely, the worse the access to knowledge generated elsewhere is, and the smaller the production elasticity of global knowledge is in comparison to that of mobile labor. An empirical implication is that we should observe a steady deglomeration process in the course of declining communication cost, increasing access to knowledge and increasing importance of global knowledge as a production factor.

The chapter also evaluates dynamic market equilibria with respect to allocative efficiency. It is shown that the market solution tends to be under-agglomerated, except for parameter constellations generating particularly low agglomeration forces. If agglomeration forces are low enough, no agglomeration emerges, and this is also socially desirable. For higher agglomeration forces, however, concentration becomes desirable though the market may not bring it about or brings it about to an insufficient degree only. This conclusion contradicts commonly held wisdom in regional policy supporting equalization between regions and aiming at a shift of resources from richer to poorer regions. This type of cohesion policy might be justified from a distributional point of view. It favors owners of immobile factors in the periphery and may therefore also support political stability. But, in the light of our theoretical results, it cannot be justified for efficiency reasons.

References

Baldwin, R. (2001), "Core-periphery model with forward-looking expectations", *Regional Science and Urban Economics*, Vol. 31, pp. 21–49.

Baldwin, R. and R. Forslid (1997), "Trade liberalization and endogenous growth: a q-theory approach", *Journal of International Economics*, Vol. 50, pp. 497–517.

Baldwin, R. and R. Forslid (2000), "The core-periphery model and endogenous growth: stabilizing and destabilizing integration", *Economica*, Vol. 67, pp. 307–324.

Barro, R. and X. Sala-i-Martin (1995), *Economic Growth*, New York: McGraw-Hill.

Fujita, M. and J.-F. Thisse (2002), *Economics of Agglomeration: Cities, Industrial Location, and Regional Growth*, Cambridge: Cambridge University Press.

Hirschman, A.O. (1958), *The Strategy of Development*, New Haven, CT: Yale University Press.

Jaffe, A., M. Trajtenberg and R. Henderson (1993), "Geographic localization of knowledge spillovers as evidenced by patent citations", *Quarterly Journal of Economics*, Vol. 108, pp. 577–598.

Kaldor, N. (1970), "The case of regional policies", *Scottish Journal of Political Economy*, Vol. 18, pp. 337–348.

Martin, P. and G. Ottaviano (2001), "Growth and agglomeration", *International Economic Review*, Vol. 42, pp. 947–968.

Myrdal, G. (1957), *Economic Theory and Underdeveloped Regions*, London: Duckworth.

Romer, P.M. (1990), "Endogenous technological change", *Journal of Political Economy*, Vol. 98, pp. S71–S102.

Walz, U. (1996), "Transport costs, intermediate goods, and localized growth", *Regional Science and Urban Economics*, Vol. 26, pp. 671–695.

Walz, U. (1999), *Dynamics of Regional Integration*, Heidelberg: Physica.

Urban Dynamics and Growth, CEA, vol. 266
Roberta Capello and Peter Nijkamp (editors)

CHAPTER 20

Innovation and the Growth of Cities

Zoltan J. Acs

Max Plank Institute, Jena, Germany
University of Baltimore, Merrick School of Business, 1420 N. Charles Street, Baltimore, MD 21201, USA

Abstract

An important problem in urban economics and cities is divergent growth rates leading to income inequality. This chapter offers a way to address the problem of divergent growth rates. Specifically, we look at the role of heterogeneity vs. specialization as an urban structure, endogenous technical change, and the importance that entrepreneurship and innovation play in the variety of local goods and the heterogeneity of the local economy both of which can mitigate the lack of convergence. Finally, an outline for a new model of regional economic growth is presented.

Keywords: innovation, knowledge spillovers, cities, entrepreneurship

JEL classifications: O40, R11, M13, C8

20.1. Introduction

> It would not be surprising if it turns out that the market-size effects emphasized by the current generation of new geography models are a less important source of agglomeration, at least at the level of urban areas, than other kinds of external economies. It is, for example, a well-documented empirical regularity that both plants and firms in large cities tend to be smaller than those in small cities; this suggests that big cites maybe sustained by increasing returns that are due to thick labor markets, or to localized knowledge spillovers, rather than those that emerge from the interaction of transport costs and scale economies at the plant level. (Krugman, 1998, p. 172)

In the modern economy, knowledge-producing inputs are not evenly distributed across space and therefore regions may not grow at the same rate (Nijkamp and Stough, 2000). The empirical work on convergence (notably the development of the notion of conditional β-convergence) was primarily stimulated by improved data series and provides a more rigorous method of quantifying relative spatial economic performance. The initial findings of Barro and Sala-I-Martin (1992) indicated that regional convergence over recent years had taken place, albeit at a very slow pace. Convergence, one could hypothesize, depends on investment in knowledge creation and on the systematic exploitation of knowledge by entrepreneurs.

The premise of this chapter is that entrepreneurship and innovation is an important local activity that translates new knowledge into start-ups that sustain the growth of cities through thick labor markets or localized knowledge spillovers (Acs, 2000, 2002). The central focus of this chapter is on the performance of cities, both success or failure, and how innovative entrepreneurs lead to a 'variety of local goods' and 'heterogeneity of the local economy' – both of which reduce the divergence of growth rates. Section 20.2 examines the relationship between heterogeneity and specialization. Section 20.3 discusses endogenous technical change. Section 20.4 examines the importance of entrepreneurship and innovation to the growth of cities. Section 20.5 outlines the direction for a new model of regional economic growth.

20.2. Heterogeneity vs. specialization

The importance of location to economic growth may seem paradoxical in the world of instant communications. However, as has been pointed out by Lucas (1988, 1993) and Black and Henderson (1999), it is localized information and knowledge spillovers, presumably through personal face-to-face contacts, that make cities the engines of economic growth. Cities grow faster than rural areas and cites are more innovative than rural areas.

Despite the general consensus that knowledge spillovers within a given location stimulate employment growth, there is little consensus as to exactly how this occurs. What type of economic activity will promote positive externalities and, therefore, economic growth? This question is important given the debate in the literature about the nature of economic activity and how it affects economic growth. The Marshall–Arrow–Romer (MAR) externality concerns knowledge spillovers between firms in an industry. Arrow (1962) presented an early formalization; the paper by Romer (1986) is a recent and influential statement. Applied to cities by Marshall (1890), this view says that the concentration of an industry in a city facilitates knowledge spillovers between firms and, therefore, the growth of that industry.

The MAR model formalizes the insight that the concentration of an industry in a city promotes knowledge spillovers between firms and therefore would facilitate employment growth in a city industry. An important assumption is that knowledge externalities with respect to firms exist, but only for firms within the same industry. Thus, the relevant unit of observation is extended from the firm to the region in the theoretical tradition of the MAR model and in subsequent empirical studies, but spillovers are limited to occur within the relevant industry. The transmission of knowledge spillovers across industries is assumed to be non-existent, or at least trivial.

These theories of externalities are extremely appealing because they try to explain simultaneously how cities form and why they grow. MAR, in particular, predicts that industries cluster geographically to absorb the knowledge spilling over between firms. In addition, they predict that regionally specialized industries grow faster because neighboring firms can learn from each other much better than geographically isolated firms.

A very different position is attributed to Jacobs (1969). Jacobs perceives information spillovers between industry clusters to be more important for the firm than within-industry information flows. Heterogeneity, not specialization, is seen as the most important regional growth factor, so Jacobs theory predicts that industries located in areas that are highly industrially diversified should grow faster.

According to Jacobs (1969), the emphasis on within-industry spillovers may be misplaced. Jacobs' idea is that the crucial externality in cities is cross-fertilization of ideas across different lines of work and industries. New York grain and cotton merchants saw the need for national and international financial transactions, and so the financial services industry was born. Rosenberg (1963) discusses the spread of machine tools across industries and describes how ideas are transmitted from one industry to another. Because cities bring people together from different walks of life, they foster transmission of ideas. Lucas (1993) emphasizes metropolitan areas as the most natural context in which the compact nature of the geographic growth facilitates personal interchange, communication and knowledge spillovers both within and across industries.

For Jacobs, the variety of industries within a geographic region promotes knowledge externalities and ultimately employment and economic growth. A common science base facilitates the exchange of existing ideas and generations of new ones across different but complementary industries. Thus, in Jacobs's view, industry diversity rather than specialization is the operative mechanism of economic growth.

Glaeser *et al.* (1992) analyze the six largest industries in each of the 170 US cities. Their results are consistent with the presence of Jacobs-type externalities. Industries will grow sluggishly in cities with high degrees of specialization. Feldman and Audretsch (1999) also test whether diversity or specialization of economic activity better promotes technological change and subsequent economic growth. They find support for the diversity thesis but little support for the specialization thesis. However, as Duranton and Puga (1999) point out in their survey of this area, the results may depend on the sector concerned. In fact, both of the above studies looked at spillovers within narrowly defined industrial sectors.

Acs *et al.* (2002b) also test the MAR hypothesis that industrial R&D spills over across regional industry clusters for 36 cities and six separate industry clusters over 4 years. Acs *et al.* estimate a model that looks at the impact of university knowledge on high-technology employment growth. They find that university research and development spills over across narrowly defined three digit industries supporting the heterogeneity hypothesis. These results suggest that risk pooling, shared infrastructure and thick labor markets are more important sources of agglomerations than knowledge spillovers (which naturally tend to be of greater value when they come from firms engaged in similar activities).

How does a local economy become heterogeneous? That is a question that has not been explored in detail. We would suggest that a city with heterogeneous firms is one where there has been lots of innovation, in the sense of creation of new products and new production processes. In this way 'heterogeneity of the local economy' and 'variety of local goods' could both be the result of a pattern of entrepreneurial innovation and the creation of new types of goods.

20.3. Endogenous technical change

In this section, we want to explore where innovation comes from. The analysis is carried out in a non-regional framework for simplicity. For a configuration of the modeling of this concept in a regional framework, see Berliant *et al.* (2002) and Lucas and Rossi-Hansberg (2002). We begin with a short survey of endogenous growth theory to bring out its strengths and weaknesses for regional analysis.[1]

The distinguishing feature of endogenous economic growth theory as compared to the neoclassical growth model is in its modeling of technological change as a result of profit-motivated investments in knowledge creation by private economic agents. Schmookler (1966) argued in great detail that it is the expected profitability of inventive activity, reflecting conditions in the relevant factor and product markets, that determines the pace and direction of industrial innovation. If scientific advances operated within the profit sector of the economy, technological progress is a subject for economic analysis. The novel formulation of technological knowledge in economic theory in Romer (1990) is the key in establishing this new and rapidly evolving field of economic growth theory. According to this formulation, technological knowledge is a non-rival, partially excludable good. This formulation of technological knowledge as a key factor in the production function results in a departure from the constant returns to scale, perfectly competitive world of the neoclassical growth theory.

Central to the neoclassical theory of economic growth as formulated in Solow (1957) is the production function. Assuming that capital does not depreciate, labor force does not grow and technology does not change over time, the production function has

---

[1] Jaffe (1989), Acs *et al.* (1992, 1994), Glaeser *et al.* (1992), Anselin *et al.* (1997, 2000) and Varga (1998) and innovation systems (e.g. Saxenian, 1994; Braczyk *et al.*, 1998; Sternberg, 1999; Acs, 2000; Fischer and Varga, 2001).

the form of

$$Y = F(K, L) \tag{20.1}$$

where Y represents aggregate production, K the capital stock and L the labor force. $F(\cdot)$ is the constant returns to scale production function. It is assumed that the capital stock grows without bounds. However, the growth rate of per-capita income is bounded. Growth rate of per-capita income is

$$g = sF_K(K, L) \tag{20.2}$$

where g is the growth rate of per capita income, s the savings rate and F_K the marginal product of capital. Equation (20.2) says that per-capita income grows as long as the marginal product of capital exceeds zero. However, assuming constant growth in the capital stock, per-capita income approaches zero. Relaxing the assumptions of stable labor force and no depreciation of capital does not change essentially the main point of the model. The condition for a sustained per-capita income growth in the long run is that, resulting from continuous capital accumulation, the marginal product of capital should not decrease below a positive lower bound.

Improvement in the state of technology is an essential force to offset the effect of capital accumulation on per capita income in the neoclassical model. Introducing technological progress in the production function it takes the form

$$Y = F(A, K, L) \tag{20.3}$$

where A stands for the state of technology. Assuming that A increases, it will increase the marginal product of capital which will lead to a higher per capita income. As a result, in steady state, the rate of technical development equals the rate of capital accumulation.

The essential role of technological progress in economic growth has been emphasized above. However, technological development remains unexplained in the neoclassical theory of economic growth. As a public good, it is considered exogenously determined although (as data show in Solow (1957) and Maddison (1987)) the major portion of economic growth can be attributed to technological change whereas capital accumulation (the main concern in the neoclassical model) explains only a fraction of it.

Primary attempts in the literature to endogenize technological progress include Arrow (1962) by introducing 'learning by doing' in technological development, Lucas (1988) by modeling human capital as the determinant factor in technical change and Romer (1986) by explicitly including research in the production function. In Arrow's formulation

$$Y_i = A(K)F(K_i, L_i) \tag{20.4}$$

the state of technology depends on the aggregate capital stock in the economy. Subscript i denotes individual firms. According to Lucas' model of endogenous technological change, spillovers resulted from human capital accumulation instead of the accumulation of physical capital increase the technological level in the economy:

$$Y_i = A(H)F(K_i, L_i) \tag{20.5}$$

where H stands for the general level of human capital in the economy. In Romer (1986) it is assumed that spillovers from private research efforts lead to the development in the public stock of knowledge. It could be written as

$$Y_i = A(R)F(R_i, K_i, L_i) \tag{20.6}$$

where R_i stands for the results of private R&D efforts by firm i and R denotes the aggregate stock of research results in the economy.

As summarized in Romer (1990), the major conceptual problem with the formulation of endogenous growth in Equations (20.4)–(20.6) is that in those models the entire stock of technological knowledge is considered to be public good. However, as evidence suggests, new technological knowledge can become partially excludable (at least for a finite amount of time) by means of patenting. Not until the formulation of monopolistic competition in Dixit and Stiglitz (1977) and Judd (1985), had modeling economic growth within an imperfectly competitive market structure becomes attainable. In Romer (1990), the approach by Judd was combined with learning by doing innovation to create the first model of endogenously determined technical change with imperfectly competing firms.

Consequently, each firm developing a new technological knowledge has some market power and earns monopoly profits on its discoveries. The 'new theory of economic growth', which follows from this first in Romer (1990), builds on a more suitable view of the available stock of technological knowledge as well as formulating the economy within the framework of imperfect competition.

At the core of the 'new' growth theory is the concept of technological knowledge as a non-rival, partially excludable good, as opposed to the neoclassical view of knowledge as an entirely public good. Knowledge is a non-rival good because it can be used by one agent without limiting its use by others. This distinguishes technology from, say a piece of capital equipment, which can only be used in one place at a time. Technology in many cases is partially excludable because it is possible to prevent its use by others to a certain extent. The excludability reflects both technological and legal considerations. Knowledge can be made partially excludable by the patent system and commercial secrecy. However, as Arrow (1962, p. 615) suggests:

> With suitable legal measures, information may become an appropriable commodity. Then the monopoly power can indeed be exerted. However, no amount of legal protection can make a thoroughly appropriable commodity of something so intangible as information. The very use of the information in any productive way is bound to reveal it, at least in part.

This partial non-excludability of knowledge suggests that industrial R&D may generate technological spillovers. According to Grossman and Helpman (1991, p. 16):

> By technological spillovers we mean that (1) firms can acquire information created by others without paying for that information in a market transaction, and (2) the creators or current owners of the information have no effective recourse, under prevailing laws, if other firms utilize information so acquired.

There are many ways in which spillovers take place; for example, the mobility of highly skilled personnel between firms represents one such mechanism. The Valley has a regional network-based industrial system that promotes learning and mutual adjustment among specialist producers of complex technologies. The region's dense social networks and open labor markets encourage entrepreneurship and experimentation resulting in knowledge spillovers (Saxenian, 1994). Innovative activity may flourish the most in environments free of bureaucratic constraints. A number of small-firm ventures have benefited from the exodus of researchers who fled thwarted by the managerial restraints as in the case of Shockley Semiconductor Labs. These small firms exploit the knowledge and experience accrued from the R&D laboratories of their previous employers.[2]

Knowledge enters production in two ways. First, newly developed technological knowledge is used in production by the firm invested in the development of this new set of technological knowledge to produce output. In this role, knowledge can be protected from being used by others in producing the same type of output. Second, knowledge increases the total stock of publicly available knowledge by spilling over to other researchers by way of studying its patent documentation (Romer, 1990). As such, it increases the productivity of creating further inventions in the research sector. The second role of knowledge in production can be formalized as

$$\mathring{A} = \delta H_A^{\lambda} A^{\varphi}, \tag{20.7}$$

where H_A stands for the number of researchers working on knowledge production in the business sector, A is the total stock of technological knowledge available at a certain point in time whereas $\mathring{A}$ is the change in technological knowledge resulting from private efforts to invest in research and development. δ, λ and φ are parameters.

Equation (20.1) presents the manner the two types of knowledge interact in the production of economically useful new technological knowledge. The particular functional form of knowledge production in Equation (20.1) is explained by the assumption of Romer (1990) that the efficiency of knowledge production is enhanced by the historically developed stock of scientific–technological knowledge. Even the same number of researchers becomes more productive if A increases over time. In the words of Grossman and Helpman (1991, p. 18), "The technological spillovers that result from commercial research may add to a pool of public knowledge, thereby lowering

[2] This section draws heavily on Acs and Varga (2002).

the cost to later generations of achieving a technological break through of some given magnitude. Such cost reductions can offset any tendency for the private returns to invention to fall as a result of increases in the number of competing technologies."

A is assumed to be perfectly accessible by everyone working in the research sector. However, as follows from the modification of Jones (1995), spillovers from the stock of codified knowledge might not be perfect. Hence the value of the aggregate codified knowledge spillovers parameter φ should be between 0 and 1. Equation (20.1) plays a central role in economic growth explanation since on the steady state growth path the rate of per capita GDP growth equals the rate of technological change ($\mathring{A}/A$).

However, not only codified but also non-codified, tacit knowledge can spillover as detailed in Section 20.2. The value of λ in Equation (20.7) reflects the extent to which tacit knowledge spills over within the research sector. Based on the literature we assume that these spillovers are influenced largely by the agglomeration of researchers as well as by the level of entrepreneurial activity in the city.

A principal assumption in the theory of endogenous growth is that for creating new sets of technological knowledge the total stock of knowledge (*A* in Equation (20.7)) and the addition to the stock of knowledge is freely accessible for anyone engaged in research. However, this assumption is not verified in the growing literature of geographic knowledge spillovers. New technological knowledge (the most valuable type of knowledge in innovation) is usually in such a tacit form that its accessibility is bounded by geographic proximity and/or by the nature and extent of the interactions among actors of an innovation system (Edquist, 1997).

Similar to the case of relaxing the neoclassical assumption of equal availability of technological opportunities in all countries of the world (Romer, 1994), a relaxation of the assumption that new knowledge *H* in Equation (20.7) is evenly distributed across space within countries seems to be also necessary. The non-excludable part of the total stock of knowledge seems rather to be correctly classified if it is assumed to have two portions: a perfectly accessible part consisting of already established knowledge elements (obtainable via scientific publications, patent applications, etc.) and a second, tacit element, accessible by interactions among actors in the innovation system.

While the first part is available without restrictions, accessibility of the second one is bounded by the nature of interactions among actors in a system of innovation.[3] Research has found that the value of φ, the rate of knowledge spillovers form the stock of codified knowledge, is less than one with a value of around 0.8 for most developed economies. Here technology commercialization could play an important role in increasing the rate at which existing knowledge is commercialized. The value of λ according to research is much smaller in both the US and in Europe, with the value in Europe being less than in the US How to increase the value of λ is an important policy question and appears to be influenced by both spatial considerations and entrepreneurship.

20.4. Entrepreneurship and innovation

In this section, we want to explore how entrepreneurship can increase the rate of knowledge spillovers in a city thereby increasing the level of heterogeneity. We define entrepreneurship as the creation of new stand-alone businesses. Even if the total stock of knowledge (in Equation (20.7)) was freely available, including the tacit and non-tacit parts, knowledge about its existence would not be. In an influential paper, Hayek (1945) pointed out that the central feature of a market economy is the partitioning of knowledge among individuals, such that no two individuals share the same knowledge or information about the economy. The key is that this knowledge is diffused in the economy and is not made available to everyone. Thus, only a few know about a particular scarcity, a new invention or a particular resource lying fallow, not being put to best use. This knowledge is typically specific to individuals' own circumstances including occupation, on the job routines, social relationships, and daily life. It is this particular knowledge, obtained in a particular knowledge base that leads to some profit making insight. The dispersion of information among different economic agents who do not have access to the same observations, interpretations or experiences has two fundamental implications for entrepreneurship.

---

[3] This is not a completed survey of an endogenous growth theory. For such surveys see, for example, Grossman and Helpman (1991), Helpman (1992), Barro and Sala-I-Martin (1992), Romer (1994), Nijkamp and Poot (1997), and Aghion and Howitt (1998).

First, opportunities for discovering or creating goods and services in the future exist precisely because of the dispersion of information. This dispersion created the opportunity in the first place. Second, the very same dispersion presents hurdles for exploiting the opportunity profitably, because of the absence or failure of current markets for future goods and services. It is therefore necessary to understand (1) how opportunities for the creation of new goods and services arise in a market economy, and (2) it is necessary to understand how and in what ways individual differences determine whether hurdles in the process of discovering, creating, and exploiting opportunities are overcome. Thus, entrepreneurship, "seeks to understand how opportunities to bring into existence 'future' goods and services are discovered, created, and exploited, by whom and with what consequences" (Shane and Venkataraman, 2000).

How do opportunities arise in the economy? In most societies, markets are imperfect, thus providing opportunities for enterprising individuals to enhance wealth by exploiting imperfections. This is most clearly articulated in the work of Kirzner (1997) where most markets are in disequilibrium. A second premise suggests that even if markets are in equilibrium, the human condition of enterprise combines with the lure of profits and advancing knowledge and technology will shift the equilibrium eventually. This premise is most often identified with Schumpeter's (1942) theory of creative destruction. Both Schumpeter and Kirzner's theories are based on the underlying assumption that change is a fact of life. And the result of this natural process is both a continuous supply of lucrative opportunities to enhance personal wealth and a continuous supply of enterprising individuals seeking such opportunities.

There are at least four classes of opportunities. The first is inefficiencies within existing markets due either to information asymmetries among market participants or to the limitations in technology to satisfy certain known but unfulfilled market needs. The second is the emergence of significant changes in social, political, demographic, and economic forces that can be exploited for economic gain that are largely outside the control of individual agents. The third source of opportunity is the accumulated stock of knowledge (A) that exists in society. The fourth source is inventions and discoveries that produce new knowledge (dA) in Equation (20.7).

It is one thing for opportunities to exist, but an entirely different matter for them to be discovered and exploited. Even new technology needs to have opportunities to exploit the new technology. Opportunity discovery is a function of the distribution of knowledge in society. Opportunities are rarely immediately obvious and/or available. They almost always have to be discovered and packaged. Thus, the nexus of opportunity and enterprising individuals is critical to entrepreneurship.

The role of specific knowledge and technical knowledge in motivating the search for profitable opportunities is critical to our understanding of what triggers the search for and exploitation of opportunities by some individuals but not others. The possession of useful knowledge varies among individuals. This variable strongly influences the search for and the decision to exploit an opportunity, and it also influences the relative success of the exploitation process.

Specific knowledge (H) by itself may only be a sufficient condition for the exercise of successful enterprise in a growth model. The ability to make the connection between specific knowledge and a commercial opportunity required a set of skills, aptitudes, insight, and circumstances that are not either uniformly or widely distributed in the population. Thus, two people with the same knowledge may put it to very different uses. It is one thing to have an insight, but an entirely different matter to profit from it. The incentive, capability, and specific behaviors needed to profit from useful knowledge or insight all vary among individuals and these differences matter for explaining and exercise of enterprise.

Bringing new products and markets into existence involves downside risk. Entrepreneurship requires making investments today without knowing what the distribution of the returns will be tomorrow. There is an uncertainty that cannot be insured against or diversified away (Knight, 1921). Individuals vary in their perception of such downside risk, and in their aptitudes and capabilities to deal with and manage them. The significant issue is that individuals vary in how they process and interpret statistical generalities and these variations may have significant but systematic impact on the decision to become an entrepreneur and the relative success of the endeavor.

While idiosyncratic insight and the ability to convert knowledge to commercial profit leads to successful enterprise, these same qualities also present the entrepreneur with problems. The process of creating products and markets implies that much of the information required by potential stakeholders – for example, technology, price, quality tastes, supply networked, distributor networks, and strategy – is not reliably available. Relevant information will only exist once the market has been successfully created. Potential stakeholders thus have to rely on the entrepreneur for information, but without the benefit of the entrepreneur's insight. In almost every project, entrepreneurs have more information about the true qualities of the project and themselves than any other parties. Because of this information asymmetry, neither buyers nor suppliers may be willing to make the necessary investment in specialized assets or formal cooperative arrangements to develop the business.

Despite the absence of current markets for future goods and services, some individuals do indeed create new markets and products. In fact, entrepreneurs are funded by venture capitalists to discover new knowledge to create future goods and services. According to Venkataraman (1997, p. 126):

> The significant point is that despite the existence of adverse selection and moral hazard problems, some individuals are able to successfully overcome these hurdles and achieve success. Thus, the ability to overcome adverse selection and moral hazard problems varies among individuals, and these differences matter for explaining successful enterprises. The interesting issue is not that such problems exist, but that in spite of them, some individuals are able to secure resources from different resource controllers, often at very favorable terms, whereby considerable risk is shifted from the entrepreneur to other stakeholders.

A critical decision for the entrepreneur is how to organize relationships with resource suppliers in order to foster the development and execution of a new business. There are often several possible institutional arrangements for creating a future product or service (such as a new firm, a franchise or license arrangement, a joint venture, or a simple contractual agreement). Why do entrepreneurs choose a particular mode? Moreover, what are the consequences of this choice on the distribution of risks and rewards among the various stakeholders? The usual assumption about the execution of entrepreneurial activity has been that most (if not all) new business creation occurs within a hierarchical framework, either as new start-ups or as new entities within an existing corporate body. However, evidence suggests that this may not be the case. Many new firms follow some different organizational form. Therefore, the most fundamental question in entrepreneurship research is, 'Why are any new entrepreneurial ventures organized as a start-up?'

In the absence of monopoly rents being earned by the incumbent firm and perfect information with no agency costs, any positive economies of scale or scope will ensure that no incentive will exist for an agent to start a new firm. If an agent had an idea for something different than is currently being practiced by the incumbent enterprise, in terms of a new product or process idea, which we will term here as an innovation, he will be presented to an incumbent enterprise. Because of the assumption of perfect information, both the firm and the agent will agree on the expected value of the innovation. However, to the degree that any economies of scale or scope exist, the expected value of implementing the innovation within the incumbent enterprise will exceed that of taking the innovation outside of the incumbent firm to start a new enterprise. Thus, the incumbent firm and the inventor of the idea would be expected to reach a bargain, splitting the value added to the firm contributed by the innovation (Audretsch, 1995).

But, of course, as Knight (1921) and others emphasized, new economic knowledge is anything but certain. Not only is new economic knowledge inherently risky, but also substantial asymmetries exist between agents and between firms. The expected value of a new idea, or innovation, is likely to be anything but unanimous between the inventor of the idea and the decision-makers of the firm confronted with proposed innovations. In fact, it is because information is uncertain that leads Knight (1921, p. 268) to argue that the primary task of the firm is to process imperfect information in order to reach a decision. According to Audretsch (1995):

> Combined with the bureaucratic organization of incumbent firms to make a decision, the asymmetry of knowledge leads to a host of agency problems, spanning incentive structures, monitoring and transaction costs. It is the existence of such agency costs, combined with asymmetric information that not only provides an incentive for agents with new ideas to start their own firms, but also at a rate that varies from industry to industry, depending upon the underlying knowledge conditions of the industry.

The degree to which incumbent firms are confronted with agency problems with respect to new knowledge and (potential) innovative activity would not be expected to be constant across industries or regions. This is because the underlying knowledge conditions vary from region to region and from industry to industry. In some industries, new knowledge generating innovative activity tends to be relatively routine and can be processed within the context of incumbent hierarchical bureaucracies. In other industries, however, innovations tend to come about in a less typical fashion and therefore tend to be rejected by the hierarchical bureaucracies of incumbent corporations. Nelson and Winter (1982) described these different underlying knowledge conditions as reflecting two distinct technological regimes – the entrepreneurial and routinized technological regimes: "an entrepreneurial regime is one that is favorable to innovative entry and unfavorable to innovative activity by established firms; a routinized regime is one in which the conditions are the other way around" (Winter, 1984, p. 297). Acs and Audretsch (1988) provided empirical evidence supporting the existence of these two distinct technological regimes.

When the underlying knowledge conditions are better characterized by the routinized technological regime, there is likely to be relatively little divergence between the evaluation of the expected value of an (potential) innovation between the inventor and the decision-making bureaucracy of the firm. Under the routinized regime there will not exist a great incentive for agents to start their own firm, or at least not for the reason of doing something differently. However, when the underlying knowledge conditions more closely adhere to the entrepreneurial regime, divergence beliefs between agent and the principal regarding the expected value an (potential) innovation is more likely to emerge. Therefore, it is under the entrepreneurial regime, where the start-up of a new firm is likely to play a more important role, presumably as a result of the motivation to appropriate the economic value of the knowledge. Due to agency problems, this cannot be transferred easily or without cost to the incumbent enterprise. As Audretsch (1995) has pointed out, "This shifts the emphasis from firms and institutions to individuals – agents with endowments of new economic knowledge."

The fundamental insight of the new growth theory is that economic growth is non-diminishing because technological knowledge is a non-rivalrous, partially excludable good. There are technological spillovers, and the profit motive ensures that entrepreneurs will continue to search out opportunities. The key feature of Austrian Economics is that the market is an entrepreneurial-driven evolutionary process. Entrepreneurship plays an important role in the discovery of knowledge and the turning of that knowledge into future goods and services through industrial innovation.

The starting point for most theories of innovation is the firm. In such theories the firm is assumed to be exogenous and its performance in generating technological change is endogenous. For example, in the most prevalent model in the literature on technological change, the knowledge production function, the firm exists exogenously and then engages in the pursuit of new knowledge as input into the process of generating innovative activity (Griliches, 1979). The most important source of new knowledge is generally considered to be R&D.

This model of innovation is questionable because in many industries small firms serve as the engine of innovation. This is startling because the bulk of industrial R&D is undertaken in the largest corporation, small enterprises account for only a minor share of R&D. Thus the knowledge production function suggests that innovative activity favor large firms. However, many smaller firms and entrepreneurs innovate (Acs and Audretsch, 1987, 1988, 1990). This leads to a fundamental question, "Where do entrepreneurs get the innovating producing inputs, that is the knowledge?" One answer is from other firms and institutions investing in R&D in the same region (Acs *et al.*, 1992, 1994; Audretsch and Feldman, 1996; Anselin *et al.*, 1997).[4]

Although the model of the knowledge production function may certainly be valid, the implicitly assumed *unit of observation* which

---

[4] Simultaneously and independently several books have appeared that try to identify the underlying processes and interconnections that govern regional innovation (DeBresson, 1996; Ratti *et al.*, 1997; Braczyk *et al.*, 1998; de la Mothe and Paquet, 1998; Acs, 2000). While these books take different approaches, rely on different methodologies, use different data, define the unit of analysis differently, they all suggest that there is something fundamental at work at the regional level. While these works are all interesting, illuminating pieces of the regional innovation puzzle, neither singularly, nor in concert, do they answer the bigger question of 'why some regions are more innovative than others and therefore grow faster."

links the knowledge inputs with the innovative outputs – at the level of the establishment or firm – may be less valid. Instead, a new literature suggests that knowledge spillovers can come from the firm or research institute producing it to a different firm commercializing that knowledge. This view is supported by theoretical models, which have focused on the role that spillovers of knowledge across firms play in generating increasing returns and ultimately economic growth (Romer, 1990).

An important theoretical development is that geography may provide a relevant unit of observation within which knowledge spillovers occur (Feldman, 1994). The theory of localization suggests that geographic proximity is needed to transmit knowledge – especially tacit knowledge.

Jaffe (1989), Jaffe *et al.* (1993), Audretsch and Feldman (1996), Anselin *et al.* (1997) and Acs *et al.* (2002a) have supported the importance of geographic proximity for knowledge spillovers in a wave of recent empirical studies. For a critical survey of the literature on spillovers see Karlsson and Manduchi (2001).

20.5. Towards a "new model of regional economic development?"

A 'spatialized' theory of technology-led regional economic growth needs to reflect three fundamental issues. First, it should provide an explanation of why knowledge-related economic activities start concentrating in certain regions and leave other regions relatively underdeveloped. Second, it needs to answer the question of how technological advance occurs and what are the key processes and institutions involved. Third, it has to present an analytical framework where the role of technological change in regional economic growth is clearly explained. In order to answer these three questions we surveyed three separate and distinct literatures: the new economic geography, the new growth theory, and the new economics of innovation. We suggest that each one of the above three approaches has its strengths and weaknesses that can be integrated to develop an appropriate model of technology-led regional economic development.

New economic geography answers the question why economic activity concentrates in certain regions but not others, but leaves out innovation and economic growth. The contribution of Krugman's theory on economic concentration is not in its elements but in the way the system was put together. It has already been well known in economic geography and regional economics that decreasing transportation costs, economies of scale, or increasing demand favor agglomeration. However, the way Krugman puts these elements together in a general equilibrium model is novel. The model provides a case for the treatment of spatial issues in the way economist are accustomed to. The model provides a technique to analyze geographical concentration of economic activities as being induced by some initial combinations of basic parameters. However, the model in its current form does not seem to be suitable for modeling technology-led regional economic growth at least for two reasons. First, Krugman's definite insistence of avoiding modeling the role of technological externalities in regional economic growth prevents the model to be applicable in innovation-led regional development since spillovers, innovation networks are in the core of this type of development as exemplified in the literature of innovation systems. Second, while the model is very strong in working out the characterization of specific initial combinations of parameters favoring geographical concentration, it is weak in actually modeling the growth process.

The new growth theory explains the causes of economic growth, but leaves out regional considerations and ignores the key processes and institutions involved in innovation. A principal assumption in the theory of endogenous growth is that for creating new sets of technological knowledge, the total stock of knowledge is freely accessible for anyone engaged in research. However, this assumption is not verified in the literature of geographic knowledge spillovers. New knowledge (potentially leading to either product of process innovations) is usually in such a tacit form that its accessibility is bounded by geographic proximity and/or by the nature and extent of the interactions among actors of an innovation system (see, e.g. Anselin *et al.*, 1997). Similar to the case of relaxing the neoclassical assumption of equal availability of technological opportunities in all countries of the world a relaxation of the assumption that knowledge is evenly distributed across space within countries seems also to be necessary. The non-excludable part of the total stock of knowledge seems rather to be correctly classified if it is assumed to have two portions: a perfectly accessible part consisting of already established knowledge elements (obtainable via scientific publications, patent applications, etc.) and a novel, tacit element, accessible by interactions among actors in the innovation system. While the first part is available without restrictions, accessibility of the second one is bounded by the nature of interactions among actors in a 'system of innovation'.

The new economics of innovation and entrepreneurship, while explaining the institutional arrangements in the innovation process, leaves out regional issues and economic growth. The systems approach is a conceptual framework that many scholars and policy makers consider useful for the analysis of innovation. Although the systems of innovation approach are not considered a formal and established theory, its development has been influenced by different theories of innovation such as interactive learning theories and evolutionary theories. In recent years, efforts have been made to percolate general theoretical and empirical observations from this literature up into a conceptual framework capable of guiding policy, loosely organized around the idea of a 'national system of innovation'. The concept of a 'national' innovation system may be problematic. Krugman has suggested that as economies become less constrained by national frontiers (as globalization intensifies), they become more geographically specialized. Important elements of the process of innovation tend to become regional rather than national. Some of the largest corporations are weakening their ties with their home country and are spreading their innovation activities to source different regional systems of innovation. Regional networks of firms are creating new forms of learning and production. These changes are important and challenge the traditional role of 'national systems of innovation'.

The specific combination of the Krugmanian theory of initial conditions for spatial concentration of economic activities with the Romerian theory of endogenous economic growth complemented with a systematic representation of interactions among the actors of Nelson's innovation system could be a way of developing an appropriate model of technology-led regional economic development. The central element of the model could be the 'regional knowledge production equation' distilled from the predominantly empirical literature of innovation networks as presented in the literature of the new economics of innovation. In the traditional model of the knowledge production function, firms exogenously exist and then engage in the pursuit of new economic knowledge as an input into the process of generating innovative activity. As suggested by Audretsch (1995, p.48) we "...propose shifting the unit of observation away from exogenously assumed firms to individuals – agents confronted with new knowledge and the decision whether and how to act upon that new knowledge." This equation would facilitate the presence of knowledge in the Krugmanian economic geography model. Here the analytical technique for deriving initial conditions of spatial concentration can be adapted to come up with the preconditions for the emergence of knowledge-induced agglomerations. Together with other parameters of the model threshold values of knowledge may be calculated following the technique developed by Krugman.

Finally, to actually model the equilibrium path of regional economic growth induced by the threshold values of knowledge and other regional parameters, the combined framework of the new economic geography and the new economics of innovation can be complemented with the Romerian analytics of economic growth.

References

Acs, Z.J. (ed.) (2000), *Regional Innovation, Knowledge and Global Change*, London: Pinter Publishers.

Acs, Z.J. (2002), *Innovation and the Growth of Cities*, Chelthenam: Edward Elgar.

Acs, Z.J. and D.B. Audretsch (1987), "Innovation market structure and firm size", *Review of Economics and Statistics*, Vol. 69, pp. 567–575.

Acs, Z.J. and D.B. Audretsch (1988), "Innovation in large and small firms: an empirical analysis", *American Economic Review*, Vol. 78, pp. 678–690.

Acs, Z.J. and D.B. Audretsch (1990), *Innovation and Small Firms*, Cambridge: MIT Press.

Acs, Z.J. and A. Varga (2002), "Geography, endogenous growth and innovation", *International Regional Science Review*, Vol. 25(1), pp. 132–148.

Acs, Z.J., D.B. Audretsch and M. Feldman (1992), "Real effects of academic research: comment", *American Economic Review*, Vol. 81, pp. 363–367.

Acs, Z.J., D.B. Audretsch and M. Feldman (1994), "R&D spillovers and recipient firm size", *The Review of Economics and Statistics*, Vol. 76, pp. 336–340.

Acs, Z.J., L. Anselin and A. Varga (2002a), "Patents and innovation counts as measures of regional production of new knowledge", *Research Policy*, Vol. 31, pp. 1069–1085.

Acs, Z.J., F.R. FitzRoy and I. Smith (2002b), "High technology employment and R&D in cities: heterogeneity vs. specialization", *The Annals of Regional Science*, Vol. 36, pp. 373–386.

Aghion, P. and P. Howitt (1998), *Endogenous Growth Theory*, Cambridge: MIT Press.

Anselin, L., A. Varga and Z.J. Acs (1997), "Local geographic spillovers between university research high-technology innovations", *Journal of Urban Economics*, Vol. 42, pp. 422–448.

Anselin, L., A. Varga and Z.J. Acs (2000), "Geographical spillovers and university research: a spatial econometric approach", *Growth and Change*, Vol. 31, pp. 435–443.

Arrow, K. (1962), "Economic welfare and the allocation of resources for invention", in: Richard Nelson, editor, *The Rate and Direction of Inventive Activity*, pp. 609–626, Princeton: Princeton University Press.

Audretsch, D.B. (1995), *Innovation and Industry Evolution*, Cambridge: MIT Press.

Audretsch, D.B. and M.P. Feldman (1996), "Knowledge spillovers and the geography of innovation and production", *American Economic Review*, Vol. 86, pp. 630–640.

Barro, R.J. and X. Sala-I-Martin (1992), "Convergence", *Journal of Political Economy*, Vol. 100, pp. 223–251.

Berliant, M., S.K. Peng and P. Wang (2002), "Production externalities and urban configurations", *Journal of Economic Theory*, Vol. 104, pp. 275–303.

Black, D. and V. Henderson (1999), "A theory of urban growth", *Journal of Political Economy*, Vol. 107, pp. 252–284.

Braczyk, H.J., P. Cooke and M. Heidenreich (1998), "Regional innovation systems", in: H.J. Braczyk, P. Cooke and M. Heidenreich, editors, *The Role of Governances in a Globalized World*, London: UCL Press.

DeBresson, C. (1996), *Economic Interdependence and Innovative Activity*, Cheltenham, UK: Edward Elgar.

de la Mothe, J. and G. Paquet (1998), *Local and Regional Systems of Innovation*, Boston: Kluwer.

Dixit, A. and J. Stiglitz (1977), "Monopolistic competition and optimum product diversity", *American Economic Review*, Vol. 67, pp. 297–308.

Duranton, G. and G. Puga (1999), Diversity and specialization in cities: why, where and when does it matter?, *Center for Economic Performance, Discussion Paper 433, August*, London: London School of Economics and Political Science.

Edquist, C. (1997), *Systems of Innovation*, London: Cassell.

Feldman, M. (1994), *The Geography of Innovation*, Boston: Kluwer Academic Publishers.

Feldman, M.P. and D.B. Audretsch (1999), "Innovation in cities: science-based diversity, specialization and localized competition", *European Economic Review*, Vol. 43, pp. 409–429.

Fischer, M. and A. Varga (2002), "Technological innovation and interfirm cooperation. An exploratory analysis using survey data from manufacturing firms in the metropolitan region Vienna", *International Journal of Technology Management*, Vol. 24(7/8), pp. 724–742.

Glaeser, E.L., H.D. Kallal, J.A. Scheinkman and A. Shleifer (1992), "Growth in cities", *Journal of Political Economy*, Vol. 100, pp. 1126–1152.

Griliches, Z. (1979), "Issues in assessing the contribution of R&D to productivity growth", *Bell Journal of Economics*, Vol. 10, pp. 92–116.

Grossman, G. and E. Helpman (1991), *Innovation and Growth in the Global Economy*, Cambridge: MIT Press.

Hayek, F.A. (1945), "The use of knowledge in society", *American Economic Review*, Vol. 35, pp. 519–530.

Helpman, E. (1992), "Endogenous macroeconomic growth theory", *European Economic Review*, Vol. 36, pp. 237–267.

Jacobs, J. (1969), *The Economy of Cities*, New York: Random House.

Jaffe, A.B. (1989), "Real effects of academic research", *American Economic Review*, Vol. 79, pp. 957–970.

Jaffe, A.B., M. Trajtenberg and R. Henderson (1993), "Geographic localization of knowledge spillovers as evidenced by patent citations", *Quarterly Journal of Economics*, pp. 577–598.

Jones, C. (1995), "R&D based models of economic growth", *Journal of Political Economy*, Vol. 103, pp. 759–784.

Judd, K. (1985), "On the performance of patents", *Econometrica*, pp. 567–586.

Karlsson and Manduchi (2001), in: M.M. Fischer and J. Frohlich, editors, *Knowledge, Complexity and Innovation Systems*, Heidelberg, Germany: Springer, pp. 101–120.

Kirzner, I.M.D. (1997), "Entrepreneurial discovery and the competitive market process", *The Journal of Economic Literature*, Vol. 35, pp. 60–85.

Knight, F.H. (1921), *Discovery and the Capitalist Process*, Chicago: University of Chicago Press.

Krugman, P. (1998), "Space: the final frontier", *The Journal of Economic Perspectives*, Vol. 12, pp. 161–174.

Lucas, R.E. (1988), "On the mechanics of economic development", *Journal of Monetary Economics*, Vol. 22, pp. 1–42.

Lucas, R.E. (1993), "Making a miracle", *Econometrica*, Vol. 61, pp. 251–272.

Lucas, R.E. and E. Rossi-Hansberg (2002), "On the internal structure of cities", *Econometrica*, Vol. 70, pp. 1445–1476.

Maddison, A. (1987), "Growth and slowdown in advanced capitalist economies", *Journal of Economic Literature*, Vol. 25, pp. 469–698.

Marshall, A. (1961), *Principles of Economics*, London: Macmillan, Original publication in 1890.

Nelson, R. and S. Winter (1982), *An Evolution Theory of Economic Change*, Cambridge, MA: Belknap Press.

Nijkamp, P. and J. Poot (1997), Endogenous technological change, long run growth and spatial interdependence: a survey, *Innovative Behavior in Time and Space*, Heidelberg, Germany: Springer, pp. 213–238.

Nijkamp, P. and R. Stough (2000), "Growth and change", *A Journal of Urban and Regional Policy*, Vol. 31, pp. 451–454.

Ratti, R., A. Bramanti and R. Gordon (1997), *The Dynamics of Innovative Regions*, Aldershot, UK: Ashgate.

Romer, P.M. (1986), "Increasing returns and long run growth", *Journal of Political Economy*, Vol. 94, pp. 1002–1037.

Romer, P.M. (1990), "Endogenous technical change", *Journal of Political Economy*, Vol. 98, pp. S72–S102.

Romer, P.M. (1994), "The origins of endogenous growth", *Journal of Economic Perspectives*, Vol. 8, pp. 3–22.

Rosenberg, N. (1963), "Technological change in the machine tool industry, 1840–1910", *Journal of Economic History*, Vol. 23, pp. 414–443.

Saxenian, A.L. (1994), *Regional Advantage: Culture and Competition in Silicon Valley and Route 128*, Cambridge: Harvard University Press.

Schmookler, J. (1966), *Inventions and Economic Growth*, Cambridge: Harvard University Press.

Schumpeter, J.A. (1942), *Capitalism, Socialism and Democracy*, New York: Harpers Collins.

Shane, S. and S. Venkataraman (2000), "The promise of entrepreneurship as a field of research", *Academy of Management Review*.

Solow, R. (1957), "Technical change in an aggregate model of economic growth", *International Economic Review*, Vol. 6, pp. 18–31.

Sternberg, R. (1999), "Innovation linkages and proximity", *Regional Studies*, Vol. 33, pp. 529–540.

Varga, A. (1998), University Research and Regional Innovation: A Spatial Econometric Analysis of Academic Technology Transfers, Boston: Kluwer.

Venkataraman, S. (1997), "The distinctive domain of entrepreneurship research in, in advances in entrepreneurship", *Firm Emergence and Growth*, Vol. 3, pp. 119–138.

Winter, S.G. (1984), "Schumpeterian competition in alternative technological regimes", *Journal of Economic Behavior and Organization*, Vol. 5, pp. 287–320.

Urban Dynamics and Growth, CEA, vol. 266
Roberta Capello and Peter Nijkamp (editors)

CHAPTER 21

Cities and Business

Roger R. Stough and Rajendra Kulkarni
School of Public Policy, George Mason University, Fairfax, VA, USA

Abstract

Recent trends and evolving relationships between business activity and urban development are examined in this chapter. The analyses described place considerable emphasis on the impact of globalization, the rise of information and computer technology (ICT) and changing political systems and ideologies on both business development and process, and urbanization. In short, an argument that much of urban growth over the past decade or two has been technology driven in nature. Consequently, the examination of the scale and geographic distribution of technology intensive business activity form the first part of the chapter. It is noted and importantly so, that the very large majority (about 90 percent) of such activity is located in urban areas due to agglomeration economies and other proximal benefits of such locations. The chapter also examines the impacts globalization, technology and political dynamics have had on business process and in turn how this has contributed to the locational dyanmics of commercial activity in urbanized areas. Some elements of urban and commercial location theory are reviewed and assessed in the context of this latter assessment along with emergent trends such as the rise of edge cites as new commercial centers "on the edge" and acceleration of so-called "urban sprawl". Next, the rapid rise in importance of enterprise development or entrepreneurship policies and programs as part of the basic matrix of urban economic development efforts is recognized and assessed. The chapter concludes with a summary and conclusion that cities will increasingly make large investment in policies and programs designed to produce proportional increases in successful growth ventures and thus the sustainability of their economies.

Keywords: information and computer technology (ICT), edge cities, decentralization, urban theory, business process, land use, multi-centered polis, entrepreneurship, enterprise development, entrepreneurial fountain, policy

JEL classifications: R00, R11

21.1. Introduction

Cities, from the emergent villages of ancient civilizations to the present, have always been centers of business activity. What is important about this today is that global society including people and organizations such as businesses is much more urban than it was just a couple of decades ago. In the past, urbanization accelerated during the industrial revolution as people moved from farms to cities to find work in new factories. This city building trend broadened and deepened as the industrial period unfolded and ran its course. However, with the rise of the knowledge age, importance of the services sector and concomitant de-industrialization beginning in the 1960s and 1970s, the pace of urbanization increased and with it a further concentration of business in urbanized areas. The purpose of this chapter is to identify and examine the forces that have accelerated urbanization and the concentration of business activity in urban and metropolitan areas. Of the various forces considered the rise of information and communications technology (ICT) is fundamental and thus receives more attention than other factors in this analysis.

The unfolding of ICT during the past decade or two, the defining and generic technology of the knowledge age, has resulted in more than just the accelerated concentration of business activity in urban places. It has been a fundamental force in changed metropolitan land use patterns and with this, where people live and work, changes in business processes, labor force demand, industrial structure, the location of business in metropolitan space, capital formation, urban and regional economic development policy, entrepreneurship and enterprise creation, and urban transport to identify just a few of the effects. Further, with globalization and the increased importance of nation building (Fukuyama, 2004), and with a concomitant emphasis placed on market principles, entrepreneurship, business formation and growth processes have become increasingly important elements of urban economic development policy. In short, profound effects have stemmed from first the ICT driven knowledge age and also from several other forces such as globalization, state building and increased emphasis on markets in the development process. The extent of these changes and their affects create a problem for writing this chapter as it is not possible to identify or address all of them adequately. The chapter thus focuses on several primary topics.

The first substantive part of the chapter examines the scale and geographic distribution of technology intensive business activity (primarily ICT) in the metropolitan areas of the US. This shows that almost all of this type of business activity is concentrated in metropolitan areas and that it is not evenly distributed across or among metropolitan areas. The next part investigates and describes changes that have occurred in business process and other elements of the business enterprise, supporting institutions and infrastructure. This is an important platform that helps explain why business location patterns have changed within urban space. The third primary part of the paper focuses analysis on patterns of business location change in urban areas. Some elements of urban theory are reviewed at the beginning of this discussion to help position the profound changes that have occurred and are occurring in the context of classical and contemporary thinking about the location of businesses. Description and analysis of location patterns are then presented. The fourth and final topic examines the recent growth of interest in entrepreneurship, enterprise and company development as a central feature of contemporary urban economic development policy. This is important because it is a recent but powerful yet underappreciated development that may considerably influence the number and rate at which businesses are formed and develop, and thus create economic value (investment, jobs, earnings and wealth). Conclusions and discussion follow these analyses. This chapter attempts to examine, in part, the impact of technological and social/political change have had on business processes and location over the past few decades. But these changes have been broad and deep and consequently, the analysis and discussion are selective. Hopefully, the reader will find value in the topics addressed.

21.2. The location of technology intensive business

Economic activity today is increasingly driven by the technical conversion of the economic base of all regions, states and nations. While awareness of this conversion is highly visible in places like the Silicon Valley and in the Route 128 concentration in the Boston Massachusetts region it is also occurring in places like Montana, Arkansas, Utah, China and India to name a few. The difference between the more visible vs. the more remote regions is in the pace of change, when started, and degree. This transformation from the old to the new technologically intense economy is examined in detail in the next part of the chapter. Here, technology employment level is used to examine some of the location patterns of technology intensive businesses. Technology employment levels are estimated by summing employment in technology intensive economic sectors as defined by engineering employment and R&D expenditure levels. Estimates are made using a modification of the Armington index of technology intensity (see Stough *et al.*, 1998 for a description of the modified index, and Stough and Kulkarni (2001) for sources of the data presented below).

Technology intensive sectors of the US economy accounted for 10.1% of the US labor force in the late 1980s. Since then technical employment has increased but surprisingly not sufficiently enough to even maintain its 10% plus portion of the labor force. Today less than 10% of the US labor force is in technically intensive economic sectors. Yet technical employment levels have remained almost constant since then. There are several reasons for this. First, the estimation method used imperfectly captures new technology employment that has occurred via the spread into new sectors up and down the value delivery chain as these sectors were not included. Another reason is due to a considerable net substitution of technology and capital for labor in the production process. Despite putting a brake of sorts on new technology employment it has resulted in increased productivity. With these caveats it is not inconsistent to argue that the economy has undergone rapid and perhaps even radical technical change.

In 1998 the large majority of technology intensive employment was achieved in metropolitan areas (90.3%). It has remained nearly at this level since then. By 2000, 50 of the 321 US metropolitan areas accounted for about 70% of all technology employment in the US. In short, the large majority of technology employment is concentrated in urban areas and more specifically in the large metropolitan areas.

A comparison of data for technology employment in some well-known technology metropolitan statistical areas of the US (see Stough and Kulkarni, 2001) shows the wide range of scale and growth patterns among well-known urban regional technology centers. For example, the Silicon Valley with nearly 400,000 technology employees grew at a rate of 2.2% per year between 1988 and 2001. Austin Texas with 101,000 technology employees grew at 10.4%. Admittedly it is easier to grow at 10% plus given a small base like Austin had in 1988 of 41,000. Boston, an older technology region like the Silicon Valley grew at 2.4%. The Washington region is interesting in that it is a large technology region, and an older economy yet it grew at a rate of 5.9%. The point of this highly selective assessment is to show that the experience in technology growth dynamics varies considerably from place to place and that despite the fact that scale seems to matter it does not explain all the variation, e.g. the Washington case.

Figures 21.1–21.3 are presented to help further expose the variation in the patterns of technology employment among the metropolitan areas of the US. Figure 21.1 shows the levels of technology intensive employment in US metropolitan areas at the end of the 1990s. The major concentrations of more than 250,000 are found in Boston, Washington, DC, Chicago, Orange County (in Los Angeles) and San Jose. This suggests that while scale is important other factors must also be at work given the failure of several large metropolitan areas to show up, e.g. New York, Philadelphia, Atlanta, Detroit.

Figure 21.2 shows in map form the changes in technology employment between the late 1980s and the late 1990s. The growth

Figure 21.1. Technology employment in US metropolitan areas 1998

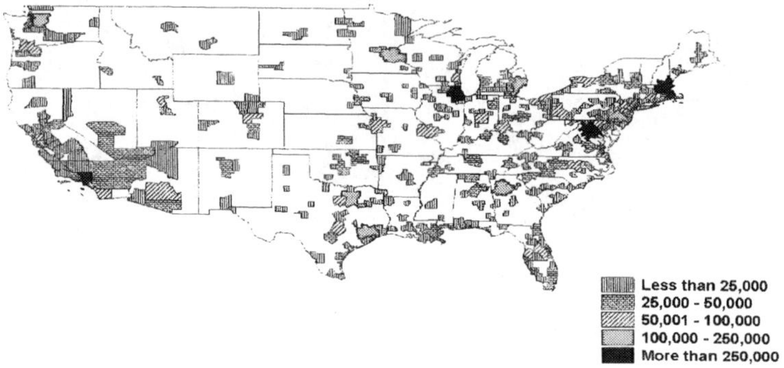

Source: U.S. Bureau of Census: County Business Patterns 1988–1998.

Figure 21.2. Technology employment change in US metropolitan areas between 1988 and 1998

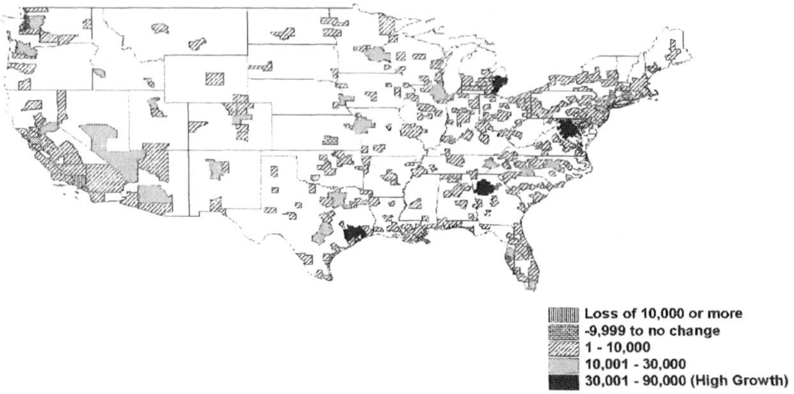

Source: U.S. Bureau of Census: County Business Patterns 1988–1998.

pattern is quite different. The largest absolute growth occurred in Washington, DC, Atlanta, Detroit, Dallas, Houston, Austin, Seattle and a county north of the Silicon Valley. Among the seven metro areas with the largest number of new technology jobs, Washington, as one of the 10 largest gainers was unique. None of the other 10 largest metro areas created 10,000 or more new technology jobs. The Texas trio is impressive in that each of these areas had more than 10,000 new technology employees meaning that a very significant agglomeration has been occurring there. Detroit's increases were

Figure 21.3. Percent change in technology employment in US metropolitan areas between 1988 and 1998

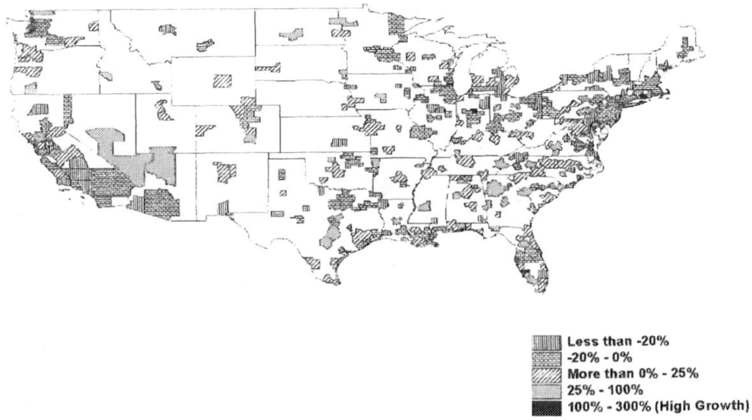

Source: U.S. Bureau of Census: County Business Patterns 1988–1998.

tied to increased demand for technical inputs in the automobile industry in that region. For Seattle, Microsoft and its supply chain demands and expansion in the aviation and aerospace industry there drove much of the growth.

The three metro areas with the greatest technology employment losses were Los Angeles, San Jose and Boston. Despite these large losses, these areas were all among the top five areas with the largest number of technology jobs in the late 1990s. That some large technology regions experienced large losses should not come as a surprise: a slight relative decrease of a few percentage points on a large base easily translates into a large absolute loss. This is the reason that regions with large relative increases and decreases are usually smaller regions.

Analyses of the relative importance of technology to the industrial mix of metropolitan economies (Figure 21.3) showed that the only large metropolitan areas that have more than 15% of their employment in technology intensive sectors are San Jose, Seattle and Washington, DC. All other regions with a large concentration are small to medium sized metropolitan areas, e.g. Kokomo, IN or Huntsville, AL. No major metropolitan area showed large gains or losses in the ratio of technology employment to total employment.

This analysis provides some insight into where technology intensive business is located within the metropolitan landscape of the US. The patterns or trends identified may be summarized as follows. First, the large majority of technology employment is located in metropolitan areas. Urban technology employment tends to be significantly correlated with size indicating that urban agglomeration economies are important location determinants. This is consistent with the analysis above that identified reasons why businesses are concentrating in outer parts of metropolitan areas in general and in edge city complexes in particular. Further, size of the urban area seems to be more important than general geographical location, e.g. east coast or west coast. Nonetheless, the larger concentrations of technology employment and thus business occur along the two coasts and across the Midwestern Manufacturing belt. Finally, it is important to note the finding that the importance of a metropolitan region's technology performance varies with the measure used (absolute number; change in number and relative to employment base) and with other endogenous attributes ranging from historical circumstance (Huntsville, Al) to amenity base (Florida, 2002), etc.

21.3. Changes in business context and operations

As a consequence of the ICT driven knowledge age and other changes such as globalization and stronger commitment to market principles the business context has been altered considerably over the earlier industrial era. Today business is not only more technical, especially in terms of the use of information technology, but increasingly is classified as technology intensive. Structurally, businesses are less vertically organized with responsibility increasingly pushed out to the so-called 'shop floor' whereby technology enhanced teams assume much of the responsibility that was formerly provided by middle management. Also, today most business operations are more narrowly defined with all but functions of greatest core competency outsourced to specialized providers or the producer services. For example, auto manufacturers throughout the world outsource almost all components of their vehicles except the drive train and engine, although even the engine in some cases, e.g. Cadillac. But, to convey the importance of this trend even further, many businesses outsource accounting, marketing, logistics, manufacture of component parts, personnel management, design, engineering, etc. In this part of the chapter we describe and explain how the context of business operations has changed in recent decades. Comparison attributes or dimensions over which this change has occurred are presented in Table 21.1.

On an economy wide basis commercial and other organizations have removed layers of middle level management to enable more horizontal organization thereby reducing transaction costs and gaining production and distribution efficiency. This has been necessary because the scope of competition today is global not national or regional and because businesses have much more information, and are more geographically mobile. Consequently, markets are more dynamic or unstable. From the perspective of the role of government, it has increasingly shirked its role as provider of services and goods to outsourcing that role to the private sector in order to gain efficiencies and focus more on policy development and

Table 21.1. Attributes of old and new economy

Issue	Old Economy (Fordist)	New Economy (Neo-Fordist)
Economy-wide characteristics		
Organizational form	Vertically integrated	Horizontal networks
Scope of competition	National	Global
Markets	Stable	Volatile
Competition among sub-national	Medium	High
Geographic mobility of business	Low	High
Role of government	Provider	Steer/row/end
Production characteristics		
Resource orientation	Material resources	Information and knowledge resources
Relation with other firms	Independent ventures	Alliance and collaboration
Source of competitive advantage	Agglomeration economies	Innovation, quality, time to market and cost
Primary source of productivity	Mechanization	Digitization
Growth driver	Capital/labor/land	Innovation, invention and knowledge
Role of research and innovation in the economy	Low moderate	High
Production methodology	Mass production	Flexible production
Role of government	Infrastructure provider	Privatization
Labor and workforce characteristics		
Labor–management relations	Adversarial	Collaborative
Skills	Job-specific skills	Global learning skills and cross-training
Requisite education	Task specialization	Lifelong learning and learning by doing
Policy goal	Jobs	Higher wages and incomes (productivity)
Infrastructure characteristics		
Form	Hard (physical)	Soft (information and organizations)
Transport	Miles of highway	Travel time reduction via application of IT
Power	Standard generation plant	Linked power grid (co-generation)
Organizational flow	Highly regulated	Deregulation
Telecommunication	Miles of copper wire	Wireless and fiber
Learning	Talking head	Distance learning

For more extended discussion, see Jin and Stough (1998).

design. In short, the overall environment within which business must operate has changed.

With these ICT driven changes in the environment of commercial activity it is no surprise that production process has also undergone transformation. Production methodology has changed from a mass production to flexible production model that is increasingly evolving an ability to 'mass produce' highly customized products and services. Production today is less focused on material resources and more oriented toward information and knowledge inputs. Increasingly the value of products and services is dominated by the knowledge imbedded in them. Consequently, innovation, invention and knowledge are becoming relatively more important than capital and labor in the input mix. In this new information rich context firms must build alliances, collaborate and cooperate with other organizations, often including competitors as well, to maintain competitiveness. The irony of the 'new economy' environment is that while competition is greater firms, at the same time, need to cooperate with competitors to learn about new innovations, build supply chain relations, and thereby reduce costs and time to market. While the primary source of productivity in the past was mechanization it is more focused today on digitization that controls production processes and helps ensure standardization of products and quality. Because the world of the 21st century is knowledge rich there is a heightened awareness and need for increasingly more research and innovation and thus increasing demand for technology transfer. As noted above the role of government has also changed from goods and services provision to managing and overseeing the privatization of these goods and services despite the fact that it may be leading to inefficiencies through a growing lack of accountability (Fritschler, 2004).

In the changing environment of the new economy labor and workforce roles have also changed. Where the 20th century labor movement took on an adversarial air with business, the 21st century economy cannot thrive in a context where labor and management are at odds because of huge related transaction costs. Reducing transaction costs is the *sin qua non* of the knowledge age and thus labor–management relations continue to move toward a more collaborative and partnership model. Out of this approach comes a view that workers with global learning skills capable of cross training and functioning are rewarded. In this context lifelong learning becomes a critical behavior for sustained high wage employment. Finally, from an economic development policy perspective the goal of the industrial era of producing jobs has been replaced by a goal of producing high wage and high income jobs that go along with high levels of worker productivity.

Finally, this assessment would be remiss if public sector infrastructure considerations were not examined. This is important because infrastructure (hard and soft) is a critical input for business success that is outside the control of individual businesses. In the industrial age physical infrastructure in the form of roads, railroads, airports, water supply and treatment systems, etc. was the critical supply side contribution of the public sector. These infrastructure components are still important. However, to be competitive these systems must be enhanced with information technology and systems control processes. Thus, concern in the areas of transport, for example, are not just with miles of highways or size of airports but with travel quality and time management via application of technology, for example, ITS (Stough, 2001). Further, the relative importance of hard or physical infrastructure has ebbed and is being replaced, in part, by soft infrastructure that includes enhanced labor quality via education and institutional restructuring to capitalize on transaction cost savings that can be achieved when organization productivity is amplified with information technology, for example. Thus, the services of the field of business process reengineering are in great demand.

Edge cities and more peripheral locations in metropolitan areas have become the major attractors of economic base business activity in the knowledge economy because, for the most part, these locations accommodate business needs better than core city or rural locations. Rural locations are less attractive because they cannot provide the location and urban agglomeration economies, i.e. economies that derive from proximity of competitors, producer services or suppliers, and support services such as education and information. Core city locations are less attractive because of a series of negative externalities or diseconomies including congestion, vertical or high-rise buildings vs. campus style structures, old infrastructure, higher taxes and crime, poor access to high quality labor. In short, outer city locations generally offer the best supporting environment.

21.4. Land use and business location patterns

Postindustrial urbanization has been of a different character than at all previous times with rapid and continuing decentralization of activity to and beyond the periphery of the development frontier. This metropolitan spread or sprawl, as some have called it, stems from several sources including technology, weak government institutions and land use laws, rising incomes and perhaps even American preferences for city access but country living. As such it is not surprising that businesses, especially newly established ones and producer services firms, have increasingly located in outer city locations such as edge cities (Garreau, 1990) and even satellite city locations (Stough, 1995). So an important theme is the decentralization of business location decisions within an intraurban context. Theory provides a partial explanation for the decentralization trend as discussed below.

21.4.1. Urban land use theory and use patterns

There is a robust body of theory regarding the location of economic activity in and among urban areas. Land rent theory originating with the work of von Thunen and Ricardo in the 19th century laid the foundation for modern urban land rent theory (Alonso, 1964) whereby the central most accessible location is viewed as the highest rent producing land with rent paying ability decreasing with distance from the center. As a consequence, most cities have evolved around a hub and spoke transportation infrastructure with higher rent paying activities located nearer the urban core. That is why high-rise physical structures and business activity capable of paying considerable rents have tended historically to emerge and grow at the geographic center of urban areas.

It is ironic, given the argument presented in the classical and neo-classical land rent literature and the considerable empirical support for it, that the core-dominated view of the city is increasingly at odds with reality. Metropolitan regions are rapidly expanding geographically as they become reorganized around multiple centers or edge cities (Garreau, 1990), the contemporary expression of the multinuclei concept (Harris, 1943). The multinuclei view of urban structure sees it expanding spatially as nodes of activity emerge along major crossroads in more peripheral areas. This activity tends to evolve in accordance with central place theory (Christaller, 1966) where initially only low-level goods and services are produced at outlying nodes. As growth continues, goods and services of increasingly higher levels are offered because of growing demand for them from positively reinforced expansion that spills over into the surrounding local periphery from the agglomeration or concentration of activities (Gordon and Richardson, 1996). Today many metropolitan areas are dominated by multiple centers that rival the historic central geographic core as centers of business activity. In the US where edge cities have evolved largely by market forces and not by plans per se, institutional coordination and management are problematic because they often evolve across multiple local government organizations. Many American cities have weak region wide institutions for managing development across political jurisdictions. In countries that have stronger metropolitan-wide planning and management institutions edge cities are often planned. For example, in China edge city type developments are organized around special economic incentives, e.g. free trade zones, high technology development parks and residential and retail complexes. Similar developments are emerging in the major cities of India as well as in other emergent national economic systems.

The rise of the multicentered polis has altered the historic pattern of demand for urban transportation as core dominated hub and spoke systems of transport that were developed in the past are ill equipped to meet the growing demand for travel between and among the new edge cities on the part of businesses and individuals. Thus, it is not surprising that the most rapid increase in demand for transport and related services for the past decade has been for cross-region travel. Metropolitan regions are having great difficulty coping with the associated increase in traffic congestion.

21.4.2. Urban decentralization and business activity

Metropolitan decentralization has long been a characteristic of the American metropolis (Wardwell and Brown, 1980; Stanback, 1991;

Nelson *et al.*, 1995). However, a rapid acceleration in the spread of metropolitan regions occurred over the past two or so decades in the US as well as in other countries such as Australia, countries of the European Union, Japan, China and India to mention a few. The geographic area of developed parts of metropolitan areas in the US has, in many cases, nearly doubled in the past 20 years. For example, the US Office of Management and Budget definition of the Washington Metropolitan region includes three state level jurisdictions (Maryland, Virginia and West Virginia) and the District of Columbia – a federal district. This region has 50% more counties in 2000 than in 1980, and grew from 3200 to 5000 square miles in area. The Washington region is not atypical as other metropolitan regions such as Atlanta Georgia, Dallas Texas, Houston Texas, Phoenix Arizona, Denver Colorado and a host of California metropolitan regions have experienced similar growth.

Metropolitan decentralization or sprawl has been driven by a diverse set of forces including the rise of the knowledge and technology intensive economy of the late 20th century (Stough *et al.*, 1998; Williams and Stimson, 2001). Among the non-technology forces contributing to these rapid spread effects are factors such as lower land costs on the periphery, extensive highway systems lowering transportation costs to outer city locations, residential preferences of Americans for the 'marriage of town and country' living styles and the vision of a Jeffersonian rural lifestyle, deteriorating conditions in central cities and finally a set of government policies that provide subsidies ranging from tax advantages to depreciation allowances to implicit subsidies in the form of building regulations and policies that discourage efforts to reuse older urban (and suburban) land (Ewing, 1994; US Office of Technology Assessment, 1995). Further, social issues related to spatial segregation by race and poverty may also be important factors (Bollens, 1988; Rusk, 1994). However, these forces have been present for many decades. So what has changed? The rapid development and ever quickening evolution of a new generic technology in the form of information and computer technology (ICT) is making a continuously changing and ever more spatially dispersed metropolitan economy and region not only possible but a reality (Niles, 1991; Kellerman, 1993; US Department of Transportation, 1993; Grantham and Nichols, 1994–1995).

Over the past decade or so the merging of computer and information technology has created vast increases in the ability of individuals and organizations to communicate. Local and wide-area networks and the Internet coupled with new and emerging high-speed wire and wireless, and large capacity telecommunications infrastructure holds the potential to connect all persons and organizations to each other almost in real time. Illustrative of the impact of this trend and critical to understanding metropolitan spread effects is that an estimated 10.56% of the US workforce telecommutes (works remotely) from home or near home at least 1 day a week (Illegems and Verbeke, 2003, p. 22). Similar patterns are emerging in the realms of 'telelearning', 'teleshopping' and 'telebanking'. As a result, the cost of physically moving across metropolitan space is reduced through substitution of virtual connections for commuter trips. For example, if the need for commuting to a central work place drops from five to three or 4 days a week due to remote working, as it has for the 10 million or so teleworkers referenced above, living farther out from the work place in more rural residential settings, for which many represents a piece of the 'American dream', is attractive. In short, ICT creates a reduction in work trip friction and thereby contributes to spread effects (Stough *et al.*, 2003). Other ways that ICT contributes to reduced friction of metropolitan travel is via smart roads and intelligent transportation systems (ITS) that enhance the productivity or performance of existing infrastructure and the vehicles that use it (Stough, 2001). In short, ICT is increasingly contributing to the substitution of communication for trip taking and more efficient movement, and thus reducing time cost of travel and in turn metropolitan spread effects.

21.4.3. Edge cities and the structure of business activity

However, the above discussion only partly explains the structural nature and mechanics of the recent era of rapid decentralization. The evolution of edge cities is central. Edge cities emerged on the periphery of metropolitan areas in the 1980s and 1990s. They are located at or near the confluence of major transport arteries and near former bedroom communities that housed white collar and highly educated workers who commuted to work in the center city. Such attributes were important for new emergent and often technology intensive enterprises because of a need for access to other infrastructure, e.g. airports, business services, etc. to high quality workers like those found in the bedroom communities, and to relatively inexpensive land that could only be found in peripheral locations.

Edge cities are large and diverse knowledge age urban concentrations that have appeared on the urban landscape over the past two decades or so and have become the standard form of American urban place according to Garreau (1990). Edge cities, in many ways, play the same role as the old traditional downtown centers did regarding jobs, shopping, entertainment, services and housing and even community. However, these new cities on the edge are often sprawling, somewhat unsystematic, are peculiar to look at, and often spill over political boundaries.

Some of the major attributes of edge cities are after Garreau (1990):

- Cater primarily to commercial office buildings (the workplaces of the knowledge age);
- Contain most of the commercial office and retail development that occurred during the various stages of urban growth experienced during the past three decades;
- Have a population base that is dominated by white-collar workers;
- Offer a variety of goods and services as well as entertainment and restaurants;
- Are perceived as one place, and an end destination for mixed use no matter how sprawling they may be; and,
- Rarely have formal political government with elected political officials.

Edge cities have at least 25,000 jobs and several million square feet of commercial office space, at least one million square feet of retail space, a dominant middle class and highly educated population, and are home to many technology intensive companies (Stough *et al.*, 1998). Edge cities are an expression of one of the most profound changes in land use experienced in the recent past. They are where most new businesses have located over the past 20 years and have thus become new urban–industrial complexes in areas that a few years ago were located on or beyond the margins of metropolitan development. Edge cities in the US have created highly 'urbanized counties in proximity to, but increasingly functionally independent from, the central cities of many metropolitan regions' (Stough *et al.*, 1998, p. 63). Edge cities are important because they represent a new urban form that has been pivotal in the changing land use reality. As such they are central to the sprawling pattern of land use and thus to increased tension between land use development, policy and transportation investment patterns. This observation is important because it explains why most of the urban economic base industry of the information age is generally locating in suburban-outlying areas and more particularly in edge city type developments.

21.5. Entrepreneurship and enterprise development in cities

Cities in the US as well as many other parts of the world have experienced a surge of interest in the formation of new businesses as a central part of their economic development strategy. Strong emphasis on enterprise development as a platform for economic development is an important new focus in the relationship between cities and business and thus is examined here. It should not be surprising that much of this entrepreneurial activity is in or proximal to outlying centers such as edge cities or other planned developments such as special trade or technology development zones in developing countries.

21.5.1. Interest in enterprise development has been increasing

In the US some 600,000–800,000 new jobs have been created annually over the past decade by new ventures (National Commission on Entrepreneurship – NCOE, 2004). The approximately 6 million small businesses in the US create the majority of these new jobs. Moreover, a large part of these jobs are created by small businesses that experience rapid growth as they move from start-ups, to pre-growth to growth companies. The NCOE (2004) report also concludes that since World War II entrepreneurs have been responsible for 67% of the inventions and 95% of all radical innovations. Further, High shows that entrepreneurial enterprises in manufacturing industries contribute 45% of the total value added by manufacturing between 1990 and 1999 (High, 2004). In short, small businesses and the processes that create them contribute greatly both quantitatively and qualitatively to the national economy (Acs, 1999). As these facts have become more highly publicized there has been a surge of interest in how countries and regions can adopt policies that will support entrepreneurship and enterprise development not just in the US but also in the EU and Asia and for that matter globally. For example, China has adopted a large-scale technology business incubation program with tens of thousands of such ventures already being developed in the cities of that country. The large majority of new company formation takes place in cities not just in the US but also in other developed and developing countries as referenced in China and the European Union (Commission of the European Communities, 2003). However, there are a variety of factors responsible for variation in new firm formation rates (Armington and Acs, 2002), including exogenous and endogenous ones.

21.5.2. Reasons for growth in enterprise development

Entrepreneurship is the way the economy and society take advantage of new wealth-creating opportunities that arise daily from constant change. However, when change accelerates, as it does during periods of rapid technological change such as during the rise of the industrial revolution or the rise of the ICT or knowledge age, entrepreneurship and enterprise development also increase. This is because technological change creates a new pool of exploitable possibilities and opportunities. Thus, it is not surprising that the recent period of rapid innovation in ICT has been one of increasing entrepreneurship. Nor is it surprising that much of this entrepreneurial activity is occurring in cities and more specifically in edge cites. That this has occurred primarily in cities is because that is where knowledge and human capital is concentrated and thus where innovation spillovers that often drive new company formation are most concentrated (Acs, 2002) as described below.

Technological change is not the only reason for the surge of interest in new firm formation. The breakup of the Soviet Union in 1989 signaled the end of a belief in the viability of centrally controlled economies (Fukuyama, 1995). This, in turn, led to a widely held belief that at the minimum a more market oriented economy was required. The basis for this arose earlier during the 1980s with increased reliance on liberalization and privatization policies in many of the more market oriented economies, e.g. New Zealand, Great Britain, Canada and Australia – including many of the lagging economies in Asia, Africa and Latin America. These political economy trends, amplified by intense technological change during the last two decades of the 20th century, interacted to create a powerful context that begged for enterprise development policies. In this context privatization became a global policy phenomenon along with a belief in the need for increased enterprise and firm development (Gomez-Ibanez and Meyer, 1993; Rao, 1998; Hodge, 2000). Thus, in addition to privatization, policies and programs to promote entrepreneurship and enterprise development also expanded. As evidence of this, it is interesting that US universities, during the 1980s and 1990s, formed 100 new entrepreneurship programs or 66% of all such entrepreneurship programs ever formed. Further, it is not surprising that since 1980 Fortune 500 companies have lost 5 million jobs while the US economy added 34 million new jobs through new ventures, start-ups or small businesses.

Globalization, partly driven by the ICT revolution that increased the availability of exploitable information, has also been a major contributor to the expansion of entrepreneurship. Audretsch (2001, p. 4) and others have observed, however, that globalization would not have become the pervasive force it has if only driven by the expansion and maturing of ICT. Political change in other parts of the world such as Eastern and Central Europe, China, India and Vietnam resulted in new stability in formerly largely inaccessible markets. As this 'opening up' occurred, access to significantly lower cost but qualified labor increased.

The response to this 'opening up' or access to lower wage labor in higher wage developed countries took two forms. One was to accelerate the substitution of capital in general and more specifically the substitution of technology for labor in an effort to retain the higher end of the more traditional manufacturing activities. This of course resulted in some job loss due to productivity enhancement but not nearly as much as when operations moved offshore. Nevertheless, global wage differentials are so great that capital substitution has been only partly effective. This is the basic concern behind recent WTO demonstrations and the rapidly emerging 'off-shoring of jobs' issue in the US and other developed countries.

The developed countries response to an inability to compete on the basis of wage cost and increasingly on the basis of capital substitution, involved shifting economic activity to high wage and high employment industries (Audretsch, 2001). In so doing, the competitive base shifted from factor cost inputs to technology and knowledge inputs. This is another major reason for the global trend in the increasing adoption of a firm formation and entrepreneurship-oriented policy on the part of developing countries. Thus, ICT and globalization are driving high wage countries and regions to focus economic development policy on the creation of even higher wage jobs and the maintenance of an environment that achieves competitiveness via innovation and high-end technology activities. Such enterprises have the ability to continuously innovate their products which in turn delays or even denies the progression of the product development cycle and thus makes it difficult if not impossible for low wage cost competitors to model or reengineer such products. Thus, continuous innovation products and companies form the foundation of competitiveness in developed countries and thus another reason for the expansion of enterprise development as an economic development strategy.

However, this begs in part the question of why there is also a surge of interest in creating technology intensive enterprise development programs in wage competitive countries like China as witnessed by the large-scale technology incubation programs that are in play in many cities there (Stough, 2003a,b). China's concern with learning how to compete on a continuous innovation basis stems in part from the fact that wages are rising in first tier cities like Beijing, Shanghai and Shenzhen making it increasingly difficult for firms there to compete with more wage competitive second level cities and cities located more to the interior. There is also long range concern that as development occurs and wages rise, China will increasingly have to compete with developed countries and to do this will require an ability to build and sustain continuous innovation enterprises. Thus, changed and changing competitive conditions globally are also driving the expansion of enterprise development initiatives in cities in both developed and developing countries.

Faced with an environment composed of a new generic technology (ICT) and associated knowledge that is enabling a host of new innovations, globalization of information and access to information partly as a result of the ICT revolution, opening up of new markets due to an acceptance of a more market oriented philosophy and increased but transformed competitive conditions it is not surprising that regions and nations have increasingly adopted stronger firm formation policies. Theory (Schumpeter, 1936, 1942; Kirzner, 1973) and experience (Birch, 1979) argue for and support a conclusion that the majority of new jobs are created in small companies and start-up ventures. While there is some debate on this (Harrison, 1994), it is difficult to argue against the evidence that the highest rates of entrepreneurship, i.e. new ventures and job growth, occur in metropolitan areas with higher levels of technology employment in the US over the past decade (Stough and Kulkarni, 2001). This provides another reason for the increasing attention that is being focused on venture formation policies.

21.5.3. Enterprise development: approaches in cities

A variety of arguments have been offered for why regions and nations are increasingly turning to enterprise development policies and programs as ways to increase jobs, wages and wealth. While it is clear that a host of macro and microeconomic policies and factors are important for achieving such goals the reasons offered above provide a rationale for why emphasis is increasingly focused on enterprise development policies. The purpose here is to provide a framework for conceptualizing how cities are approaching the challenge of supporting enterprise development.

Given any city or nation for that matter, it can be assumed that there is a stock of individuals who are either contemplating forming a company or are operating a small company. Figure 21.4 presents or models this pool of entrepreneurs as a triangle called the entrepreneurial fountain. The fountain mimics the entrepreneurial process in that its base is composed of those who are thinking about starting a company and is subsequently partitioned into smaller and smaller parts as one moves on up to existing businesses, pre-growth companies, growth companies and those that have moved out the top

Figure 21.4. Entrepreneurial fountain: static model

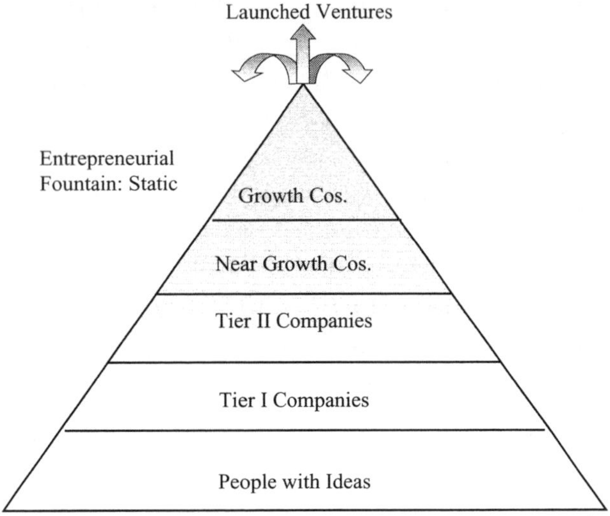

of the fountain onto sustained growth. Figure 21.5 shows the dynamic version of the fountain and recognizes that various forces ranging from macroeconomic effects and policy to endogenous efforts will result in its contraction or expansion. The economic development policy goals are to make the fountain as big as possible and to maximize the number of ventures that exit the fountain as

Figure 21.5. Entrepreneurial fountain: dynamic model

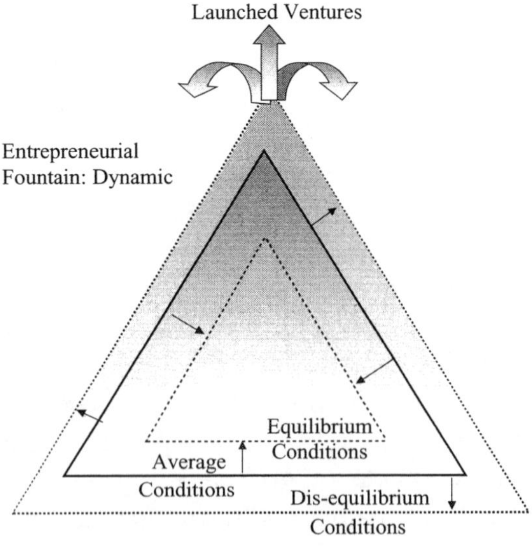

they move into sustained growth. These two policy objectives and related programs if successful will, *ceteris paribus*, create a powerful engine for job creation and sustained economic development of cities or nations.

The fountain may be viewed as a way to segment the enterprise development policy context. So policies and associated programs aimed at the base of the fountain are focused on company formation and include technical assistance for business start-ups. An example is provided by the US Small Business Administration's Small Business Development Center (SBDC) program. This is a national program with offices or centers located in major cities (metropolitan areas) in the US. SBDC centers provide free assistance for anyone who asks for help in starting a business. SBDCs also provide assistance to existing small businesses requesting assistance to help them grow, solve a business problem, enter into the export market, etc. The SBDC program provides basic assistance primarily at the first two levels or segments of the fountain. The rationale for the subsidy to support this program is that information on company formation and growth is highly imperfect and not all individuals or small businesses have the resources or ability to acquire or access appropriate information. In short, the policy and related program are justified on equity grounds on the one hand and on economic development grounds on the other.

Programs like the SBDC provide a source of basic information and assistance but are unable to provide in depth and sustained business development assistance that is required by companies that are beyond the Tier 1 level like Tier 2 companies or near growth companies. Tier 2 companies can be loosely defined as having less than \$3–4 million in annual revenue, are facing scalability issues, have basic operating capital and have a full time professional management team in place or if not in place available to be tapped. From most perspectives Tier 2 companies are viable businesses. Thus, it is not possible to rationalize public assistance to these ventures on the basis of equity. Yet, it is Tier 2 companies that move on to near growth status and then growth status that produce large numbers of jobs and thus from the perspective of job and wealth creation beg for focused assistance to increase the probability of reaching the upper echelons of the entrepreneurial fountain. While it is presumed that private sector providers, e.g. venture capital firms, provide the assistance that these firms need such assistance is more often for near growth and growth companies, not Tier 2 companies. With the exception of a few Tier 2 companies, there is thus a gap in the focused assistance that is needed to propel Tier 2 companies into sustained growth. From the perspective of the pubic sector this means that fewer Tier 2 companies reach growth status and thus less job and wealth creation occurs. All of this suggests a role for the public sector in assisting Tier 2 companies.

There are some public support programs for Tier 2 and more advanced companies but these are specialized in nature, e.g. Small Business Investment Research grants to assist companies in the further development of pre-competitive technologies, or highly dependent on local initiative, e.g. local and state government support of incubators. But even with such support the availability of incubators in the US is minimal and does not provide options for more than a small sliver of the potential Tier 2 market. In China, on the other hand, large technology incubation programs exist. For example, Qingdao, on the coast of China north of Shanghai with a population of about 4 million, supports a technology incubator with more than 600 ventures in residence. In Wuhan, with a population of more than 12 million, there are some 26 technology incubation centers supporting several thousand Tier 2 type ventures that are attempting to become growth companies. In short, there appears to be a gap in support for Tier 2 ventures in many parts of the world and cities (and nations) that choose to provide public support to these firms will likely reap powerful competitiveness benefits in the form of sustained job and wealth creation. This, rather than equity and information imperfections, provides the basis for Tier 2 public subsidies. However, steering public funding toward Tier 2 companies will be a value conflict problem in strong market based economies because this will be seen as subsidizing with public funds viable business entities.

21.5.4. General observations and conclusions: enterprise development and cities

In this part of the chapter a partially recognized trend, entrepreneurship and enterprise development, has been introduced and described. There are a number of powerful technological, competitive, ideological and political forces driving this development. Further, most cities and nations have made investments in public policy initiatives to support this trend in an effort to achieve the job and wealth benefits that are suggested. While public subsidy policies exist and have powerful support rationales for company formation and assistance to fledgling businesses, policies to support ventures that are minimally viable but facing a low probability of successfully reaching sustained growth are not prevalent in most market dominated economies. Consequently, such programs are not so prevalent. But there is an economic development rationale for publicly subsidizing the development of such Tier 2 companies in that a city (country for that matter) could reap significant additional job growth and wealth creation through the use of such subsidies. All of this of course assumes that appropriate procedures and skills exist to not only select Tier 2 companies with a high probability of success but also to assist them in reaching sustained growth. Despite these generally unanswered questions it is likely that more and more cities and countries will expand programs of this nature over the coming decade.

21.6. Discussion, conclusions and policy implications

The chapter opened with an observation that urban growth intensified in the late 20th century thus accelerating a trend that has been underway since the industrial revolution. More importantly, however, there has been a shift to increased decentralization or sprawl in metropolitan areas. This spreading out of cities has been accompanied with the rise of the polycentric city and its associated multiple nodes or edge cities. Business while becoming more concentrated in urban areas over the past decade or so has done so more intensely in edge cities that in some cases rival their core city in scale and scope. The rise of the ICT industry has perhaps more than any other development characterized the demise of the industrial era and ushered in the knowledge or information age. The chapter shows the concentration of the development of this industry in urban areas and the fact that the size of city is not the only factor determining the scale and scope of ICT business activity in a metropolitan region. Finally, the emerging trend of increased entrepreneurship and enterprise development lying at the core of contemporary urban economic development was examined.

These topics are not the only ones that this chapter could have addressed. Trends such as the great increase of the producer services as new economic base activities in urban areas could have been reviewed. There is however a large and well-developed literature on this topic (Daniels and Bryson, 1998). The chapter could also have examined the tremendous changes that have occurred in urban logistics and reverse logistics (see Brewer *et al.*, 2001). This could have been a compliment to an examination of traffic and congestion as constraints to business operations in cities. Again, these topics have received considerable attention in the literature as illustrated, for example, in Black (2004). Further, there has been no discussion of environmental issues and how these are impacting business decisions and location in cities. Again, a rich body of literature exists on this topic (Hensher and Button, 2003). Finally, globalization and outsourcing (outplacing) of jobs and operations are important developments of the last decade or so but have here only received marginal recognition.

Much of the discussion in this chapter has been descriptive and interpretive in nature. Policy insights and conclusions have been based on the same interpretive style of analysis. At the same time the chapter has with one exception, enterprise development policy, simply identified policy relevant issues such as managing the polycentric city, associated cross-jurisdictional management and policy issues, and the huge problem of coordinating policy and programs in such a diffused political context. Other policy related topics that might have been considered are land use sprawl, technology economic development policy and urban competitiveness although this is addressed a bit in the part of the chapter on enterprise development policy.

Entrepreneurship and enterprise development were given special attention in this chapter because such policies are in an early stage of development in urban areas and they are supported by powerful forces thus they are highly relevant. Consequently, significant attention was devoted to this topic to help the reader develop hypotheses about this trend and how it may impact businesses, competitiveness and business formation in cities. The conclusion is that cities are likely to make large investments in policies and programs that will not only lead to the creation of more businesses but the creation and development of more growth ventures. It is in this way that cities may improve their ability to achieve sustained economic development in the 21st century.

References

Acs, Z. (ed.) (1999), *Are Small Firms Important?*, p. 175, Boston: Kluwer.

Acs, Z. (2002), *Innovation and the Growth of Cities*, Cheltenham: Edward Elgar.

Alonso, W. (1964), *Location and Land Use: Toward a General Theory of Land Rent*, Cambridge, MA: Harvard University Press.

Armington, C. and Z. Acs (2002), "The determinants of regional variation in new firm formation, regional studies", *JRSA*, Vol. 36(1), pp. 33–46.

Audretsch, D. (2001), Entrepreneurship policy and the strategic management of places, *Proceedings of Conference on Entrepreneurship Policy: What's Government got to do with it?*, Cambridge: Kennedy School, Harvard University.

Birch, D. (1979), *The Job Creation Process*, Cambridge, MA: MIT Program on Neighborhood and Regional Change.

Black, W. (2004), *Transportation: A Geographical Analysis*, New York: Guildford Press.

Bollens, S.A. (1988), "Municipal decline and inequality in American suburban rings: 1960–1980", *Regional Studies*, Vol. 22, pp. 277–285.

Brewer, A.M., K.J. Button and D.A. Hensher (eds.) (2001), *Handbook of Logistics and Supply Chain Management*, p. 545, New York: Pergamon.

Christaller, W. (1966), *Central Places in Southern Germany*, translation by C.W. Baskin, Englewood Cliffs, NJ: Prentice-Hall Inc.

Commission of the European Communities (2003), *Green Paper: Entrepreneurship in Europe*, Brussels: European Commission.

Daniels, P. and J.R. Brysen (eds.) (1998), *Service Industries in the Global Economy: Vol. 1: Service Theories and Service Employment*; *Vol. 2: Services, Globalization and Economic Development*, p. 1746.

Ewing, R. (1994), "Characteristics, causes, and effects of sprawl: a literature review", *Environmental and Urban Issues*, pp. 1–15, Winter.

Florida, R. (2002), *The Rise of the Creative Class: And How it is Transforming Work, Leisure, Community and Everyday Life*, New York: Basic Books.

Fritschler, L. (2004), *Privatization and Accountability*, Working paper, Arlington, VA: School of Public Policy, George Mason University, Fairfax, VA.

Fukuyama, F. (1995), *The Social Values and the Creation of Prosperity*, New York: Free Press.

Fukuyama, F. (2004), *State Building: Governance and World Order in the 21st Century*, Ithaca, NY: Cornell University Press.

Garreau, J. (1990), *Edge City: Life on the New Frontier*, New York, NY: Doubleday.

Gomez-Ibanez, J.A. and J.R. Meyer (1993), *Going Private: The International Experience Transportation Privatization*, Washington, DC: The Brookings Institution.

Gordon, P. and H.W. Richardson (1996), "Beyond polycentricity: the dispersed metropolis, Los Angeles 1970–1990", *Journal of the American Planning Association*, Vol. 62, pp. 289–295.

Grantham, C.E. and L.D. Nichols (1994), "Distributed work: learning to manage at a distance", *The Public Manager*, pp. 31–34, Winter.

Harris, C.D. (1943), "A functional classification of cities in the United States", *Geographical Review*, Vol. 33, pp. 85–99, January.

Harrison, B. (1994), *Lean and Mean: The Changing Landscape of Corporate Power in the Age of Flexibility*, New York: Basic Books, p. 324.

Hensher, D.A. and K.J. Button (eds.) (2003), *Handbook of Transport and the Environment*, Boston: Elsevier.

High, J. (2004), "The roles of entrepreneurship in economic growth", in: H.L.F.P. de Groot, P. Nijkamp and R.R. Stough, editors, *Entrepreneurship and Regional Economic Development: A Spatial Perspective*, Cheltenham: Edward Elgar.

Hodge, G.A. (2000), *Privatization: An International Review of Performance*, Boulder, CO: Westview Press.

Illegems, V. and A. Verbeke (2003), *Moving Towards the Virtual Workplace: Managerial and Societal Perspectives on Telework*, Cheltenham: Edward Elgar.

Jin, D. and R.R. Stough (1998), "Learning and learning capability in the Fordist and Post-Fordist age: an integrative framework", *Environmental and Planning*, Vol. 30, pp. 1255–1278.

Kellerman, A. (1993), *Telecommunications and Geography*, New York, NY: Halsted Press.

Kirzner, I.M. (1973), *Competition and Entrepreneurship*, Chicago: University of Chicago Press.

National Commission on Entrepreneurship (NCOE) (2004), *Fact Sheet*, http://www.publicforuminstitute.org/nde/news/facts.htm.

Nelson, A.C., W.J. Drummond and D.S. Sawicki (1995), "Exurban industrialization: implications for economic development policy", *Economic Development Quarterly*, Vol. 9(2).

Niles, J.M. (1991), "Telecommuting and urban sprawl", *Transportation*, Vol. 18, pp. 411–432.

Rao, C.P. (1998), *Globalization, Privatization and Free Market Economy*, Westport, CT: Quorum Press.

Rusk, D. (1994), *Cities Without Suburbs*, Baltimore, MD: Johns Hopkins University Press.

Schumpeter, J.A. (1936), *The Theory of Economic Development: An Inquiry into Profits, Capital, Credit and the Business Cycle*, Cambridge, MA: Harvard University Press.

Schumpeter, J.A. (1942), *Capitalism, Socialism and Democracy*, New York: Harper & Row.

Stanback, T. (1991), *New Suburbanization: Challenge to the Inner City*, Boulder, CO: Westview Press.

Stough, R.R. (1995), "Technology will spur satellite cities, more sprawl", *The Edge City News*, Vol. 3(4).

Stough, R.R. (2001), *Intelligent Transportation Systems: Cases and Policies*, Cheltenham: Edward Elgar.

Stough, R.R. (2003a), "Strategic management of places and policy", *Annals of Regional Science*, Vol. 37(1), pp. 179–201.

Stough, R.R. (2003b), "The rise of global enterprise development: patterns in China and India", *The Journal of Indian Management and Strategy*, Vol. 7(6), April–June Edition.

Stough, R.R. and R. Kulkarni (2001), "Planning issues and the new generation technology economy: comparative regional analysis and the case of the U.S. national capital region", in: J.F. Williams and R. Stimson, editors, *International Review of Comparative Public Policy, International Urban Planning Settings: Lessons of Success*, Vol. 12, New York: Elsevier Science, pp 395–430.

Stough, R.R., K.E. Haynes and H.S. Campbell, Jr. (1998), "Small business entrepreneurship in the high technology services sector: an assessment for edge cities of the U.S. national capital region", *Small Business Economics*, Vol. 10, pp. 61–74.

Stough, R.R., Y. Higano, K. Button and P. Nijkamp (2003), *Transport and Information Systems*, Cheltenham: Edward Elgar.

US Department of Transportation (1993), *Transportation Implications of Telecommuting*, Washington, DC: US Department of Transportation.

US Office of Technology Assessment (1995), *The Technological Reshaping of Metropolitan America [OTA-ETI-643]*, Washington, DC: US Government Printing Office.

Wardwell, J.M. and D.L. Brown (1980), "Population redistribution in the United States during the 1970s", pp. 1–35, in: D.L. Brown and J.M. Wardwell, editors, *New Directions in Urban–Rural Migration*, New York: Academic Press.

Williams, J.F. and R. Stimson (2001). International Urban settings: innovation in planning approaches, pp. 1–17, in: J.R. Williams and R. Stimson, editors, *International Review of Comparative Public Policy, International Urban Planning Settings: Lessons of Success*, Vol. 12, New York: Elsevier Science Ltd.

PART 6

Urban Policy

Urban Dynamics and Growth, CEA, vol. 266
Roberta Capello and Peter Nijkamp (editors)

CHAPTER 22

Strengthening Municipal Fiscal Autonomy Through Intergovernmental Transfers

Peter Friedrich[a], Joanna Gwiazda[b] and Chang Woon Nam[c]

[a]University of the Federal Armed Forces Munich, Werner-Heisenberg-Weg 39, 85579, Neubiberg, Germany
[b]Universtiy of Agriculture, UI. Orlat Lwowkowskich 38/51, 02-495 Warsaw, Poland
[c]Ifo Institute for Economic Research and CESifo, Poschinger Str. 5, 81679, Munich, Germany

Abstract

With reference to the four selected European countries the ways to protect local fiscal autonomy are discussed in the framework of the vertical fiscal equalisation system. In particular, the application of the principle of parallel fiscal development between a state and its municipalities is examined, which can be adopted as a benchmark when determining the intergovernmental unconditional grants.

Keywords: municipal finance, fiscal autonomy and decentralisation, fiscal equalisation, intergovernmental transfers, Europe

JEL classifications: H7, H2, H4, H6, H8

22.1. Introduction

Worldwide a large number of municipalities suffer from fiscal stress due to various social, political and economic reasons. Local governments in various continents have been experiencing concentration of population on big cities and industrialisation processes (e.g. in India and China), transformation processes in post-socialist countries, political instability (e.g. in African countries and Russia), underdevelopment and organisational weaknesses (e.g. in South America and Africa) as well as epidemic diseases and natural problems (in Africa) or economic recessions (in more advanced countries). There is also an increasing global competition among municipalities (Bahl and Linn, 1994; Shah, 1994; Winkler, 1994; Ma, 1997; Fukasaku and de Mello, 1999; Council of Europe, 2000; Nam and Pasche, 2001; Matinez-Vazquez and Alm, 2003; Trasberg, 2003). As the conditions in the various countries are different, we focus on some European countries that operate within a similar social and economic framework but show differences in the structure of the public sector. We consider European countries with unitary government system such as Britain and Poland besides the federally structured Germany and Switzerland. Three countries co-operate within the European Union (EU), while Switzerland is a non-EU member. Poland and eastern Germany have undergone a transition from a socialist-style plan economy to a democratic market-oriented system. In these countries municipal finance through public debt, local fees, profits from municipal enterprises, sales of property has already reached natural or institutional limits. The fiscal autonomy of municipalities has rapidly weakened since 1990, as municipal expenditures continue to rise while local revenues decrease. Adapting the system of intergovernmental fiscal relations to the new circumstances is a challenge (Döring and Stahl, 1999; Borodo, 2003; Frey and Schlategger, 2003; Friedrich *et al.*, 2003).

All four countries mentioned above have implemented tax reforms; their municipalities depend heavily on vertical grants and were subject to municipal restructuring policies (Table 22.1). Municipalities in western Europe have all been faced with increased regional and global competition as well as EU policies. They have reduced local infrastructure investment and have had to increase expenditures for social aid. With the exception of Britain, where the fiscal autonomy is quite low on the local level, the financial situation of municipalities has deteriorated. Additional problems have emerged in Poland and Germany because of transformation necessities. In all the selected countries there is an urgent need to protect municipalities financially and to ensure their fiscal autonomy

Table 22.1. ***Major determinants of local finance development in the investigated European countries in the 1990s***

	Tax Reform	Large Intergovernmental Grants on the Local Level	Increase in Financial Burden of Municipalities	Increase in Social Expenditures	Decrease in Local Investment	Transformation	EU-Policy and Competition	Maastricht Budget Consolidation	Restructuring of Localities
Germany	X	X	X	X	X	X	X	X	X
Switzerland	X	X	X	X	X		X	X	X
Britain	X	X		X	X		X	X	X
Poland	X	X	X			X			X

Source: Friedrich *et al.* (2003).

from the intervention of the higher government level in order to enable the self-governing of local activities.

An improvement of fiscal autonomy for local governments must be achieved within the existing institutional arrangements. For this reason, the conventional literature on the normative theory of local public finance dealing with the welfare maximising assignments of tasks and financial means to government levels appears to be less helpful (Musgrave, 1959; Oates, 1972). Consequently, the discussion on optimal city size (Tullock, 1969; Miezkowski, 1987; Oates, 1999), contracting (Fisher, 1996), the formation of special districts and of functional overlapping competing jurisdictions (Frey and Eichenberger, 1999; Detig *et al.*, 2002; De Spindler, 2003) does not deliver information required for the improvement of municipal fiscal autonomy. Normative literature on intergovernmental grants dealing with the political effects of grants (Bradford and Oates, 1971; Fisher, 1979) and the so-called fly-paper effect (Hines and Thaler, 1995) does not adequately consider the measures to defend fiscal autonomy. Also those references on interactions among municipalities following the Tiebout (1956) approach in determining equilibrium among municipalities (Bewley, 1981; Wildasin, 1986; Scotchmer, 1994) or related to the vertical and horizontal tax competition among jurisdictions (Hamilton, 1975; McLure, 1986; Keen, 1998; Friedrich and Feng, 2002) or dealing with public choice models on local public goods (Brueckner, 2003) do not focus on ways for enhancing local fiscal autonomy aimed at decreasing financial stress. By contrast research findings shown in Pola (1996), Drennan and Netzer (1997), Zimmermann (1999), Hedkamp (2000), Dafflon (2002) and Rehm and Matern (2003), etc. address to the actual problems and suggest several ways to rectify the financial situation of municipalities. However, only some of them lead to a higher fiscal autonomy. Weaker legal requirements to increase municipal public *debt* can enhance municipal autonomy in the short run. On the other hand, credits have to be paid back and interest payments reduce the scope of financing expenditures.

The *sale of municipal real estate, firms, assets and other forms of property* may lead to a short-term liquidity effect. If these revenues are used to pay back local debts the financial situation may become healthy. However, if this local property is necessary for public production, the situation can be even worsened, especially if expensive, profit-oriented private services are procured to provide the public services. A decrease of future autonomy can also be a consequence, because of lacking revenues from the property and higher dependencies on private production and markets as well as the lack of own know-how in the production of some local services.

In countries like Germany, Poland, Switzerland and Britain *municipal fees* have been increased considerably, which led to an increase in revenues from fees. As long as profits were yielded, municipal fiscal autonomy was widened. However, higher fee revenues that stem from higher sales on the basis of given fees require higher production, thus neutralising the higher turnover by the higher total costs. In some countries there are laws that stipulate cost coverage of local firms but do not allow profit making. Therefore, an increase in fees only temporarily eases fiscal stress. More benefits may result from transferring municipalities' rights to private economic units against concession payments. Such rights include rights to provide energy, to organise passenger traffic, to use the municipal territory for storage, to organise markets, to use urban land for manufacturing, housing, gas lines, electricity and communication lines, etc.

Municipal fiscal autonomy can be enhanced if municipalities are able to apply (and expand) own *local taxes and rates* leading to higher tax revenues. In the four European countries there is an ongoing debate concerning tax reforms to provide them sufficient tax receipts. As the tax systems in these countries are different, the recommendations to change municipal taxation will differ extensively from one to another. Yet the popular taxation principles that could be well applied universally include (1) the non-business cycle sensitive tax basis, (2) the broad tax basis aimed at encompassing many citizens in taxation, (3) the taxes levied on those economic units that receive infrastructure services from the municipality, (4) the taxing of non-migrating tax objects, etc. (Oates, 1972; Musgrave and Musgrave, 1980; King, 1984; Paugam, 1999).

Generally, municipal autonomy is widened by a higher scope of self-government, e.g. decision making on a variety of tasks. There is an improved chance of maximising municipal welfare, given the financial and fiscal restrictions. The more tasks the municipalities are obliged to fulfil without receiving funds, the more restricted their autonomy will be. A farther-reaching increase in municipal autonomy requires that the revenues are also expanded. This can be arranged through an extension of *intergovernmental grants*. In general, conditional and unconditional grants exist.[1] The conditional grants can be provided according to the principle of connection between the assignment of local activities and their finance. In this sense the unconditional grants do not enhance autonomy much. The system of unconditional grants should also be changed in favour of higher autonomy of the municipalities by introducing particular rules of allocation of funds to grants to municipalities.

Therefore, we concentrate on the grant systems with respect to application of the connection principle, the shaping of conditional grants and introducing a principle of parallelism into the allocation process of unconditional intergovernmental transfers. Also relevant to the four countries and others we refer to Germany as an example that shows a rather sophisticated and legally regulated grant system.

22.2. *Fiscal equalisation to protect municipalities by conditional grants*

22.2.1. *Some fiscal issues surrounding the principle of connection*[2]

In countries like Germany, Switzerland, Poland and Britain the issue of protecting municipalities against the downward shift of tasks

---

[1] Intergovernmental transfers are aimed at rectifying not only the vertical imbalance caused by the unequal own tax revenues and expenditures of different levels of governments but also the horizontal imbalance which is led by the different fiscal capacities among same level jurisdictions. Although the local expenditure needs appear to be hardly measured in an objective way, the role of transfers becomes more crucial for those deficit jurisdictions on the sub-national level, especially when their increasing expenditures cannot be financed by borrowing or they lack direct access to capital markets. In the cases of existing externalities on other jurisdictions, the central government also needs to financially support sub-national authorities in order to guarantee the provision of certain public services on the local level like pollution control, inter-regional highways, etc. (Davis and Lucker, 1982; Ali *et al.*, 1993; Boadway and Hobson, 1993; Hyman, 1993; Rosen, 1995; Dahlby, 1996). Furthermore, the amount of grants should vary with the local expenditure needs and inversely with local fiscal capacity, while their distribution must be transparent and fair. More importantly, an effective transfer system should neither encourage overspending nor weaken tax collection efforts on the sub-national level (Gage and Mandell, 1990; Bahl and Linn, 1994; Jones and Cullis, 1994; Shah, 1994; Winkler, 1994; Oates, 1999; Nam *et al.*, 2001).

[2] The connection principle is to protect the local government tasks against crowding-out by state tasks and federal and EU tasks assigned to the state (Henneke and Vorholz, 2002). There

from higher levels of governments while leaving the fiscal burden to municipalities has become increasingly popular (Henneke, 2003).[3] The German federal constitution (Article 28, Paragraph 2) assigns a number of tasks to local authorities as well as the responsibility to execute certain functions. These tasks comprise activities of self-government where local authorities are free to choose as well as obligatory self-government where they can freely perform the respective tasks. With other public tasks they are mainly executing functions of other government levels. Within these fields of actions the scope of decision making of local authorities is not totally protected by the constitution (Klein, 2003).

In the state constitutions the connection principle is stipulated. A state should reallocate public activities to municipalities only if it offers sufficient financial resources to them required to execute these new local functions (Trapp, 1997, p. 185; Henneke, 2003, p. 142). In Germany there are debates on the extent of legal obligations to provide the appropriate financial means according to the connection

---

are also delineation and assignment of public tasks to the government level according to the subsidiarity principle to determine what public tasks should be performed by which level of government. However, there is no strict evaluation in the sense that given the stock of public tasks, state tasks are more important than municipal ones (Arnold and Geske, 1988, p. 11; Zimmermann, 1999, p. 73), although some tasks seem to be more important defined as the so-called common public tasks (Article 91a and b GG (Gemeinschaftsaufgaben)) and jointly financed programs (Article 104a GG) developed. These activities are planned, financed, and in part executed jointly by federal, state and municipal administrations. A high value and importance of local tasks signal the sole competency of local government for local affairs (Article 28 Paragraph 2 GG). In contrast, stipulations referring to intergovernmental fiscal relation to taxation show the powerful positions of the federation and states over the municipalities (Kirchhof, 2002), although local authorities spend most of the public investment expenditures. With some tasks, a mix of competencies of federation, states and municipalities exists. If considering the fundamental split of tasks laid down in the federal constitution there is no clear-cut allocation of tasks to the federation and states (Hohrmann, 1967, p. 180) and tasks are not all assigned (Trapp, 1997, p. 129), with respect to the legal ranking federal law (indirectly also European law) enjoys a higher priority than a state law and the latter over municipal law.

[3] In Germany the finance of public activities and performance of tasks does not follow the power to formulate laws but primarily the obligation to execute laws (Article 104 a, Paragraph 1 of the constitution). Finance is connected with the power of administration. Therefore, financial obligations are mostly with the states, which are to administer public functions. In cases of the federal administration (Article 84 Paragraph 1 of the constitution) the central government must finance the carrying out of a task and not the states or the municipalities.

principle (Schneider, 1998, p. 3759; Kirchhof, 2002; Klein, 2003). As there are three tiers of government, the question is whether the federal government is able to transfer tasks to municipalities leaving the financial burden to them. Article 104a of the constitution does not protect municipalities from the transfer of tasks that create additional financial burdens for them (Trapp, 1997, p. 217). Examples are social assistance for adults, youth and children (Bull and Welti, 1996; Henneke, 2003). Although local governments are safeguarded by Article 28 of the constitution, and municipalities can apply to a state constitutional court, they are legally unprotected if the respective state constitutional court does not bring the case to the Federal Constitutional Court (Klein, 2003, p. 4). Therefore, the federal government can give municipalities unfunded mandates. An open problem is how to treat these mandates in laws on intergovernmental fiscal relations (Grundlach, 2000).

A simple solution appears to be forcing the higher level government to cover the costs of administering public functions, as stipulated in Article 104 of the constitution. Although the states are responsible for executing most of the EU and federal laws, they can only claim that part of the costs that results from the administrative activities of the states. The difficulty arises in determining the amount of relevant costs (Trapp, 1997, p. 244; Schoch, 2000). Are municipalities free to influence the amount of costs because they have the autonomy to perform tasks, or should higher levels of government be able to determine an amount they are willing to pay, such as standard costs? The first possibility may lead to higher administrative costs, since local management can determine costs such as to maximise its own welfare or utility. The realisation of the second alternative may cause low payments, as higher tiers of governments maximise their utility by leaving the political and financial difficulties to the municipalities. A debate on fair costs to be met can be expected in this context. Appropriate rules must be formulated to obtain acceptable solutions. Some kind of bargaining Nash solutions may also emerge. If the higher rank of government is powerful, the principal-agent approaches can serve to fix the amount of cost covering. A municipality may execute the task according to a minimum utility constraint, and the task performance by the municipality will lead to different levels of utility of the higher level government. The payment to cover costs reduces the utility of the higher rank of government. The transfer to municipalities is to maximise higher government net utility while ensuring efficient performance of the municipalities. In a deterministic case the optimal amount of payment must equalise the marginal net benefit of the higher level government with the marginal utility of the low-level government. Special solutions result if risk is involved. According to the kind of risks, constant payments or payments that vary with the level of performance are optimal. To identify cost and its coverage, the incentive contract theory may also be applied. There may be an appropriate split of possible cost savings that causes the effective performance by the municipality and relatively low financial transfers by the higher rank municipality.

A further approach may be to compensate municipalities by stipulating a larger scope for taxation (Blankart and Borck, 2004). In this framework questions arise as to which kinds of taxes are appropriate for this purpose. Should municipalities be able to enlarge the tax base of business taxes or should their contribution from the business tax paid to higher level governments be reduced? Should the restrictions on rates of land and business taxes be simpler? If a new tax is allowed, should it be an addition to income taxation or a further turnover tax (Bull and Welti, 1996; Zimmermann, 1999)? Can new municipal taxes be introduced on packages, animals other than dogs, or on electronic communication? Severe problems of conformity with the general system of taxation may arise. If the states formulate requirements for local taxation, these problems can be reduced. On the one hand, this policy may give incentives for minimising costs for public functions transferred, but the tax receipts may not be large enough to cover the costs for fulfilling the task.

Another possibility to ensure cost-saving task performance can be the coverage of expenditures in terms of fees collected for respective services. This guarantees that services related to the transferred task are delivered to economic units that are able to pay. Increases of already existing fees may conflict with existing state laws that stipulate the principles of public fee determination, such as average cost pricing. Moreover, a non-lump sum treatment of services, etc. can be introduced.

Compensation by shifting federal or state property located in the respective municipality or fungible assets such as bonds, etc., to the municipalities carrying out additional tasks can be considered. The incidence of performing new tasks can be compensated by lowering the conditions required for borrowings by easing the cash-flow requirements for raising municipal credits by the state's municipal budget control office. State banks can provide credit to municipalities in a more easy way. This solution appears inadequate as the states or the central government will reallocate the debt burden to municipalities. This policy is similar to the local policy of switching their future borrowings to their municipal firms.

Another principle related to the connection principle is that of participation in political decision making when public tasks are reallocated or new activities are shifted to municipalities (Henneke, 2003). Therefore, municipalities should also be involved in the state or federal legislation process.[4] Yet, it appears to be difficult to expand the participation of municipalities. Law making has become more complicated in parliaments, and new bureaucratic routines, planning and participation schemes are required for government actions.

Furthermore, compensation for financial burdens related to new tasks may be considered in the framework of fiscal equalisation in terms of intergovernmental grants. Conditional grants may be adapted to meet additional municipal financial requirements, encouraging reforms of the vertical fiscal equalisation system between a state and its municipalities.

22.2.2. Conditional grants

There are several types of conditional grants in the four investigated countries (Friedrich *et al.*, 2003). We considered conditional grants provided to execute a local activity, which is originally a function of a central state or a sub-state but transmitted to municipalities. Other conditional grants are to support specified municipal tasks (Arnold and Geske, 1988; Smith, 2003). For their specification no general theory for optimal conditional grants exists. A normative approach relates to the consumer's net-benefit maximisation in the framework of a principal-agent relation. A higher level

---

[4] This type of approach supporting the application of the connection principle seems to be important because of the EU and federal policies on competencies, autonomy and financial flexibility of municipalities.

jurisdiction – the 'principal' – maximises net benefits, considering that a part of the net benefits must be allocated to the 'agent' municipality to ensure local efforts to achieve the best performance of activities. Under the certainty condition, the marginal net benefit of higher level jurisdictions must equal the marginal net benefit allocated to the assigned municipality in analogy to Gravelle and Rees (1992). If no necessities for participation of the agent in net benefit exist, the optimal conditional grant is to maximise the sum of both net benefits. Conditional grants designed in accordance with a more general support scheme cannot be adapted to individual municipalities to shape an optimal one. Therefore, when the conditions of optimal grants are considered, including the reaction of municipalities within a principal-agent relation, one may assume a representative municipality to fix the rules.[5] In such schemes risk can also be introduced. If the risk is associated with difficulties in detecting the output level, it makes sense to offer municipalities the same level of goal realisation. If grant provision to the municipality depends on an upper level decision maker who is not able to identify non-observable different efforts of the municipality, then an increasing participation should be allowed to the municipality. Therefore, different sizes of conditional grants result from this principal-agent situation (Coutry and Marschk, 2003; Grout and Stevens, 2003), when the different municipal activity levels are to be achieved. If the high-level government is able to provide different and individual amounts of support or if it has the power to decide individually on the size of conditional grants, the usual principal-agent relation changes to a game between principal and agent as shown below.

The state government may have an utility function, which depends on the output X of a municipal project and on the size of a conditional grant F, e.g. $-g_{FL^*}F + g_{XL^*}X$ where g_{FL} and g_{XL} form utility weights. The utility function U_L is depicted in

---

[5] Other goals like welfare maximisation may also be applied for the jurisdictions. If both the state and the municipality attempt to maximise employment, then total employment may be better maximised by the higher rank jurisdiction. However, a minimum employment level must be guaranteed to the municipality to ensure its efforts to perform policies which allow overall high employment within the state. The minimal employment has to be then gradually increased according to higher efforts made (or to be made). Again marginal total employment must equal to marginal minimal municipal employment.

Figure 22.1. Possible negotiation solutions

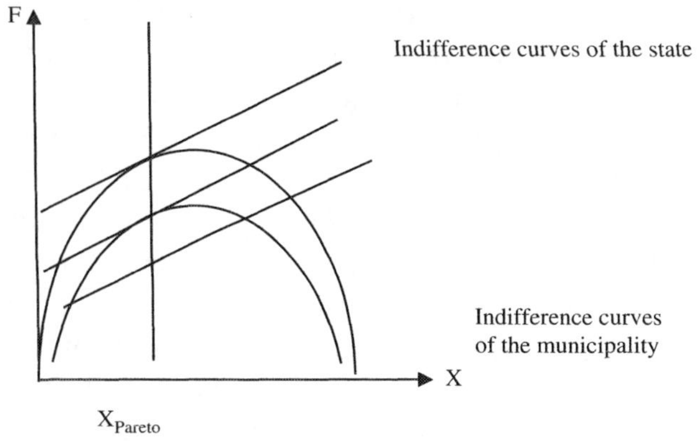

Equation (22.1). Figure 22.1 shows a set of indifference curves, which shift to the northeast with growing X. As F becomes larger the utility level of the indifference curve obtained by the state will be lower. The municipal utility U_G is measured in terms of project size X and the amount F of the conditional grant. The utility function U_G is shown by Equation (22.2) and its shape in Figure 22.1. On one hand, it increases with a higher output, reaches the top and decreases afterwards. On the other hand, it grows with the size of the

Figure 22.2. Nash solution

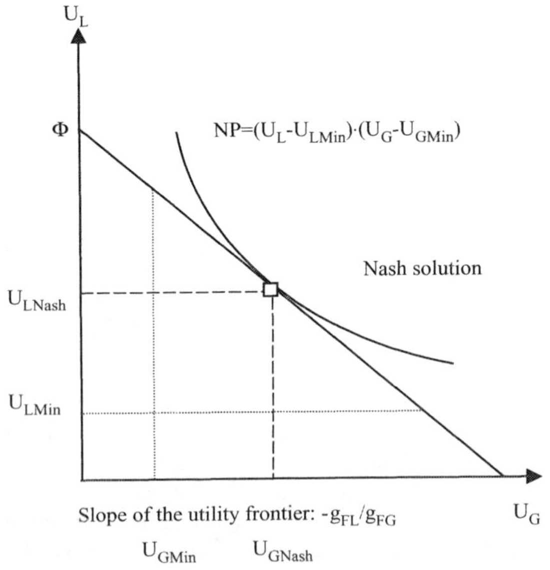

conditional grant. Therefore, there is a set of indifference curves with different utility levels positively correlated to the conditional grant. This is also depicted in Figure 22.1. If the state and the municipality start the negotiation on the size of the conditional grants, Pareto-optimal points exist at the points of tangency of the indifference curves of the state and the municipality. There is a range of Pareto-optimal possible solutions related to the utility levels of the state and municipality. They symbolise a set of possible negotiation solutions. They are depicted in Figure 22.2 and its minimum utility is multiplied by the utility change of the municipality minus municipal minimum utility. This expression is maximised considering the Pareto-optimal condition mentioned, as in Equations (22.5)–(22.9) in Box 1. The derivative to the conditional grant F delivers the Nash conditional grant F_{nash} and the utilities $U_{L\text{nash}}$ and $U_{G\text{nash}}$, as in Equation (22.8). The output X of the municipal project is determined through Equation (22.5).

Box 1.
Negotiation of conditional grant according to Nash

Utility of state:

$$\text{Max!}U_L = g_{XL}X - g_{FL}F \tag{22.1}$$

where g_{XL} is the value of activity or investment unit X to the state and g_{FL} the value of one unit of grant F to the state.

Utility of municipality:

$$\text{Max!}U_G = (a - bX)X + g_{FG}F \tag{22.2}$$

where a, b are the parameters of evaluation of activities X and g_{FG} the value of one unit of grant F to the municipality.

For both actors a set of indifference curves exist. Solutions of negotiations are related to a sequence of tangency points of indifference curves of the state and municipality.

For an indifference curve of the state holds:

$$\mathrm{d}U_L = \frac{\partial U_L}{\partial X}\mathrm{d}X + \frac{\partial U_L}{\partial F}\mathrm{d}F = g_{XL}\mathrm{d}X - g_{FL}\mathrm{d}F \overset{!}{=} 0 \tag{22.3}$$

and for that of the municipality analogously:

$$\mathrm{d}U_G = \frac{\partial U_G}{\partial X}\mathrm{d}X + \frac{\partial U_G}{\partial F}\mathrm{d}F$$

$$= (a - 2bX)\mathrm{d}X + g_{FG}\mathrm{d}F \overset{!}{=} 0 \qquad (22.4)$$

The conditions (22.3) and (22.4) denote the identity of the Pareto-solution:

$$\frac{\mathrm{d}F}{\mathrm{d}X} = \frac{g_{XL}}{g_{FL}} = -\frac{a - 2bX}{g_{FG}}, \qquad \text{or}$$

$$X_{\text{Pareto}} = \frac{g_{XL}\dfrac{g_{FG}}{g_{FL}} + a}{2b} \qquad (22.5)$$

The utilities along the identity of the Pareto-solution are given by

$$U_L = g_{XL}X_{\text{Pareto}} - g_{FL}F,$$

$$U_G = (a - bX_{\text{Pareto}})X_{\text{Pareto}} + g_{FG}F, \qquad (22.6)$$

while $X_{\text{Pareto}} = \dfrac{g_{XL}\dfrac{g_{FG}}{g_{FL}} + a}{2b} = \text{constant}$

After the substitution the grant F in Equation (22.6), we have the following frontier of the utility distribution between the state and the municipality:

$$U_L = -\frac{g_{FL}}{g_{FG}}U_G + \left[g_{XL} + \frac{g_{FL}}{g_{FG}}(a - bX_{\text{Pareto}})\right]X_{\text{Pareto}}$$

$$= -\frac{g_{FL}}{g_{FG}}U_G + \Phi, \qquad (22.7)$$

while $\Phi = \dfrac{g_{FL}}{g_{FG}} \dfrac{\left(g_{XL}\dfrac{g_{FG}}{g_{FL}} + a\right)^2}{4b}$

To derive the negotiation solution we maximise the Nash product (NP) under the constraint of the utility frontier (22.7):

$$\text{Max!}NP = (U_L - U_{L\text{Min}})(U_G - U_{G\text{Min}}) \quad \text{subject to}$$

$$U_L = -\frac{g_{FL}}{g_{FG}} U_G + \Phi$$

where $U_{L\text{Min}}$ is the minimal utility level of the state and $U_{G\text{Min}}$ the minimal utility level of the municipality.

Using the Lagrange method, we obtain the Nash solution (22.8):

$$L = (U_L - U_{L\text{Min}})(U_G - U_{G\text{min}}) + \lambda\left(\Phi - U_L - \frac{g_{FL}}{g_{FG}} U_G\right) \tag{22.8}$$

where L is Lagrange function and λ the Lagrange multiplier.

Hence

$$\frac{\partial L}{\partial U_L} = U_G - U_{G\text{Min}} - \lambda = 0, \qquad \lambda = U_G - U_{G\text{Min}}$$

$$\frac{\partial L}{\partial U_G} = U_L - U_{L\text{Min}} - \lambda \frac{g_{FL}}{g_{FG}} = 0,$$

$$\lambda = \frac{g_{FG}}{g_{FL}}(U_L - U_{L\text{Min}})$$

$$\frac{\partial L}{\partial \lambda} = \Phi - U_L - \frac{g_{FL}}{g_{FG}} U_G = 0$$

Consequently

$$U_{L\text{Nash}} = \frac{U_{L\text{Min}} - \frac{g_{FL}}{g_{FG}} U_{G\text{Min}} + \Phi}{2}$$

$$\text{F} = \frac{U_{G\text{Min}}}{2g_{FG}} - \frac{U_{L\text{Min}}}{2g_{FL}} + \frac{\left(g_{XL}\cdot\frac{g_{FG}}{g_{FL}} + a\right)\cdot\left(3g_{XL}\cdot\frac{g_{FG}}{g_{FL}} - a\right)}{8b\cdot g_{FG}} \tag{22.9}$$

$$U_{G\text{Nash}} = \frac{U_{G\text{Min}} - \frac{g_{FG}}{g_{FL}} U_{L\text{Min}} + \frac{g_{FG}}{g_{FL}} \Phi}{2}$$

In the interval

$$-g_{XL}\frac{g_{FG}}{g_{FL}} < a < 3g_{XL}\frac{g_{FG}}{g_{FL}}$$

the last part of Equation (22.9) is positive. If the parameter b rises, the grant F decreases. The maximum of F will be reached at $a = g_{XL}(g_{FG}/g_{FL})$. Therefore, the grant increases in case of $a < g_{XL}(g_{FG}/g_{FL})$, and it descends, when the parameter of evaluation of X a decreases beyond $g_{XL}(g_{FG}/g_{FL})$. In regard to a price-demand function as $a - bX$, when the municipal activities X expand, the state is willing to support these, as long as the net advantage $g_{XL}U_L - g_{FL}F$ increases. If the willingness of payment for the municipal activities is sufficiently large, then the financial situation of the municipality is stable with or without the state grant. In this case the grant F will be lower.

With respect to conditional grants there are four policies to protect municipalities from losing autonomy in self-government and finance. Apart from relating conditional grants more strongly to the connection principle as a first approach, the power structure and the resource allocation between the state and municipalities could be altered. Instead of principal-agent situations where the high power of the state is expressed through simple principal-agent model formulations, the state should be forced to negotiate conditional grants in a second attempt, thus, increasing municipalities' financial autonomy. More decision power could be given to municipalities in regional and urban planning and/or by municipal participation in formulating public utility functions with respect to conditional grants and the activities and investments concerned, e.g. through participation in decision making of municipalities on the state level. Another, third, approach would be to establish rules for co-operative decision making between the state and municipalities on some projects and public activities. A fourth policy would be to apply a general rule of parallel development of fiscal capacities between the state and municipalities for determining the amounts of unconditional grants in fiscal vertical equalisation.

22.3. Principle of parallelism to prevent fiscal autonomy through unconditional grants

22.3.1. Definition of the principle of parallelism

The principle of parallel development of fiscal capacity between a state and its municipalities comprises a guideline to determine unconditional grants from the state to the municipalities (Nam *et al.*, 2001). In Germany, it is legally implemented in the Free State of Saxony (§2 SächsFAG, law of fiscal equalisation). The state of Brandenburg tries to follow this principle as well (Grundlach, 2000, p. 10). According to this principle, the total amount of the state grants to municipalities is annually fixed, however, in a far limited way. There should be a parallel development of the municipalities' disposable income from local taxes plus the provided intergovernmental transfers by the state and the disposable income from the tax income of the state and the grants from the federal government minus the above-mentioned grants from the state to the municipalities (Nam and Pasche, 2001, p. 11). Essential for the delineation of this principle is the term parallelism. Does parallelism require a one-to-one relation or can another relation serve as well?

The one-to-one relation implies that the public tasks financed related to the fiscal capacities are of same importance regardless of the state and the municipalities manifested in an utility function concerning the state tasks and the municipal tasks. A parallel equal development of utility changes, however, does not necessarily lead to a one-to-one relation in changes of expenditures and revenues. Such utility functions do not yet exist. In reality, there are no binding political statements or juridical rules stipulating, in federal or state constitutions, that state tasks are of the same importance as municipal ones, although there exists a responsibility of a state to finance and influence the local tasks and a guarantee of self-administration of municipalities (Kirchhof, 2002) – to some extent, suggesting the priority of state tasks against municipality tasks.

Moreover, a one-to-one relation implies that the size of tasks of both levels of the government is the same. In terms of expenditures this obviously does not hold, as mentioned above, with respect to investment outlays. If future-oriented investments are of high priority they should also be considered as tasks. Then we might end up with another kind of parallelism. The relation may be three- or two-to-one favouring municipalities. A priori, one cannot be explicit on the size of this relation. A public body, which has the power to determine such a relation according to constitution, must make decisions on this matter.

In Germany, state governments and state parliaments create equalisation funds to be distributed to municipalities. The fund is collected from the state share of joint taxes, grants from the central government, revenues from state taxes, contributions from municipalities as well as from the finances from the horizontal fiscal relations between the states. These sources are fixed within a state law concerning the intergovernmental fiscal relations between a state and its municipalities. As long as the sources and the law do not change, there is a relation between the fiscal capacity of the state and the volume of the equalisation funds. However, no direct link to the total fiscal capacities of the municipalities exists. The fiscal capacity of municipalities is considered with respect to the distribution of the equalisation funds, but not when determining the size of the equalisation funds. Normally, state decision makers do not consider a parallel development of fiscal capacity.[6] The kind of parallelism chosen will depend on the aims of state politicians and municipality-oriented party members in the parliament.

Other difficulties stem from the definitions of fiscal capacities on the state and municipal level. The definition should express disposable income of the state and that of municipalities. Disposable income may include the cash flow that is at the disposal of state government.[7] Therefore, the question is whether disposable state income should be defined as all revenues minus inevitable expenditures, as is the case with cash flow. One related problem is whether grants to the municipalities are parts of these inevitable

---

[6] There is a principal-agent problem between the state and the municipalities involved which is solved and negotiated in formulating the state law of intergovernmental fiscal equalisation.

[7] Because many of the revenues are already blocked by expenditures that cannot be changed and there are juridical obligations unchangeable in the short run (or invariable), expenditures are necessary to be executed by federal (and EU) laws. For governments in Germany such a measure is called as 'free top' (freie Spitze and freie Spanne in Bavaria).

expenditures. As states must allow municipalities to participate in their tax receipts, there must be some inevitable grants to the municipalities. However, the share of these grants is not fixed by the constitution.

What revenues should be included in disposable income? A state receives financial revenues from shared taxes, purpose-oriented, conditional federal grants, unconditional grants through vertical fiscal equalisation between the federation and states, grants received through the horizontal equalisation among the states and contributions from municipalities.[8] Additionally a state has revenues from fees, sanctions, borrowings, sales of state property, profits of state enterprises, etc. What part of these revenues is disposable income? In the literature on intergovernmental fiscal relations, authors tend to refer to fiscal capacity considering only with regard to tax revenues instead of disposable income, including the items suggested above.[9]

Does a constant relation between the two indicators imply an appropriate form of parallel relationship? In the case of an agreement achieved about the size of parallelism (see Section 22.3.2.1), the self-administrative tasks of the state government and

---

[8] For example, municipalities have to transfer a part of the business tax receipts to the state.

[9] One solution to define financial municipal disposability is again to refer to the cash flow (free top = freie Spitze and freie Spanne) of the municipalities, including all the revenues mentioned above minus inevitable existing expenditure needs. As both levels of government can raise public credits and should be responsible for their own projects and tasks of self-administration, an exclusion of public debt makes sense. Service fees are considered primarily according to the benefit principles of the users or the cost covering principle. Therefore, they are not at the government's disposal and should be deducted. Profits from state enterprises reflecting the willingness to pay of clients for service should be not considered, either. The sale of public property increases the disposable revenues within the period of sale. Only if the receipts are used to decrease public debt in terms of lowering interest payments and capital service does the disposable income increase. Conditional grants refer to special services or projects related to task performance of high priority to the central state. It reflects again a willingness to pay and cost covering. Unconditional grants from other governments increase the municipal financial scope of action, thus increasing available fiscal means. Sanctions from the EU for non-fulfilment of the Maastricht criteria have to be deducted insofar the state is responsible. Therefore, it makes a sense to restrict fiscal capacity to those revenues that are without any equivalent value related to task performance. This reasoning calls for the deduction of grants to municipalities to finance municipal tasks. Therefore, disposable income means fiscal capacity from sources without services in exchange, etc.

municipal governments may not develop in the same way.[10] According to the game between the state between the municipalities and the empirical findings of Friedrich *et al.* (2003) a higher priority should be given to municipal self-administrative tasks to compensate private, European, national or state policy failures. Municipalities are also confronted with effects of natural disasters, war, population developments as well as economic developments induced by external trade, currency exchange rates, EU regulations, etc., which call for changing the size of parallelism. In general, the principle of equal fiscal development may conflict with the goal achievements of the state, individual municipalities and/or all municipalities as a group.[11] Hopefully the application of the parallel development principle will allow a satisfactory goal fulfilment of the state and municipal decision makers.

In some states the equalisation funds are split between unconditional grants and investment grants. The finances dedicated to investments, i.e. conditional grants, have to be separated from the equalisation funds. A stable form of parallelism implies that some necessary variations of grants have to take place through conditional grants. This can lead to an additional loss of autonomy of municipalities. If the principle of connection is applied when shifting tasks between different levels of government, conditional grants should be used for compensations to avoid impacts on the application of the principle of parallel fiscal development. The protagonists of the parallelism principle assume implicitly that the competition among municipalities is not influenced in an undesired way.

The specification of the principle of parallelism can be interpreted as a result of contract negotiations according to the Nash solution between the state and all municipalities, if the municipalities as a whole act as one player. A parallelism of relation one results out of a Nash solution if the minimum utility restrictions of the players do not exist and a symmetric utility distribution of the negotiators prevails with respect to disposable fiscal capacity. At least an

---

[10] According to development of the economy and social conditions the tasks of state, e.g. science, education or internal security, have become more important than those of municipalities, such as fair grounds or historical monument conservation.

[11] Regional, urban and environmental planning should not be restricted.

approximate solution should be found. In the case of the minimum utilities as guaranteed by the constitution or other utility distributions, the game may lead to another parallelism relation.

For a constant relation of parallelism, the selection of the starting relation is rather difficult. One attempt is to find a new split solution through negotiations and the other one is to select a base year in which the redistribution is considered to be fair and satisfactory. As politicians disagree on the adequate allocation of financial means on state and municipalities (Karrenberg and Münstermann, 1999, p. 207), no ideal solution will be identified. However, a relation might be chosen comprising a year where the fiscal stress of municipalities and the state was relatively low. When choosing the parallelism, the effects of parallelism should be taken into account.

22.3.2. *Analysis of the principle of parallelism*

22.3.2.1. *Basic model for one type of municipalities*

To analyse the effects of the principle of fiscal development, we refer to a simplified vertical fiscal equalisation system between a German state and its municipalities. A first type of parallelism named vertical parallelism deals with the parallel fiscal development between a state and its municipalities. A second type tackles the horizontal parallelism between municipalities or categories of municipalities through vertical unconditioned grants.

We turn first to the vertical parallelism. An expenditure need indicator is defined as $VE_i b_i GB$. It refers primarily to the weighted size of population VE_i of a municipality.[12] An amount per capita GB[13] is multiplied by the adjusted size of population and weighted again with a factor b_i, which shows the importance of different

---

[12] There is a major and additional weighting system (Haupt- und Nebenansatz) to determine a need indicator related to population. With the major weighting the number of inhabitants are multiplied by a factor larger than one, which increases with the population size of municipalities. The additional weighting adds changes to this population indicator depending on the number of pupils, central place functions, etc. An adjusted number of inhabitants (veredelte Einwohnerzahl) results.

[13] GB shows the basic needs per inhabitant, which is the same state-wide. GB is fixed by the state through simulation in such a way as to distribute an amount fixed by the state in such a way that municipalities where needs indicator receive are higher than tax indicators receive grants and the fixed amount (Schlüsselmasse) becomes exhausted.

towns i.[14] Moreover, a tax capacity indicator $SK_{i,t}$ is taken into account.[15] The unconditional grant $SZ_{i,t}$ is calculated by multiplying the difference between the indicators mentioned above by an equalisation ratio AS. This is shown in Equation (22.10).

$$SZ_{i,t} = AS(VE_i b_i GB - SK_{i,t}) \tag{22.10}$$

where $SZ_{i,t}$ refers to the transfers from the state to a municipality i ($i \in (1, \ldots, n)$) for a given year t (*Schlüsselzuweisungen*), AS the equalisation ratio (*Ausgleichsatz*), VE_i the weighted number of inhabitant of the municipality i (*Gesamtansatz*), GB the basic need per capita (*Grundbetrag*), $SK_{i,t}$ the tax capacity of the municipality i at the year t (*Steuerkraftmesszahl*) and b_i the factor of horizontal parallelism related to groups of municipalities.

The total sum of grants paid by the state to municipalities amounts to SZ_t as suggested in Equation (22.11). This equation is also used to determine the basic need per capita GB as shown by solving Equation (22.11) for GB, which leads to Equation (22.12). Inserting the result concerning GB into Equation (22.10) delivers the amount of unconditional grants paid to municipality i in period t (see Equation (22.13)).

$$SZ_t := \sum_j SZ_{j,t} = AS\left(\sum_j (VE_j b_j) GB - \sum_j SK_{j,t}\right) \tag{22.11}$$

$$(j = 1, \ldots, n)$$

and

$$GB = \frac{\dfrac{SZ_t}{AS} + \sum_j SK_{j,t}}{\sum_j (VE_j b_j)} \tag{22.12}$$

---

[14] This matters in the concept of horizontal parallelism through vertical unconditional grant allocation. The factor b_i is related to groups of municipalities. In most German states b_i, ($i = 1, \ldots, n$) turns out to be 1. To abstract from problems of horizontal competition we assume $b_i = 1$ for $i = 1, \ldots, n$ in our analysis of the vertical parallelism.

[15] The tax capacity index refers to individual tax capacity indices of local taxes (e.g. business tax, real estate tax, municipal share of income tax). These tax capacities are calculated by using state-wide normative tax rates.

Hence

$$SZ_{i,t} = AS\left(VE_i b_i \frac{\frac{SZ_t}{AS} + \sum_j SK_{j,t}}{\sum_j (VE_j b_j)} - SK_{i,t} \right) \tag{22.13}$$

Parallel development concerns municipalities' revenues from local taxes EG_t and the provided intergovernmental transfers (by the state) SZ_t, on one hand, and the state revenues from the (exclusive and shared) taxes and the grants from the federal government (to the state) EL_t minus the grants from the state to municipalities SZ_t, on the other. The size of the intergovernmental transfers is fixed in the period of 0 ($t = 0$) at a certain percentage share of the income of the state.

$$\frac{EG_t + SZ_t}{EG_{t-1} + SZ_{t-1}} = \frac{EL_t - SZ_t}{EL_{t-1} - SZ_{t-1}} \tag{22.14}$$

where EG_t the tax income of municipalities, EL_t the tax income of the state and intergovernmental transfers (from other states and the federal government) to the state at the fiscal year t).
Consequently

$$\frac{EG_t + SZ_t}{EL_t - SZ_t} = \frac{EG_{t-1} + SZ_{t-1}}{EL_{t-1} - SZ_{t-1}} = \cdots = \frac{EG_0 + SZ_0}{EL_0 - SZ_0} \tag{22.15}$$

Re-arranging equation (22.14) yields:

$$SZ_t = EL_t\left(\frac{EG_{t-1} + SZ_{t-1}}{EG_{t-1} + EL_{t-1}} \right) - EG_t\left(\frac{EL_{t-1} - SZ_{t-1}}{EG_{t-1} + EL_{t-1}} \right) \tag{22.16}$$

According to the principle of parallel development of fiscal capacity between state and municipalities one finds for period t

$$\begin{aligned} EG_t + SZ_t &= (EG_{t-1} + SZ_{t-1})\left(\frac{EL_t + EG_t}{EL_{t-1} + EG_{t-1}} \right) \\ &= \cdots = (EG_0 + SZ_0)\left(\frac{EL_t + EG_t}{EL_0 + EG_0} \right) \end{aligned} \tag{22.17}$$

$$\begin{aligned} EL_t - SZ_t &= (EL_{t-1} - SZ_{t-1})\left(\frac{EL_t + EG_t}{EL_{t-1} + EG_{t-1}} \right) \\ &= \cdots = (EL_0 - SZ_0)\left(\frac{EL_t + EG_t}{EL_0 + EG_0} \right) \end{aligned} \tag{22.18}$$

We re-arrange Equations (22.17) and (22.18) for the period $t-1$:

$$EG_{t-1} + SZ_{t-1} = (EG_0 + SZ_0)\left(\frac{EL_{t-1} + EG_{t-1}}{EL_0 + EG_0}\right),$$

$$EL_{t-1} - SZ_{t-1} = (EL_0 - SZ_0)\left(\frac{EL_{t-1} + EG_{t-1}}{EL_0 + EG_0}\right)$$

and insert these expressions into Equation (22.16), thus obtaining Equation (22.19).

Therefore, the parallelism is expressed by $(EG_0 + SZ_0)/(EL_0 - SZ_0)$, which is termed the size of parallelism. By considering Equation (22.14) one finds:

$$SZ_t = EL_t\left(\frac{EG_0 + SZ_0}{EL_0 + EG_0}\right) - EG_t\left(\frac{EL_0 - SZ_0}{EL_0 + EG_0}\right) \tag{22.19}$$

Turning to the unconditional grants of one municipality we insert Equation (22.18) into Equation (22.13), then

$$SZ_{i,t} = AS\left(VE_i b_i \frac{\dfrac{\left(EL_t\left(\dfrac{EG_0 + SZ_0}{EL_0 - SZ_0}\right) - EG_t\right)\dfrac{EL_0 - SZ_0}{EL_0 + EG_0}}{AS} + \sum_j SK_{j,t}}{\sum_j (VE_j b_j)} - SK_{i,t}\right) \tag{22.20}$$

Equation (22.20) shows the way in which the principle of parallelism is introduced into the model of vertical fiscal equalisation between the state and municipalities – here with one municipality.

22.3.2.2. *Consideration of income and population changes*

How would changes of tax revenues alter the provision of unconditional grants to that municipality? This question is tackled under the following assumptions. An income change $YG_{i,t}$ in municipality i in period t changes local tax revenues of municipality i in period t by $EG_{i,t}(YG_{i,t})$. This leads to a variation of the tax capacity indicator of that community $SK_{i,t}(EG_{i,t}(YG_{i,t}))$. However, the income change in municipality i may lead to income variations $\sum_j YG_{j,t}$ in j other communities as well. This causes a change in revenues of the state $EL_t(\sum_j YG_{j,t})$ and a j change in the tax capacities of all respective municipalities $\sum_j SK_{j,t}(EG_{j,t}(YG_{j,t}))$. Moreover, we assume that incomes vary with the population change $EGi,t := EGi,t(YGi,t(VEi,t))$ and $EL_t := EL_t(\sum_j YG_{j,t}(VE_{j,t}))$. Local revenues depend on income achieved in this community. Moreover, the state revenues are related to incomes in all municipalities, which are functions of population size.

The formula of unconditional grants of one municipality i (Equation (22.20)) turns as follows:

$$SZ_{i,t} = \left(\frac{EG_0 + SZ_0}{EL_0 - SZ_0} \frac{VE_{i,t} EL_t\left(\sum_j YG_{j,t}(VE_{j,t})\right)}{\sum_j VE_{j,t}} - \frac{VE_{i,t} \sum_j EG_{j,t}(YG_{j,t}(VE_{j,t}))}{\sum_j VE_{j,t}} \right) \times \frac{EL_0 - SZ_0}{EG_0 + EL_0} + AS \frac{VE_{i,t} \sum_j SK_{j,t}(EG_{j,t}(YG_{j,t}(VE_{j,t})))}{\sum_j VE_{j,t}} - AS \cdot SK_{i,t}(EG_{i,t}(YG_{i,t}(VE_{i,t}))) \qquad (22.21)$$

Differentiating Equation (22.21) with respect to the weighted population size $VE_{i,t}$ of the ith municipality leads to the following first-order condition:

$$\frac{\partial SZ_{i,t}}{\partial VE_{i,t}} = \left(\underbrace{\frac{EG_0 + SZ_0}{EL_0 - SZ_0}}_{\text{Constant of parallelism}} \frac{\left(1 EL_t\left(\sum_j YG_{j,t}(VE_{j,t})\right) + VE_{i,t} \frac{\partial EL_t}{\partial YG_{i,t}} \frac{\partial YG_{i,t}}{\partial VE_{i,t}}\right) \sum_j VE_{j,t} - \left(VE_{i,t} EL_t\left(\sum_j YG_{j,t}(VE_{j,t})\right)\right) 1}{\left(\sum_j VE_{j,t}\right)^2} + \frac{\left(1 \sum_j EG_{j,t}(YG_{j,t}(VE_{j,t})) + VE_{i,t} \frac{\partial EG_{i,t}}{\partial YG_{i,t}} \frac{\partial YG_{i,t}}{\partial VE_{i,t}}\right) \sum_j VE_{j,t} - \left(VE_{i,t} \sum_j EG_{j,t}(YG_{j,t}(VE_{j,t}))\right) 1}{\left(\sum_j VE_{j,t}\right)^2} \right) \frac{EL_0 - SZ_0}{EG_0 + EL_0} + AS \frac{\left(1 \sum_j SK_{j,t}(EG_{j,t}(YG_{j,t}(VE_{j,t}))) + VE_{i,t} \frac{\partial SK_{i,t}}{\partial EG_{i,t}} \frac{\partial EG_{i,t}}{\partial YG_{i,t}} \frac{\partial YG_{i,t}}{\partial VE_{i,t}}\right) \sum_j VE_{j,t} - \left(VE_{i,t} \sum_j SK_{j,t}(EG_{j,t}(YG_{j,t}(VE_{j,t})))\right) 1}{\left(\sum_j VE_{j,t}\right)^2} - AS \frac{\partial SK_{i,t}}{\partial EG_{i,t}} \frac{\partial EG_{i,t}}{\partial YG_{i,t}} \frac{\partial YG_{i,t}}{\partial VE_{i,t}} \qquad (22.22)$$

An increase in the weighted number of inhabitants of a municipality leads to a reduction of unconditional grants only if the condition stated below related to the parallelism size is fulfilled:

$$
\frac{\partial SZ_{i.t}}{\partial VE_{i.t}} < 0 \Leftrightarrow 0 < \underbrace{\frac{EG_0 + SZ_0}{EL_0 - SZ_0}}_{\text{Constant of parallelism}}
$$

$$
< \underbrace{\frac{\left(\sum_j EG_{j.t}(YG_{j.t}(VE_{j.t})) + VE_{i.t}\frac{\partial EG_{i.t}}{\partial YG_{i.t}}\frac{\partial YG_{i.t}}{\partial VE_{i.t}}\right)\sum_j VE_{j.t} - VE_{i.t}\sum_j EG_{j.t}(YG_{j.t}(VE_{j.t}))}{\left(EL_t\left(\sum_j YG_{j.t}(VE_{j.t})\right) + VE_{i.t}\frac{\partial EL_t}{\partial YG_{i.t}}\frac{\partial YG_{i.t}}{\partial VE_{i.t}}\right)\sum_j VE_{j.t} - VE_{i.t}EL_t\left(\sum_j YG_{j.t}(VE_{j.t})\right)}}_{>0}
$$

$$
- AS\frac{EG_0 + EL_0}{EL_0 - SZ_0}
$$

$$
\times \left(\underbrace{\frac{\sum_j SK_{j.t}(EG_{j.t}(YG_{j.t}(VE_{j.t})))\sum_j VE_{j.t} - VE_{i.t}\sum_j SK_{j.t}(EG_{j.t}(YG_{j.t}(VE_{j.t})))}{\left(EL_t\left(\sum_j YG_{j.t}(VE_{j.t})\right) + VE_{i.t}\frac{\partial EL_t}{\partial YG_{i.t}}\frac{\partial YG_{i.t}}{\partial VE_{i.t}}\right)\sum_j VE_{j.t} - VE_{i.t}EL_t\left(\sum_j YG_{j.t}(VE_{j.t})\right)}}_{>0} \right.
$$

$$
\left. + \underbrace{\frac{VE_{i.t}\frac{\partial SK_{i.t}}{\partial EG_{i.t}}\frac{\partial EG_{i.t}}{\partial YG_{i.t}}\frac{\partial YG_{i.t}}{\partial VE_{i.t}}\sum_j VE_{j.t} - \frac{\partial SK_{i.t}}{\partial EG_{i.t}}\frac{\partial EG_{i.t}}{\partial YG_{i.t}}\frac{\partial YG_{i.t}}{\partial VE_{i.t}}\left(\sum_j VE_{j.t}\right)^2}{\left(EL_t\left(\sum_j YG_{j.t}(VE_{j.t})\right) + VE_{i.t}\frac{\partial EL_t}{\partial YG_{i.t}}\frac{\partial YG_{i.t}}{\partial VE_{i.t}}\right)\sum_j VE_{j.t} - VE_{i.t}EL_t\left(\sum_j YG_{j.t}(VE_{j.t})\right)}}_{<0} \right)
\tag{22.23}
$$

This condition depicts two effects. On one hand, an increase of the weighted number of inhabitants leads to higher unconditional grants. On the other hand, income increases caused by the expansion of population size lead to income growth of firms and households. Consequently tax capacity grows reducing the payments of unconditional grants. The larger one of the two effects determines whether the amount of unconditional grants grows or shrinks, and whether fiscal autonomy is expanded or not.

A more complicated case occurs when a change of income in one municipality causes changes in income in other municipalities as well, which depend on the original income variation of municipality k.

This is expressed by conditions $EG_{i,t} := EG_{i,t}(YG_{i,t}(YG_{k,t}))$ and $EL_t := EL_t(\sum_j YG_{j,t}(YG_{k,t}))$. Again the unconditional grants are differentiated in terms of the income change $YG_{k,t}$. An extended derivation is made and re-arranged as shown in Equation (22.24).

$$\frac{\partial SZ_{i,t}}{\partial YG_{k,t}} = \frac{VE_i}{\sum_j VE_j}\left(\sum_j \frac{\partial EL_t}{\partial YG_{j,t}}\frac{\partial YG_{j,t}}{\partial YG_{k,t}}\frac{EG_0+SZ_0}{EL_0-SZ_0} - \sum_j \frac{\partial EG_{j,t}}{\partial YG_{j,t}}\frac{\partial YG_{j,t}}{\partial YG_{k,t}}\right)\frac{EL_0-SZ_0}{EL_0+EG_0}$$
$$+AS\left(\frac{VE_i}{\sum_j VE_j}\sum_j \frac{\partial SK_{j,t}}{\partial EG_{j,t}}\frac{\partial EG_{j,t}}{\partial YG_{j,t}}\frac{\partial YG_{j,t}}{\partial YG_{k,t}} - \frac{\partial SK_{i,t}}{\partial EG_{i,t}}\frac{\partial EG_{i,t}}{\partial YG_{i,t}}\frac{\partial YG_{i,t}}{\partial YG_{k,t}}\right) \tag{22.24}$$

We obtain a similar condition as gained before. The interpretation is also similar to that above. However, sums of revenue changes of state and municipalities play a role in this case. A further expansion considers that the population varies due to the income changes, e.g. through migration, etc. The basic relation for the unconditional grants now becomes more complicated. After differentiation and re-arrangement of terms one obtains

$$0 < \frac{EG_0+SZ_0}{EL_0-SZ_0}$$
$$< \underbrace{\frac{\left(\frac{\partial VE_{i,t}}{\partial YG_{i,t}}\sum_j EG_{j,t}(YG_{j,t}) + VE_{i,t}(YG_{i,t})\frac{\partial EG_{i,t}}{\partial YG_{i,t}}\right)\sum_j VE_{j,t}(YG_{j,t}) - \left(VE_{i,t}(YG_{i,t})\sum_j EG_{j,t}(YG_{j,t})\right)\frac{\partial VE_{i,t}}{\partial YG_{i,t}}}{\left(\frac{\partial VE_{i,t}}{\partial YG_{i,t}}EL_t\left(\sum_j YG_{j,t}\right) + VE_{i,t}(YG_{i,t})\frac{\partial EL_t}{\partial YG_{i,t}}\right)\sum_j VE_{j,t}(YG_{j,t}) - \left(VE_{i,t}(YG_{i,t})EL_t\left(\sum_j YG_{j,t}\right)\right)\frac{\partial VE_{i,t}}{\partial YG_{i,t}}}}_{>0} - AS\frac{E_0}{EL_{0,v}}$$
$$\times\left(\underbrace{\frac{\frac{\partial VE_{i,t}}{\partial YG_{i,t}}\sum_j SK_{j,t}(EG_{j,t}(YG_{j,t}))\sum_j VE_{j,t}(YG_{j,t}) - \left(VE_{i,t}(YG_{i,t})\sum_j SK_{j,t}(EG_{j,t}(YG_{j,t}))\right)\frac{\partial VE_{i,t}}{\partial YG_{i,t}}}{\left(\frac{\partial VE_{i,t}}{\partial YG_{i,t}}EL_t\left(\sum_j YG_{j,t}\right) + VE_{i,t}(YG_{i,t})\frac{\partial EL_t}{\partial YG_{i,t}}\right)\sum_j VE_{j,t}(YG_{j,t}) - \left(VE_{i,t}(YG_{i,t})EL_t\left(\sum_j YG_{j,t}\right)\right)\frac{\partial VE_{i,t}}{\partial YG_{i,t}}}}_{>0}\right.$$
$$\left.+\underbrace{\frac{VE_{i,t}(YG_{i,t})\frac{\partial SK_{i,t}}{\partial EG_{i,t}}\frac{\partial EG_{i,t}}{\partial YG_{i,t}}\sum_j VE_{j,t}(YG_{j,t}) - \frac{\partial SK_{i,t}}{\partial EG_{i,t}}\frac{\partial EG_{i,t}}{\partial YG_{i,t}}\left(\sum_j VE_{j,t}(YG_{j,t})\right)^2}{\left(\frac{\partial VE_{i,t}}{\partial YG_{i,t}}EL_t\left(\sum_j YG_{j,t}\right) + VE_{i,t}(YG_{i,t})\frac{\partial EL_t}{\partial YG_{i,t}}\right)\sum_j VE_{j,t}(YG_{j,t}) - \left(VE_{i,t}(YG_{i,t})EL_t\left(\sum_j YG_{j,t}\right)\right)\frac{\partial VE_{i,t}}{\partial YG_{i,t}}}}_{<0}\right) \tag{22.25}$$

This indicates two effects. The expressions, which have to be larger than the parallelism size, comprise three terms. The first two terms are positive but the last one is negative. However, the second one is to be subtracted from the first term. Therefore, two terms decrease the expression, which in total has to be larger than the parallelism size if the municipality is going to lose unconditional grants from an income change. A higher income can lead to an increase in population size (e.g. more jobs, more pupils and families). This fact will again lead to an increase of the needs indicator and the growth of unconditional grants; however, the increase in income raises tax capacities of municipalities, which has a counter effect on unconditional grants. Which of the terms determines the fiscal autonomy depends on the sizes and distributions of the effects.

If only income in one municipality varies the condition (22.23) reduces to

$$\frac{\partial SZ_{i,t}}{\partial YG_{i,t}} < 0 \Leftrightarrow 0 < \underbrace{\frac{EG_0 + SZ_0}{EL_0 - SZ_0}}_{\text{Constant of parallelism}} < \underbrace{\frac{\dfrac{\partial EG_{i,t}}{\partial YG_{i,t}}}{\dfrac{\partial EL_t}{\partial YG_{i,t}}}}_{>0} + \underbrace{\frac{-AS\left(\dfrac{VE_i}{\sum_j VE_j}\dfrac{\partial SK_{i,t}}{\partial EG_{i,t}}\dfrac{\partial EG_{i,t}}{\partial YG_{i,t}} - \dfrac{\partial SK_{i,t}}{\partial EG_{i,t}}\dfrac{\partial EG_{i,t}}{\partial YG_{i,t}}\right)}{\dfrac{VE_i}{\sum_j VE_j}\dfrac{EL_0 - SZ_0}{EL_0 + EG_0}\dfrac{\partial EL_t}{\partial YG_{i,t}}}}_{>0} \tag{22.26}$$

In order to avoid a reduction in unconditional grants, that means lowering fiscal autonomy, a high degree of parallelism and a high positive reaction of state revenues would be necessary if the first term on the right-hand side is positive.

The change of unconditional grants refers only to one municipality, although the size of population for other municipalities varies as well. Normally, as a consequence of changing numbers of inhabitants the total state population undergoes changes as also the total amount of unconditional grants. This is expressed in Equation (22.27) where the individual population changes and their effects on the total amount of unconditional grants are depicted.

$$\mathrm{d}SZ_{i,t} = \frac{\partial SZ_{1,t}}{\partial VE_{1,t}}\mathrm{d}VE_{1,t} + \cdots + \frac{\partial SZ_{n,t}}{\partial VE_{n,t}}\mathrm{d}VE_{n,t} \tag{22.27}$$

Equation (22.28) shows how the unconditional grants vary if the total population changes.

$$\frac{\mathrm{d}SZ_{i,t}}{\mathrm{d}VE_t} = \frac{\partial SZ_{1,t}}{\partial VE_{1,t}}\frac{\mathrm{d}VE_{1,t}}{\mathrm{d}VE_t} + \cdots + \frac{\partial SZ_{n,t}}{\partial VE_{n,t}}\frac{\mathrm{d}VE_{n,t}}{\mathrm{d}VE_t}, \tag{22.28}$$

where

$$VE_t = \sum_j VE_{j,t}$$

The effects are greatly influenced by the share of population change in the individual municipalities. Again the derivatives found above play a role. According to the degree of parallelism some of the derivatives show negative or positive values. Thus the degree of parallelism is decisive for the distribution and absolute change of total unconditional grants.

22.3.2.3. *Effects of changing parallelism*

The effects of varying the parallelism can be considered by introducing a variable α. It shows the level of unconditional grants in the basic year, thus determining the degree of parallelism. The relation (22.12) for the unconditional grants changes to Equation (22.29).

$$SZ_{i,t} = AS\left(VE_{i,t}(YG_{i,t})\frac{\dfrac{\left(EL_t\left(\sum_j YG_{j,t}\right)\dfrac{EG_0+\alpha SZ_0}{EL_0-\alpha SZ_0} - \sum_j EG_{j,t}(YG_{j,t})\right)\dfrac{EL_0-\alpha SZ_0}{EL_0+EG_0}}{AS} + \sum_j SK_{j,t}(EG_{j,t}(YG_{j,t}))}{\sum_j VE_{j,t}(YG_{j,t})} - SK_{i,t}(EG_{i,t}(YG_{i,t}))\right) \tag{22.29}$$

What happens to the unconditional grants, if α is increased? Differentiation with respect to α yields Equation (22.30):

$$\frac{\partial SZ_{i,t}}{\partial \alpha} = \frac{VE_{i,t}(YG_{i,t})}{\sum_j VE_{j,t}(YG_{j,t})} \left(EL_t \left(\sum_j YG_{j,t} \right) + \sum_j EG_{j,t}(YG_{j,t}) \right) \times \frac{SZ_0}{EL_0 + EG_0} > 0 \tag{22.30}$$

As the first-order condition is positive, the unconditional grants grow if the parallelism is changed in favour of municipalities and decreases if the state takes a bigger share of revenues by lowering α.

22.3.2.4. *Model considering two types of municipalities*

We express horizontal parallelism[16] by a constant b_i not equal to one and a number of inhabitants not normalised. The size of the different b_i determines how the unconditional grants are allocated to different types of municipalities. For the sake of simplicity we consider two types of municipalities: (1) district-free town authorities (*kreisfreie Städte*) such as large cities and counties, differ from (2) municipalities, which belong to districts (*kreisangehörige Gemeinden*).

The horizontal parallel development through vertical unconditional grants between the group of district-free towns (group 1) and

---

[16] When considering the horizontal parallelism through vertical unconditional grants we can take into account this principle in several ways. One is to keep b_i at value 1, but to consider different categories of municipalities within VE_i. This is the usual way that the German states tackle this problem. Through formula (22.3) a normalised GB, which is equal for all municipalities is found, which depends in determining its size on the weighting system and the kind of vertical parallelism. This approach comprises horizontal parallelism as mentioned. The relations of importance of municipalities among each other are fixed through the artificial number of inhabitants. From $SZ_{i,t} = AS(VE_i b_i GB - SK_{i,t})$ follows $(SZ_{it} + SK_{it})/VE_i = GB$ if b_i and AS equal 1. As GB is the same for all municipalities, the relation of unconditional grants and tax capacity per normalised inhabitant is the same for all municipalities. Therefore, also the relation shown in Equation (22.31) holds and turns out with value 1.

the group of municipalities belonging to districts (group 2) requires that the following relation holds:

$$\frac{\dfrac{SK_{1,t}}{VE_{1,t}} + \dfrac{SZ_{1,t}}{VE_{1,t}}}{\dfrac{SK_{1,0}}{VE_{1,0}} + \dfrac{SZ_{1,0}}{VE_{1,0}}} = \frac{\dfrac{SK_{2,t}}{VE_{2,t}} + \dfrac{SZ_{2,t}}{VE_{2,t}}}{\dfrac{SK_{2,0}}{VE_{2,0}} + \dfrac{SZ_{2,0}}{VE_{2,0}}} \tag{22.31}$$

$SK_{1,t}$ shows the current (exogenous) tax capacity of municipalities belonging to districts at year t. $SZ_{1,t}$ expresses the current down-flow unconditional grants to the group of municipalities belonging to districts, which is determined according to their administrative rank, $SK_{2,t}$ symbolises the current (exogenous) tax capacity of district-free towns at year t, and $SZ_{1,t}$ the current down-flow of unconditional grants to the district-free group of district-free towns. $VE_{1,t}$ ($VE_{2,t}$) depicts the number of inhabitants at year t in the group of municipalities belonging to district-free towns.

When the conditions (22.10) and (22.31) are satisfied and b_1/b_2 is constant, then the term

$$\frac{\dfrac{SK_{1,t}}{VE_{1,t}} + \dfrac{SZ_{1,t}}{VE_{1,t}}}{\dfrac{SK_{2,t}}{VE_{2,t}} + \dfrac{SZ_{2,t}}{VE_{2,t}}} = \frac{b_1}{b_2} \tag{22.32}$$

expresses the relation of horizontal parallelism between the municipality groups 1 and 2. The additional condition

$$SZ_{1,t} + SZ_{2,t} = SZ_t \tag{22.33}$$

should also be satisfied for all periods, where SZ_t means the total sum of inter-governmental unconditional grants from the state to municipalities at t, as before. This condition ensures that the vertical parallelism is considered together with the horizontal parallelism.

Integrating Equation (22.32) into Equation (22.33) we can endogenously determine the size of down-flow unconditional grants to the individual municipality groups (or in the case of one municipality of type 1 and one of type 2) at t:

$$SZ_{1,t} = \frac{\frac{b_1}{b_2} VE_{1,t} SZ_t - \left(VE_{2,t} SK_{1,t} - \frac{b_1}{b_2} VE_{1,t} SK_{2,t} \right)}{\frac{b_1}{b_2} VE_{1,t} + VE_{2,t}},$$
$$SZ_{2,t} = \frac{VE_{2,t} SZ_t + \left(VE_{2,t} SK_{1,t} - \frac{b_1}{b_2} VE_{1,t} SK_{2,t} \right)}{\frac{b_1}{b_2} VE_{1,t} + VE_{2,t}} \tag{22.34}$$

What happens to the unconditional grants, if b_1/b_2 is increased? Differentiation with respect to b_1/b_2 yields:

$$\frac{\partial SZ_{1,t}}{\partial \left(\frac{b_1}{b_2} \right)} = \frac{(VE_{1,t} SZ_t + VE_{1,t} SK_{2,t}) \left(\frac{b_1}{b_2} VE_{1,t} + VE_{2,t} \right) - \left(\frac{b_1}{b_2} VE_{1,t} SZ_t - VE_{2,t} SK_{1,t} + \frac{b_1}{b_2} VE_{1,t} SK_{2,t} \right) VE_{1,t}}{\left(\frac{b_1}{b_2} VE_{1,t} + VE_{2,t} \right)^2}$$
$$= \frac{VE_{1,t} VE_{2,t} (SZ_t + SK_{1,t} + SK_{2,t})}{\left(\frac{b_1}{b_2} VE_{1,t} + VE_{2,t} \right)^2} > 0 \tag{22.35}$$

$$\frac{\partial SZ_{2,t}}{\partial \left(\frac{b_1}{b_2} \right)} = \frac{(-VE_{1,t} SK_{2,t}) \left(\frac{b_1}{b_2} VE_{1,t} + VE_{2,t} \right) - \left(VE_{2,t} SZ_t + VE_{2,t} SK_{1,t} - \frac{b_1}{b_2} VE_{1,t} SK_{2,t} \right) VE_{1,t}}{\left(\frac{b_1}{b_2} VE_{1,t} + VE_{2,t} \right)^2}$$
$$= \frac{-VE_{1,t} VE_{2,t} (SZ_t + SK_{1,t} + SK_{2,t})}{\left(\frac{b_1}{b_2} VE_{1,t} + VE_{2,t} \right)^2} < 0 \tag{22.36}$$

The unconditional grants vary in favour of type of the municipality, which benefits from a high consideration in parallelism through its factor b. If we analyse the effects of vertical and horizontal parallelism on fiscal autonomy (the amount of unconditional grants), all relevant equations are expanded by an expression $b_i/\sum b_j$ considering horizontal parallelism as well. The conclusions about changes vary in size but the direction of changes remains same. The parallelism affects municipal fiscal autonomy differently.

22.4. Conclusions

The development of local public finance during the last decade demonstrates that (1) central government interventions with fiscal consequences have intensified, (2) territorial and administrative reforms caused financial restructuring burdens, (3) tax reforms disturbed municipal finance, (4) western European countries rapidly increased local expenditures for social tasks, (5) municipalities tried budget consolidation, and (6) municipalities experienced a reduction in fiscal autonomy. In addition, transformation played a role in countries like Poland and Germany.

In order to ensure fiscal autonomy some serious attempts appear to be necessary to protect local governments within intergovernmental fiscal relations in addition to the tax reforms favouring municipalities. In this context the connection principle is examined which aims at preventing municipalities from shifting tasks from higher to lower level governments without providing sufficient financial means. Although the fiscal compensation can be made in terms of new taxes, fees, transfer of property, simplifying the credit restrictions, etc., the most appropriate one seems to be conditional grants.

In many countries conditional current and investment grants are important. Therefore, the fiscal autonomy of municipalities can be effectively protected through the specification of conditional grants, although higher level governments control the grant conditions. Goals of higher government levels are achieved in a principal-agent situation where the higher government functions as the principal and the municipal government as the agent. The usual principal-agent solutions will be found as long as the higher government level can keep the municipality at its minimum utility level (e.g. the minimum level of goal realisation). From a negotiation model between a higher level government and a municipality as players, a Nash solution can also be found that demonstrates the size and conditions of the conditional grants. Thus political recommendations are formulated to increase municipal power with respect to conditional grants, when planning and co-operative decision making.

The implementation of the principle of fiscal parallelism between higher level governments and municipalities with respect to unconditional grants appears to be another option to safeguard local fiscal autonomy. Yet, it seems to be rather problematic to find the optimal relation between the development of fiscal capacity of a state and that of the municipalities. Further weaknesses concern the size, constancy and determination of the parallelism relation in the case of unforeseen developments, etc. The effects of such kind of parallelism on the fiscal autonomy of a municipality have not yet been discovered. Therefore, a model of intergovernmental fiscal relation through unconditional grants is extended under the consideration of parallelism. For example, it must be determined whether local fiscal autonomy is enhanced if tax revenues vary under the condition of a parallelism change. With a large size of parallelism and a considerably high increase in state revenues, the amount of unconditional grants and fiscal autonomy increase. In a similar context, various additional conditions are also elaborated for the cases of changing population size, income and the parallelism relation itself. The kinds and size of considering the parallelism depend on the interplay between population, income tax revenue movements on the municipality and state level and on the size of the parallelism. This is true for unconditional grants adopted for the vertical fiscal equalisation as well as for the horizontal one through the vertically provided unconditional grants made under the consideration of priorities for some groups of municipalities.

References

Ali, A.I., C. Lerme and R.A. Nakosteen (1993), "Assessment of intergovernmental revenue transfers", *Socio-economic Planning Science*, Vol. 27, pp. 109–118.

Arnold, V. and O.E. Geske (1988), *Öffentliche Finanzwirtschaft*, Munich: Vahlen.

Bahl, R. and J. Linn (1994), "Fiscal decentralisation and intergovernmental transfers in less developed countries", *Publius: The Journal of Federalism*, Vol. 24, pp. 1–19.

Bewley, T.F. (1981), "A critique on Tiebout's theory of local expenditures", *Econometrica*, Vol. 49, pp. 713–740.

Blankart, C.B. and R. Borck (2004), *Local Public Finance: A Survey, Discussion Paper, Economic Series, 154*, Berlin: Wirtschaftswissenschaftliche Fakultät der Humboldt-Universität Berlin.

Boadway, R.W. and A.R. Hobson (1993), *Intergovernmental Fiscal Relations in Canada, Canadian Tax Paper, 96*, Ottawa: Canadian Tax Paper Foundation.

Borodo, A. (2003), "Aktuelle Probleme der Kommunalfinanzen Polens angesichts der Reformen von 1990 und 1999", pp. 39–55, in: P. Friedrich and J.W. Tkaczynski, editors, *Business Promotion Activities in Poland and Germany (Wirtschaftsförderung in Polen und Deutschland)*, Torun: TNOiK.

Bradford, D.F. and W.E. Oates (1971), "The analysis of revenue sharing in a new approach to collective fiscal decisions", *Quarterly Journal of Economics*, Vol. 85, pp. 416–439.

Brueckner, J.K. (2003), "Strategic interaction among governments: an overview of empirical studies", *International Regional Science Review*, Vol. 26, pp. 175–188.

Bull, H.P. and F. Welti (1996), "Schwachstellen der geltenden Finanzverfassung", *Neue verwaltungswissenschaftliche Zeitschrift*, Vol. 16, pp. 836–846.

Council of Europe (2000), *Effects on Financial Autonomy of Local and Regional Authorities Resulting from the Limits Set at European Level on National Public Debt, Series of Local and Regional Authorities in Europe 71*, Strasbourg: Council of Europe Publishing.

Coutry, P. and G. Marschk (2003), "Dynamics of performance measurement systems", *Oxford Review of Economic Policy*, Vol. 19, pp. 268–284.

Dafflon, B. (ed.) (2002), "Local public finance in Europe", *Balancing the Budget and Controlling Debt*, Cheltenham: Edward Elgar.

Dahlby, B.G. (1996), "Fiscal externalities and the design of intergovernmental grants", *International Tax and Public Finance*, Vol. 3, pp. 397–412.

Davis, A. and R. Lucker (1982), "The rich-state–poor-state problem in a federal system", *National Tax Journal*, Vol. 35, pp. 337–363.

De Spindler, J. (2003), *FOCJ-Ein Konzept zur Neuordnung der Zusammenarbeit öffentlichrechtlicher Gebietskörperschaften*, Bern: Hauptverlag.

Detig, S.X., Feng and P. Friedrich (2002), "FOCJ als Grundlage Förderinstitutionen, Aufbau-Ost und Bevölkerung-Ost", pp. 82–115, in: P. Friedrich, editor, *Öffentliche Unternehmen im Standortwettbewerb für den Aufbau Ost, Diskussion Paper 41*, Munich: (Lehrstuhl für Finanzwissenschaft) University of Federal Armed Forces.

Döring, T. and D. Stahl (1999), *Räumliche Aspekte der föderalen Aufgabenverteilung, der Finanzverfassung und der Subventionspolitik in der Bundesrepublik Deutschland*, Hannover: Akademie für Raumforschung und Landesplanung.

Drennan, M.P. and D. Netzer (eds.) (1997), *Readings in State and Local Public Finance*, Oxford: Blackwell.

Fisher, R.C. (1979), "The theoretical view of revenue sharing grants", *National Tax Journal*, Vol. 32, pp. 173–184.

Fisher, R.C. (1996), *State and Local Public Finance*, Chicago: Irwin.

Frey, B.S. and R. Eichenberger (1999), *The New Democratic Federalism for Europe, Functional, Overlapping and Competing Jurisdictions*, Cheltenham: Edward Elgar.

Frey, R.L. and A. Schlategger (2003), "Finanzausgleich und Föderalismus: zur Neugestaltung der föderalen Ausgleichsbeziehungen am Beispiel der Schweiz", *Perspektiven der Wirtschaftspolitik*, Vol. 4, pp. 239–258.

Friedrich, P. and X. Feng (2002), "The role of public institutions in regional competition", pp. 79–113, in: G. Atalik and M.M. Fischer, editors, *Regional Development Reconsidered*, Berlin: Springer.

Friedrich, P., J. Gwiazda and C.W. Nam (2003), *Development of Local Public Finance in Europe*, *CESifo Working Paper 1107*, Munich.

Fukasaku, K. and L.R. de Mello (eds.) (1999), *Fiscal Decentralisation in Emerging Economies*, Paris: OECD Development Centre.

Gage, R.W. and M.P. Mandell (1990), *Strategies for Managing Intergovernmental Policies and Networks*, New York: Praeger.

Gravelle, H. and R. Rees (1992), *Microeconomics*, London: Longman.

Grout, P.A. and M. Stevens (2003), "The assessment: financing and managing public services", *Oxford Review of Economic Policy*, Vol. 19, pp. 215–234.

Grundlach, U. (2000), "Die Kommunen im steuer- und abgabenstaat", *Landes- und Kommunalverwaltung*, Vol. 10, pp. 7–13.

Hamilton, B.W. (1975), "Zoning and property taxation in a system of local governments", *Urban Studies*, Vol. 12, pp. 205–211.

Hedkamp, G. (2000), *Report by the Congress of Local and Regional Authorities (CLRAE) in Europe on Local Finance in Europe*, Strasbourg: Council of Europe Publishing.

Henneke, H.G. (2003), "Darf der Bund bestellen, wenn er bezahlt, oder soll der Bund bezahlen müssen, wenn er bestellen darf", *Landkreistag*, Vol. 73, pp. 137–148.

Henneke, H.G. and I. Vorholz (2002), "Anwendungsrelevanz des strikten Konnexitätsprinzip am Beispiel des AG-BSHG Bbg", *Landkreis Verband*, Vol. 12, pp. 297–304.

Hines, J.R. and R.H. Thaler (1995), "Anomalies: the flypaper effect", *Journal of Economic Perspectives*, Vol. 9, pp. 217–226.

Hohrmann, F. (1967), *Bundesgesetzliche Organisation landesunmittelbarer Selbstverwaltungskörperschaften*, Berlin: Duncker & Humblot.

Hyman, D.N. (1993), *Public Finance: A Contemporary Application of Theory to Policy*, 4th edition, Fort Worth: The Dryden Press, Harcourt Brace College Publishers.

Jones, P.R. and J.G. Cullis (1994), "Bureaucracy and intergovernmental grants: a comment", *Kyklos*, Vol. 47, pp. 437–448.

Karrenberg, H. and E. Münstermann (1999), *Gemeindefinanzbericht, Der Städtetag*, Stuttgart: Kohlhammer.

Keen, M. (1998), "Vertical tax externalities in the theory of fiscal federalism", *International Monetary Fund Staff Papers*, Vol. 45, pp. 454–485.

King, D.N. (1984), *Fiscal Tiers: The Economics of Multi-Level Government*, London: Allen & Unwin.

Kirchhof, P. (2002), "Die Reform der kommunalen Finanzausstattung", *Neue juristische Wochenzeitschrift*, Vol. 55, pp. 1549–1553.

Klein, O. (2003), *Das verfassungsrechtliche Konnexitätsprinzip*, Potsdam: Kommunalwissenschaftliches Institut der Universität Potsdam.

Ma, J. (1997), *Intergovernmental Fiscal Transfer: A Comparison of Nine Countries (Cases of the United States, Canada, the United Kingdom, Australia, Germany, Japan, Korea, India and Indonesia), The World Bank Policy Research Working Paper 1822,* Washington DC.

Matinez-Vazquez, J. and J. Alm (eds.) (2003), *Public Finance in Developing and Transitional Countries: Essays in Honour of Richard Bird*, Cheltenham: Edward Elgar.

McLure, C.E. (1986), "Tax competition: is what's good for the private goose also good for the public gander?", *National Tax Journal*, Vol. 34, pp. 341–348.

Miezkowski, P. (1987), "Urban economics", pp. 756–762, in: J. Eatwell, M. Milgate and P. Newman, editors, *The New Palgrave Dictionary of Economics*, London: McMillan.

Musgrave, R.A. (1959), *The Theory of Public Finance*, New York: McGraw-Hill.

Musgrave, R.A. and P.B. Musgrave (1980), *Public Finance in Theory and Practice*, New York: McGraw-Hill.

Nam, C.W. and R. Pasche (2001), "Municipal finance in Poland, the Slovak Republic, the Czech Republic and Hungary: institutional framework and recent development", *MOCT-MOST Economic Policy in Transitional Economies*, Vol. 11, pp. 113–134.

Nam, C.W., R. Pasche and M. Steinherr (2001), "The principles of parallel development of state and municipalities as useful benchmarks for the determination of intergovernmental grants in Germany", *European Planning Studies*, Vol. 9, pp. 525–537.

Oates, W.E. (1972), *Fiscal Federalism*, New York: Harcourt Brace Jovanovich.

Oates, W.E. (1999), "An essay on fiscal federalism", *Journal of Economic Literature*, Vol. 37, pp. 1120–1149.

Paugam, A. (1999), *Ad Valorem Property Taxation and Transition Economics, ECSIN Working Paper 9*, Washington, DC: World Bank.

Pola, G. (ed.) (1996), *Development in Local Government Finance: Theory and Policy*, Cheltenham: Edward Elgar.

Rehm, H. and S. Matern (2003), *Kommunale Finanzwirtschaft*, Frankfurt am Main: Lang.

Rosen, H.S. (1995), *Public Finance*, 5th edition, Homewood, IL: Richard D. Irwin.

Schneider, H.P. (1998), "Nehmen ist seliger als Geben, Oder: Wieviel Föderalismus verträgt der Bundesstaat", *Neue juristische Wochenzeitschrift*, Vol. 51, pp. 3757–3759.

Schoch, F. (2000), "Die Dogmatik zum finanzverfassungsrechtlichen Schutz der kommunalen Selbstverwaltung", *Archiv für Kommunalwissenschaften*, Vol. 39, pp. 225–242.

Scotchmer, S. (1994), "Public goods and the invisible hand", pp. 93–119, in: J.M. Quigley and E. Smolensky, editors, *Modern Public Finance 4*, Amsterdam: Elsevier.

Shah, A. (1994), *The Reform of Intergovernmental Fiscal Relations in Developing and Emerging Market Economies, The World Bank Policy and Research Series 23*, Washington, DC.

Smith, P. (2003), "Formula funding of public services: an economic analysis", *Oxford Review of Economic Policy*, Vol. 19, pp. 301–322.

Tiebout, C.M. (1956), "A pure theory of local expenditures", *Journal of Political Economy*, Vol. 64, pp. 416–424.

Trapp, G. (1997), *Das Veranlassungsprinzip in der Finanzverfassung der Bundesrepublik Deutschland*, Berlin: Duncker & Humblot.

Trasberg, V. (2003), "Local governments' fiscal situation in the Baltic states", pp. 38–52, in: J. Mønnesland, editor, *Regional Public Finances*, London: Pion.

Tullock, G. (1969), "Federalism: problems of scale", *Public Choice*, Vol. 6, pp. 19–29.

Wildasin, D.E. (1986), *Urban Public Finance*, Amsterdam: Harwood Academic Publishers.

Winkler, D.R. (1994), *The Design and Administration of Intergovernmental Transfers: Fiscal Decentralisation in Latin America, The World Bank Discussion Papers 235*, Washington, DC.

Zimmermann, H. (1999), *Kommunalfinanzen*, Baden-Baden: Nomos.

Urban Dynamics and Growth, CEA, vol. 266
Roberta Capello and Peter Nijkamp (editors)

CHAPTER 23

Urban Quality of Life and Public Policy: A Survey

Gordon Mulligan[a], John Carruthers[b] and Meagan Cahill[a]

[a]University of Arizona, Tucson, AZ 85721, USA
[b]Mundy Associates LLC, Seattle, WA 98109, USA

Abstract

This chapter assesses the recent multidisciplinary research on urban quality of life. Amenities, or site-specific benefits, drive the relocation of households and firms at both the interurban and intraurban scales. Consequently, cities or neighborhoods can be rank-ordered by their levels of amenities. Quality of life is known to affect business location decisions and to play an increasing role in local economic development plans. Urban deprivation is also addressed and its relationships with crime are briefly explored. Using the perspective of real estate capitalization, the debate between growth control and growth management is summarized.

Keywords: amenities, hedonic models, business location, local economic development, deprivation, land-use planning, growth control, growth management

JEL classifications: D10, D20, H30, I30

23.1. Introduction

Assessing quality of life (QOL) and determining its effects on human behavior are increasingly important topics in the social sciences (Dissart and Deller, 2000). Here we broadly interpret QOL as the satisfaction that a person receives from surrounding human and physical conditions. These conditions, which are scale-dependent, affect both household and firm behavior. QOL, used interchangeably with other terms like human or social well-being, is known both to influence and constrain a person's life opportunities (Smith, 1977, 1994).

Special attention is given to amenities, which are site- or region-specific goods and services that make localities attractive (or unattractive) for living and working (Diamond and Tolley, 1982; Power, 1996). These enter into the consumption or production decisions of households and firms and therefore play a crucial role in their location choices. Amenities were once considered to be largely natural; however, human-made amenities are now of greater interest for reasons of public policy. Economic agents have different levels of access to amenities depending upon the locations they occupy and the resources they command.

Some thoughtful reviews of the urban QOL literature are already available. Those by Bartik and Smith (1987) and Gyourko *et al.* (1999) are especially well regarded for their analytical insights about the diverse roles of amenities in urban life. However, in both cases the empirical material is largely confined to the American experience. Moreover, in both cases the perspectives and methodologies are drawn heavily from economics and the approaches of the other social sciences, especially geography and planning, are mostly ignored. A clear need presently exists for a multidisciplinary summary of the growing international literature on urban QOL.

In the first half, amenities are discussed at the *interurban* scale. Different approaches are used to rank-order cities by their QOL or economic health. Specific amenities (crime, pollution) are then discussed and their impacts on urban wage and growth models are assessed. Hedonic models capturing household migration and firm relocation are next examined. The role of amenities in business location decisions and the emergence of QOL concerns in local economic development (LED) initiatives are then discussed. In the second half, various issues are examined at the *intraurban* scale. Neighborhood or community deprivation is summarized before brief attention is given to its relationships with urban crime. A much longer discussion ensues about urban land use, beginning with the notion of amenity capitalization in the real estate market. As before, hedonic models are examined that allow households and firms to relocate, but now in response to local differences in amenities. Metropolitan land-use policies are then summarized from both the supply- and demand-side perspectives. The chapter concludes with a summary of the debate raging between the advocates of growth control and the advocates of growth management.

23.2. Interurban scale

23.2.1. City rankings

23.2.1.1. Indicator analysis

The seminal study of US cities is Thorndike (1939) who assessed QOL using a wide variety of single-variable indicators grouped into six categories. Hadden and Borgatta (1965) and Berry and Kasarda (1977) then followed with multivariate analyses that uncovered similar geographic patterns. Places with the highest socioeconomic scores were usually wealthy suburbs of large, coastal cities; cities with the highest deprivation scores were often smaller places found throughout the nation's South, especially near to and along the US–Mexico border.

Evidently the study by Liu (1976), which is now badly dated, remains the most popular of this genre. He rated more than 240 metropolitan areas in 1970 using five general categories of well-being – economic health, political performance, environmental conditions, health and education, and social concerns. The raw data were standardized and added so that cities, once assigned to three size-groups, could be designated as substandard, adequate, good, excellent, or outstanding. Critics have focused mainly on the arbitrariness of his methodology, including the choice of variables and the implicit weighting scheme.

Since the 1980s indicator analysis has generally fallen into disfavor among US academics, but the approach has been widely embraced by the various media. A series of six catalogs published between 1981 and 2000, called the *Places Rated Almanac*, is especially well known. US (and now Canadian) city rankings are quoted widely and the *Almanac's* copious data are regularly used in more sophisticated academic studies (see below). In the most recent edition, data are provided for more than 350 metropolitan areas across nine general categories and a variety of subcategories

(Savageau and D'Agostino, 2000). The most highly ranked cities tend to be large in population size and frequently occupy coastal locations. Canadian cities compare well against their American counterparts, presumably because they have fewer weaknesses in crime, public transportation, and health care. But Landis and Sawicki (1988) severely criticized findings of the 1985 *Almanac*, suggesting that place-wide factors are actually secondary to personal considerations in determining one's QOL. After evaluating the various categories for reliability, double counting, ecological fallacy, and other concerns, they largely dismissed the rankings as not being very relevant to planners. And the *Almanacs* do not assess the incidence of poverty (Madden, 2003), which is known to vary from one city to the next. In any case, Becker *et al.* (1987) and Burnell and Galster (1992) have argued that the *Almanac* methodology biases QOL scores in favor of larger cities and each recommends caution when using the rankings.

Other media attention has focused on QOL in America's smaller cities (Thomas, 1990; Heubusch, 1997). In fact, nearly 565 micropolitan areas, or *emerging* metropolitan areas, have just been officially recognized by the US Census Bureau (Plane *et al.*, 2002). Each of these county-based areas focuses on a high-density population cluster and ranges in size between 10,000 and 50,000 persons. An unofficial sample of these places has been shown to exhibit remarkable economic diversity (Vias *et al.*, 2002), although the underlying reasons for their continued growth remain unclear (see below). College and university communities tend to dominate the micropolitan QOL rankings.

Indicator-based studies seem to be more highly regarded by academics outside of the US. In a thoughtful series of papers, Rogerson *et al.* (1988) have examined Britain's largest cities and Morris *et al.* (1989) have examined Britain's intermediate-sized cities. Crime, health care, pollution, and the like were first identified as the main concerns and then, based on a national opinion survey, weights were computed so that age or race groups could express different preferences (Rogerson *et al.*, 1989). But only the means and not the errors are disclosed for these weights, so we have little idea about the dimensional variability in these preferences. Crime, of both the violent and non-violent variety, seems to have an especially large effect on the final rankings. Cities in the periphery performed well, thereby raising questions about the conventional view of Britain's North–South divide, at least in the 1980s. However, some highly ranked cities performed very poorly on critical indicators like crime and wage levels. This was believed to be problematic for development prospects in the periphery because professional respondents and footloose firms usually rank these two factors among the highest on their list of concerns (see below).

However, even survey-informed indicator studies are plagued by several methodological problems. First, measurement of the weights presents a serious issue. Evidence suggests that QOL preferences are not universal; instead they depend upon, among other things, one's gender, socioeconomic status, and position in the life cycle. Consequently, as any society's wealth or demography changes, its overall preferences will drift as well. This confounds interregional cross-sectional comparisons and casts suspicion on the findings of national longitudinal studies. Second, even the best indicator studies typically fail to present choices or alternatives to respondents. Happiness is very difficult to gauge and paradoxes emerge regarding the importance of relative vs. absolute income. Amenities compensate for income and therefore must be addressed similarly. Two persons facing the same crime rate likely will rank the importance of crime differently if one faces a high rate while living in a low-rate region but the other faces a low rate while living in a high-rate region. Finally, preferences can shift due to habituation as recently relocated or impacted persons adjust to their new circumstances. This research stream clearly needs more input from various disciplines, especially psychology and sociology.

23.2.1.2. Discriminant analysis

Informed by earlier work (Hall and Hay, 1980), this approach has been widely applied to the study of European cities in order to classify places according to their various problems, such as inequality and deprivation. Cheshire *et al.* (1986) focused on over 100 of the largest metropolitan areas in the EEC countries, using the functional urban region (FUR) as the appropriate geographic unit. FURs typically have a population exceeding one-third of a million persons, are defined in terms of employment cores and commuting hinterlands, and tend to be relatively self-contained units. After some experimentation QOL variables were adopted in four broad areas – per capita income, unemployment, net migration, and travel demand. Actual scores were identified for 29 cities (the training set) deemed unhealthy or healthy by EEC experts. A stepwise discriminant function was next estimated, scores were generated, and cities were rank-ordered on economic health. Cities with severe problems were largely those specializing in textiles and the heavy industries of the 19th century, but port cities and places engaged in automobile manufacturing also performed poorly. Scores were generated at three different points in time during 1971–1984 and FURs were classified into four broad types: deteriorating at a decreasing rate, deteriorating at an increasing rate, improving at a decreasing rate, or improving at an increasing rate.

Cheshire (1990) then revisited the topic and estimated a new discriminant function using more observations. The new rank positions were compared to the earlier ones and regional patterns were noted in the rank changes. The change in the health score, 1971–1988, was next estimated by a multiple regression model. Larger places generally outperformed smaller places and economic integration was shown to be beneficial. Dependence on either port- or coal-based activities, or the wider economy on agriculture, was shown to be unhealthy. Case histories were devoted to success stories and inevitable comparisons were made between national capitals. Improvement was evident in the cities of the Netherlands, Britain, and Belgium; deterioration was evident in the cities of Spain and France.

Later findings are available in Cheshire (1995) and Cheshire and Carbonaro (1996). In the first paper, new evidence is disclosed of a break with the past. Significant recentralization is shown to have taken place in many large cities, generally throughout northern Europe but especially in Germany. These settlement changes appear very similar to those reported earlier for US metropolitan areas. Comparable reasons are given, including the decline of manufacturing and the rise of high-level services, changes in household composition and rising female labor force participation, and the increasing importance of urban amenities and public transport. A positive feedback model, of the Myrdal or Kaldor variety, is briefly sketched to summarize these recent growth patterns. In the second paper, interest turns to estimating 1979–1990 growth rates in per capita income. A research and development variable, indicative of urban knowledge spillovers, proves highly significant in explaining differential rates among the FURs. Evidence is cited that highly skilled workers locate to enjoy either the general amenities of larger cities or the particular amenities of smaller cities. Policies to promote human capital growth in lagging areas will doubtless promote convergence in city growth rates across Europe (Tondl, 2001).

A more recent application of discriminant analysis can be found in Stimson *et al.* (2001) who examine community opportunity and vulnerability across Australia's metropolitan regions. Using an array of variables, 240 different statistical local areas (SLAs) are analyzed during the time period 1986–1996. Groups of similar communities are generated through hierarchical clustering and then stepwise discriminant procedures are used for parsimonious identification. Groups range from the most advantaged places (e.g. large-city core areas and their affluent suburbs) with high income opportunity to the most disadvantaged places (e.g. old extractive and manufacturing centers) with high income vulnerability. Human capital/affluence, or socioeconomic status, is by far the most prominent discriminating function. Australia, like other postindustrial societies, is experiencing polarization across many dimensions, including human capital. As in Europe and America, that nation's most highly educated and highly motivated individuals increasingly prefer to work and live in high-amenity locations, thereby reinforcing those disparities already existing between healthy and unhealthy places.

This stream of research is somewhat reminiscent of the numerous data-driven ecological studies that were widely carried out in urban geography during the 1960s and 1970s. Nevertheless, methodological improvement is seen on two fronts. First, more attention has certainly been given to identifying the attributes of problematic or vulnerable areas. And second, Cheshire's work at least addresses the very broad causal mechanisms underlying changes in city health. However, other research is needed that more fully incorporates geographic dependency, economic interdependency, fast behavioral processes, and slow institutional processes into interurban health models. Moreover, it remains unclear how useful these models are for predicting socioeconomic vulnerability.

23.2.1.3. Hedonic analysis

Although inspired by earlier work, urban hedonics really began with Rosen (1974, 1979) who imputed QOL differences among US metropolitan areas by using micro-level wage data. The novel contribution was to estimate implicit prices for location-specific amenities by regressing wages on QOL indicators, while controlling for household differences like education and race. Roback (1982, 1988) then extended this work by paying more attention to the production decisions of firms. She was concerned that land and labor markets were not fully articulated in earlier studies and argued that cross-city variations in housing prices must be examined because housing prices are known to rise with endowed amenities. Following Rosen, a number of regressions were run to explore the effects of different combinations of amenities. Cold weather, total snowfall, and frequency of cloudy days were all shown to have positive estimates, indicating they were disamenities. Clear, sunny days had positive estimates and were an amenity. Crime and particulate levels were also disamenities but their influence was not always significant. Regressions were next run on average residential site prices controlling for the amenities and various local characteristics (including density, growth, and unemployment). City-specific implicit prices for housing were then calculated. The marginal prices of amenities were then related to annual earnings with the adjustment made for housing costs. For example, the average person was shown to be willing to pay nearly an extra $70 per year for an additional sunny day. These prices were then used to impute differential QOL levels. Those cities with the highest QOL, including Los Angeles and San Francisco, were simply those places where aggregate imputed prices were the greatest. Four competing but correlated QOL indicators were introduced and the results were compared to those of Liu (1976). The two rankings were substantially different, presumably because Liu used somewhat different criteria and had entirely arbitrary weights. It is worth pointing out that both Rosen and Roback expressed caution about using their point estimates to make definitive city-to-city QOL comparisons.

Blomquist *et al.* (1988) then analyzed the variation of amenities both *within* and *across* urban areas. The observations were more numerous (some 250 counties across 185 cities), amenities were now at a more localized level, housing rent was measured for households and not for cross-city aggregates, and both wage and rent equations were updated to 1980. Signs on the wage, rent, and city-size equations depend on alternative assumptions regarding the effects of amenities on households and firms, and the effects of agglomeration. The model and its comparative statics are more complex but more realistic: now a change in the amenities of one county can induce a change in city size that in turn affects the behavior of firms in other contiguous counties. Renters and owners of housing were both considered by transforming house value to monthly rent. Climate (six variables), violent crime, teacher–pupil ratios in schools, pollution (six variables), and central city or coastal locations were included as 16 amenities. Housing and wage hedonic variables were estimated for both households and workers using a Box-Cox search procedure. Linearized estimates for the 16 components of overall QOL are disclosed. County-level units were then ranked by premiums above or below the average county. For example, individuals living in the worst-crime county were compensated a combined total of \$1600 in both markets over those individuals living in the county having the average rate of crime. The city-wide average and within-city variation of these premiums are both shown for a number of very large places. For instance, residents of metropolitan Chicago faced an average premium of −\$823, which ranged between −\$79 and −\$979 across four counties; but residents in metropolitan Denver faced an average premium of \$1198, which ranged between \$609 and \$2097 across five counties. An overall QOL ranking was first generated and then alternative rankings followed for three subsets of amenities. Significant differences occur in the various rankings (see above). Those places with the highest overall QOL were either small- or medium-sized in population and tended to be found in the Sunbelt. A less technical presentation is available in Berger *et al.* (1987), where overall city rankings are compared to those of Liu (1976) and the 1981 *Places Rated Almanac*. There is very little correlation among the three sets of rankings.

The next important paper is by Gyourko and Tracy (1991) who argue that local finance conditions can also generate compensating differences across urban land and labor markets. Worker utility now depends upon wages net of income taxes, land rentals including property taxes, the non-land cost of living including sales taxes, and the amenity and public goods package. Firm profits are modeled similarly. The amenity-public service package enters into the firm's indirect profit function. Annual housing expenditures and weekly wages are estimated using six climate variables, five other amenity variables (crime, education, pollution, city size, and coastal location), seven fiscal variables (hospital beds, fire rating, cost of living, three different tax rates, and the incidence of commuting to nearby cities), and public unionization. All 11 amenities are highly significant in both equations; the same is largely true for the fiscal variables. The implicit prices of the service proxies have the anticipated signs, the impact of public safety and health occur through the wage equation, and higher income taxes are perceived to be bad – with most of the compensation coming through the land market. The relative importance of amenities, taxes and public services, and unions in determining overall QOL compensation is determined. Amenities explain 39–43% of the variation in housing prices and 20–23% of the variation in wages, fiscal conditions explain 12–16% of the variation in housing prices and 20–21% of the variation in wages, but unionization has no discernible effect on either hedonic equation. City rankings are shown using four alternative models: random effects only, random effects combined with group effects, OLS with all fiscal variables, and OLS with no taxes and unions. Attempts are made to adjust the rankings for different household size; for example, San Francisco is shown to have a negative subsidy for retired persons but a positive subsidy for single earners. The authors are troubled that the random effects models have much higher standard errors than the OLS models, indicating that it is problematic to differentiate among closely ranked cities (see below). This issue of accuracy is revisited in Gyourko (1991).

Stover and Leven (1992) return to a narrow but important technical issue. Responding to a short discussion in Blomquist *et al.* (1988), the sensitivity of QOL rankings to functional form is analyzed. The full price for each amenity is now based on both direct and indirect impacts in land rentals, where the latter reflects feedback from the labor market. Estimates for the housing equation are similar but estimates for the wage equation shift substantially. However, the implicit price for each amenity is different from the earlier study because two equations are now estimated – a wage (housing) premium is used to adjust the housing (wage) estimates. As a result city rankings shift, in some cases considerably. Urban counties are rated in terms of quality, cost, and value. It is troubling that there exists so little correlation between the new QOL rankings and those given by in Blomquist *et al.* Caution is recommended in using one-dimensional rankings.

Kahn (1995) introduces a relatively straightforward revealed preference approach that does not require so much data on city-specific attributes. Several econometric advantages are suggested: reducing collinearity between many city-specific attributes (e.g. air quality and climate), minimizing endogeneity issues for some amenities (especially crime), and avoiding problems with excluded variables (especially local public goods). Kahn first estimates a 'price vector' for each city and then uses these to predict the behaviors of all agents. This price vector is based on city-specific employee wage (controlled for schooling and occupation) and housing rental (controlled for age and size) estimates. He then gauges the percentage of people in city *i* who, by moving to city *j*, can attain higher predicted wages and lower predicted rentals. QOL in city *j* is correlated negatively with this percentage. Using slightly different data sets, QOL in five major US cities is estimated in 1980 and 1990. San Francisco (Houston) is shown to have a high (low) QOL because a low (high) percentage of persons can raise wages and lower rentals by moving there. However, it is unclear whether this method can be easily adapted to a much larger size distribution of places for, as the author admits, households having different abilities are probably not entirely indifferent between cities.

Giannias (1997, 1998) proposes a somewhat different kind of hedonic model that allows one to identify how specific changes in variables will lead to shifts in QOL rankings. This model captures variation across census tracts in a small group of competing cities – the first study examines five large places in the US Midwest and the second study examines Canada's 13 largest cities. The idea is that residents are presented with a QOL distribution (in crime, climate, pollution, and housing) across two geographic scales. Two sets of estimates (ranks) are given; in one, housing attributes (i.e. size, age) vary across the cities but in the other attributes are held constant. In the Canadian example, rankings are dominated by three non-coastal, medium-sized cities. Again, it is unclear if this model can be successfully applied across a wide spectrum of population size classes.

Gyourko *et al.* (1999), in providing a summary and technical critique of most recent hedonic studies, point out that city-ranking studies now appear to be plagued by a lengthy list of methodological problems. They suggest this entire line of research is at a crossroads because large city-specific errors in the underlying wage and housing specifications (that were not apparent in the early OLS regression estimations) cannot be overcome with existing micro and macro data. The result is that the precision of QOL rankings will always be less than both researchers and policymakers would like. In fact they echo some of the same concerns raised earlier by Becker *et al.* (1987) for indicator-based studies. And both parties agree that differences are probably only valid between places in the upper and lower tails of the QOL distribution. The authors recommend that future research should target the intracity scale where more geographic detail regarding households and firms is available. In recent times, though, other researchers have adopted the hedonic approach to address more specific issues.

23.2.2. Other hedonic issues

23.2.2.1. Wage gaps

The literature on intercity wage differentials was revitalized during the 1990s. While this interest seems to have been mostly confined to the US, some attention was directed to the experiences of other nations (Eaton and Eckstein, 1997). These studies attempt to explain why nominal wages vary so much across places of different population size in regional or national settlement systems.

Beeson and Eberts (1989), using 1980 micro data, extended the standard model to include housing production and local non-traded goods. Quality-adjusted wage and housing expenditure equations were estimated. The former used controls for various occupations and for education, experience, employment status, gender, race, marital status, and unionization; the latter used controls for ownership, location, size, age, lot size, and a variety of attributes. Rent and wage differentials (i.e. residuals) were estimated for 35 metropolitan areas and then wages were adjusted for housing costs. The wage gap among cities was then decomposed into amenity (creating 40% of the difference) and productivity (60%) components. In many high-wage cities high productivity values for firms outweighed the low amenity values for households; however, in many low-wage cities low productivity values outweighed high amenity values. A few cities, like New York and San Diego, had very high wages because workers were productive *and* the city provided valued amenities. Beeson (1991) later revisited the data, giving special attention to the returns to education. Each additional year of schooling increased a city's hourly earnings by an average of 3.7%. Even though elasticities were generally low in the North Central region and high in the South, a broad Sunbelt–Snowbelt distinction in returns to schooling was seen as too crude. Various amenities (1981 *Places Rated Almanac*) were shown to have important effects on the estimates of wage returns to education. For example, high student–teacher ratios and high levels of crime unambiguously lowered a city's returns to schooling.

Rauch (1993) then differentiated between individual-level and city-level effects of both education and experience in the wage and rent equations. For individuals, various regressions suggested that personal education was 1.8 times as important as experience in affecting wages. But across cities, education was about seven times as important as experience in affecting wages. Largely unsuccessful attempts were made to relate city-level average education to other factors (e.g. federal research and development monies). And, using both wage and rental equations, education externalities were estimated to be about 10 times as important as experience externalities in affecting recent urban (total factor) productivity in the US.

Glaeser and Maré (2001), following Glaeser (1998), have claimed that the urban wage premium is actually comprised of two separate effects: the wage level and wage growth. Using four large data sets of urban workers, taken mostly from the 1980s and 1990s, they discounted the notion that the urban wage premium is related to the superior ability of cities to attract highly capable persons. Urban residents were shown to be not that much better endowed with human capital attributes than non-urban residents. Furthermore, once nominal wages were standardized for differential living costs there seemed to be little relationship between (real) wages and city size. One data set suggested that workers coming to a city experience an immediate wage gain and those who leave experience a significant wage loss, thus confirming the wage-level hypothesis. But two other data sets endorsed the alternative hypothesis, where city dwellers accumulate human capital or find more suitable jobs over relatively long periods of time. In this case, real wages grow steadily over time for urban workers, and when they leave cities they do not suffer severe wage declines. In any case there is plenty of evidence that age-earnings profiles are steeper in large metropolitan areas than elsewhere.

More recently, Adamson *et al.* (2002) have examined the relationships between education and city size in more detail. Like Glaeser and Mare, they make use of the National Longitudinal Survey of Youth, a panel data set having respondents aged between 23 and 36 years during the time period 1988–1993. Both the labor-supply and labor-demand equations are specified and equated at the margin, thereby allowing the urban scale (or density) effect to be associated with both positive marginal firm productivity and favorable household amenities. Two unresolved empirical questions are addressed: is there a skill bias to the relative wage gap and, if so, are the underlying mechanisms due more to supply factors or to demand factors in the labor market? The analysis looks at the full array of US metropolitan areas in 1990 but separates out the eight largest places as a special category. Statistical controls are included for worker experience and tenure and for various city attributes (including industry composition, six climate amenities, and four topographic types). Metropolitan workers are shown to have a nominal wage advantage over nonmetropolitan workers with respect to education, but this edge is shown to be smaller for well-educated persons. Skill-biased (consumption) amenities appear to overwhelm skill-biased agglomeration (production) economies in all sizes of cities. While the scale-related productivity advantages of large cities might eventually decline with size, these places will continue to enjoy skill-related productivity advantages because high-skilled workers prefer to reside in these large places (see below).

23.2.2.2. Wage curves

It is now commonly accepted that workers residing in low-amenity areas (or working in low-amenity industries) must be compensated with higher wages than workers living in high-amenity areas (working in high-amenity industries). Questions then arise about the relationship between the local unemployment rate and the local

QOL. One interesting perspective follows from the work of Blanchflower and Oswald (1994, 1995), who have examined the relationship between local (or industry) wages and local (industry) unemployment by analyzing micro data taken from 12 nations. Surprisingly, they uncover a downward-sloping wage curve that is nearly universal where the unemployment elasticity of pay is approximately -0.1. This finding runs counter to Harris-Todaro orthodoxy, which states that local wages are positively related to local unemployment. The authors address this finding by discussing the operation of local labor markets, recognizing the important role of unemployment insurance and public support programs in a world where labor mobility is not perfectly free. Presumably, their argument also holds true for the role of place-specific social capital, as discussed by Bolton (1992, 2002), Putnam (2000), and others.

Once migration is seen to be costly, firms and workers must view any move as an investment decision where expected net returns at the existing location are calculated and compared to returns at alternative locations (see below). So persons living in high-amenity areas will be more willing to accept periods of unemployment compared to workers living in low-amenity areas. This in turn suggests that regions with high levels of amenities should experience both lower wages and higher unemployment (Deller and Tsai, 1998). Recently, Deller and Zhing (2003) have adopted this perspective in a novel analysis of the effects of amenities across US counties. They estimate an amenity-adjusted wage curve, after introducing a wide array of natural and human-made features and controlling for schooling, access to medical care, poverty, local government expenditures, and demographic composition. The standard model is confirmed for all US counties in 1990, although evidence indicates the existence of some spatial dependency in the estimates. But the introduction of amenities appears to reverse the underlying relationship in the wage curve, which was an unexpected result. The approach seems promising for uncovering how QOL is related to wage levels and unemployment rates in cities of varying location, size, and complexity.

23.2.2.3. Urban growth

This literature begins with Glaeser *et al.* (1992), which leads into Glaeser (1994) and Glaeser *et al.* (1995). The last paper examines the diverse growth forces underlying some 200 large US cities (municipalities and metropolitan areas) over a recent 30-year period. The intention is to differentiate population growth experiences into various components – location, initial size and income, earlier growth, economic and demographic features, size and nature of government, and education of the labor force. The theory is informed by the work of Romer (1986) and Lucas (1988), on disembodied knowledge and human capital. The methodology follows the convergence modeling of Barro (1991) and Barro and Sala-i-Martin (1992). Individuals are assumed to have free migration across cities, so that constant utilities across space are achieved at a point in time. Utility equals wages adjusted by a QOL index. A negative correlation between initial wages and wage growth (i.e. convergence) occurs if technology improves more slowly in advanced cities, or if in-migration of labor causes wages in high-wage cities to decline. QOL for potential migrants is assumed to decline in both the level and growth rate of city population. During the study period, Western and Southern cities grew the fastest, Central and Northeastern cities grew the slowest, and all cities growing faster in 1950–1960 continued to grow faster after 1960. Furthermore, changes tended to mirror the fortunes of urban economic bases – industrial cities relatively declined in both population and per capita income. Cities with higher initial levels of schooling experienced faster subsequent income growth; a one standard-deviation rise in median years of education raised income 2.78% over the 30-year time period.

Simon (1998) gives more precise attention to the relationship between human capital and urban employment growth. His hypothesis is that cities with high concentrations of highly educated persons can absorb, transmit, and implement knowledge, and thereby be more productive. These cities should also generate more localized knowledge spillovers. More than 300 metropolitan areas in the US are examined between 1940 and 1986. The theoretical model emphasizes the role of the service sector in attracting new industries as traditional manufacturing activities decline. Particular attention is paid to the proportions of persons with different degrees. High school diplomas are shown to be important for employment growth in every decade, but especially in the 1950s. However, between 1970 and 1986 college degrees seem about twice as important as high school diplomas in affecting urban employment growth. Evidence is given that residential differences in human capital partly explain the differential employment growth rates recently experienced by central cities and their outlying counties.

Glaeser and Shapiro (2003) have recently revisited population growth in urban America. Recent growth was shown to be much like earlier growth – the correlation between population changes in the 1980s and 1990s was a remarkable 65%. Dense cities continued to decline (outside of New York and Chicago) and automobile-oriented cities grew at the expense of cities favoring public transportation. Special attention was given to the role of climate. During the 1990s, high January temperatures had a positive but diminished effect on growth and high July temperatures failed to deter growth, as shown in the remarkable population expansion of desert communities like Phoenix and Las Vegas. Rainfall had a significant and negative effect on growth in the 1980s but not in the 1990s. Initial human capital had an impressive influence on a city's subsequent population growth. However, a California effect was detected, where cities apparently grew for different reasons than other US cities during the 1990s.

Using a less technical perspective, Glaeser *et al.* (2001b) focus specifically on QOL issues in the contemporary city. They identify four general types of urban amenities: variety in consumer goods and services, aesthetics and physical setting, quality of public schools, and accessibility (Glaeser, 1998; Quigley, 1998). Some goods, like professional sports teams, only flourish in large urban markets because they enjoy substantial scale economies. Consumption value might also be related to a valuable stock of buildings or an historic district. The importance of human capital in urban productivity and creativity has of course already been studied extensively (see above). Ease of transportation not only facilitates the exchange of goods and services, but high densities improve the mixing of young singles and thereby affect the efficiency of the marriage market (Costa and Kahn, 2000). Population growth in US counties between 1977 and 1995 is estimated in a simple regression model that includes eight amenities. Dry, temperate climates and coastal locations are shown again to be very important growth determinants, but restaurants and live performance venues (used as initial conditions) also have a significant effect on growth. But not all consumer goods seem to matter because the more basic goods clearly have non-spatial substitutes. The authors also argue that urban amenities have played a comparable role in the recent growth experienced by large cities in France and England. A simple amenity index is created by regressing housing prices on per capita income, where the residuals reflect city-specific demands for local amenities. High-amenity cities are found largely in California (except for Honolulu) and low-amenity cities are found mostly in parts of the East and the Midwest. Recent population growth is then regressed against this amenity index and a strong linear relationship is apparent.

23.2.2.4. Crime

Crime is a topic of increasing concern in the urban QOL literature. In the US, for example, violent crime rates in large cities are about four times greater than in small cities, and are about seven times greater than in rural areas. This is a little surprising given that per capita expenditures on police and victim precaution are both much higher in large places.

Recently, Cullen and Levitt (1999) have provided a powerful case for the connection between crime and urban flight in the US. They argue that if the costs of crime are completely capitalized into property values, then rising crime in the short run places costs on property holders but does not necessarily encourage their movement to the suburbs or to another city. But, if these crime increases prevail over the long run, fixed costs of upkeep on housing might become exorbitant or the housing units themselves might become severely crowded or even abandoned. Sustained levels of rising crime eventually lead to city depopulation if disamenities are not fully capitalized. Their paper involves an array of regressions that relate *changes* in city population to *changes* in per capita crime rates. In the simplest model, 10-year changes in per capita crime rates are regressed against the logarithm of change in city population. More complicated models remove the city-fixed effects and control for amenities (climate, education), homeownership, age, and income. Other data sets shed further light on appropriate lags and the role of personal characteristics. Invariably, though, a change of one standard deviation in the city crime rate (representing a 10% shift in crime) induced a decline in city population of approximately 1%. Household movement in response to crime patterns was then addressed during the late 1970s. Net migration increased rapidly with the education of the household head – households with some college education had migration rates 50% higher than those comprised of high-school dropouts. Households with children were also much more likely to move. But crime-related out-migrants generally remained within the same metropolitan area, by moving to the suburbs, instead of departing entirely. Apparently households leave metropolitan areas largely for economic or job-related reasons but they tend to move within metropolitan areas in response to crime.

Glaeser and Sacerdote (1999) adopt Becker's rational behavior model and explore the relationship between big-city crime and population from a somewhat different perspective. They begin by going over some familiar ground – 1989 crime data showing that victimization rates are 2.3 times as great in large places (1,000,000 or more persons) as in small places (fewer than 10,000 persons). The authors go on to decompose this observed connection between city size and crime into three separate categories – the higher pecuniary returns, the lower arrest probabilities, and the greater numbers of crime-prone individuals in urban areas. A compelling case is made that location-specific attributes can affect each of these three categories. The analysis begins with estimates of the elasticity of reported *serious* crime with respect to city size. When adjusted for underreporting, this elasticity rises to 0.24. The crime deterrence elasticity, adjusted for city size, is shown to be somewhere between 0.02 and 0.05 (large cities spend more per capita on police but the probability of an arrest is lower). Next, the relationship between the returns to crime and city size is explored. Density-related issues like greater information flows, more victims per square mile, and the proximity of criminals and victims are briefly discussed. Based on the average financial loss ($500) per crime, a benchmark elasticity of 0.11 for city size is estimated. The authors conclude that the elasticity for returns to crime, adjusted for city size, is somewhere between 0.03 and 0.08. Finally, adjustments are made for demographic differences among cities. The crime-to-city size elasticity of 0.24 is then decomposed accordingly. Depending upon the degree to which per-person deterrence actually works, 8.33–20.8% of urban crime is due to diminished size-related arrest rates, 13.33–33.33% is due to size-related pecuniary returns to crime, and 29.2% is due to the demographic composition of the city population. Between 16.67 and 49.14% of all urban crimes in the US cannot be attributed to city population size.

23.2.2.5. Pollution

A very extensive literature is now devoted to the incidence and implications of air and water pollution in cities. Much of the seminal work in the US is summarized in Berry and Horton (1974) but a prodigious amount of research has been undertaken on both themes since that time. Given that pollution is considered a disamenity in many QOL studies, we only offer a few more comments at this juncture. It is clear that some pollutants (e.g. sulfur dioxide, ozone) are either weakly correlated or not correlated to city size (even here there is conflicting evidence), simply because they are not affixed to the locales that are responsible for their creation. However, the density of particulate matter in the atmosphere seems to be weakly related to city size. In fact, Glaeser (1998) estimates that the pollution difference between a US metropolitan area of 500,000 persons and one of 5,000,000 persons is 10.4 $\mu g/m^3$. It is worthwhile to note that the rate of increase of particulate matter with city size fell dramatically between 1980 and 1990.

Smith and Huang (1995), after accumulating the results of many willingness-to-pay studies of particulate pollution in US cities, performed a meta-analysis on the various estimates between 1967 and 1988. The median estimate of annual cost was about $38 per person but the mean estimate was $185 per person. This suggests that outliers are very important in affecting any summary statistics regarding particulate matter. Furthermore, in a study of four specific cities, the authors evaluated the benefits arising from a policy that would reduce the maximum concentration of particulate matter to national standards. The results of three popular methods for estimating benefits were shown. While the rank-ordering of annual benefits was shown to be the same in each case, the actual estimates of these benefits shifted wildly – for instance, the three competing estimates were $781, $254, and $76 (times millions of 1982–1984 dollars) for all the residents of Los Angeles. Apparently, these different models reflect varying considerations of residents and are in part driven by local variation in both demography and house-ownership.

Kahn (1997, 2002) has also recently examined the economic effects of vehicle emissions in US cities. In the second paper, he develops a hedonic model for housing prices during 1986–1994 in Southern California's Anaheim and San Bernardino counties. San Bernardino quality-adjusted homes were shown to fall in price relative to Anaheim homes during the study period. San Bernardino residents also enjoyed lower levels of smog (ozone), so at least in this labor market lower housing prices and better air quality were not mutually exclusive. One benefit of the nation's Clean Air Act might be that urban redevelopment will proceed in some areas once characterized by high levels of pollution.

Millimet and Slottje (2002) have examined how environmental quality varies with such socioeconomic attributes as race or income and other local attributes such as political activism. Their main objective was to assess how the degree of geographic inequality in per capita emissions varies with environmental compliance costs in the US. Using the broad categories of air, land, water, and underground releases, evidence is given for considerable variation both across and within states. Of the four types, air pollution is distributed the most uniformly. Moreover, consistent with the environmental justice literature, there is a significant positive correlation existing between toxic releases and both black population share and female population share. Even though environmental quality apparently did not deteriorate when federal controls passed back to the states during the 1980s, the authors conclude that any new, uniform national guidelines will not adequately diminish today's geographic inequities in the distribution of pollution. Instead, governments at all levels must target those locales exhibiting high levels of pollution. Unfortunately, the paper does not shed much light on how these environmental disparities are allocated across the population size distribution of cities.

Nevertheless, Glaeser (1998) remains very upbeat about the results of recent pollution abatement for future QOL in US cities, and argues that overall pollution costs now seem very small when compared to housing, commuting, entertainment, and other costs in the urban environment. He claims that just as government action and technological advances in earlier times eliminated many of the age-old health problems of urbanities, recent government initiatives and technological improvements have largely eliminated the mid-20th century pollution gap that once differentiated the nation's big cities from its small towns and rural areas.

23.2.3. *Population and employment relocation*

23.2.3.1. Adjustment models

There is a short but growing literature that analyzes how amenities *simultaneously* affect population and employment growth across cities. In the postindustrial societies a consensus now exists that employment change not only drives population change (the traditional view) but that population change also drives employment change. In these rich societies, individuals are mobile, live longer, are able to access their pensions or public transfers from any location, and often own two or more homes in different regions. The standard approach is to use the partial adjustment model pioneered by Steinnes and Fisher (1974), which is estimated by two-stage least-squares procedures.

The key contribution is Carlino and Mills (1987) who demonstrated that different amenities (including climate and crime) and public policy instruments differentially affected county-level growth in the US during the 1970s. A follow-up study by Clark and Murphy (1996) reached similar conclusions for growth during the 1980s. Dual causality was apparent but somewhat weak in both cases. Growth in the Sunbelt counties was also apparent in both studies; however, the metropolitan decentralization of the 1970s was not so apparent during the 1980s. Amenities and demographic characteristics were found to be more important in driving population growth but public policy variables were shown to have a greater effect on employment growth. However, policymakers were warned that they might be too optimistic in thinking that their instruments can effectively direct or influence local and regional economic development. A more recent US study by Duffy-Deno (1998) found that federal wilderness designations had no apparent effect on either population or employment change in the diverse counties of the eight-state Mountain region. The results, for the 1980s, are somewhat surprising given that so many persons, of working and retirement ages alike, claim to value nearness to the American West's many recreation areas. A related study (Duffy-Deno, 1997) indicates that state parks per acre of county land had a positive but very small effect on changes in population and employment densities during the same decade.

The most interesting recent work examines the presence and direction of linkages between rural peripheries and urban cores in metropolitan areas (Barkley *et al.*, 1996; Henry *et al.*, 1997). Their approach, which adapts Boarnet (1994) to a different geographic scale, focuses on the growth experiences of nearly 270 counties in three southern US states during the 1980s. Each county is attached to one of eight different functional economic areas and each is classified as being one of three types – core, fringe, or hinterland. Potential collinearity problems are reduced by factor analyzing a long list of amenities and fiscal variables; two groups of variables likely to impact business- and residential-location decisions are then identified. A mix of spillover and backwash effects from urban core and fringe areas to their rural hinterlands is uncovered. Rural-area population grew faster in those FEAs having slow-growth cores and fast-growth fringes. Urban population is shown to spread to nearby rural areas, but evidence of urban employment spread effects is weak at best. Cities and rural areas can position themselves competitively by improving local amenities (especially in infrastructure and education) and adopting prudent public policies. New residents seem more prone to avoid localized pockets of poverty than do new firms. The health of economies in rural areas is shown to depend intimately on the health of economies in large, proximate urban areas.

More recently, the approach has been extended to examine city–region relationships in both Denmark and France (Henry *et al.*, 1999). The 1985–1993 Danish data are organized around 275 (204 rural) municipalities and 46 functional economic areas; the 1982–1990 French data are organized around 11,170 (3515 rural) communes or municipalities in six regions. The Danish amenities include green space and school quality, the French amenities include school quality, infrastructure, and human capital, and both studies control for the spatial distributions of wealth and poverty. The known US results are basically endorsed. Rural population growth takes place near large urban centers experiencing high rates of employment growth. Rural employment growth is especially high near urban places experiencing rapid population growth. Overall the results suggest that a new typology of extended-city growth is needed that accounts for complex spread and backwash effects among central, intermediate, and peripheral zones.

In another paper, Glavac *et al.* (1998) examined population and employment growth during the 1980s in 219 (unofficial) micropolitan areas in the US. By 1990, more than 15 million persons (6.5% of the national total) were living in these emerging metropolitan areas. Bicausality was analyzed for both levels and densities, recognizing that many Western cities are automobile oriented and many Western counties are very large. Estimates are given for five different amenities (temperature extremes, overall crime rate, hospital beds and doctors per capita, and presence of recreation facilities) and for government spending on amenities (including health and recreation). A wide variety of demographic, economic, and social conditions were included as controls. The results were somewhat more encouraging for densities than for levels, but most amenities were not seen as significant factors in micropolitan growth. Crime was shown to affect population growth but in an unexpected direction – it seems to be an unfortunate byproduct of rapid population growth. Employment change was seen as important in driving population change, and not vice versa, during the decade. In the US, where the processes driving population and employment change are very complex, it remains somewhat unclear how these various processes play out across the different population size classes of cities. This research should be updated using the official (and somewhat longer) list of micropolitan areas that was identified for the 2000 US Census.

Urban land absorption would seem to be a very promising topic for interurban adjustment modeling. This is a topic of great concern to public policymakers everywhere (see below). Long-run estimates could be made of household- vs. job-related land consumption, holding constant a variety of initial conditions. A much clearer picture might result regarding the degree to which natural and human-created amenities differentially influence the land absorption rates of different cities. Moreover, through three-stage regression, it might actually become possible to endogenize a composite QOL variable in the estimation. So QOL would become not only an input to – but also an output of – the diverse land-consumption decisions of city-based agents.

23.2.3.2. Migration

The standard economic model for capturing migration is neoclassical, where rational and well-informed workers adjust to differences between regional labor markets by moving. Labor is expected to flow from low-wage, labor-surplus regions to high-wage, labor-shortage regions until wages equilibrate (or at least converge). This approach, sometimes called the *disequilibrium* model, is also associated with the human capital model pioneered by Sjaastad (1962), where households invest in migration only if they gauge the move is profitable in the long run. Moving is not free, and households decide to relocate only when they expect higher net benefits at the alternative location. But studies show that other factors, including personal or life cycle characteristics and family ties, play an important role in the decision to migrate. A persuasive alternative economic approach to modeling human migration has become popular in recent times where households are assumed to reveal their preferences for site- and region-specific amenities and public goods by relocating. As Tiebout (1956) argued, households can vote with their feet and individually seek a satisfactory tradeoff between income and QOL. When society as a whole does this, and individuals are compensated for lower wages by better amenities or public goods, the space-economy adjusts to a *spatial equilibrium* where no individual has any further incentive to relocate.

In this second stream of research, Cebula and Vedder (1973), Graves (1976, 1979, 1980, 1983), and Graves and Linneman (1979) provide the seminal contributions. Intercity income differentials do not always lead to migration because households are sometimes compensated for these differences by amenities. So changed demands for amenities result largely from either changed prices or incomes across all locations, and demographic characteristics like age, gender, and race are potentially important as underlying shifters. Gross migration data are usually preferred to net migration data because they more accurately capture household decision-making. In Graves (1980), the migration of both whites and nonwhites among US cities during 1965–1970 was examined to determine the specific effects of various climate variables. In the case of white males, the spatial equilibrium model outperformed the traditional disequilibrium model for both in- and out-migration across seven different age groups. Stage in the life cycle appeared particularly important; for example, among retirees humidity was an inferior good but warmth was a normal good. However, the signs on initial income and unemployment were unexpected. Building upon the ideas of Diamond and Tolley (1982), Graves (1983) then recast his argument by introducing contract rent as a surrogate for the host of amenities that might affect intercity migration. This composite-amenity approach reduced a number of econometric problems that seemed responsible for unanticipated signs. Rent was shown to have a very strong effect on net migration rates for all age groups of whites during the 1960s. The effect of rents on nonwhite migration was not as compelling. The important finding was that amenity bundles are superior goods – at least for whites – and that, in a world of ever-rising incomes, households will move toward high-rent locations that provide these various amenities in abundance.

Herzog and Schlottmann (1986) then used this approach to model out-migration and accordingly rank US cities. They began by borrowing data on urban QOL from the *Places Rated Almanac* by Boyer and Savageau (1981). Persons with a metropolitan residence in 1975 were examined to see which factors appeared to influence their migration over the ensuing 5-year period. Census micro data were used to control the estimates for the characteristics of some 7400 persons who were thought to be 'at risk' to the effects of urban amenities. Logit estimates of out-migration, based on movers and stayers, indicated that only four QOL factors significantly affected relocation. With their standardized weights these were housing (0.290), crime (0.226), education (0.186), and recreation (0.298). The remaining five *Almanac* factors – climate, health care, transportation, arts, and economics – all failed to be significant. Age, education, and family size were shown to selectively influence the out-movement of households and the rate of out-migration was much lower for blacks than for whites. Cities were ranked according to the four weights: of the 50 highest-ranked places, 20 were small, 18 were medium sized, and 12 were large. A surprising number of cities in the nation's northern industrial tier were ranked highly, presumably because stayers who were employed in those places were content with low housing prices and low crime rates.

Fueled by a series of mixed findings (e.g. Porell, 1982; Greenwood and Hunt, 1989; Schachter and Althaus, 1989), Clark and Cosgrove (1991) explored two important issues in more depth. First, how do US households compare marginal benefits and marginal costs in calculating a welfare-gain function? And second, does an optimal distance for migration result? Their underlying intent was to test the relative merits of the disequilibrium and equilibrium schools. A wage-opportunity equation was first estimated for some 6600 movers during 1975–1980. Both human capital and amenity variables were shown to be important in relocation. Workers living in rainy cities, cities with high murder rates, or cities with high levels of air pollution were compensated with higher wages. Cities with expensive housing also offered higher wages. A variety of cultural and recreational amenities, while not significant, did have the correct signs. A reduced-form distance equation was then estimated across the sub-sample of workers who moved between cities. High-educated householders and male householders moved greater distances than low-educated householders or female householders, and singles moved somewhat greater distances than other categories of householders. Favorable climatic amenities, especially improvements in sunshine and reductions in temperature variation, tended to increase the distance moved. Some site-specific amenities were insignificant, presumably because households could move within metropolitan areas to seek compensation (see above). An important finding was that householders were willing to move greater distances when economic opportunities were greater. Clark and Hunter (1992) then revisited the study by examining the intercounty movement of white males during 1970–1980, but included more detail on the life stages of movers and the characteristics of the origin and destination counties. Expected employment growth was very important for migrants when they were in the working-age cohorts; however, this variable was not important for migrants aged 60 or more. Fiscal preferences were also shown to shift over an individual's life cycle; for instance, older cohorts were especially reluctant to move to states with relatively high state inheritance or estate taxes. Middle-aged migrants sought areas with low housing values but older migrants preferred locations with high housing values. A general finding was that the people place age-specific values on amenities.

The most recent work on the equilibrium vs. disequilibrium debate is apparently that of Mueser and Graves (1995), who build on the work of Greenwood and Hunt (1989). A theoretical section outlines a model having optimizing firms and households, where equilibrium implies a mix of wages, rents, and amenities so that firms earn zero profits and households have a common level of utility. Disequilibrium occurs when agents face various adjustment costs, so that both firms and migrants must calculate current and future benefits over some time horizon. The thrust of the empirical work is to show how *shifts* in location profitability and desirability induce the in- and out-movement of firms and households between cities. The net migration of workers among metropolitan areas is examined over three decades – the 1950s, 1960s, and 1970s. The model works best in the first decade. Amenities, the main driver of movement between counties, enjoy a remarkably steady effect over the three decades of the study, thus endorsing Greenwood *et al.* (1991). Settlement patterns and demography are also important but the role of employment-related variables is lower than expected. In an alternative model, the amenity- and job-related effects rank first and second in importance. In terms of amenities, the same areas of the nation tended to be desirable to migrants over the entire 30-year period but, in terms of employment opportunities, areas that experienced growth in the early years tended to experience a reversal of growth in the later years. Evidently, the time period is very important when estimating intercity migration – in some years profit-shifting variables drive job-related movement but in other years utility-shifting variables drive amenity-related movement.

The alternative revealed-preference approach recently advocated by Douglas (1997) should also be noted. Building on Douglas and Wall (1993), he designs a random utility model of location choice for US states in 1970, 1980, and 1990. The model is based on the notion that distance and other symmetric fixed factors, such as cultural opportunities and economic performance, will not affect the sign of the expected net migration flow between two states. A new estimator of a state's relative QOL advantage over any other state is generated by examining the balance of net migration standardized by each state's population size. An algorithm is designed to perform all pair-wise state comparisons, rank the states by overall won–loss scores, and resolve ties. Kendall's coefficient of consistence is used to test for the incidence of intransitivities in the pair-wise comparisons. Results from three somewhat different ranking methodologies are disclosed. Rankings for a cross-migration model are stable over time, but rankings from the net in-migration and net flow probability models (the new model) are much more volatile. Presumably, many of the more extreme shifts are due to the boom-and-bust fortunes of the energy-producing states over the 20-year period of the study. Sunbelt states are clearly preferred in the three rankings, but it is unclear how the more urban states perform relative to the more rural states. It will be interesting to see if these three models can be successfully applied to US counties using Census 2000 results, where it remains unclear how severe the transitivity issue will be at this other level of geographic aggregation.

23.2.4. Industrial and business location

The transformation of the advanced economies from an industrial to a postindustrial base quite obviously has important implications for cities. Extensive research over the past 30 years has both clarified and extolled the city's productive virtues of size, density, diversity, knowledge creation, and knowledge spillovers (Jacobs, 1969, 1984; Beeson, 1992; Quigley, 1998). Glaeser *et al.* (2001a,b) recently argue that the future of cities depends in part on whether they can maintain their productive advantage in a world that is increasingly characterized by the ascent of service industries and the ever-diminishing costs for exchanging information and goods. But attention must also be given to the role of cities as centers of consumption (see above). QOL not only affects the growth fortunes of cities directly (through the demands of households) and but also indirectly (through the demands of high-skill employees). These two factors are implicitly recognized in one recent ranking of high-tech US cities, where older and more diversified centers fared very well in comparison to smaller and more specialized centers (Chapple *et al.*, 2004).

Disparate conclusions have been reached about factors influencing industrial location and relocation (Blair and Premus, 1987; Gottlieb, 1994). However, the literature seems roughly polarized around the varying importance of two broad groups of factors: economic attributes (Weberian or neo-Weberian criteria) vs. non-economic attributes (amenity or QOL criteria). Classic industrial location theory suggests that firms locate to minimize costs or to maximize profits, and investment is typically seen as a long-term decision (Smith, 1971). But in recent times the urban landscape has witnessed a dramatic shift in manufacturing from Fordist-type arrangements to flexible arrangements (with batch production of high-value, low-weight products) and has seen the rapid ascent of producer services (Piore and Sabel, 1984; Beyers, 1992). So concern for issues like the transportation costs of weighty inputs and market proximity for outputs has become less important, and concern for issues like quality of the labor force and the access of highly paid professionals to recreational and cultural opportunities has become more important. Unfortunately, our sense is that the relationships between firm location and amenities have been explored much more in rural areas than in cities (Beale and Johnson, 1998; Marcouiller *et al.*, 2002). Deller *et al.* (2001) is especially notable for designing a regional adjustment model that captures the diversity of amenities in America's more than 2200 rural counties (see above).

Analytical studies of US business location decisions appear to begin with Bartik (1985). A conditional logit model is used to model new branch plant openings across the US states during the 1970s. The author is aware of some deficiencies of his study, including its geographic scale. The effects of existing manufacturing, unionization, state taxes, and public services are all shown to be important determinants of openings. In Bartik (1988) the effect of state differences in environmental regulations on the location of Fortune 500 branch plants is analyzed. For the study period 1972–1978 he uses two measures of state water pollution and four measures of state air pollution. The study does not find any discernible effect of variation in regulations on plant location across all industries; however, the estimates are not precise enough to rule out regulatory effects on the most highly polluting industries. Then, in Bartik (1989), micro data are used to assess how different state-level attributes affect small business start-up decisions in 19 industries.

A wide list of independent variables is used, including market demand features, factor prices, taxes, demographic attributes, financial characteristics, and the level of spending for numerous public goods. The results suggest that some public goods encourage small business starts, especially fire protection and local school spending. But public welfare spending has a significant negative effect on starts, probably because higher welfare spending reduces the availability of unskilled labor. Factor costs have surprisingly small effects on variation in state-level business starts, but the education level of the workforce has a surprisingly large positive effect. Bartik argues that this last influence not only reflects better quality state-level labor but it also reflects the greater ability of well-educated persons to assume entrepreneurial risks. Obviously, other QOL variables could be included and those factors creating urban vs. rural differences in business start-ups could be identified.

More recently, Gottlieb (1995) examines 1990 employment in one high-tech industry, engineering and management services (SIC 87), across 365 contiguous municipalities in 13 counties in northern New Jersey. This area is well served by transportation, has a diverse workforce, skilled professionals, and is within New York's commuting shed. Here municipalities have autonomy in land-use regulations and compete in the provision of services. The motivation for the study is clarification of one fundamental question: do amenities attract residents that then choose to work in firms or do amenities attract firms that then hire amenity-responsive workers? Many elite firms apparently locate in high-amenity areas not only to tap into an existing labor force but also to recruit new workers seeking amenity compensations. Both employment density and employment proportion are regressed against a number of business variables (mostly related to accessibility) and amenities, and estimates are controlled for racial composition. A novel aspect of the study is the use of spatial weights to capture the effects of contiguous counties on observed counties; distance-decay parameters are adopted from an earlier study in Los Angeles by Scott (1992) and are sometimes adjusted. One of the few unambiguous findings is that municipalities having above-average violent crime rates also have a deficiency of high-tech employment. The distance-weighted amenities, which capture conditions outside each municipality, have somewhat disappointing effects; nevertheless people seem to be avoiding property crime and toxic pollution in surrounding municipalities when they choose where to work. The main conclusion is that avoidance of disamenities overrides the attraction of amenities.

A different perspective on the relationship between firm location and QOL is provided by the literature on environmental regulations. Here the recent survey article by Jeppeson *et al.* (2002) is instructive. Data are gathered from 11 US studies carried out between 1963 and 1994, and a meta-analysis is performed on 368 observations (each of which includes numerous new plant location decisions in various manufacturing industries). The research is carried out in the context of studies reaching mixed conclusions regarding the influence of environmental regulations on the location of industrial activity. Not surprisingly, the estimated influence of environmental regulations is higher when the geographic area of study is smaller. Possibly this is related to the fact that smaller areas (counties) are more homogenous with respect to taxes, climate, and labor market attributes. Wages clearly have a negative effect on new plant start-ups even when controlling for regional tax differences and the particular industry's ability to achieve scale economies. Presumably, the higher wages of metropolitan cores tend to direct much new plant investment (in polluting and non-polluting industries alike) either to metropolitan peripheries or to nonmetropolitan areas. Unfortunately, though, county-level attainment status is not controlled in the study for county population size, so it is difficult to conjecture accurately how regulations are distributing industries among large cities, small cities, and rural areas. Moreover, better geographic data might allow this stream of research to be merged with that on multi-zone regional adjustment models (see above).

A very different approach to understanding how urban amenities affect business locations has been pioneered by Florida (2002a–c). He focuses almost exclusively on the urban region's openness to creativity and diversity in the workforce. Motivated in part by traditional approaches (Foster, 1977; Blair and Premus, 1987; Gottlieb, 1994), but more by novel approaches (Hannigan, 1998; Costa and Kahn, 2000) that see cities as entertainment machines and efficient markets for acquiring partners, Florida explores the geography of human talent. Talent, which is human capital (specifically, workers with bachelor's degrees or higher), tends to flow to places that are diverse, tolerant, and cool; thus talent eventually assumes a very concentrated geographic pattern over time. Diversity is captured by the proportion of households that are bohemian or gay, and coolness is measured by the incidence of cultural and nightlife amenities. In fact, a bohemian index – based on creative occupations in the arts – is introduced in Florida (2002a) as a substitute for more traditional measures of local cultural assets. Following earlier work in Florida and Gates (2001), path analysis is used in Florida (2002b) to clarify the interrelations existing between talent, diversity, high-technology industry, and worker income. The study focuses on the 50 largest metropolitan areas in the US in 1990. Cities having highly educated populations tend to rank high in both diversity and coolness. Talented people making location choices report a preference for places having racial, ethnic, and lifestyle diversity. Furthermore, high human-capital workers change jobs frequently and thus favor locations having thick labor markets. Three types of amenities are used in regressions estimating the incidence of talent across the 50 cities. Data are taken from the *Places Rated Almanac* (Boyer and Savageau, 1989) and the tech-pole index created by the Milken Institute (DeVol *et al.*, 1999). Culture is found to be positive but insignificant, climate is negative but insignificant, and recreation is negative and significant. Florida (2002b) suggests that these somewhat disappointing results probably follow from a deficiency in the data because it is widely known that talented people seek out places with active outdoor recreation or a lively music scene.

Gertler *et al.* (2002) extend this approach in a study of urban Canada. The bohemian index and mosaic index (percentage of the population that is foreign-born) are combined with the talent index and tech-pole index. Again, city–regions with large concentrations of technology-intensive employment are found to be places that either attract or retain talent effectively. High-technology employment is strongly related to the bohemian index and moderately related to the mosaic index. Some interesting US–Canadian comparisons follow, where the authors note that the US has a higher population proportion with a university degree and Canada has a higher foreign-born population proportion. If anything, the basic relationships of Florida's model are stronger in Canada than in the US. Policy recommendations regarding cultural diversity and immigration follow; unfortunately, very little is specifically said about the role of urban amenities in this study.

23.2.5. Local economic development and planning

Some 10 years ago Kusmin (1994) reviewed many of the most important papers in the local development literature. The studies, which varied a lot in scope and intent, included industry-level, sector-level, and region-wide analyses. The characteristics believed to affect growth were found to be numerous, but advantageous location-specific amenities were often listed with favorable taxes, high public expenditures, good education, strong market demand, and excellent access to transportation. But, as noted earlier for business location studies, very few useful generalizations were reached. Since that time LED has become very fashionable (Blakely, 1994). Strong forces like deregulation, enhanced trade, cheaper communications, and the globalization of finance have shifted the interest of many academics away from the problems of the poorer countries to the concerns of peripheral regions in the richer countries. The rapid growth of local agencies and initiatives has become an international phenomenon, partly fueled by the rise of neo-liberal ideology. In many cases, too, local and national economies have become decoupled from each other. Economic agents and local governments have reduced their reliance on national governments, and frequently have devised their own local initiatives (with mixed success) to improve the economic chances of their localities in the face of intensified international competition.

In response, Wong (1996, 1998, 2002) has launched an impressive series of studies on LED. These are focused on the UK but include many international examples. In Wong (1996), she reviews a very extensive literature, highlighting studies of infrastructure, inward investment, high technology sites, and the like. The causal relationship between local development and QOL is seen to be controversial and, at best, very tenuous. Favorable perceptions of local QOL are often seen more as the result of a rising industry (including its newness and highly educated workforce) rather than a result of either the location decision or the success of the business. In fact, local successes are often believed to unravel over the longer run with rising congestion and pollution and higher costs for housing. Several observers have in fact recommended that fast-growing communities constantly monitor their local QOL trends in order to ensure long-term community health. In Wong (1998), she explores how policymakers perceive the importance of different factors for LED. Two different regions are examined, and a case study approach is used with in-depth interviews of practitioners. Seven tangible (i.e. traditional) and four intangible factors are identified. The latter group includes QOL (desirability of a place adjusted for cost of living), business culture, community identity (image from without), and institutional capacity. Respondents believe that traditional factors must be satisfied before issues like QOL are considered by relocating firms. Finally, in Wong (2002), she identifies 29 different indicators to capture these 11 broader factors. These are meant to serve as the actual inputs to practitioners defining policy problems and informing policy formulation. The desirability of the place is assessed by various data: insurance rates, housing prices, age-specific mortality rates, local tax rates, secondary school student performances, and areas designated to have outstanding natural beauty. Community identity is measured by comparing volumes of in- and out-commuters to core areas. The indicators are compiled for some 360 local authority districts in England and are analyzed by multivariate techniques. The first factor, called the big-city syndrome, explains much of the variance and highlights polarization in regional development: some areas have high earning power and short commutes but others have high housing prices, high crime risks, and low community identity. The next factor, buoyant suburbia, differentiates the affluent areas of London's suburbs and the Home Counties from the poorer areas in the periphery. The third factor highlights various amenities and reflects a desirable living environment. As expected, places performing strongly on this dimension are generally found in rural townships and scenic coastal areas, especially in the West Country. Overall, different local trajectories are recommended for successful economic development. But, overall, Wong's findings tend to reinforce the key roles of hard, traditional factors like location advantage and quality of the workforce. Special concerns are raised about preserving a sustainable living environment in the highly populated South East given postindustrial society's increased reliance on goods movement and commuting.

Another approach to facilitating local or regional economic development strategies is advocated by Roberts and Stimson (1998). Like Wong, they are concerned that standard analytical techniques, including input–output and shift-share analyses, increasingly fail to capture many of the factors that influence local growth and regional competitiveness. Non-quantifiable factors like social capital, QOL, entrepreneurship, and community attitudes toward change must be incorporated into strategic planning processes. Moreover, core competencies – the critical mass of regional skills, resources, and technology – must be assessed. They advocate an approach called multi-sector qualitative analysis (MSQA), which has its roots in structural analysis. QOL (included as part of human resource development, along with skills, wages, and labor relations) is viewed as one of 34 core competence criteria used to assess 25 industries in the Far North region of Queensland, Australia. QOL criteria are gauged to be very important for industries like communications, tourism, and some types of manufacturing, but not so important for industries like mining and fishing. In addition, a risk factor index is created for regions that incorporates environmental (loss of quality, disease) and community-level (attitudes to industry) parameters. The technique seems very promising for urban applications if practitioners want to address both tangible and non-tangible factors in formulating LED strategies.

More recently still, Mathur (1999) has outlined a human capital-based strategy for regional economic development. He begins by summarizing the well-known argument that human capital, which constitutes the source of knowledge and therefore technical change, overcomes the limitations imposed by diminishing returns to labor and capital. Then he argues that there is an important role for policymakers in correcting discrepancies between the private rewards and social benefits of human capital formation and knowledge accumulation – subsidies are needed for general education, technical training, and research and development.

Many analysts and practitioners have advocated urban sustainability in recent times. While the term sustainability remains vague and seems to contradict the very notion of a spatial division of labor, higher density urban forms are often claimed to promote greater social equity. In response, Burton (2000) has recently examined a group of small- to intermediate-sized cities in the UK. The expected attributes of compactness include better access to private and public facilities, poorer access to green space, better job accessibility with shorter commutes, better public transport, greater opportunities for walking and cycling, reduced living space, reduced crime, lower levels of social segregation, but poorer health and lower affordable housing. Fourteen indicators are used to measure compactness and another 44 indicators are used to represent the various social equity effects. Compactness is concluded to have a strong negative effect on some aspects of urban equity (e.g. crime and access to green space) and a strong positive effect on other aspects (e.g. access to facilities). Burton ranks the cities by their desirability in each category.

23.3. Intraurban scale

23.3.1. Deprivation

23.3.1.1. Definitions and assessments

Like the term QOL there are many competing definitions of *deprivation* and very little consensus exists about its exact meaning. Herbert (1975), in emphasizing its spatial nature, viewed deprivation as a level of QOL that is below that of the majority in society and recognized that it involves economic hardship and inadequate access to resources. Townsend (1987, 1993) later sought to distinguish deprivation from poverty by arguing that relatively disadvantaged states can exist along both material and social dimensions. At the same time, Wilson's (1987) exposition of the underclass – multiply disadvantaged inner-city persons – became the new standard for research on poverty and deprivation in US cities. Inner-city flight leaves neighborhoods with people who lack basic skills and qualities of leadership, who frequently engage in aberrant or even criminal behavior, and who are often dependent upon public support for their livelihood. Bursik (1986), for one, has also argued that crime, especially delinquency, can tip and then accelerate the dynamics of neighborhood change. The economic plight of many inner-city neighborhoods has been depicted by Orfield (2002), who traces the spiral into severe deprivation as private business disinvests, tax bases shrink, and local governments remove much-needed public services.

But the best known work on deprivation is that of Pacione (1986, 1989, 1995), who provided an entire decade's worth of research in this genre, mainly in Scotland. The earliest of these studies used a multi-scale methodology to examine QOL in Glasgow. Indicators of deprivation were applied at the macroscale (the entire city), the subarea scale (Easterhouse, a severely deprived area within the city), and the microscale (a local area within Easterhouse). At the macroscale, Pacione performed a multivariate analysis, using more than 50 variables. A general deprivation factor was extracted, which he used to map out deprivation across the entire city. The subarea study included only descriptive measures but, at the microscale, regression was used to determine those measures of neighborhood quality that contributed most to self-reported levels of deprivation. According to residents, a multiple deprivation measure was the most significant variable in explaining own QOL, as determined through other survey research.

Pacione (1989) then revisited the topic in a study of deprivation in Scotland's four largest cities: Glasgow, Edinburgh, Aberdeen, and Dundee. He now found two distinct types of deprivation – physical (or material) and social. But the two types displayed very different spatial patterns. Physical deprivation, epitomized by the deterioration of housing, was generally found in inner-city areas, where poor, private rental housing abounds. Social deprivation, however, was found more in outlying areas, where public housing developments provided decent housing, but where overcrowding, unemployment, and poverty were the norm. This work highlighted the spread of deprivation outward from the inner city. Pacione's (1995) most recent work reinforces this observation that severe deprivation is no longer confined to inner-city areas.

This approach has been widely adopted elsewhere. Broadway and Snyder (1989), who examined Wichita, Kansas during the 1970s, tested for evidence of convergence, divergence, and stability in relative deprivation of that city's census tracts. Areas largely populated with poor and unemployed households (Wilson's underclass) suffered from divergence, or higher levels of relative deprivation, despite a decade of being targeted with public programs. Boal (1998) has recently shown that Belfast's well-known geographic patterns of ethnic segregation are highly correlated with deprivation, especially in those inner-city areas that have changed little since the late 1970s. Apparently both Protestant and Catholic groups have experienced extreme, concentrated deprivation as a result of similar structural processes. He contends that social organization can be a mitigating factor in such circumstances. Using paired neighborhoods in Scottish cities, Atkinson and Kintrea (2001) have explored the degree to which the negative life chances of people are compounded because they happen to live in areas that are recognized as being deprived. Some support is provided for the notion that deprivation is a structural problem. More recently Witten *et al.* (2003) have developed an index of deprivation for two municipalities in the Auckland area of New Zealand. This index, applied at the meshblock level, measures the availability and quality of different types of private and public services. Higher accessibility scores were apparent for central areas and along transportation corridors, but pockets of deprivation were noticeable throughout each municipality. The new approach is recommended in order that area-based poverty and deprivation policies can be better informed.

A number of studies have addressed the usual methodological concerns of area-based QOL research. Fieldhouse and Tye (1996) compared individual-based and small area-based deprivation indices in Britain. Although some amount of ecological fallacy doubtless plagues all area-based studies, the evidence here suggests that the problem may not always be so bad. Nevertheless, individual-level studies are still recommended for distributing programs and funding. Harris and Longley (2002) recently revisited this scale issue in Britain and found that deprivation varied quite a lot across enumeration districts, thereby indicating that heterogeneity in deprivation existed at the ward level. They recommended using alternative lifestyle data to supplement the construction of deprivation indices at fine geographic scales. Kearns *et al.* (2000) have addressed the issue of measuring deprivation with more timely non-census data. Proxy measures (e.g. measuring crime rates by insurance premiums) are shown to compare favorably to census-based measures of deprivation in Scotland.

But most of the recent work on deprivation has focused on Canadian cities. Broadway (1989, 1992) contributed several early pieces. In the first of these he directly compared US and Canadian cities and uncovered something unexpected – Canadian cities had more inner-city deprivation but US cities had higher levels of multiple deprivation. However, broad cross-national conclusions could not be reached as only three pairs of observations were used. In the second paper he focused solely on different-sized Canadian cities and found that (like in Scotland) deprivation was definitely spreading from inner to outer zones. But he also found that inner cities were not declining because large numbers of well-educated persons had been enticed back to reside in downtown areas. Using unemployment as a single measure of deprivation, Broadway and Jesty (1998) next gauged the degree to which inner-city deprivation was converging or diverging in that nation's 22 largest cities.

A more recent study has looked at the geographic association of immigrants and deprivation in Canada's three largest metropolitan areas – Toronto, Montreal, and Vancouver (Ley and Smith, 2000). Various measures of deprivation did not overlap to a significant degree, a result that stood in contrast to that of an earlier US study (Hughes, 1990). The association between immigrants and deprivation in these Canadian cities was found to be modestly positive and, moreover, was stable over time.

Langlois and Kitchen (2001) next attempted to develop a standard method for identifying multiple deprivation areas. Studying deprivation in Montreal at the census tract level, six different dimensions of deprivation were addressed – demographic, income, education, language, housing, and employment. Five factors were extracted in a multivariate analysis and the factor scores were used to create a general deprivation index (GDI). By mapping the GDI scores, the authors were able to show the variable intensity of multiple deprivation in different parts of the city. Surprisingly, the most severe deprivation was not found in the inner city, although that area was deprived, but in a locale west of the city center.

Kitchen (2001) has also demonstrated the utility of a longitudinal approach. Examining the inner city of Montreal in 1986, 1991, and 1996, he revisited Broadway's approach of examining increases, decreases, and stability in deprivation. But now the analysis is decidedly multivariate – 14 different measures of deprivation were collapsed into three general factors for each of the three census years. As in other studies, the multiple deprivation experienced in various parts of the inner city was seen to be diffusing outwards to some of Montreal's inner suburbs. But Madden (2003), who focuses more narrowly on poverty shifts experienced in 27 large US cities during 1970–1990, found that distress relatively worsened only in the inner suburbs of Northeastern and Midwestern cities. Clearly much more cross-national, historical work is needed on urban poverty and deprivation.

GIS technologies increasingly will be used in many sorts of QOL studies (see above). Researchers are already re-examining the entire issue of accessibility in order to assess how overall proximity to diverse opportunities – for work, education, shopping, health, and recreation – directly affects personal health (Witten *et al.*, 2003). Other studies are examining the relationships existing between health levels and urban lifestyles, assessing among other things how transportation infrastructure affects longevity (Handy *et al.*, 2002; Boarnet *et al.*, 2003). Doubtless analysts increasingly will focus on the complex role of geographic choice or variety in enhancing the health of urban residents.

23.3.1.2. Crime

Crime and deprivation have been connected in a variety of ways. In the narrowest sense, deviant or criminal behavior is tied to poverty or economic deprivation (Messner and Tardiff, 1986). But in the wider sense, such behavior can be associated with other forms of deprivation as reflected in family disruption, neighborhood instability, and extreme racial or ethnic heterogeneity (Sampson and Groves, 1989; Bursik and Grasmik, 1993). One especially popular theme, at least in the US, is that the norms of behavior in socially and economically isolated areas are very different from those in mainstream society, and often develop and adapt to support criminal behavior (Wilson, 1987, 1991).

Herbert (1975) included crime measures in his early deprivation study of Cardiff. A strong correlation was shown to exist between levels of delinquency and those areas considered to be generally deprived. Exploring matters further, he showed that in these deprived areas children tended to have lower levels of educational attainment and that they were often physically punished by their parents. Residents often exhibited a blurring of right and wrong and some forms of behavior were entirely accepted when they would be considered delinquent in less deprived neighborhoods. More recently, Ley and Smith (2000) have briefly examined the spatial association between crime and multiple deprivation in their study of immigrants in urban Canada.

Many recent ecological studies have focused on the role of economic disadvantage in contributing to high crime rates. Hagan's (1994) work is especially well known for his views on neighborhood-based processes of capital disinvestment. Alternate norms and attitudes develop in areas of concentrated poverty and reinforce the downward spiral of the area into worsening deprivation, both social and material. Three main factors are believed to contribute most strongly to the high levels of crime and delinquency in areas of severe deprivation – residential segregation, racial inequality, and the geographic concentration of poverty. The combination of little schooling, limited job experience, and an arrest record means that little or no social capital exists to assist youths in pursuing a legitimate career path. Hagan concludes by saying that funds currently used for prisons and imprisonment might be better spent investing in areas of concentrated poverty, specifically creating better job opportunities for residents. In a closely related piece, Hagan and Petersen (1995) have suggested that US policies designed to improve the economic situation of inner-city neighborhoods have often instead served to perversely concentrate poverty. They point out that only in the areas with the highest concentrations of deprivation do blacks have homicide levels higher than whites, suggesting that the relationship between blacks and crime is the result of economic status and not of race.

Krivo and Petersen (1996) have examined whether extremely deprived neighborhoods have unusually high levels of crime when compared to other deprived neighborhoods. Drawing heavily on Wilson's ideas regarding social isolation, they suggest that severely disadvantaged communities simply cannot maintain basic institutions or local sources of social control. In other words, the social environment of very disadvantaged areas is structurally different from that of more advantaged areas and these qualitatively different features of the social environment are responsible for unusually high levels of crime. Regression analysis is performed for census tracts in Columbus, Ohio using an index of disadvantage that includes four measures – poverty, family disruption, male joblessness, and occupational composition. The results show that regardless of the dominant racial makeup of an area, extremely disadvantaged neighborhoods have a qualitatively higher level of crime than neighborhoods that are less disadvantaged. The authors contend that their findings support the thesis of Sampson and Wilson (1995) that sources of crime are deeply rooted in the structural differences among urban communities.

Shihadeh and Ousey (1998) have given a persuasive analysis of the relationship existing between deprivation and homicide in urban America. The authors focus on structural changes to most inner-city labor markets that occurred in the 1980s and 1990s. Typically, low-wage, low-skill production and distribution jobs were replaced with administration, information processing, and high-skill service positions in many revitalized downtown areas. Most entry-level jobs in heavy industry moved to peripheral areas that were distant from the concentrations of the lower skilled urban poor. This created greater joblessness and poverty among inner-city residents and eventually contributed to higher rates of violence. Educational opportunities also changed with this restructuring of the inner-city job market, leaving many residents even less equipped to compete for the types of jobs that were growing in the inner city. The result is that widespread deprivation becomes so embedded structurally that conventional norms are delegitimized and communities can no longer cultivate any allegiance to mainstream institutions. The policy prescription of the authors is a simple one – crime rates will only come down after society addresses the economic marginalization of the inner-city poor.

More recently, Clear *et al.* (2001) have discussed public control, in the form of incarceration, as contributing both to crime and deprivation in urban communities. The authors first point to the sharp geographic divisions of income and wealth in US cities, where residential segregation along ethnic, racial, and class lines is prevalent. Then the authors claim that high rates of incarceration work to increase levels of crime in disadvantaged areas by disrupting existing networks of private and parochial social control and destabilizing the legitimacy of existing means of formal social control. Two communities in Florida are analyzed in some detail. While some positive benefits of incarceration are recognized (e.g. reducing the fear of crime among residents), very pervasive negative effects arise when so many offenders are removed from the community. Apparently, a tipping point is reached when 1.5% of an area's total population is removed; beyond that threshold, the community's relationship with future crime seems assured. The financial difficulties that must be borne by the families left behind play a large role in the matter. Severely deprived neighborhoods, already confronted with high degrees of poverty, cannot absorb additional deficits. Moreover, existing family networks are severely disrupted and this eventually has serious social and economic consequences.

Similar opinions are offered by Dreier *et al.* (2001) who set themselves apart from many economists now studying the incidence of crime in the US. They claim that issues like access to health care, employment opportunities, quality of retail services, and the incidence of crime are all connected to form an intricate web of activities. Severe deprivation not only results from but also leads to structural failure in institutions as urban communities descend into a vicious cycle of increased resource inequality and they become progressively isolated from other parts of the city. The deprivation-crime problem cannot easily be corrected but requires complex and comprehensive policies.

23.3.2. Growth and planning

23.3.2.1. Capitalization of amenities

The meaning of capitalization in the QOL context is straightforward: real estate prices and rents reflect the value of non-market amenities, specifically environmental attributes and public goods and services. Wages are also affected but present interest is confined to property markets (see above). As discussed earlier, the availability of these attributes varies across space, thereby affecting both intraregional and interregional patterns of migration and development. This influence is explained through the concept of *compensating differentials* – people are willing to pay higher prices and rents for housing or accept lower wages for working in order to enjoy living in areas having rich endowments of non-market amenities. Similarly, people are willing to pay more to live in areas that remain free from crime or signs of poverty. Given that this is the case, desirable attributes are positively capitalized into the prices and rents of real estate and are negatively capitalized into wages;

undesirable attributes are capitalized the other way around. In short, other things being equal, people are willing to pay more, often while earning less, to live in places that are attractive for their endowments of non-market amenities. The availability and quality of these amenities may be controlled or rigorously enforced through land-use planning practices at the local, regional, and national scales.

From this point of departure, nearly all land-use policies may be viewed as market constraints affecting both the supply of, and demand for, space at a given location. Although the market for land, like other commodities, is affected by each of these factors simultaneously, it is different in the sense that the supply of *place* is perfectly inelastic (Fischel, 1985; Logan and Molotch, 1987). This may not seem noteworthy at first blush, given that the world's supply of land is, for all practical purposes, perfectly elastic – there is still plenty of room if a person is not discriminating about where to live. But, as arguments above made clear, people do care about where they live. If they did not, QOL would not matter. This point is hard to overemphasize in understanding the role of capitalization. Without this specific characteristic – inelastic supply – real estate values would not respond nearly so sharply to any local variations in the quality of environmental amenities, public goods and services, and neighborhoods.

A wide literature now addresses the capitalization of environmental amenities. Landscape diversity, the availability of natural open space, and proximity to farmland, scenic views, or wetlands all have been shown to increase land prices (Geoghegan *et al.*, 1997; Orford, 1999; Mahan *et al.*, 2000; Hardie *et al.*, 2001; Johnston *et al.*, 2001; Bastian *et al.*, 2002; Smith *et al.*, 2002). Thus, residents of both urban and rural locales benefit directly from the preservation of these land uses. Evidence also exists that these same features are responsible for people relocating from older, built-up areas to newer, low-density areas (Nelson, 1986a; Nelson and Deuker, 1990; Nelson and Sanchez, 1999). Moreover, the development patterns that subsequently arise in and around outlying exurbs are particularly sensitive to those highly valued landscape features (Esparza and Carruthers, 2000; Wu, 2001; Irwin and Bockstael, 2002; Vias and Carruthers, 2003). The strong draw of natural amenities, at both the intra- and inter-regional scales, suggests that environmental preservation may be necessary for sustainable economic development, at least in certain locales (Power, 1996; Power and Barrett, 2001). As Brueckner (2000) points out, one of the main drivers of urban sprawl is the failure to account for the full cost of open space – residents occupy more land than they would if they had to pay for the benefits that land provides to society as a whole.

Meanwhile, spatial variation in the availability, cost, and quality of local public goods and services also has a significant impact on residential property values. Early evidence of this was developed by Oates (1969), who tested the well-known Tiebout hypothesis by examining the effects that property taxes and spending on public schools had on real estate prices in the New York metropolitan region. The results of his analysis indicated that taxes were negatively capitalized into property values, but school quality was positively capitalized, and that the two roughly offset one another. This research remains notable to this day for directly testing the Tiebout framework. Since that time, numerous other studies have examined the capitalization of public services and almost all have found that this process plays a major role in shaping both intraurban and interurban property markets. One hedonic price model has recently shown that proximity to water and sewer services accounts for nearly $2000 of the value of a parcel of undeveloped land (Knaap and Nelson, 1992). Other models indicate that high-quality school districts are positively capitalized into the price of urban housing (Goodman and Thibodeau, 1998; Clark and Herrin, 2000; Downes and Zabel, 2002).

Within the Tiebout framework it is axiomatic that metropolitan fragmentation induces homogeneity among the constituent communities because taste and, more specifically, income are driving factors. In a perfectly fragmented region, for example, each individual would be their own mayor, with the ability to tailor their community to their own particular preferences (Carruthers and Ulfarsson, 2002). This basic premise has been validated by research illustrating that fragmentation leads to greater homogeneity by way of division along socioeconomic, racial, or ethnic lines (Dowding *et al.*, 1994; Heikkila, 1996; Hoyt and Rosenthal, 1997). Although the Tiebout hypothesis is one of the deepest and broadest theoretical frameworks ever advanced in the social sciences, unfortunately it lacks any real agency, leaving it unable to explain why urban property markets work the way they do in any meaningful way. Invisible municipal managers, not explicit local governments, are assumed to offer up a set of public goods and services leading to the *appearance* of competition among communities, as they seek to attract residents by offering the best combination of public goods and services at the lowest possible price.

Fischel's (2001) extension of the model, the Homevoter hypothesis, corrects for this shortcoming by introducing the deliberate, value-maximizing behavior of homeowners. Because the good and bad practices of local governments do eventually end up being capitalized into residential property values, homeowners retain a direct interest in ensuring that services are provided both efficiently and at a high level of quality. From this perspective, municipalities closely resemble traditional corporations, where homeowners are the shareholders and the overall goal is to maximize the value of assets through public policy decisions. Accountability comes into play as people vote, commonly with their feet, for those officials who keep homeowners' best interests in mind when they make decisions that affect residential property values (Brueckner and Joo, 1991).

The ability of local governments to accomplish this is largely embodied in their capability to regulate the type, amount, and pace of development that occurs within their boundaries (Fischel, 2001). Because fragmentation moves communities towards greater homogeneity, the smaller local governments are, the more narrowly their goals may be defined, and the better is their ability to fulfill the specific preferences of their residents (Hadden and Borgatta, 1965; Fischel, 1985, 1999; Pogodzinski and Sass, 1994). In general, people favor single-family housing, automobile ownership, low-rise work places, and environments free from the signs of poverty, all of which may be rigorously enforced through land-use regulations (Downs, 1994; Ladd, 1998). These same factors also help to ensure the highest possible quality of public service provision, because a tightly regulated land market secures enduring property values and reinforces socioeconomic homogeneity. Homeowners in particular may resist increased growth rates because they anticipate declines in service quality or increased taxes, both of which are negatively capitalized into real estate values (Logan and Molotch, 1987; Ladd, 1994; Brasington, 2002). Of course, from the perspective of the poor such practices can be viewed as being exclusionary (Ihlanfeldt, 2004).

Finally, property taxes are negatively capitalized because they increase the cost of holding a given property (Fischel, 2001). They also affect the spatial pattern of development, by encouraging both producers (developers) and consumers (households) to use more land (Gihring, 1999). At the local level, property taxes affect growth patterns by encouraging builders to reduce the intensity of development because the incidence of taxes on undeveloped land vs. improvements is heavily weighted towards the latter (Brueckner, 2002). At the same time, at least in the US, the federal tax write-off on mortgage interest, estimated by some to be as large as $1.6 billion annually (Gyourko and Sinai, 2001), effectively increases people's ability to consume more housing than they would without the benefit. Since this policy provides greater benefits to high-income households – who can afford to purchase bigger houses on larger lots – it further reinforces their motivations for locating in more spacious outlying areas, especially in the presence of binding land-use regulations (Voith and Gyorko, 2002).

23.3.2.2. Land-use planning

Obviously local QOL may be preserved through land-use policies that protect environmental amenities and work to secure enduring property values. In practice, both of these objectives may be met through land-use regulation, although the specific supply- and demand-side effects on the land market can be very difficult to disentangle. In fact, there seems to be three basic ways in which land-use controls can affect the desirability or ability of people to reside in a given place. First, agents might simply limit the supply of housing or restrict the amount of land that is available for development. Second, agents might increase the demand for land or housing (of various types) by forcing buyers to consume more than they otherwise would. And, related to this, agents can actually enhance the demand for land or housing by making a place and its environs more attractive to residents. A wide literature is devoted to these supply- and demand-side issues; see, for example, Fischel, 1990; Pogodzinski and Sass, 1990, 1991, 1994; Potepan, 1996; Knaap, 1998; Mayer and Somerville, 2000; Pendall, 2000; Malpezzi, 2002.

23.3.2.2.1. Supply constraints. As supply constraints, land-use policies affect property markets by limiting the ability of people (and firms) to locate in a given place. The simplest illustration of this is American zoning, which separates land uses – by type and intensity – into designated zones. Consider, for example, a 1 square mile municipality that designates its entire area for single-family housing, on 1 acre lots. Disregarding the space needed for roadways, drainage, and other public right-of-ways, this community has limited its size to exactly 640 households. Adding what Pogodzinski and Sass (1994) label *characteristics* zoning, which dictates the physical attributes of development, also defines the type (e.g. age, income, size) of households that may locate within the jurisdiction's boundaries. In the face of constant or rising demand, such restrictions cause housing values to rise, and people to be displaced – either directly, by being priced out of the market, or indirectly, by choosing to live in a more affordable location. Either way, this approach to land-use regulation leads to lower densities and higher real estate values than would otherwise be expected as people locate in nearby communities (Fischel, 1985, 1990, 1999). In this case, the land-use restriction is capitalized into the price of property by constraining the supply, described by the amount, style, and intensity, of developable space.

The spillover effects occur as a result of what Landis (1986, 1992) describes as a porous land market, and there is considerable empirical evidence that people and firms are displaced to less regulated areas (Shen, 1996; Levine, 1999), often all the way to the unincorporated urban fringe (Carruthers, 2003). No one community can alter the outcome of regional growth (Downs, 1999) so the situation presents something of a prisoner's dilemma, as jurisdictions remain focused on their own well-being, even if the region-wide outcomes this produces are undesirable (Bollens, 1993). For example, in a study of the 25 largest metropolitan areas in the United States, Pendall (1999) finds that land-use policies adopted explicitly to limit local growth rates actually contribute to urban sprawl, by creating significantly lower densities. A survey conducted as part of that research reveals that a significant proportion of jurisdictions located in those metropolitan areas have established zoning codes permitting fewer than eight housing units per acre as their maximum allowable density (Pendall *et al.*, 2002). The cumulative effect can become quite problematic, because communities adopt growth controls – including development moratoria, low-density zoning, and permit caps – in an effort to avoid increased growth rates brought on by other communities doing the same thing (Glickfield and Levine, 1992; Brueckner, 1998, 2003). In sum, there is substantial evidence that locally enacted growth controls – beginning, but not ending, with low-density zoning – create severe price effects and, in turn, significant spatial impacts as growth is directed to less-regulated areas (Carruthers and Ulfarsson, 2002; Ulfarsson and Carruthers, 2003).

Examples of regional and national policies that constrain supply are the urban growth boundaries (UGBs) found around many cities in the United States, including Portland, Oregon and Seattle, Washington, and the Greenbelts found around London, England, and throughout South Korea. UGBs directly limit the amount of space available for development, with the explicit goal of raising land prices, thereby forcing developers to substitute factors of production, building up, rather than out, and using less land but more costly construction plans and materials. Knaap (1985) also finds that UGBs influence *expectations* about when land will be put into urban use; the value of land outside cities is highly discounted because it will not be available for development until the city boundary is expanded. In addition, hedonic models reveal that UGBs impose location-specific price effects by raising (lowering) the value of residential property located just inside (outside) the boundary, essentially adding a price bump at the far edge of the traditional distance gradient (Knaap and Nelson, 1992). This occurs because residents located just inside the boundary view the city's edge as a greenbelt-like amenity but farmers located just outside the boundary view the city's edge as a disamenity, due to the inherent conflicts between urban and agricultural land uses (Nelson, 1986b, 1992).

Ultimately, the region-wide effect of a UGB depends on how strictly it constrains the land market and how well it is complemented with other policy instruments. If the boundary includes too much vacant land, the price effects will be minimal; however, if the boundary is drawn too tightly, it will create excessively high land values (Brueckner, 2000). In Oregon, where most evaluations of UGBs have been carried out, they have for the most part been found to be successful at separating urban from rural land uses and constraining urban sprawl. Even so, while most urban development remains inside of the boundaries, overall densities are much lower than desired, and development often exhibits patterns characteristic of sprawl (Weitz and Moore, 1998). Kline and Alig (1999) illustrate the difficulties involved in containing urban development by using a probit model to simulate land market processes in the Portland region. They find that parcels located within UGBs and subject to state-approved land-use plans are more likely to be developed than those that are not, and that some development leaks out of the UGB as a consequence of the practice not being complemented by other land-use policies. Recent thinking suggests that the effectiveness of UGBs could be improved by coordinating uses more carefully with infrastructure investments (Ding *et al.*, 1999; Knaap *et al.*, 2001) and by giving more attention to how growth boundaries affect the actual supply of developable space (Knaap and Hopkins, 2001).

Greenbelts are similar to, but qualitatively different from, UGBs: instead of separating urban and rural areas from one another with a line, they do so with open space that cannot be developed. The original and perhaps most famous greenbelt was implemented in London, England in 1938. Inspired by Howard's (1898) garden cities concept, the idea was to separate the industrial urban core from outlying rural areas in an effort to promote healthier living through access to clean, natural open space. Since that time, greenbelts have been put in place in major cities worldwide, including Ottawa, Canada; Moscow, Russia; New Delhi, India; and Seoul, South Korea (Lee, 1999). The latter is particularly notable because it is part of a nationwide system – enacted through the so-called *First National Spatial Development Plan*, in 1971 – surrounding all major cities in South Korea (Kahng, 1988).

Like UGBs in the United States, the Korean greenbelt system, and especially the one surrounding Seoul, has been extensively analyzed and often criticized from a benefit–cost standpoint. At issue here is the relationship between the types of amenities produced by greenbelts and their effect on the supply of land in the interior urban core (Lee and Fujita, 1997). Lee and Linneman (1998) find that, from an economic standpoint, the efficiency of Seoul's greenbelt has changed through time. Specifically, using a hedonic approach, they illustrate that the marginal value of accessibility to the greenbelt rose through the 1970s, after it was first implemented, but steadily declined during the 1980s, as a result of congestion effects imposed by its restrictions on developable land at the interior. The analysis concludes that greenbelts are worthwhile until they put undue pressure on the housing market, raising prices to unduly high levels, or on interior open space, leading to losses of undeveloped land, which would not have occurred without the greenbelt. Lee (1999) again criticizes the Seoul greenbelt for the congestion effects that it creates but notes that this could be mitigated by allowing for greater flexibility in its designation – an argument similar to the one presented by Knaap and Hopkins (2001) with respect to American UGBs. Finally, Cho (2002) calls for widespread reform in the Korean approach to growth management, arguing for relaxation of limitations on development in general and greenbelt boundaries in particular. All of these criticisms stem from the extreme difficulties involved in balancing the benefits of greenbelts with the strict constraints that they impose on regional land markets (Lee and Fujita, 1997).

Despite the numerous criticisms of land-use policies as supply constraints, their adoption is likely to continue to expand throughout the United States, Europe, and Asia. Although land-use controls have many drawbacks, they also produce numerous benefits, by making areas more attractive to residents. Even as land-use policies restrict the supply of housing and other developments, they raise people's demand for living in specific locations by making them more attractive, and they secure a more stable tax base for the provision of the highest quality public services.

23.3.2.2.2. Demand effects. As noted, land-use policies affect demand in two different ways: by forcing residents to consume more land or housing than they otherwise would, and by making places better to live in (Lillydahl and Singell, 1987; Knaap, 1998; Dawkins and Nelson, 2002). The first of these effects is very straightforward and is linked primarily to the kind of low-density zoning commonly used in communities throughout the United States. Specifically, since zoning dictates the intensity and style of development (in addition to type), residents and firms may literally be forced to consume more space than they otherwise would, if they wish to occupy a given area. Fischel (1985, 1990) points to this as a sign of the inefficiency of land-use controls, indicating that the kind of displacement discussed above would not occur if they were truly efficient – in other words, if their benefits perfectly offset their costs. The key point to recognize with respect to this type of demand-side effect is that it does not (necessarily) occur as a result of consumer choice. On the other hand, some observers (McMillen and McDonald, 1991; Pogodzinski and Sass, 1994) have raised the proposition of *endogenous* zoning, meaning that land-use controls follow the market and land values are thereby deliberately set by local governments through their regulatory activities.

This proposition points to the more complicated of the two demand-side effects, where land-use regulations increase the desirability of communities as places to live. If this is the case, residents will willingly pay more to live in a given location as a result of the amenity benefits associated with land-use regulation (Carruthers and Ulfarsson, 2002; Ulfarsson and Carruthers, 2003). These include the types of environmental features and the public goods and services discussed earlier. Not only do growth controls work to preserve natural amenities, they also secure enduring real estate values, by acting as a form of insurance against unwanted land uses (Fischel, 1999). Moreover, by raising property values – through both supply- and demand-side effects – land-use regulations enable communities to provide high-quality public services, while keeping tax rates relatively low. The tragedy of the kind of urban decline discussed above, for example, is that it is a self-reinforcing cycle, where property values fall, public service provision declines, and taxes rise, placing ever greater pressures on affluent residents to relocate (Orfield, 1997, 2002; Rusk, 1999). All of this relates back to the discussion on capitalization; although there is a strong theoretical base for understanding location-specific demand (Tiebout, 1956; Fischel, 2001), much empirical work remains to be done on this topic.

23.3.2.3. Growth control vs. growth management

The preceding discussion points to four distinct objectives for land-use policies: maintaining property values, shaping a compact urban form, ensuring cost-effective public service provision, and mitigating the socioeconomic inequalities that arise from patterns of urban and regional development. They also enable policies aimed at improving urban QOL to be delineated into two separable categories: growth control and growth management. The first of these involves slowing or restricting the amount, character, or pace of local development, while the other entails accommodating growth in a way that is consistent with overall public health, safety, and welfare (Nelson and Peterman, 2000; Carruthers, 2002a,b).

23.3.2.3.1. Growth control. As already mentioned, land-use policies are generally implemented with the expectation that they will raise property values, but this does not necessarily mean that such policies must work to the detriment of low-income people. The spillover effects described above happen as a result of individuals (or firms) being priced out of local property markets – often as a deliberate exercise of parochialism (Downs, 1999). Mayer and Somerville (2000) specifically describe local land-use regulation as being acrimonious, arguing that it deliberately raises the cost of new construction by adding delays and uncertainty to the development process. Their analysis of new construction illustrates that US metropolitan areas with high levels of land-use regulation have up to 45% fewer starts and significantly lower price elasticities of demand for housing than less regulated markets. Urban sprawl results as people who cannot afford to live in more regulated interior spaces are forced outward to the urban fringe. Here housing is less expensive due to its relative distance from important nodes of activity (such as the CBD and other subcenters) and because outlying land often remains comparatively free from regulation (Daniels, 1999). There is increasing evidence that urban sprawl actually increases the ability of blacks and other low-income minorities to own housing (Kahn, 2001).

Meanwhile, sprawl per se has evolved into one of the most vexing problems faced by urban and regional policymakers in the United States (Downs, 1994; Burchell, 1998; Daniels, 1999; GAO, 1999; Rusk, 1999; Fulton *et al.*, 2001; Glaeser *et al.*, 2001a; Orfield, 2002; Squires, 2002; Lang, 2003). Within the context of growth controls, the problem is perpetuated by the difficulty of separating the negative consequences of far-flung, low-density development patterns from their positive implications. Local governments represent the medium through which people express their collective residential consumer preferences, so policies tend to be inward looking, even if the result is an outcome that is undesirable for the region at large. A cliché among urban planners is that the two things residents hate most are density and sprawl. Even so, it remains unclear whether the less compact (more sprawling) development patterns produced by local growth controls, including low-density zoning, really do produce net benefits for the public at large. In particular, they seem to perpetuate the cycle of inner-city deprivation described above (Pendall, 2000) and often produce a pattern of development that tends to raise the cost of public services to unacceptably high levels.

The latter dilemma unfolds as follows. Although a robust tax base works to secure high-quality public goods provision, restrictions on the character or pace of growth are not the only way to achieve this result. In particular, contrary to popular opinion, high-density areas pay for themselves through their correspondingly high property values *and* they cost less to deliver services to by creating economies of scale or scope (Knaap and Nelson, 1992). Recent empirical evidence suggests that the cost of providing most forms of services – capital facilities, roadways, sewage and trash collection, housing and community development, police and fire protection, schools and education, parks, and libraries – declines with density and increases with the spatial extent of urbanized land (Carruthers and Ulfarsson, 2003). A site-based analysis by Speir and Stephenson (2002) reveals similar findings for public water and sewer costs. These findings contradict those of an earlier oft-cited analysis by Ladd (1992) that used a questionable measure of density, while not controlling for property values, to illustrate that high-density areas are more expensive to serve. While further work needs to be done on this topic, it should be clear that land-use policies used explicitly to control growth at the local level are not the only means to ensure high-quality, cost-effective service provision.

Finally, local growth controls do little to ameliorate the proliferation of urban pathologies, including crime and concentrated poverty – in fact, they may even contribute to these problems through their exclusionary consequences (Downs, 1994, 1999; Orfield, 1997, 2002; Rusk, 1999). Since no single jurisdiction can stem the tide of regional growth and change, problems associated with urban deprivation are better dealt with at the supra-local scale. Urban blight moves systematically across metropolitan areas by the process of filtering in the real estate market – the primary means of providing affordable housing in the United States and many other nations. This process involves a decrease in both housing services and occupant income, so that older structures are passed down through the market much in the same way used automobiles are (O'Sullivan, 2003). In this way, the filtering of housing is a natural part of the evolution of metropolitan housing markets: units in once affluent areas age and often become dilapidated, which causes their value to fall and enables lower income individuals to move in. Unless other mediating factors intervene, such as the desire of young affluents or those with alternative lifestyles to locate in older core space (Florida, 2002a), the cycle becomes self-reinforcing, with decline and deterioration systematically making its way across the urban landscape.

23.3.2.3.2. Growth management. Compared to growth control, growth management entails *accommodating* growth in a way that meets a predefined set of normative criteria (Carruthers, 2002a). One problem that planning faces almost worldwide, however, is the lack of some well-defined standard for what constitutes the ideal urban environment (Talen and Ellis, 2002), including various patterns of development conforming to this ideal (Galster *et al.*, 2001). The closest that academic research has come to identifying this is Lynch's (1981) famed *Theory of Good City Form*, which suggests that development patterns may be evaluated on the basis of five dimensions (vitality, sense, fit, access, and control) and two meta-criteria (efficiency and justice). In combination, these social use values describe how well the urban environment serves the needs of its populace and promotes or preserves their QOL. Ultimately, they suggest that an ideal urban form is one that is dynamic and responsive to the needs of its residents – in short, one that produces net benefits for the public at large and that may be continually adapted to minimize the appearance of negative externalities. This points back to some of the key objectives of planning noted above: maintaining the use *and* exchange value of property, shaping a compact urban form, ensuring cost-effective public service provision, and mitigating the socioeconomic inequalities that arise from land-development processes. All of these things may be accomplished through growth management, which generally involves supra-local intervention in the processes of urban and regional development (Bollens, 1993).

In the US, management is typically enforced through regional- and state-level planning authorities. In the first case, regional planning organizations control the flow of federal transportation dollars (as mandated by the ISTEA and subsequent TEA21 legislation), thereby exerting influence over the path of growth through careful infrastructure planning. Some regional planning organizations, such as Portland's *Metro*, Seattle's *Puget Sound Regional Council*, and San Diego's *Association of Governments*, exert even greater control by involving themselves directly in local planning practices. This may involve, for example, coordinating the planning activities of adjacent jurisdictions or helping to ensure that each community in the metropolitan area provides its fair share of affordable housing (Carruthers, 2002a). Other regional planning organizations, such as Minneapolis-St. Paul's *Metropolitan Council*, coordinate tax-base sharing programs in a direct effort to prevent aging communities from sliding into decline (Orfield, 1997). Empirical evidence suggests that there are measurable benefits to be gained from such interjurisdictional cooperation (Haughwout, 1999). Meanwhile, in the second instance, state governments increasingly require localities to produce land-use plans, often based on specific criteria for how growth is to be managed (Burby and May, 1997). Beyond planning mandates, numerous states, including Florida, Oregon, Vermont, and Washington, have adopted formal growth management programs requiring not only that communities prepare land-use plans but that they also use specific policy instruments, such as UGBs or concurrency requirements to implement them (Bollens, 1992; Gale, 1992). The relative success of these programs rests on how well they are designed to meet the needs of urban residents and each state's overall commitment to enforcing mandated planning practices (Carruthers, 2002b).

In addition to the greenbelts discussed above, a good non-US example of a centralized planning effort aimed at improving QOL is the Dutch government's effort to integrate environmental sustainability with land-use planning (Miller and de Roo, 1999). The Netherlands has advanced a national policy that involves mapping spatial patterns of pollution within urban areas, in an effort to create livable cities without the extreme separation of incompatible land uses – such as industrial and residential – that is imposed by American-style zoning. For example, *integrated environmental* zoning is used to minimize the negative externalities associated with manufacturing and other activities that may negatively affect the livability of people residing in adjacent neighborhoods (Miller and de Roo, 1996). This pragmatic approach is critical, not only to the QOL of its citizens, but also to the nation's overall economic viability. Due to its inherent scarcity, all land must be used as efficiently as possible to accommodate the various residential, commercial, and other land-use needs of the nation.

In sum, centralized land-use planning can work to produce many of the same basic goals that local planning efforts seek to accomplish, with the added benefit of equity among residents. But rising incomes, which enable people to consume more space, and falling commuting costs, which enable people to travel further between home and work, continue to work against any alteration of today's suburbanization trends (Mieszkowski and Mills, 1993). However, growth management stands as a future option to ensure a high QOL for all urban residents. The benefits of growth management operate through the same process of property capitalization that was discussed earlier but this comprehensive approach might well generate higher private plus social benefits to all residents than the more locally-oriented approaches to land-use control. Ultimately, though, a great deal of additional empirical research is needed to link policy goals with tangible QOL benefits.

23.4. Concluding remarks

This chapter has summarized the recent multidisciplinary literature addressing the complex relationships existing between cities and QOL. Hedonic models have been emphasized but other perspectives have been included. Natural amenities like climate and topography remain important in household migration and are partially responsible for the high housing costs of some cities. However, fiscal prudence, cultural and lifestyle tolerance, and the responsible management of key human-made amenities – especially crime, education, and land use – are increasingly seen as being critical for the continued success of cities. In order to be competitive in a global, high-tech economy, firms must be able to attract high human-capital workers. But these people prefer to live in large cities with broad QOL appeal or smaller places with specific QOL appeal. These same persons avoid areas of high crime, locally if not regionally, and they want their children to be educated in high-quality school districts. Housing costs are bid up accordingly and high taxes ensure the provision of high-quality public goods and services. Especially in large urban areas, these same people tolerate a wide diversity of lifestyles and, increasingly, they demand an orderly and aesthetically pleasing urban landscape. With non-interventionist state- and national-level public policies, and political fragmentation in metropolitan areas, existing resource and life-opportunity gaps between the most advantaged and the most deprived will only widen in our largest cities.

References

Adamson, D.W., D.E. Clark and M.D. Partridge (2002), *Do Urban Agglomeration Effects and Household Amenities have a Skill Bias*? Paper presented at the annual meeting of the Western Regional Science Association, Monterey, CA.

Atkinson, R. and K. Kintrea (2001), "Disentangling area effects: evidence from deprived and non-deprived neighborhoods", *Urban Studies*, Vol. 38, pp. 2277–2298.

Barkley, D.L., M.S. Henry and S. Bao (1996), "Identifying spread vs. backwash effects in regional economic areas: a density function approach", *Land Economics*, Vol. 72, pp. 334–357.

Barro, R. (1991), "Economic growth in a cross-section of countries", *Quarterly Journal of Economics*, Vol. 106, pp. 407–444.

Barro, R. and X. Sala-i-Martin (1992), "Convergence", *Journal of Political Economy*, Vol. 100, pp. 223–251.

Bartik, T.J. (1985), "Business location decisions in the United States: estimates of the effects of unionization, taxes, and other characteristics of states", *Journal of Business and Economic Statistics*, Vol. 3, pp. 14–22.

Bartik, T.J. (1988), "The effects of environmental regulation on business location in the United States", *Growth and Change*, Vol. 19(3), pp. 22–44.

Bartik, T.J. (1989), "Small business start-ups in the United States: estimates of the effects of characteristics of states", *Southern Economic Journal*, Vol. 55, pp. 1004–1018.

Bartik, T.J. and V.K. Smith (1987), "Urban amenities and public policy", pp. 1207–1254, in: E. Mills, editor, *Handbook of Regional and Urban Economics: Volume 2, Urban Economics*, Amsterdam: North-Holland.

Bastian, C., D. McLeod, M. Germino, W. Reiners and B. Blasko (2002), "Environmental amenities and agricultural land values: a hedonic model using geographic information systems data", *Ecological Economics*, Vol. 40, pp. 337–349.

Beale, C.L. and K.M. Johnson (1998), "The identification of recreational counties in nonmetropolitan areas of the USA", *Population Research and Policy Review*, Vol. 17, pp. 37–53.

Becker, R.A., L. Denby, R. McGill and A.R. Wilks (1987), "Analysis of data from the *Places Rated Almanac*", *The American Statistician*, Vol. 41, pp. 169–186.

Beeson, P.E. (1991), "Amenities and regional differences in returns to worker characteristics", *Journal of Urban Economics*, Vol. 30, pp. 224–241.

Beeson, P.E. (1992), "Agglomeration economies and productivity growth", pp. 19–35, in: E.S. Mills and J.F. MacDonald, editors, *Sources of Metropolitan Growth*, New Brunswick, NJ: Center for Urban Policy Research.

Beeson, P.E. and R.W. Eberts (1989), "Identifying productivity and amenity effects in interurban wage differentials", *Review of Economics and Statistics*, Vol. 71, pp. 443–452.

Berger, M.C., G.C. Blomquist and W. Waldner (1987), "A revealed-preference ranking of quality of life for metropolitan areas", *Social Science Quarterly*, Vol. 68, pp. 761–778.

Berry, B.J.L. and F.E. Horton (1974), *Urban Environmental Management*, Englewood Cliffs, NJ: Prentice-Hall.

Berry, B.J.L. and J.D. Kasarda (1977), *Contemporary Urban Ecology*, New York: Macmillan.

Beyers, W.B. (1992), "Producer services and metropolitan growth and development", pp. 125–158, in: E.S. Mills and J.F. MacDonald, editors, *Sources of Metropolitan Growth*, New Brunswick, NJ: Center for Urban Policy Research.

Blair, J.P. and R. Premus (1987), "Major factors in industrial location: a review", *Economic Development Quarterly*, Vol. 1, pp. 72–85.

Blakely, E.J. (1994), *Planning Local Economic Development: Theory and Practice*, 2nd edition, California: Sage.

Blanchflower, D.G. and A.J. Oswald (1994), *The Wage Curve*, Cambridge, MA: MIT Press.

Blanchflower, D.G. and A.J. Oswald (1995), "An introduction to the wage curve", *Journal of Economic Perspectives*, Vol. 9, pp. 153–167.

Blomquist, G.C., M.C. Berger and J.P. Hoehn (1988), "New estimates of quality of life in urban areas", *American Economic Review*, Vol. 78, pp. 89–107.

Boal, F.W. (1998), "Exclusion and inclusion: segregation and deprivation in Belfast", in: S. Musterd and W. Ostendorf, editors, *Urban Segregation and the Welfare State: Inequality and Exclusion in Western Cities*, London: Routledge.

Boarnet, M.G. (1994), "An empirical model of intrametropolitan population and employment growth", *Papers in Regional Science*, Vol. 73, pp. 135–153.

Boarnet, M.G., M. Greenwald and T.E. McMillan (2003), Planning and health promotion: quantifying links between land use, walking, and physical activity. Unpublished manuscript, 39 pp.

Bollens, S. (1992), "State growth management: intergovernmental frameworks and policy objectives", *Journal of the American Planning Association*, Vol. 58, pp. 454–466.

Bollens, S. (1993), "Restructuring land use governance", *Journal of Planning Literature*, Vol. 7, pp. 211–226.

Bolton, R.E. (1992), "'Place prosperity vs people prosperity' revisited: an old issue with a new angle", *Urban Studies*, Vol. 29, pp. 185–203.

Bolton, R.E. (2002), "Place surplus, exit, voice and loyalty", pp. 469–488, in: B. Johansson, C. Karlsson and R. Stough, editors, *Regional Policies and Comparative Advantage*, Cheltenham: Edward Elgar.

Boyer, R. and D. Savageau (1981), *Places Rated Almanac*, Chicago: Rand McNally.

Boyer, R. and D. Savageau (1989), *Places Rated Almanac*, New York: Prentice-Hall.

Brasington, D.M. (2002), "Edge versus center: finding common ground in the capitalization debate", *Journal of Urban Economics*, Vol. 52, pp. 524–541.

Broadway, M.J. (1989), "A comparison of patterns of urban deprivation between Canadian and U.S. cities", *Social Indicators Research*, Vol. 21, pp. 531–551.

Broadway, M.J. (1992), "Differences in inner-city deprivation: an analysis of seven Canadian cities", *The Canadian Geographer*, Vol. 36, pp. 189–196.

Broadway, M.J. and G. Jesty (1998), "Are Canadian inner cities becoming more dissimilar? An analysis of urban deprivation indicators", *Urban Studies*, Vol. 35, pp. 1423–1438.

Broadway, M.J. and S.D. Snyder (1989), "The persistence of urban deprivation: the example of Wichita, Kansas in the 1970s", *Growth and Change*, Vol. 20, pp. 50–61.

Brueckner, J. (1998), "Testing for strategic interaction among local governments: the case of growth controls", *Journal of Urban Economics*, Vol. 44, pp. 467–483.

Brueckner, J. (2000), "Urban sprawl: diagnosis and remedies", *International Regional Science Review*, Vol. 23, pp. 160–171.

Brueckner, J. (2002), "Property taxation and urban sprawl", in: W. Oates, editor, *Property Taxation and Local Government Finance*, Cambridge, MA: Lincoln Institute.

Brueckner, J. (2003), "Strategic interaction among local governments: an overview of empirical studies", *International Regional Science Review*, Vol. 26, pp. 175–188.

Brueckner, J.K. and M. Joo (1991), "Voting with capitalization", *Regional Science and Urban Economics*, Vol. 21, pp. 453–467.

Burby, R. and P. May (1997), *Making Governments Plan*, Baltimore: Johns Hopkins University Press.

Burchell, R.W. (1998), *The Costs of Sprawl Revisited*, Washington: National Academy Press.

Burnell, J.D. and G. Galster (1992), "Quality-of-life measurements and urban size: an empirical note", *Urban Studies*, Vol. 29, pp. 727–735.

Bursik, R.J., Jr. (1986), "Delinquency rates as sources of ecological change", pp. 61–74, in: J.M. Byrne and R.J. Sampson, editors, *The Social Ecology of Crime*, New York: Springer.

Bursik, R.J. and H.G. Grasmik (1993), *Neighborhoods and Crime: Dimensions of Effective Community Control*, New York: Lexington Books.

Burton, E. (2000), "The compact city: just or just compact? A preliminary analysis", *Urban Studies*, Vol. 37, pp. 1969–2001.

Carlino, G.A. and E.S. Mills (1987), "The determinants of county growth", *Journal of Regional Science*, Vol. 27, pp. 39–54.

Carruthers, J. (2002a), "Evaluating the effectiveness of regulatory growth management programs: an analytical framework", *Journal of Planning Education and Research*, Vol. 21, pp. 406–420.

Carruthers, J. (2002b), "The impacts of state growth management programmes: a comparative analysis", *Urban Studies*, Vol. 39, pp. 1959–1982.

Carruthers, J.I. (2003), "Growth at the fringe: the influence of political fragmentation in United States metropolitan areas", *Papers in Regional Science*, Vol. 84, pp. 475–499.

Carruthers, J. and G.F. Ulfarsson (2002), "Fragmentation and sprawl: evidence from interregional analysis", *Growth and Change*, Vol. 33, pp. 312–340.

Carruthers, J. and G.F. Ulfarsson (2003), "Urban sprawl and the cost of public services", *Environment and Planning B*, Vol. 30, pp. 503–522.

Cebula, R.J. and R.E. Vedder (1973), "A note on migration, economic opportunity, and the quality of life", *Journal of Regional Science*, Vol. 13, pp. 205–211.

Chapple, K., A. Markusen, G. Schrock, D. Yamamoto and P. Yu (2004), "Gauging metropolitan "high-tech" and "i-tech" activity", *Economic Development Quarterly*, Vol. 18, pp. 10–29.

Cheshire, P. (1990), "Explaining the recent performance of the European Community's major urban regions", *Urban Studies*, Vol. 27, pp. 311–333.

Cheshire, P. (1995), "A new phase of urban development in Western Europe? The evidence for the 1980s", *Urban Studies*, Vol. 32, pp. 1045–1063.

Cheshire, P. and G. Carbonaro (1996), "Urban economic growth in Europe: testing theory and policy prescriptions", *Urban Studies*, Vol. 33, pp. 1111–1128.

Cheshire, P., G. Carbonaro and D. Hay (1986), "Problems of urban decline and growth in EEC countries: or measuring degrees of elephantness", *Urban Studies*, Vol. 23, pp. 131–149.

Cho, C. (2002), "Korean growth management programs: issues, problems, and possible reforms", *Land Use Policy*, Vol. 19, pp. 12–27.

Clark, D.E. and J.C. Cosgrove (1991), "Amenities versus labor market opportunities: choosing the optimal distance to move", *Journal of Regional Science*, Vol. 31, pp. 311–328.

Clark, D.E. and W.E. Herrin (2000), "The impact of public school attributes on home sale prices in California", *Growth and Change*, Vol. 31, pp. 385–407.

Clark, D.E. and W.J. Hunter (1992), "The impact of economic opportunity, amenities and fiscal factors on age-specific migration rates", *Journal of Regional Science*, Vol. 32, pp. 349–365.

Clark, D.E. and C.A. Murphy (1996), "Countywide employment and population growth: an analysis of the 1980s", *Journal of Regional Science*, Vol. 36, pp. 235–256.

Clear, T.R., D.R. Rose and J.A. Ryder (2001), "Incarceration and the community: the problem of removing and returning offenders", *Crime and Delinquency*, Vol. 47, pp. 335–351.

Costa, D.L. and M.E. Kahn (2000), "Power couples: changes in the locational choice of the college educated, 1940–1990", *Quarterly Journal of Economics*, Vol. 115, pp. 1287–1315.

Cullen, J.B. and S.D. Levitt (1999), "Crime, urban flight, and the consequences for cities", *The Review of Economics and Statistics*, Vol. 81, pp. 159–169.

Daniels, T. (1999), *When City and Country Collide: Managing Growth in the Metropolitan Fringe*, Washington: Island Press.

Dawkins, C. and A. Nelson (2002), "Urban containment policies and housing prices: an international comparison with implications for future research", *Land Use Policy*, Vol. 19, pp. 1–12.

Deller, S.C. and T.-H. Tsai (1998), "An examination of the wage curve: a research note", *Journal of Regional Analysis and Policy*, Vol. 28, pp. 3–12.

Deller, S.C. and N. Zhing (2003), *The Wage Curve, Natural Amenities and Recreational Opportunities*. Paper presented at the annual meeting of the Western Regional Science Association, Rio Rico, AZ.

Deller, S.C., T.-H. Tsai, D.W. Marcouiller and D.B.K. English (2001), "The role of amenities and quality-of-life in rural economic growth", *American Journal of Agricultural Economics*, Vol. 83, pp. 352–365.

DeVol, R.C., P. Wong, J. Catapano and G. Robitshek (1999), *America's High-Tech Economy: Growth, Development, and Risks for Metropolitan Areas*, Santa Monica, CA: Milken Institute.

Diamond, D.B. and G.S. Tolley (eds.) (1982), *The Economics of Urban Amenities*, New York: Academic Press.

Ding, C., G. Knaap and L. Hopkins (1999), "Managing urban growth with urban growth boundaries: a theoretical analysis", *Journal of Urban Economics*, Vol. 46, pp. 53–68.

Dissart, J.-C. and S.C. Deller (2000), "Quality of life in the planning literature", *Journal of Planning Literature*, Vol. 1, pp. 36–61.

Douglas, S. (1997), "Estimating relative standard of living in the United States using cross-migration data", *Journal of Regional Science*, Vol. 37, pp. 411–436.

Douglas, S.M. and H.J. Wall (1993), "'Voting with your feet' and the quality of life index: a simple non-parametric approach applied to Canada", *Economics Letters*, Vol. 42, pp. 229–236.

Dowding, K., P. John and S. Biggs (1994), "Tiebout: a survey of the empirical literature", *Urban Studies*, Vol. 31, pp. 767–798.

Downes, T. and J. Zabel (2002), "The impact of school characteristics on house prices: Chicago 1987–1991", *Journal of Urban Economics*, Vol. 52, pp. 1–25.

Downs, A. (1994), *New Visions for Metropolitan America*, Cambridge, MA: Lincoln Institute.

Downs, A. (1999), "Some realities about sprawl and decline", *Housing Policy Debate*, Vol. 10, pp. 955–974.

Dreier, P., J. Mollenkopf and T. Swanstrom (2001), *Place Matters: Metropolitics for the Twenty-First Century*, Lawrence, KS: University Press of Kansas.

Duffy-Deno, K.T. (1997), "The effect of state parks on the county economies of the West", *Journal of Leisure Research*, Vol. 29, pp. 201–224.

Duffy-Deno, K.T. (1998), "The effect of federal wilderness on county growth in the intermountain western United States", *Journal of Regional Science*, Vol. 38, pp. 235–256.

Eaton, J. and Z. Eckstein (1997), "Cities and growth: theory and evidence from France and Japan", *Regional Science and Urban Economics*, Vol. 27, pp. 443–474.

Esparza, A. and J. Carruthers (2000), "Land use planning and exurbanization in the rural Mountain West: evidence from Arizona", *Journal of Planning Education and Research*, Vol. 20, pp. 26–39.

Fieldhouse, E.A. and R. Tye (1996), "Deprived people or deprived places? Exploring the ecological fallacy in studies of deprivation with the samples of anonymised records", *Environment and Planning A*, Vol. 28, pp. 237–259.

Fischel, W. (1985), *The Economics of Zoning Laws: A Property Rights Approach to American Land Use Controls*, Baltimore: Johns Hopkins University Press.

Fischel, W. (1990), *Do Growth Controls Matter? A Review of Empirical Evidence on the Effectiveness and Efficiency of Local Government Land Use Regulation*, Cambridge, MA: Lincoln Institute.

Fischel, W. (1999), "Does the American way of zoning cause the suburbs of metropolitan areas to be too spread out?", in: A. Altshuler, W. Morrill, H. Wolman and F. Mitchel, editors, *Governance and Opportunity in Metropolitan America*, Washington: National Academy Press.

Fischel, W.A. (2001), *The Homevoter Hypothesis: How Home Values Influence Local Government Taxation, School Finance, and Land Use Policies*, Cambridge, MA: Harvard University Press.

Florida, R. (2002a), "Bohemia and economic geography", *Journal of Economic Geography*, Vol. 2, pp. 55–71.

Florida, R. (2002b), "The economic geography of talent", *Annals of the Association of American Geographers*, Vol. 92, pp. 743–755.

Florida, R. (2002c), *The Rise of the Creative Class; and How it's Transforming Work, Leisure, and Everyday Life*, New York: Basic Books.

Florida, R. and G. Gates (2001), *Technology and Tolerance: The Importance of Diversity to High-Tech Growth*, Washington: Brookings Institute.

Foster, R. (1977), "Economic and quality of life factors in industrial location decisions", *Social Indicators Research*, Vol. 4, pp. 247–265.

Fulton, W., R. Pendall, M. Nguyen and A. Harrison (2001), *Who Sprawls the Most? How Growth Patterns Differ Across the U.S*, Washington: Brookings Institute.

Gale, D. (1992), "Eight state-sponsored growth management programs: a comparative analysis", *Journal of the American Planning Association*, Vol. 58, pp. 425–439.

Galster, G., R. Hanson, M. Ratcliffe, H. Wolman, S. Coleman and J. Freihage (2001), "Wrestling sprawl to the ground: defining and measuring an elusive concept", *Housing Policy Debate*, Vol. 12, pp. 681–717.

GAO (1999), *Community Development: Extent of Federal Influence on "Urban Sprawl" is Unclear*, Washington: United States General Accounting Office.

Geoghegan, J., L. Wainger and N. Bockstael (1997), "Spatial landscape indices in a hedonic framework: an ecological economics analysis using GIS", *Ecological Economics*, Vol. 23, pp. 251–264.

Gertler, M., R. Florida, G. Gates and T. Vinodrai (2002), *Competing on Creativity*, Toronto: Ontario Ministry of Enterprise.

Giannias, D.A. (1997), "Quality of life structural analysis", *Journal of Environmental Management*, Vol. 49, pp. 157–166.

Giannias, D.A. (1998), "A quality of life based ranking of Canadian cities", *Urban Studies*, Vol. 35, pp. 2241–2251.

Gihring, T.A. (1999), "Incentive property taxation: a potential tool for urban growth management", *Journal of the American Planning Association*, Vol. 65, pp. 62–79.

Glaeser, E.L. (1994), "Cities, information, and economic growth", *Cityscape*, Vol. 1, pp. 9–47.

Glaeser, E.L. (1998), "Are cities dying?", *Journal of Economic Perspectives*, Vol. 12, pp. 139–160.

Glaeser, E.L. and D.C. Maré (2001), "Cities and skills", *Journal of Labor Economics*, Vol. 19, pp. 316–342.

Glaeser, E.L. and B. Sacerdote (1999), "Why is there more crime in cities?", *Journal of Political Economy*, Vol. 107, pp. S225–S258.

Glaeser, E.L. and J.M. Shapiro (2003), "Urban growth rates in the 1990s: is city living back?", *Journal of Regional Science*, Vol. 43, pp. 139–165.

Glaeser, E.L., H. Kallal, J. Scheinkman and A. Shleifer (1992), "Growth in cities", *Journal of Political Economy*, Vol. 100, pp. 1126–1152.

Glaeser, E.L., J.A. Scheinkman and A. Shleifer (1995), "Economic growth in a cross-section of cities", *Journal of Monetary Economics*, Vol. 36, pp. 117–143.

Glaeser, E.L., M. Kahn and C. Chu (2001a), *Job Sprawl: Employment Location in U.S. Metropolitan Areas*, Washington: Brookings Institute.

Glaeser, E.L., J. Kolko and A. Saiz (2001b), "Consumer city", *Journal of Economic Geography*, Vol. 1, pp. 27–50.

Glavac, S.M., A.C. Vias and G.F. Mulligan (1998), "Population and employment interactions in the growth of United States micropolitan centers", *Urban Geography*, Vol. 19, pp. 632–656.

Glickfield, M. and N. Levine (1992), *Regional Growth Local Reaction: The Enactment and Effects of Local Growth Control and Management Measures in California*, Cambridge, MA: Lincoln Institute.

Goodman, A.C. and T.G. Thibodeau (1998), "Housing market segmentation", *Journal of Housing Economics*, Vol. 7, pp. 121–143.

Gottlieb, P.D. (1994), "Amenities as an economic development tool: is there enough evidence?", *Economic Development Quarterly*, Vol. 8, pp. 270–285.

Gottlieb, P.D. (1995), "Residential amenities, firm location and economic development", *Urban Studies*, Vol. 32, pp. 1413–1436.

Graves, P.E. (1976), "A reexamination of migration, economic opportunity, and the quality of life", *Journal of Regional Science*, Vol. 16, pp. 107–112.

Graves, P.E. (1979), "A life-cycle empirical analysis of migration and climate by race", *Journal of Urban Economics*, Vol. 29, pp. 135–147.

Graves, P.E. (1980), "Migration and climate", *Journal of Regional Science*, Vol. 20, pp. 227–237.

Graves, P.E. (1983), "Migration with a composite amenity: the role of rents", *Journal of Regional Science*, Vol. 23, pp. 541–546.

Graves, P.E. and P. Linneman (1979), "Household migration: theoretical and empirical results", *Journal of Urban Economics*, Vol. 6, pp. 383–404.

Greenwood, M.J. and G.L. Hunt (1989), "Jobs versus amenities in the analysis of metropolitan migration", *Journal of Urban Economics*, Vol. 25, pp. 1–16.

Greenwood, M.J., G.L. Hunt, D.S. Rickman and G.I. Treyz (1991), "Migration, regional equilibrium, and the estimation of compensating differentials", *American Economic Review*, Vol. 81, pp. 1382–1390.

Gyourko, J. (1991), "How accurate are quality-of-life rankings across cities?", *Federal Reserve Bank of Philadelphia Business Review*, pp. 3–14.

Gyourko, J. and T. Sinai (2001), *The Spatial Distribution of Housing-Related Tax Benefits in the United States*, Washington: Brookings Institute.

Gyourko, J. and J. Tracy (1991), "The structure of local public finance and the quality of life", *Journal of Political Economy*, Vol. 99, pp. 774–806.

Gyourko, J., M. Kahn and J. Tracy (1999), "Quality of life and environmental conditions", in: P. Cheshire and E.S. Mills, editors, *Handbook of Regional and Urban Economics: Volume 3, Applied Urban Economics*, pp. 1413–1454, Amsterdam: North-Holland.

Hadden, J.K. and E.F. Borgatta (1965), *American Cities: Their Social Characteristics*, Chicago: Rand McNally.

Hagan, J. (1994), "The new sociology of crime and inequality in America", *Studies on Crime and Crime Prevention*, Vol. 3, pp. 7–23.

Hagan, J. and R.D. Petersen (1995), "Criminal inequality in America: patterns and consequences", in: J. Hagan and R. Petersen, editors, *Crime and Inequality*, Stanford, CA: Stanford University Press.

Hall, P. and D. Hay (1980), *Growth Centres in the European Urban System*, London: Heinemann Educational.

Handy, S., M.G. Boarnet, R. Ewing and R.E. Killingsworth (2002), "How the built environment influences physical activity: views from urban planning", *American Journal of Preventative Medicine*, Vol. 23, pp. 64–73.

Hannigan, J. (1998), *The Fantasy City: Pleasure and Profit in the Postmodern Metropolis*, London: Routledge.

Hardie, I., T. Narayan and B. Gardner (2001), "The joint influence of agricultural and nonfarm factors on real estate values: an application to the Mid-Atlantic Region", *American Journal of Agricultural Economics*, Vol. 83, pp. 120–132.

Harris, R.J. and P.A. Longley (2002), "Creating small area measures of urban deprivation", *Environment and Planning A*, Vol. 34, pp. 1073–1093.

Haughwout, A. (1999), "Regional fiscal cooperation in metropolitan areas: an exploration", *Journal of Policy Analysis and Management*, Vol. 18, pp. 579–600.

Heikkila, E. (1996), "Are municipalities Tieboutian clubs?", *Regional Science and Urban Economics*, Vol. 26, pp. 203–226.

Henry, M.S., D.L. Barkley and S. Bao (1997), "The hinterland's stake in metropolitan growth: evidence from selected southern regions", *Journal of Regional Science*, Vol. 37, pp. 479–501.

Henry, M.S., B. Schmitt, K. Kristensen, D.L. Barkley and S. Bao (1999), "Extending Carlino-Mills to examine urban size and growth impacts on proximate rural areas", *Growth and Change*, Vol. 30, pp. 526–548.

Herbert, D.T. (1975), "Urban deprivation: definition, measurement, and spatial qualities", *Geographical Journal*, Vol. 141, pp. 362–372.

Herzog, H.W., Jr. and A.M. Schlottmann (1986), "The metro rating game: what can be learned from the recent migrants?", *Growth and Change*, Vol. 17, pp. 37–50.

Heubusch, K. (1997), *The New Rating Guide to America's Small Cities*, Amherst, NY: Prometheus Books.

Howard, E. (1898), *Tomorrow: A Peaceful Path to Real Reform*, London: Faber and Faber.

Hoyt, W. and S. Rosenthal (1997), "Household location and Tiebout: do families sort according to preferences for locational amenities?", *Journal of Urban Economics*, Vol. 42, pp. 159–178.

Hughes, M. (1990), "Formation of the impacted ghetto: evidence from large metropolitan areas", *Urban Geography*, Vol. 11, pp. 265–284.

Ihlanfeldt, K.R. (2004), "Exclusionary land-use regulations within suburban communities: a review of the evidence and policy prescriptions", *Urban Studies*, Vol. 41, pp. 261–283.

Irwin, E. and N. Bockstael (2002), "Interacting agents, spatial externalities and the evolution of residential land use patterns", *Journal of Economic Geography*, Vol. 2, pp. 31–54.

Jacobs, J. (1969), *The Economy of Cities*, New York: Random House.

Jacobs, J. (1984), *Cities and the Wealth of Nations*, New York: Random House.

Jeppeson, T., J.A. List and H. Folmer (2002), "Environmental regulations and new plant location decisions: evidence from a meta-analysis", *Journal of Regional Science*, Vol. 42, pp. 19–49.

Johnston, R., J. Opaluch, T. Grigalunas and M. Mazzotta (2001), "Estimating amenity benefits of coastal farmland", *Growth and Change*, Vol. 32, pp. 305–325.

Kahn, M. (1995), "A revealed preference approach to ranking city quality of life", *Journal of Urban Economics*, Vol. 38, pp. 221–235.

Kahn, M. (1997), "Particulate pollution trends in the United States", *Regional Science and Urban Economics*, Vol. 27, pp. 87–107.

Kahn, M. (2001), "Does sprawl reduce the black/white housing consumption gap?", *Housing Policy Debate*, Vol. 12, pp. 77–86.

Kahn, M. (2002), "Smog reduction's impact on California county growth", *Journal of Regional Science*, Vol. 40, pp. 565–582.

Kahng, B. (1988), "Towards the integration of development and conservation in Seoul, Republic of Korea", *Regional Development Dialogue*, Vol. 9, pp. 108–132.

Kearns, A., K. Gibb and D. Mackay (2000), "Area deprivation in Scotland: a new assessment", *Urban Studies*, Vol. 37, pp. 1535–1559.

Kitchen, P. (2001), "An approach to measuring urban deprivation change: the example of East Montreal and the Montreal Urban community, 1986–96", *Environment and Planning A*, Vol. 33, pp. 1901–1921.

Kline, J. and R. Alig (1999), "Does land use planning slow the conversion of forest and farm lands?", *Growth and Change*, Vol. 30, pp. 3–22.

Knaap, G. (1985), "The price effects of urban growth boundaries in metropolitan Portland, Oregon", *Land Economics*, Vol. 61, pp. 26–35.

Knaap, G. (1998), "The determinants of residential property values: implications for metropolitan planning", *Journal of Planning Literature*, Vol. 12, pp. 267–282.

Knaap, G. and L. Hopkins (2001), "The inventory approach to urban growth boundaries", *Journal of the American Planning Association*, Vol. 67, pp. 314–326.

Knaap, G. and A. Nelson (1992), *The Regulated Landscape: Lessons on State Land Use Planning from Oregon*, Cambridge, MA: Lincoln Institute.

Knaap, G., C. Ding and L. Hopkins (2001), "Managing growth for efficient use of public infrastructure: toward a theory of concurrency", *International Regional Science Review*, Vol. 24, pp. 328–343.

Krivo, L.J. and R. Petersen (1996), "Extremely disadvantaged neighborhoods and urban crime", *Social Forces*, Vol. 75, pp. 619–650.

Kusmin, L. (1994), *Factors Associated with the Growth of Local and Regional Economics: A Review of Selected Empirical Literature*, Washington: U.S. Department of Agriculture.

Ladd, H. (1992), "Population growth, density, and the costs of providing public services", *Urban Studies*, Vol. 29, pp. 273–295.

Ladd, H.F. (1994), "Fiscal impacts of local population growth: a conceptual and empirical analysis", *Regional Science and Urban Economics*, Vol. 24, pp. 661–686.

Ladd, H.F. (1998), "Land use regulation as a fiscal tool", in: H. Ladd, editor, *Local Government Tax and Land Use Policies in the United States: Understanding the Links*, Cambridge, MA: Lincoln Institute.

Landis, J. (1986), "Land regulation and the price of new housing: lessons from three California cities", *Journal of the American Planning Association*, Vol. 52, pp. 9–21.

Landis, J. (1992), "Do growth controls work? A new assessment", *Journal of the American Planning Association*, Vol. 58, pp. 489–508.

Landis, J.D. and D.S. Sawicki (1988), "A planner's guide to the *Places Rated Almanac*", *Journal of the American Planning Association*, Vol. 54, pp. 336–346.

Lang, R.E. (2003), *Edgeless Cities: Exploring the Elusive Metropolis*, Washington: Brookings Institute.

Langlois, A. and P. Kitchen (2001), "Identifying and measuring dimensions of urban deprivation in Montreal: an analysis of the 1996 Census data", *Urban Studies*, Vol. 38, pp. 119–139.

Lee, C. (1999), "An intertemporal efficiency test of a greenbelt: assessing the economic impacts of Seoul's greenbelt", *Journal of Planning Education and Research*, Vol. 19, pp. 41–52.

Lee, C. and M. Fujita (1997), "Efficient configuration of a greenbelt: theoretical modeling of greenbelt amenity", *Environment and Planning A*, Vol. 29, pp. 1999–2017.

Lee, C. and P. Linneman (1998), "Dynamics of the greenbelt amenity effect on the land market – the case of Seoul's greenbelt", *Real Estate Economics*, Vol. 26, pp. 102–129.

Levine, N. (1999), "The effects of local growth controls on regional housing and population redistribution in California", *Urban Studies*, Vol. 36, pp. 2047–2068.

Ley, D. and H. Smith (2000), "Relations between deprivation and immigrant groups in large Canadian cities", *Urban Studies*, Vol. 37, pp. 37–62.

Lillydahl, J. and L. Singell (1987), "The effects of growth management on the housing market: a review of the theoretical and empirical evidence", *Journal of Urban Affairs*, Vol. 9, pp. 63–77.

Liu, B.-C. (1976), *Quality of Life Indicators in U.S. Metropolitan Areas: A Statistical Analysis*, New York: Praeger.

Logan, J. and H. Molotch (1987), *Urban Fortunes: The Political Economy of Place*, Berkeley, CA: University of California Press.

Lucas, R. (1988), "On the mechanics of economic development", *Journal of Monetary Economics*, Vol. 22, pp. 3–42.

Lynch, K. (1981), *Theory of Good City Form*, Cambridge, MA: MIT Press.

Madden, J.F. (2003), "The changing spatial concentration of income and poverty among suburbs of large metropolitan areas", *Urban Studies*, Vol. 40, pp. 481–503.

Mahan, B., S. Polasky and R. Adams (2000), "Valuing urban wetlands: a property price approach", *Land Economics*, Vol. 76, pp. 110–113.

Malpezzi, S. (2002), "Urban regulation, the "new economy", and housing prices", *Housing Policy Debate*, Vol. 13, pp. 323–349.

Marcouiller, D.W., J.G. Clendenning and R. Kedzior (2002), "Natural amenity-led development and rural planning", *Journal of Planning Literature*, Vol. 16, pp. 515–542.

Mathur, V. (1999), "Human capital-based strategy for regional economic development", *Economic Development Quarterly*, Vol. 13, pp. 203–216.

Mayer, C. and T. Somerville (2000), "Land use regulation and new construction", *Regional Science and Urban Economics*, Vol. 30, pp. 639–662.

McMillen, D. and J. McDonald (1991), "Urban land value functions with endogenous zoning", *Journal of Urban Economics*, Vol. 29, pp. 14–27.

Messner, S.F. and K. Tardiff (1986), "Economic inequality and levels of homicide: an analysis of urban neighborhoods", *Criminology*, Vol. 24, pp. 297–316.

Mieszkowski, P. and E. Mills (1993), "The causes of metropolitan suburbanization", *Journal of Economic Perspectives*, Vol. 7, pp. 135–147.

Miller, D. and G. de Roo (1996), "Integrated environmental zoning: an innovative Dutch approach to measuring and managing environmental spillovers in urban regions", *Journal of the American Planning Association*, Vol. 62, pp. 373–380.

Miller, D. and G. de Roo (1999), *Integrating City Planning and Environmental Improvement: Practical Strategies for Sustainable Urban Development*, London: Ashgate.

Millimet, D.L. and D. Slottje (2002), "Environmental compliance costs and the distribution of emissions in the U.S", *Journal of Regional Science*, Vol. 42, pp. 87–105.

Morris, A., A. Findlay, R. Paddison and R. Rogerson (1989), "Urban quality of life and the North–South divide", *Town and Country Planning*, Vol. 58, pp. 207–211.

Mueser, P.R. and P.E. Graves (1995), "Examining the role of economic opportunity and amenities in explaining population redistribution", *Journal of Urban Economics*, Vol. 37, pp. 176–200.

Nelson, A. (1986a), "Toward a theory of the American rural residential land market", *Journal of Rural Studies*, Vol. 2, pp. 309–319.

Nelson, A. (1986b), "Using land markets to evaluate urban containment programs", *Journal of the American Planning Association*, Vol. 52, pp. 156–171.

Nelson, A. (1992), "Preserving prime farmland in the face of urbanization: lessons from Oregon", *Journal of the American Planning Association*, Vol. 58, pp. 467–488.

Nelson, A. and K. Deuker (1990), "The exurbanization of America and its planning policy implications", *Journal of Planning Education and Research*, Vol. 9, pp. 91–100.

Nelson, A. and D. Peterman (2000), "Does growth management matter? The effect of growth management on economic performance", *Journal of Planning Education and Research*, Vol. 19, pp. 277–285.

Nelson, A. and T. Sanchez (1999), "Debunking the exurban myth: a comparison of suburban households", *Housing Policy Debate*, Vol. 10, pp. 689–709.

Oates, W. (1969), "The effects of property taxes and local public spending on property value: an empirical study of tax capitalization and the Tiebout hypothesis", *Journal of Political Economy*, Vol. 77, pp. 957–971.

Orfield, M. (1997), *Metropolitics: A Regional Agenda for Community and Stability*, Washington: Brookings Institute.

Orfield, M. (2002), *American Metropolitics: The New Suburban Reality*, Washington: Brookings Institute.

Orford, S. (1999), *Valuing the Built Environment: GIS and House Price Analysis*, London: Ashgate.

O'Sullivan, A. (2003), *Urban Economics*, Boston: McGraw-Hill.

Pacione, M. (1986), "Quality of life in Glasgow: an applied geographical analysis", *Environment and Planning A*, Vol. 18, pp. 1499–1520.

Pacione, M. (1989), "The urban crisis: poverty and deprivation in the Scottish city", *Scottish Geographical Magazine*, Vol. 105, pp. 101–115.

Pacione, M. (1995), "The geography of multiple deprivation in Scotland", *Applied Geography*, Vol. 15, pp. 115–133.

Pendall, R. (1999), "Do land use controls cause sprawl?", *Environment and Planning B*, Vol. 26, pp. 555–571.

Pendall, R. (2000), "Local land use regulation and the chain of exclusion", *Journal of the American Planning Association*, Vol. 66, pp. 125–142.

Pendall, R., J. Martin and W. Fulton (2002), *Holding the Line: Urban Containment in the United States*, Washington: Brookings Institute.

Piore, M.J. and C.F. Sabel (1984), *The Second Industrial Divide*, New York: Basic Books.

Plane, D.A., C. Henrie and M. Perry (2002), *Migration Across the Micropolitan/Metropolitan Spectrum*. Paper presented at the annual meeting of the Southern Regional Science Association, Arlington, VA.

Pogodzinski, J. and T. Sass (1990), "The economic theory of zoning: a critical review", *Land Economics*, Vol. 66, pp. 294–314.

Pogodzinski, J. and T. Sass (1991), "Measuring the effects of municipal zoning regulations: a survey", *Urban Studies*, Vol. 28, pp. 597–621.

Pogodzinski, J. and T. Sass (1994), "The theory and estimation of endogenous zoning", *Regional Science and Urban Economics*, Vol. 24, pp. 601–630.

Porell, F.W. (1982), "Intermetropolitan migration and quality of life", *Journal of Regional Science*, Vol. 22, pp. 137–158.

Potepan, M. (1996), "Explaining intermetropolitan variation in housing prices, rents, and land prices", *Real Estate Economics*, Vol. 24, pp. 219–245.

Power, T.M. (1996), *Lost Landscapes and Failed Economies*, Washington: Island Press.

Power, T.M. and R. Barrett (2001), *Post-Cowboy Economics: Pay and Prosperity in the New American West*, Washington: Island Press.

Putnam, R.D. (2000), *Bowling Alone*, New York: McGraw-Hill.

Quigley, J.M. (1998), "Urban diversity and economic growth", *Journal of Economic Perspectives*, Vol. 12, pp. 127–138.

Rauch, J.E. (1993), "Productivity gains from geographic concentration of human capital: evidence from the cities", *Journal of Urban Economics*, Vol. 34, pp. 380–400.

Roback, J. (1982), "Wages, rents, and the quality of life", *Journal of Political Economy*, Vol. 90, pp. 1257–1278.

Roback, J. (1988), "Wages, rents, and amenities: differences among workers and regions", *Economic Inquiry*, Vol. 26, pp. 23–41.

Roberts, B. and R.J. Stimson (1998), "Multi-sectoral qualitative analysis: a tool for assessing the competitiveness of regions and formulating strategies for economic development", *Annals of Regional Science*, Vol. 32, pp. 469–494.

Rogerson, R., A. Findlay and A. Morris (1988), "The best cities to live in", *Town and Country Planning*, Vol. 57, pp. 270–273.

Rogerson, R.J., A.M. Findlay, A.S. Morris and M.G. Coombes (1989), "Indicators of quality of life: some methodological issues", *Environment and Planning A*, Vol. 21, pp. 1655–1666.

Romer, P. (1986), "Increasing returns and long-run growth", *Journal of Political Economy*, Vol. 98, pp. S71–S101.

Rosen, S. (1974), "Hedonic prices and implicit markets: product differentiation in pure competition", *Journal of Political Economy*, Vol. 82, pp. 34–55.

Rosen, S. (1979), "Wage-based indexes of urban quality of life", pp. 74–104, in: P. Mieszkowski and M. Straszheim, editors, *Current Issues in Urban Economics*, Baltimore: Johns Hopkins Press.

Rusk, D. (1999), *Inside Game, Outside Game: Winning Strategies for Saving Urban America*, Washington: Brookings Institute.

Sampson, R.J. and W.B. Groves (1989), "Community structure and crime: testing social-disorganization theory", *American Journal of Sociology*, Vol. 94, pp. 774–802.

Sampson, R.J. and W.J. Wilson (1995), "Toward a theory of race, crime, and urban inequality", in: J. Hagan and R. Petersen, editors, *Crime and Inequality*, Stanford, CA: Stanford University Press.

Savageau, D. and R. D'Agostino (2000), *Places Rated Almanac*, New York: Hungry Minds.

Schachter, J. and P.G. Althaus (1989), "An equilibrium model of gross migration", *Journal of Regional Science*, Vol. 29, pp. 143–160.

Scott, A.J. (1992), "The spatial organization of a local labor market: employment and residential patterns in a cohort of engineering and scientific workers", *Growth and Change*, Vol. 23, pp. 94–115.

Shen, Q. (1996), "Spatial impacts of locally enacted growth controls: the San Francisco Bay Region in the 1980s", *Environment and Planning B*, Vol. 23, pp. 61–91.

Shihadeh, E.S. and G.C. Ousey (1998), "Industrial restructuring and violence: the link between entry-level jobs, economic deprivation, and black and white homicide", *Social Forces*, Vol. 77, pp. 185–206.

Simon, C.J. (1998), "Human capital and metropolitan growth", *Journal of Urban Economics*, Vol. 43, pp. 223–243.

Sjaastad, L.A. (1962), "The costs and returns of human migration", *Journal of Political Economy*, Vol. 70, pp. 80–93.

Smith, D.M. (1971), *Industrial Location*, New York: Wiley.

Smith, D.M. (1977), *Human Geography: A Welfare Approach*, London: Edward Arnold.

Smith, D.M. (1994), *Geography and Social Justice*, Cambridge, MA: Blackwell.

Smith, V.K. and J.C. Huang (1995), "Can hedonic models value air quality? A meta-analysis", *Journal of Political Economy*, Vol. 103, pp. 209–227.

Smith, K., C. Poulos and H. Kim (2002), "Treating open space as an urban amenity", *Resource and Energy Economics*, Vol. 24, pp. 109–119.

Speir, C. and K. Stephenson (2002), "Does sprawl cost us all? Isolating the effects of housing patterns on public water and sewer costs", *Journal of the American Planning Association*, Vol. 69, pp. 56–70.

Squires, G.D. (ed.) (2002), *Urban Sprawl: Causes, Consequences, and Policy Responses*, Washington: Urban Institute.

Steinnes, D.N. and W.D. Fisher (1974), "An econometric model of intraurban location", *Journal of Regional Science*, Vol. 14, pp. 65–80.

Stimson, R., S. Baum, P. Mullins and K. O'Connor (2001), "A typology of community opportunity and vulnerability in metropolitan Australia", *Papers in Regional Science*, Vol. 80, pp. 45–66.

Stover, M.E. and C.L. Leven (1992), "Methodological issues in the determination of the quality of life in urban areas", *Urban Studies*, Vol. 29, pp. 737–754.

Talen, E. and C. Ellis (2002), "Beyond relativism: reclaiming the search for good city form", *Journal of Planning Education and Research*, Vol. 22, pp. 36–49.

Thomas, G.S. (1990), *The Rating Guide to Life in America's Small Cities*, Buffalo, NY: Prometheus Books.

Thorndike, E.L. (1939), *Your City*, New York: Harcourt, Brace, & Company.

Tiebout, C.M. (1956), "A pure theory of local expenditures", *Journal of Political Economy*, Vol. 64, pp. 416–424.

Tondl, G. (2001), *Convergence after Divergence? Regional Growth in Europe*, New York: Springer.

Townsend, P. (1987), "Deprivation", *Journal of Social Policy*, Vol. 16, pp. 125–146.

Townsend, P. (1993), *The International Analysis of Poverty*, New York: Harvester Wheatsheaf.

Ulfarsson, G.F. and J.I. Carruthers (2003), *The Cycle of Fragmentation and Sprawl: A Conceptual Framework and Empirical Model*, Seattle, WA: Mundy Associates LLC.

Vias, A.C. and J.I. Carruthers (2003), *Economic Growth and Land Use Change in the Rocky Mountain West, 1982–1997*, Storrs, CN: Department of Geography, University of Connecticut.

Vias, A.C., G.F. Mulligan and A. Molin (2002), "Economic structure and socioeconomic change in America's micropolitan areas, 1970–97", *Social Science Journal*, Vol. 39, pp. 399–417.

Voith, R. and J. Gyorko (2002), "Capitalization of federal taxes, the relative price of housing, and urban form: density and sorting effects", *Regional Science and Urban Economics*, Vol. 32, pp. 673–690.

Weitz, J. and T. Moore (1998), "Development inside urban growth boundaries: Oregon's empirical evidence of continuous urban form", *Journal of the American Planning Association*, Vol. 64, pp. 424–444.

Wilson, W.J. (1987), *The Truly Disadvantaged*, Chicago: University of Chicago Press.

Wilson, W.J. (1991), "Studying inner-city social dislocations: the challenge of public agenda research", *American Sociological Review*, Vol. 56, pp. 1–14.

Witten, K., D. Exeter and A. Field (2003), "The quality of urban environments: mapping variation in access to community resources", *Urban Studies*, Vol. 40, pp. 161–177.

Wong, C. (1996), *What is Local Development? A Conceptual Overview, Occasional Paper 49*, Manchester UK: Department of Planning and Landscape, University of Manchester.

Wong, C. (1998), "Determining factors for local economic development: the perception of practitioners in the North West and Eastern regions of the UK", *Regional Studies*, Vol. 32, pp. 707–720.

Wong, C. (2002), "Developing indicators to inform local economic development in England", *Urban Studies*, Vol. 39, pp. 1833–1863.

Wu, J. (2001), "Environmental amenities and the spatial pattern of urban sprawl", *American Journal of Agricultural Economics*, Vol. 83, pp. 691–697.

Urban Dynamics and Growth, CEA, vol. 266
Roberta Capello and Peter Nijkamp (editors)

CHAPTER 24

Policy Issues in the Urban South☆

Manie Geyer

Urban and Regional Planning, University of Potchefstroom, Potchefstroom, South Africa

Abstract

The process of globalisation has gone through different phases of evolution since the 1960s and has affected different parts of the South differently during each phase. This chapter starts with characteristics of the current phase of global regionalisation and how it impacts economically on different parts of the developing world. It explains what advantages and disadvantages the neo-liberal economic development approach holds for the South and how different parts of the South are responding to them. It shows what challenges the lagging South face in its quest to reconnect to the global economy. The chapter then moves on to the concept of sustainable urban development and how the market could be made more accessible to large parts of the lagging urban South within the framework of sustainable development. It analyses the current structure of the informal urban economic sector and demonstrates how vertical integration could be achieved between the formal and informal urban economic sectors in the urban South. Finally, it looks at different models of sustainable urban development and what consequences each holds for economic development in the urban South.

Keywords: globalisation, urbanisation, economic development, sustainable development, formal/informal sector

JEL classifications: F02, O17, O21, Q56

☆ This chapter partially draws from a paper that was presented by the author at a conference on globalisation that was held by the University of Southern California at the Rockefeller Foundation International Conference Centre, Bellagio, Italy on August 21, 2002.

24.1. Introduction

When one seriously considers the development trends in the urban South it is hard to find areas in which one could generalise because of the area's great economic, social and cultural diversity. Consequently, this chapter will begin by looking at the impact that globalisation had on the economic conditions in the South. It identifies economic factors that have led to the phenomenon of global regionalisation and the different roles regions in the South have been playing in global core-peripheral economic relationships over the past two decades. It classifies developing countries into four groups and shows how global federalism has impacted on the way in which developing economies have been moulded over the past three decades. Matters that will be addressed in this part of the chapter include: regional economics, global competition in the developing world and demographic trends. It will go on to discuss the impact of these factors on divisions of labour in the South and on migration trends.

Looking at the urban South against the backdrop of global economic trends, the chapter secondly deals with the concept of economic growth and the limitations posed by human capital in the developing world on economic sustainability, environmental sustainable development and sustainable urbanisation. The chapter looks at how the market principle has to be adapted to tie in with current economic structures of the urban South. It analyses the informal urban economic structure in the developing world and shows how elements of it can be reconciled with the traditional formal urban economic sector in the urban South. Finally, it discusses different approaches to urban sustainability and ways in which urban development policies in the South can be reconciled with current options in urban sustainability.

24.2. The South in global terms

24.2.1. Changing global divisions of labour

Generally, globalisation is associated with the relative ease with which goods and information cross international boundary lines across the globe. It integrates economies around the world and reduces social differences between nations. The instantaneous transmission of images and information to once secluded corners of the globe causes the world to 'shrink'. A global localism is being created, and as a result, local views and behaviour are being influenced by people's perceptions and behaviour elsewhere in the world (Harvey, 1989; Giddens, 1990; Allen and Thompson, 1997; Cyr, 2001). The apparent decreasing differences in tastes and needs between nations increase market sizes and the interdependence between countries. Although elements of globalisation or the extent of its impact have been questioned before (Douglas and Wind, 1987; Beinart, 1997), globalisation is by and large being regarded as a growing force in the development of economies all over the world. The corporate world plays a key role in driving the process. There is a rapid growth in offshore financial markets, an explosion in mergers and acquisitions, nationally and internationally, and all the time corporations are becoming less dependent on one nation's economy alone (Emmerij, 1992).

In the process the fragmented global economy of the Cold War era is being replaced by a rapidly integrating global economy, especially in economic layers of higher technological sophistication. However, social and cultural differences between nations remain a limiting factor. While global differences are often drawn in spatial-economic terms, the divide in social-cultural terms has become more marked in certain circles in recent years (Smith, 1995; Breathnach, 2000). According to some, large parts of the developing world have been, and are still being disadvantaged by the process of globalisation (Amin, 2001). Resistance is consequently growing against the social, religious and economic influences of the West, particularly amongst Muslim and African nations. Parts of the Muslin world visibly resist the growing global homogenisation, while sections of Africa accentuate the continent's need to find its own 'auto-centeredness', its own cultural version of an African 'renaissance' (Kaya, 2001; Sihlongonyane, 2001; Geyer, 2002; Tsheola, 2002).

In the original global division of labour, manufacturing was concentrated in the core regions of the world while the periphery was lagging behind. Up until the 1960s, core-peripheral relationships were largely exploitative – the North providing manufactured products to the South in exchange for primary products. Towards the end of the 1960s globalisation entered a new phase. During this period of regional restructuring economic changes were brought about by global shifts in manufacturing. While the North maintained an advanced form of global corporate dominance (Graham *et al.*, 1988), production processes were selectively moved from high labour cost core areas to low-wage countries with quality labour and good infrastructure (Frobel *et al.*, 1980; O'Loughlin, 1989). This resulted in a new international division of labour, the dispersion of production processes from core areas to what Wallerstein (1974) calls semi-peripheral areas, causing the replication of core labour markets in such peripheral regions (Sassen, 1991). In the process, the global economic space was transformed, from the original First, Second and Third World divide to one where distinct areas of growth in the Third World can be distinguished from the lagging Third World.

Since the 1990s, the world has witnessed yet another wave of spatial economic changes. After the demise of the Soviet Block, new arrangements of political economic associations have been evolving. The mixture of post- and neo-colonial relationships that were forged during the 1960s, which resulted in what appeared to be a fragmented arrangement of economic relationships between core and peripheral regions (Poon, 1997), are now becoming a more organised arrangement of economic super blocks. Outsourcing, international industrial integration and FDI channelling (Kakabadse and Kakabadse, 2002; Kotabe and Murray, 2004), which increasingly occur within a regional context (Allen and Thompson, 1997; Feenstra and Hanson, 1997), play an important role in this. On one hand, there are the more or less clearly defined core-peripheral divisions of the Pacific Rim centring around Japan, the Americas around the United States, and Central and Western Europe around the Anglo-Franco-German core (Emmerij, 1992; Vernon, 1996; Lipietz, 1997). On the other hand, a new underlying strategic residual force seems to be emerging, causing (members of) the Russian Federation, the Middle East/North African bloc and Central Asia, to be slowly gravitating towards one another. Links between the potential members of this emerging block are still tentative, very loose and unstructured. Some associations are of a historic political, religious, and/or cultural nature, others are based on similar resources and natural conditions, and others on strategic global geo-political interests. In the process, a new global division is being created. It consists of (i) the core regions in the North that contain most of the headquarters of enterprises with a global reach, (ii) the inner periphery consisting of the newly industrialised countries of South East Asia and the 'near-industrialised' or 'transitional' economies of Central and Eastern Europe, Latin America, and China, and (iii) the outer periphery. The latter, which consists of 'middle and low' income countries can be loosely subdivided into three groups: (i) Third World countries such as South Africa and Botswana, (ii) Fourth World countries such as Ghana, Tanzania, Zambia, and Uganda, and (iii) the Fifth World countries such as Liberia, Somalia, Sudan, Zaire, and Zimbabwe. The latter are symbols of political–economic, demographic, social and environmental stress, over population, crime and the erosion of borders (Shaw, 1994).

Broadly speaking some cross-country studies of per capita incomes seem to support this broad division. The 'twin-peak' hypothesis of the wealth of nations suggests a 'club'-like distribution of income groups with peaks or clusters of countries at the high and low-income ends of the distribution, respectively, while middle-income countries tend to converge either to the high or the low-income ends of the scale – trends that seem to become more cemented over time (Quah, 1993). Although more studies that followed confirmed this hypothesis (Durlauf and Johnson 1995; Quah, 1996, 2001; Paap and van Dijk, 1998), others suggest convergence to a single peak with a prolonged increase in polarisation and inequality over time (Galor, 1996; Kremer *et al.*, 2001).

24.2.2. Policies that caused economic change

The current phenomenon of global regionalisation has been brought about by a whole sequence of fundamental policy changes in the developed world over the post-Second World War period. First, the financial control structures that were brought about by the Great Depression of the 1930s and the Second World War were beginning to be dismantled during the late-1960s and 1970s. In the process, capital became increasingly liberated while the First World cities lost their manufacturing industries. New York lost 35% of its manufacturing employment in the 1970s and London around two-thirds over a slightly longer period (Harris, 1995). Increasingly, transnational companies started to locate in what is now known as 'newly industrialising countries' because production costs were lower there and the investment climate in those countries was more accommodating (Chan, 1996). This was a period characterised by a change in the structure of industry, from monopoly capitalism to flexible accumulation, and from a nation-state focus to a transnational focus in capitalism (O'Loughlin, 1989; Robinson, 1996). Locally, privatisation of public institutions started to gain popularity, but it was only during the 1980s that the process really gained momentum.

In response to the recessions of the early- and late-1970s the developing world resorted to protectionism in order to protect its local businesses. In their drive towards greater financial independence from colonial powers, many developing nations resorted to policies of subsidised and uncompetitive import substitution. In a climate of protectionism some domestic industries flourished, many of them operating at well below minimum efficient scale. State-controlled regulation and planning became a regular feature in those economies. As a result of insufficient monitoring measures and control, debt in the developing world escalated (Roberts, 1978). Although the increasing number of debt reschedulings over the past 20 years (Callaghy, 1997) show that debt is still a major inhibiting economic factor in the developing world, some of these countries have started to reverse their protectionist policies since the late-1980s. They are now beginning to turn to more liberal policies of less government control and market-driven economies focussed on the promotion of export industries (Harris, 1995; Bhagwati, 1997; Black and Mitchell, 2002). According to Kim (2001) this is a clear indication of the reality of global 'economic' and 'political determinism'.

24.2.3. The impact of foreign direct investment

The flow of foreign direct investment (FDI) is a very commonly used parameter for the measurement of economic relations between nations. FDI is usually affected by the political and institutional stability, and demand, supply and cost conditions in a country. Due to less favourable trade-offs between the potential returns on investments and direct and indirect cost factors such as restricting tax, tariff, quota and surcharge structures, developing countries are generally not as popular as investment destinations, as the developed world. It is, therefore, not surprising that, for many years, more FDI has been flowing to the North than to the South, and that the proportion that has flowed to the North has been increasing relative to the South (Allen and Thompson, 1997; Cyr, 2001). In 1967, 67% of the world's FDI stock was located in the North, a percentage that has increased to 81 by 1991. By the beginning of the 1990s the FDI stock in the US alone exceeded that of the entire developing world, while the stock in the UK exceeded that of Asia (Koechlin, 1995). Since the later part of the 1990s the situation has changed somewhat, however. During the period, overseas development assistance (ODA) for the South stood at around $50–55 billion, while FDI increased from $25 billion to around $200 billion (Holliday *et al.*, 2002). In 1997 the developing world received approximately 42% of all FDI (Loots, 2002). Most of it has been flowing to the Pacific Rim, Central Europe and Central and South America, however, which makes FDI regionally biased, and 'invested interests' and 'cooperating competition' important factors in globalisation.

Part of the reason why most FDI still flows to the developed world while the bulk of the remainder flows to a relatively small selection of developing countries in the inner periphery is because comparative advantages remain a major factor in international trade (Poon, 1997). Contrary to the past when comparative advantages were largely interpreted in terms of the availability of natural endowments at a particular location, it is increasingly becoming a strategically constructed phenomenon through industrial targeting and networking (Vernon, 1996). Transaction costs are another restricting factor in the dissemination of goods and services, despite decreasing friction of distance, especially in the exchange of information. It is still in the interest of transactors to move to areas where agglomeration advantages exist (Parr, 2002) and where incidents of transactions are globally higher (Storper and Scott, 1995) while transportation costs are kept low. Intra-industry trade, i.e. the internalising of markets trough vertical integration, is the third important factor that contributes to FDI convergence. A fourth factor is the concentration of high-quality labour in the North and in parts of the South. The current high-performance global economy requires high-order thinking skills and the North and the Pacific Rim have been able to attract significant amounts of FDI because they have invested massively in education over many years. This has boosted their economic performances, while the inward looking economic policies of (especially) Sub-Saharan Africa and its low quality of labour have made it a high risk area for investment (Marshal, 1995). Africa's share in the world population, for instance, increased from around 9.7% in 1970 to almost 13% in 2000, its GDP declined from around 3.4 to 2.3%, its exports from a little under 5 to 2%, and its FDI from 5.75 to 1% over the same period (OECD, 2002). This is a clear indication how disconnected Africa has become from the rest of the world economy over the years.

However, the importance of FDI as a development factor should not be overestimated. In reality, it usually only forms a relatively small percentage of domestic investment in developed countries, i.e. around 5% from 1960 to 1991 (Koechlin, 1995). Also, FDI as a percentage of local investment overall does not seem to be increasing dramatically over time. According to Bairoch (Beinart, 1997), as a percentage of GDP, FDI of the developed world in 1914 and 1993 was roughly on the same level, i.e. approximately 11%, while FDI inflows of the world as a whole stood at 2.2% in 1998 (World Bank, 2000).

24.2.4. Mega cities of the South

As a result of the economic changes that have occurred across the globe over the past five decades, major shifts in global urbanisation have also occurred over the period. In the early 1950s New York City, London and Tokyo had populations of around 8 million at the time, while the populations of large cities in the South such as Buenos Aires, Sâo Paulo, Mexico City, and Cairo varied between 1.5 and 5 million (Merriam, 1962). Since then, the world's gravity point in mega city development has shifted dramatically towards the developing world. In 1984, 34 cities had populations greater than 5 million, only nine of them were cities in the developed world (Fox, 1984). Of all the people who lived in cities with populations greater than 5 million in 1950, more than 50% were Europeans and Americans. Today, less than a quarter of the people in such cities are from those regions, and it is expected that this percentage will

Table 24.1. Population and urban growth in Africa

	Total Population (millions)			Average Annual Growth (% of Total)			Urban Population (% of Total)			Average Annual Growth (% of Urban)		
	1980	1990	2000	1980	1990	2000	1980	1990	2000	1980	1990	2000
Sub-Sahara	380.7	508.3	659.8	2.9	2.9	2.7	23.0	28.1	34.5	5.0	5.0	4.8
North	88.4	114.1	138.0	2.6	2.6	2.0	44.6	49.2	53.9	3.5	3.6	2.9
All	469.1	622.4	797.8	2.8	2.9	2.6	27.1	32.0	37.9	4.5	4.6	4.3

World Bank, 2002.

decrease to around 10% over the next decade. South America and Asia, especially India, China and parts of the Pacific Rim are all experiencing urban explosions, and based on the current low levels of urbanisation in low-income countries, especially Africa (see Tables 24.1 and 24.2) there is a great potential that this explosion of the urban South could continue well into the 21st century.

However, the economic growth in the South had not been nearly as dramatic as the population shift over the same period. Low-income economies are still largely agricultural driven and their per capita income have remained low (Table 24.3). In fact, apart from cities that are very well endowed by natural resources, such as cities near oil fields in the Middle East, only the economies of a relatively small number of cities in the Pacific Rim have shown significant progress since the 1970s – mostly as a result of the worldwide shift in post-Fordist manufacturing. In recent years, however, more FDI has been flowing from developed countries to developing countries nearby than before. In the process, countries in the Pacific Rim are now becoming stronger spatio-economically associated with Japan, Central Europe with Western Europe, and Central America with the USA. Mega cities in the Pacific Rim that form part of the 'flying geese', manage to attract industries that are no longer viable in the South East Asian urban core (Marcotullio, 2001), while cities such as Mexico and those to the north of it benefited significantly from NAFTA. Cities in the global outer periphery has been lagging behind ever since.

For this reason the mega cities of the South play significantly different roles as global players compared to mega cities in the North. Global and world cities in the North serve as apexes in the networks of urban systems in the globalising economy, while cities

Table 24.2. ***Urban population growth in global regions***

Countries	Population (millions)		Population Growth		Urban Population (% of Total)			% Population in Primary City		% Population in Cities > 1 million		
	1980	1998	1980–1998	1998–2015	1970	1980	1998	1980	1995	1980	1995	2015
Low income	2527	3536	1.9	1.3	18	22	30	13	13	7	10	13
Middle income	1114	1474	1.6	1.0	49	56	65	22	21	18	22	24
High income	789	886	0.6	0.3	73	75	77	17	17	30	32	33
East Asia/Pacific	1398	1817	1.5	0.8	19	22	34	12	9	9	12	15
C/E Europe/Central Asia	426	475	0.6	0.1	52	59	66	15	15	14	16	18
Latin America/Caribbean	360	502	1.8	1.3	57	65	75	27	25	24	28	28
North Africa/Middle East	174	286	2.8	1.8	41	48	57	31	27	17	21	24
South Asia	903	1305	2.0	1.5	19	22	28	9	11	6	10	13
Sub-Sahara	381	627	2.8	2.2	19	23	33	28	29	5	8	12

World Bank, 2000.

Table 24.3. Size of the economies of global regions

Countries (1998)	GNP 1998 (Bn$)	1997–1998 % ($)	GNP/c 1998	1997–1998 %	GPP 1998 ($)	GNP/c % of Labour		Agri-labour Trade % of GDP	
						1970	1990	1970	1998
Low income	1842	3.5	520	1.8	2170	75	68	12	46
Middle income	4401	−0.1	2990	−1.3	5990	40	28	30	56
High income	22592	1.4	25480	0.9	23420	11	5	29	44
East Asia/Pacific	1802	−1.5	990	−2.6	3280	76	68	24	75
C/E Europe/Central Asia	1044	−0.4	2200	−0.5	5510	33	23	–	71
Latin America/Caribbean	1933	2.1	3860	0.5	6340	41	25	20	32
North Africa/Middle East	581	3.7	2030	1.6	4630	50	35	–	53
South Asia	560	5.7	430	3.7	1940	71	63	12	29
Sub-Sahara	323	2.2	510	−0.4	1440	78	68	47	59

World Bank, 2000.

in the South are facing significant disadvantages. The distinguishing trademark of global cities is their ability to attract skilled labour and high value-added activities with a global reach. They are globally recognised nodes of creative thinking, innovation, information, fashion and culture, which provide specialised expertise and services and house multinational companies and substantial sources of investment capital. They act as the 'command and control centres' of the global economy (Sassen, 1991; 1997; Knox, 1994; Beaverstock *et al.*, 1999). Despite attempts to find a more comprehensive theoretical framework for the understanding of the concept of global cities, such as understanding it in the context of larger networks of globalised urban centres (Smith and Timberlake, 1995; Geyer, 1998; Jones, 2002) global cities continue to be distinguished from other centres in terms of the number of command and control functions that are located in them (Derudder *et al.*, 2003). As a consequence, mega cities in the South and North are often analysed on equal terms.

There are several reasons why this approach to mega city research is inappropriate for mega city analysis in the South. One is the fact that the spotlight falls mainly on the impact of the global reach of technologically sophisticated multinationals, the number of Fortune 500 firms located in mega cities, and the amount of FDI that is attracted by them. In the process a wide range of issues related to the impact of globalisation on layers of less economic sophistication in such cities tends to be blurred out (Geyer, 2003). An economic picture is portrayed in which the 'techno-literati' of the world (Golding, 1996) take centre stage in an exclusive economic environment, while a large percentage of the 'techno-illiterati' that live in the 'information-ghettos' of the world (cities) is being marginalised. Globalisation not only impacts directly on the survival of the lagging economies of the inner and outer peripheral regions of the world, but also impacts directly on the marginal sector inside mega cities (Friedmann and Wolff, 1982; Graham and Marvin, 1996; Graham, 1999; Warf, 2000; Geyer, 2002). Just as the global competitiveness of global core regions leads to an increase in global underdevelopment in certain global peripheral regions at the macro-scale, globalisation also tends to cause an increase in inequality and polarisation at the micro-urban scale, both in the developed and developing world. According to Sassen (1991) it causes a middle-classlessness in those societies, the proliferation of highly skilled professions followed by the 'casualisation' and 'informalisation' of a wide range of other economic activities. This leads to hourglass-shaped urban economies in the North (Hamnett, 1996) and pear-shaped mega city economies in the South.

Considering differences in the cityness of cities in advanced countries in the West compared to those in advanced and developing countries outside the West, the question should be asked how universally acceptable the application of Western definitions of cities are outside the West (King, 1997). Based upon this, the question could further be asked how applicable global city rosters are when they are compiled in terms of narrowly selected ranges of criteria. Linked to this, difficulties arise when global city rosters are compiled while disregarding the fact that global cities in advanced countries are by and large embedded in larger core regions of equal, if not greater, cumulative global significance. These cities and extensive urbanised regions around them have become spatially integrated to such an extent, that, what holds true for individual global cities in terms of capital accumulation and access to markets and resources globally, also holds true for the urban networks around them. In contrast, stand-alone mega cities (some of them emerging global cities) in developing countries are often embedded in economically less developed areas with comparatively scarce infrastructure and skilled labour (Geyer, 1998). These areas are unable to provide significant support to the modern economic activities that are clustered in the mega cities. As a result, they cause a steep decline in global significance with increasing distance from the apex of such emerging global cities in contrast to global cities in global core regions that enjoy significant economic support from associated urban agglomerations in their vicinity.

From an international development perspective, core regions in the North are not the only areas that are advantaged by international economic forces of cumulative causation. Emerging global cities in the South present similar advantages to prospective multi-national developers, albeit at lower levels of intensity and scale than those in the North. They serve as popular FDI destinations with agglomeration advantages far outstripping those of other areas in their vicinity. Serving as gateways to large economically less developed regions in the developing world, certain strategically located cities in the South have already become nodes of global significance. Others are nodes of sub-continental and even continental significance, or have the potential of becoming such nodes in the future (Geyer, 1998). Just like their counterparts in the North, these nodes in the South also attract large numbers of international migrants, most of them usually unskilled and originating from the region. Scores of migrants from Sub-Saharan Africa migrate to South Africa, many of them ending up in the inner urban areas of the country's mega cities. Similarly, a sizable proportion of the people from Asia and the Pacific Rim migrate to cities such as Singapore (Alden, 1996), and people in South America to the large urban agglomerations in that region. Those that can compete for employment migrate to countries in the North. These two migration streams are indicative of the process of international differential urbanisation (Geyer, 1998), i.e. higher qualified migrants trying to find a new life in the North, while less qualified labourers who cannot make it to the North migrate from more remote peripheral areas of the South to the most vibrant cities in the region (Geyer, 2003).

Closely associated with the problem of global block formation and the role that mega cities in the South play as migration and FDI destinations in their regions, is the problem of choosing the most appropriate economic development approach.

24.3. Development frameworks of the past

24.3.1. Neo-liberalism

Opposing views on whether the emphasis in economic development should fall on local or the export industry, are not new. Soon after the Second World War ended the debate intensified on the potential of the 'basic' or 'urban building' versus the 'non-basic' or 'urban serving' sector to bring about economic growth (Alexander, 1954; North, 1955; Tiebout, 1963; Leven, 1966). Since then the intensity of the debate on the development of the South has not at all diminished; only the focus has slightly shifted. This is clearly demonstrated by the way in which the advantages and disadvantages of neo-liberal economic development policies are still being debated in the literature. As a result of the successes that were recorded by the newly industrialised countries of the Pacific Rim over the past three decades in attracting FDI and creating economic growth through industrial development, the established global network of big business and international funding organisations still by and large regards the main elements of neo-liberalism as the most appropriate way of achieving economic development.

Proponents of neo-liberalism support the expansion of a social responsible free market system and the stimulation of private initiative. First, they believe that the reduction in financial control measures paves the way to FDIs. Protectionism, they argue, does not breed competitiveness and is not conducive to the improvement in productivity levels in the local businesses sector. Indications are that when markets are opened, the range of products broadens, the quality of products tends to improve overall, and with it also labour productivity. Second, because not all public services are always provided efficiently by the public sector, proponents of neo-liberal economics believe that, ideally, the privatisation of most public services would result in more effective management of public services and create a climate that is conducive to the growth of the private enterprise. When privatisation involves the selling of public assets it also serves as a symbolic countermove against the growth of government. However, due to high management costs to make assets sellable and the usual capital market thinness that characterises developing economies, their payoff is often not great (Bienen and Waterbury, 1989). Third, they believe that the curbing of domestic demand would reduce imports and generally improve balance of payment conditions in countries. Ways in which this can be achieved is by allowing the price of basic goods and services to rise through the elimination of subsidies and by tightening interest rates. Fourth, they believe that increasing exports would earn the necessary foreign exchange for such countries to pay off their debt. Exports are normally stimulated when hurdles to the inflow of FDI, such as tariffs, quotas and surcharges are removed. Finally, they believe that, in the long term, education and training produce a more productive labour force (Lipschutz, 1991).

Within this development framework, a country's institutional setup, its natural resource base, and the quality of its labour force which allows it to meet ever changing technology and skills requirements are key factors that determine the country's ability to compete in the international market. The OECD has been working hard to develop capabilities to conduct objective and accurate industrial development policy benchmarks. It currently uses indicators such as R&D infrastructure, educational profiles of labour forces, corporate governance structures, employment regulations, labour costs, taxation, and energy and telecommunication infrastructure as benchmark indicators (United Nations, 1999). Although attempts are now being made by the lagging South to create a more attractive economic environment for international investments (see NEPAD, 2001) many countries in this part of South still have a long way to go before reasonable international competitiveness becomes a reality (Geyer, 2003).

24.3.2. Criticism of the lagging South

Critics of neo-economic liberalism, on the other hand, believe that the approach has not delivered the desired results. Despite faster gains in income and life expectancy and literacy improvements in low- and middle-income countries compared to high-income countries, the ratio of the average income of the richest to the poorest country in the world has increased from 30:1 in 1960 to 60:1 today (MMSD, 2002). Worldwide the inequality in wealth and power is growing. According to Robinson (1996, p. 22) this amounts to permanent 'structural violence' against the world's majority. People in the South who have no access to health, clean water and sanitation and who live in poverty count in the billions. The population of the region grows exponentially because more than eight out of 10 new births now occur in the developing world (Alden, 1996).

These critics believe that a number of factors associated with economic neo-liberalism keep on marginalising lagging economies (Lipschutz, 1991; Amin, 2001; Kaya, 2001; Sihlongonyane, 2001; Tsheola, 2002). The first is the fact that poor countries have accumulated heavy debt burdens over time and the burden is growing all the time. Debt often exceeds the capabilities of marginalised countries to overcome them in the near term, resulting in endless cycles of debt refinancing and a continuing flow of capital from developing countries to pay off debt to industrialised countries. Second, there is a continued emphasis on the exporting of products that do not have large markets, or whose markets are declining in economically advanced countries. In many cases there is not enough specialisation in export products between developing countries, resulting in the oversupply of certain product lines and consequently the depressing of prices of those products on the international market. Although the focus on the export market tends to improve competitiveness and productivity locally, the urban poor in developing countries do not necessarily benefit directly from it (Lipschutz, 1991). To enable them to compete internationally, production processes have to be internationally competitive, hence the shift in focus, normally, from labour-intensive to capital-intensive production processes. In the process the lowly skilled labour force loses ground, the labour force becomes stratified and inequity deepens (Aguilar, 1997).

There are often indirect side effects to the sequence of depressing factors at the local urban level. First, excessive premature urbanisation often occurs in developing countries. When too many people migrate to large urban agglomerations in lagging countries that do not have the ability to successfully compete for employment in the urban market, over-urbanisation occurs. Large numbers of such people often live in squalid conditions with little hope that they will ever be able to improve their lot significantly over the short term under current economic conditions (Mukhertji, 2002). However, compared to rural poverty in developing countries, indicators generally are that urban poverty seems to be the lesser of the two evils.

As the urban population becomes stratified, high-wage workers with a national and international orientation operate within the formal urban economic environment, while the lowly skilled workers lag behind (Aguilar, 1997). Due to increasing capital mobility, massive shifts in capital could occur over a relatively short period of time within the urban South when urban decay sets in. Because cities are the most dynamic component of the space economy of the developing world (Harris, 1995), such shifts in capital could have devastating effects on cities or parts of cities. Business closures could occur causing crises in urban labour markets there. This has happened in many of the metropolitan and intermediate-sized cities of South Africa in recent years. Soon after the political transition, black migrants started infiltrating parts of central areas of major cities. This infiltration was accompanied by the invasion of those areas by the informal economic sector, followed by an increase in crime and urban decay and the subsequent relocation of formal businesses to areas where conditions are more conducive to formal business activities (Geyer, 2003).

24.4. Making markets work in the South

24.4.1. Pillars in the Southern economic markets

As was stated before, the focus in the neo-liberal economic development approach is on economic growth by means of procuring appropriate technology, increasing labour productivity, creating investment opportunities by removing obstacles of industrial development and international trade, and mobilising capital. This approach has been constantly criticised by the lagging South (Kaya, 2001), because many developing nations in Africa and South America have been lagging behind in economic terms, and are continuously losing ground, despite the successes that have been achieved with neo-liberalism in the Pacific Rim over the last four decades. To overcome deprivation in the South more and more voices have been going up for sustainable development (Hall, 1996). With this development framework in mind, the emphasis is shifting from conventional neo-liberal economic thinking as the principle vehicle to bring about economic growth, to ways in which new global market principles could be used to complement the economic strengths of the South. Two issues are of importance in this regard. First, how neo-liberal economics could be adapted to suit the circumstances of the lagging global periphery? Second, how elements of the new flexible global economic environment, in which capital can shift from one production branch to another and from factory production to informal subcontracting, could be exploited to supplement current approaches of development in the South?

From a corporate perspective two groups of factors impact on sustainable development. Making markets work for all is one of the key principles in the creation of wealth and life improvement. Well-structured, open, and competitive markets that take cognisance of inequities, inaccessibilities, and injustices are still one of the best ways of advancing economically lagging communities. The correlation between national scores on the Index of Economic Freedom and the Human Development Index clearly shows that governments that try to fulfil the role of the private sector keep their citizens poor (Holliday *et al.*, 2002). One sure way of prolonging inequities globally is by maintaining protectionist policies, especially in the North. At the same time, liberated and unobstructed markets do not guarantee accessibility to all. To make it work for all, especially in the South, lagging groups should be allowed easier access than before. Equitable markets create wealth, increase competition, foster innovation, create creativity and ultimately improve people's quality of life. Creating streamlined, equitable and accessible markets is the responsibility of both the private and the public sector.

Second, the financial and political regimes of developing economies are equally important in creating a sound economic environment in which business can prosper because FDI tends to flow to areas where investments are safe, where labour is productive, and where trade-distortions and financial obstacles are the least. The marginal South does not have a particularly good record in this regard. However, more FDI is also not a short-term solution because foreign investment has a long-term horizon and is often of the kind that is unlikely to provide sufficient employment to the unskilled and semi-skilled masses in the cities to radically change conditions for them. Creating a business environment that will accommodate the entire labour force of the South remains a challenge. This implies that the urban South will have to learn to live with an economic dichotomy, i.e. a formal sector that normally does not sufficiently provide employment to a large component of the lowly skilled members of the labour force, and the informal sector that seems to be the only viable option for a large percentage of the latter. Reducing the gap between the formal and informal sectors and building partnerships between the two sectors in order to increase the ability of the urban populations of the South to improve their living standard, should be strived for all the time.

24.4.2. Building new market structures in the urban South

One of the reasons why the urban South is struggling socially and economically is because there are much more inappropriately qualified people living in the cities than their formal economies can carry. Providing employment to large numbers of people who cannot compete for jobs in the informal sector becomes a constant struggle.

To put the scale of the problem in context, roughly 1 million jobs have to be created to keep pace with rising urban populations in Central Latin America alone. In comparison, 2 million jobs were created in the United States per annum as long ago as the 1970s, and that in an economy 15 times greater than the combined economies of the region at the time (Fox, 1984).

As was said before, creating a new Southern blend of market economy may offer new hope. In countries with long traditions of informal urban activities, smaller enterprises and independent formal and informal producers are often linked to large domestic firms and even multinationals (Geyer, 1988, 1989; Aguilar, 1997). One area in which Southern urban markets could be exploited more vigorously lies in (i) recognising the inherent structure of the informal urban sector, (ii) expanding and exploiting certain elements of it, and (iii) finding ways in which linkages between the formal and informal urban sectors could be maximised to the benefit of Southern economic society as a whole. One of the reasons why many of the unskilled people, who migrate from rural areas to overcrowded cities in the South, enter the informal urban market is because entering it normally does not require skills, especially at the entry level. However, over-exploitation of the informal economic sector in the urban South resulting in large numbers of vendors offering similar goods of similar quality at the same location, and at cutthroat prices, leaving most of them with barely enough income to survive, is a persisting problem (Penouil, 1981). On the other hand, the formal economic sectors of the urban South, especially those in the deep periphery of the globe not only have to battle against the odds in the global market place, they also often have to cope with informal activities that often undermine their viability (Brown and Connell, 1993) as well as growing tax demands of capital hungry local governments to foot their swelling social overhead bills. As a result of a combination of these factors the emphasis in the lagging developing countries of the world increasingly falls on the lower levels of businesses shown in Figure 24.1. The figure distinguishes between the sizes and types of businesses, their sectoral reach, levels of sophistication, locational preferences, and preferred ways of communication within the business sector as whole, and at the same time shows the (potential) overlap between the informal, formal and intellectual business

Figure 24.1. Anatomy of the business sector in the urban South

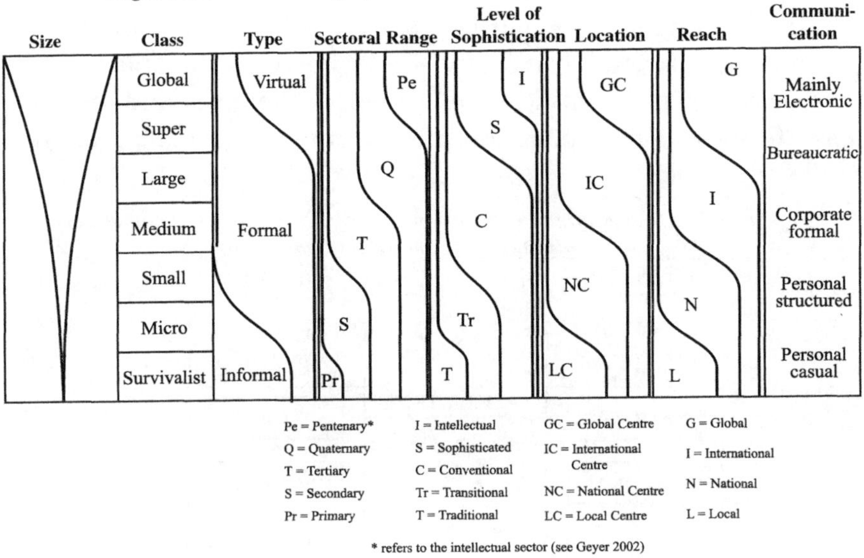

sectors from the primary to the pentanary[1] sector in the urban South. The poor quality of the environment with which the informal sector is often associated is one of the major factors that threaten the potential symbiosis between the two sectors in the urban South. This is a matter that deserves serious attention.

Indications are that the small and informal business sector in the urban South is becoming more stratified (Figure 24.2). Traditional informal economic activities, i.e. activities with a large local cultural content, such as local traditional medicine practices, the selling of traditional cuisine, the manufacturing of traditional household goods and equipment, form the lower stratum of the structure. The largest proportion of demand for goods and products in this category lies in urban areas where local population groups are concentrated. The transitional informal economic layer contains businesses that provide goods and services that have some traditional features but are adapted to incorporate more advanced elements and techniques,

[1] The pentanary sector refers to economic activities that predominantly serve the human imagination and intellect. The sector mostly lies within the new information and communication economic realm and includes the bulk of the latest computerised edutainment, religitainment, infotainment, and computainment products (Geyer, 2002).

such as the structured multiplied or even mass production of (stylised) traditional arts and crafts, or the production of modern household goods produced from locally obtained materials. While the largest market for this kind of product lies in the central business areas of cities, their production processes occur in industrial areas

Figure 24.2. Model on the integration of formal and informal commerce and industry in the urban South

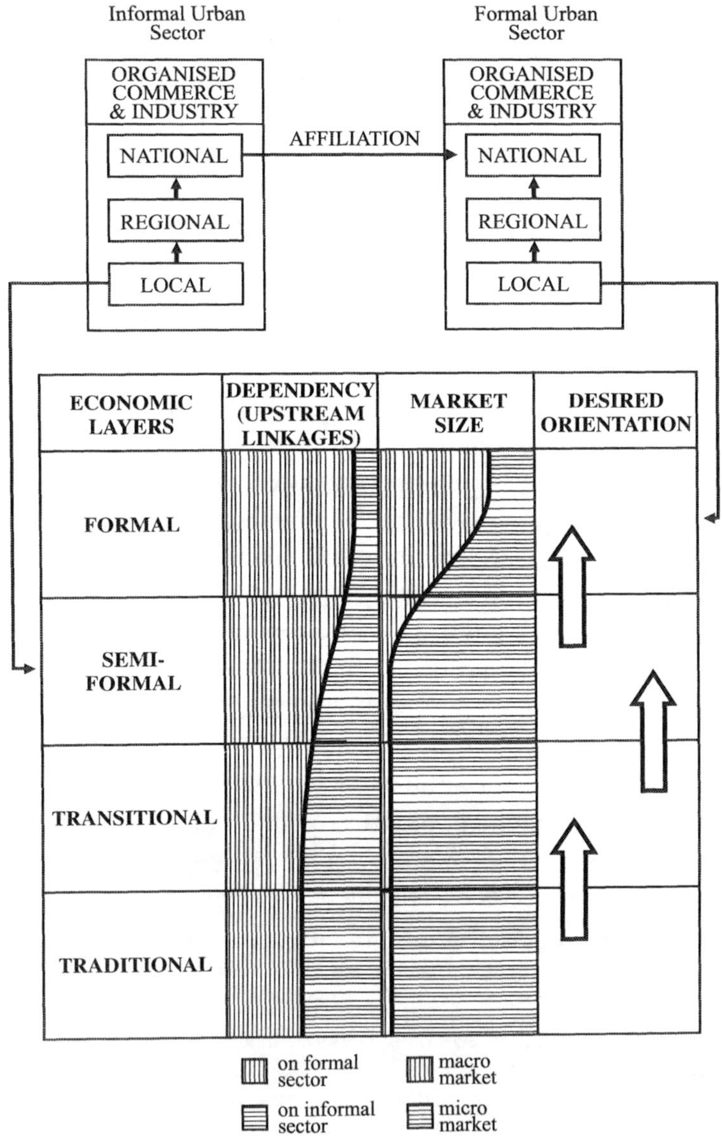

and even at home. The upper semi-formal layer consists of the selling of goods and services that are obtained from the formal sector, or that are privately produced from materials that are obtained from the formal sector. The products and services that are offered in this layer could range greatly in terms of their level of technical sophistication, from locally manufactured goods produced for local consumption only, to products that are marketed globally. Although upstream and downstream linkages are sometimes found between the formal sector and each of the three informal strata, the linkages between the formal and the semi-formal sectors are often the most promising for immediate exploitation and expansion.

The model also allows for international links with the local formal and informal sectors. Deteriorating conditions in the South often result in an increase in emigration to the North, legally or clandestine (Morris, 1997). Although this phenomenon holds negative consequences for the urban South, because of the brain drain it causes, it also holds potential for an increase in international remittances because expatriates tend to keep in contact with locals (Afshar, 1998). Initially, remittances are mostly of a financial nature but over the longer run the proportion of remittances in kind often tends to increase (Brown and Connell, 1993). Potentially, this holds benefits for both the initial remitter and the benefactor, especially when remittances in kind become a two-way stream, i.e. when products from developed countries are sold by the benefactor in developing countries and vice versa, to the benefit of both the parties.

This model (Figures 24.1 and 24.2) provides a comprehensive framework that enables one to conceptually integrate the entire range of formal and informal activities in the urban South (and North), from the smallest to the largest, and from the traditional to the most sophisticated businesses. It enables one to mix formal and informal economic activities under different development conditions in the urban South within a new market framework.

24.5. New markets and urban sustainability

While transforming and streamlining the informal sector in the urban South are aimed at increasing its viability at the bottom end of the market economy, the issue of wasteful practices in the (upper) formal sector of the economy also needs attention. Finding the balance between economic, social and environmental sustainability has never been easy. If too much focus is put in policies on economic growth, environmental and social activists frown upon it. If too much focus is placed on environmental sustainability, economic growth could be jeopardised to such an extent that living standards are compromised. For this reason three anchors are necessary in sustainable development: economic, social and environmental sustainability. In 1987, the WCED resolved that: “Sustainable development seeks to meet the needs and aspirations of the present without compromising the ability to meet those of the future. Far from requiring the cessation of economic growth, it recognises that the problems of poverty and underdevelopment cannot be solved unless we have a new era of growth in which developing countries play a large role and reap large benefits” (Holliday *et al.*, 2002, p. 13).

Since the introduction of the concept of urban ecology in California in 1975 (Roseland, 1997), the concept of urban sustainability has become a potentially important factor in the formulation of urban development policy. Essentially, it deals with two issues: the global decline in supply or availability of natural resources and the growing demand for it. In free market economics, the terms ‘demand’ and ‘need’ are often equated with people’s ability or willingness to pay for goods and services, i.e. if someone has a desire for a product or service and the ability and willingness to pay for it, then effective demand or need has been realised (Heilbroner, 1970; Samuelson, 1970). In a train of cause and effect this has led to excessive growth in demand in the wealthier parts of the world, which causes the unnecessary expending of resources, which in turn leads to an alarming narrowing of the gap between demand and supply (see Figure 24.3), and at the same time an increase in waste streams.

As a consequence of the worldwide realisation of the long-term detrimental effect of ‘wasteful’ demand, a distinction is increasingly being demanded between the concepts of ‘demand’, ‘needs’ and ‘basic’ or ‘essential needs’ (MMSD, 2002). These demands are being repeated from one world forum to the next, and cumulatively, they are becoming the driving force behind the realisation of the potential trade-offs between maximum economic growth on one hand, and economic growth tied to social well being and a sustainable environment on the other. In sustainable development

Figure 24.3. The desired effect of sustainable development on supply and demand curves

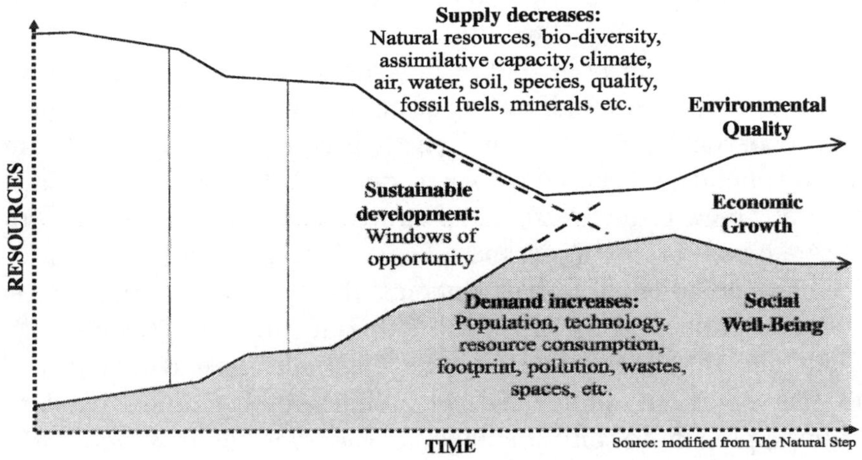

terms the overall objective is to minimise the difference between the latter and the former. To achieve this goal, a number of factors on the demand and the supply side should be addressed. On the demand side, high fertility rates in the South should be reduced, intellectual capacities expanded, wasteful consumption worldwide should be minimised, technologies that intensify resource consumption should be improved, and the ecological footprint of urban settlements in terms of the range and quantities of pollution, wastes, and space should be reduced. On the supply side, the mileage of natural resources should be stretched as far as possible, bio-diversities protected, and assimilative capacities increased.

An important area of research in the field of urban development lies in how economic development can be achieved while minimising its ecological footprint. This shift in focus has been necessitated by the general proclivity of cities (and nations) to reduce their potential damaging influence on the local environment by importing resources (which cause waste streams and the depletion of resources upstream) while exporting locally generated wastes and pollution downstream (Haughton, 1997). In the past, unbridled capitalism often led to cost transfers as a means to maximise profits locally, nationally and internationally. The question could be asked how many of the shifts of messy industries from developed to developing countries occurred during the period of industrial restructuring, only because of more favourable labour economics, and what role easier environmental policies of 'grow-first-clean-later' (Marcotullio, 2001) played in the process. Eco-sensitive urban policies are now being experimented with to attempt to reduce externalities in current urban practices. They range from eco-centric ('hard' or 'deep green') policies that are unsympathetic to economic expansion, to anthropocentric ('soft' or 'light green') policies that are aimed at balancing the need for environmental preservation and economic development.

Four categories of urban policies that potentially impact on economic policy, have been identified (Haughton, 1997): the 'self-reliant' city, the 'redesigning' city, the 'externally dependent' city, and the 'fair shares' city policy. The self-reliant city is an intrusive deep green approach to city management. It internalises economic development by emphasising local economic development within the confines of the 'bioregion' – a region that is economically and politically defined in terms of the natural boundaries of ecosystems. Although application of policies in this group is possible at the micro-scale, the location of existing cities not conforming to natural bioregions, or extending over more than one bioregion, makes application of the policy within such a framework difficult, if not impossible. Also, the distance that is created in this model between human activities and natural ecosystems poses a difficulty because the human-versus-nature mindset in ecology is rapidly changing. As a science, ecology deals with the relationship between organisms and their environment, and therefore human activities within the urban environment make humans very much part of the ecosystem, economically, socially and biologically (Pickett *et al.*, 1997; Jensen, 1998; Parlange, 1998). The impact of social and economic networks inside and between urban nodes and their hinterlands, on the natural environment, forms part of the plethora of factors that needs to be taken into account when urban economic development policy is considered.

Redesigning city policy refers to approaches that are aimed at reducing resource consumption and waste streams. The objective is to work with nature and to change the environmentally damaging practices of people in cities. One way of achieving this is through energy savings by compacting the city. While some social overhead capital savings seem possible in this model, significant savings in transport costs ultimately are not guaranteed, since the compacting of cities and the mixing of land uses do not necessarily reduce home-to-workplace travelling. In fact, the densification and mixing of land uses may eventually exacerbate travelling problems overall because land uses that were concentrated in particular areas previously, potentially improving the viability of mass and rapid transit systems, will now be more dispersed throughout the city. From an environmental sustainable point of view, the approach also defeats the objective of creating more livable urban space because compacting city spaces would necessarily result in increasing person-land ratios. Ultimately, it will reduce open spaces, make cities more 'urban', and increase pressure on social overhead capital. Mixing and densifying land uses will also put pressure on personal and social space, making it more difficult for people to recuperate psychologically. This could ultimately increase tension between individuals and communities within cities (Geyer, 2001) – something that we have learned from urban practices in Europe during the Industrial Revolution.

Externally dependent city policy is a market-centred approach to environmentally sustainable urban development. It seeks to modify market mechanisms to benefit both the environment and the society at large, without necessarily terminally damaging market systems. "Market reform is essential if sustained development is to be achieved, but this reform must be geographically sensitised, as well as linked to strong social justice programmes and environmental standards setting to ensure that both local and global environmental carrying capacities are respected" (Haughton, 1997, p. 192). Based on the classic Kutznets (1955) study that pointed to a possible divergence in secular income with economic growth, an underlying understanding of this urban model is a Kutznets-like relationship in economic growth and environmental sustainability – as economic development occurs environmental problems are increasingly overcome (Grossman and Krueger, 1995). An important disincentive against the excessive use of resources could be the realistic pricing of resources and services. Of particular interest in this model is how the cost of externalities could be built into the marketing mechanism, making polluters pay for their actions, especially in cases where wealth expands the locus of environmental challenges from local to regional and eventually to global proportions, as has been found in Asia (Marcotullio, 2001). One of the latest developments in this field is the generation of the Dow Jones Sustainability Index system (DJSI World), the first attempt at benchmarking the financial performance of sustainability of corporations on a global basis (Holliday *et al.*, 2002).

Fair shares city policy contains elements of all three previous categories. The difference lies in how it targets problems of spatial political, environmental, social and economic disequilibrium, i.e. what are the needs of people and how to balance the benefits of some against the needs of others. The unequal distribution of resources and ways in which to manage the exporting of externalities are important issues in this category.

24.6. Conclusion

Since the 1970s the industrialised world have experienced significant economic changes which changed the production conditions of entrepreneurs. The economic restructuring that followed impacted significantly on the urban environment of both the developed and developing world. The populations of large urban agglomerations in the developed world started to stagnate and even decline (Fielding, 1989; Geyer and Kontuly, 1996; Geyer, 1996) while new mega cities started to develop in the developing world. In the newly industrialising countries global shifts in industrialisation have contributed greatly to urban growth, while rural poverty remained an important factor in urban growth in the lagging corners of the world. In the process, global block formation continues to favour developing countries that are geographically closer to core regions in the North while the global outer periphery remains relatively isolated.

In the early 1990s a European Commission (EC) identified six factors that are expected to have a major impact on the economic potential of cities in the future. They are: (i) a diverse economic base, especially in the high value-added sectors, (ii) human capital that enables the exploitation of high-technology sectors, (iii) educational institutions that would provide a steady flow of skilled workers, (iv) a high quality of life to attract and retain a highly skilled workforce, (v) good transport and communication networks, and (vi) the institutional capacity to exploit potential (Alden, 1996). A key issue that is highlighted in the EC report is the ability of the central, regional and local government sectors to form partnerships with the private sector and local communities. These features are rarely seen in the urban South. Countries that have attracted industries first during the post-Fordist era often followed a develop-first-clean-later approach, while cities in the global outer periphery, such as in Africa, simply fall short in all the departments. It could be said that while informationalisation occurs in the economies of the urban North and industrialising South (Hall, 1996), informalisation has been occurring in the economies of the lagging urban South.

Based on these views, maybe one should accept the fact that the techno-economic dichotomy of the developed world and parts of the developing world will not be overcome, even in the longer term, unless ways could be found in which local economic development in lagging countries could be speeded up. As long as the gap between development conditions in the North and South remains as wide as it currently is, focusing intentionally on the vertical integration of the formal and informal urban economic sectors in the urban South seems to be one of the methods in which the gap could be narrowed. Recognising the inherent structure of the informal urban sector and narrowing the gap between the upper, technologically more sophisticated layer of the informal sector and the formal urban sector is one way of making conventional market forces more accessible to the lagging sector of the urban South. In the process, lessons that have been learned about the long-term effect of unsustainable urban economic practices in the industrialising South should be firmly kept in mind.

References

Afshar, F. (1998), "Balancing global city with global village", *Habitat*, Vol. 22, pp. 375–387.

Aguilar, A.G. (1997), "The urban labor market in Mexico: global change, informality, and social polarization", *Urban Geography*, Vol. 18, pp. 106–134.

Alden, J. (1996), "Urban development strategies: the challenge of global to local change for strategic responses", *Habitat International*, Vol. 20, pp. 553–566.

Alexander, J.W. (1954), "The basic-nonbasic concept of urban economic functions", *Economic Geography*, Vol. 30, pp. 246–261.

Allen, J. and G. Thompson (1997), "Think global, then think again – economic globalization in context", *Area*, Vol. 29, pp. 213–227.

Amin, S. (2001), "Africa: living on the fringe?", *Africa Insight*, Vol. 31, pp. 3–7.

Beaverstock, J.V., R.G. Smith and P.J. Taylor (1999), "A roster of world cities", *Cities*, Vol. 16, pp. 445–458.

Beinart, P. (1997), "An illusion of our time", *The New Republic*, Vol. 217, p. 2024.

Bhagwati, J. (1997), "The global age: from a sceptical South to a fearful North", *The World Economy*, Vol. 20, pp. 259–283.

Bienen, H. and J. Waterbury (1989), "The political economy of privatisation in developing countries", *World Development*, Vol. 17, pp. 617–632.

Black, A. and S. Mitchell (2002), "Policy in the South African motor industry: goals, incentives, and outcomes", *The South African Journal of Economics*, Vol. 70, pp. 1273–1296.

Breathnach, P. (2000), "Globalisation, information technology and the emergence of niche transnational cities: the growth of the call centre sector in Dublin", *Geoforum*, Vol. 31, pp. 477–485.

Brown, R.P.C. and J. Connell (1993), "The global flea market: migration, remittances and the informal economy in Tonga", *Development and Change*, Vol. 24, pp. 611–647.

Callaghy, T.M. (1997), "Globalization and marginalization: debt and the international underclass", *Current History*, November, pp. 392–396.

Chan, R.C.K. (1996), "Urban development strategy in an era of global competition", *Habitat International*, Vol. 20, pp. 509–523.

Cyr, A.I. (2001), "Guides to globalization", *Orbis*, Vol. 45, pp. 295–305.

Derudder, B., P.J. Taylor, F. Witlock and G. Catalano (2003), "Hierarchical tendencies and regional patterns in the world city network: a global urban analysis of 234 cities", *Regional Studies*, Vol. 37, pp. 875–886.

Douglas, S.P. and Y. Wind (1987), "The myth of globalization", *The Colombian Journal of World Business*, Vol. 22, pp. 19–29.

Durlauf, S.N. and P. Johnson (1995), "Multiple regimes and cross-country growth behavior", *Journal of Applied Econometrics*, Vol. 10, pp. 365–384.

Emmerij, L. (1992), "Globalization, regionalization and world trade", *The Colombia Journal of World Business*, Vol. 27, pp. 6–13.

Feenstra, R.C. and G.H. Hanson (1997), "Foreign direct investment and relative wages: Evidence from Mexico's maquiladoras", *Journal of International Economics*, Vol. 42, pp. 371–393.

Fielding, A.J. (1989), "Migration and urbanization in Western Europe", *The Geographical Journal*, Vol. 155, pp. 60–69.

Fox, R.W. (1984), "The world's urban explosion", *National Geographic*, Vol. 166, pp. 179–182.

Friedmann, J. and G. Wolff (1982), "World city formation: an agenda for research action", *International Journal of Urban and Regional Research*, Vol. 5, pp. 309–343.

Frobel, F., J. Heinrichs and O. Kreya (1980), *The New International Division of Labor*, Cambridge, UK: Cambridge University Press.

Galor, O. (1996), "Convergence? Inferences from theoretical models", *The Economic Journal*, Vol. 106, pp. 1056–1069.

Geyer, H.S. (1988), "On urbanization in South Africa", *The South African Journal of Economics*, Vol. 56, pp. 154–172.

Geyer, H.S. (1989), "The integration of the formal and informal urban sectors in South Africa", *Development Southern Africa*, Vol. 6, pp. 29–42.

Geyer, H.S. (1996), "Expanding the theoretical foundation of differential urbanization", *Tijdschrift voor Economische en Sociale Geografie*, Vol. 87, pp. 44–59.

Geyer, H.S. (1998), "Differential urbanization and international migration: an urban systems approach", pp. 161–184, in: C. Gorter, P. Nijkamp and J. Poot, editors, *Crossing Borders: Regional and Urban Perspectives on International Migration*, Ashgate: Aldershot.

Geyer, H.S. (2001), "Development planning transition in South Africa", pp. 143–152, in: H.C. Marais, Y. Methien, N.S. Jansen van Rensburg, M.P. Maaga, G.F. de Wet and C.J. Coetzee, editors, *Sustainable Social Development: Critical Dimensions*, Pretoria: Network Publishers.

Geyer, H.S. (2002), "The urban future", pp. 67–86, in: H.S. Geyer, editor, *International Handbook of Urban Systems: Studies of Urbanization and Migration in Advanced and Developing Countries*, Cheltenham, Aldershot, UK: Edward Elgar.

Geyer, H.S. (2003), "South Africa in the global context: the view from above and below", *Annals of Regional Science*, Vol. 37, pp. 407–420.

Geyer, H.S. and T.M. Kontuly (1996), "A theoretical foundation for the concept of differential urbanization", pp. 161–184, in: N. Hansen, K.J. Button and P. Nijkamp, editors, *Modern Classics in Regional Science: Regional Policy and Regional Integration*, Vol. 6, Cheltenham: Edward Elgar.

Giddens, A. (1990), *Consquences of modernity*, Stanford, CA: Stanford University Press.

Golding, P. (1996), "World wide wedge: division and contradiction in the global information infrastructure", *Monthly Review*, Vol. 48, pp. 70–85.

Graham, S. (1999), "Global grids of glass: on cities, telecommunications and planetary urban networks", *Urban Studies*, Vol. 36, pp. 929–949.

Graham, S. and S. Marvin (1996), *Telecommunications and the City: Electronic Spaces, Urban Places*, London: Routledge.

Graham, J., K. Gibson, R. Horvath and D.M. Shakow (1988), "Restructuring in U.S. manufacturing: the decline of monopoly capitalism", *Annals of the Association of American Geographers*, Vol. 78, pp. 473–490.

Grossman, G.M. and A.B. Krueger (1995), "Economic growth and the environment", *The Quarterly Journal of Economics*, (110), pp. 353–377.

Hall, P. (1996), "The global city", *International Social Science Journal*, Vol. 48, pp. 15–23.

Hamnett, C. (1996), "Social polarization, economic restructuring and welfare state regimes", *Urban Studies*, Vol. 33, pp. 1407–1430.

Harris, N. (1995), "Bombay in a global economy: structural adjustment and the role of cities", *Cities*, Vol. 12, pp. 175–184.

Harvey, D. (1989), *The Condition of Post-modernity*, Oxford: Basil Blackwell.

Haughton, G. (1997), "Developing sustainable urban development models", *Cities*, Vol. 14, pp. 189–195.

Heilbroner, R. (1970), *Understanding Macro Economics*, Englewood Cliffs, NJ: Prentice-Hall.

Holliday, C.O., S. Schidheiny and P. Watts (2002), *Walking the Talk: The Business Case for Sustainable Development*, San Francisco: Berrett-Koehler Publishers.

Jensen, M.N. (1998), "Ecologists go to town: investigations in Baltimore and Phoenix forge a new ecology of cities", *Science News*, Vol. 153, pp. 219–221.

Jones, A. (2002), "The global city, misconceived: the myth of 'global management' in transnational sevice firms", *Geoforum*, Vol. 33, pp. 335–350.

Kakabadse, A. and N. Kakabadse (2002), "Trends in outsourcing: contrasting USA an Europe", *European Management Journal*, Vol. 20, pp. 189–198.

Kaya, H.O. (2001), "The political economy of Africa and continental challenges in the new millennium", in: E. Maloka and E. le Roux, editors, *Africa in the New Millennium: Challenges and Prospects*, Pretoria: Africa Institute of SA.

Kim, J. (2001), "Economic integration of major industrialized areas: an empirical tracking of the continued trend", *Technological Forecasting and Social Change*, Vol. 67, pp. 187–202.

King, A. (1997), "Re-presenting world cities: cultural theory/social practice", pp. 215–231, in: P.L. Knox and P.J. Taylor, editors, *World Cities in a World-System*, Cambridge, UK: Cambridge University Press.

Knox, P.L. (1994), *Urbanization: An Introduction to Urban Geography*, London: Prentice-Hall.

Koechlin, T. (1995), "The globalization of investment", *Contemporary Economic Policy*, Vol. 13, pp. 92–100.

Kotabe, M. and J.Y. Murray (2004), "Global sourcing strategy and sustainable competitive advantage", *Industrial Marketing Management*, Vol. 33, pp. 7–14.

Kremer, M., A. Onatski and J. Stock (2001), "Searching for prosperity", *Carnegie–Rochester Conference Series on Public Policy*, Vol. 55, pp. 275–303.

Kutznets, S. (1955), "Economic growth and income inequality", *The American Economic Review*, Vol. 45, pp. 1–28.

Leven, C.L. (1966), *The economic base and regional growth, Center for Agricultural and Economic Development, Research and Education for Regional and Area Development*, Ames, IA: Iowa State University Press, pp. 79–94.

Lipietz, A. (1997), "The post-Fordist world: labour relations, international hierarchy and global ecology", *Review of International Political Economy*, Vol. 4, pp. 1–41.

Lipschutz, R.D. (1991), "One world or many? Global sustainable economic development in the 21st century", *Bulletin of Peace Proposals*, Vol. 22, pp. 189–198.

Loots, E. (2002), "Globalisation, emerging markets and the South African economy", *The South African Journal of Economics*, Vol. 70, pp. 263–286.

Marcotullio, P.J. (2001), "Asian urban sustainability in the era of globalization", *Habitat International*, Vol. 25, pp. 577–598.

Marshal, R. (1995), "The global job crisis", *Foreign Policy*, (100), pp. 50–68.

Merriam (1962), *Webster's Geographical Dictionary*, Menasha, WI: The Collegiate Press.

MMSD (2002), *Breaking New Ground, Mining, Minerals, and Sustainable Development*, London: Earthscan Publications.

Morris, L. (1997), "Globalization, migration and the nation-state: the path to a post-national Europe?", *British Journal of Sociology*, Vol. 48, pp. 192–209.

Mukhertji, S. (2002), "Urbanization and Migration in India: A Different Scene", pp. 525–562, in: H.S. Geyer, editor, *International Handbook of Urban Systems: Studies of Urbanization and Migration in Advanced and Developing Countries*, Cheltenham, Aldershot, UK: Edward Elgar.

NEPAD (2001), *New Partnership for African Development*. www.polity.org.za/misc/mapomega.html.

North, D.C. (1955), "Location theory and regional economic growth", *Journal of Political Economy*, Vol. 63, pp. 243–258.

OECD (2002), *African Economic Outlook*, African Development Bank and Development Centre of the Organization for Economic Co-operation and Development.

O'Loughlin, J.P. (1989), "Political geography: coping with global restructuring", *Progress in Human Geography*, Vol. 13, pp. 412–426.

Paap, R. and H.K. van Dijk (1998), "Distribution and mobility of wealth of nations", *European Economic Review*, Vol. 42, pp. 1269–1293.

Parlange, M. (1998), "The city as ecosystem: urban long-term ecological research projects aim to put the pieces together", *American Institute of Biological Sciences*, Vol. 48, pp. 581–585.

Parr, J.B. (2002), "Missing elements in the analysis of agglomeration economies", *International Regional Science Review*, Vol. 25, pp. 151–168.

Penouil, M. (1981), "Ivory Coast: an adaptive center-down approach in transition", pp. 305–328, in: W.B. Stöhr and D.R.F. Taylor, editors, *Development from Above or Below? The Dialectics of Regional Planning in Developing Countries*, Chichester: Wiley.

Pickett, S.T.A., W.R. Burch, Jr., S.E. Dalton, T.W. Foresman, J.M. Grove and R. Rowntree (1997), "A conceptual framework for the study of human ecosystems in urban areas", *Urban Ecosystems*, Vol. 1, pp. 185–201.

Poon, J.P. (1997), "the cosmopolitanization of trade regions: global trends and implications, 1965–1990", *Economic Geography*, Vol. 73, pp. 390–404.

Quah, D. (1993), "Empirical cross-section dynamics in economic growth", *European Economic Review*, Vol. 37(2/3), pp. 426–434.

Quah, D. (1996), "Convergence empirics across economies with (some) capital mobility", *Journal of Economic Growth*, Vol. 1(1), pp. 95–124.

Quah, D. (2001), "Searching for prosperity: a comment", *Carnegie–Rochester Conference Series on Public Policy*, Vol. 55, pp. 305–319.

Robinson, W.I. (1996), "Globalisation: nine theses on our epoch", *Race and Class*, Vol. 38, pp. 13–31.

Roberts, B. (1978), *Cities of Peasants: The Political Economy of Urbanization in the Third World*, London: Edward Arnold.

Roseland, M. (1997), "Dimensions of the eco-city", *Cities*, Vol. 14, pp. 197–202.

Samuelson, P. (1970), *Macro Economics*, New York: McGraw Hill.

Sassen, S. (1991), *The Global City*, Princeton, NJ: Princeton University Press.

Sassen, S. (1997), "Cities in the global economy", *International Journal of Urban Sciences*, Vol. 1, pp. 11–31.

Shaw, M. (1994), "The state of globalization: towards a theory of state transition", *Review of International Political Economy*, Vol. 4, pp. 497–513.

Sihlongonyane, M. (2001), "Eurocentric values are playing a role in the underdevelopment of African societies", *Africa Insight*, Vol. 30, pp. 18–25.

Smith, A.S. (1995), *Nations and Nationalism in a Global Era*, Cambridge: Polity Press.

Smith, M. and M. Timberlake (1995), "Cities in global matrices: toward mapping the world-system's city system", pp. 79–96, in: P. Knox and P. Taylor, editors, *World Cities in a World System*, Cambridge: Cambridge University Press.

Storper, M. and A.J. Scott (1995), "The wealth of regions: market forces and policy imperatives in local and global context", *Futures*, Vol. 27, pp. 505–526.

The Natural Step (2003), www.naturalstep.ca/articles/TNSFframeworkGuidebook.pdf.

Tiebout, C.M. (1963), "The urban economic base", pp. 105–109, in: H.M. Mayer and C.F. Kohn, editors, *Readings in Urban Geography*, Chicago: University of Chicago Press.

Tsheola, J. (2002), "South Africa's form of globalisation: a continental posture paradox for insertion and dependence", *Political Geography*, Vol. 21, pp. 789–811.

United Nations (1999), *Competition and The World Economy: Comparing Industrial Development Policies in the Developing and Transition Economies, United Nations Industrial Development Organisation*, Cheltenham, UK: Edward Elgar.

Vernon, R. (1996), "Passing through regionalism: the transition to global markets", *The World Economy*, Vol. 19, pp. 621–633.

Wallerstein, I. (1974), "The rise and future demise of the world capitalist system: Concepts for comparative analyses", *Comparative Studies in Society and History*, Vol. 16, pp. 387–415.

Warf, B. (2000), "New York: the Big Apple in the 1990s", *Geoforum*, Vol. 31, pp. 487–499.

World Bank (2000), *World Development Indicators*, Washington: World Bank.

Urban Dynamics and Growth, CEA, vol. 266
Roberta Capello and Peter Nijkamp (editors)

CHAPTER 25

Urban Policy in a Global Economy

Åke E. Andersson[a], Lata Chatterjee[b] and T.R. Lakshmanan[b]

[a]Department of Economics, Jönköping International Business School, Jönköping, Sweden
[b]Center for Transportation Studies, Boston University, Boston, MA 02215, USA

Abstract

Contemporary globalization is characterized by a global production system driven by two factors: the activities of global network corporations trying to capture economies of scale in knowledge, economies of scope in their corporate networks, and favorable factor prices, and by the growing demand for consumption variety attendant on rising income. Global corporations use cities as organizational commodities to maximize returns on capital, initiating a worldwide urban economic competition. The paper highlights the rise of the entrepreunerial city with an increasing role for urban economic policy and a redefined role for traditional urban public goods provision. It discusses the policy and institutional innovations to support the entrepreneurial city. The resulting spatial restructuring of urban activities increeases efficiency but also widens inequality and polarization in the urban fabric.

Keywords: trading regimes, global network corporations, intercity competition, entrepreneurial city, urban policy and institutional innovations, new urban spatial order

JEL classifications: R50, R58

25.1. Globalization: underlying processes, urban consequences, and policy implications

Globalization is not a new phenomenon, but one whose context, underlying processes, and manifest spatial forms have evolved over time.[1] The nature of contemporary globalization is twofold. It means, first, an increasing role of exports and imports in the global economy and its constituent countries and regions. This is evident in the growth of world trade at more than twice the rate of growth of world product in the last two decades.

Second, globalization also means the following: an increasing spatial extension of trade, growing economic interdependencies as manifest in a globally organized system of production and (knowledge, materials, and financial) flows, and a consequent spatial specificity. This is indicated by the fact that world trade, while still concentrated in the so-called Triad area (North America, Europe, and Japan) has been expanding rapidly in the last quarter century between the Triad and newly industrialized countries, particularly in East Asia. This vast increase in the scale, scope, and speed of global economic interactions – in terms of flows of goods, services, capital and knowledge – has, in turn, promoted a complex worldwide division of labor and a spatial organization of production (Ohmae, 1990). Such processes of globalization have a distinct spatiality in the sense that they lead to specific geographical patterns. This derives from the fact that global operations of monitoring, coordination and control carried out by global network firms (as elaborated below) is centered in urban areas. It is in these urban areas where one can observe clearly the phenomena associated with globalization forces – the evolving structure of employment, the development of vast new real estate, new forms of urban governance, and the partitioning of the urban area into parallel residential and business areas (UNCHS, 2002; Lakshmanan and Chatterjee, 2003).

25.1.1. Evolution of globalization processes, urban patterns and policy domains

A fuller understanding of the processes underlying contemporary globalization, their manifestation in the forms and functioning

---

[1] The old Silk Route, the trade of Italy and the Hanseatic League of Northern Europe from 1100 AD, the extensive world trade after the discovery of the New World, and the global trading system organized by the Colonial Powers in the Industrial Era are earlier examples of economic and cultural linking of diverse societies across very long distances (Andersson, 1986; Foltz, 2000).

of urban areas, and the consequent agenda being set for urban policy can result from a quick survey of key factors that differentiate the present era of globalization from those of the immediate preceding eras. Table 25.1 offers such a capsule summary of the forces of globalization, the resulting urban structures, and the consequent urban policy domain of the contemporary globalization as contrasted with the earlier Industrial and Mercantile eras.[2]

In each era, the forces propelling globalization were twofold: a set of physical technologies and *infrastructures*, and a set of non-physical infrastructures (Lakshmanan, 1993). In the Mercantile phase, the most important innovation was the new sailing technology, first introduced with the deep-sea-going Caravelle and later by the more *economically* efficient ship called the Flute, invented by the Dutch ship builders. These inventions and innovations made trans-oceanic transportation possible and opened up Europe for long-distance international trade with the Americas and by seaways to Asia (Hugill, 1995). Trade was further stimulated by new means of payment. The arrival of precious metals from the Americas allowed transactions at a much larger scale than ever before. The creation of new financial institutions supported by the city administration of Amsterdam further facilitated the expansion of world trade. Slowly, but steadily, the focus of world trade was shifted towards the north of Europe and by 1650 the capital of world trade had been established in Amsterdam, which will hold that position for over 100 years thereafter (Braudel, 1992). This logistical revolution carried with it a new phase of urbanization as a mirror image of increasing division of labor and world trade. This time urbanization was more in terms of growing size of the most important commercial cities rather than in an increasing number of towns (Andersson, 1986).

By the end of the 18th century the role of Amsterdam and the Netherlands as the focal point of the world economy was fading, giving way for an increasing role of Great Britain as the new center of world trade and the first focal point of the next logistical or industrial

---

[2] There is an extensive literature on cross-cultural trade in earlier periods. Curtin (1984) surveys cross-cultural trade from early times to mercantile era. Pomeranz and Topik (2000) provide a broad view of trade development since 1400, and discuss the interrelationships between culture, society, and the world economy.

Table 25.1. Evolution of globalization forces, urban patterns and urban policy domain

Key Attributes	Globalization Era		
	Mercantile Era	Industrial Era	Contemporary Era
Technologies and underlying processes	New transport technology; sailing ships; *Caraville*; *Flute*, sextant, etc.	Steam power; railroad; steamships; machine fabrication	New transport and communication technologies; knowledge-rich technologies of production
Non-material infrastructure	Cartography; new means of payment: precious metals and financial innovations of Amsterdam facilitating trade	Exploiting economies of scale; vertical integration of production; factory systems; assembly line; labor unions; property rights; central bank; currency; monetary policies; compulsory minimum education	Exploitation of economies of scope; open trade regime institutions; new logistical innovations; facilitating flows of goods; services, capital and knowledge
Spatial patterns; urban forms; and functions	Increasing urbanization arriving with new division of labor; increasing size of major cities; e.g. Amsterdam, London, etc.	Massive urbanization; rise of new factory towns; increase in average size of towns. Massive problems of urban areas; *problems of cities* (housing, infrastructure; spatial organization); *problems in cities* (unemployment, health, welfare, education, etc.)	New urban regions competing globally for economic activities; relatively fast changes in economic fortunes of many cities causing local dislocations; rise of large global urban regions or corridors around major cities
Urban policy domain	NA	*National public policy. domain.* Keynesian fiscal and monetary policies. *Redistributive public goods* (health, welfare, education and environment), *Urban public policy domain. Distributive public goods* (water supply, sewer, transport and energy infrastructure), etc.	Urban policy is the joint work *of public, private and civil society sectors*; increased focus on urban economic development, a mix of (a) supply side policies to attract globally mobile capital and (b) entrepreneurial demand side strategies supporting urban endogenous economic growth

revolution (Braudel, 1992). With the industrial revolution, division of labor had become a deeper concept, in the sense that it was now openly accepted that conditions of production as well as of transportation and transactions could be influenced by the capitalists involved in trade (Hobsbawm 1996). The key new technologies of this era were steam power, railroads, steamships, and machine fabrication (Hugill, 1995). The supportive organizational developments came from the massive increase in the division of labor, exploitation of economies of scale, vertical integration of production and the development of the assembly line. In addition, in every industrializing country the nation states have played a significant role as providers of basic material and non-material infrastructure. Postal and telephone networks and railroads were established, connecting factory towns and raw material provision centers with each other and the world market, served by major deep-sea ports. Every industrializing country provided the firms with the financial infrastructure in terms of a national central bank, reliable currency and monetary policies. A system of free basic education was established and distributed to most regions. National defense systems, police forces and other institutions provided law and order and protection of property rights (Andersson, 1986; Lakshmanan, 1993).

The trading decision problem in this era was: Can the price in the region of import be sufficiently larger than the marginal cost of production to compensate for the minimized transaction and transport costs? This formulation of the problem implies that the prices of goods in different regions must be seen as functions of the quantities produced and traded:

$$\text{Is } \rho_i - c_j(x_j) > \tau_{ji}(x) + t_{ji}(x)?$$

where $c_j(x_j)$ is the marginal cost of production of a given good in region j, $\tau_{ji}(x)$ the marginal cost of transaction when the given good is traded from region j to region i, $t_{ji}(x)$ the marginal cost of transporting the given good from region j to region i, and x the vector of production of the different goods produced in the regions.

A general equilibrium requires that

$$\rho_i^c - c_j^c(x_j) \leq \tau_{ji}^c(x) + t_{ji}^c(x)$$

for all goods c and all i and j, with an inequality, if no trade in that good and the given regions i and j is profitable.

The fine division and massive deployment of labor to exploit economies of scale in production and the development of assembly line associated with the industrial era has led to distinct spatial patterns – *in the form of a massive urbanization, and specific urban forms and functioning*. Not only did the average size of towns and cities increase, but with growing industrialization new factory towns were also established in every industrializing country.

The arrival of new technologies of fossil energy use based manufacturing into these factory towns influenced their physical, economic, and quality of life characteristics. The owners of capital and the means of production were the principal beneficiaries in terms of wealth accumulation. The workers providing the labor for this wealth creation drew low wages and often lived under miserable conditions. As this spatial form of industrial towns developed over the national landscape, two types of problems emerged. First, the rising income inequalities in early industrialization yielded widespread poverty, and low levels of education and health among the industrial workers. Since these workers lived in cities, these national problems of poverty and welfare became *problems in cities* (Lakshmanan and Chatterjee, 1977).

While massive investments of capital took place in machinery and buildings to support productive activities in these towns, complementary investments on urban infrastructure (water supply, sewers, transportation, energy, etc.) necessary to support basic quality of life in the dense urban environments lagged far behind – leading to extensive slums. The resulting problems – inherent in urban living and organization, with its high densities, shared services, and externalities – represent the second class of problems, namely the *problems of cities* (Lakshmanan and Chatterjee, 1977).

These sets of problems of the industrial era yielded over time a broad policy domain (an arena where decision makers develop and implement policy solutions). As Table 25.1 indicates, public sector actors inhabited this policy domain in the industrial era. The *national level* public policy actors offered solutions: (a) Keynesian type (guided capitalism) national fiscal and monetary policies to promote national and urban employment, and (b) for *problems in cities* in the form of *redistributive public goods* (health, welfare, education, and environment). The *urban level* public sector, for its part, provided a set of *distributive collective goods* (water supply, sewer, transport, and energy infrastructure), to address *problems of cities*.

It is important to note that urban policy in the industrial era globalization was dominated by the public sector – though the precursors for many such initiatives were in the civil society such as the Settlement House Movement in the late 19th and early 20th centuries. As we will note below, the advent of contemporary globalization processes in the last two decades or more has made possible by a weaker role for public policy. In reality, contemporary urban policy is largely the outcome of partnerships between the public, private, and civil society sectors. This recent evolution of the urban policy domain and actors reflects powerful processes, changes in ideology and institutions, and spatial and urban consequences of the new globalization. We turn next to a discussion of such changes and how they define the agenda for today's urban policy.

25.1.2. Contemporary global network corporations, demand for variety and urban consequences

Contemporary globalization is driven by a combination of new transport and communication technologies, knowledge-rich technologies of production, new open trade regime institutions, neo-liberal ideologies, new logistical innovations facilitating flows of goods, services, capital and knowledge, etc. (Drucker, 1990; Ohmae, 1990; Castells, 2000). Many countries have now come to this stage of post-industrial development (Bell, 1999). This stage is no different in terms of advantages of spatial division of labor. But it is different from earlier stages of economic transformation in one important respect. The dependency on an expanding supply of natural resources and thus of land has drastically weakened. Simultaneously, there has been a shift towards creativity and the use of knowledge and information as major resources for economic decision making (Jacobs, 1970; Bell, 1999). This has implied an increasing emphasis on accessibility by communication for the transmission of information and accessibility to other forms of knowledge by personal transportation (Andersson, 1986).

Further, current notions of liberalization and deregulation are now slowly but steadily weakening the nation states institutions. The free flow of information and the advantages of international diversification of financial portfolios have eroded the role of monetary policies by the free flow of financial capital. Traditional Keynesian fiscal policy is losing its efficiency in the same process. Policies generating large government budget deficits regularly generate counteracting reactions in the financial markets in the form of increasing rates of interest and decreasing value of the currency.

Also as a provider of infrastructure, the nation state has lost its fundamental role. With the dramatic increase of real income per capita, combined with a dramatic rise of levels of knowledge in all regions of industrialized countries, the regional dependency on centralized provisions of funds for most types of infrastructure has decreased while privatization trends are stronger here.[3] A similar loss of nation state responsibilities is with the development of the networks for long-distance transportation.[4] It is worth noting that while the above statement is true in general, the case of the newly industrializing countries is somewhat different. In these countries which are telescoping the industrial and the post industrial eras, the role of the public sector in creating and enabling infrastructure for transportation and communication is still vital.

A major agent of this current economic transformation is the global network corporation (e.g. Ikea, Ford, GM, Microsoft, or Pfizer). The economic rationality of these global network corporations is the simultaneous reaping of the advantages of

1. economies of scale in knowledge,
2. economies of scope in the use of the corporate network and

---

[3] Meanwhile, much of the new infrastructure is being provided at levels higher than the nation state – i.e. by continental authorities like the European Union or multinational communication network providers. However, some types of infrastructure, which earlier for financial reasons had to be provided by national funds, like universities and research institutions, are now funded at the regional level. One example of such a shift is Germany. Other examples are given by the restructuring of higher education and research in Japan and other East Asian countries.

[4] At the early stages of industrialization, nation states invested heavily in railroad networks and deep-sea ports, especially in Western Europe. Currently the airline networks with their airports are primarily invested in by private companies or regional governments. Most of the deep-sea ports of Western Europe are being privatized or regionalized. The nation state is losing importance for the regional and the multi-national levels, and the supra-national level institutions are becoming important.

3. the enormous variation in local market conditions and the corresponding variance in local prices and wage rates.

The impact on profitability by these strategic factors can be shown with the following corporate profit maximization model.

In this model, we assume that the net price, after deduction of transportation and transaction costs, of a given product sold in a given region is determined by the corporate-wide availability of knowledge. The role of knowledge varies between different markets. Creativity might play a great role in some design-sensitive markets, while other markets are less interested in such characteristics. Similarly, some markets with a highly sophisticated infrastructure might place a high value on more sophisticated products than less developed markets. This can be reflected by an assumption that the net price of the product in a given market depends on the use of the knowledge stock available to the corporation. The second aspect of the optimization model is the assumption that the volume of output per period of production is dependent both on the capacity of the corporate network as a whole and the use of local resources. A corporate network normally consists of a financial network, a marketing and sales network, a logistical network and an information network.

The local resources are the conventional input of a standard production function, e.g. local labor and its capital equipment. For analytical simplification knowledge is assumed to enter the price functions only, while network resources are assumed to enter the production functions only. Both the price functions and the production functions are assumed to be concave and differentiable everywhere. The optimization model is summarized in the following equation.

The network corporation profit maximization problem:

$$\underset{(K,N,\ell)}{\text{Maximize}}\, P = \sum_i \rho_i(K) q_i(N, L_i) - \rho K - \lambda N - \sum_i \omega_i L_i$$

where P is the corporation-wide profit, ρ_i the price of product of region i, net of transaction and transport cost, K the level of corporation-wide knowledge stock influencing the price in region i, q_i the quantity of production in region i, N the logistical network capacity of the corporation in terms of finances, information, transportation, etc., L_i the local input in region i, ρ the unit cost of expanding corporation-wide knowledge stock,

λ the unit cost of expanding network capacity, and ω_i the unit cost of local input in region i.

The ρ_i – and q_i – functions are assumed to be concave and differentiable everywhere.

The conditions of maximum profit are thus:

$$\frac{\partial P}{\partial K} = \sum_i \frac{\partial p_i}{\partial K} q_i^* - \rho = 0$$

$$\frac{\partial P}{\partial N} = \sum_i \frac{\partial q_i}{\partial N} \rho_i^* - \lambda = 0$$

$$\frac{\partial P}{\partial L_i} = \frac{\partial q_i}{\partial L_i} \rho_i^* - \omega_i = 0$$

Define : $\sum_i q_i \equiv Q; \ \alpha_i \equiv \frac{q_i}{Q}; \ \sum_i \rho_i = \rho; \ \beta_i \equiv \frac{\rho_i}{\rho};$

Thus:

$$\sum_i \alpha_i \frac{\partial \rho_i}{\partial K} = \frac{\rho}{Q} \qquad \text{[Economies of scale in knowledge]}$$

$$\sum_i \beta_i \frac{\partial q_i}{\partial N} = \frac{\lambda}{\rho} \qquad \text{[Economies of scope in network]}$$

This model can, of course, be made more realistic (and complicated) by introducing a further differentiation of the product space, the knowledge and network characteristics of the local resources. However, these complications would add little to the explanation of the rationales behind the formation of global network corporations in the post-industrial economy with its increasing role of creative, logistic, and communicative capacities.

These network corporations, with their global extension and dependency on efficient logistics, are able to simultaneously exploit economies of scale and scope at the corporate level, while maintaining production units in many urban regions around the world.

In this context, global capital in these network corporations uses cities and urban regions as an *organizational commodity* to maximize returns on their capital. Two factors govern the choice of particular production regions or cities, and thus the global geography of production. First, while this global distribution of production centers is driven by the need to reduce factor costs, a key factor is the choice of production locations with sufficient infrastructure capacity in terms of integrator terminals with reliable global accessibility. This favors locations not only in the Triad area (North America, Europe, and Japan), but also locations in newly industrializing countries which have invested and developed their global accessibility. A second factor pertains to the fact that the global network corporations derive their economic advantage from corporate-wide exploitation of knowledge in terms of design and other qualitative characteristics of their products. Locating creative and efficient knowledge centers in regions with comparative advantage in terms of higher education, research and development facilities in relevant scientific technological fields is thus a key strategic variable in these corporations. Thus global network corporations seek a combination of infrastructure investments for improved global accessibility and the provision of knowledge. This explains the rapid growth of multinational firms in towns and cities in large metropolitan corridors around global cities, e.g. the corridor connecting Cambridge, London, Heathrow, Reading, and Oxford.

25.1.3. Demand for variety

While the emphasis so far has been on global network corporations and urban growth stimulated from the production side, recent developments on the consumption side also generate incentives for growth and evolution. These derive from the growing demand for consumption variety attendant on rising income. A new analytical framework (driven by *scale economies*) has emerged – called the 'new economic geography' – which addresses just such situations and provides a host of new insights into the spatial configuration of economic activities.

It develops a new framework for trade by exploiting the analytical breakthrough of Dixit and Stiglitz (1977) who incorporated scale economies into a general equilibrium model assuming a monopolistically competitive market structure (Krugman, 1991, 1999). In this model, product variety is the critical component of competition so that all firms produce distinct but substitutable goods. Consumers' utility functions are defined in such a way that they prefer to consume a variety of goods rather than to concentrate their production on a small number of goods. Thus goods are imperfect substitutes. This means each firm has some degree of monopoly power and can therefore set its price above its marginal cost. The cost structure for each firm includes a fixed component and a constant marginal cost, which results in a downward sloping average cost function indicative of scale economies.

By opening up trade, producers in each region are able to reach broader markets for their unique goods, allowing them to move down their average cost curves and earn greater profits. Naturally this market expansion effect is limited by interregional transportation costs, and any reduction in transportation costs yields increased trade benefits. In the long run, however, more firms – each providing a unique product variety – will enter any market where profits are being earned. The presence of more firms shifts the demand curves of all preexisting firms downward until excess profits are exhausted. Thus it is entry of new firms that brings about an equilibrium.

This view has a number of advantages. For one thing, its emphasis on product differentiation as opposed to direct price competition among producers of perfectly substitutable goods is in keeping with the long-term trend away from commodity production towards highly differentiated and specialized goods. Growth in consumer utility in recent decades has been due not only to the quantity of goods consumed, but also to the ever-increasing variety of goods – especially consumer electronics and other categories of goods where constant product innovations define the competitive environment. Furthermore, the fastest economic growth has occurred in consumer and producer services, which are also highly differentiated.

Another advantage is that by adopting imperfect competition and scale economies, this view is able to tackle a whole range of explicitly spatial phenomena, such as agglomeration and persistent regional differences in wages, which mainstream economic theory has largely ignored. Finally, the new view provides theoretical underpinnings to a variety of observations about contemporary urbanization. For example, the strategy by which some urban regions attain competitive advantage on a global scale by specializing in the production of one or a few high value added commodities, which has been observed by Porter (1990), is consistent with results on agglomeration.

Indeed the notion of demand for variety in the context of economic and urban growth needs to be framed in a larger context than that of goods and services. For example, urban areas attract (and have done so in the past) a variety of persons and views, promote their disciplined interactions, leading over time to technical and organizational innovations and creativity (Jacobs, 1970).[5] Indeed, the demand for variety in the broad sense is ultimately a search for new ideas, techniques, and organizational and institutional innovations (a more probable outcome in urban agglomerations) for promoting growth and development (Andersson, 1986; Sassen, 2000). Florida (2002) suggests the rise of a varied new creative class in contemporary society (numbering in the tens of millions), whose function is to generate new ideas and technologies, which transform work, leisure and other aspects of contemporary life. As the global cities serve to some degree as extra-national platforms for global capital, they also serve as magnets for attracting a broad range of knowledgeable and creative people from around the world – including dynamic minorities, immigrants, and refugees. Multiculturalism is as much part of large cities as international finance (Lakshmanan *et al.*, 2000; UNCHS, 2002).

25.1.4. Urban consequences

As noted below, global cities such as London or New York are not passive observers of the technological and organizational changes initiated by global network corporations and the demand for diversity, but actively create their own environments (Savitch, 1988; Lakshmanan *et al.*, 2000). These cities are creative actors who endogenize their own growth. Utilizing their own resources, they adapt old areas to new uses, mixing them in numerous permutations and combinations, and pyramiding one asset on another until they have reinvented themselves.

However, smaller urban areas, less endowed in the combination of global accessibility and knowledge stocks, fare differently in the competition for production locations in the global production system. Multinational corporations, driven by the need to reduce factor costs, distribute the various stages of production among countries and cities

[5] Rosenberg and Birdzill (1986) suggest (in their book, *How the West got Rich*) that the creativity sustained in Europe over many centuries derive from that society's ability to create and maintain the following conditions: diversity of groups and views, their autonomy and experimentation as a method of going forward.

to take advantage of the spatial differences in factor costs. The ability of an urban region to participate in this global division of labor depends upon its ability to engage in this spatial competition by offering such cost advantages. This process of competition among urban areas to serve as production sites is, however, dynamic. The global distribution of production activities itself changes over time as factor costs change among countries and regions, and component production activities get shifted among countries. In such a context, the geography of production activities is constantly evolving. What is produced, and how and where it is produced undergo frequent change.

These accelerating changes in the scale, composition, and location of production activities in the urban areas worldwide have major consequences. These urban consequences of global forces in the post-industrial era are in terms of changes in the fortunes of industries, cities and urban regions, and in the life chances of urban residents. Such changes in urban *economic performance and its volatility* become a matter of policy concern for the relevant private and public economic actors in the affected cities. The urban economic consequences of globalization thus enter the urban policy agenda.

The urban policy actors, as they attempt to respond to these economic shifts, sense a *major change in the contexts or the environment* in which individuals, firms, and urban public sector entities exist and operate (representing a new 'urban world' for these actors).[6] As the technical, economic, and social forces of globalization

---

[6] There is a significant business and social science literature on this new 'action contexts' of such private sector actors (firms) and city and state governments, and the broad range of economic and social consequences of such actions. The consequences of such globalization are extensively debated. One group emphasizes, in the neo-liberal vein, the inevitability of globalization and the overall positive economic, and technological gains deriving from the globalization market forces; the other group notes the very uneven incidence of such gains worldwide on households, neighborhoods, cities, and countries, and the powerlessness of the regional or local public sector in mediating these impacts. There is, however, a growing third view that the impacts of globalization are more complex than is allowed in the above two views. The corporate economy, society, and the state and urban public sectors are all changing, but talk of 'decline and withering away' of state and urban governments is premature. There is an increasing role for public policy (complementary to other social actors) to help fashion desirable economic and social outcomes from globalization forces. This third nuanced view of globalization impacts is beginning to attract scholarly attention in social sciences, and urban planning (e.g. Cerney, 2000; Clarke and Gaile, 1998; Savitch, 1988; Palan and Abbott 1996).

create a new global geography of competitive advantage and restructured webs of power, the contexts in which urban governments exist and function are altered. In an open and integrated world economy where global network corporations depend on production chains that span several countries, urban areas compete to attract mobile economic activities in an increasingly competitive global economic environment. In turn, the politicians and bureaucrats who populate the different urban governments have an increasing incentive: (a) to promote their urban area's competitive advantage for specific production and service sectors, and (b) to enable the private sector in their localities to innovate and produce goods and services for the global markets.

25.2. Emerging urban policy domains and strategies

The central reality of today's globalization is the increasing competition between cities as well as among nations (Porter, 1998). Today's post-industrial city not only represents itself but also the hopes of its nation. Success in adaptation to the new competitive global environment entails for the city advantages in terms of jobs, income, and tax revenue. Since successful cities in this global competition acquire economic prowess, and leadership, there is a premium on developing successful adaptation strategies. Such strategies must be fashioned to be consistent with the context of emerging political and institutional trends. Two such broad political and contextual developments which influence the development of urban policy are noteworthy.

25.2.1. Increasing role for urban economic policy

The first such broad development is the changing *ideology* of development and how it impacts on the economic policy role of urban areas. The last two decades or more when a globally organized production system was evolving, was also a period of growing neo-liberal ideologies. National governments have been shedding their economic policy roles with a 'privatization of the public sphere', in the form of deregulation, liberalization, muting of macroeconomic policy roles, and an emphasis on freer markets and voluntarism. What was accepted previously as belonging to the public sphere becomes essentially 'private phenomena' in the wider global context. As the economic policy role at the national level becomes weakened, urban areas engaged in global competition assume *an increasing economic policy role*, as explicit spatial policies are required to help market capitalism function effectively in the ongoing economic competition among cities worldwide.

In contemporary capitalism, private entrepreneurs face additional economic crises that manifest at the urban scale. Cities and localities in cities face multiple problems resulting from an imbalance between demand and supply of goods and services such as transportation infrastructure, land developed previously for industry, housing of inadequate quality, lagging skills, rising crime rates, etc. The creation of a new physical urban environment – transport facilities and terminals, office towers, shopping malls, universities, hospitals, research centers, etc. – has to be done smoothly while synchronizing a vast set of transactions between private, public, and civil society actors. There is need for coordinating facilities, securing of finances and working out a new system of laws and rules. The neutral role of the intermediary is no longer adequate for the urban public sector. The private entrepreneurs at the urban level are incapable of solving the basic production problem of securing capital and land at the right place, at the right time and at the right cost, in order to lower their fixed and operating costs and to maintain profitability in a global competition environment. Whether it be Times Square (New York), Docklands (London) or Le Defense (Paris), there is public–private partnership: public acquisition of land, partial public ownership, concessions, leases, etc. (Savitch, 1988). Thus, the increasing economic policy role of urban government (when national economic policy is weakening) represents the *current phase of guided capitalism*, as active public intervention is required to ensure that the market for these spatially determined goods function in an orderly way.

This assumption of active economic policy role by city governments has an important implication. The ideology of market primacy associated with globalization modifies and transforms the traditional roles, tasks, and activities of urban governments. Changes occur in two types of public goods provided to urban residents in a national welfare state framework of the earlier era – *redistributive public goods*, and *distributive public goods*. In the contemporary globally oriented urban economy, new ideologies of liberalization, and privatization of governmental activities, combine with the strong imperative to retain competitive conditions in a variety of different and fast changing international markets for the city's goods and services. The result is an undermining of the traditional urban government's *ideology, identity, and role* – in the process modifying its functions, tasks, and activities. In other words, the way the urban politicians and bureaucrats view, pursue, and deliver 'public goods' is being transformed. As national level resources in general, and those available for *redistributive* public goods (health and welfare services, education, environmental protection, etc.) at the urban level are dropping significantly, the provision of these goods is declining significantly. As cities had to pick up more of the burden for these services through their local tax base, richer cities had an advantage, thereby augmenting income inequalities by service inequalities among cities.

At the same time, the urban *distributive* public goods (garbage collection, water supply and quality services, transport and energy infrastructure) are being organized for market or quasi-market provision.

25.2.2. Emerging institutions and policy strategies

As urban governments begin to pursue international competitiveness as a major goal, the stakeholders (to whom the urban political actors respond) expand beyond the urban households, resident firms and civil society to *include* the economic interests in the global economy and the complex linkages and interdependencies thereof. In this context, the boundaries between international private actors and functions on one hand, and urban actors and roles on the other, intersect and influence one another at many levels. The consequences, as elaborated below, are threefold:

- the emergence of new aims and roles for the politicians and bureaucrats in urban government,
- the redefinition (if not erosion) of some of their earlier public goods functions, and
- the emergence of new institutions or urban governance structures.

As the urban politicians and bureaucrats develop their expanded economic policy role, they create new institutions for governance that bring together various stakeholders in urban development. Typically, they take the form of different partnerships between urban public sector, private sector, and urban non-profit sector. Partnerships are just illustrative of the wide range of institutional innovations which permit the assumption and implementation of new urban policy missions. The more successful the region, the greater the variety and depth of institutions, with considerable redundancy in the institutional structure (Grabher, 1993). The institutional variety an urban region can offer provides a constantly changing portfolio of potential organizational solutions to problems of collective response to changing economic circumstances (Amin and Thrift, 1994). Indeed, we highlight in Section 25.2.3 this institutional diversity as we describe the urban policy strategies being pursued by urban areas in US and elsewhere.

25.2.3. Changing urban policy orientation and innovations

As urban governments have tried in recent years to reduce their vulnerability to 'shocks' from globalization forces and stabilize the level of good jobs for residents and their own revenue base, they have experimented with different strategies and policy instruments to increase local employment and income. This urban vulnerability and policy response has a longer history in the US (where the contemporary globalization arrived earlier) than in Europe. The combination of such forces – outmigration of firms (seeking lower factor costs), increasing participation of US in international markets, and the real declines of national aid to cities – led initially to a set of supply side policies on the part of cities to attract mobile capital to themselves. As elaborated below, this approach turned out to be ineffective and wasteful.

Obliged to depend more on their local resources and increasingly aware of the limitations of the supply side policies to attract mobile capital, urban governments did not adopt the risk aversive stances of political decision units with limited resources. Instead, they turned towards an entrepreneurial strategy, exposing their revenues to some risk and incurring some opportunity costs in order to create endogenous growth. Towards this end, the entrepreneurial urban areas have taken over the last two decades a variety of innovative steps to support their local firms in generating employment. These innovations are of two kinds: new policy instruments and new institutional arrangements for developing and implementing the policies to support generation of local economic growth. We explore next how the entrepreneurial city has expanded its policy domain with a range of new instruments which, while pragmatic and non-ideological, express the ascendance of market feasibility criteria over social criteria, and which redefine the city's responsibilities as a public developer. Further, we describe how the entrepreneurial city has enlarged its institutional space for policy implementation, with an emphasis on public partnerships with the private and non-profit or civil society sector.

25.3. *The American entrepreneurial city: policies, institutions, and new spatial order*

The entrepreneurial city functions much like a Schumpeterian entrepreneur, but in partnership with private sector economic agents. It seeks to identify market opportunities for private economic actors whose exploitation of those opportunities can serve the city's public objectives. The city government becomes a risk taker, a promoter of city businesses' global competitiveness, an enabler in finding new markets, and a catalyst in forming private–public partnerships in testing and developing new technology. Table 25.2 provides an overview of the city's recently changing policy domain – an arena where political actors attempt to develop and implement problem solutions. The policy domain has expanded as the economic contexts have changed and the orientations and scope of policies of urban areas have adopted increasingly entrepreneurial postures. It is worth noting that once policies are adopted there develops supporting constituencies, making termination of older policies difficult for political concerns. While the tool kit of policy instruments expands, older instruments remain to some degree, providing a layering of approaches. One can recognize three phases in this policy evolution (Table 25.2):

- *Phase 1.* Supply side policies and strategies to attract mobile capital,
- *Phase 2.* A transitional phase towards entrepreneurial strategies, and

Table 25.2. ***Evolution of the urban policy domain in the global economy***

Phase 1[a] (Supply Side Policies)	Phase 2[b] (Transition to Entrepreneurial Approach)	Phase 3[b] (Entrepreneurial Policies)
Attract firms to the city through locational incentives	Emphasize local capacity building for market expansion using federal grants	Stimulate new enterprise development by shaping market opportunities and through leverage of public funds to increase private investment Channel investment for physical, institutional and human capital expansion
Based on the concept of spatial comparative advantage	Based on the concept of enabling strategies and removal of barriers in labor, capital and land markets	Based on concepts of public, private partnerships and coproduction Based on concept of knowledge and network economies

[a]The actual time period of the phases varies between cities. Some innovative cities such as New York and Baltimore started earlier. Later cities acted as imitative public entrepreneurs.
[b]New objectives did not replace earlier ones. These objectives were added on and prioritized.

- *Phase 3*. Entrepreneurial policies to promote endogenous growth of human, physical, and social capital in the urban area.

25.3.1. Phase 1: policies and strategies to attract mobile capital: supply side

Guided by supply side view of the economy, these cities attempted to create for mobile firms an advantageous factor price structure in the urban area. The policy instruments used were typically capital subsidies in the form of assistance in site selection and preparation, low-interest financing (tax credits, abatements, exemptions, deferments, industrial revenue bonds, etc.) and expenditures for labor training (Clarke and Gaile, 1998). The aim was to offer lower costs for business in one's city compared to others. Over time this approach lost appeal for several reasons. As this policy approach of bidding for businesses became widespread among cities, it could not pass muster on grounds of effectiveness or efficiency (Blair and Premus, 1987). No net national economic growth occurred, and the firms enjoyed rents. Cities were left with underused infrastructures (which were built with government bonds), leaving them with spatial rigidities of fixed capital, while firms enjoyed the benefits of capital mobility. How appropriate is the use of local public funds for subsidies to firms, which might have made investments anyway? Changing economic conditions and a growing new understanding of urban economic growth process reduced the appeal of supply side policies of inducing firms to locate in one's area (Premus *et al.*, 1985).

25.3.2. Phase 2: a transitional phase towards entrepreneurial strategies

These strategies derive from an understanding that markets have failed and public intervention is required to reduce market imperfections. The policies (a) lower barriers to the creation and expansion of small firms and the level of entrepreneurial activity in the urban area, and (b) risk public funds and invest them to help local private firms engage in value creating activities. The overall thrust of the urban public sector in this approach is to promote the flexibility and adaptive capability of the urban private sector, and help it adapt to the 'shocks' from globalization processes. Utilizing funds provided by federal government programs such as Economic Development Administration (Business Loans, Adjustment Assistance), and Department of Housing and Urban Development (Community Development Block Grant programs), the cities learnt and acquired experience with market-based strategies.

The institutional innovation to leverage the city's funds was public–private partnerships which take many forms: public ownership and private rental as in industrial parks, public tax expenditures for revitalizing urban neighborhoods, small business grants, funding community development corporations, equity participation in development projects. Indeed, all sectors were mobilized as in waterfront development schemes in many cities. Rigid laws and codes were modified, preferring performance-based criteria over formula-based entitlement criteria. Adaptive use of abandoned structures, like the Piano Factory in Boston as housing and center for artists, adaptive reuse of schools for high end condominiums were developed either by the private sector or with partnerships between the private and civil sectors like Community Developments Corporations (CDC). Security and crime prevention became important and neighborhoods began to partner with the police departments through Neighborhood Watch schemes. Chambers of commerce with their urban boosterism, venture capitalists looking to support innovative firms in their local communities are all ingredients of the entrepreneurial city at this stage. Universities partnered through R&D activities, research and consulting jointly with the public sector. As a result of this multipronged and multifaceted approach firms begin to cluster in attractive/entrepreneurial cities where the quality of life issues for their employees are critical for retaining qualified labor. Employees of IT firms, and other tech sectors are attracted by exciting neighborhoods, plays, music venues and such.

Thus, both production and consumption attractors emerge as important ingredients in urban revitalization in a period of stiff competition. New York and Cleveland, for example, provide good examples of the synergies between the public, private, and civil sectors in production and consumption activities. In an increasingly knowledge-intensive economy, growth and nurturing of human capital is a critical element of entrepreneurial strategies.

25.3.3. Phase 3: entrepreneurial policies to promote endogenous growth in the urban area

A wide range of instruments are used to shape the structure of local market opportunities, to leverage public and private funds to stimulate new enterprises and local economic growth, and to use contingent partnerships with private sector and civil society – while exposing their own resources to risk. There are two interactive elements in these partnerships. There is a hierarchical element and a cross-sectoral element. There is a growing partnership between the three levels of government: federal, state, and local to increase the competitive advantage of urban areas. Currently, the highly competitive environment has compelled the adoption of entrepreneurial practices for basic survival *in all sectors.*

In Table 25.3 selected illustrations of the variety of institutional forms developed by entrepreneurial cities are provided. For each institutional form innovated, the table identifies the objectives sought, the types of policy instruments (e.g. financial and fiscal), and the physical and administrative infrastructure used to promote the sought objectives. For example, take the institutional form of a local development council and a business incubator which represents

Table 25.3. ***Entrepreneurial city's objectives, institutions, instruments and infrastructure***

Institutional Forms	Types of Instruments[a]		Infrastructure		Objectives
	Financial	Fiscal	Physical	Administrative	
Community Development Corporation (CDC)	Community bloc development grants, other project grants, venture capital pools	Tax abatement from the city government	Capital improvement in streets and neighborhoods	Streamlining licensing permitting	Neighborhood revitalization and employment creation
Local development councils and business incubators	Use of various debt instruments such as revenue bonds, venture capital pools	Tax increment financing	Industrial parks on Greenfield and Brownfield sites	Skilling, networking	Seed and nurture startup companies; attract high-tech firms
Empowerment zones and enterprise communities. EZ/EC (Partnership between Fed/State, city government and private sector	Federal grants for capital promotion through leveraging	Property improvement tax abatement; sales tax exemptions	Public transit provision; worker training	Creation of business technology centers; promote and market trade associations	Retain existing industry; attract new investment; foster local entrepreneurship
Rebuild Boston Energy Initiative (Partnership of city government, utility companies, community organizations and energy consultation	Funding for energy companies	Income tax credit	Retrofitting buildings; consulting services for new construction	Linkage to business owners via Boston Office of Business Development	Energy conservation through upgrading, retrofitting, and consulting for new construction, reduce factor costs, of firms, attract new investment

[a]A selected and illustrative list. Not all instruments were used by all cities. The package varied between cities and in time within a city.

a partnership between actors in the private, public, and university sectors, aiming to promote start-up companies that can offer new goods and services in the global markets. Towards this end, the incubator wants to create infrastructure – industrial parks on Greenfield and Brownfield sites and flexible and sophisticated human capital. The urban public sector financial instruments to develop this infrastructure capital will be drawn from debt instruments such as revenue bonds and venture capital pools. Tax increment financing is the fiscal tool in this case. The specific forms derive from the context. Globalization affects cities differentially, but the responses of the city varies by the local aims of ameriolization, local, social, and political mobilization and institutional history.

25.3.4. A new spatial order in the entrepreneurial city

The deployment of the policies and institutional innovations noted above have enabled the entrepreneurial cities in the US and other OECD countries to restructure their economic activities and adapt and grow in the context of globalization. While the forces of globalization and creative urban adaptation and restructuring have created *aggregate growth benefits*, the incidence of these benefits is very uneven. Mirroring a key aspect of globalization, there is widening inequality or polarization among cities and within cities (Harvey, 1985; Lash and Urry, 2000; Marcuse and Kampen, 2000; Friedmann, 2002). The economic and social disparities one observes between core and peripheral regions on the world scale are typically reproduced *within cities* (UNCHS, 2002). Affluent districts and vital business districts coexist with dilapidated neighborhoods in cities, with little interaction among them.

While cities in the industrial and earlier eras have shown functional, cultural, and status divisions, the widening inequality of incomes and polarization in contemporary globalization has further increased the differentiation within urban areas, hardening the lines between the areas, sometimes walling off the rich from the poor. There is increased fragmentation of cities – physically, economically, and socially (Marcuse and Kampen, 2000).

This widening inequality has a strong spatial dimension in the emerging urban areas. The link between inequality and contemporary urban space is not the simple division of urban society into a rich and a poor part, as observed in 19th century industrial era. It is today a deeply divided, partitioned, or quartered city, reflecting the extent and forms of emerging income and asset inequalities in urban areas (Friedmann, 2002). The increasing mobility of capital, the growing span of control that is possible have led to a great concentration of the wealth at the top and a widening gap between the rich, the affluent, and the poor. Indeed, this polarization has several dimensions: an increase in the relative numbers of the rich and the poor; second, a widening of the wealth distance between them; and third, a greater differentiation of the people who are in between the rich and the poor. Consequently, there are multiple classes, and a sharper differentiation among them. Such differences are reflected in the patterns of segregation of people and land uses in the emerging city (Marcuse and Kampen, 2000; Beauregard and Haila, 2000; UNCHS, 2002).

The resulting partitioning of the city is a fivefold formation of parallel residential and business districts (with limited interactions among them):

- the luxury district and controlling district (citadels of wealth and business);
- the gentrified district and the district of the advanced services;
- the suburban district and the district of direct production;
- the tenement district and the district of unskilled work; and
- the abandoned district and the residual district.

Little policy attention has been directed at the resultant welfare inequities across urban space. The focus of urban entrepreneurial policy actors has been on creating aggregate growth benefits at the urban level, neglecting distributional issues. In view of the powerful forces underlying the process of intercity competition, it is likely that these unfavorable consequences in terms of urban living may worsen in the future – creating thereby worsening livelihood chances and unsustainable cities. The current focus on growth and the neglect of distributional and *intraurban* quality of life issues are likely to be unsustainable in the continuing age of globalization. Very likely, urban entrepreneurs will generate an agenda for future urban policy and move towards redressing the current imbalance and promoting equity issues. Such future policy efforts are likely to build upon on the experience and learning from current civil society initiatives (e.g. low income units in mixed housing development).

References

Andersson, A. (1986), "Presidential address: the four logistical revolutions", *Papers of the Regional Science Association*, Vol. 59, pp. 1–12.

Amin, A. and N. Thrift (eds.) (1994), *Globalization, Institutions and Regional Development in Europe*, Oxford: Oxford University Press.

Beauregard, R.A. and A. Haila (2000), "The unavoidable continuities of the city", pp. 22–36, in: P. Marcuse and R.V. Kampen, editors, *Globalizing Cities: A New Spatial Order?*, London: Blackwell.

Bell, D. (1999), *The Coming of Post-industrial Society: A Venture in Social Forcasting*, New York: Basic Books.

Blair, J.P. and R. Premus (1987), "Major factors in industrial location: a review", *Economic Development Quarterly*, Vol. 1(72), p. 8.

Braudel, F. (1992), *Perspective of the World Civilization: Civilization and Capitalism in 15–18th Century*, Berkeley, CA: University of California Press.

Castells, M. (2000), *The Rise of Network Society*, 2nd edition, London: Blackwell.

Cerney, P.G. (2000), "Structuring the political arena: public goods, states and the governance in a globalizing world", pp. 21–35, in: R. Palen, editor, *Global Political Economy*, London: Routledge.

Clarke, S. and G. Gaile (1998), *The Work of Cities*, Minneapolis: The University of Minneapolis Press.

Curtin, P. (1984), *Cross-Cultural Trade in History*, London: Cambridge University Press.

Dixit, A. and J.E. Stiglitz (1977), "Monopolistic competition and optimum product diversity", *American Economic Review*, Vol. 67(3), pp. 297–308.

Drucker, P. (1990), *The New Realities*, New York: Mandarin.

Florida, R. (2002), *Rise of the Creative Class: How it is Transforming Work, Leisure, Community and Everyday Life*, New York: Basic Books.

Foltz, R. (2000), *Religions of the Silk Road: Overland Trade and Cultural Exchange from Antiquity and the Fifteenth Century*, Basingstoke, UK: Macmillan.

Friedmann, J. (2002), *The Prospect of Cities*, Minneapolis: University of Minnesota Press.

Grabher, G. (1993), *In Praise of Waste: Redundancy in Regional Development*, Berlin: Edition Sigma.

Harvey, D. (1985), *Urbanization of Capital: Studies in the History and Theory of Capitalistic Urbanization*, Baltimore, MD: Johns Hopkins University Press.

Hobsbawm, E. (1996), *The Age of Capital: 1848–1875*, New York: Vintage.

Hugill, P.J. (1995), *World Trade Since 1431: Geography, Technology, and Capitalism*, Baltimore, MD: Johns Hopkins Press.

Jacobs, J. (1970), *The Economy of Cities*, New York: Vintage Books.

Krugman, P.K. (1991), *Geography and Trade*, Cambridge, MA: MIT Press.

Krugman, P.K. (1999), "The role of geography in development", *International Regional Science Review*, Vol. 22(2), pp. 142–161.

Lakshmanan, T.R. (1993), "Social change induced by technology: promotion and resistance", in: Nordal Ackerman, editor, *The Necessity of Friction*, Berlin: Springer, Reissued as paperback, Westview Press, 1999.

Lakshmanan, T.R. and L. Chatterjee (1977), *Urbanization and Environmental Quality, Resource Paper RP 77-1*, Washington, DC: Association of American Geographers.

Lakshmanan, T.R. and L. Chatterjee (2003), The entrepreneurial city in the global marketplace, paper prepared for the International Workshop on Modern Entrepreneurship, Regional Development and Policy: Dynamic and Evolutionary Perspectives at Tinbergen Institute, Amsterdam, May 23–24.

Lakshmanan, T.R., D.E. Andersson, L. Chatterjee and K. Sasaki (2000), *Three Global Cities: New York, London, and Tokyo, Gateways to the Global Economy*, Cheltenham, UK: Edward Elgar, pp. 49–82.

Lash, S. and J. Urry (2000), *The End of Capitalism*, Madison, WI: University of Wisconsin Press.

Marcuse, P. and R.V. Kampen (2000), "Conclusion: a changed spatial order", pp. 249–275, in: P. Marcuse and R.V. Kampen, editors, *Globalizing Cities: A New Spatial Order*, London: Blackwell.

Ohmae, K. (1990), *The Borderless World: Power and Strategy in the Interlinked Economy*, New York: Harper Business.

Palan, R. and J. Abbott (1996), *State Strategie in the Global Political Economy*, London: Pinter.

Pomeranz, K. and S. Topik (eds.) (2000), *The World Trade Created: Culture, Society and the World Economy*, New York: M.E. Sharpe.

Porter, M. (1990), *The Competitive Advantage of Nations*, New York: Free Press.

Porter, M. (1998), *On Competition*, Cambridge, MA: Harvard University Press.

Premus, R., C. Bradford, G. Krumbhaar and W. Schact (1985), The US Climate for Entrepreneurship and Innovation, A report prepared by the Joint Economic Committee US Congress, Washington, DC.

Rosenberg, N. and L.E. Birdzill, Jr. (1986), *How the West Got Rich*, New York: Basic Books.

Sassen, S. (2000), *Cities in a World Economy*, 2nd edition, California: Pine Forgr/ Sage.

Savitch, H.V. (1988), *Post-Industrial Cities: Politics and Planning in New York, Paris and London*, Princeton, NJ: Princeton University Press.

UNCHS (United Nations Centre for Human Settlements) (2002), *Cities in a Globalizing World, Earthscan*, London: Earthscan.

INDEX

The manufacturer's authorised representative in the EU for product safety is Easy Access System Europe – Mustamäe tee 50, 10621 Tallinn, Estonia, easproject.com, +372 5696 8939, gpsr.requests@easproject.com

Printed and bound by CPI Group (UK) Ltd, Croydon, CR0 4YY
15/09/2026
02217231-0001